Assessing Revolutionary and Insurgent Strategies

I0095705

CASEBOOK ON INSURGENCY
AND REVOLUTIONARY WARFARE
VOLUME II: 1962–2009

Paul J. Tompkins Jr., USASOC Project Lead

Chuck Crossett, Editor

United States Army Special Operations Command

and

The Johns Hopkins University/Applied Physics Laboratory

National Security Analysis Department

This publication is a work of the United States Government in accordance with Title 17, United States Code, sections 101 and 105.

Published by:

The United States Army Special Operations Command

Fort Bragg, North Carolina

27 April 2012

Second Printing 2013

Reproduction in whole or in part is permitted for any purpose of the US government.

Nonmateriel research on special warfare is performed in support of the requirements stated by the US Army Special Operations Command, Department of the Army.

Comments correcting errors of fact and opinion, filling or indicating gaps of information, and suggesting other changes that may be appropriate should be addressed to:

United States Army Special Operations Command

G-3X, Special Programs Division

2929 Desert Storm Drive

Fort Bragg, NC 28310

When citing this book, please refer to the version available at www.soc.mil.

All ARIS products are available from USASOC at www.soc.mil under the ARIS link.

Published by Conflict Research Group.

First published by USASOC in 2013

ISBN: 978-1-925907-29-2

ASSESSING REVOLUTIONARY AND INSURGENT STRATEGIES

The Assessing Revolutionary and Insurgent Strategies (ARIS) series consists of a set of case studies and research conducted for the US Army Special Operations Command by the National Security Analysis Department of The Johns Hopkins University Applied Physics Laboratory.

The purpose of the ARIS series is to produce a collection of academically rigorous yet operationally relevant research materials to develop and illustrate a common understanding of insurgency and revolution. This research, intended to form a bedrock body of knowledge for members of the Special Forces, will allow users to distill vast amounts of material from a wide array of campaigns and extract relevant lessons, thereby enabling the development of future doctrine, professional education, and training.

From its inception, ARIS has been focused on exploring historical and current revolutions and insurgencies for the purpose of identifying emerging trends in operational designs and patterns. ARIS encompasses research and studies on the general characteristics of revolutionary movements and insurgencies and examines unique adaptations by specific organizations or groups to overcome various environmental and contextual challenges.

The ARIS series follows in the tradition of research conducted by the Special Operations Research Office (SORO) of American University in the 1950s and 1960s, by adding new research to that body of work and in several instances releasing updated editions of original SORO studies.

VOLUMES IN THE ARIS SERIES

Casebook on Insurgency and Revolutionary Warfare, Volume I: 1933–1962 (Rev. Ed.)
Casebook on Insurgency and Revolutionary Warfare, Volume II: 1962–2009
Undergrounds in Insurgent, Revolutionary, and Resistance Warfare (2nd Ed.)
Human Factors Considerations of Undergrounds in Insurgencies (2nd Ed.)
Irregular Warfare Annotated Bibliography

FUTURE STUDIES

The Legal Status of Participants in Irregular Warfare
Case Studies in Insurgency and Revolutionary Warfare—Colombia (1964–2009)
Case Studies in Insurgency and Revolutionary Warfare—Sri Lanka (1976–2009)

SORO STUDIES

Case Study in Guerrilla War: Greece During World War II (pub.1961)
Case Studies in Insurgency and Revolutionary Warfare: Algeria 1954–1962 (pub. 1963)
Case Studies in Insurgency and Revolutionary Warfare: Cuba 1953–1959 (pub. 1963)
Case Study in Insurgency and Revolutionary Warfare: Guatemala 1944–1954 (pub. 1964)
Case Studies in Insurgency and Revolutionary Warfare: Vietnam 1941–1954 (pub. 1964)

In a rare spare moment during a training exercise, the Operational Detachment-Alpha (ODA) Team Sergeant took an old book down from the shelf and tossed it into the young Green Beret's lap. "Read and learn." The book on human factors considerations in insurgencies was already more than twenty years old and very out of vogue. But the younger sergeant soon became engrossed and took other forgotten revolution-related texts off the shelf, including the 1962 *Casebook on Insurgency and Revolutionary Warfare,* which described the organization of undergrounds and the motivations and behaviors of revolutionaries. He became a student of the history of unconventional warfare and soon championed its revival as a teaching subject for the US Army Special Forces. When his country faced pop-up resistance in Iraq and tenacious guerrilla bands in Afghanistan during the mid-2000s, his vision of modernizing the research and reintroducing it into standard education and training took hold.

This second volume owes its creation to the vision of that young Green Beret, Paul Tompkins, and to the challenge that his sergeant, Ed Brody, threw into his lap.

FOREWORD

Unconventional Warfare is the core mission and organizing principle for US Army Special Forces. The Army is the only military organization specifically trained and organized to wage Unconventional Warfare. From their inception, Special Forces and Army Special Operations Forces were largely focused on developing regional, cultural, and language skills in recognition of the singular importance of the human dimensions of war among the people. We have consistently recognized the importance of dedicating intellectual efforts to better understand the nature of our environment, the motivations and behavior of our enemies. Investment in our human capital is an essential part of developing and maintaining sufficient capability to conduct Unconventional Warfare or Unconventional Warfare-related operations in sensitive environments or conditions.

In the 1960s, our predecessors had the Special Operations Research Office (SORO) at American University produce a collection of case studies on insurgent movements; these studies characterized the motivations and behaviors of revolutionaries and insurgents. The book provided rich reading and study for generations of scholars, Green Berets, and other practitioners and is still a relevant part of our professional literature today. That investment informed our tactics and operations and set the tone for how US Army Special Operations practiced irregular warfare.

Today we again find ourselves facing a dynamic, agile, and flexible enemy whose motivations and behaviors have changed since our historic studies. Our challenge is to understand today's very capable, intelligent, and adaptable enemy and to understand that enemy's relationship to relevant populations. We partnered with Johns Hopkins scholars to build on the foundations of our historic case studies to produce a new case-study series to help us better understand the characteristics of the modern operational environment.

I strongly encourage the men and women of Army Special Operations, the joint Special Operations community, and anyone whose professional interest encompasses unconventional warfare and irregular warfare to make these studies a fundamental part of their professional reading and development. The understanding and successful practice of Unconventional Warfare and Irregular Warfare demands our best intellectual appreciation and application as much as it does excellence within our tactical skill sets.

Strength and Honor,

Lieutenant General John F. Mulholland
US Army, Commanding General

On behalf of the John F. Kennedy Special Warfare Center (SWC) and School, I am proud to present this collection of twenty-three outstanding case studies of insurgencies and revolutions as a survey of modern unconventional warfare. These case studies are a valuable reference for those who study contemporary conflict and for those who practice it, and they are a significant contribution to the body of knowledge on unconventional warfare.

At SWC, our mission is to prepare soldiers for unconventional warfare. We have pursued the singular goal of excellence in all areas of unconventional operations since our inception, and we have established ourselves as the Center of Excellence for Unconventional Warfare. Today, as US Army Special Forces (SF) move into the twenty-first-century operating environment, our ability to think creatively and critically to find solutions to the complex situations we will face becomes even more significant. The operating environment of today requires the traditional skills that have been a mainstay of SF operations since our founding. It also demands that these skills be updated to meet the new demands of an agile and adaptable enemy.

Green Berets employ both a direct and an indirect approach to countering modern threats. At times there is value in and need for the direct approach. For SF, however, the enduring results, and our legacy, live with the indirect approach whereby we enable partners and populations to combat extremist organizations themselves and we contribute to their capabilities through advising and training. It is in this environment that these updated case studies will have the most value.

These chapters offer an opportunity to capture and learn important lessons from history. In them, you can inform your own actions and decisions by better understanding the steps that led to the 1979 Iranian Revolution. By developing an appreciation for the cultural and political frustrations in the Middle East, you will better understand the conditions that brought about modern-day militant Islam. By taking the time to become a student of the history of revolutionaries and insurgencies, you will better be able to carry out the indirect missions of advising and training our partners and allies.

Brigadier General Bennet S. Sacolick
USAJFKSWCS Commanding General

"How do you set about raising an insurrection?" inquired the King.

"Why," replied Coppenole, "the thing is not at all difficult. There are a hundred ways. In the first place the city must be discontented. That is not a rare circumstance. And then the character of the inhabitants. Those of Ghent are disposed to rebellion. They always love the son of the reigning prince, but never the prince himself. Well, suppose some morning someone comes into my shop and says to me: 'Father Coppenole, there is this, and there is that; the maid of Flanders is determined to save her ministers; the high bailiff has doubled the tax for grinding corn'—or anything else for that matter. That's all we need. I leave my work, and I go out into the street and shout, 'To arms!' There is always some cask or barrel lying around. I leap up on it, and cry out the first words that come to mind what I have upon my heart, and when one belongs to the people, sire, one always has something upon the heart. Then everyone assembles, they shout, they ring the alarm bell, they arm themselves with weapons taken from the soldiers, the market people join them, and they go to work. And this will always be the way, while there are lords in the seigneuries, bourgeois in the cities, and peasants in the country."

"And against whom do you rebel?" inquired the King. "Against your bailiffs? Against your liege lords?"

"Sometimes one, sometimes the other; sometimes it is against the Duke."

> —Victor Hugo, *The Hunchback of Notre Dame*, Book X, Chapter V, trans. Catherine Liu (New York: Random House, 2002).

PREFACE

This Casebook provides a summary of twenty-three insurgencies and revolutions; the goal of the book is to introduce the reader to modern-style irregular and unconventional warfare, as well as to act as an informational resource on these particular cases. While not trying to provide an in-depth analysis of any case, our intent was to provide enough background material and description of the revolution to allow comparisons and analysis of broader ideas and insights across this broad spectrum of cases. If further study is desired, each case contains a detailed bibliography that points toward what we found to be the most helpful and insightful sources.

All cases in this book are presented in a standardized format, a research framework, making it easy to compare various aspects of revolutionary warfare. The Methodology section will define what each section of the framework provides and our justification for its inclusion.

All of the sources used in preparation of this Casebook are unclassified and for the most part are secondary rather than primary sources. Where we could, we used primary sources to describe the objectives of the revolution and to give a sense of the perspective of the revolutionary or another participant or observer. Our limitation to unclassified sources allows a much wider distribution of these case studies, while hindering the inclusion of revealing or perhaps more accurate information. We have endeavored to use sources that we believe to be reliable and accurate.

These studies are also meant to be strictly neutral in terms of bias toward the revolution or those to whom the revolution is directed. We sought to balance any interpretive bias in our sources and in the case presentation so that it may be studied without any indication by the author of moral, ethical, or other judgment.

TABLE OF CONTENTS

INTRODUCTION

HISTORY

In 1962, the Special Operations Research Office (SORO) published the first *Casebook on Insurgency and Revolutionary Warfare.*[1] The volume was written under contract to the US Army, which funded a number of social science projects at SORO and similar organizations.[2] SORO was part of American University, which hired a team of social scientists expressly to conduct research on Army contracts. During the 1950s through the mid-1960s, SORO sociologists, historians, psychologists, anthropologists, and former military personnel provided to the Army descriptions of worldwide cultures, social movements, and regional political conditions. These teams also analyzed the effects of propaganda and psychological operations as well as the roles of the military in developing countries, and they provided large bibliographies of unclassified materials related to counterinsurgency and unconventional warfare. There was fair hope during those years that the social sciences would prove as useful as physics and engineering had been to the Armed Forces during World War II and the start of the Cold War.[3]

The SORO Casebook (considered to be Volume I) was both an introductory piece for students of insurgency and unconventional warfare, as well as an initial piece of background research for the next few years of the organization's study. It contained summaries of twenty-three revolutions and insurgencies across the world that occurred during the mid-twentieth century (1927–1962). The Army wished to understand the processes of violent social change in order to be able to cope directly or indirectly, through assistance and advice, with revolutionary actions. The Army also was, at times, a participant in efforts to bring about change. Volume I, therefore, was written to extend the Army's knowledge of how revolutions are born, grow, succeed, or fail. The cases summarized in Volume I are listed in Table 1.

---

[1] Paul A. Jureidini, Norman A. La Clarite, Bert H. Cooper, and William A. Lybrand, *Casebook on Insurgency and Revolutionary Warfare: 23 Summary Accounts* (Special Operations Research Office, Washington, DC, 1962).

[2] The Johns Hopkins University had the Operations Research Office (ORO), which provided more of an mathematical and engineering analysis role to the US Army from 1948–1961, although it too had psychologists, historians, and other social scientists involved in studying psychological warfare under contract. The George Washington University stood up the Human Resources Research Office (HumRRO) in 1951 to provide human factors and human resources research to the Army.

[3] The motivations of the SORO social scientists and their perspectives on the future of their disciplines as related to pragmatic use in policy and analysis can be seen in Joy Elizabeth Rhode's unpublished dissertation, *The Social Scientists' War: Expertise in a Cold War Nation* (University of Pennsylvania, 2007).

Table 1: Contents of Volume I published in 1962

After the publication of the SORO Casebook and a few single volumes of expanded treatments of individual revolutions, SORO started research into the covert organizations that fed and operated within the movements or groups: the undergrounds. Studies on the roles and functions provided by various undergrounds and then the motivations and behaviors of the people performing those functions led SORO to an expanded understanding of how insurgencies developed in the early stages and uncovered various factors that could lead to their success or failure. The research path soon grew into the more exotic realm of trying to predict the likelihood that a country was "ripe" for a revolution. This research was called Project Camelot and received a great deal of attention within the international media and diplomatic corps. This attention led to the demise of SORO as an organization and also presaged a dwindling amount of social science research sponsored by the military. Although the usefulness of the original SORO Casebook and its follow-on research is not disputed, insurgencies and revolutions have now modernized and transformed to the point where a second volume is necessary. The student of modern revolutions needs a new set of studies.

PURPOSE OF THE CASEBOOK

This Casebook is intended to provide a foundation for common understanding on the topic of insurgency and revolution. This foundation will allow readers to distill vast amounts of material from a wide array of campaigns and extract relevant lessons, thereby enabling the development of future doctrine, professional education, and training. Volume I was primarily focused on the communist threat. Although Volume I has served the Special Operations community exceptionally well in the past, the current fundamentalist Islamic threat is significantly different from its communist and Maoist predecessors, thereby warranting new analysis.

This Casebook summarizes twenty-three revolutions that have occurred since 1962. Many of these revolutions are still active—some in a steady-state violent conflict, others in decline, still others possibly approaching a complete resurgence. We have therefore attempted to bound each revolution in time and place in order to present a digestible study. For example, at the time of the writing of this Casebook, the US fight against Al Qaeda is ongoing and of prime importance to the US Army and other US government organizations. We have chosen, however, to study the group from its formation up to the point at which it was dislodged from its primary bases in Afghanistan at the end of 2001. This provides a marked transformation of the nature of the conflict between the movement and other nations, as well as a point of an organizational and functional transformation by the group in order to maintain its operations. It might be possible to write another case study about the post-2001 Al Qaeda movement, but the one included in this Casebook suffices for instruction into current styles of warfare.

Each case presents a background of physical, cultural, social, economic, and political factors that are relevant and important to understand the revolution. The nature of the revolution is explored through its objectives, its leadership and organization; its operations, communications, and interactions with the surrounding population; and the government's response. Then a short section describes how the revolution changed the environment, as well as how it changed the movement itself. Each case is focused almost entirely on the revolutionary movement. This Casebook is not meant to provide a complete historical account of the revolution from all sides. It is meant to explain, explore, and investigate the revolutionary movement and the surroundings that affected it. It is therefore "red-centric."

ORGANIZATION OF THE CASEBOOK

This Casebook has five sections that divide the twenty-three studies by the predominant motivation behind each revolution. This is not to suggest, however, that there is only a single motivating factor behind each or any of these revolutions. The summaries presented, in fact, highlight the myriad of factors that contribute to the creation of organizations, the participation by various social groups, and the eventual outcome that satisfies one side, both, or neither. But in order to present some structure for analysis and discussion of similarities and differences, categorizing the various events herein by the most evident or persuasive cause of the revolt seems the most beneficial.[4]

The first section deals with revolutions that desire to greatly **modify the type of government**. They include:

1. New People's Army (NPA)
2. Fuerzas Armada Revolucionarias de Colombia (FARC)
3. Sendero Luminoso (Shining Path)

---

[4] In the original SORO Casebook, cases were divided into geographic regions. Geographic areas of interest can be a factor in the last fifty years as well, especially in regard to similarities in either the cultural or physical environment. But the rapidity of information dissemination has narrowed the spectrum of regional differences in tactics, resources, or even ideological nuance. The original SORO cases were also indicative of the era in which they occurred, most being motivated strongly by the desire for more socialistic or communistic government systems or policies.

4. 1979 Iranian Revolution

5. Frente Farabundo Martí para la Liberación Nacional (FMLN)

6. Karen National Liberation Army (KNLA)

The second section describes revolutions where **identity or ethnic issues** are prime motivations for the warfare:

7. Liberation Tigers of Tamil Eelam (LTTE)

8. Palestine Liberation Organization (PLO)

9. Hutu-Tutsi genocides

10. Kosovo Liberation Army (KLA)

11. Provisional Irish Republican Army (PIRA)

The desire to **drive out a foreign power** from their area constitutes the third section, with the cases:

12. Afghan Mujahidin

13. Vietcong

14. Chechen Revolution

15. Hizbollah

16. Hizbul Mujahedeen

The fourth section deals with the pressing rise of revolutions based upon **religious fundamentalism**:

17. Egyptian Islamic Jihad (EIJ)

18. Taliban

19. Al Qaeda

The last section covers **issues of modernization or reform**, including:

20. Movement for the Emancipation of the Niger Delta (MEND)

21. Revolutionary United Front (RUF)

22. Orange Revolution of Ukraine

23. Solidarity

RESEARCH FRAMEWORK

The outline of each study presented in this Casebook has been standardized to allow the reader to compare particular aspects or factors across cases, as well as to ensure that the research conducted to write the case was inclusive of all-important factors or conditions.

Each case has four major sections. The first is a SYNOPSIS of the case, which summarizes key facts for a reader not acquainted with the revolution. For cases that are ongoing as of the writing of this book, this section will also provide a description of the time frame chosen for the study.

The second section describes the ENVIRONMENT in which the revolution takes place. This is meant to provide contextual information, such as the physical terrain and geography, as well as socioeconomic and cultural factors that could influence or determine how the revolution occurred. This section also describes historical information that is important for highlighting motivational issues, such as long-lasting grievances, historical events that provide a narrative, or even social/cultural clashes that may color the events of today. Each revolution is reacting to a particular governing and political environment, which is described to explore the issues, organizations, or policies that bare upon the desired changes of the revolutionary movement. Finally, any weaknesses in the entire environment are described, as well as particular catalysts or events that directly surround or lead to the start of the revolutionary movement itself.

The third section explains the FORM of the revolution and describes the important CHARACTERISTICS of the movement. Its objectives and goals are delineated, both as originally conceived and any evolution or alteration that may occur during the time covered. The leadership and organizational structure are described, as well as the means and methodology of communications, both within the movement and with external actors. A large subsection covers the operational methods of the movement, usually covering events chronologically and placing them in the context of how the organization tries to achieve its goals, such as through the use of violence or protests. The means by which the movement recruits participants is summarized, identifying motivations for joining and the process by which recruits are made members of a functioning organization. Another subsection describes how the revolutionary movement sustains itself through resupply, finances, and logistics. Obtaining legitimacy is an additional important theme for the success of any revolutionary movement, so we also describe how the movement presents itself to the general population, the opposing government, and external actors that allows it to be taken seriously as a legitimate actor within the political realm. Any external support that is crucial to the success of the movement is then listed, followed by a brief section on how the government responded to the methods and actions of the movement.[5]

The final section summarizes the aftermath of the revolution itself, again covering the multiple environments around which the event took place and how the revolution attempted to change, or did change, any political, economic, or social conditions. The changes to the government itself are explained, as well as the changes that happened to the movement, whether it disappeared, became the ruling government, or perhaps even became a legitimate political player at the national level.

Each case concludes with a comparatively robust bibliography that encompasses the key sources used by the author(s) and all citations. The standard outline for each case is as follows:

SYNOPSIS

TIMELINE

THE ENVIRONMENT OF THE REVOLUTION

    Physical environment

    Cultural and demographic environment

    Socioeconomic environment

    Historical factors

    Governing environment

    Weaknesses of the prerevolutionary environment and catalysts

FORM AND CHARACTERISTICS OF THE REVOLUTION

    Objectives and goals

    Leadership and organizational structure

    Communications

    Methods of action and violence

    Methods of recruitment

    Methods of sustainment

    Methods of obtaining legitimacy

    External support

    Countermeasures taken by the government

---

[5] The purpose of this Casebook is to focus on the factors which contribute to the emergence and execution of a revolutionary movement within a given environment, not the study of the counter-insurgency efforts employed by a government. As such, the discussion on how a government responded to the methods and actions of the movement is kept somewhat brief but still included since it is a contextual aspect of the growth and transformation of a revolutionary movement.

SHORT- AND LONG-TERM EFFECTS

TYPES OF CASES INCLUDED

To understand how we selected these cases, it is important to understand how we define the terms *insurgency* and *revolutionary warfare*. For the purposes of this study, we define a *revolution* to be:

> An attempt to modify the existing political system at least partially through unconstitutional or illegal use of force or protest.

Insurgency or *revolutionary warfare*, then, is used to describe the means by which a revolution is attempted or achieved. Throughout this Casebook, we use the terms *insurgency* and *revolutionary warfare* interchangeably. *Insurgent* and *revolutionary* are therefore interchangeable as well.

The aforementioned definition of a revolution describes the desired end state (a modification of the existing political system), as well as the means to achieving that end state (unconstitutional or illegal use of force or protest). Therefore, a revolution does not necessarily encompass the usurpation of power by the insurgents but can be a modification of the existing system according to the group's demands. The revolutionary must also include some unconstitutional or illegal actions to achieve that end state. Here we take our cue from the original SORO Casebook in which the authors carefully point out that even though the government the revolution is directed against may have been unconstitutionally situated in the first place, this is irrelevant to our definition. The revolution may in fact be an attempt to restore the legitimate and constitutional system. For the purposes of this research, the de facto government's laws are what constitute legality, whether in violation of the country's previous constitution, international law, or other. Therefore, even the former power trying to reseat its constitutional system of laws against a dictatorship or despot is an insurgency in our parlance.

We have, however, had to depart from the previous Casebook in the matter of means. The original SORO study required that a revolution had to involve the use of, or the threat of use of, force. *Force* was open violence, guerrilla warfare, or even civil war within the military. Mere propaganda, protest, strikes, or even passive resistance did not meet the criteria in the previous study. For modern revolutionary warfare, we have decided to include examples of revolutions begotten by either entirely or predominately nonviolent means. The strikes and underground propaganda activities by Solidarity in Poland constitute a viable revolution by our criteria, as does the Orange Revolution. These "velvet" revolutions can teach the student of modern warfare how mass protests, coordinated propaganda and information campaigns, and coercive yet nonviolent means can topple an entrenched governmental system. The revolutionary must decide whether a nonviolent campaign could be more effective by understanding its pros and cons and also recent successes and failures of such attempts. Therefore, this Casebook includes examples of both violent and nonviolent revolutionary warfare.

Our selection of cases for inclusion was determined by three primary factors. First, we wished to contain ourselves as much as possible with discrete cases where the revolution could be delineated from surrounding chaos or other events and could be said to have reached a major transition point in the success or failure of the movement. Because of current high-interest revolutionary movements, however, we have included some cases that are difficult to define independently and discretely. Our justification is that these cases can illuminate various aspects of the form and characteristics of modern insurgency.

We also wished to select cases that represent a diversity across multiple criteria. The cases included span major geographic regions of the world, cover each decade from the 1960s to 2000s, cover at least five different primary motivations for their existence, and proffer a wide range of outcomes from complete success to near annihilation. We felt such a wide range of cases allowed for more interesting comparisons and analysis of variation.

Our final criterion for inclusion in this Casebook was that the revolution had to have sufficient source material in the unclassified academic domain as to allow us to complete each section of the research framework. Therefore, campaigns that were more minor or quickly unsuccessful have been given little consideration. We caution any reader from using this selected set of cases to be statistically similar to the entire set of insurgencies during this time period, for we have not attempted to cover the range of movement *size* or campaign *duration*.

An accompanying set of larger studies have been developed to provide the reader with a more in-depth treatment of six cases summarized in this volume. These larger studies are found in separate volumes and cover the following cases: Taliban, FARC, LTTE, Viet Cong, Hizbollah, and the Provisional IRA.

METHODOLOGICAL NOTES

A few methodological explanations should be stated. First, we are indebted to the reviewers who graciously reviewed and commented on drafts of individual cases at our request. Cases were given to noted subject-matter experts in academia and the military and these experts were asked to review the cases for accuracy and relevance and to ensure that all important issues were treated. While we tried our best to incorporate and interpret all of their suggestions, any remaining errors and omissions are purely the responsibility of our team.

We selected sources that provided the most authoritative and unbiased research we could obtain. We favored secondary accounts by reliable and trustworthy historians or political scientists, and we used primary sources as much as possible to understand motivations, objectives, and behaviors of the participants. We were limited by language and classification; hence, all of our sources were unclassified and almost all were in English.

SECTION I

....

REVOLUTION TO MODIFY
THE TYPE OF GOVERNMENT

GENERAL DISCUSSION

The first type of rebellion is perhaps best understood from an academic viewpoint and constitutes the majority of the cases studied in the original SORO Casebook. The desire to change the style of a nation's government has been a cause for revolution dating back millennia and encompasses the more celebrated revolutions in history, such as the American revolt in the eighteenth century, the ensuing French overthrow of their monarchy, the pursuit of Lenin and Mao to establish communist states, and so forth.

The desire to overthrow and replace an entire system of government requires two crucial factors: (1) an existing type of government that does not meet the needs of the population, or at least does not meet the needs of powerful segments of the population, and (2) an ideologically mature alternative government to be proposed and rallied around. The first factor requires that the populace, or again important segments, such as the political elite, the merchant class, blue-collar workers, academia, etc., be unhappy in some fashion with their present lot. Their displeasure may be rooted in socioeconomic conditions, such as an economic depression, lack of political influence or representation, and large disparities between upper and lower economic classes, or it may be displeasure with the actions of the government, such as excessive taxation, harsh security measures such as curfews and martial law, or even dramatic changes in policy such as the introduction of modern economic measures or restriction of goods and services once provided. This motivated segment of the population becomes the recruitment base for the fomenting revolution, and support networks often operate purely within those social groups throughout the revolution, rarely expanding into other segments unless the revolution truly becomes a populist wave. In many examples this group is a well-educated segment of the population, and their relative disillusionment or disenfranchisement with the government is the motivating factor.

The second factor for a revolution to modify the government type is an ideologically different government system that has either worked in other nations or has reached a level of theoretical maturity such that it can convince a large group of people that it would be successful. In most of the cases, both in the earlier Special Operations Research Office (SORO) Casebook, as well as in this section, the ideological form of government is a well-established socialist construct, usually traceable to some form of pure Marxism or its main variants where the population would benefit from a less-capitalistic economic system to one that is more communal and centrally governed. The Communist revolutions of the mid-twentieth century have ceased to be an effective rallying cry for most of the geographic regions, but three Central and South American revolutions and one from the Philippines are used in this section to describe the modern socialist-style revolution and the adaptations that have been made since the heyday of Che Guevara and Fidel Castro. The other two revolutions, 1979 Iran and Burma, are descriptions of how revolt can occur against a powerful dictatorial regime. Iran provides a fascinating example of how three different segments of the population each had their own choice for an alternative governance but worked together to overthrow the monarchy of the Shah. The cunning and sheer popularity of the Ayatollah Khomeini allowed him to implement his vision of Islamic-based governance over the constitutionalist middle-class and Marxist university students.

It is evident that the combination of the two factors allows one to easily see these types of revolutions in terms of class-based warfare. The discontented are often responding to conditions related to their economic or political prowess. The resonance of a new government type can often appeal within a specific class, such as the intellectual environs of the university for some socialist concepts, the peasant-based approach of Maoism, or the exploited workers approach of Marxism. We have tried to point out these class issues as they have been used within the revolution itself, as well as the analysis from this point of view in historical works on these cases. However, we also emphasize that the local conditions and contexts of each of these cases ensure that they play out in very unique and sometimes surprising ways. The Fuerzas Armada Revolucionarias de Colombia (FARC) has had to modify its message and purpose to adjust to changing conditions, while the New People's Army (NPA) failed to do so, each deviating from the standard communist revolutionary template that was prevalent in the 1960s. A class-based viewpoint allows for good analysis but cannot give a student the complete picture of how a revolution occurs (or does not occur).

NEW PEOPLE'S ARMY (NPA)

Ron Buikema and Matt Burger

SYNOPSIS

The New People's Army (NPA) insurgency in the Republic of the Philippines became a major threat to the security and stability of the government, particularly during the tenure of Presidents Ferdinand Marcos and Corazon Aquino. The NPA effectively mobilized support from the economically disadvantaged, who believed that the government had either done nothing to support their economic situation or that it had become an enemy to the common man, through its corruption, greed, and economic abuse of the masses. The NPA was successful in establishing operations in almost all of the 73 Philippine provinces, conducting extensive urban operations and by the mid-1980s actually controlling approximately 12% of the *barangays*[1] nationwide.[2] NPA's success was due to the establishment and execution of an effective national strategy that strained civil–military relations in addition to the slow, reactive, and uncoordinated response by the Philippine government to the insurgency.

This study covers the NPA from its inception in 1969 until 2009 but focuses on the period of 1969–1992, highlighting the formation, strategic planning, operational execution, continued growth and "high-water mark" of the late 1980s; then into a steady decline in popular support and capabilities in the early 1990s, resulting from political infighting, increased counterinsurgency actions by the Philippine government, and the Second Great Rectification Movement. The NPA was still an active insurgency at the time of this writing, with a presence in numerous Philippine provinces.

---

[1] *Barangays* refer to the smallest administrative organization within the Philippine government. The Spanish equivalent is a *barrio*.

[2] Larry A. Niksch, *Insurgency and Counterinsurgency in the Philippines*, Prepared for the Committee on Foreign Relations United States Senate by The Foreign Affairs and National Defense Division-Congressional Research Service, Library of Congress (Washington, DC: US Government Printing Office, 1985).

TIMELINE

1969	NPA formed in Central Luzon. Limited tactical ambush capability only.
1972	Marcos declares martial law. New recruits flock to join NPA, provide support to the NPA movement, or both; movement widely seen as the only option to the corrupt Marcos regime. NPA expands influence to provinces throughout the country.
1986	Corazon Aquino elected president. Path to democratic reforms initiated (NPA fails to support the election). Professionalization of the armed forces campaign instituted.
1988	First rectification purge commences as a result of failed policies post-Marcos, internal power struggles, and lack of strategic direction.
1988, 1989	NPA high-water mark, approximately 25,000 combatants, company-sized attacks against Philippine security forces, complex urban operations in Manila, attacks on government infrastructure and hard targets, political assassinations.
1992	US bases, including Clark Air Base and Subic Bay Naval Station, close after failure to renegotiate status of forces agreement (SOFA) and massive damage caused by Mount Pinatubo volcano eruption.
1992–1998	Second rectification purge, Jose Maria Sison's attempt to regain control—thousands of NPA leaders killed.
2009	NPA remains active at the small-unit level in several provinces, focusing principally on propaganda activities. Political assassinations, kidnappings, and ambushes of security forces remain common.

THE ENVIRONMENT OF THE REVOLUTION

PHYSICAL ENVIRONMENT

The Philippines are made up of more than 7,100 islands in Southeast Asia, with a land mass slightly larger than the state of Arizona. The archipelago has a tropical climate. The Philippines have a number of active volcanoes, and key terrain throughout the islands is dominated by both active and dormant volcanic mountains. Terrain composition includes mountains and valleys, with channelized terrain and micro-terrain features, as well as limited lines of communication, particularly on the smaller, less-populated islands. The NPA cited terrain as a factor in their protracted war strategy. Terrain also limited the NPA's ability to amass forces and centralize infrastructure, becoming both an advantage and disadvantage to their cause.[3]

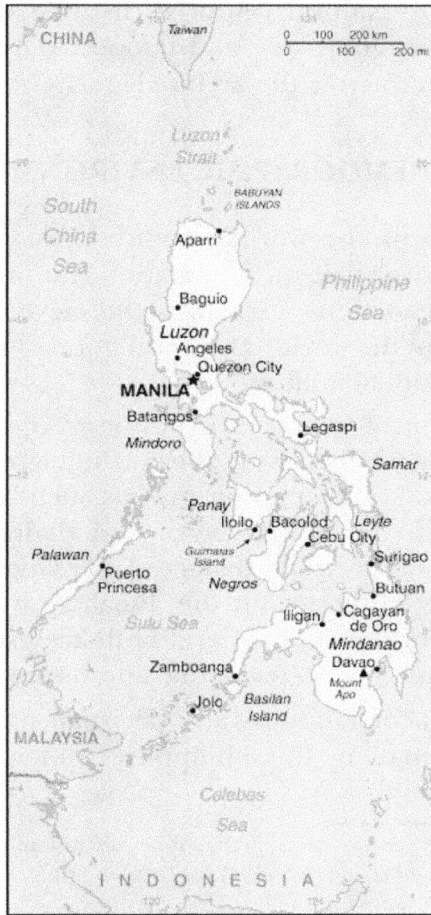

Figure 1. Map of the Philippines.[4]

[3] Gregg Jones, *The Red Revolution: Inside the Philippine Guerrilla Movement* (Boulder, CO: Westview, 1989); Michael J. Montesano, "The Philippines in 2003: Troubles, None of Them New," *Asian Survey* 44, no. 1 (January–February 2004): 93–101; Gary Hawes, "Theories of Peasant Revolution: A Critique and Contribution from the Philippines," *World Politics* 42, no. 2 (January 1990): 261–298; William Chapman, *Inside the Philippine Revolution* (New York: W.W. Norton, 1987); Ava Patricia C. Avila, *Midlife Crisis of the Philippine Red Movement* (Singapore: S. Rajaratnam School of International Studies, NTU, 2008); Maria J. Stephan and Erica Chenoweth, "Why Civil Resistance Works; the Strategic Logic of Nonviolent Conflict," *International Security* 33, no. 1 (2008): 7–44; Dev Nathan, "Armed Struggle in Philippines," *Economic and Political Weekly* 22, no. 51 (December 19, 1987): 2201–2203; F. A. Mediansky, "The New People's Army: A Nationwide Insurgency in the Philippines," *Contemporary Southeast Asia* 8, no. 1 (1986): 1–17.

[4] Central Intelligence Agency, "Philippines," *The World Factbook*, accessed March 15, 2011, https://www.cia.gov/library/publications/the-world-factbook/maps/maptemplate_rp.html.

Approximately 19% of the land is arable, with the majority of agricultural activity being subsistence farming.[5] Lines of communication vary from province to province, with dirt roads and paths predominating in rural and economically impoverished areas. Traffic between islands is routinely conducted via ferry and small boats, called *bancas*. Natural resources include timber, petroleum, nickel, cobalt, silver, gold, salt, and copper.[6] The NPA operates in nearly every province across the archipelago.

CULTURAL AND DEMOGRAPHIC ENVIRONMENT

The Philippines is the world's twelfth most populated country (92 million), with over half the population residing on the island of Luzon. Manila, the nation's capital, is the eleventh most populous metropolitan area in the world. The population of the Greater Manila Metro Area is around 20 million.[7]

Filipinos are a mix of several Asian ethnic groups but can be broadly categorized as Malayo-Polynesian. These ethnic groups include tribal, nontribal, aboriginal, and migrant groups such as Chinese.[8] There are more than 180 native languages and dialects spoken in the Philippines. Filipino, an urban dialect of the language spoke by the Philippines largest ethnic group the Tagalog, and English are the official languages[9] of the country, and both are used in government, education, entertainment, news media, and business.

More than 90% of Filipinos are Christian;[10] of those, 80% are Roman Catholic, making the Philippines one of only two major

---

[5] See *Socioeconomic Environment* section for a fuller discussion.

[6] See Central Intelligence Agency, *The World Factbook*, http://www.cia.gov/library/publications/the-world-factbook.

[7] Census Bureau of the Philippines (2009).

[8] Ibid.; Rodolfo Severino, Lorraine Carlos Salazar, and Institute of Southeast Asian Studies, *Whither the Philippines in the 21st Century?* (Singapore: Konrad Adenauer Stiftung, Institute of Southeast Asian Studies, 2007).

[9] Bicolano (3.7 million speakers), Cebuano (20 million), Ilocano (7.7 million), Hiligaynon (Ilonggo) (7 million), Kapampangan (2.9 million), Pangasinan (2.4 million), and Waray-Waray (3.1 million) are also recognized as official languages by the constitution, while Spanish and Arabic are recognized as auxiliary languages. Census Bureau of the Philippines.

[10] The remaining 10% are Protestants or nontraditional Christians belonging to denominations including the Philippine Independent Church, Iglesia ni Christo, The Church of Jesus Christ of Latter-Day Saints (Mormon), the Seventh Day Adventists, the United Church of Christ, and the Orthodox Church. About 5% of the population is Muslim, the majority of whom are members of the Moro people living in western Mindanao, Palawan, and the Sulu Archipelago. Particularly among tribal groups, Animism, Shamanism, and folk religions are still practiced, while Buddhism and Taoism are practiced among ethnic Chinese communities. Other minor religions include Baha'i, Hinduism,

Catholic nations in Asia.[11] The Catholic Church has remained a major force in political and social life in the Philippines.[12]

Philippine culture is a cornucopia of Eastern and Western influences. Eastern influences have come from China and Malaysia, while the colonial legacies of Spain and the United States have left a distinctly Western mark on Philippine culture.[13] Concurrently, the dominant use of English, as well as American music, film, and television, is a lasting legacy of the presence of the United States in the Philippines. The Philippines has a Western education and university system. Philippine culture is marked by an emphasis on family life and loyalty, a trait that has been reinforced by the Catholic Church.[14]

SOCIOECONOMIC ENVIRONMENT

Four forces worked to stifle the socioeconomic development of the Philippines during the second half of the twentieth century. These forces included: (1) the corruption of the government of President Ferdinand Marcos (1965–1986); (2) the failure to develop an export economy; (3) the persistent insurgent threat, which made the Philippines an unattractive place for foreign economic investment; and (4) agricultural feudalism. Much of the rural Philippines is defined by large *haciendas* (plantations).[15] Tenant farmers work (manual labor) for planter families who, in most instances, live either in Manila or the province's town.[16] The plight of these farmers became a central cause and base of support for the NPA, with demands for reduced land rents and the elimination of usury (i.e., high interest

---

Judaism, and atheism/no religion. US Department of State, *Philippines: International Religious Freedom Report 2008* (2008).

[11] East Timor is the only other majority Roman Catholic country in Asia.

[12] Kathleen M. Nadeau, *Liberation Theology in the Philippines: Faith in a Revolution* (Westport, CT: Greenwood Publishing Group, 2002); Robert Youngblood, "Structural Imperialism: An Analysis of the Catholic Bishops' Conference of the Philippines," *Comparative Political Studies* 15, no. 1 (1982): 29–56; Robert L. Youngblood, "The Corazon Aquino 'Miracle' and the Philippine Churches," *Asian Survey* 27, no. 12 (December 1987): 1240–1255.

[13] Paul A. Rodell, *Culture and Customs of the Philippines* (Westport, CT: Greenwood Press, 2002).

[14] Nadeau, *Liberation Theology in the Philippines*.

[15] Some *haciendas* are as small as 20 hectares and other as large as 1,000 hectares; most, however, are between 20 and 150 hectares (50–100 acres). Rosanne Rutten, "High-Cost Activism and the Worker Household: Interests, Commitment, and the Costs of Revolutionary Activism in a Philippine Plantation Region," *Theory and Society* 29, no. 2 (April 2000): 215–252.

[16] As of 2007 about 40% of Filipinos earned less than $2 a day. Andrew Marshall, "The War with no End," *Time Magazine* (January 25, 2007); Severino, Salazar, and Institute of Southeast Asian Studies, *Whither the Philippines*.

charged by landowners to peasant farmers); increased agricultural productivity via community cooperation; the formation of peasant associations to advocate for worker rights; the assurance of fair prices for agriculture produce; and the opposition of the illegal takeover of peasant-owned land by wealthy planter families.[17] The NPA demonstrated both empathy for the plight of the peasant farming community, as well as an option for their situation—revolution and overthrow of the government.

The NPA persisted in the Philippines not because of a broad commitment to a Communist or Maoist ideology, but because of the reality of poverty, government corruption, and unemployment. Until the late 1960s, the Philippines enjoyed the second-highest standard of living among Asian nations, second only to Japan. Yet by the 1980s, it had been surpassed by South Korea, Taiwan, Hong Kong, Singapore, Malaysia, Thailand, and Indonesia. During the 1970s and 1980s, these other nations intentionally industrialized and created an export economy, yet this type of development was stifled in the Philippines because of (1) corruption by President Marcos; (2) broader government corruption; (3) government encouragement of import substitution rather than an export economy;[18] and (4) persistent pressure from unions, leading to an exit of foreign capital.[19]

It was not until the Philippines emerged from the Marcos regime in the 1990s that serious development began again. By then, however,

---

[17] By 1988, however, the National Democratic Front (NDF), an alliance of left-leaning organizations that serve as the political front for the CPP-NPA, noting the inadequacy of these former demands, advocated the wholesale seizure and redistribution of land along Maoist lines. Jones, *The Red Revolution*; Peter R. Kann, "The Philippines without Democracy," *Foreign Affairs* 52, no. 3 (April 1974): 612–632; David Rosenberg, "Communism in the Philippines," *Problem of Communism* 33, no. 5: 24–46 (1984); Robert A. Manning, "The Philippines in Crisis," *Foreign Affairs* 63, no. 2 (Winter 1984): 392–410.

[18] For instance, during the late 1950s the Philippines passed the Minimum Wage Law, which deterred US and other foreign companies from locating in the Philippines, preferring Taiwan and Hong Kong, which had no such statutory guarantees. Antonio C. Abaya, "Defeating the Communists III," *Manila Standard* (February 15, 2007); Antonio C. Abaya, "Defeating the Communists II," *Manila Standard* (February 8, 2007); A. M. Balisacan and Hal Hill, *The Philippine Economy: Development, Policies, and Challenges* (Oxford: Oxford University Press, 2003).

[19] Abaya, "Defeating the Communists III"; Abaya, "Defeating the Communists II"; Severino, Salazar, and Institute of Southeast Asian Studies, *Whither the Philippines*; Balisacan and Hill, *The Philippine Economy*; Misagh Parsa, *States, Ideologies, and Social Revolutions: A Comparative Analysis of Iran, Nicaragua, and the Philippines* (Cambridge: Cambridge University Press, 2000); Ben Reid, *Philippine Left: Political Crisis and Social Change* (Manila, Philippine: Journal of Contemporary Asia Publishers, 2000); Carlos H. Conde, "Peace Effort with Philippine Rebels Breaks Down," *The International Herald Tribune* (August 5, 2005); Carlos H. Conde, "Philippines Insurgency Takes Toll; Economic Strategy could be Undercut," *The International Herald Tribune* (November 23, 2005).

the Philippines had fallen far behind their neighbors in terms of export dollars, particularly with the rise of China as a major exporter.[20]

HISTORICAL FACTORS

Several historical forces coalesced to create the NPA as a persistent and pervasive threat in the Philippines. These factors included: (1) the prior existence of an armed insurgency (i.e., the communist Hukbalahap or Huk); (2) an educated leadership rooted in the left-wing university student movement of the late 1960s; (3) economic factors (i.e., a large peasant population); (4) political factors (i.e., government corruption and autocracy);[21] and (5) popular support.

The Hukbalahap (Huk) developed as an armed resistance to the Japanese occupation of the Philippines. After the war the group continued its communist-based guerrilla campaign over issues of unequal land ownership. With five provinces under Huk control by 1950, the Philippine government launched a vigorous military campaign. In 1954, the Huk leader Luis Taruc voluntarily surrendered, after which the movement died out. The grievances that had motivated the Huk, however, remained.[22]

During the late 1960s, Philippine universities across the country were rocked by mass student protests dubbed the "First Quarter Storm." Amid political and cultural turmoil, the Communist Party of Philippines (CPP)-Mao Tse-Tung Thought (MTT) was formed by middle-class undergraduates in December 1968. Most of these protests were centered on the University of the Philippines, where a young academic, Jose Maria Sison, the primary founder of the new party, asserted that as a "semi-feudal, semi-colonial country" the Philippines was ideal for revolution.[23] Sison included US imperialism (e.g., the war in Vietnam), US colonialism (e.g., perceived US interference in domestic politics), rising oil prices, social and economic injustice, and government corruption in his list of grievances against the people.[24]

---

[20] 2005 export totals by country: China ($850 billion), Japan ($700 billion), South Korea ($288.2 billion), Singapore ($204.8 billion), Taiwan ($189.4 billion, Malaysia ($147.1 billion), Thailand ($105.8 billion), Indonesia ($83.6 billion), Philippines ($41.3 billion); Antonio C. Abaya, "Defeating the Communists," *Manila Standard* (February 1, 2007); Gregory C. Chow, *China's Economic Transformation* (Malden, MA: Blackwell Publishers, 2007).

[21] See *Governing Environment* for a fuller discussion.

[22] Jones, *The Red Revolution*; Mediansky, *The New People's Army*, 1–17.

[23] Jose Maria Sison, *Philippine Society and Revolution* (Manila, Philippines: Pulang Tala, 1971); Jose Maria Sison, *Struggle for National Democracy* (Quezon City, Philippines: Progressive Publications, 1967).

[24] Kann, "The Philippines without Democracy"; Rosenberg, "Communism in the Philippines"; Manning, "The Philippines in Crisis."

Sison had broken with the older Communist Party of the Philippines (PKP), which was based on a Soviet/Leninist model rather than Maoism. Sison believed the older party had become complacent. Influenced by the Great Proletarian Cultural Revolution, which was rocking China, Sison merged elements of Maoism and new nationalism to advocate a national democratic revolution along Maoist lines. The NPA was created from the remnants of the Huk, whose leader Bernabe Buscayno (also known as Commander Dante), had also become disenchanted with the PKP. Sison took the leadership of CPP, while Buscayno headed the NPA.[25]

Following the Maoist model, Buscayno initiated a "Protracted People's War,"[26] with a small band of guerrillas armed with only ten rifles. Politically, Sison believed that in a society where the middle class included both urban bourgeoisie and rural landowners, an alliance with this class was unimportant. He favored capturing the cities from the countryside via the protracted people's war.[27] Initially, the NPA was nearly destroyed by the Armed Forces of the Philippines. The NPA regrouped and established a revolutionary base in Isabela Province. Its ranks soon swelled after the violent crackdown of anti-government forces by the Marcos regime in the early 1970s.[28] This is exemplified by the 1970 defection of Lieutenant Victor Corpuz, who, disenchanted with the corrupt and undemocratic Marcos regime, took several automatic weapons from the armory of the Philippine Military Academy and joined the rebels.

The growth of liberation theology in the Philippines' Catholic Church paralleled that of the NPA. The movement, which mixes Christian theology and Marxist ideology, was considered a threat by both the government and the church.[29] The state began to persecute the fledgling Christian Base Communities (CBC),[30] accusing them of being a front for the NPA. In addition to being self-reliant communities

---

[25] Mediansky, *The New People's Army*; Jose P. Magno Jr. and A. James Gregor, "Insurgency and Counterinsurgency in the Philippines," *Asian Survey* 26, no. 5 (May 1986): 501–517; Alexander R. Magno, "A Nation Reborn," vol. 9 of *Kasaysayan: The Story of the Filipino People* (Manila, Philippines: Asia Publishing Co., 1998); David Wurfel, *Filipino Politics: Development and Decay* (Ithaca, NY: Cornell University Press, 1988); Jones, *The Red Revolution*.

[26] The Protracted People's War is a Maoist strategy for a long-term revolutionary struggle in which enemy forces are drawn into rural areas where they can be slowly decimated by mobile guerrilla fighters supported by the local population. The strategy has been advocated and employed frequently throughout the twentieth century, namely in the Cuban Revolution, the Vietnam War, the Sendero Luminoso in Peru, the FARC-ELN in Colombia, the Maoist in Nepal, and numerous others.

[27] Ibid.

[28] Ibid.; Wurfel, *Filipino Politics*; Mediansky, *The New People's Army*.

[29] Nadeau, *Liberation Theology in the Philippines*; Youngblood, "Structural Imperialism."

[30] Christian Base Communities were organized by local Catholic clergy as self-reliant religion communes as a response to onerous working and living conditions on *haciendas*.

of worship, the CBC served as mechanisms for organizing peasants to voice their grievances against the planters and the state.[31] Numerous priests, nuns, and lay leaders were harassed, imprisoned, and murdered by the military.[32] These oppressive tactics, along with the broader injustices under martial law, drove some clergy to join the insurgency,[33] while concurrently cultivating less direct support among the broader church.[34]

A related event through which the NPA gained popular support was the formation of labor unions among the *haciendas*, after the declaration of martial law. The fledgling peasant organizations were transformed into chapters of the newly formed militant labor union, the National Federation of Sugar Workers (NFSW). The NFSW led to broad popular support among hacienda workers for the NPA, providing organization and a clear articulation of their grievances.[35]

GOVERNING ENVIRONMENT

Having liberated the Philippines at the conclusion of the World War II, the United States reestablished an independent democratic nation. In 1969, Ferdinand Marcos was reelected to a second term, a first in Philippine history. Initially, Marcos undertook massive public works projects, such as the Pan-Philippine Highway, and also tried to tackle corruption.[36] In 1972, after the seizure of weapons from China bound for the NPA and the bombing of an opposition political rally in Manila in 1971, Marcos declared martial law.[37] This allowed him to circumvent the legislature, while his regime became increasingly autocratic and corrupt.

With continued US support for his regime, Marcos centralized power in the executive branch and began to accumulate personal wealth by embezzling funds from state monopolies, US aid, and

---

[31] Rutten, "High-Cost Activism and the Worker Household."

[32] David Kowalewski, "Cultism, Insurgency, and Vigilantism in the Philippines," *Sociological Analysis* 52, no. 3: 241–253 (1991).

[33] Father Conrado Balwag formed the Cordillera People's Liberation Army, which allied itself with the NPA against the Marcos regime. More than 50 priests joined the NPA directly, creating the moniker "New Priests Army." Nadeau, *Liberation Theology in the Philippines*.

[34] Rosenberg, "Communism in the Philippines."

[35] Rutten, "High-Cost Activism and the Worker Household."

[36] Wurfel, *Filipino Politics*.

[37] As a major US ally in a region beset with communists threats, such as China, North Korea, and Indochina, successive US administrations supported Marcos, particularly as he couched his autocracy as a necessary means to defeat the communists. Olle Törnquist, "Communists and Democracy in the Philippines," *Economic and Political Weekly* 26, no. 27/28 (July 6–13, 1991): 1683–1691.

loans from international financial institutions.[38] Accusing political opposition of aiding the communists, Marcos seized their assets and had many imprisoned. Leaders of mainstream opposition parties were either arrested or silenced, stifling an organized aboveground opposition to the regime.[39] The closed political environment forced political opposition into a cool alliance with the NPA.[40] This alliance served to bolster both the communist's domestic and international financial support and its legitimacy.[41]

Although the NPA continued to gain strength during this period, the Marcos government launched several military campaigns against them. Between 1972 and 1977, fighting against the Moro National Liberation Front (MNLF) in Mindanao was considered a greater immediate threat to the security of the regime—critical security resources were fighting the MNLF, not the NPA. After 1977, as the NPA's strength exploded, the regime intensified operations against it. In most instances, however, the government did just enough to keep the insurgency at bay. Although there were notable victories, such as the capture of NPA leaders Victor Corpuz and Bernabe Buscayno, as well as CPP-NPA founder Jose Maria Sison, any real movement to eliminate the NPA was infeasible due to a lack of political will in the Marcos regime, exacerbated by widespread corruption in the government and armed forces.[42]

WEAKNESSES OF THE PREREVOLUTIONARY ENVIRONMENT AND CATALYSTS

One catalyst for the NPA was clearly the establishment of permanent US military installations at Subic Bay and Clark. Numerous anti-US, anti-Philippine government demonstrations were held in Manila during the 1965–1966 time frame. The bases were regarded as a symbol of repression.

---

[38] Stephan and Chenoweth, *Why Civil Resistance Works*, 7–44.

[39] Ibid.

[40] The NDF served as the umbrella through which organizations forced underground by martial law, such as the Patriotic Movement of New Women and Christians for National Liberation, could unite in opposition to Marcos.

[41] The CPP exploited the human rights violations by the Marcos regime to recruit, as well as to gain support financially and otherwise, internationally and among the Philippine middle class. By 1977, human rights abuses were widespread, as evident by the annual averages of 30 disappearances and 50 summary executions, as well as the more than 70,000 Filipinos arrested for political reasons since the imposition of martial law. Andrew Tian Huat Tan, ed., *A Handbook of Terrorism and Insurgency in Southeast Asia* (Northampton, MA: Edward Elgar Publishing, 2007); Gareth Porter, *The Politics of Counterinsurgency in the Philippines: Military and Political Options* (Honolulu: University of Hawaii, 1987).

[42] Wurfel, *Filipino Politics*; Magno, "A Nation Reborn."

The Chinese Cultural Revolution also served as a catalyst for the CPP-NPA. Student groups formed "Serve the People Brigades" and dispersed to the countryside, sharing snippets of Mao doctrine with farmers while learning more about the lives of the "peasants." Gregg Jones, in *Red Revolution*, notes "For idealistic students, the experience was one more step in the transition from 'bourgeois nationalists' to Marxist revolutionaries."[43]

The single most destructive, pervasive, and enduring characteristic of the government of the Philippines was corruption. Political corruption reached a zenith during the regime of President Ferdinand Marcos (1965–1986). Marcos's policies were considered repressive by the large working class and peasants. These policies (including the resulting corruption) served as the major catalyst of the insurgency. However, when the Marcos regime was removed from power, a weakness of the insurgency was recognized. The peasant and working classes began, over time, to see peaceful opportunities for economic advancement within the existing political process. The most powerful recruiting tool of the insurgency, Ferdinand Marcos, was gone.

Another catalyst, US presence via Clark Air Base and Subic Bay Naval Station, was removed with the destructive eruption of Mount Pinatubo during June 1991. While talks were ongoing between the US and Philippine governments to renegotiate the SOFA and usage of military facilities, the damage to both installations, particularly Clark Air Base, meant that rebuilding costs were economically unacceptable to the US government. Suddenly, the US presence was leaving the Republic of the Philippines, and with that another founding principle of the NPA was removed.

FORM AND CHARACTERISTICS OF THE REVOLUTION

OBJECTIVES AND GOALS

As the NPA Commander-in-Chief, Commander Dante worked closely with Sison and the CPP leadership to establish military objectives and goals that supported the overall political objectives. Sison had envisioned an NPA army based on the Chinese model, and Maoist military strategy was mirrored as much as possible. The resulting military strategy was built on establishing a main army on Luzon, while smaller units would be established on other islands in order to

---

[43] Jones, *The Red Revolution.*

disperse Philippine security forces as much as possible, denying them the ability to amass forces. Sison described the protracted people's war in three stages: (1) establish a strategic defense in order to build up forces and establish basic military capabilities; (2) reach parity with government forces; and (3) conduct offensive operations in order to attack isolated security forces. By 1985, the NPA believed that they were only two years away from establishing a strategic stalemate; and that a stalemate had already been achieved on the operationally important stronghold of Mindanao.[44]

LEADERSHIP AND ORGANIZATIONAL STRUCTURE

The insurgency consisted of three basic structures: the National Democratic Front (NDF) formed in 1973, the CPP formed in 1968, and the NPA formed in 1969. Sison described the communist movement as a Warrior with a Sword (i.e., the NPA) that strikes blows against the enemy and a shield (i.e., the NDF) to protect the movement from enemy blows.[45] The CPP was composed primarily of urban, middle-class university-educated intellectuals. Its leadership was highly centralized, with a small group virtually running the CPP by decree.

The NPA is the armed wing of the CPP and is supervised through the Central Committee and Military Commission of the CPP, which monitors and directs local NPA activity to ensure adherence to the party.[46] While NPA units are relatively autonomous, they do operate

---

[44] The NPA also adopted Mao's vision of "iron discipline," and a code of conduct was enacted that established the following: Three Main Rules of Discipline—Obey orders in all your actions. Do not take a single needle or piece of thread from the masses. Turn in everything captured; Eight Points of Attention—Speak politely. Pay fairly for what you buy. Return everything you borrow. Pay for everything you damage. Do not hit or swear at people. Do not damage crops. Do not take liberties with women. Do not ill-treat captives. Teodoro Agoncillo, *A Short History of the Philippines* (New York: Mentor Publishing, 1975); Niksch, *Insurgency and Counterinsurgency in the Philippines*. The CPP-NPA strategy worked initially, with the rapid growth far exceeding projected estimates. The Congressional Research Service noted in a brief for the Senate Foreign Relations Committee during 1985 that "The CPP has established supportive links with people in the towns and barangays, as traditional support mechanisms for people have eroded. It offers status and a new kind of group security and camaraderie to people like the unemployed and displaced, who otherwise face bleak prospects. It has convinced sizeable numbers of Filipinos that the government at various levels is responsible for their individual problems and grievances or that the government institutions will do nothing to help them. It apparently has persuaded many that they can act effectively against the institutions and circumstances hurting them if they join the insurgents." Ibid. The insurgent leadership would have heartily agreed with the Congressional Research Service assessment.

[45] Abaya, "Defeating the Communists."

[46] Tan, *Handbook of Terrorism and Insurgency in Southeast Asia.*

under a strict code of conduct prescribed over the years by the CPP central committees, as well as Sison himself.[47]

Burgeoning numbers and a series of tactical victories prompted the CPP-NPA leadership to begin moving from the strategic defensive stage to the strategic offensive stage in 1981. In 1983, numerous small irregular guerrilla bands were formed into five companies. In 1985, the number of companies increased to fourteen, with two battalions in Samar and Northern Luzon, respectively.[48] This move, however, demonstrated the reality that the NPA remained inferior to the Philippines armed forces.[49] The increased unit size meant greater visibility and financial cost, leading to a series of military defeats.[50] Since the late 1980s, the NPA has favored small mobile units. Among these small units are the "sparrow death squad" units, which are trained and employed in attacks on urban police and military installations in order to capture weapons; these units also conduct assassinations and punitive killings for counter-revolutionary activities, such as failure to pay revolutionary taxes or cooperation with government forces.

No one has been a greater guiding force in the ideological and strategic development of the organization than Jose Maria Sison.[51]

---

[47] Every aspect of NPA is regulated. Robert Francis Garcia, *To Suffer Thy Comrades: How the Revolution Decimated its Own* (Quezon City, Philippines: Anvil Publishing, 2001); Hawes, "Theories of Peasant Revolution"; Chapman, *Inside the Philippine Revolution*. For instance, NPA members are discouraged from fraternizing with the opposite sex. Request for marriage must be made to party officials, premarital sex is forbidden, and couples who do request marriage must engage in a multiple-year courtship process. Should a marriage be permitted, both parties must have an equal commitment to the movement and undergo a Party Marriage, by which they pledge to place their primary commitment to the movement above that to their spouse. These rules are set down in a CPP document "On the Proletarian Relationship of the Sexes." Marshall, "The War with no End"; Rutten, "High-Cost Activism and the Worker Household." Even daily life is strictly prescribed. NPA members rise at 4 a.m. for military drills, including martial arts training. Every day follows a specific schedule including military and medical training, basic education, and indoctrination in the Maoist ideology. Weekends are typically reserved for recreation and food production. Alcohol is banned. Recreation times may consist of fighters gathering with a guitar to sing revolutionary songs, including the NPA's anthem. Garcia, *To Suffer Thy Comrades*; Jones, *The Red Revolution*; Chapman, *Inside the Philippine Revolution*.

[48] Tan, *Handbook of Terrorism and Insurgency in Southeast Asia.*

[49] Ibid.

[50] Sison himself admitted the failure of this move in "Reaffirm Our Basic Principles and Rectify Error" (*Kasarinlan: Philippine Journal of Third World Studies* 8, no. 1 [1992]: 96–157). Written under the name Armando Liwanag, one of at least two pseudonyms, Sison claimed that the enlargement of NPA units was premature, noting that absence of secure bases from which to launch large operations. However, there was also a political element at play. Facing challenges to his leadership in the late 1980s and early 1990s, Sison employed bloody purges to break up these large NPA formations, which were perceived as a challenge to his leadership. Severino, Salazar, and Institute of Southeast Asian Studies, *Whither the Philippines.*

[51] Although there are other notable leaders including former NPA head Bernabe Buscayno, who like Sison, was arrested in 1976 and released under the amnesty program

Sison is not as charismatic as Fidel Castro or as intellectual as Ho Chi Minh or Mao Tse-Tung;[52] rather, his leadership was more about persistent, ruthless, and quixotic adherence to Maoist orthodoxy. The Philippine revolution has been propelled more by chance and circumstance than by powerful personalities, and historical accident seems to have played a large role.[53] After founding and leading the CPP, Sison was imprisoned by the Marcos regime in 1976. The fall of Marcos led to Sison's release under the amnesty program of President Corazon Aquino (1986–1992). After a leadership struggle in the late 1980s and early 1990s, Sison was firmly reentrenched as CPP head.[54]

In 1986, with the loss of US support, President Marcos was forced to resign and leave the country following a failed attempt to nullify democratic elections.[55] A new democratic regime, under President Corazon Aquino, came to power. While the CPP and NPA called for a boycott of the election, moderate anti-Marcos opposition, as well as many local NPA insurgents, overwhelmingly rejected this course. The CPP-NPA thus rejected the popular democratic uprising that had toppled the Marcos regime, leaving them on the sidelines of the most significant political change in the Philippines since World War II. This proved to be a strategic error with far-reaching impact, as their popularity among the working and peasant classes began to steadily wane. NPA opposition to Marcos was responsible for the unprecedented support the movement had enjoyed during the 1980s. Now at the peak of its military and organizational strength, the movement had marginalized itself. Blamed for the mistake, NPA Commander-in-Chief, Rodolfo Salas,[56] resigned and was demoted amid unparalleled debate within the movement's leadership. Many CPP-NPA leaders called for reconciliation with the popular new

---

of President Corazon Aquino. He left the movement following his release. Others include Rodolfo Salas, former CPP chairman, and Romulo Kintanar, former NPA head, who led the movement during the imprisonment of Buscayno and Sison, only to be imprisoned themselves under President Aquino. Today, apart from Sison, the most visible leader is Luis Jalandoni, who since the 1970s has led the NDF from the Netherlands, particularly its efforts in Europe to garner foreign support.

[52] Chapman, *Inside the Philippine Revolution.*

[53] Ibid.

[54] Since 2002, Sison has been listed as a "person supporting terrorism" by both the United States and European Union (EU). In 2007, an EU court ordered his name be removed from the list and reversed actions by several government members that had frozen his assets. Concurrently, there has been an ongoing effort since 2006, by the administration of President Arroyo, to seek Sison's prosecution in the Netherlands for his connection with the assassinations of several former CPP-NPA leaders, including Romulo Kintanar in 2003.

[55] See *Methods of Obtaining Legitimacy and Changes in Government* for a fuller discussion.

[56] Salas (CPP Chairman 1977–1986 and NPA Commander-in-Chief 1976–1986) was replaced briefly as CPP Chairman by Benito Tiamzon and afterward by Sison, following Sison's release from prison.

Aquino government, including dismantling the armed opposition and participating openly in the democratic process. The result was a long power struggle and period of decline for the NPA.

In 1985, following a series of military defeats and leadership arrests in Mindanao, party leaders launched Operation Kampanyang Ahos (KAHOS), an internal purge operation. Because of a lack of strong social bonds and the perceived failure to provide education in the movement's ideology, many new recruits were suspected of being government agents.[57] Paranoia over the so-called "deep penetration agents" led NPA to force party members, NPA fighters, and local supporters into makeshift camps, where interrogations quickly descended into torture and killings.[58] The purge claimed the lives of 900 people.

During 1988, in the wake of the NPA's failure to move from the strategic defensive to the strategic offensive stage of the revolution, another purge took place in Luzon where at least 121 party leaders were killed.[59]

The geographic proximity of this later purge to Manila increased its visibility among the general population. The damage to party morale and legitimacy, as well as to their public image, was immeasurable.[60] These purges effectively eliminated the next generation of potential leaders from the NPA, either by death or defection.

The NPA not only witnessed the defection of members to the democratic process, but it also saw numerous armed breakaway factions, most notably the Revolutionary Proletarian Army-Alex Buscayno Brigade (RPA-ABB).[61] In the wake of these leadership struggles and defections, as well as continued failure to deal with the strategic errors of the 1980s, the Second Great Rectification Movement was initiated

---

[57] Jones, *The Red Revolution*; Garcia, *To Suffer Thy Comrades*; Patricio N. Abinales, ed., *The Revolution Falters: The Left in Philippine Politics After 1986* (New York: Cornell Southeast Asia Program Publications, 1996).; R. J. May and Francisco Nemenzo, *The Philippines After Marcos* (New York: St. Martin's Press, 1985).

[58] Joel Rocamora, "The Left in the Philippines: Learning from the People, Learning from each Other," (Colombo, Sri Lanka: Transnational Institute, March 25, 2000).

[59] Although Kintanar escaped after only a few months in prison. Reid, *Philippine Left*; Garcia, *To Suffer Thy Comrades*; Abinales, *The Revolution Falters*; Joel Rocamora, *Breaking Through: The Struggle in the Communist Party of the Philippines* (Quezon City, Philippines: Anvil Publishing, 1994); Rocamora, "The Left in the Philippines."

[60] The cohesiveness of NPA units and sustaining the armed conflict depended on comradeship (i.e., trusting your comrade with your life), which suffered greatly because of the purges.

[61] The separate defections of the Manila-Rizal and Central Mindanao Regional Commission and the Negros Island party Committee in 1993 are two notable examples. The Negros Island Party Committee has 1,800 CPP members, four NPA companies, a popular base of 36,000, and 570 high-powered rifles, while the Manila-Rizal group included an urban guerrilla RPA-ABB unit. *Tan, Handbook of Terrorism and Insurgency in Southeast Asia.*

in 1992 and completed in 1998. More widespread than the smaller purges of the late 1980s, the Second Rectification led to the killing of thousands of members accused of being "deep penetration agents."[62] Yet in reality, these purges were an effort by the Sison faction to regain control of the CPP-NPA from the autonomous regional commission and large NPA formations, such as the RPA-ABB, which had risen to power during the late Marcos era.[63]

The NPA actually started changing as an organization as early as the 1970s.[64] During the Sison imprisonment under the Marcos regime, innovative young leaders arising from the student movement of the late 1960s took over key leadership positions. These leaders relaxed the rigid commitment to Maoist orthodoxy, preferring decentralization, the formation of larger guerrilla formations, variation of tactics (e.g., emphasis on nonarmed tactics such as political mobilization and protests, particularly in urban areas), and greater allowance for internal debate. These innovations, particularly the increased allowances of internal debate, directly precipitated the first internal purges.[65]

In the wake of the election of 1986, the NPA entered a long period of decline marked by diminished public image and relevance, the defection of those who rejected continued armed struggle, and the internal purges and leadership struggles. During the Marcos era, the urban mass movement had provided highly educated middle-class leaders capable of providing financial and logistical support for rural fighters. After 1986, many of those key leaders were no longer in the organization.

COMMUNICATIONS

Insurgent leaders used routine, face-to-face meetings with their subordinates, complemented by the use of couriers, from one tactical command to another. The NPA became newsworthy by the 1970s, and numerous national newspapers covered official statements from CPP and NPA leadership, which demonstrated legitimacy while also supporting their recruiting and support goals.

---

[62] Like the first series of purges, the Second Rectification severely damaged the public image of the CPP-NPA. Peace Advocates for Truth, Healing and Justice (PATH) was formed as a support group for survivors and the families of victims. In addition, the NPA has offered apologies and compensation to affected communities. Garcia, *To Suffer Thy Comrades.*

[63] Severino, Salazar, and Institute of Southeast Asian Studies, *Whither the Philippines.*

[64] Ibid.; Reid, *Philippine Left,* 210; Magno, "A Nation Reborn"; Abinales, *The Revolution Falters.*

[65] Severino, Salazar, and Institute of Southeast Asian Studies, *Whither the Philippines.*

At the column level (the equivalent of a rifle company, generally organized as 80–100 combatants), the NPA was very much a field command, with forces using tactical communications equipment captured or stolen from the Philippine security forces. Shortwave radios were also monitored for news on the movement of Philippine forces. Tactical communications were not sophisticated. Messages delivered by courier often took weeks, and even months, to get from one unit to another.[66] Tactical radios, however, largely adapted by the NPA in the 1980s, provided extensive improvements that even rivaled the capabilities of the Armed Forces of the Philippines.[67]

METHODS OF ACTION AND VIOLENCE

The NPA understood that a protracted war across thousands of islands would be extremely difficult to coordinate. As such, Jose Sison directed the establishment of autonomous fronts on each of the major islands. Sison was convinced that this would cause the Philippine security forces to widely disperse their own forces in response, thereby avoiding a massive assault at any time against the NPA. Sison called this "centralized leadership, decentralized operations."[68] This strategy called for great autonomy and self-sufficiency at the operational level and also meant that the NPA units would be responsible for recruitment, political education, and propaganda, as well as military action. Gregg Jones noted, "The policy of decentralized operations proved to be a masterstroke that enabled the NPA to adapt to the Philippines' complex matrix of ethnic and linguistic diversity, which was so great that even adjacent barrios were sometimes cleaved by custom and language. Armed with the flexibility to discard unworkable tactics and experiment with new ones, leaders of the various fronts patiently developed the inept insurgency into a national movement of vast potential."[69]

During the early 2000s, the NPA remained capable of limited tactical operations, but their current composition of small guerrilla units is not suited to engage government troops unless several units are temporarily combined for a "tactical offensive," typically against undermanned and outgunned police outposts. This reality and the preoccupation with fund-raising explains the significant drop

---

[66] Jones, *The Red Revolution.*
[67] Ibid.
[68] Ibid.
[69] Ibid.

in the number of NPA encounters with government forces during the decade.[70]

METHODS OF RECRUITMENT

Factors that have influenced recruitment include government corruption and incompetence (notably during the Marcos regime), government and military human rights violations, and grievous socioeconomic disparities.[71] In part, the ability to recruit in rural communities can be attributed to the ability of the NPA to function as a separate state, providing peace, order, and social justice in places where the government is unable or unwilling to do so.[72] During the early 1980s, the communist insurgency grew rapidly as a result of increased dissatisfaction with the Marcos government's corruption and human rights abuses.[73] Many, even middle-class urban moderates, supported the NPA as the only force capable of challenging the Marcos dictatorship.[74]

---

[70] Severino, Salazar, and Institute of Southeast Asian Studies, *Whither the Philippines*; Magno, "A Nation Reborn."

[71] Marshall, "The War with no End"; Severino, Salazar, and Institute of Southeast Asian Studies, *Whither the Philippines*; Nadeau, *Liberation Theology in the Philippines*; Youngblood, "Structural Imperialism"; Avila, *Midlife Crisis of the Philippine Red Movement*; Mayand Nemenzo, *The Philippines After Marcos*; A. Hicken, "The Philippines in 2007: Ballots, Budgets, and Bribes," *Asian Survey* 48, no. 1 (January/February 2008): 75; Sheila S. Coronel, "The Philippines in 2006: Democracy and its Discontents," *Asian Survey* 47, no. 1 (January/February 2007): 175; Magno, "A Nation Reborn"; Magno and Gregor, "Insurgency and Counterinsurgency in the Philippines."

[72] Tan, *Handbook of Terrorism and Insurgency in Southeast Asia.*

[73] In 1983 the NPA exercised control over 2–3% of the nation's villages with 6,000 full-time fighters. By 1986 that number had grown to 22,500 full-time fighters and controlled over 40% of the nation's villages. Ibid. This growth is what prompted CPP-NPA leadership to begin the transition to larger units in preparation for the move from the strategic defensive stage to the strategic offensive stage of the Protracted People's War. Reid, *Philippine Left*; Rocamora, *Breaking Through: The Struggle in the Communist Party of the Philippines.* (Quezon City, Philippines: Anvil, 1994); Chapman, *Inside the Philippine Revolution*; Jones, *The Red Revolution*; Wurfel, *Filipino Politics.*

[74] Reasons for NPA growth under martial law include (1) the politicization of students and young urban professionals during the protest movements of the 1960s and 1970s and their subsequent radicalization when legal means of opposition were foreclosed; (2) political repression and human rights violations; (3) militarization of the countryside; (4) a general absence of social and political justice; (5) declining economic prosperity generally and growing economic disparities due to the oil crisis, the cronyism and monopolies of the Marcos government, and the erosion of effective governance and governing legitimacy leading to uncertainty among the business and middle classes. Tan, *Handbook of Terrorism and Insurgency in Southeast Asia;* Törnquist, "Communists and Democracy in the Philippines"; Jones, *The Red Revolution;* Abinales, *The Revolution Falters;* Chapman, *Inside the Philippine Revolution;* May and Nemenzo, *the Philippines After Marcos;* Mediansky, *The New People's Army;* Jose P. Magno Jr. and A. James Gregor, "Insurgency and Counterinsurgency in the Philippines"; Reid, *Philippine Left.*

Marcos' unpopularity was the most effective recruiting tool of the NPA. With his ouster, the NPA was thrown into disarray.[75] The highly popular President Aquino caused a huge drop in recruitment, particularly among the urban middle class, students, and labor unions.[76]

Initially, NPA mobilizers did not seek commitment to the Maoist ideology; rather, they appealed directly to self-interest, which most Filipino peasants equated with the responsibility of mothers and fathers for the family finances.[77] Operating through labor unions, such as the NFSW, the NPA used coercion to pressure *hacienda* landlords for small concessions such as higher wages and the use of some plantation land for subsistence farming. The workers came to view the NPA as their only means for securing concessions from the landowners and believed the planters would renege on concessions without the guerrilla coercion.[78] The NPA ensured worker protection from planter and state repression.

The NPA framed commitment to one's family interests above that of "service to the people"[79] in negative terms in an effort to erode the cultural centrality of family loyalty. The NPA employed seminars, song, dances, and plays, performed by their youth groups, to emphasize the need for sacrifice in order to achieve the ultimate liberation of the people. Commitment to communist ideology was socialized through existing family and social networks, which the NPA had thoroughly infiltrated. NPA activists organized new community groups based on gender and generation, not on family units.[80] These groups helped to shift loyalty from the household to the communist cause. The NPA would earnestly ask parents for permission to recruit a child as a full-time fighter, emphasizing that their sacrifice would provide a better future for their children and grandchildren.[81]

Among the youth, the NPA employed peer-pressure tactics to encourage recruitment.[82] By far, the youth group was the most effective

---

[75] Jones, *The Red Revolution*; Reid, *Philippine Left*; Garcia, *To Suffer Thy Comrades*; Magno, "A Nation Reborn"; Abinales, *The Revolution Falters*; Parsa, *States, Ideologies, and Social Revolutions*.

[76] Abinales, *The Revolution Falters*; Parsa, *States, Ideologies, and Social Revolutions*.

[77] Abinales, *The Revolution Falters*.

[78] Rutten, "High-Cost Activism and the Worker Household."

[79] Terms such as *personal enteres* (personal interests), *burgis* (bourgeois emphasis on the family accumulation of power of wealth), and *pyudal* (the feudal or authoritarian personal relationships, including that of the family) were all employed by recruiters.

[80] Ibid.

[81] This appealed directly to the deeply held Filipino cultural values to work hard and sacrifice so that future generations will have better lives. Ibid.; Rodell, *Culture and Customs of the Philippines*.

[82] Abinales, *The Revolution Falters*.

means by which the NPA redirected loyalties away from the family.[83] Poverty-stricken teenage children of the *hacienda* workers were quick to join the youth groups that provided peer acceptance, excitement, camaraderie, and importance, while increasing the teenagers' commitment to the movement.

With the fall of Marcos, NPA resources to provide household-oriented goods as well as to foster commitment to the movement's ideology were lost. Renewed counterinsurgency efforts, as well as the presence of paramilitary forces,[84] restricted NPA movements in the lowland *haciendas*. NPA guerrillas and recruiters could no longer move freely among the *hacienda* workers. The NPA's capacity to coerce planter concession and protect workers was diminished.[85]

METHODS OF SUSTAINMENT

Logistically, the NPA was designed from the start to be self-sufficient. As the insurgency became more complex, however, the logistics requirements became greater. During 1971, the CPP established a permanent delegation in Beijing to coordinate support from the Chinese government. Some shipments of arms were likely received and then offloaded by small *banca* boats, but these shipments were quite limited.[86] The NPA, more often, became proficient at obtaining Philippine security weapons and supplies from multiple means including raids on combat outposts, recovery of material and supplies from combat, and even paying for arms and supplies on the black market. The principle of self-sustainment, however, did continue. Over time, safe havens were established in hundreds of areas, where the local populace could be depended on for logistics support from food, to medical supplies, to batteries.

Gregg Jones spent extensive time in the Philippines with NPA forces from 1984 to 1989. In *Red Revolution*, he commented on life and activity within an NPA encampment:

---

[83] Rutten, "High-Cost Activism and the Worker Household."

[84] See *Changes in Government Policy* for a fuller discussion.

[85] Under increased police surveillance the youth groups also dissolved. Ibid.

[86] See Jones, *The Red Revolution* for a description of the Chinese support, including attempted deliveries of Chinese-provided materiels by motor vessel *Karagatan* and motor vessel *Andrea*.

In contrast to most Philippines armed forces installations I had visited, the NPA camps were usually beehives of activity; couriers coming and going throughout the day, porters arriving with supplies, and the constant construction of new shelters and better facilities. By 5:30 or 6:00 A.M., the camps were alive with *kasamas* [workers] doing chores like fetching water, sweeping the packed mud floors of their shelters, and doing laundry. Life inside the guerrilla zone was largely self-contained, and news from the outside world was sometimes limited. A *kasama* arriving from the outside world took pains to buy one or two newspapers, which were devoured at every stop along the trail by Party workers and literate guerrillas even as the papers grew steadily more out of date. Often, the rebels read articles to illiterate peasants. An effective, if sometimes slow, underground mail system had developed in the communist zones. Travelers coming from outside destinations usually carried letters from friends and loved ones of comrades living inside the zone. Anyone departing from a camp or from peasant houses that were popular rest stops along the jungle trails was handed letters folded into tiny "chiclets" and wrapped in clear tape."[87]

METHODS OF OBTAINING LEGITIMACY

The NPA made ardent efforts to bolster its legitimacy in the public's eyes both domestically and internationally. Seeing itself as the rightful government of the Philippines, the CPP-NPA has typically observed human rights and international human rights laws.[88] To that end, the NPA has not generally engaged in random bombing operations, with the notable exception of the 1971 bombing of the Liberal Party rally in Plaza Miranda, which killed a number of civilians and precipitated the declaration of martial law.[89] In addition, the CPP-NPA initially

---

[87] Ibid.

[88] For instance, while the CPP-NPA used homemade command-detonated land mines to ambush Philippines Armed Forces soldiers, it refrained from mining large tracts of land with self-detonating mines. Tan, *Handbook of Terrorism and Insurgency in Southeast Asia.*

[89] Although many at the time accused Marcos himself of orchestrating the bombing as an excuse to declare martial law, more recent evidence suggests that Sison was the bombing's chief architect. Ibid. Reid, *Philippine Left*; Törnquist, "Communists and Democracy in the Philippines"; Garcia, *To Suffer Thy Comrades*; Hawes, "Theories of Peasant Revolution"; Chapman, *Inside the Philippine Revolution*, 288; Avila, *Midlife Crisis of the Philippine Red Movement*; Parsa, *States, Ideologies, and Social Revolutions.*

tried to practice what they preached, by focusing on the plight of the rural poor and opposing the Marcos regime, winning mass rural and urban bases, as well as support from anti-Marcos human rights activists abroad.[90]

Yet several factors have damaged the legitimacy of the NPA in recent decades including: (1) the 1986 election and subsequent internal struggles (i.e., purges), (2) the emphasis on violence over the plight of the workers, and (3) the increase of financial extortion as a means of funding the revolution. Finding itself in a more hostile environment, particularly with the advent of the global war on terrorism in the wake of 9/11, it has become increasingly difficult for the NPA to maintain a distinction between revolutionaries and terrorists in the eyes of most observers.

No other event precipitated the decline of the NPA more than the failure to adjust to the changing realities in the Philippines during and in the immediate wake of the election of 1986.[91] The communists' emphasis on armed conflict over unarmed protests, rural over urban, and revolution over democratic organization allowed it to thrive under the conditions created by the repressive Marcos regime. Even support from the urban middle class exploded as an avenue to oppose Marcos through large unarmed protests, particularly after the assassination of opposition leader Ninoy Aquino in 1983. The CPP called for a boycott of the snap election of February 1986, dubbing it a bourgeois process.[92] The NPA sat on the sidelines while the discounted unarmed urban mass movement ousted Marcos and restored democracy. By the CPP's own admission[93] it had made a grave tactical error, standing by while hundreds of thousands of its own supporters toppled the Marcos regime, "because it . . . decided that their particular political action did not after all fit the party's strategic framework."[94] Given the strength of the NPA (i.e., tens of thousands of armed fighters, as well as mass popular support movements in the rural and urban areas), they should have been the primary revolutionary force that

---

[90]   Rutten, "High-Cost Activism and the Worker Household."

[91]   See *Changes in Government* for a fuller discussion. Reid, *Philippine Left.*; Törnquist, "Communists and Democracy in the Philippines"; Magno, "A Nation Reborn"; Rocamora, "The Left in the Philippines"; Rocamora, *Breaking Through*; Abinales, *The Revolution Falters.*

[92]   Stephan and Chenoweth, "Why Civil Resistance Works."

[93]   The CPP noted that "when the aroused and militant masses moved spontaneously but resolutely to out the hated grime last February 22–25, the Party and its forces were not there to lead them. In large measure, the Party and its forces were on the sidelines, unable to lead or influence the hundreds of thousands of people who moved with amazing speed and decisiveness to overthrow the regime." Rocamora, *Breaking Through.*

[94]   Rocamora, "The Left in the Philippines."

ousted Marcos.[95] Yet, the revolution was dominated by middle-class moderates, not armed communist radicals. The reality that Marcos was toppled by middle-class moderates, through peaceful democratic means—and not by armed communist guerrillas—shifted popular support away from the CPP-NPA.[96]

The NPA failure to capitalize on the election of 1986 can be traced to "the party's flawed understanding of Philippine politics and society and its overemphasis on a rural-based and militarist protracted people's war strategy."[97] Reduced to the position of irrelevant bystanders during the election and in the subsequent transition to democracy, the CPP-NPA rapidly lost legitimacy. Despite the use of terror via the murder of local officials, villages under NPA control did not support candidates selected by the CPP in the congressional elections of May 1987. In the midst of the rapid decline in popular support, Jose Maria Sison publicly supported the killing of unarmed student protesters in Tiananmen Square in 1989—another in a series of political blunders by the NPA leadership.

There seems to be a direct correlation between the declining resources of the NPA and their declining legitimacy. As resources declined and the NPA spent more of its time securing resources through the use of terror and extortion, it further alienated itself from the people.[98]

EXTERNAL SUPPORT

The bulk of external support for the NPA has not been derived from the communist countries such as China, but from nongovernmental organizations (NGOs) within European pro-Western democracies. Given its Maoist ideology, the NPA turned initially to China for support in the early 1970s. The Chinese made two failed attempts to land weapons in the early 1970s, and the weapons were seized by the Marcos government and used as a pretext for the declaration of martial law.[99] From the mid-1970s onward, Chinese government officials backed the Marcos regime, breaking ties with the CPP-NPA.[100]

---

[95] Reid, *Philippine Left*; Hawes, "Theories of Peasant Revolution," 261–298.; Parsa, *States, Ideologies, and Social Revolutions*.

[96] Reid, *Philippine Left*.; Hawes, "Theories of Peasant Revolution," 261–298; Parsa, *States, Ideologies, and Social Revolutions*.

[97] Reid, *Philippine Left*.

[98] Severino, Salazar, and Institute of Southeast Asian Studies, *Whither the Philippines*.

[99] Wurfel, *Filipino Politics*; Joshua Kurlantzick, *Charm Offensive: How China's Soft Power is Transforming the World* (New Haven, CT: Yale University Press, 2007).

[100] Marcos dispatched his wife Imelda to Beijing in 1974, where she purportedly wooed Mao Tse-Tung himself, who remarked, "I like Mrs. Marcos because she is so natural and that

Other attempts to secure arms purchased from overseas sources in the 1980s and early 1990s were largely unsuccessful.[101]

The main source of overseas funding for the NPA was from humanitarian organizations, including a number of European churches, and radical groups in Europe. Touting its position as the only viable opposition to the human rights abuses of the Marcos regime, the communists, through their public face of the NDF, had established support networks in more than 25 countries, attracting the support of numerous internationally recognized human rights organizations by 1987.[102] These organizations remained a major source of support even after the fall of Marcos.[103] NPA strategy was to divert resources from the NGOs through aboveground institutions run by NPA supporters under the auspices of rural aid and development.[104]

COUNTERMEASURES TAKEN BY THE GOVERNMENT

As the NPA continued to grow in force and capability during the early 1970s, President Marcos declared martial law on September 22, 1972. Domestic unrest and the threat posed by the NPA to national security were only two of several reasons touted for the declaration, but the NPA saw it as an opportunity. Marcos and the government were clearly painted as the enemy of the people.

At the time martial law was declared, Philippine security forces were severely limited in capabilities, having received no counterinsurgency training. Additionally, resources were dispersed from Luzon to the north and from Mindanao to the south, where the Marcos administration was also dealing with the MNLF insurgency. In the key early years of 1969–1972, security forces did almost nothing to counter or confront the NPA. Promotion in the officer corps was based on political patronage, not leadership abilities. Morale was very low. Thomas Marks, in *Maoist Insurgency Since Vietnam*, states that "Companies which at full strength should have had more than 150 personnel were mentioned in numerous operational reports as

---

is perfection." Marshall, "The War with no End"; Kurlantzick, *Charm Offensive*.

[101] Jones, *The Red Revolution*; Chapman, *Inside the Philippine Revolution*; Florante Solmerin, "NPA Rejects Truce, Slays 2 Policemen in New Raid," *Manila Standard* (December 23, 2005).

[102] The CPP-NPA enjoyed particularly strong support in the Netherlands and Germany, one among many reasons self-exiled leaders chose the Netherlands as a base of operations. Severino, Salazar, and Institute of Southeast Asian Studies, *Whither the Philippines*; Törnquist, "Communists and Democracy in the Philippines"; Reid, *Philippine Left*; Magno, "A Nation Reborn."

[103] Törnquist, "Communists and Democracy in the Philippines."

[104] Severino, Salazar, and Institute of Southeast Asian Studies, *Whither the Philippines*

putting but 70–80 men into the field. . . . training, for all practical purposes, vanished. Units were formed, taught basics, then deployed to the field. . . . In supply, modern infantry arms and ammunition were frequently not available."[105]

The war was seen as a battalion war, to be waged at the tactical level, *barangay* to *barangay*. Over time, however, the government forces became more tactically proficient, and their improved military capabilities coincided with political changes as well, including the cessation of martial law, increased violence by the NPA (and its associated negative perception by the people), and improved understanding on the linkage of the people to the counterinsurgency campaign by Philippine military leadership. The government began to reclaim areas over time, amass forces, and improve intelligence sources and methods. Command and control was also improved, enabling better protection of local infrastructure. A gradual yet noticeable shift occurred: with the NPA no longer applying their points of discipline and attention, they were suddenly seen as the oppressors, not the liberators.

Marked changes in leadership within the Philippine military also resulted in operational changes. One interesting player was Victor Corpuz. Corpuz, a Class of 1967 Philippine Military Academy graduate, ". . . created a sensation by defecting to the NPA. Six years later, however, disillusioned with the NPA, he returned to the government fold, only to be imprisoned for ten years. Released when Marcos was ousted from power . . . he became the central force in radically reorienting Philippine counterinsurgency strategy away from its fruitless emphasis upon military operations. Instead, the weight of effort went to socioeconomic-political development."[106] Lieutenant Colonel Corpuz became a key adviser to the Ministry of National Defense Counterinsurgency Study Group, touted with developing a change in course for the military response. These changes included ending the "search and destroy" methods that had actually worked to strengthen the ties between the NPA and local villages, changing the structure and composition of security organizations (including the police and military forces) to conduce counterinsurgency operations, and increasing coordination of strategy to operations.[107]

---

[105] Thomas A. Marks, *Maoist Insurgency since Vietnam* (Portland, OR: Frank Cass, 1996).

[106] Ibid.

[107] Tan, *Handbook of Terrorism and Insurgency.*

SHORT- AND LONG-TERM EFFECTS

CHANGES IN GOVERNMENT

Unquestionably, the most significant change in government was the fall of Marcos following the election of 1986. Despite a communist boycott, Marcos lost the election and nullified the results. The "people power" revolution had done what the CPP-NPA could not.

The election of President Aquino brought hope for reconciliation with the NPA. A cease-fire was announced in December 1986, which was followed by negotiations and an amnesty program[108] for communists who rejected the armed conflict. Yet, in January 1987, negotiation collapsed and violence resumed. With the NPA at its greatest strength, 25,000 full-time fighters, incidents of violence reached their highest point to date, even compared with the later years of the Marcos regime.[109]

With the renewal of violence, the Aquino administration increasingly favored a military solution over negotiation. Another important factor during this period was US involvement in the Philippines. With the end of the Cold War, the strategic value of the Philippines to the United States began to diminish. In 1989, the NPA offered a unilateral cease-fire if the government would refuse to renew the lease agreement for the US Naval Base Subic Bay and Clark Air Base. President Aquino declined the offer, but the Philippine Senate voted not to renew and the United States withdrew all military forces in 1991.

The ability of the current administration of President Gloria Macapagal-Arroyo (2001–present) to deal with the NPA has been impacted by a renewed US military presence in the Philippines, the global political and economic realities, and the continued tenuousness of Philippine democracy. The presence of Islamic extremists, including the Moro Islamic Liberation Front, the Abu Sayyaf Group, and the Rajah Sulaiman Group, the latter two having connections with Jemaah Islamiyah and Al Qaeda, has led to renewed US military involvement in the Philippines. Concurrently, the NPA has become an anachronism to the end of the Cold War alliances and

---

[108] As part of the goodwill gestures of the Aquino government, former CPP chairman Jose Maria Sison and former NPA leader Bernabe Buscayno were released from prison.

[109] The assassination of government officials by the sparrow units of the NPA grew at an alarming rate during this period. Kowalewski, "Cultism, Insurgency, and Vigilantism in the Philippines," 241–253; Chapman, *Inside the Philippine Revolution*; James Clad, "Betting on Violence," *Far Eastern Economic Review* 138, no. 51: 35–41 (1987); Nathan, "Armed Struggle in Philippines."

domination of market forces with the advent of globalization, notably in the transformation of China into a market-driven economy.[110]

CHANGES IN POLICY

Since the fall of Marcos, successive Philippine governments have favored a development-based approach, along with a tacit commitment to peace talks and persistent military offensives to address the communist insurgency. The development-based approach recognizes that a military solution alone will not work because the root cause of support for the NPA lies in genuine socioeconomic grievances that must be addressed.[111] Progress on this front, however, has remained elusive as the insurgency makes investing in business in the poorest areas of the Philippines unattractive. The NPA has killed or driven away potential entrepreneurs and imposed revolutionary taxes on those who stay. Administrations since Marcos have employed a combination of economic development, military pressure, and peace offers to produce a slow but steady decline in NPA strength and support.

With the failure of peace talks and the NPA at its greatest strength, pressure from conservative military and political elites led an intensified and sophisticated counterinsurgency program, with unprecedented levels of violence.[112] This campaign led to the destruction of NPA bases, as well as the capture of more than 100 leading national and regional leaders. Between 1986 and 1990, the conflict created more than 1.3 million internal refugees. The NPA was in clear decline because of forces of history, the mistakes surrounding the 1986 election, changing economic realities, and internal conflict (i.e., purges, disagreement of mission, and leadership struggles).

The policy of the government during both the presidencies of Fidel V. Ramos and Joseph Estrada included attempts at peace talks, as well as a military campaign to crush the NPA.[113] In 1992, the government of President Ramos recognized the CPP as a legitimate political party, in a bid to rekindle peace talks. Although President Estrada was initially conciliatory toward the CPP-NPA, he took an increasingly hard-line approach, particularly after the NPA kidnapped a senior

---

[110]   Severino, Salazar, and Institute of Southeast Asian Studies, *Whither the Philippines*.

[111]   Ibid.; Magno, "A Nation Reborn"; R. T. Naylor, *Wages of Crime: Black Markets, Illegal Finance, and the Underworld Economy* (Ithaca, NY: Cornell University Press, 2004).

[112]   Tan, *Handbook of Terrorism and Insurgency in Southeast Asia*.

[113]   Ibid.

officer of the Philippines armed forces.[114] Like negotiations between rebels and governments in many other countries, those between the government of the Philippines and the NPA have been cyclical, derailed by "unacceptable" hostile action or the failure of one side or both to "perfectly" meet the terms of an agreement. Thus, a cycle of peace talks, derailment, and renewed violence continues.

Since 2001, the military has come under scrutiny for the illegal detention, abduction, assassination, and disappearance of more than 800 left-leaning activists, lawyers, union leaders, journalists, and clergy as part of its counterinsurgency operations.[115] In a widely circulated PowerPoint presentation entitled "Knowing the Enemy," the Philippines armed forces identified numerous aboveground organizations as being fronts for the CPP-NPA and suffering from communist infiltration. In 2006, Amnesty International released a scathing report describing a "pattern of politically targeted extra-judicial executions taking place within the broader context of a continuing counter-insurgency campaign."[116] These killings were also accompanied by an intensified campaign to prosecute public officials accused of supporting the NPA.[117] These incidents damaged the legitimacy of both the civilian government and the military.

Along with the NPA's own self-destruction following the fall of Marcos, the two most important factors precipitating their decline have been (1) the continued growth and modernization of the Philippine economy[118] and (2) the availability of democratic means to effect reform.[119] The chaos within the NPA meant that those who

---

[114] Even after the release of the abducted officer, both the government and the CPP-NPA refused to return to the negotiating table. The renewal of the mutual defense treaty with the United States and the associated ratification of the Visiting Forces Agreements led the communists to refuse to resume negotiations until the end of Estrada's presidency. Ibid. See Footnote 55.

[115] Philip Alston, *Promotion and Protection of all Human Rights, Civil, Political, Economic, Social And Cultural Rights, Including the Right to Development* (Human Rights Council, 2008).

[116] See Amnesty International's report *Philippines: Political Killings, Human Rights, and the Peace Process* (August 15, 2006). The report also accused the Arroyo government of allowing the killings in order to win support from the military, while concurrently eliminating political opponents.

[117] This has included the attempted prosecution of congressmen from left-wing political parties for crimes relating to their time as former NPA members or for their support of the organization, including former NDF negotiator, Satur Ocampo. Many on the left have seen the extra-judicial killings and the intensified prosecution of left-wing political leaders as an attempt by the Arroyo administration to eliminate democratic opposition.

[118] Economic growth has resulted both from market forces such as globalization and from government policy, although government policy has attempted to target economic growth in ways that benefit the poor Filipinos, from whose ranks the NPA draws its support, with some success.

[119] Abinales, *The Revolution Falters*; Törnquist, "Communists and Democracy in the Philippines"; Rocamora, *Breaking Through*.

had organized in the final years of the Marcos regime turned not to the communists for the restoration of democracy, but to grassroots organizations and NGOs, whose growth exploded in the post-Marcos period.[120] The reform and democratization work of these groups was facilitated by the passage of the Local Government Code in 1991, which gave significant fiscal and legislative autonomy to local jurisdictions.[121] Not only did this granting of authority to local jurisdictions decrease political corruption, it also normalized democratic participation in small rural communities, which had been fertile recruiting grounds for the NPA.[122] The institutionalization of political and economic reforms to address the governance and socioeconomic issues may prove the most efficacious means of destroying the NPA.

CHANGES IN THE REVOLUTIONARY MOVEMENT

Time may have worked against the NPA. Thomas Marks noted "nearly 30 years of formal struggle in the case of the CPP—the original grievances, whatever their form, have declined dramatically in salience."[123] Maintaining momentum over an extended period of time posed challenges for the NPA. Stagnation resulted from a populace that became as disenchanted with the NPA as they had with the government. As the Marcos regime was ousted, democracy took hold under the tutelage of President Corazon Aquino. Suddenly, the government was taking back the very causes that the CPP had previously claimed—freedom and democracy. Commenting on the sudden loss of supporters after the end of the Marcos regime and the beginning of the Aquino era, one CPP leader noted, "We have found that some whom we thought we had convinced to rationally understand the structural problems were in fact only anti-Marcos

---

[120] These organizations also provided an outlet for those disenchanted with the CPP-NPA to continue the work of progressive reform through legal means. Ibid.

[121] This allowed local governments to work with grassroots organizations to enact real and visible reform and development at the local level. These efforts were aided by international organizations such as the US Agency for International Development. Rocamora, "The Left in the Philippines."

[122] Before the implementation of the Local Government Code, provincial towns and cities had been the center of political life, as smaller villages had no self-governance. This reality meant that rural politics was controlled from the towns and political power was concentrated in the hands of political families led by either local landowners or businessmen. The political culture created by this arrangement fostered nepotism, corruption, violence, and lack of transparency, as politicians would trade jobs and other services in exchange for favors or support for their political family. With the devolvement of government authority (paid leadership, government appropriations, limited taxation and legislative power, etc.) to the village level, local politics was no longer dependent on the patronage of political families in the towns. Ibid.

[123] Marks, *Maoist Insurgency since Vietnam.*

and anti-military. We had tried to teach them that the problems were not caused by evil men but by an unjust system. Some obviously did not understand . . . To that extent the victory of Cory [Aquino] is a real dilemma for us."[124] However, it was more than a dilemma—it was the culminating point for the NPA. Moreover, real changes in the socioeconomic conditions were noted.[125] The result was that the NPA initiated even more violent action, particularly from the 1987–1990 time frame. These actions included urban violence, ambushes, increased use of improvised explosive devices (IEDs), and assassinations of political targets. The populace, however, only grew further alienated from the NPA's cause. While the Philippine security forces improved their professionalism, adopted a sound counterinsurgency strategy, and focused on protection and respect of the populace, the CPP-NPA became "more bandit than rebel."[126] The NPA had lost popular support and their political ethos. As they lost the ability to conduct sustained offensive operations and continue a protracted war, the NPA modified their tactics, limiting operations to raids and political assassinations, while also increasing their emphasis on nonkinetic action, specifically information operations. The NPA has been using many means to get their word out to the people, including the Internet, the radio, pamphlets, and rallies. Their message, however, is no longer well received, as the working class has observed democratic action taking hold. The government is no longer seen as the root of evil and the basis for the economic malaise. The NPA today has lost the support of their former base, and they have never recovered from the loss of their greatest unifier, Ferdinand Marcos.

OTHER EFFECTS

Logistically, the NPA established a complex intra-island and inter-island network, tied to small boats, called *bancas. Bancas* are used extensively for fishing and legitimate island trade, and it was almost impossible to distinguish an NPA *banca* from another *banca.* Moreover, the Philippine security forces lacked the brown water navy and patrol boat structure to effectively secure the thousands of miles of navigable waterways. During the three decades of active insurgent operations by the NPA, only limited logistics shipments were ever interdicted by

---

[124] Chapman, *Inside the Philippine Revolution.*

[125] Measurable socioeconomic change did not happen overnight. New policies, international support, and a recognition of human rights, military professionalism, and democracy did set the conditions for long-term improvement.

[126] Marks, *Maoist Insurgency since Vietnam.*

Philippine security forces. Waterways were effectively conceded to the NPA.

The NPA first conducted urban operations in Manila during the 1970s, but by the 1980s, they had specially trained brigades focused on urban operations, for the purpose of forcing Armed Forces of the Philippines soldiers to be held in the capital. Urban operations became hit squads, targeting politicians and others. They also maintained conventional capabilities, assaulting military bases in Manila with column-sized (approximately 100 combatants) or greater-strength troops.

BIBLIOGRAPHY

Abaya, Antonio C. "Defeating the Communists." *Manila Standard,* February 1, 2007.

————. "Defeating the Communists II." *Manila Standard,* February 8, 2007.

————. "Defeating the Communists III." *Manila Standard,* February 15, 2007.

Abinales, Patricio N., ed. *The Revolution Falters: The Left in Philippine Politics After 1986.* New York: Cornell Southeast Asia Program Publications, 1996.

Agoncillo, Teodoro. *A Short History of the Philippines.* New York: Mentor Publishing, 1975.

Alston, Philip. *Promotion and Protection of all Human Rights, Civil, Political, Economic, Social and Cultural Rights, Including the Right to Development.* Human Rights Council, 2008.

Amnesty International. *Philippines: Political Killings, Human Rights, and the Peace Process.* August 15, 2006.

Avila, Ava Patricia C. *Midlife Crisis of the Philippine Red Movement.* Singapore: S. Rajaratnam School of International Studies, NTU, 2008.

Balisacan, A. M., and Hal Hill. *The Philippine Economy: Development, Policies, and Challenges.* New York: Oxford University Press, 2003.

Census Bureau of the Philippines. Manila: Census Bureau of the Philippines, 2009.

Chapman, William. *Inside the Philippine Revolution.* New York: W.W. Norton, 1987.

Chow, Gregory C. *China's Economic Transformation.* Malden, MA: Blackwell Publishers, 2007.

Clad, James. "Betting on Violence." *Far Eastern Economic Review* 138, no. 51: 35–41 (1987).

Conde, Carlos H. "Peace Effort with Philippine Rebels Breaks Down." *The International Herald Tribune,* August 5, 2005.

———. "Philippines Insurgency Takes Toll; Economic Strategy could be Undercut." *The International Herald Tribune,* November 23, 2005.

Coronel, Sheila S. "The Philippines in 2006: Democracy and its Discontents." *Asian Survey* 47, no. 1 (January/February 2007): 175.

Garcia, Robert Francis. *To Suffer Thy Comrades: How the Revolution Decimated its Own.* Quezon, Philippines: Anvil Publishing, 2001.

Hawes, Gary. "Theories of Peasant Revolution: A Critique and Contribution from the Philippines." *World Politics* 42, no. 2 (January 1990): 261–298.

Hicken, A. "The Philippines in 2007: Ballots, Budgets, and Bribes." *Asian Survey* 48, no. 1 (January/February 2008): 75.

Jones, Gregg. *The Red Revolution: Inside the Philippine Guerrilla Movement.* Boulder, CO: Westview, 1989.

Kann, Peter R. "The Philippines without Democracy." *Foreign Affairs* 52, no. 3 (April 1974): 612–632.

Kowalewski, David. "Cultism, Insurgency, and Vigilantism in the Philippines." *Sociological Analysis* 52, no. 3 (1991): 241–253.

Kurlantzick, Joshua. *Charm Offensive: How China's Soft Power is Transforming the World.* New Haven, CT: Yale University Press, 2007.

Liwanag, Armando. "Reaffirm Our Basic Principles and Rectify Errors." *Kasarinlan: Philippine Journal of Third World Studies* 8, no. 1 (1992): 96–157.

Magno, Alexander R. "A Nation Reborn." Vol. 9 of *Kasaysayan: The Story of the Filipino People.* Manila, Philippines: Asia Publishing Co., 1998.

Magno, Jose P. Jr., and A. James Gregor. "Insurgency and Counterinsurgency in the Philippines." *Asian Survey* 26, no. 5 (May 1986): 501–517.

Manning, Robert A. "The Philippines in Crisis." *Foreign Affairs* 63, no. 2 (Winter 1984): 392–410.

Marks, Thomas A. *Maoist Insurgency since Vietnam.* Portland, OR: Frank Cass, 1996.

Marshall, Andrew. "The War with no End." *Time Magazine,* 2007.

May, R. J., and Francisco Nemenzo. *The Philippines After Marcos.* New York: St. Martin's Press, 1985.

Mediansky, F. A. "The New People's Army: A Nationwide Insurgency in the Philippines." *Contemporary Southeast Asia* 8, no. 1 (1986): 1–17.

Montesano, Michael J. "The Philippines in 2003: Troubles, None of Them New." *Asian Survey* 44, no. 1 (January–February 2004): 93–101.

Nadeau, Kathleen, M. *Liberation Theology in the Philippines: Faith in a Revolution.* Religion in the Age of Transformation Praeger Series in Political Communication. Westport, CT: Greenwood Publishing Group, 2002.

Nathan, Dev. "Armed Struggle in Philippines." *Economic and Political Weekly* 22, no. 51 (December 19, 1987): 2201–2203.

Naylor, R. T. *Wages of Crime: Black Markets, Illegal Finance, and the Underworld Economy.* Rev ed. Ithaca, NY: Cornell University Press, 2004.

Niksch, Larry A., *Insurgency and Counterinsurgency in the Philippines.* Prepared for the Committee on Foreign Relations United States Senate by The Foreign Affairs and National Defense Division-Congressional Research Service, Library of Congress. Washington, DC: US Government Printing Office, 1985.

Parsa, Misagh. *States, Ideologies, and Social Revolutions: A Comparative Analysis of Iran, Nicaragua, and the Philippines.* New York: Cambridge University Press, 2000.

Porter, Gareth. *The Politics of Counterinsurgency in the Philippines: Military and Political Options.* Honolulu: University of Hawaii, 1987.

Reid, Ben. *Philippine Left: Political Crisis and Social Change.* Manila, Philippines: Journal of Contemporary Asia Publishers, 2000.

Rocamora, Joel. *Breaking Through: The Struggle in the Communist Party of the Philippines.* Quezon City, Philippines: Anvil Publishing, 1994.

———. "The Left in the Philippines: Learning from the People, Learning from each Other." Colombo, Sri Lanka: Transnational Institute, March 25, 2000.

Rodell, Paul A. *Culture and Customs of the Philippines.* Culture and Customs of Asia. Westport, CT: Greenwood Press, 2002.

Rosenberg, David. "Communism in the Philippines." *Problem of Communism* 33, no. 5: 24–46 (1984).

Rutten, Rosanne. "High-Cost Activism and the Worker Household: Interests, Commitment, and the Costs of Revolutionary Activism in a Philippine Plantation Region." *Theory and Society* 29, no. 2 (April 2000): 215–252.

Severino, Rodolfo, Lorraine Carlos Salazar, and Institute of Southeast Asian Studies. *Whither the Philippines in the 21st Century?* Singapore: Konrad Adenauer Stiftung, Institute of Southeast Asian Studies, 2007.

Sison, Jose Maria. *Philippine Society and Revolution.* Manila, Philippines: Pulang Tala, 1971.

———. *Struggle for National Democracy.* Quezon City, Philippines: Progressive Publications, 1967.

Solmerin, Florante. "NPA Rejects Truce, Slays 2 Policemen in New Raid." *Manila Standard,* December 23, 2005.

Stephan, Maria J., and Erica Chenoweth. "Why Civil Resistance Works; the Strategic Logic of Nonviolent Conflict." *International Security* 33, no. 1 (2008): 7–44.

Tan, Andrew Tian Huat, ed. *A Handbook of Terrorism and Insurgency in Southeast Asia.* Northampton, MA: Edward Elgar Publishing, 2007.

Törnquist, Olle. "Communists and Democracy in the Philippines." *Economic and Political Weekly* 26, no. 27/28 (July 6–13, 1991): 1683–1691.

US Department of State. *Philippines: International Religious Freedom Report 2008,* 2008.

Wurfel, David. *Filipino Politics: Development and Decay.* Politics and International Relations of Southeast Asia. Ithaca, NY: Cornell University Press, 1988.

Youngblood, Robert. "Structural Imperialism: An Analysis of the Catholic Bishops' Conference of the Philippines." *Comparative Political Studies* 15, no. 1 (1982): 29–56.

Youngblood, Robert L. "The Corazon Aquino 'Miracle' and the Philippine Churches." *Asian Survey* 27, no. 12 (December 1987): 1240–1255.

FUERZAS ARMADAS REVOLUCIONARIAS DE COLOMBIA (FARC)

Ron Buikema and Matt Burger

SYNOPSIS

In the mid-1960s, the *Fuerzas Armadas Revolucionarias de Colombia* (Revolutionary Armed Forces of Colombia), or FARC, and the *Ejército de Liberación Nacional* (National Liberation Army), or ELN, were formed in part as a consequence of the seventeen-year period of violence known as *La Violencia* (The Violence), and between 1946 and 1965, an estimated 200,000 people were killed. During this time, politics polarized further between the Conservatives, who favored the government, and Liberals, who supported the peasant class. Violent atrocities fell along familial lines and working-class groups organized armed self-defense squads. In order to establish themselves more firmly as political forces, the FARC recruited members from among the rural, working class, and the ELN found a base among the student movements of the universities. The ELN also found an unusual ally in the Catholic Church, with its liberation theology. As their numbers and political and military power grew in the 1980s, both organizations tried to suggest reasonable worker reforms to the government; however, the government's continued opposition strengthened the popularity of the FARC and the ELN, resulting in fresh recruits. As of 2009, both the FARC and the ELN maintain a political and military presence in the country and continue to recruit among young people at both public and private universities.

Some of the FARC's most notorious military campaigns include the 1998 ambush and destruction of the Colombian Army's 52nd Counter-Guerrilla Battalion, the 1998 kidnapping and murder of police officers from the Colombian National Police antinarcotics base at Miraflores, and continuing kidnapping and drug trafficking. However, the FARC received a serious blow in 1999 when the European Union (EU) and the US government (in conjunction with the Colombian government) launched Plan Colombia, which was designed to strengthen the Colombian government, integrate isolated areas, and enforce current laws. With the election of Alvaro Uribe in 2002, further steps were taken to develop a new national security strategy, reduce insurgent violence, and improve the economic status of the country.

TIMELINE

1966	FARC founded.
1978	Colombian government begins counternarcotics security actions.
1982	President Betancur grants amnesty to insurgents.
1984	Autodefensas Unidas de Colombia–United Self-Defense Forces of Colombia (AUC) paramilitary group forms.
1986	AUC begins murder campaign against Patriotic Union Party politicians; violence from insurgent groups also increases.
1989	Presidential candidates from both major political parties are assassinated.
1993	Pablo Escobar, Medellín drug network leader, is killed.
1998	President Pastrana is elected and cedes *Despeje* to the FARC as a demilitarized zone—an area the size of Switzerland.
2000	Plan Colombia is approved, with multi-billion-dollar investment from the United States, Colombia, EU, and the greater international community.
2002	Uribe elected president, enacts aggressive policy changes against insurgent groups, state of emergency declared.
2003	AUC commences disarmament after successful negotiations with the government.
2004	US government announces first decline in 30 years of drug production activity in Colombia (2002–2004 time frame).
2005	Tensions increase with Venezuela over allegations that they provide support to FARC.
2006	President Uribe elected to second presidential term, based on a platform of continued actions against insurgent groups and drug traffickers.

THE ENVIRONMENT OF THE REVOLUTION

PHYSICAL ENVIRONMENT

Figure 1. Maps of Colombia.[1]

---

[1]   Central Intelligence Agency, "Colombia," *The World Factbook*, accessed November 2, 2009, https://www.cia.gov/library/publications/the-world-factbook/maps/maptemplate_co.html; https://www.cia.gov/library/publications/the-world-factbook/maps/co_largelocator_template.html.

Colombia is the only South American country with access to both the Atlantic and Pacific Oceans and is considered one of the most biodiverse countries in the world. Land boundaries include Brazil, Ecuador, Panama, Peru, and Venezuela. The country has more than 5,000 miles of coastline, and more than 6,000 miles of navigable rivers. In area, Colombia is approximately twice the size of the US state of Texas. Colombia's terrain varies from mountains that extend to 17,000 feet, to plateaus, plains, and ranges that continue to the Amazon Basin.[2] Colombia is rich with natural resources, including oil, natural gas, and coal and is prone to natural disasters, including volcanic eruptions and earthquakes.[3]

CULTURAL AND DEMOGRAPHIC ENVIRONMENT

Colombia's population is approximately 43 million people. It is diverse socially, culturally, and economically—from the working peasant class in rural parts of the country to an educated and upwardly mobile middle class and upper class, living in or moving to the urban areas—in a style and manner not dissimilar to some American cities. The working and peasant class continue to work the land, as it is still culturally believed that the earth gives man strength.[4] Quality of life has degraded because of economic stagnation, limited international investment, and continued violence and the threat of violence from insurgent groups, narco-traffickers, and common criminals. Additionally, intellectual and economic flight from Colombia to other countries, principally by those who had the means to leave and live abroad, has been a continuing problem since the 1980s. Overall, approximately 1 million Colombians, 2.5% of the population, have been displaced, either internally or externally, because of ongoing violence.[5] Continued violence has taken a serious toll on the people at both macro and micro levels. Colombia has one of the highest homicide rates of any county in the world—50 times higher than a typical European country.[6] In 1995, 58% of municipalities in Colombia reported some type of insurgent presence in their local area.[7] Violence

---

[2] Stephen P. Weiler, "Colombia: Gateway to Defeating Transnational Hell in the Western Hemisphere" (master's thesis, US Army War College, 2004), 6.

[3] Central Intelligence Agency, "Colombia," *The World Factbook*, accessed November 2, 2009, https://www.cia.gov/library/publications/the-world-factbook/geos/co.html.

[4] Stephen Gudeman and Alberto Rivera, *Conversations in Colombia: The Domestic Economy in Life and Text* (Cambridge: Cambridge University Press, 1990).

[5] Marcelo Giugale, Oliver Lafourcade, and Connie Luff, *Colombia: The Economic Foundation of Peace* (Washington, DC: World Bank Publications, 2002), 39.

[6] Ibid., 36.

[7] Ibid., 37.

has increased in both rural and urban areas, with youth and ethnic minorities being most affected. The population is composed of 58% Mestizo, 20% Caucasian, 14% Mulatto, 4% Black, and 4% mixed Black-Amerindian or Amerindian. Ninety percent describe themselves as Roman Catholic. Spanish is the official and predominant language throughout the country. Seventy-four percent of the populace live in an urban environment, with a rate of urbanization estimated at 1.7% annually.[8]

SOCIOECONOMIC ENVIRONMENT

More than 54% of the Colombian populace live in poverty. Many live on less than $2 per day. During the 1990s, the country faced a long recession, with negative growth in the agricultural sector. As a result, rural employment decreased while poverty rates continued to increase.[9] Colombia's major trading partners include Brazil, Mexico, Venezuela, and the United States. The United States sees long-term economic development as a means of supporting stability in Colombia. International investments have been focused on the natural resources sector, particularly oil, gas, and chemical manufacturing.[10]

Because of the long-term presence of multiple armed insurgent groups that have been active within Colombia for decades, there is a general social-cultural construct regarding the perception of insurgencies. Perez describes the popular response to the FARC/ELN insurgencies by three distinct phases, or time periods:

> The attitude of the Colombian people can be divided into three periods: Indifference, Coexistence, and Rejection. These periods have been determined according to the level of impact by illegal groups on the civilian population. Indifference was the attitude during the 1960's and 1970's when guerrilla groups were emerging and expanding, as their actions were generally concentrated in isolated regions of the national territory. The framework of the Cold War helped lead the people to perceive communist groups as messianic organizations, particularly among some sectors of the poor, students, and leftist parties. In the urban areas, the problem was seen as a peripheral matter, and

---

[8] Central Intelligence Agency, "Colombia," *The World Factbook.*

[9] Ricardo Vargas, *The Revolutionary Armed Forces of Colombia (FARC) and the Illicit Drug Trade* (The Netherlands: Transnational Institute, 1999).

[10] Weiler, "Colombia: Gateway to Defeating Transnational Hell," 3.

people acted as would a neutral spectator watching a football game. The results of the struggle were remote from their interests; only peasants in the countryside were directly affected by the violent situation.

Coexistence came in the 1980's; at that time guerrilla groups achieved considerable ability to disrupt the country. Their illegal actions approached the main cities, far beyond the peasant population. Then, farmers, ranchers, businessmen, industrialists and landowners became targets of guerrillas, who asked them for economical support and that they not denounce the guerrillas to the authorities for their criminal activities. To combat the situation, most of the affected people tried to obtain the guerrillas' consent by paying extortion money, ransoms, and supporting them with logistical activities. Silence with the authorities on guerrilla movements was an extra "charge." Some people decided to confront the problem by creating and supporting self-defense groups to counter the growing threat.

Rejection started in the 1990's when indiscriminate terrorist actions spread throughout Colombia and the civilian population was the focus of the attacks. This environment convinced the people that guerrillas posed a significant threat . . . The feelings of civil society changed from disinterest to a decisive desire to support an initiative that confronted this threat once and for all.[11]

Violence throughout the country has increased as economic disparity throughout the country has also increased. "In 1975, an urban family earned 1.5 times more than a rural family, but 20 years later it earns 4.5 times more."[12] Continued violence has created an unfavorable investment environment while also taking a measurable toll on society in terms of increased military and security costs, loss of life, and damage to economic infrastructure (e.g., oil pipelines, a frequent target of the ELN). Violence and insurgent activities add up to an estimated cost of 18% of the gross domestic product (GDP), as estimated by the World Bank and other organizations.[13] Economic loss

---

[11] William F. Perez, "An Effective Strategy for Colombia: A Potential End to the Current Crisis" (master's thesis, US Army War College, 2004), 2–3.

[12] Giugale, Lafourcade, and Luff, *Colombia: The Economic Foundation*, 42.

[13] Ibid., 45.

includes that of farmers and landowners unable to even gain access to their land because it is occupied and even used by insurgent groups or narco-traffickers. In real terms, "Colombia's annual GDP growth fell from an average of 5% between 1950 and 1980 to 3% between 1980 and 2000."[14] This long-term decline is attributed to a continued fall in production, largely due to the scale of violence via insurgencies, crime, and narco-trafficking activities.

HISTORICAL FACTORS

Colombia has a history of armed insurgency, beginning in the 1920s. Traditional friction points included land use and perceived abuses as well as working conditions of the agricultural/peasant class. In the southern area where coffee is produced, armed resistance erupted during the 1930s, with the government reacting to the resistance movements with overwhelming force, leading to organized, armed resistance from the left that lasted for more than a decade.[15] By the 1960s, the peasant class had formed armed self-defense organizations.[16]

In 1948, a political uprising began that led to a period called *La Violencia* (the Violence). Today, the period of 1946–1965 is generally referred to as *La Violencia*, and it remains an emotional issue for many Colombians. In 1945, the Conservative Party won the presidency and national election. Liberals, however, coalesced around a new populist charismatic leader, Jorge Eliécer Gaitán. Gaitán was assassinated in 1948, and Conservatives were quickly blamed. Resulting violence, initiated by the Liberals, burned much of the capital of Bogota. Peasant groups armed and organized, following a community self-defense model. This process was referred to as "armed colonization."[17] They also considered new social constructs at the local level, including socialism. Armed defense and local socialist politics became the central tenets of a growing insurgent movement. The country was polarized into conservative and liberal spheres of influence. Stories of violent atrocities are repeated through familial lines to this day, with "good guys" and "bad guys" related to the family association with either liberal or conservative sides. Estimates of the total killed over the resulting seventeen years vary greatly, from 100,000 to 300,000, with 200,000 generally accepted as the approximate number of

---

[14] Ibid., 46.

[15] Vargas, *FARC and the Illicit Drug Trade.*

[16] Eduardo Pizarro, *Las FARC (1949–1966): De la Autodefensa a la Combinación de Todas las Formas de Lucha (Sociologia y Politica)* (Bogota: Tercer Mundo Editores, 1992).

[17] Vargas, *FARC and the Illicit Drug Trade,* 2.

deaths. Violence and polarization did not end in the 1960s. Rather, it transitioned for the Liberals into either support for the FARC or ELN, while the Conservatives continued to rally around the government, which was seen as the defender of the status quo.[18] In 1964, the FARC "declared its intention to use the armed struggle as part of a political strategy to seize national power."[19] The FARC was already establishing bases of support in several geographic areas by the mid-1960s.[20]

Colombia's political system continued to be largely exclusionary during the 1950s and 1960s. A power-sharing agreement adopted in the 1950s among the two major political parties largely left power in the hands of the elite. Any threat to the status quo was thwarted, frequently by declaring states of emergency, providing justification for security forces to respond aggressively to any leftist organization.[21]

During the 1950s, the Amazon Basin began to be colonized by peasants, following their displacement as a result of *La Violencia*. Coca proved to be the crop most hospitable to the environment, with easier production and greater potential for profit than any competitive cash crop. Coca would continue to flourish as the crop of choice for much of south and southeast Colombia.[22]

GOVERNING ENVIRONMENT

The Colombian government is a functioning democracy, with the elected president serving as head of state and commander of all military forces. The president is elected every four years, with no option of reelection.[23] Perez has stated that the change of government every four years leads to a lack of continuity and that this situation is exploited by insurgent groups, who have no such mandate— they maintain the same leadership and objectives year after year. The result, since the inception of the FARC and ELN, has been a continuing, ever-present shift in policies by administrations. Actions have included various states of emergency, offers of amnesty, mediated peace initiatives, and extensive, offensive military operations. This continuing flux of policy, Perez states, has caused a lack of confidence in the Colombian people on national resolve and ability to counter

---

[18] Steffen W. Schmidt. "La Violencia Revisited: The Clientelist Bases of Political Violence in Colombia," *Journal of Latin American Studies* 6, no. I (1974): 97–111; Vargas, *FARC and the Illicit Drug Trade.*

[19] Vargas, *FARC and the Illicit Drug Trade*, 1.

[20] Ibid.

[21] Ibid.

[22] Ibid.

[23] President Uribe effectively changed the law in 2005, leading to reelection in 2006.

the various insurgent organizations; and he also notes that the state's ability to function effectively has been undermined by this continuous shift in policies.[24]

Colombia's governing body as it exists today is not how it functioned during most of the twentieth century. Prior to 1958, the country was ruled by a military dictatorship. From 1958 to 1986, conservatives refused to participate in the government, although it was formally considered a coalition government. "After thirty years, the country was transformed socially, economically and demographically with little political change. The country's population doubled, and it became far younger, better educated, and more urban."[25]

Since 1983, Colombia instituted a policy of decentralization, transferring revenue from a national value-added tax (VAT) to municipalities and regional governments. As a result, funds were increased in areas of health and education. Decentralization was implemented in order to encourage greater participation in government.

During 1991, constitutional reform was enacted. Reforms did improve the legitimacy of the government by instituting improvement in the check-balance system, strengthening the position of the president, and improving the national electoral voting system.[26] Additionally, democratic participation increased with the enactment of reforms. Gamboa notes "A democratic institution is emerging . . . after a long period of ignoring or even denying the need for one, and it is incapable of resolving the problems produced by its negation."[27]

Election abuses and violence have been widespread throughout Colombia since at least the 1980s. The right-wing paramilitary Autodefensas Unidas de Colombia–United Self-Defense Forces of Colombia (AUC) has conducted active targeting and death-squad operations against left-wing candidates, while the FARC and the ELN have also actively dissuaded people from either voting or running for office.[28]

---

[24] Perez, "An Effective Strategy for Colombia," 5.

[25] Jonathan Hartlyn, *Drug Trafficking and Democracy in Colombia in the 1980s*, Working Paper no. 70 (Barcelona: Institut de Ciencies Politiques i Socials), 5, 6.

[26] Mauricio Cárdenas, Roberto Junguito, and Alberto Alesin, "Political Institutions, Policymaking Processes, and Policy Outcomes: The Case of Colombia," Research Proposal Presented by Fedesarrollo to the Inter-American Development Bank (2004).

[27] Miguel Gamboa, "Democratic Discourse and the Conflict in Colombia," *Latin American Perspectives* 116, no. 28 (2001): 93–109.

[28] Joe Foweraker and Roman Krznaric, "The Uneven Performance of Third Wave Democracies: Electoral Politics and the Imperfect Rule of Law in Latin America," *Latin American Politics and Society* 44, no. 3 (2002): 29–60.

WEAKNESSES OF THE PREREVOLUTIONARY ENVIRONMENT AND CATALYSTS

Since the 1960s, Colombia has been facing the challenge of dealing with multiple left-wing insurgent groups. The FARC formed in 1964, and the ELN formed in 1966.[29] The FARC specifically track their inception to the seventeen-year period of violence known as *La Violencia* (The Violence), which killed approximately 200,000 Colombians.

The FARC and the ELN both base their foundations on Marxist doctrine. While the FARC established a base within the rural, working class, the ELN established a foundation within the student movement of the universities. The ELN also garnered ties with the Catholic Church and liberation theology, similar to the association of the Salvadoran-based *Frente Farabundo Martí para la Liberación Nacional* (FMLN) with the Catholic Church.

The FARC represent the largest and most capable armed insurgency within Colombia. They maintain a force in both urban and rural areas of the country.[30] Although the FARC started as a predominately local movement with a rural support base, maintaining control over limited physical space in the rural southern section of Colombia, it now has national standing and is capable of influencing political and military action in vast areas of the country. Original grievances of land reform have now increased to include "charges of corruption, the perversion of capitalism, and US imperialism to its motivations."[31] During the 1960s and 1970s, the government's failure to identify the threat the organization represented, and their failure to respond aggressively, provided the FARC with several years of political and military growth among their peasant, working class, and rural base.[32]

At the start of the FARC insurgency in the 1960s, a minority of its founders stressed the need to expand the base of support beyond the peasant/working class; after all, the FARC represented just one of several armed insurgencies in the country. In addition to the ELN, the Popular Liberation Front (EPL) (founded in 1967) and 19th of April Movement (M-19, founded in 1970) were also vying for a left-wing support base. There was, in fact, competition between the Communist

---

[29] Perez, "An Effective Strategy for Colombia."

[30] James Petras, "Geopolitics of Plan Colombia," *Economic and Political Weekly* 35, no. 52/53 (2000): 4617–4623.

[31] Jennifer S. Holmes, Sheila A. G. De Piñeres, and Kevin. M. Curtin, "A Subnational Study of Insurgency: FARC Violence in the 1990s," *Studies in Conflict and Terrorism* 30, no. 3 (2007): 249–265.

[32] Vargas, *FARC and the Illicit Drug Trade.*

Party, FARC, ELN, EPL, and M-19 with regard to recruiting and their support bases. For the first decade, the FARC maintained marginal political and military capabilities. Internal heated discussions were common. Only a few fronts were operational, including five in the south, two in the central region, and one other in the north.[33]

During the 1980s, both the FARC and the ELN continued to see their political and military power grow. In 1983, with 18 military fronts active, the FARC changed their title to FARC-EP (Ejército del Pueblo, or Army of the People). Growth was attributed to continued intransigence by the government in not assuaging concerns regarding worker reforms and land rights. The FARC, as well as the ELN, became associated with viable alternatives to the government, a means of reform, and institutions that defended the causes of the worker and peasant classes. Government opposition to any reforms was fueling the causes and bringing in fresh recruits.[34]

FORM AND CHARACTERISTICS OF THE REVOLUTION

OBJECTIVES AND GOALS

Since the 1980s, the operational environment has become very complex, with the emergence of paramilitary self-defense groups and the collaboration of insurgent groups with drug cartels.[35] Objectives and goals of the FARC have transitioned many times since its inception in the 1960s. While the FARC was formed under the nexus of a Marxist–Leninist ideology, that was transitioned to a system described as Bolivarian, which includes a combination of nationalist and leftist ideals (i.e., land and social reform). Román Ortiz, of the Gutiérrez Mellado University Institute in Madrid, Spain, notes that the FARC have questioned the legitimacy of the government and presented themselves as a viable alternative. "Somehow, the Colombian guerrilla movement has gone from criticizing the legitimacy of the origin of the state to questioning its functional legitimacy. This political transformation has manifested itself in various aspects of the strategic orientation of the organization."[36] The FARC's original position and objectives were based on land and social reform. It is fair, however, to

---

[33] Ibid.

[34] Ibid.

[35] Perez, "An Effective Strategy for Colombia."

[36] Román D. Ortiz, "Insurgent Strategies in the Post-Cold War: The Case of the Revolutionary Armed Forces of Colombia," *Studies in Conflict & Terrorism* 25, no. 2 (2002): 127–143.

state that the FARC does not have a defined ideology—certainly not when compared, for example, to the Communist Party of Colombia (PCC). Saskiewicz calls this an "ideological cocktail" and states "the hybrid nature of the FARC-EP's political ideology was and continues to be reflected in the eclectic membership of the insurgency. In fact, the FARC-EP finds its unity not in its ideology, but in its strategy, which is its commitment to the 'armed struggle.' "[37]

From 1982 to 1990, the FARC adopted a strategic plan entitled "Strategic Plan for Taking Power," which was divided into three phases (offensive, government, and defense of the revolution). The plan was based on overthrowing the military and civilian government, culminating with an urban offensive in Bogota, then establishing authority and establishing a new regime that would lead the cause to the drafting of a new constitution. The plan was to cut off vast parts of the country, leading to the eventual siege on Bogota.[38] Militarily, the plan made little sense and seemed as though it could have been concocted during medieval times. Probably more significant than the adoption of the strategic plan, however, was the accompanying tactical employment of FARC combatants. Offensive action against Colombian Army forces was stressed, based on surprise, mobility, and secrecy.[39]

By 1989, after several attempts at peace talks had failed, the FARC adopted the "Bolivarian Campaign for the New Colombia," a four-phased, potentially eight-year plan. Phases included (1) building the military force, (2) increasing FARC membership and areas of action to 80 designated fronts, (3) employing new offensive action (although Bogota continued to be the military focus), and (4) creating a contingency plan for retrograde operations and reorganization, if phase 3 failed.[40] The FARC also increased their emphasis on attacking the national infrastructure; this included sabotage of electricity, telecommunications networks, and roads and bridges. The goal was to undermine the legitimacy of the government while also causing a negative economic impact.

---

[37] Paul E. Saskiewicz, "The Revolutionary Armed Forces of Colombia–People's Army (FARC-EP): Marxist-Leninist Insurgency or Criminal Enterprise?" (master's thesis, Naval Postgraduate School, 2005), 12.

[38] Ibid.

[39] Ibid., 22.

[40] Ibid.

LEADERSHIP AND ORGANIZATIONAL STRUCTURE

During the 1980s, there were several active left-wing insurgent groups in Colombia, with the decade overall being an active dynamic period of growth for multiple insurgent organizations. Both the FARC and ELN faced internal tension regarding how to grow and how to balance political growth with increasing military capabilities. At first, the FARC had applied the concept of combining legal with illegal actions—political engagement with military operations—but this concept failed. Left-wing groups organized politically under the banner of the Patriotic Union (Union Patriotica), or UP. The UP coalition had the goal of organizing a left-wing coalition political party, based on Communism. The government, however, successfully targeted UP political members through arrest and direct action, resulting in its disbandment by the 1990s. As the UP disbanded, the FARC, which had continued to organize and gain military and political capabilities, became the key insurgent force in the country, without equal in Colombia.[41]

During the early 1990s, the FARC organized into seven operational regions. Each region, which controls political and military activity, has an associated military block. These blocks include the Northern (Caribbean), Northwestern (bordering Panama), Middle Magdalena (along the Venezuelan border), Central, Eastern, Western, and Southern blocks. Each block contains subordinate elements called fronts, with five to fifteen fronts per block. The subordinate element to a front is a column, and the subordinate element of a column is a company. A FARC company, however, is much smaller than an associated US Army or Marine Corps rifle company. A column is normally associated with 100–150 combatants. The element normally associated with conducting specific tactical missions is the column. The FARC has approximately 67 specific fronts, with 17,000 total combatants.[42]

FARC leaders also recognized the need for strategic organization. As a result, the Estado Mayor Central (EMC) was formed, with the responsibility of strategic leadership for the entire organization. Additionally, formal training centers were established, providing a means of standardizing training, implementing a code of conduct,

---

[41] Vargas, *FARC and the Illicit Drug Trade.*

[42] Maria A. Vélez, "FARC-ELN: Evolución y Expansión Territorial," *Revista Desarrollo y Sociedad* no. 47, 151–225 (2001); Holmes, De Piñeres, and Curtin, "A Subnational Study of Insurgency: FARC Violence in the 1990s."

and legitimizing the group's principles internally.[43] By 1987, the FARC had established a formal military academy in Caqueta Province.

There may be some vulnerabilities and weaknesses associated with this structure, however, as noted by Ortiz:

> The problem with this structure lies in the fact that orders and resources flow in opposite directions. Obviously, the definition of primary strategic directives and the decisions regarding large-scale operations are taken by the general staff, who communicate them to the blocks, and, from there, to the fronts. However, the economic resources are collected from the base of the organization. In fact, in most cases it is the fronts that collect all the payments from drug dealers or the ransoms from the kidnappings. Part of these funds are destined for the maintenance of the unit that collects them, whereas another part is passed up to higher levels . . . for common use by the organization as a whole . . . the high degree of decentralization is very effective from a strategic point of view, since it increases the organization's flexibility . . . this structure facilitates fictionalization. If, for example, the commander of a front or a block does not agree with some decision of his superiors, he may decide to appropriate all the funds collected by his unit, instead of handing them over to his superiors. The means for gaining independence are thus very easily available and in a country like Colombia, with a high degree of political and social fragmentation, and within a guerilla organization characterized by weak ideological cohesion, such divisions are a real possibility.[44]

Ortiz notes four ways that the FARC have achieved success, where other Latin American insurgent groups have not:

> First, the FARC has reduced rigidity to its ideology in order to make its political message more attractive. Second, it has made a great effort to boost its military potential. Third, it has established independent channels of funding and arms supply. Finally, the Colombian rebels have developed a very decentralized organic structure that nevertheless maintains a sufficient degree

---

[43] Saskiewicz, "The Revolutionary Armed Forces of Colombia–People's Army."

[44] Ortiz, "Insurgent Strategies in the Post-Cold War: The Case of the Revolutionary Armed Forces of Colombia," 140.

of cohesion. These innovations have made the FARC a new model of insurgency that has managed to corner the Bogota government and destabilize a significant part of the Andean region.[45]

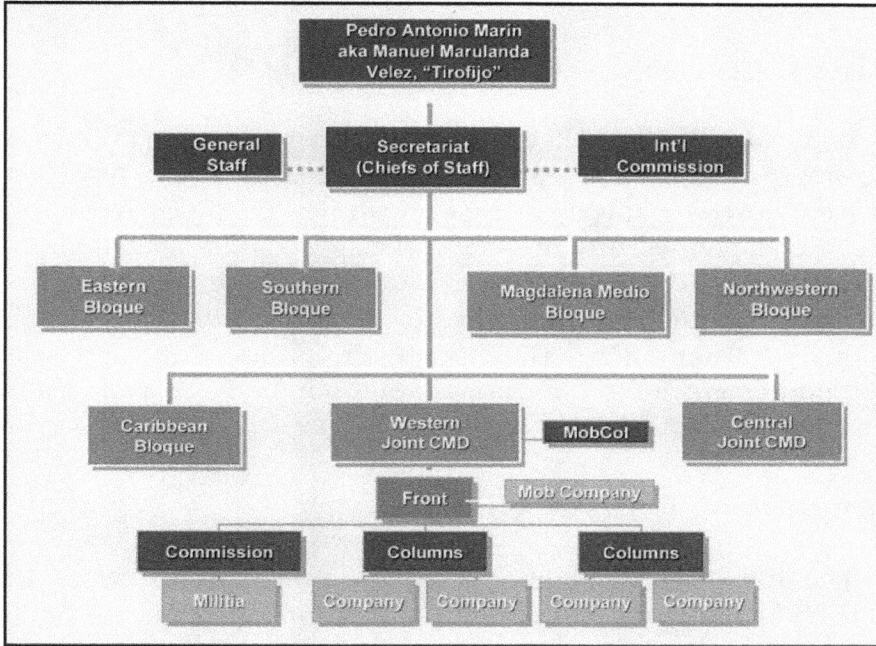

Figure 2. FARC organization structure, 1993 time frame.[46]

FARC relationships with the PCC have also transitioned over time. For the first decade, the PCC considered itself to be the controlling body of the FARC. By the 1970s, however, there were clear distinctions between formal statements of the PCC and those of the FARC, indicating the ongoing political transition of the FARC to Bolivarianism. By 1978, the FARC had established formal leadership structure, guidelines, and a secretariat that were completely separated from the PCC. FARC also took charge of its own political ideology and since that time, the PCC and FARC have maintained their own identities. In 1983, FARC leader Manuel Marulanda said "within our organization there is a little of everything. There are Marxist-Leninists too. But above all the FARC is an armed guerrilla organization which opens its doors to all political, philosophical, religious and ideological tendencies, and which brings together people with the common

[45] Ibid., 127.
[46] Darren D. Sprunk, "Transformation in the Developing World: An Analysis of Colombia's Security Transformation" (master's thesis, Naval Postgraduate School, 2004), 35.

ideal of liberating this country."[47] No one from the PCC could have ever imagined making such a statement. In 1990, influential FARC leader Jacobo Arenas died. Arenas had served for years as the political link between the FARC and PCC. When Arenas died, so did the link between the two organizations.

COMMUNICATIONS

By the 1990s, the FARC had the financial resources to purchase the best communication equipment available.[48] Tactically, VHF radios were employed organically at the company level. FARC also employed satellite telephones, the Internet, messengers, and even cell phones. During their countless raids against Colombian security forces, they also collected an abundance of weapons and communication equipment from the government, although frequently the weapons and equipment were not up to the standards that the FARC considered acceptable.

METHODS OF ACTION AND VIOLENCE

The 1990s saw a change in FARC (as well as ELN) tactics, with increased use of terrorism as a means of intimidating and controlling the greater populace. Kidnapping and extortions continued as sources of violence, but attacks on targets of infrastructure, such as electrical substations, increased. Oil pipelines were also routinely targeted, especially by the ELN. Government security forces also now faced well-organized military forces, capable of conducting offensive operations such as ambushes and even attacks on security outposts. The FARC especially were capable of striking targets in urban areas as well as in their stronghold fronts in the central and southern parts of the country. FARC and ELN forces also increased their use of assassinations of political leaders as both a means of intimidating the populace and influencing local elections.[49] Vargas noted that "Their armed capability has grown because of their ability to take military—more than political—advantage of cracks in a highly decayed regime that requires reforms of the existing socioeconomic and institutional framework.[50]

---

[47]   Ibid., 134.

[48]   Saskiewicz, "The Revolutionary Armed Forces of Colombia–People's Army."

[49]   Perez, "An Effective Strategy for Colombia."

[50]   Vargas, *FARC and the Illicit Drug Trade*, 2.

With the death of FARC leader Jacobo Arenas is 1990, the new FARC leadership shifted standing policies drastically, establishing relationships with the powerful narco-traffickers in the Amazon Basin region. Leadership saw an increasing role in drug trafficking as a means of (1) increasing revenue, (2) more closely linking their work with that of the peasant-class farmers, and (3) expanding power and influence geographically. While levels of influence in the illicit drug trade are difficult to quantify, it is estimated that from drug trafficking the FARC realized an annual profit of approximately $700 million. As a result of the expansion into the drug trade, the FARC also established "business" relationships with external organizations, including the Russian mafia and the Tijuana, Mexico drug cartel.[51]

The FARC have embraced the position as a pseudo-government, providing social services, including health services and education, to civilians in the areas they control. In so doing, they have added to their legitimacy while garnering additional support from the populace, governing where the government could not.

FARC military activities include the use of special forces for direct action missions and specialized commando-like raids making use of heavy weapons including mortars, rocket launchers, and heavy machine guns, as well as extensive use of improvised explosive devices (IEDs).[52]

By the 1990s, FARC military capabilities had grown to a level seldom seen by insurgent forces anywhere. During March 1998, the Colombian Army's 52nd Counter-Guerrilla Battalion, an elite military unit, was ambushed and effectively destroyed by the FARC, leaving sixty-two soldiers dead and forty-three taken prisoner.[53] During August of that same year, 1,200 insurgent fighters massed and attacked a Colombian National Police anti-narcotics base at Miraflores, Guaviare Province. More than one hundred police officers were kidnapped, with thirty more killed and fifty more wounded. The FARC had demonstrated the ability to amass forces and launch sophisticated military operations. They had 18,000 well-trained and dedicated fighters in the field, and the Colombian security forces were being fought on parity, or worse.[54] By 1999, the FARC launched even larger

---

[51] Frank Cilluffo, "The Threat Posed from the Convergence of Organized Crime, Drug Trafficking, and Terrorism," Testimony before the US House Committee on the Judiciary, Subcommittee on Crime, 106th Cong., 2nd Sess. (2000), 4–5.

[52] Ortiz, "Insurgent Strategies in the Post-Cold War: The Case of the Revolutionary Armed Forces of Colombia," 127–143.

[53] Saskiewicz, "The Revolutionary Armed Forces of Colombia–People's Army."

[54] Ibid.

coordinated attacks across multiple fronts—at one point having 4,000 combatants in offensive action.

METHODS OF RECRUITMENT

FARC generally recruits men and women between the ages of 15 and 30. Forced recruitment is rare and is contrary to the FARC "safety rules." Recruitment by both FARC and ELN groups includes maintaining a presence on public and private university campuses. During the 2006–2008 time frame, desertion became a concern for the FARC. Losses on the battlefield may have corresponded to an increase in the desertion rate. For many, joining the FARC represents a steady "job" that includes pay, possibly education benefits, and some social status at the local level. Maintaining popularity with the peasant and working classes in traditional stronghold areas means that the FARC and ELN have been able to maintain a steady flow of new recruits without having to expend many resources on this traditionally challenging activity.[55]

METHODS OF SUSTAINMENT

As the FARC continued to grow, particularly throughout the 1980s, new sources of revenue and resources were required to sustain the force. The FARC began selling resources, such as cattle, commercial agriculture, oil, and even gold, on the black market. The FARC sometimes served as brokers between wholesalers and retail markets, extorted from business enterprises, and, in other cases, simply stole the assets and then sold them directly on the black market. In some cases, the FARC established a presence in an area with the express interest of exploiting resources and increasing revenue.[56]

When the *Despeje* demilitarized zone was introduced by the Pastrana administration in hopes of resolving the conflict through negotiation, the FARC took advantage of the opportunity and used funds they had acquired via their illicit activities to acquire additional resources including 20,000 rifles, mortars, man-portable air defense

---

[55] Immigration and Refugee Board of Canada, *Colombia: The Recruitment Methods of the Revolutionary Armed Forces of Colombia (Fuerzas Armadas Revolucionarias De Colombia, FARC) and Government Measures to Help FARC Members Reintegrate into Civilian Society (2005–February 2008)*, COL102787.FE (April 14, 2008), http://www.unhcr.org/refworld/docid/4829b55c23.html.

[56] Saskiewicz, "The Revolutionary Armed Forces of Colombia–People's Army," 79.

weapons (MANPADs), electronic communications equipment, and even aircraft.[57]

FARC methods of sustainment continued to expand as the organization itself grew since its inception. As of 2010, FARC raised cash by extortion of businesses, including legitimate and illegitimate entities, and by kidnapping. With coca production continuing to be a major illegitimate cash crop for southern Colombia, it is estimated that the FARC taxed traffickers approximately 8–10% of the value of their coca. FARC relationships with drug traffickers were also reported as a source of friction for the insurgents, with some drug traffickers building their own military capability to protect their land and assets from any threat, whether it be the government or an insurgent group.[58] As of 2010, the FARC and ELN had aggregate revenues of approximately $900 million (US dollars) annually. Approximately $500 million of that revenue resulted from drug-related activities, with the remaining coming from kidnappings and other acts of extortion. Shifter notes that "their criminal activities help sustain a political agenda."[59] The ELN, operating generally in the northern areas, which is the source of oil, also received untold amounts of money from the oil industry. These funds may have resulted from the capture and resale of oil, extortion of direct funds from oil producers, kidnapping of oil workers, or other illegitimate activities.[60]

Ortiz attributes much of the FARC success to their logistics sophistication:

> The guerrillas have developed a supply network that combines legal and illegal operations as well as state and private suppliers. This has resulted in a logistical web that is extremely difficult to break up due to its range and diversity of connections … the FARC has had no problems in acquiring equipment essential to establishing a dense network of communications. [Additionally] there is a growing capacity for the homemade manufacture of relatively crude yet extremely effective weaponry . . . Finally, there exists the possibility of their tapping

---

[57] Ibid. Original source: United States Congress, House of Representatives, House Appropriations Committee Subcommittee on Foreign Operations, Written Statement of Major General Gary D. Speer, US Army, Acting Commander in Chief, US Southern Command, Before the 107th Cong. Senate Foreign Relations Committee, Subcommittee on Western Hemisphere, Peace Corps and Narcotics Affairs (April 24, 2002).

[58] Holmes, De Piñeres, and Curtin, "A Subnational Study of Insurgency: FARC Violence in the 1990s."

[59] Michael Shifter, "Colombia on the Brink: There Goes the Neighborhood," *Foreign Affairs* 78, no. 4 (1999): 15.

[60] Ibid.

a black arms market that, after the end of the Cold War, is capable of supplying combat equipment of unprecedented quantity and quality. In fact, the name of the Colombian guerrilla force has appeared linked to large clandestine arms deals.[61]

METHODS OF OBTAINING LEGITIMACY

Both the FARC and ELN have effectively used information operations campaigns to highlight possible human rights abuses, or even violations of the Law of Land Warfare, conducted or allegedly conducted by Colombian security forces. FARC and ELN forces have depicted themselves as the protectors of human rights, although the track records for both—including terrorist attacks and the indiscriminate killing of other noncombatants—tell a very different story. FARC specifically targeted Colombian and international media to highlight any human rights violations by the Colombian Military (COLMIL) whenever possible. Their appeals successfully drew the attention of nongovernmental organizations (NGOs). Legitimacy of the COLMIL was undermined successfully, particularly during the 1980s. As a result, however, COLMIL forces in the 1990s and 2000s gave greater attention to human rights, which ended up having the opposite effect: public perception of the military increased over time and decreased for the insurgent forces. Human rights principles became a major tenet of success for the COLMIL.[62]

EXTERNAL SUPPORT

The FARC and ELN initially maintained ties with Fidel Castro and the Cuban government for political and (possibly) military support, but details on the level of that support are limited. While Castro's revolution clearly served as an initial motivator, particularly for the FARC, those relationships were not maintained.[63] Material support from Cuba or the Soviet Union was never planned or required. FARC leaders were concerned about external political influence on the organization.[64]

---

[61] Ortiz, "Insurgent Strategies in the Post-Cold War: The Case of the Revolutionary Armed Forces of Colombia," 138.

[62] Perez and Army, "An Effective Strategy for Colombia," 10, 11.

[63] Shifter, "Colombia on the Brink: There Goes the Neighborhood."

[64] Saskiewicz, "The Revolutionary Armed Forces of Colombia–People's Army."

Venezuela and the Chavez administration have maintained an active hand regarding insurgent organizations in Colombia, principally with the FARC and ELN. Venezuelan officials, however, have continued to deny any allegations of support for these insurgent groups. US media sources (including *US News & World Report*) have published information that supports the existence of insurgent camps inside Venezuelan territory. FARC deserters have also corroborated information on support at the local level from within Venezuela. One camp, the Resumidero base, has even been described as supporting 700 people. It has been alleged that Venezuela has provided safe havens for insurgent fighters and leadership, supported training sites, and harbored command and control locations.[65] Venezuelans have also been victims of the insurgent organizations, as Venezuelan citizens have been extorted and even kidnapped for ransom, in the same way that Colombian citizens have suffered.[66] Venezuelan officials have also been associated with "materially assisting the trafficking activities of the FARC" according to the US Department of Treasury, Office of Foreign Assets Control.[67]

The FARC is also reported to use territory in Ecuador, along Colombia's southern border, for safe haven and training, as well as some illicit business activities associated with coca production. Colombian military forces conducted a raid on a FARC safe haven within Ecuador during 2008. To the northern border of Panama, FARC insurgents are routinely reported in the Darien province, including entering towns for provisions. The Panamanian government monitors FARC activities closely, occasionally capturing and arresting insurgents.[68] With the exception of the Venezuelan government, neighboring countries generally have not been accused of actively supporting the FARC and ELN.

The FARC have also received technical training and support from other terrorist and insurgent groups. Throughout most of the 1980s, the FARC and FMLN maintained relationships. FARC combatants were trained in tactical operations and large-scale raids. They may have also received technical training and support from organizations as diverse as Hizbollah, Japanese Red Army, and the Provisional Irish Republican Army.[69]

---

[65] Linda Robinson, "Terror Close to Home," *US News & World Report* (October 6, 2003).

[66] Mark P. Sullivan,"Latin America: Terrorism Issues," in *Focus on Terrorism*, vol. 9, ed. Edward V. Linden (New York: Nova Science Publishers: 2007): 301.

[67] Ibid., 2.

[68] Ibid.

[69] Ortiz, "Insurgent Strategies in the Post-Cold War: The Case of the Revolutionary Armed Forces of Colombia."

COUNTERMEASURES TAKEN BY THE GOVERNMENT

The Colombian government had a history of using the military forces to protect the status quo, going back to *La Violencia*, in the 1940s and 1950s. Perez describes this as a "militarizing spirit" within Colombian leadership.[70] He further describes the initial military response to the FARC and ELN:

> Civilian authorities considered insurgency as a problem of public order. The responsibility was left to the military, which had to deal with the issue without a coordinated national security policy. Consequently, military operations were at the core of the strategy, and there were few initiatives to carry out solutions other than military, based on a sense of distrust of political leaders toward the armed forces and the lack of leadership to generate popular support. Without a clear strategy, the intensity of conflict grew and the army was lured into the scheme of a protracted popular war, the type of warfare suited to guerrillas.[71]

By the 1990s, the military initiated sweeping reforms and modernization, largely in response to their lack of strategic success against the insurgencies.

During the first five years of the FARC buildup, the Colombian military principally focused on restructuring and modernization, under a program called "Military Forces Facing the 21st Century." The modernization plan had a goal of upgrading the military capabilities against conventional threats, as well as new insurgent threats, of which the FARC and ELN were just two of many. The modernization did allow the COLMIL to take the initiative tactically, with widespread claims of the insurgent forces conducting tactical retreats from areas previously considered strongholds.[72] Perez discussed the COLMIL operational concept:

> The Colombian military forces designed a new vision of operations with emphasis on mobility and rapid reaction, improvement of the collection and processing of information, development of an integrated communications system, and integration of air power in support of ground operations. This new operational concept is based on the implementation of new doctrine,

---

[70] Perez, "An Effective Strategy for Colombia," 3.
[71] Ibid., 3–4.
[72] Ibid., 9.

strengthening of training, improved planning efficiency, better capacity to react day and night, and an increased capability of responsive and agile air support. Mobility, mass and flexibility are the keys to the Colombian military strategy. The Air Force and Army Aviation are now able to conduct night operations using night-vision equipment that have brought major improvement in the employment of air power, integrating operations effectively with the land forces. By moving troops quickly, anywhere and anytime, the Colombian military has neutralized the guerrillas' operational and tactical advantages. Mobility means deploying forces rapidly and safely by air, avoiding guerrillas that ambush the troops and mine the roads that military convoys must cross to reach combat areas. The creation of the Rapid Deployment Force in December 1999 initiated this process. This unit, composed of three mobile brigades and a Special Forces brigade, consolidates a striking force of some 5,000 troops that can be deployed anywhere in Colombia. Despite the fact that the force still has insufficient air transport resources, it has conducted the most successful operations to date throughout the country, becoming a source of pride for the population. As a maxim of combat, before carrying out an operation, commanders in all levels must fulfill three basic elements: "accurate intelligence, excellent planning and correct leadership." This simple phrase sums up the new culture exhibited by the Colombian military that has brought a chain of continuous successes on the battlefield.[73]

During the Samper administration (1994–1998) the level of armed conflict in Colombia further intensified. The FARC demonstrated increasing military prowess and territorial control between 1995 and 1997.[74]

When President Uribe's administration assumed power in 2002, he promised a new strategy in dealing with the insurgents. It contained specific objectives that were discussed during the 1990s and the Samper administration, but never implemented. They included: the consolidation of state control through Colombia

---

[73] Ibid., 11, 12.
[74] Vargas, *FARC and the Illicit Drug Trade.*

to deny sanctuary to terrorists and perpetrators of violence; protection of the population through the increase of state presence and a corresponding reduction in violence; destruction of the illegal drug trade in Colombia to eliminate the revenues which finance terrorism and generate corruption and crime; maintenance of a deterrent military capability as a long-term guarantee of democratic sustainability; and transparent and efficient management of resources as a means to reform and improve the performance of the government.[75]

During the first year of implementation (August 2002–August 2003), rates of kidnappings, murders, and infrastructure attacks all significantly decreased, in large part due to a government security presence established in 158 towns that had previously been without. President Uribe's popularity had also increased as a result of this initial success. Morale increased in the military as the president lauded its successes. One of the most popular new programs adopted was called the "town soldiers" program (Soldados Campesinos), allowing new inductees to remain in their hometown areas as they served their country.[76]

SHORT- AND LONG-TERM EFFECTS

CHANGES IN GOVERNMENT

President Andrés Pastrana was elected in 1998—a pivotal period for the FARC insurgency. The FARC controlled large areas of southern Colombia where drug trafficking was flourishing. They represented the "law of the land" in several territories where the Colombian government had no means of exercising control. Meanwhile, the ELN continued to attack the oil pipelines in the north, negatively impacting the economic lifeblood of the government. The Colombian government needed a strong president, possessing moral courage, strategic vision, and an effective administration. Pastrana was seen as lacking on all counts. His administration was considered corrupt and ineffective. His strategy toward the insurgents, particularly the FARC, was one offering major concessions and incentives to participate in the peace

---

[75] Perez, "An Effective Strategy for Colombia," 6.

[76] Ann Mason, "Colombia's Democratic Security Agenda: Public Order in the Security Tripod." *Security Dialogue* 34, no. 4 (2003): 391; Perez, "An Effective Strategy for Colombia: A Potential End to the Current Crisis."

process. His weak position, however, made implementation of Plan Colombia and several military campaigns difficult, if not impossible. Pastrana assumed that pacification, not military confrontation, would drive the FARC to negotiate. Pastrana even ceded an area of the country as a demilitarized zone, allowing de facto control by the FARC, called the *Despeje*—an area that was the size of Switzerland. The FARC took this amazing opportunity to increase its legitimacy in the eyes of Colombians and the international community. Ambassadors from various countries were invited to the *Despeje* to meet with FARC leadership, further supporting their position as a legitimate governing body. They also used this expansive safe haven to recruit and train members, improve capabilities, and establish governing bodies in local communities. Meanwhile, production, processing, and trafficking of coca and cocaine flourished. As Plan Colombia funds were provided, corrupt government officials filled their personal coffers via embezzlement, cronyism, or facilitation of illegal contracts. Offers of support for Plan Colombia from donor countries in the EU as well as Japan began to waiver. Pastrana's leadership failure had the potential to severely impact international relations, in addition to its negative impact on the internal security of Colombia.[77]

As Alvaro Uribe was inaugurated as the new Colombian president on August 7, 2002, the FARC launched a coordinated mortar attack in the capital, killing twenty and wounding sixty people.[78] The war was already personal to Uribe; insurgents had previously killed his father.[79] With the transition from Pastrana to Uribe complete, the new administration quickly established a different approach to countering the insurgencies. Uribe had successfully run on a platform based on the government establishing effective control throughout the entire country. Upon accepting office, Uribe declared a "State of Limited Emergency," developed a new national security strategy, and took the country's political and security apparatus on a drastically different course than his predecessor. After Uribe had spent less than two years in office, the United States reported that drug trafficking in Colombia had decreased for the first time in 30 years. With assistance from the United States and partner nations under Plan Colombia, Uribe poured resources not only into an increase in end strength of the military but also into improving their capabilities through formal training, new organizations, and new equipment. He also concentrated on protecting the nation's infrastructure, especially

---

[77] Weiler, "Colombia: Gateway to Defeating Transnational Hell."

[78] Saskiewicz, "The Revolutionary Armed Forces of Colombia–People's Army."

[79] Julia E. Sweig, "What Kind of War for Colombia?" *Foreign Affairs* 81, no. 5 (September–October 2002): 122–141.

the pipelines that had been attacked time and time again by ELN insurgents. Coinciding with President Uribe's placement in office, the US Congress lifted the restrictions that stipulated that Plan Colombia funds could be used only for counternarcotics operations. Uribe now had the assets that he needed to counter the insurgents via military action.[80] Uribe also approved the execution of Plan Patriota (the Patriot Plan), which focused on hitting the FARC in their former stronghold and safe haven, the southeast portion of the country (the same area that had formerly been used as a demilitarized zone, the *Despeje*). Plan Patriota represented the largest military offensive ever conducted by the Colombian military. While the operation did not attain a strategic success for the government (nor was it intended to), it did accomplish its objectives of denying the FARC an operational safe haven and reestablishing the government forces in the offensive—a condition that Uribe planned on sustaining.

CHANGES IN POLICY

During 1999, Plan Colombia was presented as a means for the United States, other participating nations, and Colombia to confront the complexities of dealing with narco-traffickers, right-wing paramilitary organizations, and multiple insurgent groups. Plan Colombia was created during the Clinton and Pastrana administrations, with the entire "package" costing $7.5 billion over five years, funded by the Colombian and US governments, as well as the EU. It included political, social, and military facets, focused on the strengthening of Colombian government institutions, integration of isolated areas, and enforcement of the rule of law. It also focused on economic advancement steps aimed at supporting the legitimacy of the Colombian government and institutions, as well as improving social capacity. Most initial funds, however, were focused on security and support to the Colombian military, improving equipment, training, and personnel.[81] Success via Plan Colombia was not envisaged as a military "solution," but as a negotiated settlement to the fighting. Plan Colombia had seven focus areas: (1) alternative crop development, which aimed to provide small farmers with viable alternatives to raising coca and other illicit crops; (2) assistance to displaced persons; (3) the protection of human rights; (4) improvement of government capacity; (5) specific initiatives that focused on the peace process; (6)

---

[80] Saskiewicz, "The Revolutionary Armed Forces of Colombia–People's Army."

[81] Perez, "An Effective Strategy for Colombia," 7.

eradication of illicit crops; and (7) the denial of transportation of illicit crops.[82]

President Uribe developed a new national security strategy, called "Democratic Security and Defense Policy." Its five main objectives were (1) the consolidation of state control and denial of sanctuary to the insurgents, (2) protection of the populace, (3) destruction of the illegal drug trade, (4) maintenance of military capability and assurance of democratic sustainability, and (5) improvement of the performance of the government.

Uribe's strategy was bold and aggressive. He also took a number of pragmatic steps, such as increasing the percentage of GDP to the defense budget from 3.6 to 6%, drastically increasing the number of security forces (from 250,000 to 850,000), and establishing government fiscal reforms to ensure accountability and oversight. Uribe's initiative and leadership were credited with an improved security situation throughout the country, declining violence, and improvements in the economic and internal security status of the country.[83]

Perez notes:

> The problem of violence and security in Colombia increased because the state lacked leadership to integrate a strategy that addressed the political, economic, military, and social dynamics of the conflict. Despite several attempts to resolve the situation, none was able to effectively synchronize the capacity of government, people and the armed forces in a unified effort to deal with this threat. However, this situation changed in 2002 [with the election of Uribe], when a strategy was developed that finally integrated them.[84]

CHANGES IN THE REVOLUTIONARY MOVEMENT

By the 1990s, there was a changing dynamic that impacted the FARC, ELN, Colombian government, and the populace. That dynamic was the ever-increasing illicit drug trafficking in the region. Some drug traffickers built their own security forces, capable of attacking Colombian security forces one day and attacking a FARC column the next. Individual combatants applying an "entrepreneurial spirit" could be fighting with the ELN one day and providing security for a

---

[82] Weiler, "Colombia: Gateway to Defeating Transnational Hell," 9–10.
[83] Ibid., 9–12.
[84] Perez, "An Effective Strategy for Colombia," 2.

narco-trafficker the next. From the Colombian military perspective, it was not uncommon to face tactical engagements without ever knowing what group—insurgent or drug trafficker—had been encountered. Civilians were increasingly caught in the crossfire (sometimes literally), while extortions, kidnappings, and assassinations continued. Parts of Colombia were considered ungoverned.[85]

From an insurgent perspective, there were many challenges. Right-wing paramilitary groups, Colombian security forces, drug traffickers with their own paramilitary organization, and other insurgent groups all posed potential threats to security and growth.

Vargas notes:

> One of the most significant changes in the guerrilla forces in the 1990s has been their increased control over local economic resources and increased economic reserves to fuel their war machine. Guerrillas have become involved in the armed oversight of municipal budget administration, which has involved kidnapping and threatening mayors. They have also been active in gathering intelligence on resource administration at the departmental level. Kidnapping and extortion have become a major source of resources, targeting individuals such as politicians and executives from the petroleum, banana, commercial agriculture, and cattle industries. Finally, the FARC has profited tremendously from its multi-dimensional relationship with segments of the illegal drug circuit.[86]

OTHER EFFECTS

The Colombian military implemented a number of effective steps, starting in 2000, that have improved the ability to collect and process intelligence. With the implementation of Plan Colombia, forces received better intelligence training and equipment. COLMIL also established a Joint Intelligence Center, based on the US military construct, that improved processing, analysis, and dissemination of finished intelligence. COLMIL also acquired improved technical intelligence and signals intelligence sensors. The most effective improvement, however, may be the emphasis on human intelligence. With the disbandment of the *Despeje* demilitarized zone, the Uribe

---

[85] Vargas, *FARC and the Illicit Drug Trade.*
[86] Ibid.

administration ushered in a new emphasis on intelligence collection and denial of safe haven for insurgent groups. Uribe focused on keeping soldiers in their home provinces ("town soldiers program"), which proved extremely successful in improving intelligence collection. Now, soldiers who have been born and raised in an area are still serving as soldiers in the same physical environment—meaning that they know the key personalities and can rapidly identify anomalous activities that may be related to insurgent forces. Additionally, the "cooperative network" was established, touted as an organization of more than one million volunteers who report to military forces about possible insurgent activity. A cooperative network is akin to a national neighborhood watch program that is focused on insurgent activities.[87]

Drug trafficking in parts of Colombia represents the principal economic engine. Trafficking may take many forms, from a peasant farmer harvesting a few acres to a complex industrial manufacturing array managed by powerful narco-traffickers complete with their own security forces. The coca industry has continued to grow throughout the country, likely employing more than 300,000 people. FARC insurgents, particularly, benefit from production by imposing a "tax" on larger farmers, production facilities, and cocaine shippers. Cocaine production, therefore, is viewed by the government as directly facilitating insurgent operations. Complemented by US funds and policy that also fully support counternarcotics operations, the military and security forces have maintained a strategy that has focused on countering drug trafficking as much as on countering the insurgents head-on.[88]

In the same areas where the FARC and ELN operate, the AUC also are active. This composite organization of various armed right-wing paramilitary groups has autonomously launched campaigns against insurgent and narco-trafficking organizations, especially in the Amazon Basin area. The AUC have been associated with numerous human rights abuses and blatant attacks on civilians. Civilians in some areas report being targeted because they represent a potential resource to insurgents, either by illicit cooperation or by providing "taxes." The AUC complicated the security challenges of the Colombian government, particularly during their most active time frame of the 1990s–2002. They further diminished the role of the legitimacy of the government of Colombia, reinforcing the belief that some regions of the country were truly lawless.[89]

---

[87] Perez, "An Effective Strategy for Colombia," 12.

[88] Vargas, *FARC and the Illicit Drug Trade.*

[89] Ibid.

During 1990 and 1991, the FARC and ELN coordinated the largest insurgency operation in the history of the country. Operation Wasp occurred just weeks after the government had refused to allow either the FARC or ELN to participate in constituent assembly elections. Concurrently, the FARC former headquarters, La Casa Verde, had been bombed, which indicated that the government had closed the door on a negotiated settlement. The action had enraged FARC leadership. FARC and ELN attacks were carried out for several weeks, from December 1990 through February 1991. Hundreds were killed on both sides. During February, the decision to continue operations, while focusing on oil pipelines and other economic targets, was initiated. More than 650,000 barrels of fuel were spilled—an amount 2.5 times greater than that spilled by the Exxon Valdez tanker in 1989.[90] The combined offensive action of the FARC and ELN, over such a sustained period of time, indicated joint capabilities and coordination that had previously not been understood. This represented a new military challenge to the government.

BIBLIOGRAPHY

Cárdenas, Mauricio, Roberto Junguito, and Alberto Alesin. "Political Institutions, Policymaking Processes, and Policy Outcomes: The Case of Colombia." Research Proposal Presented by Fedesarrollo to the Inter-American Development Bank, 2004.

Cilluffo, Frank. "The Threat Posed from the Convergence of Organized Crime, Drug Trafficking, and Terrorism." Testimony before the US House Committee on the Judiciary, Subcommittee on Crime, 106th Cong., 2nd Sess, 2000.

Central Intelligence Agency. "Colombia." *The World Factbook.* Accessed November 2, 2009. https://www.cia.gov/library/publications/the-world-factbook/geos/co.html.

Foweraker, Joe, and Roman Krznaric. "The Uneven Performance of Third Wave Democracies: Electoral Politics and the Imperfect Rule of Law in Latin America." *Latin American Politics and Society* 44, no. 3 (2002): 29–60.

Gamboa, Miguel. "Democratic Discourse and the Conflict in Colombia." *Latin American Perspectives* 116, no. 28 (2001): 93–109.

Giugale, Marcelo, Oliver Lafourcade, and Connie Luff. *Colombia: The Economic Foundation of Peace.* Washington, DC: World Bank Publications, 2002.

---

[90] Saskiewicz, "The Revolutionary Armed Forces of Colombia–People's Army," 46, 47.

Gudeman, Stephen, and Alberto Rivera. *Conversations in Colombia: The Domestic Economy in Life and Text.* Cambridge: Cambridge University Press, 1990.

Hartlyn, Jonathan. *Drug Trafficking and Democracy in Colombia in the 1980s.* Working Paper no. 70. Barcelona: Institut de Ciencies Politiques i Socials, 1993.

Holmes, Jennifer S., Sheila A. G. De Piñeres, and Kevin M. Curtin. "A Subnational Study of Insurgency: FARC Violence in the 1990s." *Studies in Conflict and Terrorism* 30, no. 3 (2007): 249–265.

Immigration and Refugee Board of Canada. *Colombia: The Recruitment Methods of the Revolutionary Armed Forces of Colombia (Fuerzas Armadas Revolucionarias De Colombia, FARC) and Government Measures to Help FARC Members Reintegrate into Civilian Society (2005–February 2008).* COL102787.FE. April 14, 2008. Accessed March 20, 2012. http://www.unhcr.org/refworld/docid/4829b55c23.html.

Mason, Ann. "Colombia's Democratic Security Agenda: Public Order in the Security Tripod." *Security Dialogue* 34, no. 4 (2003): 391.

Ortiz, Román D. "Insurgent Strategies in the Post-Cold War: The Case of the Revolutionary Armed Forces of Colombia." *Studies in Conflict & Terrorism* 25, no. 2 (2002): 127–143.

Perez, William F. "An Effective Strategy for Colombia: A Potential End to the Current Crisis." Master's thesis, US Army War College, 2004.

Petras, James "Geopolitics of Plan Colombia." *Economic and Political Weekly* 35, no. 52/53 (2000): 4617–4623.

Pizarro, Eduardo. *Las FARC (1949–1966): De la Autodefensa a la Combinación de Todas las Formas de Lucha (Sociologia y Politica).* Bogota: Tercer Mundo Editores, 1992.

Robinson, Linda. "Terror Close to Home," *US News & World Report.* October 6, 2003.

Saskiewicz, P. E. "The Revolutionary Armed Forces of Colombia–People's Army (FARC-EP): Marxist-Leninist Insurgency or Criminal Enterprise." Master's thesis, Naval Postgraduate School, 2005.

Schmidt, Steffen W. "La Violencia Revisited: The Clientelist Bases of Political Violence in Colombia." *Journal of Latin American Studies* 6, no. 1 (1974): 97–111.

Shifter, Michael. "Colombia on the Brink: There Goes the Neighborhood." *Foreign Affairs* 78, no. 4 (1999): 14–20.

Sprunk, Darren D. "Transformation in the Developing World: An Analysis of Colombia's Security Transformation." Master's thesis, Naval Postgraduate School, 2004.

Sullivan, Mark P. "Latin America: Terrorism Issues." In *Focus on Terrorism*, edited by Edward V. Linden, 301. New York: Nova Science Publishers, 2007.

Sweig, Julia E. "What Kind of War for Colombia?" *Foreign Affairs* 81, no. 5 (September–October 2002): 122–141.

Vargas, Ricardo. *The Revolutionary Armed Forces of Colombia (FARC) and the Illicit Drug Trade*. The Netherlands: Transnational Institute, 1999.

Vélez, Maria A. "FARC-ELN: Evolución y Expansión Territorial." *Revista Desarrollo y Sociedad* no. 47, 151–225 (2001).

Weiler, Stephen P. "Colombia: Gateway to Defeating Transnational Hell in the Western Hemisphere." Master's thesis, US Army War College, 2004.

SENDERO LUMINOSO (SHINING PATH)

Ron Buikema and Matt Burger

SYNOPSIS

Sendero Luminoso, or "Shining Path," initiated a Maoist-based Peruvian insurgency that was renowned for its rapid escalation of violence and brutality throughout the countryside. It was also unique in that the insurgency actually took place *after* an agrarian reform initiative had been instituted (1968–1980). Resulting economic effects of reforms, however, were minimized, while the political implications were far-reaching.[1] Sendero Luminoso was considered the most radical communist movement in all of Latin America.

The Sendero Luminoso insurgency is characterized as a peasant revolt, although it contrasts significantly with previous insurgencies or revolts initiated in Peru, especially during the 1960s. From early on, Sendero Luminoso leadership understood the importance of grasping the sociocultural perspective of the indigenous communities, and they were dedicated to working strategically on political development in the countryside, living in the communities, learning the language, even marrying into the local communities.[2] The movement was also characterized by the charismatic leadership of one man, Abimael Guzmán, who served as the focal point for Maoist political ideals and the requirement for military action against the state. Without Guzmán, there would have been no Sendero Luminoso. He was the impetus for its founding, organization, and purpose.

This study covers Sendero Luminoso from its inception in military operations in 1980 until 1992, when Guzmán was captured and the movement hit a culminating point regarding insurgent capabilities.

---

[1]   Cynthia McClintock, "Why Peasants Rebel: The Case of Peru's Sendero Luminoso," *World Politics* 37, no. 1 (October 1984), 48–84.

[2]   Ibid.

TIMELINE

1962	Abimael Guzmán, founder of Sendero Luminoso, becomes a professor at University of San Cristóbal de Huamanga and begins activist group that eventually grows into Sendero Luminoso.
1980	Sendero Luminoso begins military attacks.
1982	Ayacucho region mostly under Sendero Luminoso control.
1984	Government declares an emergency zone.
1989	Sendero Luminoso begins siege of cities around Peru.
1990	President Fujimori elected.
1992	Fujimori suspends constitution and declares a state of emergency.
September 1992	Guzmán captured.

THE ENVIRONMENT OF THE REVOLUTION

PHYSICAL ENVIRONMENT

Peru is located in western South America, bordering the Pacific Ocean (to the west), Chile and Bolivia (to the south and southeast), Ecuador and Colombia (to the north), and Brazil (to the east). In size, Peru is slightly smaller than Alaska. Climate varies widely, from tropical to frigid (in the Andes). Terrain features include a vast coastal plain, rugged mountains of the Andes (with the highest elevation reaching more than 22,000 feet), and a lowland jungle (including the Amazon Basin). Peru is prone to natural disasters caused by earthquakes, volcanoes, flooding, and landslides. The country is ripe with natural resources, including copper, silver, gold, coal, and natural gas. Only 2.8% of the land is arable, with permanent crops accounting for only 0.47% of the land.[3]

---

[3] Central Intelligence Agency, "Peru," *The World Factbook*, accessed November 2, 2009, https://www.cia.gov/library/publications/the-world-factbook/maps/maptemplate_pe.html.

Figure 1. Map of Peru.[4]

CULTURAL AND DEMOGRAPHIC ENVIRONMENT

Peru is dominated by two cultures, distinct in some ways and interwoven in others: Amerindian, indigenous peoples descended from the conquered Incan Empire, and the predominantly white descendants of their Spanish conquerors, who retain vestiges of European culture and religion. The rural, inland, and highland regions are dominated by Amerindians, while the coastal cities are dominated by white creoles of Spanish descent. The mestizos, those of mixed ancestry, form a link between these two distinct communities. The indigenous communities have their own distinct art and culture, largely rooted in the history of their pre-Colombian civilizations, while the creole and mestizo communities are primarily rooted in Spanish traditions, with flavors of the native culture. There are also significant Asian (primarily Japanese and Chinese) immigrant communities in urban areas.[5]

The country is overwhelmingly Catholic (81.3% according to the 2007 census), although evangelical Christianity has made significant in-roads, particularly among indigenous peoples during the past 30 years, now comprising roughly 12.5% of the population (according

---

[4]   Ibid.

[5]   Orin Starn, Carlos Iván Degregori, and Robin Kirk, *The Peru Reader: History, Culture, Politics,* 2nd ed. (Durham: Duke University Press, 2005).

to the 2007 census). Although most indigenous peoples are at least nominally Catholic, significant syncretism is rooted in pre-Colombian and tribal belief systems.[6]

SOCIOECONOMIC ENVIRONMENT

Between 1960 and 1980 Peru's population increased dramatically from 9.9 million to 17.7 million, particular in urban areas, such as in Lima, where the population grew from 1.7 million to 5.5 million during the same period. Labor union membership increased fourfold during this same period, and the number of eligible voters also expanded dramatically with the elimination of literacy restrictions.[7]

In 1969, a reform-oriented military government began a land reform program to redistribute land from large *haciendas* to landless peasants. While moderately successful in some regions, the concept of private land ownership was ill suited to the community-based agricultural culture of the highland regions of Peru.[8] This issue was critical in the region of Ayacucho[9] where only a few *haciendas* existed, given the scarcity of arable land, and thus only around 10% of the peasant population was impacted by the reform.[10] Moreover, with the restoration of democracy, the 1980s witnessed the reprivatization of the land that had been allocated to community-based agricultural cooperatives by the land reform.[11]

Beginning in 1975, Peru began a precipitous economic decline that continued throughout the 1980s and was augmented by the growing devastation caused by Sendero Luminoso. Inflation by 1980 had reached between 75 and 125% and real wages began to drop, continuing throughout the 1980s, so that by 1989 real wages were

---

[6] Apocalyptic and millenarian aspects of these beliefs, which were prevalent among the indigenous peoples of the Peruvian highlands, were exploited by Sendero Luminoso to curry support. Susan Eckstein and Manuel A. Garretón Merino, *Power and Popular Protest: Latin American Social Movements,* updated and expanded ed. (Berkeley: University of California Press, 2001).

[7] David Scott Palmer, "Rebellion in Rural Peru: The Origins and Evolution of Sendero Luminoso," *Comparative Politics* 18, no. 2 (January 1986): 127–146.

[8] Kenneth M. Roberts, "Economic Crisis and the Demise of the Legal Left in Peru," *Comparative Politics* 29, no. 1 (October 1996): 69–92.

[9] It is in this region that Sendero Luminoso would first gain a foothold, under Abimael Guzmán, a professor at the University of San Cristóbal de Huamanga in Ayacucho and the founder and leader of Sendero Luminoso.

[10] Ibid.; McClintock, "Why Peasants Rebel: The Case of Peru's Sendero Luminoso."

[11] Roberts, "Economic Crisis and the Demise of the Legal Left in Peru."

only 52% of their 1970 levels.[12] This decline was felt even more acutely in the highland regions, such as Ayacucho, where the standards of living had already been much lower.[13]

Concurrently, Peru expanded access to higher education during this period, opening new universities, such as the University of San Cristóbal de Huamanga in Ayacucho. These new opportunities raised the professional expectations of middle- and working-class Peruvians. Yet in the wake of the 1975 economic depression, opportunities were scarce, particularly for graduates from a provincial university like Huamanga in the highlands. The only profession that saw hiring in sizable numbers was educators in the public schools. Thus, the ranks of teachers in Ayacucho were increasingly dominated by graduates from Huamanga by the early 1980s.[14]

HISTORICAL FACTORS

The government of Peru has long neglected the interior highland regions,[15] ignoring the economic plight of the indigenous peoples and creating an image of a distant, indifferent, and illegitimate authority.[16] The disparity between the prosperous coastal cities and the impoverished highlands crossed the racial divide between whites of European descent and the indigenous Amerindians.[17] Moreover, the grievances of the highland Indians were regularly ignored. This neglect was rooted in a long history of racism, which was a legacy of the colonial period and maintained by the creole elites after independence. In this caste system, the indigenous people, called "*la mancha india*" (The Indian Stain) by the creoles, were on the bottom, the mestizos were in the middle, and the white creoles on the top.[18] Even recent immigrants from Asian countries, such as Japan and China, enjoyed more social mobility and economic status than

---

[12] T. David Mason, "'Take Two Acres and Call Me in the Morning': Is Land Reform a Prescription for Peasant Unrest?" *The Journal of Politics* 60, no. 1 (February 1998): 199–230; Roberts, "Economic Crisis and the Demise of the Legal Left in Peru."

[13] Eckstein and Garretón Merino, *Power and Popular Protest: Latin American Social Movements.*

[14] Palmer, "Rebellion in Rural Peru: The Origins and Evolution of Sendero Luminoso."

[15] The economic impetus of the Sendero Luminoso among indigenous peoples was thus less about land and more about government service, such as free access to education, to improve their economic plight. David Post, "Political Goals of Peruvian Students: The Foundations of Legitimacy in Education," *Sociology of Education* 61, no. 3 (July 1988): 178–190.

[16] McClintock, "Why Peasants Rebel: The Case of Peru's Sendero Luminoso."

[17] Ibid.

[18] Ibid.

indigenous peoples, as evidenced by the election of Alberto Fujimori in 1990. This reality was resented by the indigenous peoples and, after the advent of Sendero Luminoso insurrection, it was reinforced by the harsh tactics of the police and military.[19]

GOVERNING ENVIRONMENT

In a move that seemingly epitomized strategic ineptness,[20] Sendero Luminoso launched its insurrection to coincide with the first democratic election with universal suffrage in Peru's history and the return of civilian government after twelve years of rule by a reformist military regime. The election included participation from political parties on the left and right, including numerous Marxist parties with bases of support in universities and labor unions.[21] Sendero Luminoso was able to flourish in an otherwise stable democratic nation primarily because of frivolity, inaction, or covert conciliation in the face of terrorist subversion "by the government [which] took a country to the edge of collapse."[22]

The 1980 election was won by President Fernando Belaúnde Terry (1980–1985), who had been ousted by the military regime twelve years earlier. As a result, Belaúnde feared the power of the military, leading him to cut the military budget, limit its intelligence capacity, and place

---

[19] C. I. Degregori, "After the Fall of Abimael Guzmán: The Limits of Sendero Luminoso," in *The Peruvian Labyrinth: Polity, Society, Economy*, ed. Maxwell A. Cameron and Philip Mauceri (University Park, PA: Pennsylvania State University Press, 1997); Orin Starn, "To Revolt Against the Revolution: War and Resistance in Peru's Andes," *Cultural Anthropology* 10, no. 4 (November 1995): 547–580; Enrique Mayer, "Peru in Deep Trouble: Mario Vargas Llosa's 'Inquest in the Andes' Reexamined," *Cultural Anthropology* 6, no. 4 (November 1991): 466–504.

[20] Scholars typically cite one or more groups' exclusion from political participation as a leading cause of Marxist insurgencies. Yet Sendero Luminoso would blossom during a period of open democratic participation. It was the rigid ideology of Sendero Luminoso that led it to reject democratic participation from the start. Linda J. Seligmann, *Between Reform & Revolution: Political Struggles in the Peruvian Andes, 1969–1991* (Stanford, CA: Stanford University Press, 1995); Robert B. Kent, "Geographical Dimensions of the Shining Path Insurgency in Peru," *Geographical Review* 83, no. 4 (October 1993): 441–454; Larry A. Niksch and Mark P. Sullivan, *Peru's Shining Path* (Washington, DC: Congressional Research Service, Library of Congress, 1993); Ronald H. Berg, "Sendero Luminoso and the Peasantry of Andahuaylas," *Journal of Interamerican Studies and World Affairs* 28, no. 4 (Winter 1986): 165–196.

[21] Orin Starn, "Maoism in the Andes: The Communist Party of Peru-Shining Path and the Refusal of History," *Journal of Latin American Studies* 27, no. 2 (May 1995): 399–421; David Scott Palmer, *The Shining Path of Peru*, 2nd ed. (New York: St. Martin's Press, 1994); Niksch and Sullivan, *Peru's Shining Path*, 35; Carol L. Graham, "The Latin American Quagmire: Beyond Debt and Democracy," *The Brookings Review* 7, no. 2 (Spring 1989): 42–47; Palmer, "Rebellion in Rural Peru: The Origins and Evolution of Sendero Luminoso."

[22] Yonah Alexander and Michael Kraft, *Evolution of U.S. Counterterrorism Policy*, vol. 1–3 (Santa Barbara, CA: Greenwood Publishing Group, 2007).

military commanders under the control of civilian bureaucrats who were unable to work together in an integrated, efficient manner.[23] Concomitantly, Belaúnde's fear of renewed military influence over the civil government also led him to ignore and downplay the growing insurgent threat for more than two years.[24] The changes crippled the ability of the military[25] to effectively counter Sendero Luminoso in its infancy as well as provided a two-year window during which the insurgency developed virtually unchallenged. In addition, Belaúnde failed to address the economic and racial issues that abetted Sendero Luminoso's growth. When Belaúnde finally did act, he sent the *sinchis* (special police units) into Ayacucho, the Sendero Luminoso epicenter. This action exposed the festering racism in Peru, as the police's inappropriately excessive use of violent and demeaning tactics toward the mestizos and indigenous peoples drove them to Sendero Luminoso's side while simultaneously providing justification for the insurgency.[26]

WEAKNESSES OF THE PREREVOLUTIONARY ENVIRONMENT AND CATALYSTS

Most of the core leadership of Sendero Luminoso became politically active during the 1960s, within the confines of the University of San Cristóbal de Huamanga in Ayacucho or the larger urban areas of the country, especially Lima.[27] The founder of the Comité Central del Partido Comunista del Perú (as Sendero Luminoso refers to itself, also the Communist Party of Peru or PCP), Abimael Guzmán, was a professor of philosophy at that university until 1963 (although the party was not founded until 1968). Guzmán was a charismatic figure and a talented public speaker, holding political discussions in his home and the surrounding area with university students, many from peasant families themselves.[28] Ayacucho was not an area chosen by accident

---

[23] Sandra Woy-Hazleton and William A. Hazleton, "Sendero Luminoso and the Future of Peruvian Democracy," *Third World Quarterly* 12, no. 2 (April 1990): 21–35; Palmer, "Rebellion in Rural Peru: The Origins and Evolution of Sendero Luminoso."

[24] James Ron, "Ideology in Context: Explaining Sendero Luminoso's Tactical Escalation," *Journal of Peace Research* 38, no. 5 (September 2001): 569–592; Gustavo Gorriti Ellenbogen, *The Shining Path: A History of the Millenarian War in Peru* (Chapel Hill: University of North Carolina Press, 1999), 290; Palmer, *The Shining Path of Peru.*

[25] In addition to limitations placed upon it by the Belaúnde administration, the military also suffered from a history of infighting and distrust between its branches as well as between it and the national police.

[26] Ibid.

[27] Cynthia McClintock, *Revolutionary Movements in Latin America: El Salvador's FMLN & Peru's Shining Path* (Washington, DC: United States Institute of Peace Press, 1998).

[28] McClintock, "Why Peasants Rebel: The Case of Peru's Sendero Luminoso."

as the center of the Sendero Luminoso movement. Founded in 1540, the city was not linked to the rest of Peru by road until 1924. For decades, Ayacucho in many respects served as a forgotten city, lacking in economic growth or greater opportunities for its populace—the city lacked roads, rail linkages, and trade and often failed to address urgent needs such as chronic water shortages.[29] During the 1960s, the area opened up to the Peace Corps and other nongovernmental organizations, with an emphasis on agrarian reform, education, and infrastructure improvements. The 1959 founding of the university was established along with these ongoing changes. The school quickly became a haven for radical political discussion focused on Marxist principles. By the mid-1960s, the university's extension services, reaching into the broader realms of the province, became an effective method to expand the radical Marxist beliefs to a broader, receptive audience.

Abimael Guzmán served as the founder, leader, and charismatic center of strategic planning for the PCP, or Sendero Luminoso. Guzmán began to study Marxist ideas through his university colleagues during the 1950s. He accepted a position as a philosophy professor in Huamanga in 1962. At that point, he had already fully embraced Marxist ideology. By 1970, Guzmán had become disenchanted with the Communist Party, principally because of their unwillingness to take up arms against the state. By this time, he had been promoted to personnel director of the university, a position that he used to hire loyalists and like-minded individuals to positions of influence in the university and the local community. He concurrently became influential with the faculty. Guzmán's increasing base of supporters was promised that they would be active participants in a "world proletarian revolution."[30] Guzmán was expelled from the university in 1975, but he merely shifted his focus from the university to the greater space of Ayacucho's society at large, particularly within the poorest provinces. Guzmán continued his ties with his supporters at the university, and new graduates, fully indoctrinated by Guzmán, were now heading to be teachers and local leaders in these same impoverished provinces. Guzmán had already become a semi-godlike personality to some.[31] Sendero Luminoso could not have become an established insurgent force without both the charismatic leadership of Guzmán and the

---

[29] Palmer, "Rebellion in Rural Peru: The Origins and Evolution of Sendero Luminoso."

[30] Starn, "Maoism in the Andes: The Communist Party of Peru-Shining Path and the Refusal of History."

[31] A. Portugal, *Voices from the War: Exploring the Motivation of Sendero Luminoso Militants* (Oxford: Center for Research on Inequality, Human Security, and Ethnicity, 2008).

relative safe havens offered within both the greater area of Ayacucho and the confines of a supportive university structure.

The prerevolutionary environment in Ayacucho, key to the introduction of Sendero Luminoso as a political/military force, was unique in that the region was geographically, socially, and economically marginalized. The Indian population that had been exploited historically was combined with political passion fueled by a charismatic leader working through the instrument of a politicized provincial university.[32]

During the early to mid-1980s, Sendero Luminoso published a few documents that described their political ideology in detail; however, leaders did not provide public statements or interviews. The few statements that were collected, including those from placards and posters, suggest that Sendero Luminoso had a skewed view of Peru, seeing it as a feudal state with landlords, akin to Maoist doctrine but not consistent with the social/cultural landscape of the country. Sendero Luminoso members spoke of Peru as an agrarian state, although even in 1980, only 10% of the country's gross domestic product (GDP) was derived from the agricultural sector.[33] The days of a feudal society with landlords, however, had long passed. This tie to an agricultural society may explain the attraction of its base— principally university students from what were generally known as the forgotten areas of Peru's Andean backcountry. Sendero Luminoso also supported the message of the indigenous population. Taylor noted political meetings in Ayacucho "where statements such as *necesitamos un gobierno de Indios* ('We need a government of Indians') and *hay que matar a los blancos y destuir las ciudades que siempre nos han explotado* ('We have to kill the whites and destroy the towns that have always exploited us') have been frequently heard."[34] Sendero Luminoso saw the outside world as a threat, with the intent of exploiting the people. Sendero Luminoso combined Maoism with "Andean millennialism," suggesting the perceived golden era of the Incan reign. The peasant class was considered the essential component of a successful Sendero Luminoso revolution. Some researchers, however, have also noted that ethnicity was not as important as class to the Sendero Luminoso cause and that the links to the call of millennialism may have represented

---

[32] Palmer, "Rebellion in Rural Peru: The Origins and Evolution of Sendero Luminoso."

[33] L. Taylor, "Maoism in the Andes: Sendero Luminoso and the Contemporary Guerilla Movement in Peru," Working Paper No. 2 (Liverpool: Institute of Latin American Studies, 1983).

[34] Ibid.

nothing more than a convenient linkage with the Andean peasant class, rather than a core for political Marxist–Leninist pragmatism.[35]

Nationally, there were noted improvements in the quality of life, although the Ayacucho region was generally excluded. Gross national product (GNP) per capita increased from $526 in 1960 to $1,294 in 1981 and literacy rates increased by more than 25% in the same period. The Marxist parties, composed of a growing body of union workers, increased from 3.6% of the voting populace in 1962 to 13.7% in 1980. In Ayacucho province, however, little improvement was noted. In 1981, only 7% of the residents had access to running water. There were only 30 doctors for more than 500,000 residents. Life expectancy was at an abysmal 44 years. Possessing 3% of the country's populace, Ayacucho received only 1% of the national expenditures.[36]

Land reform had been instituted in 1969, but the Ayacucho region did not benefit at the same rate as the rest of Peru. The national government lacked both the technical expertise and the personnel to manage implementation in the region, or failed to enact reform because of the "special challenges" faced by the local Indian communities.[37] These conditions exacerbated the perception and reality that the communities of Ayacucho were falling further behind the rest of Peru. By the 1979–1980 time frame, Sendero Luminoso represented an avenue of hope for many of the disenfranchised in the Ayacucho region.

In April 1980, Guzmán effectively declared war when he announced the following to the PCP-Sendero Luminoso: "Comrades: Our labor without guns has ended, the armed struggle has begun . . . The invincible flames of the revolution will glow, turning to lead and steel . . . There will be a great rupture and we will be the makers of the new dawn . . . We shall convert the black fire into red and the red into pure light."[38]

During May 1980, Sendero Luminoso commenced an escalation of terrorist and violent activities, beginning a new stage of the

---

[35] Gorriti Ellenbogen, *The Shining Path: A History of the Millenarian War in Peru*, 290; Orin Starn, "New Literature on Peru's Sendero Luminoso; El Surgimiento De Sendero Luminoso: Del Movimiento Por La Gratitud De La Ensenanza Al Inicio De La Lucha Armada; Que Dificil Es Ser Dios: Ideologia y Violencia Politica En Sendero Luminoso; Sendero: La Historia De La Guerra Milenaria; Juventud y Terrorismo; the Decade of Chaqwa: Peru's Internal Refugees; Violencia Politica En El Peru, 1980–88," *Latin American Research Review* 27, no. 2 (1992), 212–226.

[36] Palmer, "Rebellion in Rural Peru: The Origins and Evolution of Sendero Luminoso."

[37] Ibid.

[38] Gorriti Ellenbogen, *The Shining Path: A History of the Millenarian War in Peru* as quoted in Starn, "New Literature on Peru's Sendero Luminoso."

insurgency. Although the conditions were not ideal in 1980 for launching a protracted people's war, Guzmán believed that once the war was launched, conditions would improve over time.[39] Sendero Luminoso stole munitions and arms from security forces and attacked infrastructure targets, including communication facilities, electrical power facilities, and even embassies. Factories were also targeted, with the focus on imposing an economic price on the government and its supporters.[40] Success of these attacks resulted in a rapid increase of insurgents and supporters. From 1980 to 1983, forces increased from approximately 300 to 3,000. By 1982, the Ayacucho region was firmly under Sendero Luminoso control, administered by a Popular Committee. Growth was supported by focused recruiting on campus in Ayacucho. By 1984, the government had declared an emergency zone, encompassing thirteen provinces and 12% of the peasant population.[41]

FORM AND CHARACTERISTICS OF THE REVOLUTION

OBJECTIVES AND GOALS

According to Guzmán, the principal objective of Sendero Luminoso was power derived from "an intelligent, well-motivated, and highly disciplined organization with a . . . purposeful long-term program for gaining control of the state."[42] This organization would strategically employ and maintain this power as the state disintegrated and the organization took on that mantle.

Sendero Luminoso's ideology was Maoist. It was self-sufficient, receiving no outside funding or political/military support from any other country. There were no indications that Sendero Luminoso solicited outside funding or support during the early years of the military campaign. Guzmán kept Sendero Luminoso focused on strategic goals, based on a five-point program: (1) convert the backward areas into advanced and solid bases of revolutionary support; (2) attack the symbols of the bourgeois state; (3) generalize violence and

---

[39] Starn, "Maoism in the Andes"; Palmer, *The Shining Path of Peru*; Niksch and Sullivan, *Peru's Shining Path*; Graham, "The Latin American Quagmire: Beyond Debt and Democracy."

[40] McClintock, *Revolutionary Movements in Latin America: El Salvador's FMLN & Peru's Shining Path.*

[41] McClintock, "Why Peasants Rebel: The Case of Peru's Sendero Luminoso."

[42] Max G. Manwaring, "Peru's Sendero Luminoso: The Shining Path Beckons," *Annals of the American Academy of Political and Social Science* 541, no. 1 (1995): 157–166.

develop a guerrilla war; (4) conquer and expand the bases of support; and (5) lay siege to the cities and bring about the total collapse of the state.[43]

Sendero Luminoso's focus on the Indian populace was unique. Early on, there was little or no focus on public affairs or information operations and no attempt to build a broad base of sympathizers either internally or externally. Instead, Sendero Luminoso focused on building a sociocultural "connection" with the Indian base of support. The Indians had become disenfranchised from the government and provided the likely base of support—geographically, politically, and socially—that would be required to operationally conduct a long-term insurgent campaign.

Although Sendero Luminoso's early success was built on the strong relationships it established with the peasant class and the Indian populace of Ayacucho, these relationships also caused challenges as the organization matured. The focus on "pure native peasant communism" did not appeal to those outside of the Indian community, and Sendero Luminoso's operational base of support was never expanded. This lack of expansion limited Sendero Luminoso's operations throughout the rest of Peru and extended their lines of communication, causing a weakness that could be exploited by government security forces.

Sendero Luminoso also focused growth—both political and military—in areas where the government and security forces lacked control, support, or widespread influence. In fact, they demonstrated the ability to grow and adapt more effectively than the government throughout most of the 1980s. Sendero Luminoso eschewed international recognition or support. Unlike the *Frente Farabundo Martí para la Liberación Nacional* (FMLN) and other Latin American communist insurgencies that received international training, equipment, and political support from other states and entities, including Cuba and Nicaragua, Sendero Luminoso made no attempt to either unify other leftist and violent organizations within Peru or to seek assistance and support abroad. This self-imposed isolation may have impacted potential recruiting, command and control, logistics and financial support, and legitimacy of the cause.[44] During the 1990s, however, Sendero Luminoso became much more outwardly focused, realizing the importance of getting their message out, and they even

---

[43] Palmer, "Rebellion in Rural Peru: The Origins and Evolution of Sendero Luminoso."

[44] William A. Hazleton and Sandra Woy-Hazleton, "Sendero Luminoso and the Future of Peruvian Democracy," *Third World Quarterly* 12, no. 2 (April 1990): 21–35.

established an external website in 1996 maintained by a support group in Berkeley, California.[45]

When Sendero Luminoso shifted its focus to the urban areas during the late 1980s and early 1990s, it attempted to fix security forces in the cities, permitting the organization to continue to consolidate support in its traditional rural areas. At some point, leadership believed, the urban organizations would link up with the rural organizations for a final defeat of the government.[46]

LEADERSHIP AND ORGANIZATIONAL STRUCTURE

Abimael Guzmán was the undisputed leader of Sendero Luminoso. This physically unremarkable man, rarely seen by Sendero Luminoso members apart from the highest leadership, employed an extraordinary capacity for persuasion and organization to create a cult-like organization whose members literally revered him as a god in many cases.[47] For Sendero Luminoso members, Guzmán was shrouded in mystery, a charismatic, almost hypnotic leader who held the one true vision of the future and the means to achieve it. In their minds, he was almost superhuman and his commands were obeyed without question or hesitation. He demanded and received absolute devotion.[48] Indeed, Guzmán saw himself as a "revolutionary Moses who will lead his followers across a river of blood into the Maoist promise land of communism."[49] Sendero Luminoso believed that Peru was but the epicenter of a world revolution and ultimate victory depended on absolute obedience to Guzmán himself, the leader of the world revolution.[50]

The top leadership was primarily drawn from Guzmán's former students at Huamanga University and this group remained fairly static throughout the conflict.[51] In addition to Guzmán's wife and former

[45] Kathy Crilley, "Information Warfare: New Battlefields; Terrorists, Propaganda and the Internet," *Aslib Proceedings* 53, no. 7 (July/August 2001): 250–264.

[46] Gordon H. McCormick, *The Shining Path and the Future of Peru* (Santa Monica, CA: RAND, 1990), 58.

[47] This was especially true among some indigenous peoples who saw Guzmán as the prophesied return of one of their gods. Carlos Iván Degregori, "How Difficult it is to be God: Ideology and Political Violence in Sendero Luminoso," *Critique of Anthropology* 11, no. 3 (1991); Hazleton and Woy-Hazleton, "Sendero Luminoso and the Future of Peruvian Democracy."

[48] Degregori, "How Difficult it is to be God: Ideology and Political Violence in Sendero Luminoso"; McCormick and others, *The Shining Path and the Future of Peru.*

[49] Starn, "New Literature on Peru's Sendero Luminoso."

[50] Manwaring, "Peru's Sendero Luminoso: The Shining Path Beckons."

[51] Woy-Hazleton and Hazleton, "Sendero Luminoso and the Future of Peruvian Democracy."

wife, the leadership included Antonio Díaz Martínez, an architect of Sendero Luminoso ideology killed in a prison mutiny in 1986, and Oscar Ramírez Durand, the son of a Peruvian army general, an expert with improvised explosive devices (IEDs), and the Sendero Luminoso leader after Guzmán's arrest in 1992. Durand, however, lacked the organizational skills and charismatic personality of Guzmán, and the organization suffered from the leadership gap after 1992. In one sense, the near deification of Guzmán ensured unparalleled organizational unity and clarity of vision.[52]

The Sendero Luminoso zealots adhered to a rigid ideology that made it "savage, sectarian, and fanatical . . . compared to Pol Pot's Khmer Rouge rather than to the Sandinistas or the *Frente Farabundo Martí para la Liberación Nacional* movement (FMLN) in El Salvador."[53] Sendero Luminoso saw itself as the "last bastion of true communism in the world" and Guzmán as its last true prophet.[54] The group rejected Peru's other Marxist parties for participating in Peru's democracy, dubbing them "parliamentary cretins," as well as other communist states, such as China and the Soviet Union,[55] because Sendero Luminoso considered them slaves under the rule of "capitalist-imperialist dogs."[56] Thus, Sendero Luminoso impartially targeted Chinese, Indian, Israeli, American, and Soviet embassies for terrorist attacks.[57] Accordingly, Sendero Luminoso rejected any negotiation with the Peruvian government that they believed to be too corrupt to be redeemed;[58] rather, their ideological purity required that Peru be purified through the blood of armed struggle.

In part, the success of Sendero Luminoso is linked to its efficient organization, which indoctrinated new recruits to unquestioningly obey

---

[52] Portugal, *Voices from the War* (2008); Degregori, "After the Fall of Abimael Guzmán."

[53] Eckstein and Garretón Merino, *Power and Popular Protest: Latin American Social Movements.*

[54] US Congress, "The Shining Path After Guzman: The Threat and the International Response," Hearing before the Subcommittee on Western Hemisphere Affairs of the Committee on Foreign Affairs, House of Representatives, 102nd Cong., 2nd sess. (September 23, 1992).

[55] Examples of this ideological fanaticism include Sendero Luminoso graffiti on the walls of rural villages calling for the death the Chinese leader, Deng Xiaoping, regardless of the reality that none of the villagers had a remote idea as to the identity of this "traitor." Eckstein and Garretón Merino, *Power and Popular Protest: Latin American Social Movements.*

[56] US Congress, *The Shining Path After Guzman.*

[57] Lewis Taylor, *Shining Path: Guerrilla War in Peru's Northern Highlands, 1980–1997,* Liverpool Latin American Studies New Series 6 (Liverpool: Liverpool University Press, 2006), 232; Colin Harding, "Antonio Diaz Martinez and the Ideology of Sendero Luminoso," *Bulletin of Latin American Research* 7, no. 1 (1988), 65–73.

[58] Bernard W. Aronson, "Brutal Insurgency: Sendero Luminoso," *U.S. Department of State Dispatch* 3, no. 12 (1992); Woy-Hazleton and Hazleton, "Sendero Luminoso and the Future of Peruvian Democracy."

and sacrifice themselves in order to meet the "quota of blood" Guzmán asserted was necessary for victory.[59] Although it held several national congresses where strategy and ideology were discussed more broadly, Sendero Luminoso was highly centralized with regard to ideological, political, and strategic decision making, while tactical decision making was decentralized.[60] Local militants were organized into cells, similar to contemporary terrorist cells, and for security reasons had limited contacts outside their immediate five- to nine-member unit.[61] Even a regional commander had direct contact with no more than eight other insurgents.[62] This structure, as well as the insulation of the leadership via tactical decentralization, made counterinsurgency efforts, particularly intelligence gathering, extremely difficult. The Sendero Luminoso organization consisted of five levels: the Cupola, regional leadership, militants, activists, and sympathizers.[63] The first level, the Cupola, consisted of Guzmán and other top leaders who formed the national directorate and made all strategic and ideological decisions. The Cupola also included the twenty-five-member Central Committee, which advised the top leadership and issued directives to regional commanders and committees. The regional leadership, directly below the Cupola, consisted of the regional commanders and committees. These leaders made military and political decisions within the six regions into which Sendero Luminoso divided

---

[59] Degregori, "How Difficult it is to be God: Ideology and Political Violence in Sendero Luminoso."

[60] McClintock, *Revolutionary Movements in Latin America: El Salvador's FMLN & Peru's Shining Path*; Niksch and Sullivan, *Peru's Shining Path*; Simon Strong, *Shining Path: A Case Study in Ideological Terrorism*, Conflict Studies Series no. 260 (London: Research Institute for the Study of Conflict and Terrorism, 1993); Michael F. Brown and Eduardo Fernández, *War of Shadows: The Struggle for Utopia in the Peruvian Amazon* (Berkeley, CA: University of California Press, 1991); Barry M. Schutz and Robert O. Slater, *Revolution & Political Change in the Third World* (Boulder, CO: L. Rienner Publishers, 1990).

[61] Taylor, *Shining Path: Guerrilla War in Peru's Northern Highlands*; Stéphane Courtois, *The Black Book of Communism: Crimes, Terror, Repression* (Cambridge, MA.: Harvard University Press, 1999); Woy-Hazleton and Hazleton, "Sendero Luminoso and the Future of Peruvian Democracy."

[62] Taylor, *Shining Path: Guerrilla War in Peru's Northern Highlands*.

[63] Ibid.; John M. Bennett and Laurence Hallewell, *Sendero Luminoso in Context: An Annotated Bibliography* (Lanham, MD: Scarecrow Press, 1998); Jeffrey John Ryan, "The Dynamics of Latin American Insurgencies: 1956–1986" (Ph.D. diss., Rice University, 1989).

Peru.[64] The militants,[65] under the direct command of the regional leadership, acted as the organization's principal means of violence.[66] The activists coordinated and operated the schools established by Sendero Luminoso for indoctrination and also organized protests and disseminated propaganda. The sympathizers provided financial and material support for the militants and served as couriers of supplies and information. Both activists and sympathizers were separated from the higher levels of Sendero Luminoso and often lacked the fanatical religious devotion and had limited awareness of and contact with the broader organization.[67]

COMMUNICATIONS

Sendero Luminoso implemented a cellular tactical structure, with each cell generally composed of ten or fewer personnel and a commander. Only the leader of the cell would know other members outside of that cell, and then, only by an alias. Identities of members were carefully protected.[68] During the early 1980s, several cell commanders were purported to have received training in China and North Korea. All official communication between cells was conducted by the commander. Each cell also had two explosives specialists, a political representative, and a physical fitness instructor. All members of the cell were trained in small arms, self-defense, and first aid.[69]

Sendero Luminoso was also supported by a large base of sympathizers who were willing to pass on general information, provide safe haven, and conduct other general services. Means of communication included couriers and limited use of tactical

---

[64] In addition to the committees based on geography there was a series of support committees that crossed regional boundaries. These included the (1) Department of Organization Support, which provided logistical support for insurgent cells; (2) the Group of Popular Support, which provided intelligence and organized popular support; (3) the Department of Finance; and (4) the Department of International Relations, which worked with support groups in the United States and Europe to raise political and financial support. Palmer, *The Shining Path of Peru.*

[65] No definitive estimate of the number of fighters exists. Estimates range from as low as 5,000 (Ron, "Ideology in Context: Explaining Sendero Luminoso's Tactical Escalation"; Degregori, "After the Fall of Abimael Guzmán") to as high as more than 20,000 (James Francis Rochlin, *Vanguard Revolutionaries in Latin America: Peru, Colombia, Mexico* [Boulder, CO: L. Rienner Publishers, 2003]; McClintock, *Revolutionary Movements in Latin America: El Salvador's FMLN & Peru's Shining Path.*)

[66] Taylor, *Shining Path: Guerrilla War in Peru's Northern Highlands;* Bennett and Hallewell, *Sendero Luminoso in Context: An Annotated Bibliography;* Ryan, "The Dynamics of Latin American Insurgencies: 1956–1986."

[67] Schutz and Slater, *Revolution & Political Change in the Third World.*

[68] McClintock, "Why Peasants Rebel: The Case of Peru's Sendero Luminoso."

[69] Taylor, "Maoism in the Andes: Sendero Luminoso and the Contemporary Guerilla Movement in Peru."

radios.[70] Operational security remained paramount; therefore only cellular commanders were authorized to conduct communications knowingly with other Sendero Luminoso members. The rest of the cell had no tactical or operational knowledge and no ability to divulge valuable operational information to Peruvian security or intelligence personnel; they simply were not privy to such information at any time. This method of cellular communication emphasized operational security as a high priority but also highlighted a risk—if cell leaders were killed, captured, or compromised, the remaining members of the cell could become totally cut off, unable to function or coordinate their activities effectively.

The use of propaganda by Sendero Luminoso varied throughout the conflict. After retreating from an active campaign of indoctrination in the mid-1980s in favor of focusing on military operations, Sendero Luminoso renewed an active effort to win popular support in the late 1980s.[71] To this end, Sendero Luminoso distributed propaganda, such as pamphlets and posters; spoke to the news media, including a 1988 interview with Guzmán himself; and organized public meetings and rallies. Yet Sendero Luminoso's appeal remained limited[72] and by 1991, 68% of Peruvians labeled Sendero Luminoso a terrorist organization and the nation's primary threat, rather than its savior.[73]

METHODS OF ACTION AND VIOLENCE

Sendero Luminoso's military campaign commenced in May 1980. Initial actions included the disrupting provincial and regional elections, sabotaging infrastructure (specifically focused on electrical sabotage within Lima), firebombings, attacking police stations, conducting assassinations of political leadership, occupying towns, taking over radio stations (which were then used as propaganda tools), occupying schools, organizing general strikes, setting IEDs (especially in urban areas), and attacking economic targets, including major factories. Sendero Luminoso carried out actions of disruption and

---

[70] Sendero Luminoso's top leader used the newspaper, *El Diario*, to augment other modes of communication with the cells and to reinforce party ideology. McCormick, *The Shining Path and the Future of Peru*.

[71] Woy-Hazleton and Hazleton, "Sendero Luminoso and the Future of Peruvian Democracy."

[72] The numeric evidence of this reality was made clear when contrasting the approximately 250 supporters who attended a Sendero Luminoso rally in Lima in 1988 vs. the 30,000 people who demonstrated in support of Peru's democracy in November 1989. Ibid.

[73] Rex A. Hudson, ed. *Peru: A Country Study*, Area Handbook Series, 4th ed., vol. 550-42 (Washington, DC: US Department of Defense, 1993).

destruction, focusing on their five strategic goals and emphasizing economic targets that would delegitimize the government.[74]

Sendero Luminoso secured physical areas by actively targeting and killing government authorities, while also promising a better life for the peasantry. Additionally, business leaders and shopkeepers were frequently threatened or killed, with Sendero Luminoso also taking control of their property and then disseminating it to the local villagers.[75] This early approach, and its positive appeal to the peasants, could be associated with a Robin Hood-esque appeal.

Sendero Luminoso attempted to maintain a positive face, but they were also capable of resorting to terror and coercion. Many of the students and teachers at the universities and schools where Guzmán spoke attended the leader's speaking events because they felt obliged. Those who did not attend were threatened, at a minimum. Students and teachers did not want to run the risk of crossing the paths of Sendero Luminoso leadership, as physical and verbal aggression was considered commonplace for those who did not support the cause. Faculty members who did not support Guzmán were encouraged to resign their posts and leave the region, even under the threat of death if they chose to stay. Fear was used as the great neutralizer of the political opposition.[76]

Sendero Luminoso relied heavily on hit-and-run attacks, striking targets and then moving on. Even the "capture" of towns was principally conducted as a temporary measure, lasting at most a few days. Their intent was to spread fear and violence throughout a large area, without necessarily controlling the ground or confronting the security forces head-on.[77] These tactics proved effective in minimizing amassment of Peruvian security forces and requiring them to remain in a reactive mode, waiting for the next act of violence and then responding. In this regard, the declared emergency zone, a broad area where the

---

[74] Taylor, "Maoism in the Andes: Sendero Luminoso and the Contemporary Guerilla Movement in Peru."

[75] McClintock, "Why Peasants Rebel: The Case of Peru's Sendero Luminoso."

[76] Local government officials served as examples for uncooperative villages and towns; officials were executed while those who had elected them were forced to watch. Portugal, *Voices from the War*; Rochlin, *Vanguard Revolutionaries in Latin America: Peru, Colombia, Mexico*; Ron, "Ideology in Context: Explaining Sendero Luminoso's Tactical Escalation"; Strong, *Shining Path: A Case Study in Ideological Terrorism*; Angela Cornell and Kenneth Roberts, "Democracy, Counterinsurgency, and Human Rights: The Case of Peru," *Human Rights Quarterly* 12, no. 4 (November 1990): 529–553.

[77] The story of "people's trials," the brutal murder of those who refused to cooperate, became infamous. A Catholic priest from Ayacucho related how peasants were stripped and tied to a post in the center of the village, while the villagers, including women and children, were forced to cut a piece of flesh from the victim. This torture would continue until the victim died from blood loss or shock. US Congress, *The Shining Path After Guzman: The Threat and the International Response*.

security forces focused their activities, included large expanses where there was little if any Sendero Luminoso presence. Peruvian security forces seemed to be defending everything while protecting nothing.[78]

Sendero Luminoso influenced numerous municipal elections by stealing voting ballots (their first widely reported action was in 1980), attacking public transportation, cutting off electricity on election days, and assassinating political candidates.[79]

Since 1962, Sendero Luminoso went through several stages and conducted major activities in the pursuit of power. From 1962 to 1980, during its first, "organization" stage, Sendero Luminoso undertook doctrine and leadership development while expanding organizational relationships with peasant communities. From 1980 to 1982, Sendero Luminoso commenced its second, "offensive," stage, with bombings of public buildings and private companies and attacks on and assassinations of local public figures. Sendero Luminoso also commenced its creation of a local political vacuum during this time. Throughout the following two years, 1982 and 1983, Sendero Luminoso continued to spread the violence started in stage 2, but they also increased the level of violence, including the execution of local political leaders. In March of 1982, Sendero Luminoso attacked the Ayacucho prison and released its prisoners, and in December of 1982, the insurgency attacked the Lima electrical grid. From 1983 to 1993, Sendero Luminoso began consolidating and expanding its support bases. Base presence was expanded to 114 provinces and logistics support expansion included entry into the coca trade. Sendero Luminoso also began extorting businesses and gained control of agricultural production while implementing the isolation of Lima. The years 1989–1992 also saw Sendero Luminoso besieging cities and pushing for the collapse of the state. Sendero Luminoso increased its focus on Lima and underwent an operational shift from rural to urban areas; Guzmán was captured during this time. Since 1992, Sendero Luminoso has "prepared for a world revolution," based on a plan revealed by Guzmán from prison (during his sentencing procedures). The plan involved widening the political base and

---

[78] Starn, "New Literature on Peru's Sendero Luminoso."

[79] Sendero Luminoso employed sabotage, assassination, and other forms of terrorism to break down law and order and thereby undermine confidence in the legitimacy of the state. To that end, targets for sabotage were selected both because of the expense to repair or replace them as well as because their destruction would impact the greatest number of Peruvians. High tension towers, which brought electricity to the major urban centers, were thus favorite targets. Rochlin, *Vanguard Revolutionaries in Latin America: Peru, Colombia, Mexico*; Ron, "Ideology in Context: Explaining Sendero Luminoso's Tactical Escalation"; Strong, *Shining Path: A Case Study in Ideological Terrorism*; Cornell and Roberts, "Democracy, Counterinsurgency, and Human Rights: The Case of Peru."

further destabilizing Peru (Sendero Luminoso efforts for this stage were never executed).[80]

METHODS OF RECRUITMENT

In the early 1980s, Sendero Luminoso actively appealed for popular support among the highland peasants by holding a mandatory indoctrination session, distributing confiscated money, property, livestock, and other goods and conducting community improvement projects.[81] However, as more emphasis was placed on the military campaign, these efforts waned, although the eventual goal of winning the hearts and minds of the people remained.[82] Sendero Luminoso's broad use of violence and rigid ideology also alienated peasant leaders, compelling the organization to take the initiative in recruitment, as the absence of popular appeal resulted in a limited pool of voluntary recruits. Thus, Sendero Luminoso employed coercion and fear to gain recruits, making "targets" feel that they or their families would be killed if they refused to join.[83] The absence of government forces to protect the citizenry made recruitment by intimidation particularly successful, yet it also meant that Sendero Luminoso remained a relatively small elitist organization.[84]

One of the reasons for the success of Sendero Luminoso was its educated membership, which ensured effective leadership and organization.[85] Indeed, the majority of Sendero Luminoso recruits[86] were young middle-class university students and high school graduates,[87] mainly mestizos, who believed that Peruvian society

---

[80] Manwaring, "Peru's Sendero Luminoso: The Shining Path Beckons," 161–164.

[81] McCormick, *The Shining Path and the Future of Peru.*

[82] Portugal, *Voices from the War.*

[83] Ibid.; Woy-Hazleton and Hazleton, "Sendero Luminoso and the Future of Peruvian Democracy."

[84] Portugal, *Voices from the War.*

[85] Gorriti Ellenbogen, *The Shining Path: A History of the Millenarian War in Peru;* Steve J. Stern, *Shining and Other Paths: War and Society in Peru, 1980–1995* (Durham, NC: Duke University Press, 1998); Lewis Taylor, "Counter-Insurgency Strategy, the PCP-Sendero Luminoso and the Civil War in Peru, 1980–1996," *Bulletin of Latin American Research* 17, no. 1 (January 1998): 35–58.

[86] The breakdown of organization members is as follows: 22% were students from universities and technical schools, 19% were peasants, 16% were merchants, 10% were working class, 8% were local officials and leaders, and 6% were college professors and school teachers. Overall more than 26% of Sendero Luminoso members had some higher education and more than 46% had finished high school. Portugal, *Voice from the War.*

[87] The thousands of high school graduates who were unable to pass the highly competitive university entrance exams were a receptive audience for recruiters. Stern, *Shining and Other Paths: War and Society in Peru, 1980–1995;* Taylor, "Counter-Insurgency Strategy, the PCP-Sendero Luminoso and the Civil War in Peru, 1980–1996."

could not provide the opportunities to which they were entitled.[88] These students faced a virtually nonexistent job market; this, along with Peru's increasing economic woes through the 1980s, aggravated the impact of the Sendero Luminoso insurgency as unemployed or underemployed discontents eager to see changes were prime targets for insurgent recruitment. Sendero Luminoso agents infiltrated public universities, frequently posing as students, and co-opted legal Marxist student organizations, such as the Student Revolutionary Front, as a stage for voicing Sendero Luminoso ideology and encouraging recruitment.[89] Recruiters would target students and teachers, requesting only tacit support initially and then increasing demands on the recruit and pushing indoctrination over time. Many universities essentially became recruitment and training academies for Sendero Luminoso.[90]

A unique and disturbing aspect of Sendero Luminoso, relative to other Marxist insurgencies, is its striking similarities to a religious cult.[91] Propaganda related to Guzmán and Sendero Luminoso was marked by images of blood and death, portraying armed struggle as a "purifying fire." Guzmán himself spoke of the "quota of blood" that must be paid, where "the quota is the stamp of commitment to our revolution . . . with that blood of the people that runs in our country . . . they form lakes of blood, we form pools. The blood strengthens us."[92] Recruits were given the promise of a future paradise and they were expected to give their lives to Guzmán, who would lead them to that paradise.[93] After no less than two years of training and indoctrination, recruits went through a rite of initiation through which they were expected to "cross the river of blood"—that is, to

---

[88] Woy-Hazleton and Hazleton, "Sendero Luminoso and the Future of Peruvian Democracy."

[89] Gorriti Ellenbogen, *The Shining Path: A History of the Millenarian War in Peru;* Stern, *Shining and Other Paths: War and Society in Peru, 1980–1995;* Taylor, "Counter-Insurgency Strategy, the PCP-Sendero Luminoso and the Civil War in Peru, 1980–1996."

[90] Woy-Hazleton and Hazleton, "Sendero Luminoso and the Future of Peruvian Democracy."

[91] Starn, "Maoism in the Andes: The Communist Party of Peru-Shining Path and the Refusal of History"; Terry Whalin and Chris Woehr, *One Bright Shining Path: Faith in the Midst of Terrorism* (Wheaton, IL: Crossway Books, 1993); Degregori, "How Difficult it is to be God: Ideology and Political Violence in Sendero Luminoso."

[92] Purportedly from documents used in indoctrination sessions, seized by the Peruvian military. (See Starn, "To Revolt Against the Revolution: War and Resistance in Peru's Andes.")

[93] Starn, "Maoism in the Andes: The Communist Party of Peru-Shining Path and the Refusal of History," 399–421; Degregori, "How Difficult it is to be God: Ideology and Political Violence in Sendero Luminoso."

murder to prove they were true believers.[94] In one such incident, Sendero Luminoso insurgents removed two French tourists from a bus traveling in rural Peru and shot them. As his rite of initiation, the youngest member of the group, which included mostly members who were no older than sixteen, totally crushed the skull of one of the tourists by beating the tourist with a rock.[95] No amount of reasoned argument could cause these disciples to lose faith in Guzmán, the movement, and the realization of the Maoist promised land through military victory.[96]

All recruitment was initiated by Sendero Luminoso; volunteers were viewed with suspicion. Additionally, two current Sendero Luminoso members had to vouch for new recruits. During the first year of a less than two-year training process, recruits engaged in simple noncombat tasks, such as distributing propaganda. They also received classroom instruction on Marxist texts and guerrilla warfare,[97] as well as indoctrination in Sendero Luminoso's ideology.[98] After one or two years, recruits began military training, which sometimes included participating in acts of sabotage, such as destroying high tension towers, bridges, or other infrastructure. Recruits also engaged in physical conditioning and training in the use of small arms, explosives, combat triage, and other specialized guerrilla warfare proficiencies.[99] At the conclusion of the training period and a scrupulous investigation of the candidate's background, particularly his or her personal associations, a formal determination was made as to whether the recruit would be initiated. Recruits who were accepted took an oath before four regional leaders, hooded to protect their identity. Even after two or three years of training, a recruit possessed very limited knowledge of the organizational structure and had contact during that time with only a handful of other members, as the training cadre also employed a cell-like structure.[100]

---

[94] McClintock, *Revolutionary Movements in Latin America: El Salvador's FMLN & Peru's Shining Path*; Niksch and Sullivan, *Peru's Shining Path*; Whalin and Woehr, *One Bright Shining Path: Faith in the Midst of Terrorism*.

[95] US Congress, *The Shining Path After Guzman: The Threat and the International*.

[96] Whalin and Woehr, *One Bright Shining Path: Faith in the Midst of Terrorism*.

[97] Specifically the theory of the Maoist Protracted People's War.

[98] Taylor, *Shining Path: Guerrilla War in Peru's Northern Highlands*.

[99] Ibid.

[100] Taylor, "Maoism in the Andes: Sendero Luminoso and the Contemporary Guerilla Movement in Peru."

METHODS OF SUSTAINMENT

At the outset of the insurgency, Sendero Luminoso obtained weapons and explosives by raiding police stations and mining camps. Yet by the mid-1980s, having expanded in the primary coca-producing region of Peru, the Upper Huallaga Valley, Sendero Luminoso was able to tap the profits of the drug traffickers.[101] In order to tap the ever-burgeoning cocaine trade, Sendero Luminoso sent units into the Upper Huallaga Valley to identify and kill government enforcement agents and their supporters. Once they had taken effective control of the region, Sendero Luminoso members served as middlemen between the coca growers and the drug traffickers, reportedly receiving 10% of the sale of every kilo of coca.[102]

The arrangement benefited all sides. Sendero Luminoso provided coca growers with protection from government forces and unscrupulous drug traffickers, as well as higher prices for the sale of their coca. In this role, Sendero Luminoso sold itself as the defender of the indigenous farmers whose livelihood was threatened by the government, which wanted to replace their cash crop with some less profitable produce.[103] The drug traffickers were provided with increased coca production[104] and a secure region in which to operate, which included several airfields used by traffickers to transport raw materials to Colombia for refinement.[105] Sendero Luminoso reportedly earned between 20 and 50 million dollars annually, income that was

---

[101] Coletta Youngers and Eileen Rosin, *Drugs and Democracy in Latin America: The Impact of U.S. Policy* (Boulder, CO: L. Rienner Publishers, 2004).

[102] Ibid.; Bruce H. Kay, "Violent Opportunities: The Rise and Fall of 'King Coca' and Shining Path," *Journal of Interamerican Studies and World Affairs* 41, no. 3 (Autumn 1999): vi–127; McClintock, *Revolutionary Movements in Latin America: El Salvador's FMLN & Peru's Shining Path*; David Scott Palmer, "Peru, the Drug Business and Shining Path: Between Scylla and Charybdis?" *Journal of Interamerican Studies and World Affairs* 34, no. 3 (Autumn 1992): 65–88; Starn, "New Literature on Peru's Sendero Luminoso," 212–226; Woy-Hazleton and Hazleton, "Sendero Luminoso and the Future of Peruvian Democracy"; Cynthia McClintock, "The War on Drugs: The Peruvian Case," *Journal of Interamerican Studies and World Affairs* 30, no. 2/3 (Summer–Autumn 1988): 127–142.

[103] Palmer, "Peru, the Drug Business and Shining Path: Between Scylla and Charybdis?".

[104] Sendero Luminoso demanded a strong work ethic from farmers, forbidding diversions such as alcohol and prostitution. Youngers and Rosin, *Drugs and Democracy in Latin America: The Impact of U.S. Policy*, 414; Kay, "Violent Opportunities: The Rise and Fall of 'King Coca' and Shining Path."

[105] Youngers and Rosin, *Drugs and Democracy in Latin America: The Impact of U.S. Policy;* Kay, "Violent Opportunities: The Rise and Fall of 'King Coca' and Shining Path"; McClintock, *Revolutionary Movements in Latin America: El Salvador's FMLN & Peru's Shining Path;* Palmer, "Peru, the Drug Business and Shining Path: Between Scylla and Charybdis?"; Starn, "New Literature on Peru's Sendero Luminoso"; Woy-Hazleton and Hazleton, "Sendero Luminoso and the Future of Peruvian Democracy"; McClintock," The War on Drugs: The Peruvian Case."

used to purchase weapons[106] and pay militants.[107] Although Guzmán originally disavowed any connection between Sendero Luminoso and the ongoing drug trafficking along the Andean ridge, pragmatic considerations appeared to have triumphed. However, the perceived disconnect between Sendero Luminoso's purported ideological purity and its involvement in the black market drug trade did impact its legitimacy.[108]

Sendero Luminoso's ability to operate freely in large sections of the Huallaga Valley simplified their internal lines of communication for logistics and resupply.[109] Logistical support for the insurgency was administered via a regional leader (commissar) who led a five-person committee charged with overall operational planning and execution for each region.[110] Logistics support was generally provided by villagers, either voluntarily or through coercion, as well as a small cadre of trained and specialized logistics personnel who provided weapons and ammunition. Local villagers would routinely be directed to hide ammunition or other contraband items, with their compliance motivated either positively, by ideological and emotional support for the cause, or negatively, by fear of violence and even death.

METHODS OF OBTAINING LEGITIMACY

Sendero Luminoso's appeal for legitimacy was dynamic and varied and depended on the audience and circumstances it was targeting. Among the educated mestizos, Guzmán employed his charismatic personality and leveraged the discontent with the dearth of economic opportunities. Among the highland Indians, he appealed to vague notions of injustice, racial resentment, and economic disparity, and he highlighted the harsh consequences on the native populations by government forces in response to the Sendero Luminoso insurgency. Broadly, as a means to curry legitimacy, Sendero Luminoso cited economic disparities, opportunism, a sense of government neglect by highland mestizos and Indians tied to racial inequalities, an intentional campaign to portray the government as ineffectual, and the harsh government reprisal upon the population in response to the insurgency.

---

[106] Documents captured by the Peruvian military in 1989 revealed that Sendero Luminoso had purchased thousands of Belgian-made assault rifles from the drug traffickers. McClintock, *Revolutionary Movements in Latin America: El Salvador's FMLN & Peru's Shining Path.*

[107] Militants were paid between 250 and 1,000 per month. Profits from the drug trade also went to support the families of insurgents who had been killed. Ibid.

[108] Kay, "Violent Opportunities: The Rise and Fall of 'King Coca' and Shining Path."

[109] Manwaring, "Peru's Sendero Luminoso: The Shining Path Beckons."

[110] Ibid.

Although the agrarian land redistribution had broken up large *haciendas*, Sendero Luminoso was still able to exploit land inequalities, particularly in indigenous communities where lands that had been taken in the past had not been restored to the original owners or where agricultural cooperatives had been dismantled and privatized.[111] Sendero Luminoso even divided indigenous communities by turning landless peasants against wealthier members of the community who did own land, dubbing them "landlords" no matter how modest their holdings.

Regardless of the land issue, many highland Indians still lived in extreme poverty, facing hunger, malnutrition, and disease, as well as a government that appeared unconcerned with their plight.[112] These already onerous economic conditions deteriorated further throughout the 1980s, abetted by the devastating impact of the insurgency. Yet, Sendero Luminoso was able to convincingly level blame for these woes on the government and the capitalist system, as 70% inflation in 1982 climbed to 7,650% inflation by 1990, decimating the purchasing power of most Peruvians.[113] It was within this economic climate that Sendero Luminoso opportunistically[114] won the hearts and minds of peasant farmers and urban poor, despite the reality that these individuals had only a vague notion of Sendero Luminoso's actual ideology.[115]

Given the limited allure of land redistribution, Sendero Luminoso instead appealed to the perceptions and realities of government indifference toward the plight of the mestizos and indigenous people, as well as the related racial resentment.[116] In contrast to the Peruvian state, which failed to provide basic social services or address rampant poverty, Sendero Luminoso policed regions under its control and punished corrupt officials, establishing law and order in regions the government in Lima had long neglected. Indeed, the highland Indians had experienced a long history of brutal repression and neglect by

---

[111] Degregori, "After the Fall of Abimael Guzmán"; Starn, "Maoism in the Andes: The Communist Party of Peru-Shining Path and the Refusal of History"; Mayer, "Peru in Deep Trouble: Mario Vargas Llosa's 'Inquest in the Andes' Reexamined."

[112] McClintock, *Revolutionary Movements in Latin America: El Salvador's FMLN & Peru's Shining Path.*

[113] Sally Bowen, "The Fujimori File: Peru and its President 1990–2000," *Peru Monitor,* 2000.

[114] Sendero Luminoso also took advantage of the Peruvian government's coca eradication program, portraying the government and their US backers as threats to the livelihood of the farmers, thereby winning the farmers to their side. Kay, "Violent Opportunities: The Rise and Fall of 'King Coca' and Shining Path."

[115] Degregori, "How Difficult it is to be God: Ideology and Political Violence in Sendero Luminoso."

[116] Degregori, "After the Fall of Abimael Guzmán," 179; Starn, "Maoism in the Andes: The Communist Party of Peru-Shining Path and the Refusal of History," 399–421; Mayer, "Peru in Deep Trouble: Mario Vargas Llosa's 'Inquest in the Andes' Reexamined."

the creole elites rooted in overt racial discrimination. Thus, Sendero Luminoso exploited what US Ambassador Anthony Quainton labeled "the sense of profound cultural grievance in the Indian peoples of the highlands."[117] Guzmán exploited these grievances, at least rhetorically, posing Sendero Luminoso as the champion that would overthrow the illegitimate foreign government that had oppressed native Peruvians since the Spanish conquest and restore an authentically Peruvian democracy.[118] Guzmán asserted that the ideology of Sendero Luminoso was rooted in the structure of pre-Colombian indigenous communities and that authentic Peruvian democracy was nationalistic, popular, and Indian.[119]

Unlike past and contemporary Maoist insurgencies, Sendero Luminoso's appeal to the people was nuanced and indirect. Rather than expend resources to appeal broadly and gain a large support base, they instead concentrated on undermining the legitimacy of the government, hoping the disenchanted masses would flock to their side. Sendero Luminoso claimed to offer what the government could not or would not. The success of its insurgency in the 1980s can be attributed, in part, to a heavy-handed military response of the government, which early on failed to understand that legitimacy was the currency of the ultimate victory.[120]

The insurgents employed terrorism, assassination, and sabotage as forms of "armed propaganda" designed to call into question the effectiveness, and thus, legitimacy, of the Peruvian government.[121] Assassinations and other acts of terrorism seemed to expose an incompetent and ineffectual police force and military incapable of protecting the citizenry or themselves,[122] ironically increasing

---

[117] As cited in Peter Winn, *Americas: The Changing Face of Latin America and the Caribbean*, 3rd ed. (Berkeley, CA: University of California Press, 2006).

[118] Manwaring, "Peru's Sendero Luminoso: The Shining Path Beckons."

[119] In truth, the rigid Maoist ideology of Sendero Luminoso placed little emphasis on ethnicity, but rather focused on class. Ironically, therefore, because its leadership was drawn from university students, Sendero Luminoso's own chain of command mirrored the racial hierarchy of Peru with Amerindian fighters following creole and mestizo leaders. McClintock, "Why Peasants Rebel: The Case of Peru's Sendero Luminoso."

[120] Manwaring, "Peru's Sendero Luminoso: The Shining Path Beckons."

[121] Rochlin, *Vanguard Revolutionaries in Latin America: Peru, Colombia, Mexico*; Ron, "Ideology in Context: Explaining Sendero Luminoso's Tactical Escalation"; Strong, *Shining Path: A Case Study in Ideological Terrorism*; Cornell and Roberts, "Democracy, Counterinsurgency, and Human Rights: The Case of Peru."

[122] In May 1987, a Civil Guard base outside the town of Uchiza was attacked by the insurgents. Even though the engagement lasted for several hours, the army failed to relieve the defenders before exhaustion of their ammunition forced them to surrender. Many of those who died were executed by Sendero Luminoso. Incidents like that at Uchiza fostered public perception that the military and government were ineffectual, impotent, and easily outmaneuvered by the insurgents, serving Sendero Luminoso's aim of undermining the regime's legitimacy. Niksch and Sullivan, *Peru's Shining Path*.

perceptions of injustice among the population and building sympathy for Sendero Luminoso.[123] Concurrently, the insurgency sapped public resources for social programs that could have redressed the grievances of the population that Sendero Luminoso exploited for support. At the very least, the use of terror ensured that those who did not openly support the insurgents would not, in turn, assist the government. These tactics were also useful in undermining Peru's economy, already in the midst of a severe decline, further eroding confidence in the government. By 1990, with Sendero Luminoso at the height of its military power, the economy in a free fall exacerbated by the insurgent campaign, and the legitimacy of the regime at historic lows, the strategy appeared to be successful.[124]

Rather than contravening the affront to its legitimacy, the government aggravated feelings of injustice among the population by disregarding human rights in its response to the insurgency.[125] Sendero Luminoso had intentionally created an environment in which they knew the government would respond excessively and the military naively did just that. Areas of insurgent activity were declared emergency zones by the government, allowing for the suspension of constitutional rights and resulting in looting, torture, extrajudicial killings, and the disappearance of suspects.[126] Blinded by the racism of Peru's creole elites, the military strip-searched innocent Indian villagers and attacked grassroots organizations that advocated for the indigenous peoples—ironically the very entities that could have countered Sendero Luminoso's influence.[127] Sendero Luminoso then positioned itself as a guardian of the people against the racism and brutality of the military.[128] Even when Sendero Luminoso's violence against the indigenous people intensified after 1982 and broad support for the insurgents waned, support for the government did not increase.

---

[123] Rochlin, *Vanguard Revolutionaries in Latin America: Peru, Colombia, Mexico*, 293; Ron, "Ideology in Context: Explaining Sendero Luminoso's Tactical Escalation"; Strong, *Shining Path: A Case Study in Ideological Terrorism*; Cornell and Roberts, "Democracy, Counterinsurgency, and Human Rights: The Case of Peru."

[124] Manwaring, "Peru's Sendero Luminoso: The Shining Path Beckons."

[125] Gordon H. McCormick, *From the Sierra to the Cities: The Urban Campaign of the Shining Path* (Santa Monica, CA: RAND, 1992).

[126] Truth and Reconciliation Commission, *Final Report* (2003); Degregori, "After the Fall of Abimael Guzmán"; Brown and Fernandez, *War of Shadows: The Struggle for Utopia in the Peruvian Amazon*.

[127] Degregori, "After the Fall of Abimael Guzmán"; Brown and Fernandez, *War of Shadows: The Struggle for Utopia in the Peruvian Amazon*.

[128] Ibid. Family members of many Sendero Luminoso recruits had been abused by military or police forces.

The use of "revolutionary violence," which included massacres of peasants and assassinations, was an explicit strategy of Sendero Luminoso,[129] intended to persuade reluctant communities to support Sendero Luminoso, as well as to destroy or intimidate any political or military rivals.[130] Potential rivals included the government, other leftist insurgent groups, other Marxist parties, as well as religious and other organizations and leaders that might compete for the loyalties of the people by providing social justice advocacy and economic development.[131] With one method of recruitment by fear, parents were murdered in front of their children, and the children were then forced to eat their parents' tongues. Likewise, children of uncooperative parents would be tortured in front of their parents in an attempt to coerce the parents to support Sendero Luminoso.[132] As the insurgency grew, emphasis on the military campaign and frustration with the tacit support of the peasants led Sendero Luminoso to largely abandon the services they provided to local communities, instead favoring violent coercion as a means to win support.[133] Moreover, Sendero Luminoso actually began to attack the livelihood of peasants by closing local markets in order to undermine the capitalist system. The absence

---

[129] In an official document, Guzmán articulated Sendero Luminoso's position on human rights: "We start by not ascribing to either Universal Declaration of Human Rights or the Costa Rica Convention on Human Rights, but we have used their legal devices to unmask and denounce the old Peruvian state. . . . For us, human rights are contradictory to the rights of the people, because we base rights in man as a social product, not man as an abstract with innate rights. 'Human rights' do not exist except for the bourgeois man, a position that was at the forefront of feudalism, like liberty, equality, and fraternity were advanced for the bourgeoisie of the past. But today, since the appearance of the proletariat as an organized class in the Communist Party, with the experience of triumphant revolutions, with the construction of socialism, new democracy and the dictatorship of the proletariat, it has been proven that human rights serve the oppressor class and the exploiters who run the imperialist and landowner-bureaucratic states. Bourgeois states in general . . . Our position is very clear. We reject and condemn human rights because they are bourgeois, reactionary, counterrevolutionary rights, and are today a weapon of revisionists and imperialists, principally Yankee imperialists." Comité Central del Partido Comunista del Perú, *Sobre Las Dos Colinas: Documento De Estudio Para El Balance De La III Campaña De Impulsar Las Bases De Apoyo*, 1991.

[130] Brown and Fernandez, *War of Shadows: The Struggle for Utopia in the Peruvian Amazon.*

[131] In May 1991, Sister Irene McCormick, an Australian nun with the Catholic relief organization Caritas, was shot by Sendero Luminoso guerrillas. Local villagers were forbidden by Sendero Luminoso from moving her body for twenty-four hours. In a similar incident, María Elena Moyano, the popular mayor of one of Lima's slums, was shot because her reform programs, designed to improve the economic conditions of the urban poor, provided an alternative means of hope to the violence offered by Sendero Luminoso. As her family looked on, insurgents blew up her body with dynamite, leaving pieces more than one hundred yards away. US Congress, *The Shining Path After Guzman: The Threat and the International Response.*

[132] Ibid.

[133] Starn, Degregori, and Kirk, *The Peru Reader: History, Culture, Politics*; Bennett and Hallewell, *Sendero Luminoso in Context: An Annotated Bibliography*; Aronson, "Brutal Insurgency: Sendero Luminoso."

of visible benefits made the rigid ideology of Sendero Luminoso markedly less compelling, and the increasing use of violence and repression alienated the peasantry.[134]

Blinded by their religious devotion to Guzmán and his austere brand of Maoism, Sendero Luminoso sought to impose, through appalling violence, a system that was contrary to the culture, inclinations, needs, and aspirations of the people.[135] Although the comparably harsh tactics and racism of the military had made the people reluctant to support the government and even drove some, particularly university students, to the Sendero Luminoso cause, the military became increasingly sensitive to human rights abuses over time.[136] By being more discriminate in their counterinsurgency operations, the military won the support of the local population. Concurrently, the government of President Alan Garcia Perez (1985–1990), recognizing the need for economic reform, provided modest economic aid to the poor interior regions. This effort was dramatically expanded under President Alberto Fujimori (1990–2000).

EXTERNAL SUPPORT

Although Sendero Luminoso rejected other communist nations because of their ideological laxity, they were not completely isolated internationally.[137] Through its Department of International Relations,[138] Sendero Luminoso organized support committees abroad. Composed of expatriates, these committees operated throughout the United States and Western Europe, engaging in fund-raising and public relations in order to gain legitimacy in the eyes of the international community, as well as to cultivate political support from leftist groups and individuals.[139] Support for Sendero Luminoso came from a variety of human rights organizations, academics, and

---

[134] Ibid.

[135] Palmer, *The Shining Path of Peru*, 298.

[136] This contrasts rigid adherence of Sendero Luminoso to an ideology that demanded the use of violence, even after it became clear that this strategy was no longer fruitful. Ibid.

[137] Aronson, "Brutal Insurgency: Sendero Luminoso"; Woy-Hazleton and Hazleton, "Sendero Luminoso and the Future of Peruvian Democracy."

[138] Sendero Luminoso had offices in capitals such as London and Paris. The head of their London office, Adolfo Héctor Olaechea, created the Musical Guerrilla Army, which held concerts through England. Their songs reportedly extolled the revolution with lyrics such as "the people's blood has a beautiful aroma . . . Chairman Gonzalo, Light of the Masses . . . The blood of the armed people nourishes the armed struggle . . . Victory is ours." Olaechea also arranged for BBC reporters to do a very sympathetic report on Sendero Luminoso at the height of their power in the 1992. Strong, *Shining Path: A Case Study in Ideological Terrorism.*

[139] Philip Mauceri, *State Under Siege: Development and Policy Making in Peru* (Boulder, CO: Westview Press, 1996); Palmer, *The Shining Path of Peru.*

others with left-wing sympathies, who falsely viewed the insurgency as an indigenous peasant movement.[140]

Within Peru, a network of supposedly independent organizations developed that were, in reality, legal fronts that Sendero Luminoso used to recruit and provide logistical and financial support to its members and their families, as well as legal advocacy for captured insurgents and the Sendero Luminoso cause more broadly.[141] These organizations were populated with left-wing supporters of the movement, many of whom naively supported Sendero Luminoso, viewing it as the solution to legitimate grievances among the poor of Peru and a counter to the discriminations and repression of the military. One such front was the César Vallejo Academy, an elegant Lima ballet studio, which served as the headquarters for Sendero Luminoso's Department of Organizational Support, which managed internal communications and evaluated strategy.[142] Fronts like the César Vallejo Academy allowed Sendero Luminoso to hide its financial activities and recruitment strategies. Attempts to target these organizations[143] for offering material support for terrorism was decried by other leftist organizations, as well as domestic and international human rights groups such as Amnesty International and Human Rights Watch.[144] These organizations remained in operation until President Alberto Fujimori (1990–2000) declared a state of emergency and suspended the constitution in April 1992, which allowed the government to shut them down.

COUNTERMEASURES TAKEN BY THE GOVERNMENT

The initial government response to Sendero Luminoso was the use of specially trained police units (called *sinchis*), not the military, during the early 1980s. Sendero Luminoso was characterized as a

---

[140] Mauceri, *State Under Siege: Development and Policy Making in Peru.*

[141] Eckstein and Garretón Merino, *Power and Popular Protest: Latin American Social Movements*; Stern, *Shining and Other Paths: War and Society in Peru, 1980–1995*; William A. Hazleton and Sandra Woy-Hazleton, "Terrorism and the Marxist Left: Peru's Struggle Against Sendero Luminoso," *Studies in Conflict & Terrorism* 11, no. 6 (1988).

[142] Eduardo Toche, *ONG, Enemigos Imaginados*, 1st ed. (Lima: DESCO, Centro de Estudios y Promoción del Desarrollo, 2003).

[143] These organizations included the Movement of Classist Workers and Peasants, the Association of Democratic Lawyers, the Movement of Popular Intellectuals, the Movement of Popular Artists, Popular Aid, the Federation of Revolutionary Students, and the Neighborhood Movement. Eckstein and Garretón Merino, *Power and Popular Protest: Latin American Social Movements;* Stern, *Shining and Other Paths: War and Society in Peru, 1980–1995*; Hazleton and Woy-Hazleton, "Terrorism and the Marxist Left: Peru's Struggle Against Sendero Luminoso."

[144] Toche, *ONG, Enemigos Imaginados.*

terrorist group and was not considered a threat to the government. The *sinchis* were not properly trained or equipped for the mission, however, and in many cases, their indiscriminate violence only made a bad situation worse, further alienating the peasant populace from the government. There was little attempt to address root social and economic conditions in Ayacucho. For many, radical violence seemed to be the only response to their desperate situation.

As the government's initial response was one of confusion and misunderstanding, Sendero Luminoso was not seen as a threat to the state and they were treated more as an annoyance. By 1983, however, the government understood that more urgent steps were needed, and the Army deployed a reinforced division into Ayacucho to confront Sendero Luminoso head-on with brute force. Soldiers had received little or no training in conducting counterinsurgency operations, and their mission was ill defined beyond securing the region and destroying Sendero Luminoso. Sendero Luminoso, however, had no intention of facing the military in a *mano y mano* confrontation. While the military attempted to control the region by force, the adversary was not prepared to do the same. The result was a series of human rights violations that raised the stakes of the insurgency (such as international news coverage of the murder of eight journalists killed not by the military but by Indian peasants with rocks and machetes in January 1983). For Peruvians living in Lima and other urban areas, the lesson they learned related to how different the lives of native Indians in the Andes were from their own. Many spoke of "the two Perus." Thousands of civilians were killed, tortured, kidnapped, or simply disappeared (*las desapariciones*) during the conflict. Civilians feared both Sendero Luminoso and the military. In many cases, peasants were uncertain as to who had instigated the violent acts. By 1986, the government had dismissed 1,700 corrupt police officers, including 120 police generals and colonels. The pattern of violence, abuse, and corruption had continued unabated for years. The real threat of Sendero Luminoso, however, had grown, expanding violent acts throughout the country while successfully remaining elusive to police and military forces.[145]

When Sendero Luminoso shifted the focus of its movement from the predominately rural areas of the Andean ridge to the cities, Peruvian security forces also shifted their focus. First, they recognized that the state had the upper hand in "controlling" the media, including print, radio, and television. Peruvian authorities

---

[145] K. Theidon, "Terror's Talk: Fieldwork and War," *Dialectical Anthropology* 26, no. 1 (2001): 19–35; Susan C. Bourque and Kay B. Warren, "Democracy without Peace: The Cultural Politics of Terror in Peru," *Latin American Research Review* 24, no. 1 (1989): 7–34.

emphasized the human rights abuses of Sendero Luminoso (while minimizing their own) and touted the economic and security improvements that the government had been implementing. Second, security forces established an intelligence network throughout the urban areas, specifically within the capital of Lima. The establishment and use of this human intelligence (HUMINT) network, focused on the urban terrain, began to pay off as lower- and mid-level leaders were discovered and either killed or captured. Based on the cellular structure of the organization, security forces were able to destroy the effectiveness of an entire cell by simply focusing on the leadership, not the individual members. Because none of the other members of a cell had any contact with other cells, their effectiveness was generally gone.[146] By 1992, the government's HUMINT network netted the biggest prize of all: Guzmán. In August 1992, Sendero Luminoso announced a new military offensive, Grand Military Plan VI. Just weeks later, however, on September 12, the special antiterrorist police captured Guzmán, two Political Bureau members, and valuable Sendero Luminoso strategic documents. Within months, 19 of the 22 members of the Central Committee had also been captured, and the strategic leadership of the organization was in disarray.

One year after Guzmán's capture, in 1993, President Fujimori surprised the world when he announced at the United Nations that Guzmán was calling for peace talks with the government. The reorganized Sendero Luminoso national leadership had not anticipated this event, and they immediately denied it, claiming a ruse by the Fujimori administration. Days later, however, Guzmán appeared on television, reiterating the call for peace and negotiation with the government. Guzmán, as well as other Sendero Luminoso leaders detained in prison, began to appear regularly on television, all stating the need for peace. Although Sendero Luminoso continued to operate in the field, their morale was shaken; their leadership had been dismantled and their future direction was in question. The next year, 1994, saw a marked decline in terrorist actions to the lowest level of violence recorded since 1981.[147]

Peruvian police and military authorities demonstrated little tactical success in countering Sendero Luminoso during the early 1980s. Forces had received no counterinsurgency training. Their equipment, including communications and transportation, was completely inadequate. Attacked police posts frequently had no

---

[146] M. Elkhamri et al., *Urban Population Control in a Counterinsurgency* (Foreign Military Studies Office [FMSO], Center for Army Lessons Learned [CALL], Fort Leavenworth, KS, 2005).

[147] Degregori, "After the Fall of Abimael Guzmán."

means to call for support. The National Guard was understaffed by 17,000 personnel. Pay and morale were low.[148] The government also had no success in infiltrating the organization. The cellular structure that Sendero Luminoso had adopted from its inception had proven to be successful in maintaining operational security. In a January 1983 interview, Peruvian Minister of War General Cisneros commented, "the police force do not know where the Senderistas are, nor how many there are, nor when they are going to attack,"[149] revealing that Sendero Luminoso was making steady progress in its protracted Maoist war.

General Roberto Noel Moral was assigned the task of developing a counterterrorist strategy. He believed that Sendero Luminoso was a movement of fanatics and was viewed as a disease that had to be removed and annihilated from the state. Violence was considered the only option for dealing with the threat. Mario Vargas Llosa, president of the Commission of Inquiry into the Uchuraccay Incident (journalist killings), noted that Sendero Luminoso consisted of bloodthirsty fanatics, "detached from life and common sense . . . committed to destroying and killing."[150]

During 1986, the Peruvian military responded to prison riots in three Lima prisons by using extensive force, resulting in the deaths of 270 people who were accused of terrorist activities. At the Lurigancho prison, more than 100 prisoners were shot at close range, reportedly after they had surrendered.[151] Images of corpses were displayed extensively by the media, with new video and photographs leading almost every news broadcast and every newspaper front page. The public outcry and backlash led to an opportunity for fresh support, now including the middle class and urbanites, who became convinced that the government's actions confirmed Sendero Luminoso's claims of abuse toward its people. Building on this burgeoning support base during the September 1987 meeting of the Sendero Luminoso Central Leadership, the decision was made to move from a protracted rural campaign to an urban-based revolution. Sendero Luminoso also increased their propaganda campaign, with Guzmán conducting a series of interviews with journalists, and organized a series of marches through the capital demanding support for the Sendero Luminoso cause.

---

[148] Taylor, "Maoism in the Andes: Sendero Luminoso and the Contemporary Guerilla Movement in Peru."

[149] Ibid.

[150] Portugal, *Voices from the War.*

[151] Bourque and Warren, "Democracy without Peace: The Cultural Politics of Terror in Peru."

In response to the ever-growing threat, the Garcia government requested and received US military equipment, training, and support, commencing in 1988. This US support enabled the Peruvian security forces to initiate a major military offensive in 1989. Sendero Luminoso, however, continued to skip away, and although the military claimed great success, there was no victory. Sendero Luminoso became as elusive as ever, with their threat to security continuing almost unabated. The military, however, was accused of massive human rights abuses resulting from the torture and killing of civilians. Amnesty International widely reported that Peru had the highest number of disappearances in the world, claiming the security forces were responsible. Amnesty International also noted the abuses at the hands of Sendero Luminoso, but the damage to the legitimacy of the government was now clearly in question.[152] The actions, real or perceived, of the police and military against civilians caused resentment, convincing many that Sendero Luminoso provided a better alternative than the government. The result was a sympathetic base at the village level that was willing to provide either active support to Sendero Luminoso in the form of safe havens, intelligence, or medical and logistics support, or passive support through reduced interactions with the Army and police.[153]

In 1983, the government began relocating rural communities to areas that were more defensible, simultaneously establishing local committees to organize community defense units, called *Rondas Campesinas*.[154] The *Rondas Campesinas* (*Rondas*) were a civil defense organization originally formed to stop cattle rustlers. These Civil Defense Hamlets, also called "mounds" or "masses," were targeted by Sendero Luminoso, who dubbed them "flocks of sheep."[155] This policy alienated the indigenous villagers, who were forced from lands held by their families for as long as they could remember and served as the subsistence to their agricultural economies. Many who had owned land now became landless, and the new communities often incorporated multiple remote villages, compelling members to work together with strangers, which, given their previous isolation, naturally bred distrust. Moreover, public infrastructure was limited, of poor quality, and quickly decayed. For some of the numerous farmers who lost

---

[152] Woy-Hazleton and Hazleton, "Sendero Luminoso and the Future of Peruvian Democracy."

[153] Portugal, *Voices from the War*.

[154] Taylor, *Shining Path: Guerrilla War in Peru's Northern Highlands*; Taylor, "Counter-Insurgency Strategy, the PCP-Sendero Luminoso and the Civil War in Peru, 1980–1996"; Cornell and Roberts, "Democracy, Counterinsurgency, and Human Rights: The Case of Peru."

[155] Degregori, "How Difficult it is to be God: Ideology and Political Violence in Sendero Luminoso."

their land through the relocation, Sendero Luminoso appeared to be a natural ally against an unjust government in the fight for restoration of their land.[156] Because of the long history of racism, which raised doubts regarding the ability of armed indigenous peasants to defend themselves, as well as the fear that such groups could turn against the government, the military resisted any efforts to arm the *Rondas*.[157]

The *Rondas* provided intelligence to the National Intelligence Service while also helping to defend peasant communities that had little or no local security capabilities. Some analysts have credited the *Rondas* with forcing Sendero Luminoso out of the peasant communities and into the urban environment before they were operationally ready to do so. The *Rondas* also reestablished the positive sense of community that Sendero Luminoso had previously severed. Contrary opinions note, however, that the *Rondas* were poorly trained, lacked oversight, and committed numerous human rights violations.[158]

SHORT- AND LONG-TERM EFFECTS

CHANGES IN GOVERNMENT

In 1990, the third president to confront Sendero Luminoso, Alberto Fujimori, was installed. Fujimori radically shifted the Peruvian government's response from a national police-led counterinsurgency campaign to a military-led counterinsurgency campaign, which was acknowledged for its aggressiveness but also for its lack of respect for human rights and the rule of law. In April 1992, just five months prior to Guzmán's capture, Fujimori successfully conducted a *coup de main*, dissolving the Congress, Constitution, and judiciary. All national power rested in his hands. With the capture of Guzmán and the additional capture of 3,600 Sendero Luminoso insurgents in the following 18 months, Fujimori rode a wave of internal popularity, although much of the international community condemned his actions. His authoritarian, and even ruthless, actions resulted in a sudden quelling of the insurgency, with the government firmly on the offensive. These changes came at a price of untold violence to noncombatants, near economic ruin to the country, and a badly

[156] Taylor, *Shining Path: Guerrilla War in Peru's Northern Highlands*; Taylor, "Counter-Insurgency Strategy, the PCP-Sendero Luminoso and the Civil War in Peru, 1980–1996"; Cornell and Roberts, "Democracy, Counterinsurgency, and Human Rights: The Case of Peru."

[157] Starn, "To Revolt Against the Revolution: War and Resistance in Peru's Andes."

[158] Gerald N. Vevon, "Sendero Luminoso: A Failed Revolution In Peru?" (Carlisle Barracks, PA: US Army War College, 1998).

tarnished international reputation, resulting in Fujimori's departure for exile in Japan, followed later by his arrest and imprisonment for crimes against humanity.[159]

CHANGES IN POLICY

The Ministry of the Interior, which oversaw all national police forces, formed a counterintelligence service in the mid-1980s, called DIRCOTE and subsequently renamed DINCOTE. This organization's painstaking analysis is credited with the capture of Guzmán, which proved to be the waning moment for the Sendero Luminoso insurgency. DINCOTE's activities were completely segregated from the National Intelligence Service, as well as other ongoing military operational and intelligence activities. Their activities were focused on attacking key nodes of Sendero Luminoso, as opposed to confronting the organization from a *mano y mano* perspective, using force on force. DINCOTE focused on investigation and analysis, including tracking visitors of imprisoned Sendero Luminoso personnel, collecting intelligence from trash, and networking. During 1992, DINCOTE raided a Lima college campus, capturing Central Committee members in charge of overall logistics activities, as well as capturing computers filled with operational information on the organization. Apparently, the Central Committee members had not adhered to the same stringent operational security measures that were successfully incorporated for years by their subordinates. Even Guzmán admitted that this raid proved to be a major blow to Sendero Luminoso.

On September 12, 1992, DINCOTE agents captured and arrested Abimael Guzmán in Lima. Their analysis, and subsequent awareness, of his location was based on the discovery of medicine for psoriasis, from which Guzmán was known to suffer, cigarette stubs from Guzmán's favorite brand, and chicken bones (Guzmán enjoyed Peruvian roast chicken). With hard analysis, DINCOTE had successfully dealt a major blow to the organization, and Sendero Luminoso has never fully recovered.[160]

---

[159] E. Morón and C. Sanborn, "The Pitfalls of Policymaking in Peru: Actors, Institutions and Rules of the Game," *Universidad Del Pacífico, April (Unpublished)* (2004); Vevon, "Sendero Luminoso: A Failed Revolution in Peru?"

[160] Ibid.

CHANGES IN THE REVOLUTIONARY MOVEMENT

Without question, the capture of Abimael Guzmán by DINCOTE in 1992 served as the high-water mark for the insurgency. Although Sendero Luminoso continues to exist as a political element, it no longer has the capability to hold territory; pose a threat to the national, regional, or provincial governments; or conduct organized armed resistance against police or military forces. With the exile of Fujimori in 2000, Sendero Luminoso became a secondary, and even tertiary, priority of the government as it became more concerned about economic recovery and improving international relations than continuing a widespread offensive against the remains of a tattered organization. Sendero Luminoso continues to exist, but its capabilities are extremely limited. Guzmán served as the cohesive, charismatic element around which all else flowed. Not only was his capture a severe blow to the organization, but his apparent change of political character, encouraging negotiation and peace talks, threw the organization into a kind of political shock. Sendero Luminoso was defeated operationally by strong intelligence and analysis and the removal of one operational leader after another until a key analytic thread led to Guzmán himself.

OTHER EFFECTS

One major success for the government was the National Intelligence Service, which was formed from other intelligence organizations, principally from the armed forces, and consisted of approximately 100 personnel. The National Intelligence Service was credited with infiltrating the Sendero Luminoso cellular structure, leading to the arrest or killing of several mid-level regional leaders throughout the organization. Although the strength of the Sendero Luminoso cellular structure was operational security, this also meant that if key leaders were taken out of the organization, some operational capabilities and key personnel would have to be reconstituted, a process that could take several months, even years. The *Rondas* were a valuable source of information for the National Intelligence Service.[161]

---

[161] Ibid.

BIBLIOGRAPHY

Alexander, Yonah, and Michael Kraft. *Evolution of U.S. Counterterrorism Policy.* Vol. 1–3. Santa Barbara, CA: Greenwood Publishing Group, 2007.

Aronson, Bernard W. "Brutal Insurgency: Sendero Luminoso." *U.S. Department of State Dispatch* 3, no. 12 (1992): 236.

Bennett, John M., and Laurence Hallewell. *Sendero Luminoso in Context: An Annotated Bibliography.* Lanham, MD: Scarecrow Press, 1998.

Berg, Ronald H. "Sendero Luminoso and the Peasantry of Andahuaylas." *Journal of Interamerican Studies and World Affairs* 28, no. 4 (Winter 1986): 165–196.

Bourque, Susan C., and Kay B. Warren. "Democracy without Peace: The Cultural Politics of Terror in Peru." *Latin American Research Review* 24, no. 1 (1989): 7–34.

Bowen, Sally. *The Fujimori File: Peru and its President 1990–2000.* Lima, Peru: Monitor, 2000.

Brown, Michael F., and Eduardo Fernández. *War of Shadows: The Struggle for Utopia in the Peruvian Amazon.* Berkeley, CA: University of California Press, 1991.

Comité Central del Partido Comunista del Perú. *Sobre Las Dos Colinas: Documento De Estudio Para El Balance De La III Campaña De Impulsar Las Bases De Apoyo,* 1991.

Central Intelligence Agency. "Peru," *The World Factbook.* "The World Factbook." Accessed November 2, 2009. https://www.cia.gov/library/publications/the-world-factbook/geos/pe.html.

Cornell, Angela, and Kenneth Roberts. "Democracy, Counterinsurgency, and Human Rights: The Case of Peru." *Human Rights Quarterly* 12, no. 4 (November 1990): 529–553.

Courtois, Stéphane. *The Black Book of Communism: Crimes, Terror, Repression.* Cambridge, MA: Harvard University Press, 1999.

Crilley, Kathy. "Information Warfare: New Battlefields; Terrorists, Propaganda and the Internet." *Aslib Proceedings* 53, no. 7 (July/August 2001): 250–264.

Degregori, C. I. "After the Fall of Abimael Guzmán: The Limits of Sendero Luminoso." In *The Peruvian Labyrinth: Polity, Society, Economy,* edited by Maxwell A. Cameron and Philip Mauceri. University Park, PA: Pennsylvania State University Press, 1997.

———. "How Difficult it is to be God: Ideology and Political Violence in Sendero Luminoso." *Critique of Anthropology* 11, no. 3 (1991): 233.

Eckstein, Susan, and Manuel A. Garretón Merino. *Power and Popular Protest: Latin American Social Movements.* Updated and expanded ed. Berkeley, CA: University of California Press, 2001.

Elkhamri, M., L. W. Grau, L. King-Irani, A. S. Mitchell, and L. Tasa-Bennett. "Urban Population Control in a Counterinsurgency." Foreign Military Studies Office (FMSO), Center for Army Lessons Learned (CALL), Fort Leavenworth, KS, 2005.

Gorriti Ellenbogen, Gustavo. *The Shining Path: A History of the Millenarian War in Peru.* Latin America in translation/en traducción/em tradução. Chapel Hill, NC: University of North Carolina Press, 1999.

Graham, Carol L. "The Latin American Quagmire: Beyond Debt and Democracy." *The Brookings Review* 7, no. 2 (Spring 1989): 42–47.

Harding, Colin. "Antonio Diaz Martinez and the Ideology of Sendero Luminoso." *Bulletin of Latin American Research* 7, no. 1 (1988): 65–73.

Hazleton, William A., and Sandra Woy-Hazleton. "Terrorism and the Marxist Left: Peru's Struggle Against Sendero Luminoso." *Studies in Conflict & Terrorism* 11, no. 6 (1988): 471.

Hudson, Rex A., ed., *Peru: A Country Study.* Area Handbook Series. 4th ed. Vol. 550-42. Washington, DC: US Department of Defense, 1993.

Kay, Bruce H. "Violent Opportunities: The Rise and Fall of 'King Coca' and Shining Path." *Journal of Interamerican Studies and World Affairs* 41, no. 3 (Autumn 1999): vi–127.

Kent, Robert B. "Geographical Dimensions of the Shining Path Insurgency in Peru." *Geographical Review* 83, no. 4 (October 1993): 441–454.

Manwaring, Max G. "Peru's Sendero Luminoso: The Shining Path Beckons." *Annals of the American Academy of Political and Social Science* 541, no. 1 (1995): 157–166.

Mason, T. David. "'Take Two Acres and Call Me in the Morning': Is Land Reform a Prescription for Peasant Unrest?" *The Journal of Politics* 60, no. 1 (February 1998): 199–230.

Mauceri, Philip. *State Under Siege: Development and Policy Making in Peru.* Boulder, CO: Westview Press, 1996.

Mayer, Enrique. "Peru in Deep Trouble: Mario Vargas Llosa's 'Inquest in the Andes' Reexamined." *Cultural Anthropology* 6, no. 4 (November 1991): 466–504.

McClintock, Cynthia. *Revolutionary Movements in Latin America: El Salvador's FMLN & Peru's Shining Path.* Washington, DC: United States Institute of Peace Press, 1998.

————. "The War on Drugs: The Peruvian Case." *Journal of Interamerican Studies and World Affairs* 30, no. 2/3 (Summer–Autumn 1988): 127–142.

————. "Why Peasants Rebel: The Case of Peru's Sendero Luminoso." *World Politics* 37, no. 1 (October 1984): 48–84.

McCormick, Gordon H. *From the Sierra to the Cities: The Urban Campaign of the Shining Path.* Santa Monica, CA: RAND, 1992.

————. *The Shining Path and the Future of Peru.* Santa Monica, CA: RAND, 1990.

Morón, E., and C. Sanborn. "The Pitfalls of Policymaking in Peru: Actors, Institutions and Rules of the Game." *Universidad Del Pacífico,* April (Unpublished) 2004.

Niksch, Larry A., and Mark P. Sullivan. *Peru's Shining Path.* Washington, DC: Congressional Research Service, Library of Congress, 1993.

Palmer, David Scott. "Peru, the Drug Business and Shining Path: Between Scylla and Charybdis?" *Journal of Interamerican Studies and World Affairs* 34, no. 3 (Autumn 1992): 65–88.

————. "Rebellion in Rural Peru: The Origins and Evolution of Sendero Luminoso." *Comparative Politics* 18, no. 2 (January 1986): 127–146.

————. *The Shining Path of Peru.* 2nd ed. New York: St. Martin's Press, 1994.

Portugal, A. *Voices from the War: Exploring the Motivation of Sendero Luminoso Militants.* Oxford: Center for Research on Inequality, Human Security, and Ethnicity, 2008.

Post, David. "Political Goals of Peruvian Students: The Foundations of Legitimacy in Education." *Sociology of Education* 61, no. 3 (July 1988): 178–190.

Roberts, Kenneth M. "Economic Crisis and the Demise of the Legal Left in Peru." *Comparative Politics* 29, no. 1 (October 1996): 69–92.

Rochlin, James Francis. *Vanguard Revolutionaries in Latin America: Peru, Colombia, Mexico.* Boulder, CO: L. Rienner Publishers, 2003.

Ron, James. "Ideology in Context: Explaining Sendero Luminoso's Tactical Escalation." *Journal of Peace Research* 38, no. 5 (September 2001): 569–592.

Ryan, Jeffrey John. "The Dynamics of Latin American Insurgencies: 1956–1986." Ph.D. diss., Rice University, 1989.

Schutz, Barry M., and Robert O. Slater. *Revolution & Political Change in the Third World.* Boulder, CO: L. Rienner Publishers, 1990.

Seligmann, Linda J. *Between Reform & Revolution: Political Struggles in the Peruvian Andes, 1969–1991.* Stanford, CA: Stanford University Press, 1995.

Starn, Orin. "Maoism in the Andes: The Communist Party of Peru-Shining Path and the Refusal of History." *Journal of Latin American Studies* 27, no. 2 (May 1995): 399–421.

———. "New Literature on Peru's Sendero Luminoso; El Surgimiento De Sendero Luminoso: Del Movimiento Por La Gratitud De La Ensenanza Al Inicio De La Lucha Armada; Que Dificil Es Ser Dios: Ideologia y Violencia Politica En Sendero Luminoso; Sendero: La Historia De La Guerra Milenaria; Juventud y Terrorismo; the Decade of Chaqwa: Peru's Internal Refugees; Violencia Politica En El Peru, 1980–88." *Latin American Research Review* 27, no. 2 (1992): 212–226.

———. "To Revolt Against the Revolution: War and Resistance in Peru's Andes." *Cultural Anthropology* 10, no. 4 (November 1995): 547–580.

Starn, Orin, Carlos Iván Degregori, and Robin Kirk. *The Peru Reader: History, Culture, Politics.* Latin America Readers. 2nd ed. Durham, NC: Duke University Press, 2005.

Stern, Steve J. *Shining and Other Paths: War and Society in Peru, 1980–1995.* Latin America Otherwise. Durham NC: Duke University Press, 1998.

Strong, Simon. *Shining Path: A Case Study in Ideological Terrorism.* Conflict Studies Series no. 260. London: Research Institute for the Study of Conflict and Terrorism, 1993.

Taylor, L. "Maoism in the Andes: Sendero Luminoso and the Contemporary Guerilla Movement in Peru." Working Paper No. 2. Liverpool, UK: Institute of Latin American Studies, 1983.

Taylor, Lewis. "Counter-Insurgency Strategy, the PCP-Sendero Luminoso and the Civil War in Peru, 1980–1996." *Bulletin of Latin American Research* 17, no. 1 (January 1998): 35–58.

———. *Shining Path: Guerrilla War in Peru's Northern Highlands, 1980–1997* Liverpool Latin American Studies New Series 6. Liverpool, UK: Liverpool University Press, 2006.

Theidon, K. "Terror's Talk: Fieldwork and War." *Dialectical Anthropology* 26, no. 1 (2001): 19–35.

Toche, Eduardo. *ONG, Enemigos Imaginados.* 1st ed. Lima: DESCO, Centro de Estudios y Promoción del Desarrollo, 2003.

Truth and Reconciliation Commission. *Final Report.* 2003.

US Congress. "The Shining Path After Guzman: The Threat and the International Response." Hearing before the Subcommittee on Western Hemisphere Affairs of the Committee on Foreign Affairs, House of Representatives, 102nd Cong., 2nd sess., September 23, 1992.

Vevon, Gerald N. "Sendero Luminoso: A Failed Revolution in Peru?" Carlisle Barracks, PA: US Army War College, 1998.

Whalin, Terry, and Chris Woehr. *One Bright Shining Path: Faith in the Midst of Terrorism.* Wheaton, IL: Crossway Books, 1993.

Winn, Peter. *Americas: The Changing Face of Latin America and the Caribbean.* 3rd ed. Berkeley, CA: University of California Press, 2006.

Woy-Hazleton, Sandra, and William A. Hazleton. "Sendero Luminoso and the Future of Peruvian Democracy." *Third World Quarterly* 12, no. 2 (April 1990): 21–35.

Youngers, Coletta, and Eileen Rosin. *Drugs and Democracy in Latin America: The Impact of U.S. Policy.* Boulder, CO: L. Rienner Publishers, 2004.

1979 IRANIAN REVOLUTION

Chuck Crossett and Summer Newton

SYNOPSIS

The 1979 collapse of the Pahlavi monarchy in Iran happened quickly and was somewhat unexpected in the West. Opposition to the Shah's government started to expand early in 1978 after years of tight control over any dissent. A very loose confederation of secular politicians and Islamic fundamentalist clerics helped to stir up anger and protest, but the violent crackdowns by the police backfired as the lower and merchant classes took to the streets. Massive demonstrations overtook Tehran and Tabriz, while the exiled Ayatollah Khomeini emerged as the voice and de facto leader of the opposition through his attacks on the Shah for policies aimed at Westernizing and secularizing Iran. Strikes by the industrial workers and government employees brought the economy to a standstill and forced the Shah to step down. As 1979 began, his caretaker government crumbled in the face of popular support for the parallel government set up by Khomeini and his allies, although Khomeini soon purged those who did not desire a full Islamic government, leading to the formation of the Islamic Republic of Iran in 1982.

TIMELINE

1941	Reza Shah forced by Allies to abdicate Iranian throne and his son Mohammed Reza Pahlavi was installed as Shah.
1950–1951	National Front Party makes large gains in national elections and Mosaddeq named prime minister; legislation passed to nationalize the Iranian oil industry.
1953	Coup removes Mosaddeq and Shah appoints a US/British-approved prime minister.
1963	Shah's "White Revolution" enacted land reforms and social welfare programs.
	Ayatollah Ruhollah Khomeini gives a speech in Qom criticizing the monarchy, leading to his arrest.
1965	Khomeini was exiled to Iraq for his continued criticism of the Shah and his policies.

1977	National Front Party distributes three open letters complaining of government corruption and repression.
January 1978	Appearance of a government-vetted newspaper article attacking Khomeini's past and foreign ties, inciting protests by his followers in Qom that lead to more than 70 protestors being killed.
February 1978	Forty-day commemorations for the dead in Qom lead to further riots and protests across Iran.
Fall 1978	400 men, women, and children are killed in a fire at a movie theater in Abadan, with both the government and protestors charging the other side with arson.
	Security forces kill hundreds of protesters on "Black Friday."
	Shah requests the expulsion of Khomeini from Iraq. Khomeini moves to Paris.
	Oil workers, electrical workers, and teachers strike as protests continue.
December 1978	Shah appoints Shahpur Bakhtiar to form a new government, with the condition that the Shah would leave the country temporarily.
January 1979	Shah leaves Iran for Egypt and Khomeini announces creation of the Council of the Islamic Revolution to begin formation of an alternative government.
February 1979	Khomeini returns to a hero's welcome in Iran, Bakhtiar goes into hiding, Bazargan heads the new government at the request of Khomeini.
November 1979	Students seize the US embassy in Tehran, leading to Bazargan's downfall and solidification of the Council of the Islamic Revolution's power. American hostages are held for 444 days.

THE ENVIRONMENT OF THE REVOLUTION

PHYSICAL ENVIRONMENT

Iran, the second-largest country in the Middle East, is bordered by seven countries, including Afghanistan, Armenia, Azerbaijan, Iraq, Pakistan, Turkey, and Turkmenistan. The country, divided into 30 provinces, is demarcated to the south by the Persian Gulf; to the east by the deserts and mountains of Khurasan, Sistan, and Baluchestan; to the west by Shatt al-Arab, Iraqi marshes, and the Kurdish mountains; and to the north by the Aras River from Mt. Ararat to the Caspian Sea and by the Atrek River stretching from the Caspian Sea into Central Asia.[1] Iran has a strategic position on the Persian Gulf and the Strait of Hormuz, used for the maritime transport of crude oil.[2] Slightly smaller than the state of Alaska, Iran's total area is 1,648,195 square kilometers. Iran's climate is mostly arid or semiarid with subtropical regions along the Caspian coast. As a result, only 9.7% of the land is arable.[3] Nearly three-fifths of the country, especially the central plateau, lacks the rainfall necessary for sustainable agricultural production.[4] Iran holds the world's third-largest known oil reserves (around 10% of the world's total reserves) and second-largest natural gas reserves (around 20% of the world's total reserves).[5]

Figure 1. Map of Iran.[6]

[1] Ervand Abrahamian, *A History of Modern Iran* (New York: Cambridge University Press, 2008), 1–2.

[2] Central Intelligence Agency, "Iran," *The World Factbook*, accessed December 4, 2009, https://www.cia.gov/library/publications/the-world-factbook/geos/ir.html.

[3] Ibid.

[4] Ibid.

[5] US Energy Information Administration, "Country Analysis Briefs: Iran," accessed July 9, 2010, http://www.eia.doe.gov/cabs/Iran/pdf.pdf.

[6] Central Intelligence Agency, "Iran," *The World Factbook*, accessed December 15, 2010, https://www.cia.gov/library/publications/the-world-factbook/maps/maptemplate_ir.html.

CULTURAL AND DEMOGRAPHIC ENVIRONMENT

Iran is a populous, ethnically diverse, although religiously homogenous, society. Pre-Islamic Persia and Shi'a Islam are the predominant cultural influences, although in the period leading up to the Revolution, many Iranians interpreted these influences in new ways that complemented revolutionary ideologies and agendas.

During the latter half of the twentieth century, as Mohammed Reza Shah embarked on a modernizing program, the urban population of the country increased significantly. In 1976, the official census showed a population of 33.7 million.[7] Of that number, 47% resided in urban areas in 373 cities.[8] Tehran, the largest city, had a population of 4.5 million, or 28.6% of the urban population. Qom, a holy city important during the 1979 Revolution,[9] had a population of 247,000, or 1.6% of the urban population. Tabriz, a fairly substantial city and also the site of several important events, held 598,000 persons, or 3.8% of the urban population. Iran experienced a 2.4% average urban growth in the census year.

Persians constitute the majority of Iran, before and after the Revolution, but Iran is nevertheless an ethnically diverse, but religiously homogenous, society. Substantial populations of Azeris, Kurds, Baluchis, and Arabs all inhabit modern Iran. The Pahlavi rulers in modern Iran engaged in a concerted effort to create a unified Iranian identity based on the nation's Persian heritage, eclipsing other ethnic or tribal identities. During the early twentieth century, and in the period before the Revolution, the centralized state heavily promoted the Persian language and identity. Religious diversity is far less prevalent; the vast majority of Iranians are Shi'a Muslims, with small pockets of Sunni Muslims (usually associated with non-Persian ethnic groups), Jews, Christians, Baha'is, and Zoroastrians.

The predominant cultural influences of Iran are a mixture of pre-Islamic Persia and Shi'a Islam. Persian history dates back several millennia—the Achaemenid kings were contemporaries of ancient Athens. One of the cornerstones of Persian identity is the ancient epic poem "Shahnameh," or the "Book of Kings," written by the poet Ferdowsi in the tenth century A.D. Still read in the modern age, the "Shahnameh" is a testament to the perseverance of Persian nationalism across the millennia. However, the ways in which Persians

---

[7] Zohreh Fanni, "Cities and Urbanization in Iran After the Islamic Revolution," *Cities* 23, no. 6 (2006).

[8] A city in the Iranian census is variably defined as a dwelling place with more than 5,000 inhabitants or one with a municipality. Ibid.

[9] Referred to hereafter as "the Revolution."

have interpreted that nationalism over the ages has changed. Like Shi'a Islam, the "Shahnameh" and Persian identity evolved in the era before the Revolution. For centuries, the epic was interpreted as legitimizing the monarchy of Iran. By contrast, in the years leading up to the Revolution, the "Shahnameh" was interpreted not as praise of Iranian kingship but as a condemnation of the institution.[10]

As opposition to the Shah increased, Shi'a Islam, formerly a quietist, pious, and relatively apolitical religion, also underwent a profound transformation, developing into a comprehensive language of resistance. Previously, adherents and clerics turned their attention not to temporal affairs but to those of the afterlife "in matters of personal behavior and ethics."[11] For example, one of the most prominent symbols of Shi'ism is the holiday *Ashura* in the month of *Muharram,* commemorated to mark the day in 680 A.D. when the Imam Hussein willingly went to his martyrdom in the battle at Karbala to fulfill his divinely predetermined will.[12] In the politicized context of prerevolutionary Iran, Shi'as increasingly understood *Muharram* and *Ashura* as a struggle for social justice, a political revolution rather than submission to divine will. Of course, clerics were not universally comfortable with such interpretations. Regardless, by the time of the

---

[10] Abrahamian, *A History of Modern Iran*, 5.

[11] Ibid.

[12] Imam Hussein was the son of the third caliph, Ali, who attempted to claim the caliphship. He was killed by the Umayyad caliph Yazid in 680 A.D. The major break between Sunni and Shi'a theology concerns their differing positions as to the legitimate rulers of the Muslim world after the death of the Prophet Muhammad. Shi'as, themselves separated into various sects, hold that the familial line of Ali, the Prophet's cousin and son-in-law, held the rightful claim to the caliphship. Shi'as deny the legitimacy of the three "rightly guided" caliphs preceding Ali's rule and commemorated in Sunni theological doctrine. Later, during the Abbasid caliphate, Shi'a divided into distinct sects, each revering a different imam. The most numerous, and the predominant sect in Iran, are the Twelvers, who believe in the Twelve Imams, including the "Hidden Imam" or the "Twelfth Imam." Nikki R. Keddie and Richard Yann, *Roots of Revolution: An Interpretive History of Modern Iran* (New Haven, CT: Yale University Press, 1981), 4–9. For further information on Shi'a Islam, please see Moojan Momen, *An Introduction to Shi'i Islam: The History and Doctrines of Twelver Shi'ism* (New Haven, CT: Yale University Press, 1985); Martin S. Kramer, *Shi'ism, Resistance, and Revolution* (Boulder, CO: Westview Press, 1987); Rainer Brunner and Werner Ende, *The Twelver Shia in Modern Times: Religious Culture & Political History*, vol. 72 (Boston: Brill, 2001).

Revolution, Shi'ism looked "more like a radical ideology than a pious and conservative religion."[13]

SOCIOECONOMIC ENVIRONMENT

Several financial crises in the decades preceding the Revolution destabilized Iran, leading to large-scale riots and stiff opposition. The Shah, through use of coercive force and promised reforms, was able to maintain control of the country during the various crises, with the notable exception of the late 1970s. A cornerstone of the Shah's promised reforms after the 1960s crisis was the White Revolution, later referred to as the Shah-People Revolution, which included economic, land, and social reforms.[14] The White Revolution was the Shah's gateway to a supposed "Great Civilization," leading to a dramatic turn in fortune domestically and internationally. Although the reforms of the White Revolution did lead to improvements for some Iranians, the outcomes fell short of the Shah's megalomaniac notions.[15] Economic indicators pointed to some initial economic successes of the reforms, but the figures belied underlying endemic corruption and mismanagement that effectively derailed the efforts of the White Revolution despite the substantial influx of oil revenues in the 1970s.

---

[13] Abrahamian, *A History of Modern Iran*, 6. For further information on the relationship between Shi'ism and politics in the modern era, please see Youssef M. Choueiri, *Islamic Fundamentalism* (Washington, DC: Pinter, 1997); Juan Ricardo Cole, *Sacred Space and Holy War: The Politics, Culture and History of Shi'ite Islam* (London: I.B. Tauris, 2002); Juan Ricardo Cole and Nikki R. Keddie, *Shi'ism and Social Protest* (New Haven, CT: Yale University Press, 1986); Nikki R. Keddie, *Religion and Politics in Iran: Shi'ism from Quietism to Revolution* (New Haven, CT: Yale University Press, 1983). This is not to say that the "Islamic Revival" was an entirely new phenomenon. In some ways, it "follows a long tradition in both Iran and the Muslim world of expressing socioeconomic and cultural grievances in the only way familiar to most people—a religious idiom arraying the forces of good against the forces of evil and promising to bring justice to the oppressed." Keddie and Richard, *Roots of Revolution*, 3.

[14] The Shah referred to his extensive set of reforms as the "White Revolution" to draw distinctions between the two dominant ideological contenders in Iran, "red" Communism and "black" Islamism. He initiated a White Revolution from above to prevent a "red" revolution from below. The Shah proposed the reform package in a national referendum that was, unsurprisingly, passed by Iranian voters despite its boycott by the National Front, which argued that such reforms should be legislated by the *Majles*, not the crown. The referendum included land reform, sale of government-owned factories to finance land reform, new election laws including women's suffrage, the nationalization of forests, a national literacy corps, and profit sharing for industrial workers. Other reforms were tacked on to the original six in succeeding years. Ibid., 156.

[15] Abrahamian, *A History of Modern Iran*, 131.

The initial economic promise of the 1950s came to an untimely end when it met the serious inflation,[16] corruption, and electoral fraud of the latter half of the decade and the early 1960s. Unofficial opposition gained headway in 1960, charging the regime with electoral fraud. Afterward, amid pressures from the National Front to dissolve Parliament, charges of widespread corruption, teachers' strikes, and demonstrations that left several dead, the Shah appointed Ali Amini, an independent leader of the opposition, as Prime Minister in 1961.[17] Amini, with the encouragement of the Kennedy administration and his American advisers, initiated reforms but was consistently blocked by the Shah and eventually resigned in frustration. The Shah reversed most of Amini's reform attempts, with the exception of a watered-down land reform. Only in 1963 did the Shah, gaining cognizance that reform was needed to retain American support and strengthen his weakening political base, announce the White Revolution economic and social reforms.[18]

Despite some initial indicators to the contrary, the economic reforms initiated under the White Revolution met with little success, shadowed as they were with corruption and mismanagement. The influx of foreign exchange from oil and foreign investment did lead to an increased gross national product (GNP), and an impressive growth rate of 13%, from 1959 to 1976.[19] Likewise, reforms led to a measure of success in the industrial sector, including increases in large and small factories and infrastructure.[20] However, because of corruption and mismanagement, the Plan and Budget Organization (PO), tasked with developing and implementing economic reforms, received only 55% of the 100% of promised revenues from the newly denationalized oil industry. This development, along with a rise in inflation, led to many project cuts, and those projects that were implemented were often more showy than practical.[21] Additionally, the export of petroleum reserves, which boosted the GNP, encouraged the growth of little

---

[16] Excessive credit, little control on foreign currency, and nonessential imports contributed to inflation, hitting lower and middle classes the hardest. Keddie and Richard, *Roots of Revolution*, 151–152.

[17] Ibid., 150–155.

[18] April Summitt, "For a White Revolution: John F. Kennedy and the Shah of Iran," *Middle East Journal* 58, no. 4 (2004): 560–575. See fn.14 for a detailed description of the reform package.

[19] Manoucher Parvin and Amir Zamani, "Political Economy of Growth and Destruction: A Statistical Interpretation of the Iranian Case," *Iranian Studies* 12, no. 1/2 (1979): 43–78.

[20] Abrahamian, *A History of Modern Iran*, 133. Large factories, employing more than 500 employees, increased from less than 100 to 150, while small factories increased from 1,500 to 7,000.

[21] Keddie and Richard, *Roots of Revolution*, 148.

productive capacity as Iran engaged in only the primary stages of the production process. As a result, even though in 1972 oil revenues accounted for 28% of GNP, the oil sector employed only 0.54% of the country's population. In the mid-1970s, the Shah's policies to control inflation played no small part in mobilizing the *bazaaris* (traditional, middle-class businessman), effecting price controls and other harsh measures in an antiprofiteering campaign.[22]

Oil Revenues, 1954–1976[23]

Year	Oil Revenue ($m)
1954–1955	34.4
1956–1957	181
1958–1959	344
1960–1961	359
1962–1963	437.2
1964–1965	555.4
1966–1967	968.5
1968–1969	958.5
1970–1971	1,200
1972–1973	2,500
1973–1974	5,000
1974–1975	1,800
1975–1976	20,000

Likewise, land reform, the crown jewel of the Shah's White Revolution, having little effect on the rural poor and spurring opposition among the clergy, served to exacerbate, not mitigate, social and political tensions in Iran. The reform's core beneficiaries were big business farms and agribusinesses. Although some peasants received land, the reforms left more than 1.2 million families without the 10 hectares necessary for subsistence farming in Iran's hardscrabble landscape.[24] Moreover, the focus on large-scale, mechanized agriculture eliminated wage labor positions needed by the landless and rapidly decreased productivity.[25] As a result, during the 1970s, Iran became a net importer, rather than a net exporter, of food. The reforms also contributed to the break between the state

---

[22] Abrahamian, *A History of Modern Iran*, 151–152.
[23] Ibid., 124.
[24] Keddie and Richard, *Roots of Revolution*. See discussion, 162–163.
[25] Ibid., 166–168.

and the clergy, many of whom were landholders, and land owned by religious institutions and mosques was slated for confiscation.[26]

In addition to economic programs, the White Revolution also included extensive social reforms, including reforms in health care, education, and women's issues. Educational institutions increased threefold, and the number of health care professionals and medical facilities increased significantly. Women also benefitted from the Shah's programs, gaining the right to vote, run for office, and serve in the judiciary as both lawyers and judges. Although the head scarf was not banned, women were discouraged from donning the *hijab* in public institutions.[27]

Both the financial crisis and the reforms to mitigate its impact were met with opposition, including large-scale riots and demonstrations in the early 1960s. Some members of the clergy, including Ayatollah Khomeini and Ayatollah Shariatmadari, opposed the Shah's increasingly autocratic regime, women's reforms, and subservience to Western powers, leading to Khomeini's exile in 1963. This opposition by the clergy marked a turning point in the relationship between the state and the *ulama* as the latter had supported the Pahlavis against the secular National Front. The Shah restored order after extensive arrests of religious and nationalist opposition figures, the shooting of demonstrators, and reforms.[28] The reforms, however, rather than managing political and social tensions, had the opposite effect in the long term. Increasing inequality, failed promises, and the iron fist of the Shah's centralized state exacerbated the already tense relations between the state and society during the 1970s.

HISTORICAL FACTORS

Mohammed Reza Shah was not the first Pahlavi to sit on the Iranian throne. Decades earlier, his father, Reza Shah, captured the crown and steered Iran toward the path later adopted by his son. Like his successor, the first Pahlavi shah adopted étatist policies to modernize Iran.[29] Also like his son, Reza Shah's contemporaries did not universally regard him as a benevolent state builder. To some, he brought necessary order and discipline to the burgeoning state; for others, he brought oppression and taxation.[30] Reza Shah was

---

[26] Misagh Parsa, *Social Origins of the Iranian Revolution* (New Brunswick, NJ: Rutgers University Press, 1989), 195.

[27] Abrahamian, *A History of Modern Iran*, 134.

[28] Keddie and Richard, *Roots of Revolution*, 159.

[29] Abrahamian, *A History of Modern Iran*, 83–84.

[30] Ibid.

forced to abdicate in 1941, leading to the installation of his son on the throne. Prior to the Central Intelligence Agency (CIA)-staged coup of 1953, the Shah was a more or less a constitutional monarch, exercising power dispersed through governmental institutions and societal sectors. After the ouster of Mosaddeq's democratically elected government, which sought to nationalize the oil industry, the Shah's vision for Iran included a highly centralized state and the concentration of power in the royal palace backed by the resources and power of the United States.

Mosaddeq's popular support and rise in the Iranian government was tied to dissatisfaction with the oil agreement Iran established with the AIOC (Anglo-Iranian Oil Company)[31] during the deteriorating socioeconomic conditions of the postwar period. Oil revenues increased; AIOC's profits increased, but Iranian profits, under the agreement, remained stagnant. Opposition to the agreement became more vocal, especially among leftist parties. After a disappointing reworked agreement and a rigged election, a rainbow coalition, the National Front, formed to coordinate opposition to the Shah.[32] The National Front organized demonstrations against the Shah and the British presence in Iran. Voters ushered National Front candidates into the *Majles* in the 1950 election, with Mohammed Mosaddeq as the de facto leader.[33]

In office, the National Front pressed for reduction of the Shah's powers as well as the nationalization of the Iranian oil industry. Mosaddeq's proposal, and the *Majles*' passage, of legislation nationalizing the oil industry led the Shah to appoint him as prime minister in response to popular pressure. The leftist move antagonized British interests in the region, prompting a host of British machinations to remove Mosaddeq from power, including

---

[31] The AIOC would later become British Petroleum, an early shoot of the mammoth British oil company today, BP.

[32] Keddie and Richard, *Roots of Revolution*, 321. Nationalists, leftists, some clergy, and unaffiliated individuals, mostly from the urban lower and middle classes, formed the National Front. The strongly left-leaning, communist Tudeh Party was not part of the coalition, nor was the *Fedayin-e Islam*.

[33] In the 1940s and the early 1950s, two issues predominated Iranian politics: the transfer of political power from the royal court to the elected parliament, the *Majles*, and increasing Iran's control over the oil industry, then controlled by the British-owned AIOC. Mosaddeq's platform included strong positions in favor of both issues. Mark Gasiorowski, "The 1953 Coup d'Etat in Iran," *International Journal of Middle Eastern Studies* 19, no. 3 (1987).

a plan, dashed by the Truman administration, to invade.[34] While Mosaddeq at this juncture enjoyed a great deal of popular support, internal dissension within the National Front, partially engineered by British agents,[35] fragmented the movement. Members of the National Front and some in the military, notably General Zahedi, began to actively plot Mosaddeq's downfall. However, although several covert CIA operations were active in Iran at the time, it was not the policy of the Truman administration, or Dean Acheson's State Department, to seek the forcible removal of the Mosaddeq government.

After the election of Eisenhower, the United States changed its tune. Unlike Truman, Eisenhower supported an Iranian coup. After the policy switch, the CIA used an already existing operation, BEDAMN, involved in anti-Soviet and anti-Communist propaganda and political action, to undermine Mosaddeq's rule.[36] Along with various other measures to sow discord and faction among Mosaddeq's supporters and Iranian politicians, BEDAMN agents also reportedly bribed clergy to denounce Mosaddeq and create a "political crisis." Efforts to enlist clergy, however, were only marginally successful, as most clergy members failed to follow through with the agents' requests.[37] Several days before Mosaddeq's surrender to General Zahedi, BEDAMN agents orchestrated a pro-Tudeh demonstration to push the military and others into Zahedi's arms, which succeeded beautifully.[38] The

---

[34] The British attempted to undermine support for Mosaddeq by imposing economic sanctions on the country and engaging in military maneuvers in the region. British paratroopers were stationed in Cypress, and the cruiser *Mauritius*, and eventually four other cruisers, were deployed to the region, which held firing practice near Abadan. British land forces were also bolstered. The events evolved into a full blockade of Iranian oil exports in which major oil companies participated. Ibid., 263. British intelligence services (MI6) played a role in the CIA-led coup, developing a plan with the CIA and choosing General Zahedi to replace the Mosaddeq. Mark Gasiorowski, "The CIA Looks Back at the 1953 Coup in Iran," *Middle East Report* 216 (2000), 4.

[35] MI6 agents used the Rashidian brothers, British agents in Iran since the 1940s, to increase tension and dissension among National Front leaders. The Rashidian brothers would also join General Zahedi's plans to stage a coup. The brothers' network played an important role in overthrowing Mosaddeq. Gasiorowski, "The 1953 Coup d'Etat in Iran," 263–270.

[36] US operatives in the country had to work to secure the Shah's approval for the coup. The CIA recruited Princess Ashraf, the Shah's sister, and Colonel Norman Schwarzkopf Sr., father to General Norman Schwarzkopf, to meet with the Shah to convince him to approve the plan. However, the Shah refused to commit to the plan until he heard official announcements of British and US involvement over a special radio broadcast. Ibid., 273. Wilber's history indicates that the coup would have proceeded without the support of the Shah. Gasiorowski, "The CIA Looks Back at the 1953 Coup in Iran," 4.

[37] Ibid. One of the clergy with whom the CIA had a "firm contract" was more than likely the leader of *Fedayin-e Islam.*

[38] Tudeh members unwittingly joined the demonstration, unaware that it was organized by the CIA. The uprisings in the streets that resulted in Mosaddeq's ouster were only partially spontaneous, and violence was deliberately incited by CIA Iranian agents. Ibid.

next day, the CIA helped to incite another anti-Mosaddeq crowd that was joined by police and army units who, after wreaking havoc on government buildings and pro-Mosaddeq institutions, marched on towards Mosaddeq's house, where he later surrendered to Zahedi.[39]

The 1953 coup supplanted the last democratically elected, and democratically oriented, government in Iran. Although purely conjecture, it is likely that in the absence of the social and political tensions created by Mohammed Shah's dictatorship, the Revolution would not have occurred had Mosaddeq been allowed to stay in power.[40] Mosaddeq became something of a martyr after his forcible ouster and death under house arrest. The role of the United States in the demise of the democratically elected government contributed later to the decidedly anti-American flavor of the Revolution. After Mohammed Pahlavi regained the throne, the United States provided staunch support to the Shah's authoritarian regime, a blow to moderate Iranians who had counted on the United States to push for more democratic governance against the imperial British. British influence in Iran waned notably after the coup, making the United States the largest, most influential Western power in the country.[41]

GOVERNING ENVIRONMENT

Iran underwent tremendous political transformations in the twentieth century. Reza Shah's ambition was partly responsible. During his rule, he embarked on a concerted mission to modernize Iran. When he was forced to abdicate in 1941, the state, whose control had previously only extended into the capital, increasingly extended across the nation and into the everyday lives of more Iranians. After the 1953 coup, Mohammed Reza mimicked his father's authoritarian rule, bolstering the state's interventional capacity in society and centralizing power in his hands, but unlike Reza Shah, Mohammed Reza had the benefit of substantial revenues from the oil boom of the 1970s.

After Britain and the Allied powers deposed Reza Shah in 1941, his son, Mohammed Reza, was given the throne. The British allowed the Pahlavis to stay in power because of their special relationship with the military. Mohammed Reza was allowed to keep control over the military in return for acquiescence to the Allies' other demands. He was in office pending "good behavior," which he initially offered, although hints of

---

[39] Gasiorowski, "The 1953 Coup d'Etat in Iran."
[40] Ibid.
[41] Keddie and Richard, *Roots of Revolution*, 142.

his later megalomania were already apparent. When sworn onto the throne by the *Majles*, he swore to rule as a constitutional monarch, obeying the fundamental laws of the country. The first period of the Shah's rule, from 1941 until 1953, notably lacked the former Shah's supreme control over the bureaucracy and court patronage system, although Mohammed Reza, like his father, maintained extensive control over the military. Regardless, in distinction to his father and to his rule after the 1953 coup, governmental power was dispersed, "contested between the royal palace, the cabinet, the Majles, and the urban masses."[42]

In addition to the extensive reforms of the White Revolution, Mohammed Reza's drive toward a powerful, centralized state also included dramatic reforms and expansions of the military, bureaucracy, and court patronage system. The former was especially important because of the relative ease with which the Shah's opponents were able to marginalize his power during the National Front's triumph under the leadership of Mosaddeq. Upon entering office, the Shah purged his political enemies from government and military positions. Perhaps the most hated institution in Iran was the Shah's internal security agency, SAVAK, founded after the 1953 coup.[43] In a 1976 report, Amnesty International charged SAVAK with extensive violations of human rights, including torture.[44] SAVAK created an atmosphere of fear and distrust in Iran.

The Shah maintained tight control over the military, preventing any civilian "muddling" in martial affairs, and renamed the Ministry of Defense the Ministry of War. Upon retaking office in 1953, the Shah expanded military expenditures, increasing them more than twelvefold in 25 years, from $60 million in 1954 to $7.2 billion in 1977. During the same period, military expenditures rose from 24% to 35% of total expenditures.

By 1975, the military increased from 127,000 to 410,000, making the Shah's army the fifth largest in the world, the navy the largest in the Persian Gulf, and the air force the largest in western Asia. A 1976 report to the US Senate Committee on Foreign Relations indicates that Iran's military purchases at the time were the largest in the world,[45] with another $12 billion on order in 1978 before the Revolution.

---

[42] Abrahamian, *A History of Modern Iran*, 101.

[43] Colonel Norman Schwarzkopf Sr. was instrumental in training SAVAK along with the Israeli Mossad.

[44] Amnesty International, "The Amnesty International Report 1975–1976."

[45] Robert Mantel and Geoffrey Kemp, "US Military Sales to Iran," staff report to the Subcommittee on Foreign Assistance of the Committee on Foreign Relations, US Senate (Washington, DC: Government Printing Office, 1976).

Military Expenditures, 1954–1977. Amounts reflect 1973 prices and exchange rates.[46]

Year	Expenditures ($m)	Year	Expenditures ($m)
1954	60	1966	598
1955	64	1967	752
1956	68	1968	852
1957	203	1969	759
1958	326	1970	958
1959	364	1971	944
1960	290	1972	1,300
1961	290	1973	1,800
1962	287	1974	4,000
1963	292	1975	5,500
1964	323	1976	5,700
1965	434	1977	7,200

The Shah took a remarkably personal role in military affairs, from training to barracks, but he also took measures to prevent a military coup. He purportedly showered his officers with gifts, providing generous salaries and pensions, foreign travel, and real estate among other things. Additionally, in a move that would later harm the Shah during the Revolution, he appointed family and friends with "underwhelming personalities" to key military positions.[47] Ironically, despite the bloated military sector, the Shah was overthrown in a nearly bloodless coup, sapped of all legitimacy in the eyes of Iranian society.

Another pillar of the Shah's authoritarian government, the bureaucracy, did not suffer from neglect. The extensive social service reforms included in the White Revolution demanded a large bureaucracy. During the Shah's second stint on the throne, government ministries increased from 12 to 20, and by 1975 the state employed more than 304,000 civil servants and approximately one million white-collar and blue-collar workers, nearly half of all full-time employees.[48]

In addition to the military and the bureaucracy, the Shah's government also relied on an extensive court patronage system to maintain control. The Shah created a tax-exempt charity, the Pahlavi Foundation, and at its height, the Foundation had some $3 billion

---

[46] Ibid., 132.

[47] Ibid., 125.

[48] Ibid., 126.

in assets, with shares in 207 companies including international ones such as General Electric and Krupp. The Foundation also held the landed estates granted to Mohammed Reza from his father and the assets of the Shah's 64 family members, whose total assets amounted to approximately $20 billion. Underneath the facade, the ostensibly charitable organization was used to exert influence on the economy, as a source of funds for the royal family, and to distribute largess to the regime's supporters.[49]

Mohammed Reza used the extensive institutions to ensure his dominance of the parliament and cabinet. He gave himself the constitutional power to appoint prime ministers, and of the eight prime ministers ruling from 1955 to 1977, all, with the exception of Ali Amini in 1961–1962, were the Shah's henchman. The Shah appointed Amini as a concession to the opposition during the financial crisis in the early 1960s after public outcries over a series of fraudulent elections.[50] Amini, a favorite also of the Kennedy administration, initiated land reform and financial stabilization programs during his tenure but resigned from lack of support from the Shah and other oppositional players.[51] His tenure marked the only real opposition in the Shah's government until the Revolution. Otherwise, the vouchsafed premiers filled the parliament with their supporters, turning the *Majles* into an ineffectual, rubber-stamp institution.[52]

Prior to the crisis of 1960–1963, the Shah introduced a two-party system in response to demands for more democratic governance. However, no discernible differences were apparent in the two parties, Melliyun (Nationalist Party) and Mardom (People's Party).[53] In his autobiography, Mohammed Reza exclaimed support for the two-party system in Iran as a harbinger of liberal democracy,[54] and SAVAK assigned deputies to their party affiliations.[55] The Shah, like other regimes in the Middle East, had two ways of dealing with opposition: repression or co-optation. SAVAK was instrumental in the former, infiltrating and systematically destroying non-officially sanctioned opposition groups, especially pro-Tudeh factions. A great number

---

[49] Ibid., 127.

[50] Amini attributes less noble motivations to the Shah in the appointment of Amini, describing it as an act of cowardice and an unwillingness to confront opposition forces directly, instead foisting the distasteful task on an enemy. Parvin Merat Amini, "A Single Party State in Iran, 1975–78: The Rastakhiz Party—the Final Attempt by the Shah to Consolidate His Political Base," *Middle Eastern Studies* 38, no. 1 (2002), 133.

[51] Keddie and Richard, *Roots of Revolution*, 150–155.

[52] Abrahamian, *A History of Modern Iran*, 130.

[53] Keddie and Richard, *Roots of Revolution*, 150.

[54] Mohammed Reza Shah Pahlavi, *My Mission My Country* (London: Hutchinson, 1974).

[55] Abrahamian, *A History of Modern Iran*.

of executions, purges within the military, and new legislation against opposition organizations accompanied the Shah's return to power. Along with repression, the Shah neutralized opposition by co-opting regime critics, using tactics such as providing students with government jobs in return for support.[56]

Virtually any criticism of the Shah's regime was treasonous—in the Shah's mind, critics were either "black reactionaries" (religious reactionaries) or "red reactionaries" (Communist reactionaries).[57] The Shah loosened his stranglehold on Iranian society only in 1977 as discontent grew, in part, because of deteriorating economic conditions despite the continued influx of substantial oil revenues. Moreover, the Shah's terminal cancer and the Western focus on Iran's human rights violations, including President Carter's attention to the matter, were also potential contributors to his increased lenience in the short period before the Revolution.[58] Criticism of the regime increased, leading to massive demonstrations in Tehran, Tabriz, and Qom, prompting the Shah to declare martial law in late 1978. By the following February, the Shah had fled the country, and Tehran Radio announced "This is the voice of Iran, the voice of the Islamic Republic."[59]

WEAKNESSES OF THE PREREVOLUTIONARY ENVIRONMENT AND CATALYSTS

The extensive reforms initiated under the White Revolution, ostensibly enacted under populist themes, proved to have nearly the opposite effect, exacerbating social and political tensions. Land reform and the extensive social welfare programs were presented to the population as measures to improve the social and economic positions of Iranians across the class spectrum, obviating the attractions of a socialist revolution. Tensions attributable to the reforms came to a head in the mid-1970s. The Shah's response was the eradication of the two-party system and the establishment of a one-party state led by the Resurgence Party. The Resurgence Party, supposedly meant to repair the broken relations between state and society, only served to drive crucial sectors of society, the *bazaaris* and the *ulama*, into the only available avenue of resistance, the mosque and Khomeini.

Although the Iranian government lacks statistical data on the level of income inequality during the time period under consideration, a

---

[56] Keddie and Richard, *Roots of Revolution*, 144.

[57] Ibid., 145.

[58] Abrahamian, *A History of Modern Iran*, 157.

[59] Ibid., 162.

picture of the failure of the reforms to trickle down to those segments of society most in need is readily apparent in statistics on urban household expenditures gathered by the Central Bank of Iran.[60] From 1960 to 1970, while the expenditures of the wealthiest 20% rose from 44% to 64%, those for the lowest 20% increased a bare 1.5%. Moreover, although the White Revolution made some strides, the reforms fell far short of the Shah's promises. The infant mortality and doctor–patient rates remained some of the worst in the region, and illiteracy remained high. In addition, the White Revolution did not touch most of the countryside. The White Revolution reforms did increase the ranks of the intelligentsia and the urban working class, both of which were traditionally hostile to the Pahlavi regime.[61] Massive workers' strikes in the fall of 1978 crippled the regime.[62]

These social tensions were mirrored by political tensions in Iran. During the 1970s, the opposition to the Shah, which had initially gained steam during the 1960–1963 crisis, became increasingly vocal. The two ideologues of the Revolution, Ayatollah Khomeini and Ayatollah Shariati, exercised enormous influence on Iranian society, transforming quietist Shi'ism into a rough-and-tumble political ideology. Both Shariati and Khomeini argued for an activist Shi'ism, but whereas Shariati reinterpreted Shi'ism as a revolutionary ideology struggling against all forms of oppression,[63] Khomeini advocated a clerical populism. In his lectures, anonymously published works, and cassette tapes that famously traveled considerable distances, Khomeini advocated *velayat-e faqeh hokumat-e Islami*, or the jurists' guardianship of Islam.[64] According to *velayat-e faqeh*, senior *mojtahads* had the authority to rule the state, even in the absence of the occulted Twelfth

---

[60] According to Amini, when the rather dismal figures were reported to the Queen's Council, the Central Bank statistical staff was shuffled and required to resubmit figures. The resulting revised data grant more generous figures to the expenditures in the mid-range but still demonstrate a marked inequality between the upper, middle, and lower classes. Amini gathered the revised statistical data from M. H. Pearsan, "Income Distribution in Iran," *Iran: Past, Present, and Future*, ed. Jane W. Jacqs (New York: Aspen Institute of Humanities, 1976).

[61] Abrahamian, *A History of Modern Iran*, 140–143.

[62] Keddie and Richard, *Roots of Revolution*, 250–251.

[63] For an extended treatment of Shariati's influence, please see Ibid., 183–230, and Abrahamian, *A History of Modern Iran*, 143–146.

[64] For an extended treatment of Khomeini's works and influence, please see Keddie and Richard, *Roots of Revolution*, 183–230, and Ervand Abrahamian, *Khomeinism: Essays on the Islamic Republic* (Berkeley, CA: University of California Press, 1993).

Imam.[65] Several stillborn guerrilla organizations, one associated with the works of Shariati, *Mojahedin-e Khalq*, and another with the Tudeh Party, *Fedayin-e Khalq*, also emerged. Members of both organizations were mostly drawn from the ranks of the 177,000 university students in Iran, some 65,000 of which had studied in the United States. The radical Marxist and "socialist Shi'ism" espoused by the revolutionaries had little resonance, however, with grassroots Iranian society and the guerrilla organizations failed. It was left to Khomeini, through a bit of political wizardry, to unite the disparate opponents of the regime— *bazaaris*, urban secularists, the working class, the intelligentsia and the clergy—into a cohesive mass capable of toppling the Shah.[66] While Khomeini's disciples began openly calling for the replacement of the monarchy with a republic, the "loyal" opposition party, Mardom, unexpectedly won a series of elections after fielding candidates who were not associated with the court, unsettling the Shah and SAVAK.[67]

The unraveling of the political, social, and economic fabric of Iran during the mid-1970s played a role in the Shah's decision to replace the multiparty system with a one-party state.[68] Ostensibly to provide closer ties between the government and people in order to better realize the reforms of the White Revolution, the Shah dismantled the existing parties, replacing them with the singular Resurgence Party (*Hezb-e Rastakhiz*). In 1975, the Shah established the Resurgence Party, a statewide organization that incorporated a myriad of other state organizations, penetrating ever deeper into Iranian society. The party

---

[65] The concept of *velayat-e faqeh* was not necessarily a new one; Khomeini simply extended its application from the guidance of those not able to guide themselves (the mentally handicapped, widows, and children) to society at large. His interpretation of *velayat-e faqeh*, although based on conventional Shi'a premises, had no precedent in the Quran or in the teachings of the Twelve Imams, a fact not lost upon his followers. Abrahamian, *A History of Modern Iran*, 146, fn. 36. For a description of the tradition of the occulted Imam in Shi'ism, see fn. 11.

[66] Gilles Kepel, *Jihad: The Trail of Political Islam* (Cambridge, MA: Harvard University Press, 2002), 107–113. Several months before the events of 1979 that led to the Shah's demise, one journalist aptly noted that Khomeini's solution, although irrational and shortsighted, was embraced even among the intelligentsia and the *bazaaris* because the Shah had managed to alienate so much of Iranian society that many were "prepared to swallow" Khomeini's vision for Iran if it meant the end of the Shah. Charles Douglas-Home, "Will the Shah be Toppled from His Shaky Throne?" *Sunday Times*, November 28, 1978. Khomeini's support stemmed not only from his willingness to confront the corrupt regime but also from his denouncement of policies adversely affecting other sectors of society, such as farmers and *bazaaris*, in effect championing their cause. See discussion in Parsa, *Social Origins of the Iranian Revolution*, 216–217.

[67] Abrahamian, *A History of Modern Iran*, 147–149.

[68] In a scene worthy of Orwell's *1984*, the Shah, who had championed multiparty systems in his book, *My Mission, My Country*, had SAVAK remove copies of the book from libraries and bookstores. Ibid., 150.

increased the regime's stranglehold over the salaried middle class, the urban working class, and even rural farming co-ops.

Most importantly, however, the party extended its reach into the *bazaari* and the *ulama* sectors. *Bazaaris* were an economic group within Iranian society that conducted mostly petty trade and banking in a traditional, rather than a modernized, fashion. Although diversity existed within the class, the *bazaaris* were typically highly respectful of the clergy, following their lead in most matters. The Resurgence Party mounted a concerted attack on the *bazaaris* to control inflation and to modernize the economy, replacing the network of small shops with modern markets, dissolving the centuries-old guilds, and enforcing price controls.[69] The price controls were enforced on luckless retailers during a period of high inflation. One official described the Shah as sensitive to the problem of inflation but remarkably unwilling to curb public spending; on television the Shah announced new public programs moments after his Minister of Planning laid out the dire inflationary situation. Consequently, the Shah embarked on an antiprofiteering campaign to combat "high prices," threatening retailers with lashes and imprisonment for not adhering to price control measures.[70] One government economist recounted how he and his colleagues prepared a report for the Shah in 1972, detailing the necessity for addressing increasing economic inequalities and inflationary spending or else face inevitable "socioeconomic explosions." The Shah reportedly dismissed the charges as too negative and despairing in an almost willful and infantile disregard for empirical reality.[71] The highly integrated, collective structure of the *bazaaris* made them one of the most effective, and pivotal, members of the opposition, driven by the government's attacks on their institutions.[72]

Likewise, the Resurgence Party attempted to "nationalize" religion, proclaiming the Shah as the political and spiritual leader of the country. Various measures were taken against the "black reactionaries," including requirements for state sanctions of publication, and, in a final blow, the Shah scrapped the religious calendar, replacing it with an imperial one. Additionally, many reforms benefitted women's

---

[69] Amini, *A Single Party State in Iran, 1975–78*, 139–145. Parvin and Zamani, *Political Economy of Growth and Destruction: A Statistical Interpretation of the Iranian Case*, 43–78.

[70] SAVAK Guild Courts handed out 250,000 fines, 8,000 prison sentences, and charges against another 180,000. Few *bazaari* families escaped unscathed from the antiprofiteering campaign. Abrahamian, *A History of Modern Iran*, 152.

[71] Amini, *A Single Party State in Iran, 1975–78*. Negative reports led only to a reform of data, not to the major structural reforms required to better the economy.

[72] Parsa, *Social Origins of the Iranian Revolution*. See discussion on pp. 91–125 regarding the mobilization of the *bazaaris* and the important role they played in the Revolution.

equal status, from the establishment of a ministry of women's affairs to permitting birth control and abortions. In response, many clerics, even heretofore apolitical ones, were effectively driven into the welcoming arms of Khomeini. Some began issuing *fatwas* against the Resurgence Party, leading to their imprisonment.[73]

Ironically enough, the Shah theorized that the Resurgence Party was necessary to stabilize the regime. However, on the contrary, the Shah's one-party state succeeded in alienating nearly every sector of society while also obliterating nearly all channels for airing grievances in the political arena. Reform of the existing government looked increasingly impossible to many in the opposition, leaving revolution as the most appealing, and viable, option. With political parties, local notables, trade unions, and other collective organizations eliminated, marginalized, or under the domineering hand of the Resurgence Party, the mosque was the only institution offering a semblance of cohesion and autonomy from the state, leaving it the only avenue for mobilization, gathering, and communicating. Although various divisions and factions existed within the clergy,[74] and the clerics themselves were to play a muted role in the Revolution, the mosques as an institution were crucial.[75] The Resurgence Party, in short, severed any remaining ties between the government and society rather than repairing them, paving the way for the Revolution.

FORM AND CHARACTERISTICS OF THE REVOLUTION

OBJECTIVES AND GOALS

The opposition that rose up against the Shah came from many different directions as his iron-fisted authority and reforms grew. At least five major lines of dissent operated in the 1960s and 1970s and factored in the eventual overthrow of the Pahlavi government.

---

[73] Abrahamian, *A History of Modern Iran*, 152-153.

[74] With the obvious exception of Khomeini. Parsa, *Social Origins of the Iranian Revolution* (see discussion on pp. 189–219).

[75] The mosque's role in the Revolution extended beyond instigations to violence or protest by religious authorities. A great deal of spontaneous collective activity was a by-product of the mosque's function as a gathering place, not as an institution of religious indoctrination. For example, after groups of individuals gathered in the mosque for Friday prayers during Ramadan, the men often attacked banks and government offices on their way home despite appeals by religious leaders to refrain from violence. The month of Ramadan preceding the events of 1979 proved to be an especially boisterous one, provoking widespread rioting and repressive measures by the government, leading to the implementation of martial law, further enraging the opposition. Ibid., 210–211.

They ranged from guerrillas to religious clerics to political parties and all at their root were motivated by their opposition to the Shah's governance style and policies. A few of the groups that formed the opposition desired the immediate overthrow of the Shah from their formation, but most of the dissent started as a call for changes in policy and governmental structure.

The Islamic religious scholars (the *ulama*), for example, began to raise objections to the Shah's proposed land reform and Local Council Elections Bill in the early 1960s. The land reforms of 1961 were seen as crucial to deterring communist expansion, but Ayatollah Borujerdi wrote that any limitations on land ownership were shameful to the traditional Islamic law.[76] Borujerdi was often supportive of the Shah and was a moderate cleric; therefore, his opposition was a first crack in the relationship between the *ulama* and the monarchy. The Local Council Elections Bill created a further split as it granted suffrage to women and replaced the term "holy Quran" in the oath of office with "holy book," two actions that greatly upset the *ulama*.

As the Shah's policies were implemented, a new and more radical Ayatollah replaced the deceased Borujerdi. Ayatollah Ruhollah Mussavi Khomeini became a vocal critic of the Shah and by 1969 had declared Islam to be incompatible with the monarchy itself (not the Shah as a person, but the position). The *ulama* was a diverse enough community that some moderates still called only for changes in policy throughout most of the protest and revolutionary fervor that Khomeini and others were to utilize in the Revolution. But even by the late stages of the Shah's rule, the more orthodox Ayatollah Shariatmadari doubted that any compromise with the Shah was possible.[77]

The guerrilla groups, primarily the *Mojahedin-e Khalq* and the *Fedayin*, were an outgrowth of the bloody suppression of the 1963 riots[78] and were heavily influenced by Marxist ideology. They both thought armed struggle was required to change the government system and were primarily motivated by the repression and harsh tactics of the regime against protesters and innocent civilians.

Two political groups also had a role in the revolutionary opposition. The National Front, the nationalist party of Mosaddeq, was still alive, although fairly fractured since the coup, and was to be a key player to both the regime and the revolutionary alliance that would form. They desired a return to the Constitutional system of the past and straddled

---

[76] Mohsen M. Milani, *The Making of Iran's Islamic Revolution: From Monarchy to Islamic Republic* (Boulder, CO: Westview Press, 1994), 48.

[77] Ibid., 118

[78] Ervand Abrahamian, *Radical Islam: The Iranian Mojahedin* (London: I.B. Tauris, 1989), 85.

the line between being involved with the Shah's reorganization of the government to try to appease the growing dissent in the 1970s and playing key roles with Ayatollah Khomeini as the Revolution took hold and an alternative government was formed upon the Shah's exile. The Tudeh party, however, was not as active or as influential in the Revolution itself. During the 1950s and 1960s, the Tudeh party was fairly popular with the younger dissenters, given its Marxist ideology and propaganda. But the party was seen as a front for the Soviet Union's policies and influence in Iran, which initially was alluring to the opposition. As the Shah and the Soviet Union began to take a more conciliatory tone with each other, and as the Cold War was reaching a détente, the allure of the party became muted. SAVAK had also heavily infiltrated Tudeh and kept it under close watch, ensuring that it remained weak.[79]

LEADERSHIP AND ORGANIZATIONAL STRUCTURE

The Revolution of 1979 was almost entirely driven by popular resistance rather than armed struggle. Hence, the leadership figures who played a primary role in instigating, supporting, or otherwise influencing the fervor are more important than the exact organizational structure of the guerrilla groups or political parties. It is also important to understand the classes of people that made up the resistance, especially the religious clerics and the *bazaaris*, as well as their dissatisfaction with the Shah's regime and role in the Revolution.

The process of Westernizing and secularizing Iran had been a source of some tension with the Shi'a religious community during the earlier years of the Shah's rule, but the combination of Cold War-driven policies, the reforms of the White Revolution, and the declining influence of the *ulama* within the regime soon drove a large wedge between the Shah and the *ulama*. When Ayatollah Borujerdi, who had openly opposed the land reforms, died in March of 1961, it created a vacuum in the Shi'a hierarchy. The Shah tried to intervene in the selection process by recognizing Ayatollah Mohesen Hakim as the leader of the clerics. But the *ulama* rejected the Shah's attempt, although none of the candidates was able to ascend to the position at the time. The seat was left open, leaving all of the candidates to have a lesser, but still significant, voice. One of these candidates, Ruhollah Khomeini, held the position of Ayatollah at Qom and soon became a vocal critic of the Shah's reforms, drawing national attention as a political figure for this opposition. On June 3, 1963, Khomeini made

---

[79] Milani, *The Making of Iran's Islamic Revolution*, 76.

a speech in Qom in which he dramatically reproached the Shah: "You miserable wretch, forty-five years of your life have passed; isn't it time for you to think and reflect a little, to ponder about where all this is leading you, to learn a lesson from the experience of your father. . . . You won't be able to go on living; the nation will not allow you to continue this way."[80] The next day he was arrested. A riot broke out in Tehran during protests of Khomeini's arrest, with many killed or injured by machine-gun fire and the imposition of martial law.[81] This was the riot that led some of the secular opposition to feel that guerrilla warfare was necessary and justified.

Khomeini was released but had become a much more popular leader and was given a hero's welcome on his return to Qom. He still pressed the Shah to reform his policies, preaching that his reforms were a US conspiracy against Islam. By 1965, the regime felt that he could not be peacefully silenced and exiled him. Khomeini ended up residing in Najaf, Iraq, where, in 1970, he gave a series of lectures that were published as the *Hukumat-i Islami*, or "Islamic Government." The lectures detailed for the first time his call for the establishment of an Islamic political institution that would subordinate political power to religious criteria. It also called for the *ulama* to help bring about this Islamic state.[82] His continued opposition to the Shah from abroad cemented the Ayatollah Khomeini as the key religious figure for the opposition movement and the ideological head of the call for a new, religion-based governmental system.

A colleague of Khomeini's was Ayatollah Sayyid Mohammad Kazem Shariatmadari of Tabriz, another candidate for the seat left by the death of Borujerdi. He joined with Khomeini and others in open opposition to the Shah's modernization efforts and was instrumental in orchestrating the release of Khomeini after his arrest in 1963.[83] Together, he and Khomeini were seen as leaders in the calls for the Shah to drop his reforms, and up until the Revolution the population rarely distinguished Shariatmadari's ideas from Khomeini's.[84] Shariatmadari's views of the role of the *ulama*, however, were much more orthodox than Khomeini's. He supported the idea of the monarchy, although he eventually proclaimed that agreement

---

[80] Ruhollah Khomeini and Hamid Algar, *Islam and Revolution: Writings and Declarations of Imam Khomeini* (Berkeley, CA: Mizan Press, 1981), 179.

[81] "The official government estimate was that 20 were killed and 1,000 injured. The opposition claimed that thousands were massacred." Milani, *The Making of Iran's Islamic Revolution*, 51.

[82] Ibid., 52, and Khomeini and Algar, *Islam and Revolution*, 25.

[83] Milani, *The Making of Iran's Islamic Revolution*, 51.

[84] Nikki R. Keddie, Yann Richard, and Nikki R. Keddie, *Modern Iran: Roots and Results of Revolution* (New Haven, CT: Yale University Press, 2006), 226.

with the Shah had become impossible. The orthodox view of Shariatmadari and others claimed that the *ulama* should avoid being involved in political matters except in situations where un-Islamic laws or threats to Islam itself were being advocated by the regime. This placed him in the position of arguing on the side of Khomeini regarding the abuses and new policies of the regime, but not joining in the calls by Khomeini for active or violent protest by the faithful. He refused to support the call for strikes and protests.[85] He called for peaceful reform, and this distinction led to an eventual split between Shariatmadari and Khomeini after the Revolution.

The *ulama* in general had the same issues as the two leaders noted above, generally growing irritated with the Shah's secularization of the educational system, the introduction of coeducation, and the usurpation of traditional Islamic practice, for example, the installation of examinations for becoming part of the *ulama* and the unveiling of women. The slow erosion of both the *ulama's* power over the people and their influence within the government cemented the inclusion of much of the religious community within the opposition to the regime. The *ulama* was not homogenous in its views toward the government, however, or its opinions on how to best replace it, as shown in the differences between Khomeini's views and those of the orthodox.[86] As the Shah had closed off most areas of dissent across Iran, including the unions and the political parties, the mosques were one of the only outlets whose message was not controlled by the government.

The *bazaaris* included not only the shop owners in the traditional town bazaars but also anyone who operated any trade or manufacturing in the traditional sense (rather than more "Western" or "modern" ideas of business). The richest and more powerful of the *bazaaris* were extremely helpful in organizing and then populating massive rallies. Their ties with the *ulama* were mostly political, for they both desired continuation of the traditional ways and viewed encroaching Westernism as a threat. Their estimated control of the marketplace in 1976 was two-thirds of all domestic wholesale trade and 15% of private-sector credit. The new supermarkets, banks, and machine-made carpeting competed with their small food markets, money-lending operations, and handwoven Persian rugs. The bazaar areas, often centered on a mosque, offered an easy area for rapid communication and organization. The more notable *bazaaris* often had meetings and gatherings at their houses, providing an easy social network for the protest movement.[87]

---

[85] Ibid., 194.
[86] Ibid., 222.
[87] Ibid., 226–228.

COMMUNICATIONS

There is little evidence of secret communications between the key players during the critical years of 1978 and 1979. Some diplomatic efforts were attempted by the United States, but these efforts were often befuddled by the conflicting assessments and approaches of two key players, Ambassador William Sullivan and National Security Advisor Zbigniew Brzezinski.

The early dissatisfaction with regime policies was aired in public by the religious leaders either through open speeches or lectures. The National Front published open letters attacking specific Shah policies and called for reform. During his exile in Najaf, Khomeini's lectures were often smuggled into Iran on cassette tapes, maintaining his position as the key oppositional leader even from across the border.

When Khomeini was forced to leave Iraq by Saddam Hussein in late 1978, he gained incredible access to the Western media during his 114-day stay in Paris. Khomeini had a cadre of Western-education advisers, who helped him skillfully exploit the modern communication system that he had lacked in Najaf, and his visibility and ability to rally the population in Iran was greatly enhanced at this crucial moment.[88]

The mosques also acted as a medium for spreading dissenting messages, as the clerics used their sermons to judge the government's actions according to Islam's precepts. More than 8,400 mosques around the country, along with hundreds of community Islamic organizations, provided a means of motivating the faithful.[89] Islam provided a set of standards by which the *ulama* could argue against the Shah, and the religion also presented symbols and rituals that could be used to galvanize the resistance. Moreover, with the organization of the massive protests, Shi'a symbols and rituals afforded the means to select days for events. Commemorations are traditionally held forty days after an act of martyrdom. When the governmental forces repressed one protest with violence, another gathering would occur forty days later at the site or a nearby mosque. This forty-day cycle escalated the violence in early 1978. The religious opposition also called for massive rallies on religious holidays, which placed the Shah in a difficult position. If he called for people to stay home and enforced a curfew, the opposition could point to his secular practices and his opposition to Islam. If he allowed the rallies, they quickly turned to protests against the government.

---

[88] Milani, *The Making of Iran's Islamic Revolution*, 118.
[89] Ibid., 18.

METHODS OF ACTION AND VIOLENCE

Two particular events are notable in turning the opposition toward a full-blown protest movement. In early 1977, the National Front circulated three open letters to the Shah complaining of the prevalence of corruption and repression by his government. The government took no action to arrest or harass the authors, and this was perceived to be a sign that the government repression was weakening. This encouraged further demonstrations, which again met with little show of force by the police or SAVAK. Gradually, the opposition mobilized further, both internationally and within the student population. The government organized and held its own rallies as a show of support.[90]

In January of 1978, however, an article was published in the semiofficial newspaper *Ettela'at*, supposedly at the instigation of the Shah. In the article Khomeini was crassly attacked for having a dubious past and purportedly accepting money from the British to fight against the regime. The article instantly produced protests by the students of Khomeini in Qom, to which the police did respond, and brutally. Over two days at least seventy were killed, and some claim this event is the point at which the movement shifted from being dominated by the secular opposition to being led by the *ulama*, more particularly Khomeini. Whereas the government had been able to successfully abate any secular threat, it now faced a less manageable and more popular religious-based opposition.[91]

The cycle of forty-day commemorations started in February to honor those killed in Qom. One of the major sites was Tabriz, home of Ayatollah Shariatmadari, where police moved to block access to the mosque. Mourners were turned away; in anger, they soon ravaged the city. The symbols of dependence on the West, such as Bank-e Saderat, movie theaters, liquor stores, shops, and even the headquarters of the Women's Association, were attacked and burned. The mobs did not target people, and this approach held true throughout most of the entire Revolution. The army was called in, and they killed and arrested many and restored calm quickly.[92]

With Tabriz as a template, riots and protests soon spread to other cities around Iran, and the commemorations saw an increase in participation and potency in each cycle. SAVAK and the government struggled against preventing the protests because of their lack of coordination by any centralized control structure and their growing size. Martial law was imposed in some cities, but the growing

---

[90] Ibid., 110–111.
[91] Keddie, Richard, and Keddie, *Modern Iran*, 225.
[92] Ibid., 226; Milani, *The Making of Iran's Islamic Revolution*, 113.

opposition was gathering momentum, with the disparate antiregime groups now sensing a possibility of success. The protest gatherings were a mixture of calls for the Shah to step down, calls for reform, praises for Khomeini, and anger at the violence perpetrated by the government's forces. The protesters moved from running away from security forces to direct confrontation and conflict. There were few exchanges of gunfire. The vast majority of arms were in the hands of the government. The contests were ones of crowds versus crowd control. However, the Iranian armed forces were not trained in such operations and had little of the equipment necessary for them to accomplish the task.[93]

On August 19, 1978, a fire in an Abadan movie theater killed more than 400 men, women, and children. Although the government tried to blame the fire on the opposition, the opposition was firmly convinced that the Shah had ordered the arson to discredit the religious protests and leadership. The result of the fire was a galvanization of the revolutionary movement as well as defensive moves by the regime. The Shah replaced his premier in late August, appointing Ja'far Sharif Imami to resolve the worsening situation. Imami undertook immediate reforms to appease the diverse opposition, including shutting down casinos, nightclubs, and abolishing the ministry of women's affairs to appease the *ulama*. To mollify the secular movement, he began an anticorruption campaign, ordered punishment for the officials responsible for the killings of protesters, and granted more freedoms to political parties. The reforms were rejected by both branches of the revolution. Ayatollah Shariatmadari declared that the Shah had three months to resolve the tension between the regime and the people. The National Front called the reforms a sham and demanded the dissolution of SAVAK and the immediate release of all political prisoners. The opposition sensed weakness and opportunity.[94]

A tragedy in Tehran early the next month probably squashed any possibility of reconciliation. At the army's request, Imami imposed martial law and curfew in the capital city in anticipation of a rally in Jaleh Square scheduled for the 8th of September. The demonstrators ignored the curfew, and the army reacted violently. The number of protesters killed probably numbered in the hundreds and the date became known as "Black Friday." Over the next two months, the Shah's government tried to negotiate directly with both Khomeini and the National Front, only to be rebuffed. By November, the Shah had replaced his staff with military men and requested the government of

---

[93] Milani states that Iran's request to the United States for tear gas did not get fulfilled until November of 1978 because of delays by the State Department. Ibid., 113.

[94] Ibid., 116–117; Keddie, Richard, and Keddie, *Modern Iran*, 231.

Iraq to expel Khomeini or severely limit his activities. The Ayatollah obtained permission from France to live near Paris, thereby increasing his visibility and influence over the Revolution.[95]

Another crucial component to the success of the Revolution took place when workers and public employees seized the tumultuous period as an opportunity to strike. The economy had been harsh on blue-collar workers, as well as government employees, over the previous two years, and their plight surfaced in the fall of 1978. Oil refinery workers walked off the job first in Tehran and then across the country. Within a month, production had fallen to 28% of normal rates, causing shortages of heating oil at the start of winter for the population and massive drops in oil revenues for the government. Strikes by the electrical workers' unions led to periodic blackouts. Teachers decided to strike just as school was scheduled to start, turning students out into the streets to join the demonstrations. Students were attracted to the more militant groups, such as the *Fedayin* and *Mojahedin*, which grew rapidly, as well as the more radical religious message of Khomeini. Grassroots support for the now-visible cleric exploded, with his revolutionary ideas spreading widely and quickly in the fall and early winter.[96]

The only piece of the puzzle that remained was the alliance of the oppositional forces. This was achieved when Karim Sanjabi of the National Front met with Khomeini in Paris and left with a short declaration of agreement. It stated that the existing monarchy had no constitutional or religious legitimacy and called for a future political system in accordance with the precepts of Islam, democracy, and independence. This seemed to assuage the National Front that Khomeini would support a secular government and had no intentions of directly ruling the government.[97] It is probable that most of the secular opposition had not read his *Hukumat-i Islami*. Even the chief of the Tudeh party recognized the leadership of Khomeini in the opposition movement and called for a united front.[98] The agreement between the multiple branches foreclosed any compromise between the monarchy and the opposition. The elite and prosperous residents

---

[95] Milani, *The Making of Iran's Islamic Revolution*, 117–119; Keddie, Richard, and Keddie, *Modern Iran*, 231–232.

[96] Milani, *The Making of Iran's Islamic*, 119; Keddie, Richard, and Keddie, *Modern Iran*, 232–233.

[97] By late October, the US State Department had recognized the probable end of the Shah but predicted that his successor would come from the secular opposition or the military. Milani, *The Making of Iran's Islamic Revolution*, 120.

[98] There was no alliance between the Marxists and Khomeini, just recognition of Khomeini's leadership by the Tudehs. Khomeini viewed them as being as dangerous to Iran as the Shah. Ibid., 120 and 136, fn. 54.

of Iran recognized the beginning of the end and soon had transferred much of their wealth and families out of the country, mostly to the Western nations. The support of the Shah was collapsing quickly.[99]

Rampaging crowds were destroying Tehrani government buildings, theaters, and other targets with little intervention by the army. The army leadership and new military government seemed to be allowing things to fester to convince the Shah to order a hard crackdown. He never did.[100] Repression increased, but a brutal response was discouraged by the United States and the ailing Shah[101] did not order it. Khomeini was busy encouraging the strikes to continue and pushing for desertion within the army. Massive rallies were held to commemorate *Ashura*, drawing millions into the streets. The crowds were diverse in age, background, and class, and signs against the Shah, calling for an Islamic republic, or just praising Khomeini were equally spread across them.[102]

Figure 2. Islamic Revolution protesters in Tehran, 1979.[103]

---

[99] Keddie, Richard, and Keddie, *Modern Iran*, 233–234; Milani, *The Making of Iran's Islamic Revolution*, 120.

[100] See Milani and Keddie for discussions of the influence of US policy on the Shah's actions. The infighting and delayed reactions by the United States certainly gave the Shah pause in deciding how to respond to the protests. He often waited to see if the United States would support his actions, whether harsh or conciliatory, before implementing.

[101] The Shah was by now visibly ailing and growing weak, probably due to non-Hodgkins lymphoma; complications would cause his death in 1980.

[102] Ibid., 121–123.

[103] "File:1979 Islamic Revolution.jpg," *Wikipedia*, accessed March 11, 2011, http://en.wikipedia.org/wiki/File:1979_Islamic_Revolution.jpg.

In December 1978, the Shah turned again to the National Front to reorganize his government to appease the revolt. Many National Front members turned him down, causing him to turn to Shahpur Bakhtiar, who had been a junior member of the Mosaddeq administration,[104] to form a new government. Bakhtiar agreed on the condition that the Shah leave Iran. On December 31, it was announced that the Shah was leaving Iran, supposedly temporarily.

The possibility of a military coup concerned both the US administration and the opposition. Ayatollah Shariatmadari asked the US embassy, through an intermediary, to take measures to prevent a coup, and even Khomeini indicated that such a move by the United States would be a positive gesture between Washington and the opposition. The Carter administration sent General Robert E. "Dutch" Huyser, deputy commander of US forces in Europe, to meet with the Iranian military leadership. Huyser soon found that a coup was in the making, but the logistics were not being worked out between the five Iranian generals. The Shah had prevented them from meeting in any organized forum, and, therefore, they were dependent upon his leadership and unable to work out even simple operations by themselves.[105]

As soon as the Shah left Iran for Egypt in mid-January 1979, Ayatollah Khomeini announced the creation of a secret Council of the Islamic Revolution to coordinate the opposition formally. "It has been entrusted with the task of examining and studying conditions for the establishment of a transitional government and making all the necessary preliminary arrangements."[106] The National Front and other secular groups were denied membership, as were moderate *ulama* members who favored some type of constitutional monarchy. A fierce competition for public support emerged between the Bakhtiar government, the National Front (which had expelled Bakhtiar for his agreement with the Shah), and the revolutionaries under Khomeini. The support of the armed forces was also paramount, and the Council of the Islamic Revolution took many steps to neutralize the armed forces' support of the current government. The Council arranged meetings with the more moderate military leadership, leading them to report to Khomeini and Shariatmadari that the armed forces would not intervene in any confrontation between the revolution and the Bakhtiar government. This assertion was tested in early February when the Imperial Guard attacked a base of rebelling air force technicians.

---

[104] Bakhtiar had been deputy minister of Labor under Mosaddeq and had helped organize the Second National Front after the coup. Ibid., 45.

[105] Ibid., 127–128.

[106] Khomeini and Algar, *Islam and Revolution*, 246.

The *Mojahedin* and *Fedayin* mobilized to defend the technicians and successfully fended off the Guard. The guerrilla groups took over the garrison, seized thousands of weapons, and then started a rampage to open prisons and ransack SAVAK headquarters and military bases. The armed forces quickly declared their neutrality to prevent more bloodshed.[107]

Bakhtiar was still trying to gain support for his government and started negotiations with the Council about ending Khomeini's exile. The demands came back that Bakhtiar would have to resign and acknowledge Khomeini's leadership; these demands were refused. The pressure continued on the government to allow the Ayatollah to return. Bakhtiar even closed the Tehran airport to stall for some time but finally relented. Ayatollah Khomeini returned on February 1 to a massive welcome and made his intentions known with a speech at a cemetery to honor the martyrs of the Revolution:

> [Bakhtiar's] government and all those associated with it are illegal. If he and his colleagues persist, they will be counted as criminals who must be brought to trial. Yes, we will put them on trial. I will appoint a government, and I will give this government a punch in the mouth. . . . The government I intend to appoint is a government based on divine ordinance, and to oppose it is to deny God as well as the will of the people.[108]

A parallel government was set up within six days, and Mehdi Bazargan was asked by Khomeini to head the government. Bazargan had split from the National Front in the 1960s to lead a small independent secular opposition group that had good ties with the *ulama*. He was respected by both secular and religious leaders and was assumed to be uneager for political power. He constructed a tenuous alliance between the secular and religious opposition and had strong backing from the *bazaaris*, the middle class, and the orthodox *ulama*.[109] Bakhtiar went into hiding,[110] and soon the Bazargan government had effectively taken over Iranian affairs.

---

[107] Milani, *The Making of Iran's Islamic Revolution*, 131–132; Keddie, Richard, and Keddie, *Modern Iran*, 238.

[108] Khomeini and Algar, *Islam and Revolution*, 259.

[109] Milani, *The Making of Iran's Islamic Revolution*, 143–144; Keddie, Richard, and Keddie, *Modern Iran*, 240–243.

[110] Bakhtiar was assassinated in Paris in 1992. The Islamic Republic has denied any involvement in the shooting. Milani, *The Making of Iran's Islamic Revolution*, 230.

METHODS OF RECRUITMENT

The guerrilla groups were able to recruit from the Western-educated student population and middle classes, and the National Front drew from the educated classes, as well. The allure of joining in the protests and demonstrations often started, however, as a way to vent frustration at the government's policies and brutal tactics. The working and poor classes soon added to the demonstrations as conditions worsened and the allure of a religious figure drew some to his call.

Desertion within the armed forces became common, with many lower ranks either leaving to join the crowds or shooting their own commanders rather than firing into the demonstration. Khomeini encouraged the soldiers to join them, and as the regime's power was visibly crumbling, units soon grew apathetic or supportive of the protests themselves.[111]

METHODS OF SUSTAINMENT

Because this revolution had little material support, its sustainment was more tied to the revolutionary fervor that could be generated by the leadership of Ayatollah Khomeini, as well as the opposition to the policies and actions of the Shah. The ability for fiercely combative messages from Khomeini to reach the Iranian population certainly allowed his ascendance to leader of the entire opposition. Social networks based upon religious, familial, and marketplace relationships allowed for rapid trafficking of these messages and plans for future protests. The importance of the forty-day cycle of commemoration, as well as the decentralized control of the revolution's operations, was critical to the sustainment of the revolution.

METHODS OF OBTAINING LEGITIMACY

Two important factors allowed Khomeini to achieve a legitimacy that had eluded the National Front or other groups since Mosaddeq. First was his religious authority as an Ayatollah from one of the most respected Islamic training centers, Qom. His hard-line interpretation of Islam as a means of living and governing brought a deep respect from the faithful and his colleagues, even though they disagreed.[112]

---

[111] Parsa, *Social Origins of the Iranian Revolution*, 241–248.

[112] Khomeini makes a good case of fulfilling Weber's theory of the "charismatic leader," one whose power is legitimized by his demonstration of exemplary character, heroism, and/or sanctity, spurring loyalty and devotion among his followers.

Perhaps most important, however, was the fearless attacks Khomeini made on a figure who was seen to be invincible and all-powerful. The Shah had generated a formidable autocracy built on fear and power, and a cleric challenged this ruler without bending.

Khomeini's political savvy in building legitimacy can be demonstrated by an event that took place as the Shah was preparing to leave Iran. The oil strikes had been crippling the country, and Khomeini asked two of his advisers to go to the oil fields to talk with the workers. They immediately came back with a resolution to begin producing enough oil for domestic consumption. That Khomeini could wield such influence and decided to do so without inviting the rest of the opposition to participate showed that he was clearly in control.[113]

Khomeini's approach to the alliances formed between the secular and religious factions shows his pragmatic and Machiavellian determination to succeed. The declaration that emerged from the meeting in Paris of Khomeini and the National Front spoke of both Islam and democracy as basic principles. Once in power, Khomeini did not support any form of democracy, seeing it as counter to Islam. His willingness to agree to the declaration convinced the secular faction that he would support their form of government once the Shah was overthrown, an assumption that proved inaccurate.[114]

EXTERNAL SUPPORT

Little external support was required for the successful revolution. Both the Soviet Union and the United States did not aid the opposition in any great manner; however, the infighting and indecision within the Carter administration about how to respond may have further emboldened Khomeini. The mixed signals by the United States certainly caused part of the Shah's indecision and awkward responses as the Revolution intensified.

Student organizations abroad had some influence in bringing attention to the policies and conditions under the Shah's rule. Groups such as the Confederation of Iranian Students and the Moslem Student Association had been created in the aftermath of the 1963 riots and were the main anti-Shah propaganda machines outside Iran.[115]

---

[113] Milani, *The Making of Iran's Islamic Revolution*, 125.

[114] Keddie, Richard, and Keddie, *Modern Iran*, 233–234.

[115] Milani, *The Making of Iran's Islamic Revolution*, 62.

COUNTERMEASURES TAKEN BY THE GOVERNMENT

Previous sections have given some indication of the various measures taken by the government including martial law and curfews, attempts at riot control, and the infiltration of groups by SAVAK. The frequent shuffling of the government was also seen by the Shah as an attempt to appease the opposition, although it was primarily policies that were opposed rather than specific government appointees. The increasing liberalization of the Shah's hold on government was a key factor in the ability of the opposition to thrive in 1977 and 1978; he alternated this liberal approach with harsh repression that was unable to put the genie back into the bottle. The Shah's concessions were always a step or two behind the opposition's demands.[116]

SHORT- AND LONG-TERM EFFECTS

CHANGES IN THE ENVIRONMENT

The success of an Islamic-led revolution showed that there were viable alternatives to monarchies, democracy, or Marxist-style government, spurring a new wave of thinking within the Mideast religious communities. But the intellectual and upper and middle classes were absent for much of the post-1979 transition, with the emigration of many during the final days of the Shah's rule and many others leaving or muted by the purges of the universities by the 1980 Cultural Revolution. Khomeini established a community that dismissed "subversives" from the higher-education system, causing many universities to shut down until they could find acceptable replacement teachers and enough students to continue.[117]

CHANGES IN GOVERNMENT

Bazargan headed the Iranian government for only 10 months, resigning in November 1979 upon the seizure of the US embassy in Tehran. Radicalism was not quenched by his secular government, and militant students taking hostages in defiance of his orders showed the powerlessness of his government. Abolhassan Bani Sadr was elected Iran's first president in early 1980. His government had to deal with the hostage crisis, the Iraqi invasion of Iran, and the growing competition with the religious fundamentalists under Khomeini. He soon found

---

[116] Keddie, Richard, and Keddie, *Modern Iran*, 236.
[117] Ibid., 250.

himself powerless to settle the hostage crisis, negotiated and settled by the judicial branch, which was under fundamentalist control.[118]

Khomeini had formed the Islamic Republican Party, which grew in power and control of the government's institutions. Khomeini had also created his own forces, both the semilegal Revolutionary Guard and the more violent and underground Hizbollah. His Council of the Islamic Revolution continued to operate, sometimes in competition with the new government. Khomeini's appointees and faithful soon came to dominate the *ulama*, the judicial branches of government, and many new institutions. He ordered a purge of the university systems, which paralyzed some of the schools for almost a year while they found enough acceptable people to again teach and be students. All of these changes were viewed as a Cultural Revolution.[119]

The tensions between the Bani Sadr government and the Islamic Republican Party grew fierce, and Sadr was declared incompetent by the judicial branch. Leaked US embassy documents showed that Bani Sadr had met with a CIA agent.[120] Ayatollah Khomeini dismissed Bani Sadr, and by 1982 Khomeini was in full control of the state and its institutions.[121]

CHANGES IN POLICY

The relationship with Western governments obviously suffered greatly under the new Islamic government as opposed to the Shah's. The creation of a state built around the concept of Islamic revolution changed the dynamic of the Middle East. Its internal policies were a dramatic change from the modernist intentions of Mohammed Shah Pahlavi, returning to the more traditionalist roles of women, laws based on *shari'a*, and a rejection of anything corresponding to Western "vices." The hostage crisis marked a dramatic end to any rapprochement between the new Iranian government and the United States. Although the crisis was not instigated by Khomeini or the new government, it was prolonged in order to show the world the new limits of US power.

When the Carter administration decided to allow the Shah to enter the United States for medical treatment in October of 1979, the more militant Islamic groups viewed the move as the United States'

---

[118] Milani, *The Making of Iran's Islamic Revolution*, 178–181.

[119] Keddie, Richard and Keddie, *Modern Iran*, 250.

[120] Bani Sadr denied that he knew the man was a CIA agent but acknowledged that he was offered money as a "consulting fee," which he rejected. Milani, *The Making of Iran's Islamic Revolution*, 184.

[121] Ibid., 197.

first step toward either trying to reinstall him or destroying the Islamic revolution. The US embassy was besieged by less than 500 students, taking the personnel hostage and confiscating hundreds of documents. Both diplomatic and military efforts failed to release the Americans and helped to strengthen the position of the fundamentalists within the Iranian political environment. Islamic groups seized secular offices, implicated many military and government officials in coup attempts, and thereby increased their own hold on the government. The anger at the United States was also solidified within the population as it was used to strengthen Khomeini's position.

CHANGES IN THE REVOLUTIONARY MOVEMENT

The revolutionary movement had been a diverse assortment of Marxist, orthodox Islam, radical Islam, constitutionalist, and modernist believers that were united by their hatred of the Shah's policies and, eventually, the Shah himself. Upon Bazargan's usurpation of power at the request of Ayatollah Khomeini, however, the individual groups all clamored for legitimacy and political power. No one did this with as much dexterity as the Ayatollah. He had started from a position of leadership of the movement, but his adept takeover of the major institutions and governmental positions for his followers, and not himself, allowed him to emerge victorious.

The *Mojahedin* quickly broke with Khomeini and suffered under attacks by his Hizbollah. A brutal repression of the *Mojahedin* caused hundreds of executions and the leadership to flee the country. The group turned back to guerrilla tactics but soon found themselves decimated. By 1983, the leadership ordered remaining cells to move to Kurdistan.[122]

The National Front backed the Bani Sadr government but found themselves attacked by Khomeini. In June of 1981, just before a planned demonstration by the party in support of the government, Khomeini attacked the National Front in a speech, declaring them to be more concerned about nationalism than Islam. The rally was a bust, and Hizbollah soon destroyed their newspaper operations and headquarters. The Front quickly ceased to be a viable party. The Tudeh and *Fedayin* were allowed to operate until 1983, when accusations of coup plots and spying for the Soviet Union were declared. Their leaders were arrested and the parties declared illegal in May of 1983.[123]

---

[122] Abrahamian, *Radical Islam*, 206–223.
[123] Keddie, Richard, and Keddie, *Modern Iran*, 251–254.

BIBLIOGRAPHY

Abrahamian, Ervand. *A History of Modern Iran.* New York: Cambridge University Press, 2008.

———. *Khomeinism: Essays on the Islamic Republic.* Berkeley, CA: University of California Press, 1993.

———. *Radical Islam: The Iranian Mojahedin.* London: I.B. Tauris, 1989.

Amini, Parvin Merat. "A Single Party State in Iran, 1975–78: The Rastakhiz Party—the Final Attempt by the Shah to Consolidate His Political Base." *Middle Eastern Studies* 38, no. 1 (2002).

Amnesty International. "The Amnesty International Report 1975–1976." (1976).

"Anatomy of a Coup: The CIA in Iran," Edited by Tom Seligson, Susan Werbe, Ann Dean, and Rebecca Donahue. Documentary Film. The History Channel, 2006.

Brunner, Rainer, and Werner Ende. *The Twelver Shia in Modern Time: Religious Culture & Political History.* Boston, MA: Brill, 2001.

Central Intelligence Agency. "Iran." *The World Factbook.* https://www.cia.gov/library/publications/the-world-factbook/geos/ir.html

Choueiri, Youssef M. *Islamic Fundamentalism.* Washington, DC: Pinter, 1997.

Cole, Juan Ricardo. *Sacred Space and Holy War: The Politics, Culture and History of Shi'Ite Islam.* London: I.B. Tauris, 2002.

Cole, Juan Ricardo, and Nikki R. Keddie. *Shi'Ism and Social Protest.* New Haven, CT: Yale University Press, 1986.

Douglas-Home, Charles. "Will the Shah be Toppled from His Shaky Throne?" *Sunday Times,* November 28, 1978.

Fanni, Zohreh. "Cities and Urbanization in Iran after the Islamic Revolution." *Cities* 23, no. 6 (2006).

Gasiorowski, Mark. "The CIA Looks Back at the 1953 Coup in Iran." *Middle East Report* 216, (2000): 4–5.

———. "The 1953 Coup d'Etat in Iran." *International Journal of Middle Eastern Studies* 19, no. 3 (1987).

Gasiorowski, Mark, and Malcolm Byrne. *Mohammad Mosaddeq and the 1953 Coup in Iran.* Syracuse, NY: Syracuse University Press, 2004.

Keddie, Nikki R. *Religion and Politics in Iran: Shi'ism from Quietism to Revolution.* New Haven, CT: Yale University Press, 1983.

Keddie, Nikki R., and Yann Richard. *Roots of Revolution: An Interpretive History of Modern Iran.* New Haven, CT: Yale University Press, 1981.

Keddie, Nikki R., and Yann Richard. *Modern Iran: Roots and Results of Revolution.* New Haven, CT: Yale University Press, 2006.

Kepel, Gilles. *Jihad: The Trail of Political Islam.* Cambridge: Harvard University Press, 2002.

Khomeini, Ruhollah, and Hamid Algar. *Islam and Revolution: Writings and Declarations of Imam Khomeini.* Berkeley, CA: Mizan Press, 1981.

Kramer, Martin S. *Shi'ism, Resistance, and Revolution.* Boulder, CO: Westview Press, 1987.

Mantel, Robert, and Geoffrey Kemp. "US Military Sales to Iran." Staff report to the Subcommittee on Foreign Assistance of the Committee on Foreign Relations, US Senate. Washington, DC: Government Printing Office, 1976.

Milani, Mohsen M. *The Making of Iran's Islamic Revolution: From Monarchy to Islamic Republic.* Boulder, CO: Westview Press, 1994.

Momen, Moojan. *An Introduction to Shi'i Islam: The History and Doctrines of Twelver Shi'ism.* New Haven, CT: Yale University Press, 1985.

Pahlavi, Mohammed Reza Shah. *My Mission My Country.* London: Hutchinson, 1974.

Parsa, Misagh. *Social Origins of the Iranian Revolution.* New Brunswick, NJ: Rutgers University Press, 1989.

Parvin, Manoucher, and Amir Zamani. "Political Economy of Growth and Destruction: A Statistical Interpretation of the Iranian Case." *Iranian Studies* 12, no. 1/2 (1979): 43–78.

Reza. "A Visual Witness to Iran's Revolution." Nieman Report, Nieman Foundation for Journalism, Harvard University. Accessed July 9, 2010. http://www.nieman.harvard.edu/reportsitem. aspx?id=101470.

Pearsan, M. H. "Income Distribution in Iran," in *Iran: Past, Present, and Future,* edited by Jane W. Jacqs. New York: Aspen Institute of Humanities, 1976.

Scott, James C. *Seeing Like a State: How Certain Schemes to Improve the Human Condition have Failed.* New Haven, CT: Yale University Press, 1998.

Summitt, April. "For a White Revolution: John F. Kennedy and the Shah of Iran." *Middle East Journal* 58, no. 4 (2004): 560–575.

Weiner, Tim. "C.I.A. Destroyed Files on 1953 Iran Coup." *New York Times,* May 29, 1997.

Wilber, Donald. "CIA Clandestine Service History, 'Overthrow of Prime Minister Mossadeq of Iran, November 1952–August 1953.' " *New York Times,* 2000.

FRENTE FARABUNDO MARTI PARA LA LIBERACION NACIONAL (FMLN)

Ron Buikema and Matt Burger

SYNOPSIS

The *Frente Farabundo Martí para la Liberación Nacional* (FMLN) became a major insurgency of the 1980s and 1990s in the Central American country of El Salvador, not only threatening the legitimate government of El Salvador but also playing out on the larger stage, as it was a hot war within the construct of the Cold War. The insurgency resulted in dramatic political drama in San Salvador and Washington, DC, including emotional debates about alleged and proven human rights abuses by the Salvadoran government, as well as about the role of US military, intelligence, and political resources. Twelve years of war killed approximately 1.5% of the population, displaced another 30% (1.5 million people), and caused widespread destruction of the country's economy.[1] El Salvador became a high priority for US administrations in demonstrating resolve to defeating communism, particularly within the Western Hemisphere. US military and economic aid to El Salvador amounted to more than $6 billion over the course of the conflict.[2] Throughout two decades, the United States was increasingly involved in an insurgency that became known for human rights abuses on both sides as well as urban and rural violence on scales that had not been seen for decades within Central America. The insurgency demonstrated the ability to win on the battlefield, build a political and popular support structure, and wage a serious fight for control of El Salvador. This study covers the FMLN from its inception to the execution of the United Nations (UN)-brokered cease-fire in 1992.

---

[1]   C. D. Brockett, "El Salvador: The Long Journey from Violence to Reconciliation," *Latin American Research Review* 29, no. 3 (1994): 174–187.

[2]   T. Buergenthal, "The United Nations Truth Commission for El Salvador," *Vanderbilt Journal of Transnational Law* 27, no. 3 (1994), 497.

TIMELINE

December 1979–January 1980	The FMLN formed.
1980	National Guard attacks a crowd of demonstrators, resulting in more than 500 casualties; US ambassador reports mutilated bodies on roadsides; Archbishop Romero is assassinated, polarizing Salvadoran society; four US church women are raped and murdered by National Guard troops; José Napoleón Duarte becomes El Salvador's first civilian president in 49 years.
1981	The FMLN launches first major offensive action against military targets across the country, surprising US and Salvadoran officials with their demonstrated military capabilities; 1,000 villagers are massacred by the Salvadoran Army near the village of El Mozote.
1983	President Reagan requests increased military aid for El Salvador; the level of violence and civilian deaths continues to spiral; first US adviser is killed; peace talks between the government and the FMLN are canceled; the FMLN commences widespread use of anti-personnel mines and improvised explosive devices; UN reports that 20% of Salvadoran population is either displaced or in exile.
1984	Right-wing death squad activity continues to increase, with hundreds killed and thousands missing.
1988	The FMLN boycotts national elections; death squads remain active; the FMLN kills eight mayors.
1989	The FMLN commences the largest offensive of the war, including a major urban campaign in San Salvador. Fighting continues for thirty days. Indiscriminate aerial bombing by the Salvadoran Air Force sharply turns public opinion to the FMLN.

1990	US House of Representatives, concerned over continued reports of human rights abuses by the Salvadoran military, cuts aid by 50%. UN begins to mediate talks between the government and the FMLN. New Minister of Defense is accused in the murder of Jesuits earlier in the war, but the government takes no action against him.
1991	UN stands up the UN Observer Mission, El Salvador with a Human Rights Division. The FMLN and Salvadoran officials agree to establish the Commission on the Truth and reach significant agreements regarding constitutional reform of the armed forces, the judiciary, and the electoral system.
1992	The Peace Accords of Chapultepec are signed in Mexico City; UN Truth Commission begins investigating acts of violence and human rights abuses committed by both belligerents during the war; the FMLN demobilizes and becomes a political party.

THE ENVIRONMENT OF THE REVOLUTION

PHYSICAL ENVIRONMENT

El Salvador is slightly smaller than the state of Massachusetts and is the smallest country in Central America. It borders Guatemala and Honduras and has more than 300 kilometers of coastline in the Gulf of Fonseca (Pacific Ocean). El Salvador has a tropical climate, with rainy and dry seasons. Terrain is mostly mountainous with a narrow coastal belt and central plateau. Known as the land of volcanoes for an active string that runs east–west throughout the country, natural hazards include destructive earthquakes and some volcanic activity. Approximately 31% of the land is arable, with 11% populated by permanent crops.[3]

---

[3] Central Intelligence Agency, "El Salvador," *The World Factbook*, accessed November 2, 2009, https://www.cia.gov/library/publications/the-world-factbook/geos/es.html.

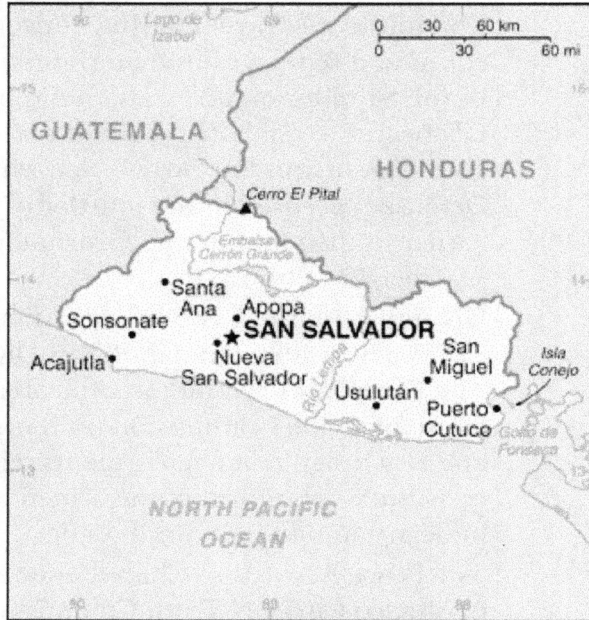

Figure 1. Map of El Salvador.[4]

CULTURAL AND DEMOGRAPHIC ENVIRONMENT

Ethnically, Salvadorans are principally mestizo (90%), with a significant Caucasian minority (9%) and only a small Amerindian (indigenous) remnant (1%). Although Spanish is the predominant language, often words from the indigenous language are incorporated, particularly in rural areas. More than half (57.1%)[5] of Salvadorans identify themselves as Roman Catholic. The Catholic Church has and continues to play a significant role in Salvadoran society, often as an unofficial representative of the peasant and working classes, even to the point of confronting political leadership from the pulpit.[6]

Before the civil war, Salvadoran culture remained essentially colonial. The country relied on the export of cash crops, especially coffee, and a small group of landed elites controlled the nation's wealth and land resources while the majority of Salvadorans were impoverished peasants. Political regimes were unstable and subject to the will of the landed elites and the military who bowed to their

---

[4] Ibid.

[5] Protestants make up roughly 27.6% of the population, and their numbers have rapidly grown since the 1980s. Those with no religious affiliation constitute 11.1% of the population, with the remainder identifying with a variety of other religions (e.g., Islam, Judaism, Mormonism, etc.). Ibid.

[6] Ibid.

bidding.[7] Moreover, the cultural ties that knitted together Salvadoran society had slowly eroded. In the late nineteenth and early twentieth centuries, indigenous people were displaced from their communities and their practice of communal land ownership was restricted as land was accumulated by the landed elites in order to maximize profits from coffee production.[8] With the ruthless suppression of a peasant uprising in 1932, indigenous culture was essentially destroyed. After World War II, the traditional patron–client relationship between the landed elites and the peasants on their plantations also began to deteriorate as the old system of mutual obligation was replaced by an extremely asymmetrical wage labor system that favored the landowners. As a result, many peasants moved to urban areas or other parts of the country to find work, severing local social ties.[9] This cultural breakdown alienated Salvadorans from each other and widened the chasm between peasants and the landed elites.

SOCIOECONOMIC ENVIRONMENT

Samuel Huntington and Francis Fukuyama noted, "Where the conditions of land ownership are equitable and provide a viable living for the peasant, revolution is unlikely. Where they are inequitable and where the peasant lives in poverty and suffering, revolution is likely, if not inevitable."[10] They succinctly described the sociocultural construct that was El Salvador during the 1970s and 1980s.

La matanza (the slaughter), the term given to a 1932 government response to a rebellion that resulted in the death of more than 10,000 Salvadorans, proved to be historically significant in influencing the sociocultural environment in El Salvador for the next sixty years.[11] To the elite right-wing segment, the event represented the downfall of communism and a popular uprising, with the clear understanding that violence would continue to be used if necessary to protect the rights of the landowners and elites. The right controlled the military, the political apparatus, and the land. For the left, the event represented the ominous threat of violence and repression by the government. However, it also served as a focal point for unifying the peasant class's growing sentiment that the status quo could not be tolerated.

---

[7] Steve Hobden, "El Salvador: Civil War, Civil Society and the State," *Civil Wars* 3, no. 2 (2000): 106–120.

[8] Elisabeth Jean Wood, *Insurgent Collective Action and Civil War in El Salvador* (Cambridge, UK: Cambridge University Press, 2003).

[9] Ibid.

[10] S. P. Huntington and F. Fukuyama, *Political Order in Changing Societies* (New Haven, CT: Yale University Press, 2006).

[11] See *Historical Factors* section for a fuller discussion.

Coffee represented one half of El Salvador's exports; the country was dubbed a coffee republic. Coffee was grown on large plantations, where peasants labored like feudal serfs for a small cadre of elites that constituted 2% of the population but owned 60% of the land.[12] These elites, which by most accounts comprised no more than about fourteen families, were economically progressive but vigorously opposed to any reforms that threatened their power, despite the potential benefits to peasant farmers. These families held extraordinary sway over the government and virtually controlled the military as their own private army.[13] Moreover, these large landowners seized land from small farmers for the production of coffee for export, forcing impoverished peasants to become laborers on the *haciendas* or move to the cities. As the 1970s drew to a close, political forces became increasingly polarized over issues related to these inequalities.[14] Between 1978 and 1982, the country was near anarchy, and the rise of the FMLN insurgency and the inadequate reforms of El Salvador's military junta[15] led to a 22% decline in the country's gross domestic product.

El Salvador is a predominately Catholic country. The religious values taught in the Catholic Church regarding suffering, unjust persecution, and even martyrdom were all themes that became associated with the plight of the poor and, eventually, with the FMLN. The Vatican Council and Medellín conferences in the 1960s resulted in the Church's emphasis on social justice, pasturing to the poor, and fighting for economic equality. In El Salvador, the result was that "a growing commitment to human rights, political democracy, and economic equality brought Christian activists into conflicts with political and economic elites, and the military forces that defended them. This meant that for the first time in Latin American history, Catholics became targets of political violence for their work "on behalf of the faith."[16]

---

[12] Jeffery M. Paige, *Coffee and Power: Revolution and the Rise of Democracy in Central America* (Cambridge, MA: Harvard University Press, 1997); Edwin G. Corr, "Societal Transformation for Peace in El Salvador," *Annals of the American Academy of Political and Social Science* 541, no. 1 (September 1995): 144–156.

[13] Paige, *Coffee and Power: Revolution and the Rise of Democracy in Central America*; Robert W. Taylor and Harry E. Vanden, "Defining Terrorism in El Salvador: 'La Matanza,'" *Annals of the American Academy of Political and Social Science* 463 (September 1982): 106–118; Leonel Gómez and Bruce Cameron, "El Salvador: The Current Danger. American Myths," *Foreign Policy*, no. 43 (Summer 1981): 71–92.

[14] Corr, "Societal Transformation for Peace in El Salvador."

[15] In 1979, in the wake of the Sandinista overthrow the conservative Somoza regime in neighboring Nicaragua, a group of reformist military officers allied the with moderate Christian Democrats seized power and formed a military junta that would rule El Salvador from 1979 to 1982.

[16] Aldo Antonio Lauria Santiago, "An Agrarian Republic: Production, Politics, and the Peasantry in El Salvador, 1740–1920" (PhD diss., University of Chicago, 1992).

HISTORICAL FACTORS

There is a long history of state violence against the peasants of El Salvador, beginning as early as the 1830s. Driven by the value of agricultural land for coffee production, the expansion of large plantations has displaced subsistence farmers, concentrating the most valuable land in the hands of wealthy landlords.[17] By the 1920s, El Salvador was essentially a feudal state, where peasants worked the land for the wealthy. With the government and military firmly under the influence of the landowners, any peasant protest was quickly repressed. The cycle of protest and repression was horrifically illustrated in the 1932 massacre branded as *la matanza* (the slaughter).[18]

In December 1931, a military junta overthrew the recently elected reformist president, Arturo Araujo. The next month, about 5,000 (mostly indigenous) people organized by the Communist Party launched an uprising centered in the western provinces of Sonsonate and Ahuachapán. The rebels took over and destroyed several town halls and killed fifteen to twenty people, including landlords, national guardsmen, and a retired general. The military government responded by killing not only the insurrection's participants and leaders, including Communist Party founder Farabundo Martí, but also huge numbers of people who had not participated in the rebellion. Ten thousand people died at the hands of the government.[19]

El Salvador was the most densely populated country in Central America by the 1960s,[20] and as many as 300,000 Salvadorans were living and working illegally in neighboring Honduras.[21] In 1967, spurred by violent clashes during a soccer match between the two countries, Honduras began expelling Salvadoran migrants. These actions combined with a border dispute erupted into a four-day war, appropriately called the *Futbol*, or Soccer, War. After a burst of Salvadoran patriotic fervor and an abortive invasion of neighboring Honduras, the war ended with a precarious cease-fire and an ongoing border dispute. The disruption of trade caused by the conflict ended

---

[17] Ibid.

[18] Paige, *Coffee and Power: Revolution and the Rise of Democracy in Central America*; Taylor and Vanden, "Defining Terrorism in El Salvador: 'La Matanza'"; Gómez and Cameron, "El Salvador: The Current Danger."

[19] Lauria Santiago, "An Agrarian Republic: Production, Politics, and the Peasantry in El Salvador, 1740–1920."

[20] Tommie Sue Montgomery, *Revolution in El Salvador: From Civil Strife to Civil Peace*, 2nd ed. (Boulder, CO: Westview Press, 1995); Wim Pelupessy and John F. Uggen, "Economic Adjustment Policies in El Salvador during the 1980s," *Latin American Perspectives* 18, no. 4 (Autumn 1991), 48–78.

[21] Paige, *Coffee and Power: Revolution and the Rise of Democracy in Central America*; Corr, "Societal Transformation for Peace in El Salvador."

a period of economic growth and prosperity, inaugurating a period of economic stagnation and decline.[22]

During the 1970s, fueled by widespread electoral fraud in 1972 and 1977, many activists believed that armed revolution provided the only path toward change. As repression continued, the country became more polarized. In 1979, bowing to Soviet pressure, the Salvadoran Communist Party broke with its traditional rejection of violence and embraced a strategy of armed resistance. At the urging of Fidel Castro and as a condition for military support from both the Soviets and Cubans, the Communist Party leader, Schafik Jorge Handal, began negotiations with the various other communist insurgent groups to form a unified organization.[23]

In addition to the Salvadoran Communist Party founded in 1930, numerous communist insurgency groups had formed throughout the 1970s with the goal of establishing a Cuban-style communist state in El Salvador.[24] The Popular Liberation Front (FPL) was founded in 1970 by Cayetano Carpio, the former Secretary-General of the Communist Party, after he was expelled for advocating a Maoist-style protracted people's war.[25] After training in Cuba and Vietnam, Carpio began a guerrilla war against the Salvadoran government. By 1979, his FPL had 50,000–80,000 members.[26] Founded in 1972 by radicalized university youth, the *Ejército Revolucionario del Pueblo* (People's Revolutionary Army, or ERP), under the leadership of Joaquín Villalobos,[27] boasted 2,000 guerrilla fighters.[28] Inclined toward mass popular protest rather than armed struggle, the 100,000-member Armed Forces of the National Resistance (FARN) was formed by moderates in the ERP in 1975.[29] The fifth organization, the Revolutionary Party of Central

---

[22] Ibid.

[23] Thomas Sheehan, "Recent Developments in El Salvador," *The Threepenny Review*, no. 16 (Winter 1984), 10–11; Robert H. Dix, "Why Revolutions Succeed & Fail," *Polity* 16, no. 3 (Spring 1984), 423–446.

[24] Cynthia McClintock, *Revolutionary Movements in Latin America: El Salvador's FMLN & Peru's Shining Path* (Washington, DC: United States Institute of Peace Press, 1998); Hugh Byrne, *El Salvador's Civil War: A Study of Revolution* (Boulder, CO: L. Rienner Publishers, 1996).

[25] Montgomery, *Revolution in El Salvador: From Civil Strife to Civil Peace*; J. Michael Waller, *The Third Current of Revolution: Inside the North America Front of El Salvador's Guerrilla War* (Lanham, VA: University Press of America, 1991).

[26] Montgomery, *Revolution in El Salvador: From Civil Strife to Civil Peace*.

[27] Joaquín Villalobos, though, had vision of a Communist state uniting all of Central America.

[28] William Bollinger, "Villalobos on 'Popular Insurrection,'" *Latin American Perspectives* 16, no. 3, (Summer 1989): 38–47.

[29] The impetus for this split was the assassination of Roque Dalton, El Salvador's leading poet and a communist supporter, at the bequest of Villalobos. James LeMoyne, "El Salvador's Forgotten War," *Foreign Affairs* 68, no. 3 (Summer 1989): 105–125.

American Workers (PRTC), was a small cadre of urban activists and terrorists that severed ties to companion organizations across Central American in 1980.

In 1980, after negotiation in Cuba between these disparate organizations, the FMLN asserted itself as the single revolutionary party in El Salvador. To support its revolutionary goals, the FMLN was able to secure economic and military aid, financing, and training from the Soviet Union, Libya, Cuba, Nicaragua, and even radical groups in the United States.[30] Although some scholars have viewed the development of the FMLN and the subsequent civil war as an inevitable grassroots union of peasant organizations, labor unions, and Christian-based communities in the wake of government oppression and rampant social injustice, this view ignores the reality that the FMLN, as well as right-wing opposition groups, was organized and mobilized by elites such as Carpio, Villalobos, Zamora, and Handal.[31] Indeed, much of the leadership of the FMLN was made up of university students who had been activists promoting economic and social justice in El Salvador and were radicalized when the military foreclosed democratic modes of political opposition.[32] These elites rejected moderation that could have prevented the civil war.

With the economy in steep decline, grievous social and economic disparities, increasing violence, greater polarization between the military and landowners who sought to maintain the status quo, an increasingly unified opposition centered on the FMLN, and a weak government unable to address these issues, El Salvador was ripe for war. The spark came in March 1980, when Roman Catholic Archbishop Oscar Romero, a vocal opponent of the social injustice and growing violence,[33] was murdered during mass by a right-wing death squad.[34] During his funeral, television news cameras captured members of the military who opened fire on unarmed mourners and demonstrators. Although these events were condemned by the United States and the international community, the ranks of those in armed opposition to

---

[30] Richard D. Newton, "The Seeds of Surrogate Warfare," *Joint Special Operations University and the Strategic Studies Department* 09, no. 3 (February 2009): 1.

[31] Yvon Grenier, *The Emergence of Insurgency in El Salvador* (Pittsburgh: University of Pittsburgh Press, 1999).

[32] McClintock, *Revolutionary Movements in Latin America: El Salvador's FMLN & Peru's Shining Path.*

[33] This included the assassination of six Catholic priests, who were targeted for their social activism.

[34] See *Governing Environment* and *Methods of Obtaining Legitimacy* sections for a fuller discussion.

the government swelled, and violence dramatically escalated. By 1981, the country was in civil war.[35]

GOVERNING ENVIRONMENT

On the eve of civil war, El Salvador had been effectively governed by the military since 1931, supported by the landed elites. The military regime openly supported the fascist ideology during World War II, yet by the 1950s, a cadre of younger reform-minded officers, led by José María Lemus, initiated joint military–civilian rule.[36] This reformist movement was opposed by the landed elites and more conservative elements of the military, who levied charges of covert communism against the government. Precipitated by the rise of Fidel Castro in Cuba in 1959, conservative military officers seized power in a coup in January 1961.[37]

By 1970, declining economic fortunes, gross economic disparities, and the absence of a viable democratic opposition to the military regime caused growing political unrest. Modeled on the Christian Democratic Party that had emerged in Chile in the 1960s and rooted in the principles of Catholic social justice, the moderate Christian Democratic Party was formed, mainly by the middle and upper classes who championed economic growth, political stability, and moderate reforms.[38] In 1972, Christian Democratic presidential candidate José Napoleón Duarte was poised for an election victory when the military declared its candidate, Colonel Arturo Molina, the winner, even though Duarte had received the majority of the popular vote. In the wake of an attempted coup by reformist military officers in support of Duarte, he was arrested, tortured, and exiled.

After 1972, the military repression proliferated, provoking a climate of revolution. In light of growing dissent, the military government formed the Democratic Nationalist Organization (ORDEN), a covert paramilitary group.[39] Concurrently, death squads, composed of

---

[35] Anna Lisa Peterson, *Martyrdom and the Politics of Religion: Progressive Catholicism in El Salvador's Civil War* (Albany: State University of New York Press, 1997); Montgomery, *Revolution in El Salvador: From Civil Strife to Civil Peace;* Michael McClintock, *State Terror and Popular Resistance in El Salvador,* vol. 1 (London: Zed Books, 1985).

[36] Grenier, *The Emergence of Insurgency in El Salvador;* Paige, *Coffee and Power: Revolution and the Rise of Democracy in Central America.*

[37] Ibid.

[38] Bruce Campbell and Arthur David Brenner, *Death Squads in Global Perspective: Murder with Deniability* (New York: St. Martin's Press, 2000); Paige, *Coffee and Power: Revolution and the Rise of Democracy in Central America;* Montgomery, *Revolution in El Salvador: From Civil Strife to Civil Peace.*

[39] Campbell and Brenner, *Death Squads in Global Perspective: Murder with Deniability.*

former and current members of the military and police, financed by the wealthy elites, and tied to the military government, specifically ORDEN, appeared. Political assassinations and disappearances became commonplace.[40] Even overt repression of dissonance increased as in July 1975 when demonstrators in San Salvador were fired upon by the military.[41]

During the elections of 1972, with democratic avenues of opposition apparently closed, armed insurgent groups, such as the FPL and ERP, committed acts of sabotage and terrorism.[42] To address the burgeoning unrest, the government enacted minor land reforms, but it refused to challenge the land monopoly of the agricultural elites and, moreover, the reforms were not enforced. Meanwhile, opposition groups were being mobilized by secular leftist revolutionaries, as well as Catholic priests who were influenced by the growing prominence of liberation theology and established subsistence farming collectives for the rural poor.[43]

By 1979, the country was in near anarchy. Mass demonstrations of 100,000 people became common, protesters virtually besieged government ministries and large businesses, and bombing became a nightly occurrence.[44] In October 1979, a coup led by reformist military officers seized power and named José Napoleón Duarte, returned from exile, as provisional president in 1980. Duarte called for the election of a Constituent Assembly that would create a new

---

[40] Aldo Lauria-Santiago and Leigh Binford, *Landscapes of Struggle: Politics, Society, and Community in El Salvador* (Pittsburgh: University of Pittsburgh Press, 2004); Paige, *Coffee and Power: Revolution and the Rise of Democracy in Central America;* Byrne, *El Salvador's Civil War: A Study of Revolution.*

[41] Robert E. White, "Preliminary Assessment of Situation in El Salvador" (US Department of State, 1980) summarizes Salvadoran society from this time frame: "The major, immediate threat to the existence of this government is the right-wing violence. In the city of San Salvador, the hired thugs of the extreme right, some of them well-trained Cuban and Nicaraguan terrorists, kill moderate left leaders and blow up government buildings. In the countryside, elements of the security forces torture and kill the campesinos, shoot up their houses and burn their crops. At least two hundred refugees, from the countryside arrive daily in the capital city. This campaign of terror is radicalizing the rural areas just as surely as Somoza's National Guard did in Nicaragua. Unfortunately, the command structure of the army and the security forces either tolerates or encourages this activity. These senior officers believe, or pretend to believe, that they are eliminating the guerillas."

[42] McClintock, *Revolutionary Movements in Latin America: El Salvador's FMLN & Peru's Shining Path;* Sewall H. Menzel, *Bullets Versus Ballots: Political Violence and Revolutionary War in El Salvador, 1979–1991* (New Brunswick: Transaction Publishers, 1994); Waller, *The Third Current of Revolution: Inside the North America Front of El Salvador's Guerrilla War.*

[43] Peterson, *Martyrdom and the Politics of Religion: Progressive Catholicism in El Salvador's Civil War;* Montgomery, *Revolution in El Salvador: From Civil Strife to Civil Peace;* McClintock, *State Terror and Popular Resistance in El Salvador.*

[44] Corr, "Societal Transformation for Peace in El Salvador."

constitution and pave the way for democratic elections in 1982.[45] Within the prevailing climate of violence, the desire for law and order gave the majority of seats in the assembly to the National Republican Alliance (ARENA) party, which represented the conservative, military, and landed interests. Its leader, Roberto D'Aubuisson, was a former intelligence officer with ties to the death squads and had purportedly ordered the assassination of Archbishop Romero.[46] Despite this image, D'Aubuisson tried to reach out to moderates and was elected president of the assembly.

On the eve of the 1982 presidential elections, El Salvador was clearly in a state of civil war. The military used helicopter gunships and indiscriminate aerial bombings to punish towns and villages that supported the insurgents. The FMLN began to expand out of its bases in the northern mountains toward the Pacific coast and the capital San Salvador. The US government sent military advisers and aid to the beleaguered government to stem the communist threat, concurrently calling for an end to death squad activity and human rights abuses. Yet the government, particularly the military, continued to repress all opposition with any means of violence it deemed necessary.

WEAKNESSES OF THE PREREVOLUTIONARY ENVIRONMENT AND CATALYSTS

When considering a true catalyst for the commencement of an insurgency in El Salvador, two specific conditions must be considered. First, there was a large disparity between the rich and poor, with a middle class that was effectively ceasing to exist. Second, political oppression, using the power of the government and military forces, created an environment that Major Chris Herrera, US Army Reserves, noted in his study "Why Choose Peace? The El Salvador Experience," as "inevitable."[47] Civil society, as described by Cathy McIlwaine

---

[45] Max G. Manwaring and Court Prisk, *El Salvador at War: An Oral History* (Washington, DC: National Defense University Press, Government Printing Office, 1988).

[46] Peterson, *Martyrdom and the Politics of Religion: Progressive Catholicism in El Salvador's Civil War*; Montgomery, *Revolution in El Salvador: From Civil Strife to Civil Peace*; McClintock, *State Terror and Popular Resistance in El Salvador*.

[47] M. Chris Herrera, "Why Choose Peace? The El Salvador Experience" (master's thesis, School of Advanced Military Studies, US Army Command and General Staff College, 2007), http://handle.dtic.mil/100.2/ADA485593.

in "Contesting Civil Society: Reflections from El Salvador," was fragmented "along political, geographical and social lines."[48]

Conditions for the masses were intolerable, with repression and abuse continuing for decades. The government recognized no bounds in the subjugation of the masses. As a result, unrest, primarily initiated by students from the national university, became commonplace after the 1972 presidential election, which was largely believed to have been stolen because of widespread fraud. "The thrust of the revolutionary program centered on the redress of real as well as perceived grievances and deprivations."[49]

One of the major grievances was economic disparity. Heritage Foundation policy analyst Jorge Salaverry noted the dire straits of the Salvadoran economy in a 1988 paper, *A Winning U.S. Policy is Needed in El Salvador*:

> The decline in El Salvador's economy since U.S. aid began to increase in 1979 is dramatic. More than 1,000,000 Salvadorans have no jobs (700,000 more than in 1978). Real wages are half their 1979 level. Last year's inflation rate was 25 percent, while in 1978 it was around 13 percent. Income per capita has declined to the level of 27 years ago. Real gross domestic product in 1987 was equivalent to only 70 percent of that in 1978. Assuming a population growth rate of 2.5 percent per year, the Salvadoran economy would have to grow at an average of 6.5 percent per year over the next ten years just to achieve in 1997 the per capita product of 1978.

Joaquín Villalobos, leader of the ERP, one of the five fighting factions of the FMLN, recognized the economic disparity as a central grievance, and noted:

> The objective conditions of poverty and the lack of a solution set a clear course toward social upheaval . . . The Salvadoran people have a tradition of organization and struggle, an ability to conspire, and have endured a wide variety of experiences. We have only to recall the

---

[48] Cathy McIlwaine, "Contesting Civil Society: Reflections from El Salvador," *Third World Quarterly* 19, no. 4 (1998): 651–672. In El Salvador in the 1970s and 1980s, the authoritarian state used military aggression to squash popular dissent, which was growing in response to deteriorating social and economic conditions. The initial targets were representatives of various civil society organizations, such as trade union leaders, community organizers, cooperatives, and, especially, church leaders. As the war escalated in the 1980s, this assault continued, taking on a distinctly geographical character.

[49] M. G. Manwaring and Court Prisk, "A Strategic View of Insurgencies: Insights from El Salvador," *Small Wars and Insurgencies* 4, no. 1 (Spring 1993): 53–72.

peasant insurrection of 1932; the patriotic uprising of
the army combined with the mass uprising in 1944;
the student struggles of 1960 and the *coup d'etat* that
followed it; the great workers' and teachers' strikes
of the 1960s; the electoral struggles of the 1970s and
the great revolutionary surge of the masses in the
1980s . . . It is no coincidence that the most complex
revolutionary popular war in Latin America has unfolded
in El Salvador. This can only be explained by the depth
of a class struggle generated by the endemic misery of a
heavily populated country lacking in resources.[50]

In "El Salvador's Forgotten War," James LeMoyne succinctly
captures the essence of the conditions that support violent unrest and
insurgency:

El Salvador is at war because it is one of the sickest
societies in Latin America. Its archaic social structure
remains basically colonial. Despite some efforts at
change, a tiny urban elite and dominating cast of army
officers essentially rule, but do not effectively govern,
an illiterate, disease-ridden and frustrated majority
of peasants and urban slum-dwellers. Order is often
imposed by violence; there is not now, nor has there
ever been, a just legal system. The rebels, in short, have
had ample cause to lead a revolution.[51]

FORM AND CHARACTERISTICS OF THE REVOLUTION

OBJECTIVES AND GOALS

The strategic goal of the FMLN was the overthrow of the repressive
regime and reinstatement of a communist political body aligned with
Cuba and the Soviet Union. It was this strategic goal that was the
impetus for continued US engagement throughout years of the FMLN
insurgency. Specific FMLN objectives included organizing and winning
broad popular support via indoctrination and widespread political
activity; coordinating and sustaining the military actions of the five
main factions across the country, including in urban operations within

---

[50] Joaquín Villalobos, "A Democratic Revolution for El Salvador," *Foreign Policy*, no. 74 (1989): 103–122.

[51] LeMoyne, "El Salvador's Forgotten War."

the capital; establishment of legitimacy with external organizations and governments; and a bleeding of government resources and personnel over the long haul, leading to the eventual collapse of the oligarchic regime. José Angel Moroni Bracamonte and David Spencer discussed the military strategy in *Strategy and Tactics of the Salvadoran FMLN Guerillas: Last Battle of the Cold War, Blueprint for Future Conflicts*:

> The objective . . . was not annihilation or the capture of large numbers of weapons, but rather to inflict a slow, steady stream of casualties on the government troops while paying the lowest cost possible in ammunition, resources, and blood. While in the short run numbers of government casualties would be low, over time the casualties would add up, and instead of being inflicted in open combat, they would be inflicted by an enemy that was rarely seen. Soldiers would never be sure when or where they would be shot at, step on a mine, or run into an ambush. Mental tension would be constant and high, severely affecting morale.[52]

Joaquín Villalobos also commented on the military attrition strategy of the FMLN:

> In this war of attrition there is an average of one ambush a day—on all the strategic roads and highways, and now in the capital and the other cities. As a result of the "wear and tear" tactics of the FMLN, the army has more than 4,500 people crippled by the war [as of 1989]. The FMLN has ground-destroyed more than 60 air force units. Likewise, hundreds of trucks and armored vehicles have been destroyed in combat. Dozens of garrisons and military installations have been burned to ashes. The FMLN has access to all roads; it has positions in thirteen of the fourteen departments. In all of the mountain ranges of the country and in the heights and volcanic ranges near the cities, there are permanently armed guerillas, and now there are urban commandos in practically all of the cities.

> By blowing up tens of bridges, including the two most important ones, the FMLN has obstructed the counterinsurgent economic plan with a strategy

---

[52] José Angel Moroni Bracamonte and David E. Spencer, *Strategy and Tactics of the Salvadoran FMLN Guerrillas: Last Battle of the Cold War, Blueprint for Future Conflicts* (Westport, CT: Praeger, 1995).

of sabotage. It has also destroyed dozens of coffee, sugarcane, and cotton installations, including all of the largest and most important ones; and it has constantly interrupted the electrical system in more than 80 percent of the country. From the military point of view, the FMLN has been able to sustain a wide offensive drive, modifying its strategy according to political circumstances.

There is no precedent in America for a destabilizing revolutionary military pressure like the one that the FMLN has carried out. If it were not for US support, the government and the army would have collapsed a long time ago. The war can be described as a game of chess in which the FMLN has constantly checked the army, which has always been saved by US support. The problem is that, under the current situation, the FMLN has the decisive piece in its hands—the masses—and it is going to use them for the checkmate.[53]

Joaquín Villalobos discussed the elements of the revolution as[54]

1. The FMLN's military offensive;
2. The insurrectional process of the masses in both the cities and the countryside;
3. Generalized repression;
4. Political disintegration of the government and the armed forces; and
5. Weakening of US policies and its instruments in El Salvador.

In the early 1980s, the FMLN advanced the dual goals of rectifying the social and economic injustice in the country and establishing a communist regime akin to those in Cuba and Nicaragua. By the late 1980s, however, FMLN leaders increasingly favored being included as a legal participant in El Salvador's democratic system.[55]

---

[53] Although Villalobos clearly wrote with an element of bravado, the FMLN was achieving operational success, effectively cutting the country in two by destroying all the main bridges that crossed the Río Lempa, the economic lifeline that was the US equivalent of the Mississippi River. The army was facing determined opposition, and the FMLN had widespread popular support in the rural northern and eastern departments of the country. Joaquín Villalobos, "Popular Insurrection: Desire or Reality?", *Latin American Perspectives* 62, no. 3 (Summer 1989): 5–37.

[54] Ibid.

[55] Corr, "Societal Transformation for Peace in El Salvador."

LEADERSHIP AND ORGANIZATIONAL STRUCTURE

Despite the pretensions of a single united opposition organization, the FMLN was, in fact, a confederate of distinct constituencies drawn together in their opposition to military repression, economic injustice, and the death squads.[56] It included the five communist groups united in Cuba in 1980, student activists, and disenfranchised Christian Democrats.[57] The membership disparity of this confederation was also mirrored in the group's agenda, where hard-line communists favored a protracted people's war ending with a Cuban-style communist state, while moderates sought an eventual peace settlement that would provide for democratic participation toward establishing a social democratic regime.[58] This division was illustrated in a bizarre incident in 1984, when a debate broke out in the FPL between its leader Cayetano Carpio, who had favored a protracted people's war since his time as Secretary-General of Communist Party, and his former mistress and second in command, Mélida Anaya, who, at the suggestion of Nicaragua and Cuba, advocated for negotiation and democratic participation.[59] The dispute ended when Carpio had Anaya murdered and then killed himself. By the late 1980s this division had become acute, although by this point both victory in the civil war and a general election victory were increasingly remote.[60]

The FMLN had no centralized military command, and the five insurgent groups that had formed the organization maintained their own political and military structures, as well as their own command and control apparatuses, making coordination difficult.[61] The leadership organ of the FMLN was the general command, which was an executive committee for coordinating joint operations and consisted of the military commanders of the five principal constituent

---

[56] Moroni Bracamonte and Spencer, *Strategy and Tactics of the Salvadoran FMLN Guerrillas: Last Battle of the Cold War, Blueprint for Future Conflicts.*

[57] McClintock, *Revolutionary Movements in Latin America: El Salvador's FMLN & Peru's Shining Path.*

[58] Ibid.

[59] Grenier, *The Emergence of Insurgency in El Salvador;* Byrne, *El Salvador's Civil War: A Study of Revolution.*

[60] Montgomery, *Revolution in El Salvador: From Civil Strife to Civil Peace.*

[61] Todd R. Greentree, *Crossroads of Intervention: Insurgency and Counterinsurgency Lessons from Central America* (Westport, CT: Praeger Security International, 2008).

groups.[62] Decisions were made by simple majority vote,[63] guided by the Leninist principle of democratic centralism.[64] An attempt to create the National Democratic Army (*Ejercito Nacional Democraticia*), a unified military command, was made in 1989 and 1990 without results.

Infighting between the five constituent groups was well known;[65] indeed, by some accounts these groups hated one other more than they hated the government of El Salvador.[66] This infighting was never exploited by the Armed Forces of El Salvador (ESAF), and none of the groups left the FMLN during the twelve-year civil war. In fact, the general command was an astute agent for coordinating strategy and military operations, as well as providing military provisions for guerrilla fighters.[67]

In 1981, the FMLN created the Political-Diplomatic Commission (CPD) to encourage international support and recognition, particularly among Western nations.[68] The CPD had agents, de facto ambassadors, in thirty-three countries, which was more agents than the government of El Salvador maintained in its diplomatic missions. FMLN's recognition by France and Mexico as a "representative political force" in 1981 revealed the fruit of these efforts.[69]

---

[62]   McClintock, *Revolutionary Movements in Latin America: El Salvador's FMLN & Peru's Shining Path*; Menzel, *Bullets Versus Ballots: Political Violence and Revolutionary War in El Salvador, 1979–1991*; Waller, *The Third Current of Revolution: Inside the North America Front of El Salvador's Guerrilla War.*

[63]   Although these commanders were equal in principle, Villalobos was counted as the "first among equals," because militarily the ERP was most adept and was the favored recipient of arms and other aid from Nicaragua and Cuba. Greentree, *Crossroads of Intervention: Insurgency and Counterinsurgency Lessons from Central America.*

[64]   Democratic centralism states that regardless of the presence of real disagreements, one party's position is decided by majority vote, and members will abide by the decision and refrain from public opposition or disunity. Thus, in principle, the FMLN has a single command, military plan, and political position, but this was not strictly observed by the constituent groups. Richard Stahler-Sholk, "El Salvador's Negotiated Transition: From Low-Intensity Conflict to Low-Intensity Democracy," *Journal of Interamerican Studies & World Affairs* 36, no. 4 (1994): 1–59; Ruben Zamora Rivas and Schafik Jorge Handal, "Proposal of the FMLN/FDR," *Latin American Perspectives* 14, no. 4, (Autumn 1987): 481–486.

[65]   Stahler-Sholk, "El Salvador's Negotiated Transition: From Low-Intensity Conflict to Low-Intensity Democracy."

[66]   Herrera, "Why Choose Peace? The El Salvador Experience."

[67]   Greentree, *Crossroads of Intervention: Insurgency and Counterinsurgency Lessons from Central America*; Richard Stahler-Sholk, "Central America: A Few Steps Backward, a Few Steps Forward," *Latin American Perspectives* 26, no. 2 (March 1999): 3–12.

[68]   Montgomery, *Revolution in El Salvador: From Civil Strife to Civil Peace.*

[69]   The inability of the FMLN to make territorial gains after 1982 meant that "the nations that had toyed with the idea of recognizing the FMLN backed off" and the recognition of France and Mexico did not materialize into real support. Moroni Bracamonte and Spencer, *Strategy and Tactics of the Salvadoran FMLN Guerrillas: Last Battle of the Cold War, Blueprint for Future Conflicts.*

After a brief retreat into their mountain sanctuaries after the failure of the 1981 offensive,[70] the guerrillas reemerged in 1982, operating throughout the countryside and also infiltrating the cities, where they carried out terrorist attacks and formed death squads to counter those on the right.[71] As a result of search and destroy missions by the ESAF, the insurgents abandoned larger formations in favor of smaller more mobile units, although they retained the ability to mobilize larger forces when needed, such as to assault an ESAF base.[72] Most of the FMLN constituent groups had decentralized command structures, allowing for semiautonomy among tactical units.[73]

COMMUNICATIONS

Tactical communications between units was generally conducted by a runner or a tactical radio. Radios were provided from Cuba and the Soviet Union or stolen from raids against the Salvadoran army. Radios became the lifeblood of tactical communications for the FMLN as early as 1982.[74]

David E. Spencer studied the tactical communications of FMLN Special Forces. He researched FMLN papers that had been captured and also conducted first-person interviews with participants. He notes how communications would work in support of a tactical engagement:

> Communications between the unit leader and his groups were done through runners. The teams did not normally carry radios onto the objective. Communications outside the objective, between the unit leader and the overall leader or supporting regular forces, were done by radio. The distances involved outside the objective made sending runners for personal contact communication prohibitive and impractical. After the team leader reported to the unit leader, this information was transmitted to the higher echelons.[75]

---

[70] See *Methods of Action and Violence* section for a fuller discussion.

[71] McClintock, *Revolutionary Movements in Latin America: El Salvador's FMLN & Peru's Shining Path.*

[72] Byrne, *El Salvador's Civil War: A Study of Revolution*; Moroni Bracamonte and Spencer, *Strategy and Tactics of the Salvadoran FMLN Guerrillas: Last Battle of the Cold War, Blueprint for Future Conflicts.*

[73] Greentree, *Crossroads of Intervention: Insurgency and Counterinsurgency Lessons from Central America*; David E. Spencer, *From Vietnam to El Salvador: The Saga of the FMLN Sappers and Other Guerrilla Special Forces in Latin America* (Westport, CT: Greenwood Publishing Group, 1996).

[74] Spencer, *From Vietnam to El Salvador: The Saga of the FMLN Sappers and Other Guerrilla Special Forces in Latin America.*

[75] Ibid.

The FMLN also developed an extensive propaganda apparatus. Propaganda efforts included pamphlets, posters, newspaper distribution, and video; but the most effective means of dissemination was the clandestine radio networks, named *Farabundo Martí* and *Radio Venceremos*. *Radio Venceremos* operated via shortwave radio and had a support structure of approximately 100 insurgents who were responsible for security, production, and program writing. Programming included the playing of motivational music, reading of the news, and even delivery of Sunday Catholic mass.[76]

METHODS OF ACTION AND VIOLENCE

The insurgency, considered a civil war by the majority of Salvadorans, became extremely violent throughout its existence, with accusations of human rights abuses, including the indiscriminate killing of noncombatants, raised by both sides against the other. Elisabeth J. Wood noted that nearly one in fifty-six Salvadorans died as a result of El Salvador's civil war, a figure comparable to that of casualties in the United States during the Civil War (one in fifty-five) and of Britain in World War I (one in fifty-seven). About two-thirds of those killed were civilians, with a few killed in the cross-fire of battle, but the great majority were killed intentionally. The war resulted in major population movements, reversing the pre-war urban to rural ratio of 40:60. Most violence took place in areas controlled or contested by insurgent forces.[77]

During the course of the war, the FMLN developed increasingly proficient tactical forces, often coordinating complex assaults on ESAF and its installations with multiple battalions. During 1989 and 1990, some of the most elite battalions of the ESAF were reportedly "chased" out of the country and into Honduras. Thomas Sheehan, in "El Salvador: The Forgotten War" captured the essence of FMLN military capabilities in his account of an engagement between the ESAF and FMLN in Suchitoto during November, 1983:

---

[76] Different methods were used to communicate the same or similar messages. The RAND Corporation, led by Christina Meyer, conducted a study for the Under Secretary of Defense for Policy entitled *Underground Voices: Insurgent Propaganda in El Salvador, Nicaragua and Peru*. The study notes that the ERP propaganda plan "details methods by which propaganda can be put to work to 'awaken the people's consciences' by promoting the view that class structure is the root of political as well as socioeconomic ills, to denounce the Salvadoran government and army, and to convince the people that the FMLN's plan is the solution to their specific woes." The FMLN even encouraged sympathizers to organize groups to listen to and discuss radio broadcasts. C. Meyer, *Underground Voices: Insurgent Propaganda in El Salvador, Nicaragua and Peru* (New York: RAND Corporation, 1991).

[77] Wood, *Insurgent Collective Action and Civil War in El Salvador*.

Well armed and numbering close to a thousand, the guerrillas planned their surprise attack on the town of Suchitoto for the pre-dawn hours of Friday, November 9. For two days they had been slipping out of the surrounding hills in groups of twenty to twenty-five—from Chalatenango to the north, Cuscatlan to the west, and Cabanas to the east—hiding out among the peasants during the day while at night edging toward their goal, the army's outpost on the south shore of the Cerron Grande lake in the center of El Salvador. Even though split up into small groups, the guerrillas belonged to three crack battalions . . . By dusk on November 8 more than thirty of these small groups, most of them wearing blue jeans and baseball caps, had converged on the outskirts of Suchitoto, and in a matter of hours they had regrouped and reunited under their battalion commanders. Using the cover of darkness, advance units . . . skipped past the sentries and entered the sleeping town. At precisely 2:00 a.m. the attack began. Using field radios, the advance units inside Suchitoto called in mortar salvos with deadly accuracy. Sixty-millimeter shells, fired from American-made M-19 mortars, crashed into the headquarters of the National Police near the town, causing twenty-six casualties in the first minutes of the battle. Advancing through the cobblestone streets of the town, guerrillas opened fire with their captured M-16s. Army units within the strongly fortified town and Civil Defense units on the periphery were pinned down for almost four hours, and they began to take heavy losses. It was after dawn before air-transported reinforcements could reach Suchitoto . . . and when they arrived, the guerrillas were ready for them. The rebels had all but surrounded the town's helicopter pad, and as the Hueys landed, the guerrillas trained their American-made M-60 machine guns and M-70 grenade launchers on the helicopters, putting two of them out of commission and badly damaging three others. The battle raged until noon, by which time the Salvadoran Air Force called in A-37 Dragonfly jets which strafed and bombed the outskirts of the town. Before withdrawing, the guerrillas damaged one of the jets. Even though Suchitoto is in the heart of territory that the Salvadoran military claims to control, the guerrilla force, again by breaking down into small company-size units, managed to slip away during the daylight hours without any effective pursuit by the Army or Air Force.[78]

During the early and mid-1980s, ESAF members were not trained or equipped to deal with an active, popular insurgency. Starting with a traditional Latin American standing army of 11,000–15,000, the force was transformed into an organization of closer to 60,000 soldiers, fully supported by the United States for training and equipping. The Reagan and George H. W. Bush administrations cited El Salvador as a top priority, engaging with the Salvadoran government for training, equipment, intelligence, and adviser support. This support included close coordination on strategic and operational planning.[79] The ESAF, however, did not always maintain tactical or operational advantage, with several offensive campaigns by the FMLN achieving success.[80]

The FMLN developed special and conventional military capabilities, which included assassinations; direct attacks on military *cuartels* (garrisons), including the Ministry of Defense; urban operations; special operations; and sophisticated attacks on infrastructure targets, from telephone poles to dams. In fact, the FMLN effectively cut off power to the capital for weeks during the late 1980s/early 1990s, degrading the morale of the populace while diminishing the legitimacy of the government. From 1980 through 1987, the FMLN caused an estimated $2 billion in direct and indirect damages to the Salvadoran economy and infrastructure. US economic aid during that same time frame was $2.1 billion.[81]

Military strategy for the FMLN shifted somewhat during 1983 and 1984. Binford notes the "shift in FMLN strategy from one of massed forced aimed at enlargement of territorial control to a mobile 'war of resistance' designed to defeat the 'low-intensity warfare' designed and financed by the United States . . . the new strategy emphasized guerrilla warfare conducted by small mobile units and involving ambushes, economic sabotage, and political action."[82] In fact, increasing military assistance and employment of revised operations caused a reaction and adjustment by the FMLN, but defeat as a result of a new government strategy was in no way imminent. Banks were

---

[78] Sheehan, "Recent Developments in El Salvador."

[79] Ibid.

[80] Manwaring and Prisk, *A Strategic View of Insurgencies: Insights from El Salvador*

[81] Sam Dillon, "Dateline El Salvador: Crisis Renewed," *Foreign Policy*, no. 73 (Winter 1988): 153–170.

[82] Leigh Binford, "Grassroots Development in Conflict Zones of Northeastern El Salvador," *Latin American Perspectives* 24, no. 2 (March 1997): 56–79.

also a significant target of the FMLN, but their goal was not to loot the banks or raise money; rather, the FMLN's goal was to hinder the economy, attacking the legitimacy of the state while also discouraging commerce at the local level.

In the wake of the failure of the 1989 offensive to topple the government, it became increasingly clear to both hard-liners who favored a protracted people's war and to moderates within the FMLN that a military victory would not be possible.[83]

METHODS OF RECRUITMENT

There was never widespread support for the communist ideology of the FMLN and support for the insurgents ranged from indifference to tacit support for pragmatic reasons.[84] In regions under FMLN control, the insurgents were able to gain support from local populations by providing immediate and promised benefits.[85] However, the guerrillas were also willing to employ violence when necessary to ensure support, which alienated the peasantry. On the other hand, particularly before 1985, the peasants viewed the government as even worse because of their perception that it placed the interests of the landed elites above those of the people, as well as its long history of egregious human rights violations.[86] These violations included collusion with the death squads, the excessively brutal treatment of peasants by the ESAF as they executed the civil war, and the ESAF's absolute disregard for property destruction and civilian causalities. The result was a climate ripe for FMLN recruitment.[87]

Among the middle class, the FMLN received recruits from popular organizations, such as the Democratic Revolutionary Front (FDR), which had been co-opted by guerrilla cells as platforms for propaganda and logistical support, as well as the national university, whose leadership staunchly promulgated an extreme leftist ideology. Moreover, the prominence of Catholic liberation theology radicalized not only rural peasants but also the urban middle class.[88]

---

[83]   Montgomery, *Revolution in El Salvador: From Civil Strife to Civil Peace.*

[84]   McClintock, *Revolutionary Movements in Latin America: El Salvador's FMLN & Peru's Shining Path.*

[85]   See *Methods of Obtaining Legitimacy* for a fuller discussion.

[86]   Corr, "Societal Transformation for Peace in El Salvador."

[87]   McClintock, *Revolutionary Movements in Latin America: El Salvador's FMLN & Peru's Shining Path,* 492; Binford, "Grassroots Development in Conflict Zones of Northeastern El Salvador."

[88]   Grenier, *The Emergence of Insurgency in El Salvador.*

In provinces under insurgent control, the FMLN served as a de facto government providing order and even some limited social services.[89] This reality, along with a dearth of economic opportunity and the fact that fighters and their families enjoyed a level of distinction, served to aid insurgent recruitment in the early 1980s.[90] With a notable exception from 1983 to 1984, the FMLN generally avoided forced conscription, believing it to be a counterproductive strategy.[91] Nonetheless, force recruitment did damage the public support for the FMLN.[92]

By 1985, the inability of the FMLN to move beyond stalemate, a shift in tactics by the insurgents that increased their own responsibility for human rights violations and civilian causalities, and the economic and political reform of the government President José Napoleón Duarte (1984–1989) began to erode public support for the FMLN.[93] Despite its continued appearance of military strength and an active campaign of recruitment, the FMLN faced growing attrition, which reduced the number of guerrilla fighters by half between 1982 and 1989.[94] Despite these setbacks throughout most of the country, the growing urban presence of the FMLN in the late 1980s did aid recruitment in cities, particularly San Salvador.[95]

METHODS OF SUSTAINMENT

Although their primary source of military funding and supplies came from the Soviet Union and its allies, particularly Nicaragua and Cuba, the FMLN did attempt to get weapons internally. The insurgents established underground factories that made weapons, ammunition (including improvised explosive devices), and supporting material. Looting from the military and even simple recovery from abandoned military, police, and civil facilities were also effective means of acquiring needed supplies. By 1983 attacks on isolated army bases had increased so much that "the guerillas could make credible claims that most of their weapons, including even mortars and other artillery pieces, came from the United States by way of captured government

---

89   Binford, "Grassroots Development in Conflict Zones of Northeastern El Salvador."
90   Ibid.
91   In contrast the ESAF was infamous for its forced conscription. One ESAF strategy was to hold a youth event, such as a dance, and then force the attendees onto buses. Ibid.
92   Moroni Bracamonte and Spencer, *Strategy and Tactics of the Salvadoran FMLN Guerrillas: Last Battle of the Cold War, Blueprint for Future Conflicts.*
93   Ibid.
94   Greentree, *Crossroads of Intervention: Insurgency and Counterinsurgency Lessons from Central America.*
95   Wood, *Insurgent Collective Action and Civil War in El Salvador.*

troops."[96] The FMLN adopted the policy of burying weapons caches throughout the countryside, avoiding the establishment of large supply depots. In that regard, a single capture by the ESAF would not significantly cripple military operations. Concurrently, the FMLN did impose "revolutionary taxes," particularly on the sale of coffee in zones that it controlled. Generally, however, taxes on individuals were not encouraged in order to maintain strong relationships with the impoverished peasant class, which were the base of the FMLN support.

METHODS OF OBTAINING LEGITIMACY

In the Salvadoran civil war, legitimacy was a scarce resource for which the government and the FMLN competed both domestically and internationally. The FMLN did not provoke a popular uprising to take away power from the government—but this was not because the majority of Salvadorans supported the government but rather because the majority did not support the insurgency.[97] The people were caught between a government that was complicit in gross human rights violations while simultaneously offering democracy and moderate economic reforms and an insurgency that promised to right the abuses of the government and establish an egalitarian utopia while accumulating its own record of abuses. They doubted that an FMLN regime would be any better than the status quo.

The right-wing death squads, which epitomized human rights violations by the government, first appeared around 1975.[98] Although officially clandestine, these death squads were composed of former and current members of the military and police, financed by members of the landed elite and the business community, and organized by right-wing elements inside and outside the government, including the covert paramilitary force known as ORDEN, the National Guard, and the Treasury Police.[99] The purpose of the death squads was to create a climate of terror that would discourage any opposition to the dominance of the military and landed elites.

---

[96] Christopher Dickey, "Central America: From Quagmire to Cauldron," *Foreign Affairs and the World* 62, no. 3 (1983).

[97] Greentree, *Crossroads of Intervention: Insurgency and Counterinsurgency Lessons from Central America.*

[98] James R. Crouch Sr., "The Decade of the Seventies in El Salvador: Prelude to Revolution" (master's thesis, US Army Command and General Staff College, 1992).

[99] Campbell and Brenner, *Death Squads in Global Perspective: Murder with Deniability;* Douglas Payne, *El Salvador Re-Emergence of "Social Cleansing" Death Squads* (Washington DC: INS Resource Information Center,1999); T. David Mason and Dale A. Krane, "The Political Economy of Death Squads: Toward a Theory of the Impact of State-Sanctioned Terror," *International Studies Quarterly* 33, no. 2 (June 1989): 175–198; Raymond Bonner, *Weakness and Deceit: U.S. Policy and El Salvador* (New York: Times Books, 1984).

To that end, the death squads carried out the assassination, torture, and abduction of thousands. The US embassy noted that 750 civilians were killed by political violence every month in 1980.[100] In 1981, 12,000 victims were attributed to the death squads, and the pattern of killings and disappearances continued through the 1980s.[101] The killings became so prolific that public service announcements were run on radio and television stations requesting that no more corpses be thrown in Lake Ilopango, the source of fresh water for San Salvador, because of concerns of contamination. To augment the terror, lists of assassination targets were announced on radio and television. Yet as the violence became ubiquitous, its perceived connection to support for the left became obfuscated and ceased to function as a deterrent. Instead, the violence encouraged support for the FMLN and diminished the legitimacy and, therefore, Salvadorans' support for the government.[102]

One cannot understate the impact of government collusion, both by its failure to stop the death squads and in some cases its direct support for them, on the competition for legitimacy between it and the FMLN. Major Roberto D'Aubuisson, the founder of the conservative ARENA party, dominated Salvadoran politics during the civil war (with the exception of the presidency of Christian Democrat José Napoleón Duarte) and was a principal organizer of death squad activity.[103] In some instances, the death squads were the legally established organs of the government and military. The Atlacatl Battalion, a counterinsurgency unit formed at the US Army School of the Americas in Panama in 1980 and trained at Fort Bragg, North Carolina, massacred hundreds in the village of El Mozote in 1981 and murdered six Jesuit priests as well as their housekeeper and her daughter in 1989.[104]

These widely publicized events and others, such as the assassination of Archbishop Oscar Romero, the rape and murder of three American nuns and a layman in 1980, the murder of four US Marines at a café in 1985, as well as endless assassinations and abductions, crippled the legitimacy of the government domestically and internationally,

[100] Corr, "Societal Transformation for Peace in El Salvador."

[101] Campbell and Brenner, *Death Squads in Global Perspective: Murder with Deniability*; Payne, *El Salvador Re-Emergence of "Social Cleansing" Death Squads*; Mason and Krane, "The Political Economy of Death Squads: Toward a Theory of the Impact of State-Sanctioned Terror"; Bonner, *Weakness and Deceit: U.S. Policy and El Salvador.*

[102] Mason and Krane, "The Political Economy of Death Squads: Toward a Theory of the Impact of State-Sanctioned Terror."

[103] Campbell and Brenner, *Death Squads in Global Perspective: Murder with Deniability*; Payne, *El Salvador Re-Emergence of "Social Cleansing" Death Squads*; Bonner, *Weakness and Deceit: U.S. Policy and El Salvador.*

[104] Leigh Binford, *The El Mozote Massacre: Anthropology and Human Rights* (Tucson: University of Arizona Press, 1996); Bonner, *Weakness and Deceit: U.S. Policy and El Salvador.*

even those elements that officially condemned these acts. This created formidable pressure on the Reagan administration to end the training and funding of the Salvadoran military.[105] Yet because of the perceived urgency of preventing the communist threat in El Salvador, the Reagan administration discounted the Salvadoran government's involvement in these activities, dismissing the purported scale and characterization of specific incidents as being FMLN propaganda.[106] Still, even apart from its connection to the death squads, the ESAF, especially the National Guard, had a deplorable human rights record. The abuse, disappearance, or killing of a close relative by the ESAF was one of the best predictors of support for the FMLN.[107]

In 1981, the FMLN launched a general offensive, designed to incite a popular uprising that would topple the government of El Salvador. Contrary to this objective and significantly before the advent of US funding and arms shipments,[108] the ESAF was able to scatter the rebels, driving them from urban bases to mountain strongholds.[109] This failure cast doubts on the ability of the rebels to achieve a military victory, if for no other reason than because the desired popular insurrection did not materialize, revealing that the insurgents had not successfully demonstrated that the FMLN was a legitimate or desired alternative to the established government.[110]

While the FMLN could not rely on the popular support that rallied the people behind the Sandinistas in Nicaragua in their successful revolution in 1979, it did make significant efforts to curry the support of the people by establishing itself as an alternative government and by providing assistance to the peasants.[111] Fighters from the FPL and ERP regularly lived in peasant villages, where they alerted the peasants to the approach of the ESAF, trained them to defend themselves, worked alongside them in the fields, furnished medical care, and began a literacy program for which it produced teaching

---

[105] Campbell and Brenner, *Death Squads in Global Perspective: Murder with Deniability*; Payne, *El Salvador Re-Emergence of "Social Cleansing" Death Squads;* Bonner, *Weakness and Deceit: U.S. Policy and El Salvador.*

[106] Ibid.

[107] Wood, *Insurgent Collective Action and Civil War in El Salvador.*

[108] Linda Robinson, *Intervention Or Neglect: The United States and Central America Beyond the 1980s* (New York: Council on Foreign Relations Press, 1991).

[109] Menzel, *Bullets Versus Ballots: Political Violence and Revolutionary War in El Salvador, 1979–1991.*

[110] McClintock, *Revolutionary Movements in Latin America: El Salvador's FMLN & Peru's Shining Path*; Menzel, *Bullets Versus Ballots: Political Violence and Revolutionary War in El Salvador, 1979–1991*; Waller, *The Third Current of Revolution: Inside the North America Front of El Salvador's Guerrilla War.*

[111] Byrne, *El Salvador's Civil War: A Study of Revolution*; Montgomery, *Revolution in El Salvador: From Civil Strife to Civil Peace*; Manwaring, "A Strategic View of Insurgencies: Insights from El Salvador."

guides and workbooks.[112] The rebels set up local elected governments and held congresses to allow peasants to express concerns to FMLN leadership. Concurrently, the FMLN worked to maintain law and order, establishing commissions to arbitrate civil disputes and banning alcohol and drug use both for the population and its guerrillas in order to prevent abuses by drunken fighters. Even if the FMLN did not enjoy the massive popular support it desired,[113] these efforts did generate positive feelings, particularly among the peasantry.[114]

Internationally, the FMLN also achieved a degree of legitimacy. Among communist states that actively supported the insurgency (i.e., the Soviet Union and its allies, such as Cuba and Nicaragua), the FMLN was recognized as the legitimate government of El Salvador.[115] Moreover, through its Political-Diplomatic Commission and by positioning itself as the legitimate alternative to the economic injustice and the egregious human rights violations of the established government, the FMLN was able to gain real and moral support throughout the Western world from human rights organizations, social justice advocates, academics, and even those in government.[116] In the wake of the 1981 general offensive, Mexico and France recognized the FMLN as a "representative political force" and called for a negotiated resolution to the conflict that would uphold that reality.[117] However, pressure from the United States and the inability of the FMLN to make real progress in the civil war decreased direct international support in the West over time.[118] Still, the international media characterized FMLN leaders, such as Jorge Schafik Handal, as state representatives when they traveled abroad.

At least officially, FMLN adhered to the UN Declaration of Human Rights and disciplined those among their ranks who perpetrated human rights violations.[119] Yet by 1985, the increasing strength and

---

[112] Wood, *Insurgent Collective Action and Civil War in El Salvador.*

[113] However, the fact that most of its supplies, with the exception of arms and munitions, came from within urban centers in El Salvador attests to the presence of sizable popular support. Montgomery, *Revolution in El Salvador: From Civil Strife to Civil Peace.*

[114] Byrne, *El Salvador's Civil War: A Study of Revolution,* 242; Montgomery. *Revolution in El Salvador: From Civil Strife to Civil Peace.*

[115] William M. LeoGrande, *Our Own Backyard: The United States in Central America, 1977–1992* (Chapel Hill, NC: University of North Carolina Press, 1998); Byrne, *El Salvador's Civil War: A Study of Revolution.*

[116] Montgomery, *Revolution in El Salvador: From Civil Strife to Civil Peace.*

[117] LeoGrande, *Our Own Backyard: The United States in Central America, 1977–1992;* Byrne, *El Salvador's Civil War: A Study of Revolution.*

[118] Moroni Bracamonte and Spencer, *Strategy and Tactics of the Salvadoran FMLN Guerrillas: Last Battle of the Cold War, Blueprint for Future Conflicts.*

[119] Greentree, *Crossroads of Intervention: Insurgency and Counterinsurgency Lessons from Central America;* Diana Villiers Negroponte, "Conflict Resolution at the End of the Cold War: The Case of El Salvador, 1989–1994" (PhD diss., Georgetown University, 2006).

stability of the ESAF and the democratic government of El Salvador, in part because of US aid, forced the FMLN to adjust its military strategy away from direct engagement with the ESAF and toward assassination, terrorism, and economic sabotage with the goal of demonstrating the incapability of the government to rule effectively, thereby undermining its legitimacy.[120] To this end the FMLN, particularly the ERP under the direction of Joaquín Villalobos, began using land mines and improvised explosive devices that led to heavy civilian causalities. These actions, along with other terrorism, did not erode the legitimacy of the government but rather eroded the legitimacy of the insurgents.[121] By the end of twelve years of war, both sides had accumulated long lists of human rights violations.

Observers of the Salvadoran civil war have noted that "the most important factor in small wars is legitimacy, the moral right to govern."[122] Between the extremes on the right and left in El Salvador was a moderate and democratic center, which was represented in the established government by the Christian Democratic Party. Whereas the radicals on the right and left set themselves outside the nation's shaky constitutional order, moderates attempted to work within it, and it was these elements in the government that won support from the United States and neighbors in Central and South America.[123] Despite its many problems, El Salvador's democracy did provide a space in which moderates could legally seek political, economic, and social reform, as well as legitimacy, in the eyes of the Salvadoran people and the international community.

Still, in 1989 when the FMLN launched its "final offensive," a peaceful resolution to the conflict seemed remote. During the 1989 offensive, the FMLN occupied numerous lower-class neighborhoods in San Salvador. In response, the ESAF conducted indiscriminate aerial bombings of these areas, producing large numbers of civilian causalities. In response to the unsuccessful offensive, the government conducted mass arrests and permitted the continued activity of the death squads, which resulted in the high-profile murder of six Jesuit priests as well as their housekeeper and her daughter.[124] For the FMLN, the improbability of a military victory seriously weakened its

---

[120]   Corr, "Societal Transformation for Peace in El Salvador."

[121]   Byrne, *El Salvador's Civil War: A Study of Revolution.*

[122]   Max G. Manwaring and William J. Olson, *Managing Contemporary Conflict: Pillars of Success* (Boulder, CO: Westview Press, 1996).

[123]   Corr, "Societal Transformation for Peace in El Salvador."

[124]   Payne, *El Salvador Re-Emergence of "Social Cleansing" Death Squads;* Peterson, *Martyrdom and the Politics of Religion: Progressive Catholicism in El Salvador's Civil War.*

position as a legitimate alternative to the established government.[125] For the government, continued gross human rights violations during the offensive and in its aftermath had exasperated even its most ardent international supporters, especially the United States, where pressure on the government mounted for a negotiated settlement.[126] The legitimacy of military victory for either side was quickly waning.

It was in this climate that moderates within the FMLN were able to shift the objective from a military victory to negotiated peace and democratic participation. Although they continued the military campaign during the negotiation process, this shift bolstered the legitimacy of the FMLN, as it provided both a public forum for airing political grievances against the government and also demonstration that the FMLN was intent on achieving real political goals. The FMLN demonstrated its commitment to the peace process by maintaining formal channels of communication and making concessions, including the creation of a truth commission on human rights, which increased its legitimacy both internationally and among the Salvadoran populace.

EXTERNAL SUPPORT

The civil war was fought not only because of the particular grievances of the internal factions within El Salvador, but also because it served as a point of conflict in the broader Cold War between the East and West. Indeed, some have argued that without the external military training, funding, and encouragement, the FMLN would not have seen the violent overthrow of the government as a feasible option.[127] It is certain that the scale of the fighting and causalities would not have been possible without such external support for both sides.[128]

The incredible amount of external support for the FMLN makes it unique among insurgent groups. The rebels received arms,

---

[125] Moroni Bracamonte and Spencer, *Strategy and Tactics of the Salvadoran FMLN Guerrillas: Last Battle of the Cold War, Blueprint for Future Conflicts.*

[126] Robinson, *Intervention Or Neglect: The United States and Central America Beyond the 1980s.*

[127] Corr, "Societal Transformation for Peace in El Salvador."

[128] The FMLN received support from the Soviet Union and its allies as well as some Latin American and Western European governments (with the latter governments providing support because of anti-American or anti-Reagan sentiments). The government of El Salvador was supported by the United States as an ally against Soviet encroachment, while the Salvadoran right, including the death squads, were funded by the landed elites. Wood, *Insurgent Collective Action and Civil War in El Salvador.*; Corr, "Societal Transformation for Peace in El Salvador"; Teresa Whitfield, *Paying the Price: Ignacio Ellacuría and the Murdered Jesuits of El Salvador* (Philadelphia: Temple University Press, 1994).

training, and other supplies primarily[129] from the Soviet Union, Cuba, and Nicaragua.[130] Concurrently, the FMLN received financial and moral support from individuals and organizations in the West, which together with funds raised internally from activities such as kidnapping and ransom, allowed the insurgents to purchase arms from abroad. For instance, a 1981 campaign in West Germany called "Arms for El Salvador" raised more than one million dollars. The insurgents also received relief and development aid from a number of nongovernmental organizations, and this aid was channeled into projects consistent with the FMLN goals, freeing up funds for military operations.[131] Arms and other supplies were smuggled by the Sandinistas through Honduras or in small boats across the Gulf of Fonseca from Nicaragua.[132] Although the scope of these shipments was exaggerated by the Reagan administration and the government of El Salvador, the shipments were nonetheless significant.[133] Cuba, for instance, supplied Soviet-produced man-portable air defense systems (MANPADS), such as the SA-7 and SA-14, and postwar disarmament of the FMLN produced more than 10,000 weapons, 74 missiles, 9,000 grenades, and four million rounds of ammunition, in addition to the 9,500 anti-personnel land mines laid by the insurgents.

---

[129] However, the FMLN also received funding, training, and other support from Angola, Algeria, Vietnam, the Palestine Liberation Organization (PLO), as well as most of the other revolutionary Marxist organizations throughout Latin America, save the Sendero Luminso in Peru. Greentree, *Crossroads of Intervention: Insurgency and Counterinsurgency Lessons from Central America.*

[130] Grenier, *The Emergence of Insurgency in El Salvador.*

[131] R. T. Naylor, "The Insurgent Economy: Black Market Operations of Guerrilla Organizations," *Crime, Law and Social Change* 20, no. 1 (1993): 13–51.

[132] Herrera, "Why Choose Peace? The El Salvador Experience."

[133] LeoGrande, *Our Own Backyard: The United States in Central America, 1977–1992*; Marvin E. Gettleman, *El Salvador: Central America in the New Cold War* (New York: Grove Press, 1987).

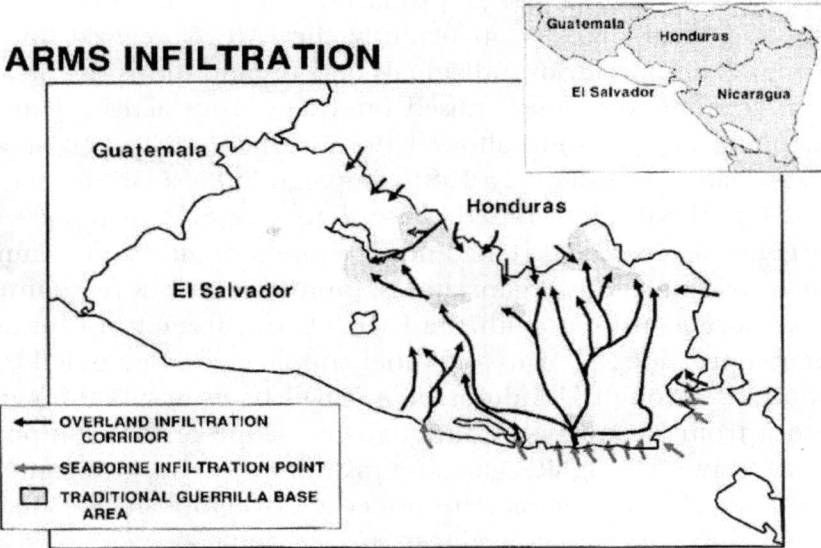

Figure 2. Arms infiltration.[134]

Arms shipments from Cuba and Nicaragua began in late 1980 and early 1981 as the FMLN prepared for the 1981 general offensive. Although the trafficking ceased temporarily, possibly as a result of the failed 1981 offensive and US pressure after clear evidence of the arms shipments emerged, the practice continued through the war.[135] Indeed, even after the peace accords in 1992 and 1993, the UN discovered an additional 15 tons of FMLN-owned weapons and equipment in Honduras and Nicaragua.[136] In addition to providing arms, both Cuba and Nicaragua provided military training and logistical, economic, and political support for the insurgency, including medical facilities for wounded FMLN fighters in Cuba.[137]

However, by the late 1980s, with relations between the East and West warming, the notion that El Salvador was a strategic front in the Cold War became increasingly anachronistic. As part of Premier Mikhail Gorbachev's efforts to improve relations with the United

---

[134] Herrera, "Why Choose Peace? The El Salvador Experience," 26.

[135] Greentree, *Crossroads of Intervention: Insurgency and Counterinsurgency Lessons from Central America*, 196; Negroponte, "Conflict Resolution at the End of the Cold War: The Case of El Salvador, 1989–1994"; Wood, *Insurgent Collective Action and Civil War in El Salvador*; LeoGrande, *Our Own Backyard: The United States in Central America, 1977–1992*.

[136] Edward J. Laurance and William H. Godnick, "Weapons Collection in Central America: El Salvador and Guatemala," in *Managing the Remnants of War: Micro-Disarmament as an Element of Peace-Building*, eds. Sami Faltas and Joseph Di Chiaro III (Germany: Nomos, 2001), 15–48.

[137] Manwaring, *A Strategic View of Insurgencies: Insights from El Salvador*.

States, Soviet aid to Cuba, and thus to the FMLN, virtually ended.[138] Although the Cubans continued to provide minimal support, the attention of Nicaragua turned increasingly to domestic affairs as the changing climate of the Cold War diminished support for advancing communism in the Americas. In 1991, democratic elections in Nicaragua brought a moderate conservative, Violeta Chamorro, to power, removing a key FMLN ally. With the collapse of the Soviet Union in that same year, it became clear that the FMLN would no longer enjoy the external support it had in the 1980s.[139]

COUNTERMEASURES TAKEN BY THE GOVERNMENT

Major Robert J. Molinari, in his 2004 thesis for the Naval War College, noted some of the challenges that the government of El Salvador, as well as the United States, faced in confronting the FMLN insurgency. Molinari notes that "the ESAF were transformed from an 11–14,000 man praetorian guard abusive of the population, to a 50–55,000 man Army able to defeat the insurgents. Further, the ESAF was transformed from a conventionally trained force overly preoccupied with its last war with Honduras, to an unconventional warfare (UW) force that combined small unit operations, intelligence, civic action, psychological operations, protection of economic infrastructure, and winning the support of the population."[140] Molinari noted the critical role the US government played in the military, as well as the political posture of El Salvador, applying a carrot and stick approach. "If ESAF corruption incidents, threats, or challenges against elected civilian leaders, or human rights violations increased—funds were decreased; if FMLN military actions or demonstrations seemed to increase—funds were increased."

At the start of the war, the ESAF consisted of "14,000 firemen, policemen, and soldiers who were neither equipped nor trained for counterinsurgency."[141] Initial responses by the government were similar to previous actions conducted during violent uprisings in the 1960s and 1970s. Notably, the government passed new antiterrorist laws and conducted mass arrests. The ESAF did not substantively change

---

[138] Terry Lynn Karl, "El Salvador's Negotiated Revolution," *Foreign Affairs* 71, no. 2 (Spring 1992): 147–164; Waller, *The Third Current of Revolution: Inside the North America Front of El Salvador's Guerrilla War.*

[139] Stahler-Sholk, *El Salvador's Negotiated Transition: From Low-Intensity Conflict to Low-Intensity Democracy.*

[140] Robert J. Molinari, "Carrots and Sticks: Questions for COCOMs Who Must Leverage National Power in Counter Insurgency Warfare" (master's thesis, Naval War College, 2004).

[141] Corr, "Societal Transformation for Peace in El Salvador."

their tactics during the course of the war. The objective remained to defeat the FMLN militarily—an objective that was never met.

SHORT- AND LONG-TERM EFFECTS

CHANGES IN GOVERNMENT

Land reform, and its supposed relationship to poverty and repression of the peasant class, was always seen as a key issue in both the cause and solution to violence and revolution in El Salvador. The government initiated two massive land programs to first assuage violence and, second, as a key facet of the UN-monitored peace reform process adopted in 1992. In 1980, the government provided 125,000 workers with land. All farms larger than 500 hectares in size were seized and reapportioned. The 1992 accords provided additional land (3.5 hectares each) to 47,500 families. During the course of the twelve-year insurgency, however, the population shifted, and this shift was inextricably linked to the repression of the peasant class, with land reform being a central tenet. First, there was a shift due to the flight from El Salvador to other countries because of the conflict; second, there was a shift from an agrarian, rural society to an urban/suburban trade-based society. Land ownership did not build wealth for the peasant class as effectively as learning an urban industrial trade.[142] In fact, land reform as a central issue for the FMLN political charter had diminished in importance, relative to wealth and social structure, over the course of the conflict.

Herrera notes that, during the Duarte presidential administration, real efforts were made at government reform. During 1982–1989, six free and fair elections were held, and the administration also focused on land reform (see the preceding paragraph). US Ambassador Passage noted, "The lesson we preached (from the Embassy) to the Salvadoran military and security forces and to those who controlled Salvador's economy, was that they had to change themselves so that the typical campesino and his family would begin to actively support the Government side rather than the guerrillas."[143] Furthermore, FMLN Commander Joaquín Villalobos noted the resulting change, stating, "The democratic changes that took place before the Peace Agreement were partial and imperfect but felt tangibly by the insurgency. This

---

[142] Mitchell A. Seligson, "Thirty Years of Transformation in the Agrarian Structure of El Salvador, 1961–1991," *Latin American Research Review* 30, no. 3 (1995): 43–74.

[143] Herrera, "Why Choose Peace? The El Salvador Experience."

gave credibility to the idea that working politically in a context of peace was more beneficial than continuing war."[144]

CHANGES IN POLICY

In anticipation of the elections for the constituent assembly, which was tasked by the junta with writing a new constitution, Major Roberto D'Aubuisson, along with other conservatives including the landed elites, business owners, and the military, formed the ARENA.[145] Although ARENA captured the majority of the seats in the assembly, its leader Roberto D'Aubuisson was not chosen as the interim president because the Reagan administration could not accept someone so closely linked with the death squads. Instead, Alvaro Magaña, a conservative banker who was viewed favorably by the ESAF, was elected president by the assembly. In the 1984 presidential elections, Christian Democrat José Napoleón Duarte was elected. Although human rights abuses continued, the new president did manage some minor agrarian and economic reforms to address the economic injustice that fueled the FMLN, despite being hampered by a failing economy and opposed by ARENA members of the legislature.[146]

Duarte favored a negotiated peace with the FMLN and, in 1986, began overtures toward the insurgents.[147] Formal negotiations hosted by the Papal Nuncio were held in 1987, leading to the Esquipulas Process, a blueprint for peace. Yet Duarte was unable to meet the many FMLN demands as he faced opposition from the ARENA-controlled legislature. Concurrently, the FMLN was also not ready to concede and used the peace negotiations as a tactical ruse rather than a sincere offer of peace.[148] In 1989, the moderate ARENA candidate Alfredo Cristiani was elected president. Cristiani also favored a negotiated peace, and as a member of the ARENA party, he was in the political position to unite moderates and conservatives behind the initiative.[149] Cristiani carried out massive economic reforms along free-market lines, ended state marketing monopolies, discouraged the ownership of farmland by large landowners or cooperatives, and privatized the

---

[144] Villalobos, "A Democratic Revolution for El Salvador."

[145] Manwaring and Prisk, *El Salvador at War: An Oral History.*

[146] LeoGrande, *Our Own Backyard: The United States in Central America, 1977–1992*; Corr, "Societal Transformation for Peace in El Salvador."

[147] Montgomery, *Revolution in El Salvador: From Civil Strife to Civil Peace*; Pratap C. Chitnis, "Observing El Salvador: The 1984 Elections," *Third World Quarterly* 6, no. 4 (October 1984): 963–980.

[148] Corr, "Societal Transformation for Peace in El Salvador."

[149] Menzel, *Bullets Versus Ballots: Political Violence and Revolutionary War in El Salvador, 1979–1991*; Karl, *El Salvador's Negotiated Revolution.*

financial sector. Most importantly, he created a national commission to begin the peace process with the FMLN.

CHANGES IN THE REVOLUTIONARY MOVEMENT

Responding to the adoption of low-intensity conflict operations with the election of José Napoleón Duarte in 1984, the FMLN organized an operational response entitled *poder de doble cara*, or "two-faced power."

> Encouraged by guerrilla political organizers, civilians decided to remain in their homes during army incursions and responded to human rights violations by sending delegations to San Salvador to complain to the media and human rights agencies. This placed additional pressure on the government to live up to its self-proclaimed democratic image. Second, a rooted civilian population freed combatants to carry out actions elsewhere in the nation; rural and urban sabotage and harassment eventually pinned the army down to defensive positions and relieved some of the pressure on zones under the control or influence of the guerrillas, thus creating even more political space for civilian organizations. The military and political sides of the 'war of resistance' reinforced one another and constituted an effective response to 'low-intensity warfare.'[150]

The government of El Salvador had not anticipated the coordinated political and military response from the FMLN. In fact, they had demonstrated the ability to adjust strategy and operational initiatives while continuing to conduct coordinated political and military campaigns. It became clear that a concerted effort, supported by the US administration, would be required to improve the capabilities of the ESAF. Additional time and resources were going to be needed, and El Salvador remained in crisis.

OTHER EFFECTS

Three factors contributed to the desire of both sides for peace, including: (1) the self-interest of the landed elites who had suffered economically because of the civil war; (2) the FMLN's recognition

---

[150] Binford, "Grassroots Development in Conflict Zones of Northeastern El Salvador."

that military victory was impossible, which led to the inclusion of more moderates in the FMLN leadership; and (3) the collapse of the Soviet Union and the end of the Cold War, which diminished for the United States the importance of defeating the FMLN militarily.[151]

During UN-brokered peace negotiations that took place during 1991 (and were implemented on January 16, 1992), the FMLN cited their lack of confidence in the Salvadoran justice system and the need for an independent body to investigate atrocities during the course of the war, noting that the FMLN "considered such action a necessary guarantee to protect the FMLN leadership and its supporters against potential government abuses once the FMLN laid down its weapons. The Parties therefore concluded that a special body would have to be established to carry out the investigations."[152]

The result was the formation of the UN Truth Commission for El Salvador, which explored actions during the course of the war between 1980 and 1991. The commission provided specific names of individuals responsible for violent acts, on both the government and FMLN sides. Both President Cristiani and FMLN leadership agreed that the "bad apples" had to be named, and that this was part of the healing process. ESAF military leadership, however, strongly opposed this, openly threatening a military coup and circulating the threat within the halls of the Ministry of Defense. Thomas Buergenthal, one of the three central commission members, noted:

> Our hope for a quantitatively balanced report could not be realized. Despite the massive wartime propaganda to the contrary, the government side had committed a substantially larger number of egregious acts than the FMLN. Moreover, some of these acts—among them the El Mozote massacre in which hundreds of innocent civilians were slaughtered—had no comparable counterparts among the crimes committed by the FMLN.

The commission recommended amnesty, not trials. When published, the commission's findings caused shock and dismay throughout the Salvadoran communities, including the large displaced communities in the areas of Washington, DC, and Los Angeles, California, that, during the course of the war, had become the de facto second- and third-largest cities with a Salvadoran population. Over time, however, the commission report served as the impetus for a national healing.

---

[151] Montgomery, *Revolution in El Salvador: From Civil Strife to Civil Peace.*
[152] Buergenthal, "United Nations Truth Commission for El Salvador."

Buergenthal noted the cathartic effect that the commission report had on the people:

> The war in El Salvador not only pitted the combatants in the armed conflict against each other, but also totally polarized the population. It became a country in which there was no room for moderation or tolerance for peaceful political debate. Political opponents were treated as enemies and acts of violence against them rationalized as necessary or denied as propaganda. Political allegiance rather than basic human decency determined one's actions and reactions to the crimes that both sides committed. El Salvador was a country in which man lived in fear, and where their next of kin often did not dare to denounce publicly what had been done to them or even speak about it lest their claims expose them to further abuse. People kept their suffering to themselves, hoping for justice—a very human instinct—but not really expecting it.[153]

Although disheartened by the Sandinista loss in democratic elections in nearby Nicaragua on the eve of the peace accords, the FMLN found the Cristiani administration's concessions acceptable. These included the establishment of a civilian police force that would include members of the FMLN, the transformation of the FMLN into a legitimate political party, and constitutional limits that restricted the military to national defense and border control. Although the FMLN did not attain the power-sharing agreement it had hoped for, both sides signed the accords in Mexico City, ending the twelve-year civil war in January 1992.[154]

The FMLN agreed to report to fifty camps located throughout the country, principally in areas that had been under FMLN control as of 1991. From these camps, fifteen UN-administered verification centers were established; at these centers, UN observers administered the demobilization of FMLN combatants. Demobilization consisted of FMLN members registering, turning in weapons, and providing personal preference regarding options for continued education, paths to learning a trade, or farming. Concurrently, the ESAF also began demobilization to 100 designated garrison locations. Although the entire demobilization process took more than 18 months to

---

[153] Ibid.

[154] Stahler-Sholk, "Central America: A Few Steps Backward, a Few Steps Forward"; Corr, "Societal Transformation for Peace in El Salvador"; Montgomery, *Revolution in El Salvador: From Civil Strife to Civil Peace.*

implement, violations were minor. Trust was the major factor used for oversight by the UN observers, with no major peacekeeping force ever deployed.[155]

BIBLIOGRAPHY

Binford, Leigh. *The El Mozote Massacre: Anthropology and Human Rights.* Tucson: University of Arizona Press, 1996.

————. "Grassroots Development in Conflict Zones of Northeastern El Salvador." *Latin American Perspectives* 24, no. 2 (March 1997): 56–79.

Bollinger, William. "Villalobos on 'Popular Insurrection.'" *Latin American Perspectives* 16, no. 3 (Summer 1989): 38–47.

Bonner, Raymond. *Weakness and Deceit: U.S. Policy and El Salvador.* New York: Times Books, 1984.

Brockett, C. D. "El Salvador: The Long Journey from Violence to Reconciliation." *Latin American Research Review* 29, no. 3 (1994): 174–187.

Buergenthal, T. "The United Nations Truth Commission for El Salvador," *Vanderbilt Journal of Transnational Law* 27, no. 3 (1994): 497.

Byrne, Hugh. El Salvador's Civil War: A Study of Revolution. Boulder, CO: L. Rienner Publishers, 1996.

Campbell, Bruce, and Arthur David Brenner. *Death Squads in Global Perspective: Murder with Deniability.* New York: St. Martin's Press, 2000.

Chitnis, Pratap C. "Observing El Salvador: The 1984 Elections." *Third World Quarterly* 6, no. 4 (October 1984): 963–980.

Central Intelligence Agency. "El Salvador." *The World Factbook.* Accessed November 2, 2009. https://www.cia.gov/library/publications/the-world-factbook/geos/es.html.

Corr, Edwin G. "Societal Transformation for Peace in El Salvador." *Annals of the American Academy of Political and Social Science* 541, no. 1 (1995): 144–156.

Crouch, James R. Sr. "The Decade of the Seventies in El Salvador: Prelude to Revolution." Master's thesis. US Army Command and General Staff College, 1997.

Dickey, Christopher. "Central America: From Quagmire to Cauldron." *Foreign Affairs and the World* 62, no. 3 (1983).

---

[155] Laurance and Godnick, "Weapons Collection in Central America: El Salvador and Guatemala."

Dillon, Sam. "Dateline El Salvador: Crisis Renewed." *Foreign Policy* no. 73 (Winter 1988): 153–170.

Dix, Robert H. "Why Revolutions Succeed & Fail." *Polity* 16, no. 3 (Spring 1984): 423–446.

Gettleman, Marvin E. *El Salvador: Central America in the New Cold War.* 1, revised and updated. New York: Grove Press, 1987.

Gómez, Leonel, and Bruce Cameron. "El Salvador: The Current Danger. American Myths." *Foreign Policy* no. 43 (1981): 71–92.

Greentree, Todd R. *Crossroads of Intervention: Insurgency and Counterinsurgency Lessons from Central America.* Westport, CT: Praeger Security International, 2008.

Grenier, Yvon. *The Emergence of Insurgency in El Salvador* University of Pittsburgh Press, 1999.

Herrera, M. Chris. "Why Choose Peace? The El Salvador Experience." Master's thesis, School of Advanced Military Studies, United States Army Command and General Staff College, 2007. http://handle.dtic.mil/100.2/ADA485593.

Hobden, Steve. "El Salvador: Civil War, Civil Society and the State." *Civil Wars* 3, no. 2 (2000): 106–120.

Huntington, S. P., and F. Fukuyama. *Political Order in Changing Societies.* New Haven, CT: Yale University Press, 2006.

Karl, Terry Lynn. "El Salvador's Negotiated Revolution." *Foreign Affairs* 71, no. 2 (Spring 1992): 147–164.

Laurance, Edward J., and William H. Godnick. "Weapons Collection in Central America: El Salvador and Guatemala." In *Managing the Remnants of War: Micro-Disarmament as an Element of Peace-Building,* edited by Sami Faltas and Joseph Di Chiaro III, 15–48. Germany: Nomos, 2001.

Lauria Santiago, Aldo Antonio. "An Agrarian Republic: Production, Politics, and the Peasantry in El Salvador, 1740–1920." PhD diss., University of Chicago, 1992.

Lauria-Santiago, Aldo, and Leigh Binford. *Landscapes of Struggle: Politics, Society, and Community in El Salvador.* Pittsburgh: University of Pittsburgh Press, 2004.

LeMoyne, James. "El Salvador's Forgotten War." *Foreign Affairs* 68, no. 3 (Summer 1989): 105–125.

LeoGrande, William M. *Our Own Backyard: The United States in Central America, 1977–1992.* Chapel Hill, NC: University of North Carolina Press, 1998.

Manwaring, M. G., and Court Prisk. "A Strategic View of Insurgencies: Insights from El Salvador." *Small Wars and Insurgencies* 4, no. 1 (Spring 1993).

Manwaring, Max G., and William J. Olson. *Managing Contemporary Conflict: Pillars of Success.* Westview Studies in Regional Security. Boulder, CO: Westview Press, 1996.

Manwaring, Max G., and Court Prisk. *El Salvador at War: An Oral History.* Washington, DC: National Defense University Press, Government Printing Office, 1988.

Mason, T. David, and Dale A. Krane. "The Political Economy of Death Squads: Toward a Theory of the Impact of State-Sanctioned Terror." *International Studies Quarterly* 33, no. 2 (June 1989): 175–198.

McClintock, Cynthia. *Revolutionary Movements in Latin America: El Salvador's FMLN & Peru's Shining Path.* Washington, DC: United States Institute of Peace Press, 1998.

McClintock, Michael. *State Terror and Popular Resistance in El Salvador.* The American Connection. Vol. 1. London: Zed Books, 1985.

McIlwaine, Cathy. "Contesting Civil Society: Reflections from El Salvador." *Third World Quarterly* 19, no. 4 (1998): 651–672.

Menzel, Sewall H. *Bullets Versus Ballots: Political Violence and Revolutionary War in El Salvador, 1979–1991.* New Brunswick: Transaction Publishers, 1994.

Meyer, C. *Underground Voices: Insurgent Propaganda in El Salvador, Nicaragua and Peru.* RAND Corporation, 1991.

Molinari, Robert J. "Carrots and Sticks: Questions for COCOMs Who Must Leverage National Power in Counter Insurgency Warfare." Master's thesis, Naval War College, 2004.

Montgomery, Tommie Sue. *Revolution in El Salvador: From Civil Strife to Civil Peace.* 2nd ed. Boulder, CO: Westview Press, 1995.

Moroni Bracamonte, José Angel, and David E. Spencer. *Strategy and Tactics of the Salvadoran FMLN Guerrillas: Last Battle of the Cold War, Blueprint for Future Conflicts.* Westport, CT: Praeger, 1995.

Naylor, R. T. "The Insurgent Economy: Black Market Operations of Guerrilla Organizations." *Crime, Law and Social Change* 20, no. 1 (1993): 13–51.

Negroponte, Diana Villiers. "Conflict Resolution at the End of the Cold War: The Case of El Salvador, 1989–1994." PhD diss., Georgetown University, 2006.

Newton, Richard D. "The Seeds of Surrogate Warfare." *Joint Special Operations University and the Strategic Studies Department* 09, no. 3 (February 2009): 1.

Paige, Jeffery M. *Coffee and Power: Revolution and the Rise of Democracy in Central America.* Cambridge: Harvard University Press, 1997.

Payne, Douglas. *El Salvador Re-Emergence of "Social Cleansing" Death Squads.* Washington DC: INS Resource Information Center, 1999.

Pelupessy, Wim, and John F. Uggen. "Economic Adjustment Policies in El Salvador during the 1980s." *Latin American Perspectives* 18, no. 4 (Autumn 1991): 48–78.

Peterson, Anna Lisa. *Martyrdom and the Politics of Religion: Progressive Catholicism in El Salvador's Civil War.* Albany: State University of New York Press, 1997.

Rivas, Ruben Zamora, and Schafik Jorge Handal. "Proposal of the FMLN/FDR." *Latin American Perspectives* 14, no. 4 (Autumn 1987): 481–486.

Robinson, Linda. *Intervention Or Neglect: The United States and Central America Beyond the 1980s.* New York: Council on Foreign Relations Press, 1991.

Salaverry, Jorge. "A Winning U.S. Policy Is Needed in El Salvador." Washington, DC: Heritage Foundation, 1988.

Seligson, Mitchell A. "Thirty Years of Transformation in the Agrarian Structure of El Salvador, 1961-1991." *Latin American Research Review* 30, no. 3 (1995): 43–74.

Sheehan, Thomas. "Recent Developments in El Salvador." *The Threepenny Review* no. 16 (Winter 1984): 10–11.

Spencer, David E. *From Vietnam to El Salvador: The Saga of the FMLN Sappers and Other Guerrilla Special Forces in Latin America,* Westport, CT: Greenwood Publishing Group, 1996.

Stahler-Sholk, Richard. "Central America: A Few Steps Backward, a Few Steps Forward." *Latin American Perspectives* 26, no. 2 (March 1999): 3–12.

———. "El Salvador's Negotiated Transition: From Low-Intensity Conflict to Low-Intensity Democracy." *Journal of Interamerican Studies & World Affairs* 36, no. 4 (1994): 1–59.

Taylor, Robert W., and Harry E. Vanden. "Defining Terrorism in El Salvador: 'La Matanza.'" *Annals of the American Academy of Political and Social Science* 463, International Terrorism (September 1982): 106–118.

Villalobos, Joaquín. "Popular Insurrection: Desire or Reality?" *Latin American Perspectives* 62, no. 3 (Summer 1989): 5–37.

Villalobos, Joaquín. "A Democratic Revolution for El Salvador." *Foreign Policy* 74 (1989): 103–122.

Waller, J. Michael. *The Third Current of Revolution: Inside the North America Front of El Salvador's Guerrilla War.* Lanham, VA: University Press of America, 1991.

White, Robert E. *Preliminary Assessment of Situation in El Salvador.* US Department of State, 1980.

Whitfield, Teresa. *Paying the Price: Ignacio Ellacuría and the Murdered Jesuits of El Salvador.* Philadelphia: Temple University Press, 1994.

Wood, Elisabeth Jean. *Insurgent Collective Action and Civil War in El Salvador.* Cambridge Studies in Comparative Politics. Cambridge: Cambridge University Press, 2003.

KAREN NATIONAL LIBERATION ARMY (KNLA)

Ron Buikema and Jason Spitaletta

SYNOPSIS

The Karen National Union (KNU) and its armed wing, the Karen National Liberation Army (KNLA), represent the longest ongoing insurgency in the world, having commenced in 1949. The Karen separatist movement dates back to the end of World War II when the British began the global relinquishment of their colonial holdings. The British-aligned Karen wanted an independent state distinct from the Japanese-aligned, and ethnically distinct, Burmans. The ensuing conflict persists to this day. This study examines the KNLA from the 1962 military coup in Burma to March 2010. Studying the KNLA presents a unique opportunity to evaluate a group that has maintained a military force with the political goal of separatism while several other groups fighting against the Burmese government have failed to achieve the same.

TIMELINE

1881	Karen National Association (KNA) formed to gain a fair representation for the Karen people in the then-British Burmese government.
1947	The KNU formed, combining the KNA, the Buddhist Karen National Association (BKNA), the Karen Central Organization (KCO), and the Karen Youth Organization (KYO).
1948	Burma was granted independence from British rule.
1949	Civil war in Burma commences.
1962	Military successfully launches coup d'etat.
1988	Democratic uprising is launched by KNLA and other ethno-nationalist groups; Burmese Army successfully thwarts the uprising.
1989	Majority of ethno-nationalist insurgencies announce cease-fire with State Peace and Development Council (SPDC), with one exception—the KNLA. Burmese Army is now focused on KNLA as principal threat to the state.
1994	KNU/KNLA headquarters is overrun by Burmese Army.

1995	KNLA becomes an all-volunteer force.
2004	KNU delegation of twenty Karen officials goes to Rangoon for talks with top junta leaders to discuss an official cease-fire agreement.
2008	KNLA deactivates several brigades and battalions.

THE ENVIRONMENT OF THE REVOLUTION

PHYSICAL ENVIRONMENT

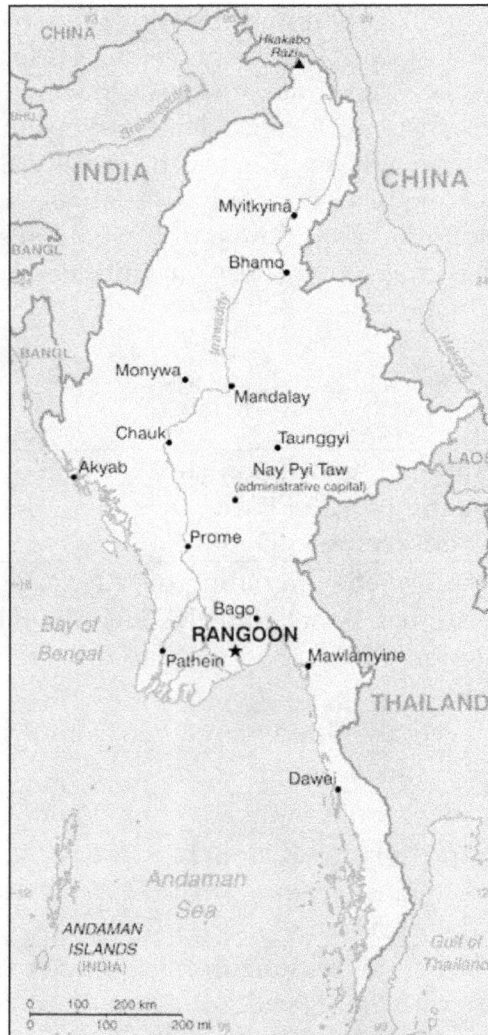

Figure 1. Map of Burma.[1]

---

[1] Central Intelligence Agency, "Burma," *The World Factbook*, https://www.cia.gov/library/publications/the-world-factbook/geos/bm.html.

CULTURAL AND DEMOGRAPHIC ENVIRONMENT

The population of Burma is estimated at fifty million, with the ethnic Burman totaling more than 50% of the nation's population. The remaining 50% is divided among at least twelve other distinct ethnic groups, including the Karen as one of the largest groups, with subcultural groups numbering many more. Burma is a country that remains a collection of fiefdoms ruled by traditional tribal chiefs, insurgents, drug lords, military regional commanders, and black marketers. Control of these remote regions shifts between them as alliances are made and broken.[2] There are more than one hundred distinct languages spoken within Burma.

The Karen ethnic group, the second largest in Burma, is a group of Indo-Chinese tribes in the easternmost province of the former British Indian Empire.[3] The term "Karen" encompasses approximately twenty-four subgroups of Karen-speaking peoples. Most of the ethnic groups are associated with specific geographic areas within the country. More than 80% of the population is characterized as rural/agrarian, with many living as subsistence farmers. Many of the Burmese ethnic groups, including the Karen, retain a strong affinity to the land, seeing it as a focus of nourishment, both physically and spiritually. The two dominant Karen groups are the Sgaw (mostly Christian and animists in the hill regions) and the Pwo (mostly lowland Buddhists) accounting for 80–85% of the population.[4] The Sgaw is the largest and most scattered group and can be found through the Irrawaddy Delta to the Arracan coast. The Pwo are concentrated in the coastal areas from Arracan to Mergui.[5] The Karen have historically been differentiated through tribal distinctions and linguistic differences; the Sgaw and the Pwo Karen speak different dialects (both of which are called "Karen") that are not easily understood by the speakers of the other tribe.[6]

The Karen reside in both plains and forested areas and predominately live in small villages. Within this village structure, important lines of differentiation include gender, age, education, civil

---

[2] E. W. Rogers, "Burma on the Brink: Complications for U.S. Policy in Burma" (master's thesis, Naval Postgraduate School, 1991).

[3] H. I. Marshall, *The Karen People of Burma: A Study in Anthropology and Ethnology* (Columbus, OH: Dodo Press, 1922).

[4] A. T. Thawnghmung, *The Karen Revolution in Burma: Diverse Voices, Uncertain Ends* (Washington DC: East-West Center, 2008), 3.

[5] Marshall, *The Karen People of Burma,* 2.

[6] H. MacLachlan, "The Don Dance: An Expression of Karen Nationalism," *Voices: The Journal of New York Folklore* 32 (2007): 26–34.

status, and religion.[7] Social and work structure, as well as commerce, is built on the village. Trade may also be conducted from one village to another village, generally within the same ethnic group. The Karen social structure is matriarchal, with elderly women generally serving as village leaders. The Chairman of the KNU is a woman. Mother figures are revered, and they are accorded both power and authority throughout the Karen society.[8]

The Burmese people have known ethnic conflict for decades. Many ethnic minorities, including the Karen, have been repeatedly and forcibly displaced from their homelands. The Karen community, estimated at five to seven million, is predominantly Buddhist, with Christian Karen representing a politically (and militarily) powerful minority within the culture. The Karen nationalist movement started in the 1940s during British colonial rule. Repression of the Karen and other ethnic groups has continued since the 1940s.[9] Karen people still refer to the Burmese nation-state as *gkaw p'yaw*, literally "land of the Burmans," and people still speak of "going down into Burma."[10] The KNLA has maintained close relationships with the villages, with most families having a member in the organization. Maintaining this local tie at the village level has ensured the cohesion of common cultural and political ties from the people and the organization.

Many displaced Karen have fled Burma for refugee camps along the Thai–Burma border. The camps were established in 1984 after the KNU's base at Me Thaw Waw was taken over by the Tatmadaw and, by 2007, housed more than 130,000 Karen.[11] Conditions in the numerous camps vary greatly ranging from pseudo-internment facilities to those where inhabitants are permitted to leave and return for employment purposes. There is also a sizable Karen diaspora residing in various countries.[12]

SOCIOECONOMIC ENVIRONMENT

Burma remains one of the poorest countries in Southeast Asia, but as recently as fifty years ago it had the strongest economy in the region.

---

[7] K. Malseed, "Networks of Noncompliance: Grassroots Resistance and Sovereignty in Militarised Burma" (presentation, Yale Agrarian Studies Colloquium, April 25, 2008), http://www.khrg.org/khrg2008/khrg08w3.pdf.

[8] Ibid.

[9] K. MacLean, "Spaces of Extraction," in *Myanmar: The State, Community, and the Environment*, eds. M. Skidmore and T. Wilson (Canberra, Australia: Asian Pacific Press, 2007).

[10] Malseed, "Networks of Noncompliance."

[11] Thawnghmung, *The Karen Revolution in Burma*, 21.

[12] Ibid., 23.

The degradation of the economy is attributed principally to how the military regime "systematically dismantled the fundamental economic institutions—effective property rights, contract enforcement, the measures that define the 'rules of the game' for efficient economic transactions—that history [reveals] are necessary for sustainable long-term growth."[13] Corruption is rampant, while investment in education, health services, and agriculture has been largely neglected. The State Peace and Development Council (SPDC) has routinely made claims of 10% gross domestic product (GDP) growth per year for a number of decades, but the growth rate is actually estimated to be much closer to 1.5–4% per year.[14] Economic growth is primarily due to the export of natural gas, which began producing revenue in 1998.

Outside of natural resources, Burma is not attracting foreign investment with the exception of China. Kokang Chinese living within Burma were granted national registration cards under an agreement with former Prime Minister General Khin Nyunt. As a result, formal and informal relations and investment opportunities with the Chinese steadily increased after approximately 2003.[15] Some foreign investment and economic growth has been noted in the tourism industry, which has focused on the restoration of historic sites and construction of tourist hotels. Logging of teak and other hardwoods produced some economic activity, although overcutting and deforestation have been noted as major problems, with an extensive portion of illegal logging activities associated with export to China.[16] Finally, export of gems, including rubies, sapphires, and jade, continues to generate revenue, with principal global exports managed via China, India, and Thailand.[17]

Socioeconomic disparities between the Karen villagers and government workers, particularly military officers, are vast. Military officers "regularly show off their high status, driving around in expensive cars, eating at expensive restaurants, promoting their children in business or educational sectors."[18] Working for the government, particularly the military, has been the preferred route for climbing the socioeconomic ladder in Burma for the past three decades.

---

[13] MacLean, *Spaces of Extraction*, 76.

[14] Ibid., 76.

[15] Ibid.

[16] D. S. Heaney, "Burma: Assessing Options for U.S. Engagement" (master's thesis, Naval Postgraduate School, 2009).

[17] The Tom Lantos Burmese JADE Act bans the import of Burmese gems to the United States.

[18] MacLean, *Spaces of Extraction*, 76, 41.

Most Burmese people, including the Karen, have no access to a formal financial institution. Instead, they rely on family members or the local community. There are thousands of informal and illicit moneylenders operating within a complex black- and grey-market economy, funding everything from small loans to legitimate business operations, to drug trafficking and armed militia activities.[19]

HISTORICAL FACTORS

The hostility between ethnic Karens and Burmans predates the precolonial period when Burman kings attempted to subjugate the Karen tribes living in the hill regions. Differences in religious beliefs, cultural practices, and agricultural methods also caused tension.[20] The British exploited these tensions and employed Karens in their conflict against the Burmans in 1853 and 1855. The Karen's aspiration for a coherent national identity and self-determination dates to 1881 with the establishment of the Karen National Association (KNA). As with many of its colonies, the British established a representative government, with the majority Burmans holding the most power. This system lasted for well over a century, and Burma was officially recognized as a country in 1937. Many Karens benefited from their support to the British through access to Western-style education. The increased exposure to Western missionaries enabled the heretofore-disconnected Karen villages to come into contact, and a concept of pan-Karen nationalism began to emerge. The idea of an independent Karen state was first proposed by Dr. San C. Po, who advocated it as a component of a federation.[21]

The colonial-era policies did little to engender mutual understanding among Burma's ethnic populations. Britain's preference for incorporating the Karen into the armed forces, along with the Karen's access to missionary education, led to their disproportionate representation in the military, police, and civil service. Because communal seats in the legislature were reserved for Karens living among Burmans in the lowland areas, the requirement to build cross-ethnic relationships between the groups was eliminated, thus fostering greater resentment.[22]

The military incarnation of the Karen insurgency traces its roots to the Karen Rifles, a regiment that served with the British Army

---

[19]   Ibid., 76.
[20]   Thawnghmung, *The Karen Revolution in Burma*, 4.
[21]   Ibid., 4.
[22]   Ibid., 5.

until 1946. They were renowned by the British for their loyalty and fighting spirit. The Karen were adept at harassing isolated Japanese outposts, gathering intelligence, and serving as guides to Allied forces. During World War II, when the Japanese occupied the region, long-term tensions between the Karen and Burma turned into open fighting. Consequently, many villages were destroyed, and massacres were committed by both the Japanese and the Burma Independence Army troops who assisted them. The army took hostage and executed approximately 2,000 Karen, prompting three months of violence between the two groups.[23] A government report later claimed that the excesses of the army and the loyalty of the Karens toward the British were the reasons for these attacks. American operatives from the Office of Strategic Services (OSS), along with British operatives of the Special Operations Executive, organized a successful guerrilla movement among the Kachin tribesmen of northern Burma, who were hostile to the Burmese and Thai ethnic groups aligned with Japan and against whom the Japanese committed a number of atrocities.

In April 1942, Major General Orde Wingate arrived in India to organize guerrillas against the Japanese in Burma. Wingate's 3,000-strong 77th Indian Infantry Brigade, the "Chindits," received extensive training at Ramgarh and then moved more than 200 miles behind Japanese lines in Burma. The Allies supported the guerrillas from Fort Hertz, the only Allied base in Burma with an airfield. The three regiments of guerrillas, the Karen Rifles, the Kachin Rifles, and the Kachin Levies, were proficient jungle fighting units but lacked the proper training and the equipment needed to effectively engage Japan's mechanized infantry and armor. Relying solely on air assets for resupply and medical evacuation, the Chindits ambushed Japanese patrols, attacked outposts and supply depots, destroyed bridges, and repeatedly cut the Myitkyina railroad for several months.[24]

Lieutenant General William J. Slim, commander of the British 14th Army, criticized Wingate's efforts, but Winston Churchill praised Wingate who later recommended that the OSS expand its guerrilla-warfare activities into Burma. The ongoing resistance of the hill tribes integrated well with the British plan (called Guerrilla Forces–Plan V) to support small units operating behind Japanese lines, leading US General Stilwell to reconstitute the Kachin Levies in 1943. The V-Force recruited the hill tribesmen and trained them to collect intelligence; provide early warnings of air attacks; recover downed Allied aircrews;

---

[23] Ibid., 5.

[24] C. H. D. Briscoe, "Kachin Rangers: Allied Guerrillas in WWII Burma," HtoiGinTwang.Over-Blog.com (blog), June 10, 2010, http://htoigintawng.over-blog.com/article-kachin-rangers-allied-guerrillas-in-ww-ii-burma-51989324.html.

conduct ambushes, reconnaissance, and flank patrols; and scout for conventional forces.[25]

The initial successes of the V-Force led Stilwell to expand his guerrilla operations, and he directed his OSS Detachment 101 to plan and execute operations against the roads and the railroad into Myitkyina in order to deny the Japanese the use of the Myitkyina airfield. By the end of 1943, Detachment 101 had established several operating bases in northern Burma. Each base detachment recruited and trained small tribal elements for base/personnel security and internal defense as well as for conducting limited sabotage and ambush operations. The guerrilla forces were equipped with modern carbines, as well as light and heavy machine guns along with ammunition and demolitions. Japanese weapons and equipment in northern Burma were outdated, and the superior firepower of the guerrilla units was critical to their success.[26]

An excerpt from Detachment 101's Presidential Unit Citation, awarded for the unit's capture of several strategic Japanese strongpoints in Burma's Central Shan States in 1945, extolled the warrior ethos of the hill tribesman. Detachment 101 and its guerrilla cadre became a highly effective strike force, continually on the offensive against the veteran Japanese 18th and 56th divisions. Although they were cited officially only by the Americans, the Karen were heavily involved throughout the China–Burma–India theater of operations: they served as levies with the British from Fort Hertz, supported Wingate's two Chindit expeditions, engaged in direct combat with the Japanese, collected intelligence, reported weather, and rescued downed Allied aircrews.[27]

After World War II, the Karen people aspired to have a Karen-majority subdivision within Burma similar to what the Shan, Kachin, and Chin peoples had been given. An August 1946 goodwill mission to London led by Saw Tha Din and Saw Ba U Gyi failed to achieve endorsement from the British government for any separatist demands. When a delegation of representatives of the Governor's Executive Council headed by Aung San was invited to London to negotiate for the Aung San-Attlee Treaty in January 1947, none of the ethnic minority members were included by the British government. The following month at the Panglong Conference, when an agreement was signed between Aung San as head of the interim Burmese government and the Shan, Kachin, and Chin leaders, the Karen were

---

[25] Ibid.

[26] Ibid.

[27] Ibid.

present only as observers. Although the situation of the Karen was discussed, nothing practical was done before the British left Burma. The 1947 Constitution, drawn without Karen participation because of their boycott of the elections to the Constituent Assembly, also failed to address the Karen question specifically and clearly, leaving it to be discussed only after independence. The Shan and Karenni states were given the right to secession after ten years, the Kachin their own state, and the Chin a special division. The Mon and Arakanese of Ministerial Burma were not given any consideration.[28]

The Karen ethno-nationalist movement grew stronger during the transition from British colonial rule when it became evident that the new independent government did not intend to grant autonomy to any specific ethnic group. In February 1947, the KNU was formed at a Karen Congress. The meeting called for a Karen state with a seaboard, an increased number of seats in the Constituent Assembly, a new ethnic census, and a continuance of Karen units in the armed forces. The March deadline passed without a reply from the British government.[29]

Burma was granted independence in January 1948, and the Karen, represented by the KNU, attempted to coexist peacefully with the Burman ethnic majority. Karen people held leading positions in both the government and the army. In the fall of 1948, the Burmese government began raising and arming irregular political militias known as *Sitwundan* outside the control of the regular army. In January 1949, some of these militias rampaged through Karen communities. In late January, the Army Chief of Staff, General Smith Dun, a Karen, was removed from office, imprisoned, and replaced by Burmese nationalist Ne Win. The Karen National Defense Organization (KNDO), which was formed in July 1947, commenced an insurgency against the government after General Dun's removal as the army chief of staff. They were supported by the defections of the Karen Rifles and the Union Military Police units that had been successfully deployed in suppressing the earlier Burmese Communist rebellions, and they came close to capturing Rangoon.[30] During the 1950s, a period of civil conflict followed, as ethnic groups throughout the country took up arms to fight for local sovereignty. In 1953, the KNU officers in the Irrawaddy Delta region established a vanguard party in the Maoist tradition. The Karen National Unity Party (KNUP) was intended to generate rural support for the separatist movements by educating

---

[28] Z. Oo and W. Min, *Assessing Burma's Ceasefire Accords* (Washington DC: East-West Center, 2007), 4.

[29] Ibid., 5–6.

[30] Ibid., 5–6.

and training the Karen in the tactics of people's warfare.[31] Many of the groups were well armed and trained from their experience under British tutelage during World War II. Since gaining independence, the Burmese government has had to focus resources and attention on defeat of numerous ethnic insurgencies. They have successfully defeated the Kuomintang and Community Party of Burma, backed by the People's Republic of China; the National League for Democracy; and the Buddhist monk "Saffron Revolution."[32]

GOVERNING ENVIRONMENT

During the constitutional period from 1948 to 1962, when Burma had a parliamentary government, the country suffered widespread conflict and continuing internal struggle. Constitutional disputes and persistent division among political and ethnic groups contributed to the democratic government's weak hold on power.

The outbreak of rebellion was catastrophic for the average Karen. Besides the many killed, wounded, or homeless, thousands of Karen civil servants, soldiers, and policemen were arrested and interned, while numerous others lost their jobs. Only in 1951 did the Burmese government start reinstating a handful of Karen into the police and, in 1952, into the military. However, the Karen community never regained its former influence in government.

In 1958, Prime Minister U Nu accepted military rule temporarily to restore political order, and the military stepped down after eighteen months. In 1962, General Ne Win led a military coup, abolishing the constitution and establishing a xenophobic military government with socialist economic policies that had devastating effects on the country's economy and business climate.[33]

The State Peace and Development Council (SPDC), previously referred to as the State Law and Order Restoration Committee (SLORC) from 1988 to 1997, have continued to rule by military force since 1962. The standing conscripted Army has a force of approximately 400,000. The Army maintains bases throughout eastern Burma, where SPDC forces have been reported to "impress civilians, women as well as men, as porters for months at a time. Hungry soldiers take villagers' crops and livestock. Refugees also report frequent beatings, rapes,

---

[31] A. T. H. Tan, ed., *A Handbook of Terrorism and Insurgency in Southeast Asia* (New York: Edward Elgar Publishing, 2007), 302.

[32] Heaney, "Burma: Assessing Options for U.S. Engagement," 35.

[33] US Department of State, "Background Note: Burma," accessed September 9, 2010, http://www.state.gov/r/pa/ei/bgn/35910.htm.

and murder."[34] The military regime controls all forms of media in the country. Any person or entity that speaks disparagingly of the government is labeled a threat, a stooge, or an enemy of the state. The government has decreed that civilians have no role in politics or government; they should be dependent on and subservient to the state without question or protest. Internet access is routinely blocked and monitored. Telephone communications are also restricted and closely monitored by the state. All documents, including books and periodicals, are subject to censure. Messages countering the military regime are strictly forbidden. The government imposes work requirements at the village level, including determining which crops will be grown and what the expected crop yield will be. Failure to meet production quotas can result in imprisonment and loss of land rights. Trade of crops and other goods from one village to another is closely controlled, frequently taxed, and strictly limited by the Army.[35]

Governance from a Karen village perspective is challenging because there are two distinct chains of command. One, the Karen governing system, is composed of a village elder, likely a woman, who oversees village activities, including work. The village elders generally build a self-sustaining village capable of providing food, shelter, and security to all of its members. The second system is the Burmese government. Even in predominantly Karen areas, there are likely nearby Burmese Army (Tatmadaw) camps, charged with ensuring that the local villages comply with government orders for work. An SPDC township authority assigns work to each village on a monthly basis; the work includes such activities as road maintenance. Soldiers may also demand work from villagers; tasks might include harvesting crops, providing building materials, or maintaining military facilities. Travel outside of the village area requires written authorization from the government. Insubordination to orders could result in punishment to the village at large, assignment to a forced labor camp, or even relocation of the entire village. Bribes are commonly paid by villagers in order to avoid forced labor.[36]

[34] Doug Bandow, "Forgotten War in a Forgotten Country," The Cato Institute, accessed September 9, 2010, http://www.cato.org/pub_display.php?pub_id=4622.

[35] US Department of State, "Background Note: Burma," accessed September 9, 2010, http://www.state.gov/r/pa/ei/bgn/35910.htm.

[36] Ibid.

Figure 2. Government poster, Mandalay, Burma.[37]

WEAKNESSES OF THE PREREVOLUTIONARY ENVIRONMENT AND CATALYSTS

On February 11, 1948, the KNU organized tens of thousands of Karen people in a peaceful demonstration demanding an autonomous Karen state. The peaceful demand made by the KNU to establish a Karen State was not considered by the Burmese government, which ultimately conducted military operations against the Karen at Insein in January 1949. The KNDO, the armed branch of the KNU, resisted the military incursion, marking the beginning of the current insurgency.

The principal catalyst of the Karen insurgency was the denial of an autonomous region of Burma for the ethnic Karen at the conclusion of World War II and the postcolonial era. Since the military takeover in 1962, the polarization of the Karen and the government results from the military regime's desire to administer a unified state and the Karen's desire to maintain ethnic, cultural, economic, and administrative autonomy. The strength, over time, of the ethno-nationalist cause has been exacerbated by the fact that the state is so weak.[38]

---

[37] "File:Myanmar-message.jpg," *Wikipedia*, accessed March 14, 2011, http://commons.wikimedia.org/wiki/File:Myanmar-message.jpg.

[38] A. Rajah, "A Nation of Intent in Burma: Karen Ethno-Nationalism, Nationalism and Narrations of Nation," *The Pacific Review* 15, no. 4 (2002): 517–537.

FORM AND CHARACTERISTICS OF THE REVOLUTION

OBJECTIVES AND GOALS

The KNU and KNLA seek to defend Karen culture and interests, and ethnic autonomy remains their goal. The KNU established its political goals in 1949, codified during the 1956 KNU Congress. They are contained in the manifesto and seek to (1) establish a Karen state with a right to self-determination; (2) establish national states for all the nationalities, with the right to self-determination; (3) establish a Federal Union with all the states having equal rights and the right to self-determination; and (4) pursue the policy of National Democracy.[39] The political goals of the KNU have undergone three stages of development corresponding to the three Burmese/Myanmar regimes. At the beginning of the parliamentary era of the Anti-Fascist People's Freedom League (1948–1962), the KNU demanded the right to secession and the inclusion in the Kawthoolei state of mixed Burmese–Karen territories in the Irrawaddy Delta. In the Burma Socialist Programme Party era (1962–1988), the goal of the KNU shifted from territorial demands to preventing the marginalization (and elimination) of ethnic minorities. This shift led to the establishment of the National Democratic Front as an alliance of ethnic minorities that sought a federal union of Burma. By the end of the 1980s, the KNU strongholds along the border with Thailand became the main refuge for students and prodemocracy activists fleeing from the SLORC, and, thus, the KNU joined the democratic goal of the exile organizations and established the Democratic Alliance of Burma. The Democratic Alliance of Burma initiated an alliance with the National League for Democracy, a combination that, in the 1990s, evolved into the National Council of the Union of Burma.[40]

Although the Karen have long sought independence from Burma, they have recently begun to call for a federal democratic government of Burma offering adequate safeguards for the various ethnic groups constituting the state.

---

[39] Ibid.

[40] M. Smith, *Burma: Insurgence and the Politics of Ethnicity* (Dhaka, Bangladesh: The University Press, 1999), 185.

LEADERSHIP AND ORGANIZATIONAL STRUCTURE

The KNU is the largest, best-organized, and most powerful and influential political/military organization in Burma.[41] The KNU and its armed wing, the KNLA, operated as a quasi-government along the Thai–Burma border (from Toungoo province in the north to Tavoy in the south) from the 1970s until the 1990s.

Given its longevity, the KNU has experienced several structural reorganizations. In 1953, Karen leaders, inspired by Maoist models, established the KNUP, assigning it the role of political leadership, while the KNU remained the mass organization with the restructured Kawthoolei Armed Forces acting as the military wing. The socialist political line of the KNUP/KNU dissatisfied the eastern units led by General Bo Mya, which, in 1968, split in order to form the Karen National United Front (KNUF). In 1975, after the dissolution of KNUP, the two factions of the Karen movement reunited under the banner of the KNU, and the combined armies become known as the KNLA. In December 1994, a significant number of Karen Buddhists left the KNU over allegations of discrimination by the Christian-dominated leadership and formed the Democratic Karen Buddhist Organization/Army (DKBO/A). Information on the KNU/KNLA provided by the DKBO/A to the Burmese army was hypothesized to be the vital intelligence that led to the capture of the KNU headquarters at Manerplaw and the stronghold at Kawmoorah in 1995. The most recent split took place in 2007 when the commander of the KNLA 7th Brigade left the KNU and established a new organization, the KNU/KNLA Peace Council, which subsequently negotiated a cease-fire agreement with the military government.[42]

The KNU was organized into seven administrative districts (Thaton, Toungoo, Nyaunglebin, Mergui-Tavoy, Papun, Dooplaya, and Paan), each of which was subdivided into townships and tracts. Each KNU district selected a committee, a chairperson, a vice-chairperson, a secretary, and departmental officers. The KNU was governed by an executive committee of eleven members, a group typically dominated by Sgaw Christians from the Delta region.[43]

The KNU districts roughly correspond with KNLA brigades, which were responsible for raising their own funds, as well as organizing, training, and equipping their members. The KNLA has a parallel command structure of seven military brigades, each headed by a brigade commander and subject to the KNU's Defense Minister. These

---

[41] Thawnghmung, *The Karen Revolution in Burma*, 25.

[42] Smith, *Burma: Insurgence and the Politics of Ethnicity*, 285–287.

[43] Thawnghmung, *The Karen Revolution in Burma*, 25–26.

parallel structures, overlapping but not entirely coinciding, afford the opportunity for distinct factions to develop within the KNU/KNLA leadership with somewhat divergent outlooks and constituencies. The KNLA is led by Saw Tamla Baw, who serves as commander-in-chief and vice-chair of the KNU. Tamla Baw was imprisoned by the Japanese forces during World War II, joined the initial Karen uprising in 1949, and has been affiliated with the KNLA since 1949. The KNLA has approximately 5,000–7,000 combatants, organized into seven brigades. These brigades include mobile battalions and village militias, and each brigade may also have a political cadre of approximately 1,000 personnel. Half of the forces are believed to be operating from within ten refugee camps maintained along the border with Thailand.[44]

In response to the Burmese military's Four-Cuts strategy,[45] the KNLA decentralized their command and control. The result was the formation of six "battle areas" with corresponding forces, generally referred to as brigades, for each area.[46]

KNLA also maintains the 101 Special Battalion, which may possess special training or equipment focused on ambush and offensive operations.[47] By 2008, some brigades and several battalions were likely deactivated because of a loss of KNLA operational capability and a lack of ability to staff all of the units. Additionally, because of Burmese military offensive activity, some traditional areas of operation (i.e., Battle Area 7) had been lost.

---

[44] Ibid., 25–26.

[45] The Four-Cuts strategy is described in detail in the *Countermeasures Taken by the Government* section. This strategy was a Tatmadaw policy intended to sever the insurgent groups' links to food, funds, intelligence, and recruits.

[46] (1) Thaton District, composed of the 1st Brigade and three subordinate battalions; (2) Toungoo District, 2nd Brigade; (3) Nyaunglebin District, 3rd Brigade; (4) Mergui-Tavoy District, no. 10 Battalion; (5) Duklay Area, 6th Brigade; (6) Karen Central Military District, 7th Brigade; (7) Pa-an District, 7th Brigade (this area was abandoned by the KNLA during 2009).

[47] Multiple radio reports from several years of reporting were used to associate geographic areas with specific units. Guy J. Pauker, *Government Responses to Armed Insurgency in Southeast Asia: A Comparative Examination of Failures and Successes and their Likely Implications for the Future* (Santa Monica, CA: RAND Corporation, 1985).

Figure 3. KNLA soldiers.[48]

COMMUNICATIONS

The KNLA has an established communications network through the Karen diaspora that is active in Thailand. This diaspora communicates freely within Thailand via the Internet and can provide logistics support and even safe haven to KNLA members once they cross the border into Thailand. Within Burma, communication is largely passed from one village to the next via courier or word of mouth.[49] This is not to say, however, that the KNLA have become antiquated in their use of technology for communications: they are proficient in their use of the Internet to respond to official SPDC pronouncements. Rather, they are disciplined in their approach, allowing the KNU to take the lead on strategic and political statements. There are no indications of friction or misunderstanding between the KNU and KNLA organizations.

KNLA and supporting organizations in Thailand have also used radio as a means of conducting information operations, both internally to the KNU/KNLA and to the greater diaspora, particularly within Thailand. Radio Kawthulay, one of the radio groups, was very popular during the 1980s and 1990s, providing casualty reports in

---

[48] KNLA Karen National Liberation Army Unofficial Facebook page, accessed March 15, 2011, http://www.facebook.com/group.php?v=wall&gid=2333008449#!/group.php?gid=2333008449&v=photos.

[49] I. Brees, "Burmese Refugee Transnationalism: What is the Effect?" *Journal of Current Southeast Asian Affairs* 28, no. 2 (2009): 23–46.

detail, even on a daily basis. Radio stations have remained a vital source of information to the Karen community.

Tactically, the KNLA have principally relied on captured VHF radios recovered during combat with the Burmese military forces. They have also recently reported having a signals intelligence capability, but this has not been confirmed.

It is unclear whether there is a centralized command and control component to the strategic communications effort on behalf of the Karen; however, from human rights groups, to prodemocracy advocates, to myriad investigative journalists the world over, there is no shortage of small articles, websites, and blogs dedicated to furthering the understanding of the Karen struggle.

METHODS OF ACTION AND VIOLENCE

The KNLA has retained many of the tactics, techniques, and procedures employed by the Karen Rifles under the direction of the US OSS and the British Special Operations Executive during World War II. The KNLA continues to collect intelligence, conduct sabotage operations, set hasty ambushes, and conduct long-range reconnaissance patrols. Early in the fighting, Karen forces overran much of Northern Burma, including towns such as Mandalay, and established strong positions outside Rangoon at Insein Township. However, lacking a port from which to receive military supplies, the Karen forces gradually withdrew to the southeast of Burma. Since that time, the KNLA has been fighting for an independent state, called Kawthoolei, to be located in eastern Burma near the border with Thailand and in other places with large Karen populations.

Following the strategy and principles enumerated by Mao Tse-Tung, the KNUP was established as the vanguard of the Karen people. KNUP cadre training programs began in late 1953. In areas of mixed Karen–Burman population, support from local Burman villages grew. Along with this support, finances were more centralized and the KAF was better prepared to conduct mobile guerrilla war, so that by 1955 KAF units were able to reoccupy many areas from which they were displaced in 1952.[50]

The KNLA and Karen splinter groups survive by avoiding direct mass engagement and refusing to cooperate with state forces. This allows the KNLA to avoid tactical attrition, further enabling them to

---

[50] Ba Saw Khin, "Fifty Years of Struggle: A Review of the Fight for Karen People's Autonomy," Life In Picture website, accessed October 10, 2010, http://www.kwekalu.net/photojournal1/soldier/story6.htm#hist.

continue both their recruiting efforts and political objectives. The KNLA are generally facing government forces that have numerical supremacy. Ambushes and hit-and-run tactics are the preferred means of military engagement. They additionally have to defend against government forces that attack villages. KNLA General Htey said "the SPDC try to fight the grass roots, our backbone, the villages," so the people "don't have the morale to support us with food or anything else."[51] KNLA carry equipment that has been smuggled into their areas of control via Thailand, acquired on the black market, or recovered on the battlefield, and they routinely employ mortars and crew-served weapons. Moreover, the KNLA does, at times, engage in direct confrontation with the Burmese military, often publicizing the Burmese casualty rates as proof of their tactical prowess and continued will to fight.

When the Burmese military adopted the Four-Cuts strategy,[52] the KNLA responded by initiating a guerrilla warfare approach. In so doing, they sacrificed territory, no longer directly engaging the military in many cases. Actions included ambushes and placement of improvised explosive devices (IEDs) along principal lines of communication. Mines were also placed as a defensive tactic to protect some villages and military camps.[53] After the loss of Manerplaw and control of the checkpoints along the lucrative trade routes on the Thai–Burma border, the KNLA regrouped into approximately 300 small bases throughout the frontier area.[54]

Although the Burmese military (or Tatmadaw) typically outnumber the KNLA guerrillas (on paper and in the field), the KNLA hold their own during confrontations. One account claims that during January through June of 2006, the KNLA lost twelve soldiers, compared to 185 Burmese military soldiers, and suffered fourteen casualties while inflicting 448.[55]

By 2009, the KNLA had lost control of most of its territorial bases. Continued army offensives, tumultuous cease-fire negotiations that never bore fruit, and infighting between the political and military elites resulted in what one observer describes as a "critical phase of its

---

[51] Bandow, "Forgotten War in a Forgotten Country."

[52] The Four-Cuts strategy is described in detail in the *Countermeasures Taken by the Government* section. It was a Tatmadaw policy intended to sever the insurgent groups' links to food, funds, intelligence, and recruits.

[53] Pauker, "Government Responses to Armed Insurgency," i.

[54] Tan, *A Handbook of Terrorism and Insurgency*, 47.

[55] "Mahn Sha, General Secretary, Karen National Union (KNU), Interview," *Jane's Intelligence Review*, accessed October 10, 2010, http://www.janes.com/articles/Janes-Intelligence-Review-2006/Interview–Mahn-Sha-General-secretary-Karen-National-Union-KNU.html.

life."[56] This tactical redeployment by the KNLA was effected to limit both Karen guerrilla and civilian casualties in the wake of offensive operations of increasing intensity by the government. Burmese forces, augmented by the breakaway DKBO/A, launched a new offensive against the KNU on June 3, 2009, resulting in a large flow of Karen refugees into Thailand.

The KNLA's 101 Special Battalion (or Special Warfare Branch), created in 2001, was charged with deep reconnaissance and sabotage missions against strategic targets. The specially trained soldiers covertly infiltrated enemy territory under the cover of darkness and with minimal radio communication; they planted timed explosives and then quickly retreated along predetermined egress routes. Simultaneously, other teams established ambush sites and employed obstacles (including IEDs and mines) to disrupt the forces giving chase. These tactics not only served as effective countermobility practices against pursuing infantry but also served to demoralize the adversary, given the high value of the chosen target and the target's distance from KNLA territory. The effectiveness of these missions, particularly the booby-trapped withdrawals, led to their adoption by the KNLA Second and Third brigades and, later, the 6th and 7th brigades, resulting in increased Tatmadaw causalities in those areas.

Since the beginning of the insurgency, the typical KNLA soldier has engaged in the same type of small-unit ambush and raid operations, with substandard equipment, inadequate supplies of food and medicine, and the knowledge that if he is seriously wounded in combat, he will likely die before reaching the nearest hospital.

METHODS OF RECRUITMENT

The KNLA recruits from the local Karen villages. Recruiting is usually voluntary, although there have been incidents of forced recruiting reported. Mutual allegations of human rights abuses are widespread about both the Tatmadaw and the KNLA, both of which have accused the other of forced conscription of child soldiers[57] in the conflict.[58]

---

[56] P. Core, "Burma/Myanmar: Challenges of a Ceasefire Accord in Karen State," *Journal of Current Southeast Asian Affairs* 28, no. 3 (2009): 95–105.

[57] Child recruitment by the Burmese military has increased in recent years as the government has expanded its armed forces. Some child soldiers have been forced to commit atrocities against civilians, including burning homes and rounding up villagers, including children, for forced labor. Joe Becker, "Children at War," *The New York Times*, October 16, 2002.

[58] Malseed, "Networks of Noncompliance."

The KNU/KNLA established an ethno-nationalist culture, with "ethno-history" taught in school and established as KNU policy. The KNU established a flag, a coat of arms, national dress, and a national anthem. By adapting culturally acceptable symbols passed down from generation to generation, the goals and objectives of the organization were reinforced and accepted, and at the same time the symbols helped to draw on the greater Karen populace for a call to service.[59]

The KNLA troop strength peaked in the early 1980s at an estimated 5,000 regulars and another 5,000 village militia. Recruits received no pay and were provided with only food and uniforms; they had to get pocket money from their own families. Despite these hardships, for many Karens, service in the KNLA has for many years been a quite respectable profession, and the troops have displayed remarkable endurance and bravery. However, the military system is flexible, and at short notice the KNLA can call up or demobilize large numbers of troops according to the state of war.[60]

Figure 4. KNLA flag.[61]

METHODS OF SUSTAINMENT

The KNLA organization and finances were ostensibly centralized; however, each brigade was largely responsible for raising its own funds and arming its own troops. This meant that the strongest brigades, the 6th and 7th, with their once very lucrative trading and customs posts along the Thai border, prospered, while the smaller ones, such as the

---

[59] Rajah, "A Nation of Intent in Burma," 517–537.

[60] Khin, "Fifty Years of Struggle."

[61] "Karen National Liberation Army flag," *Wikipedia*, accessed March 15, 2010, http://en.wikipedia.org/wiki/File:Karen_National_Liberation_Army_flag.svg.

Toungoo (the 2nd) and Nyaunglebin (the 3rd) brigades, had only meager resources.[62]

Cross-border trade and taxes constituted the main sources of revenue for the KNU, with 1962–1988 being the most profitable period.[63] The KNLA predominately received support from the Karen network of villages. Food and medical supplies were probably stored within the villages. The Karen diaspora community provided financial support, principally flowing via the Thai border, where there are an estimated 120,000 Burmese refugees, many affiliated with Karen. There was relative freedom of movement into and out of refugee camps, which facilitated provision of food, medical supplies, and other items of supply for KNLA forces. Large explosive caches have been previously discovered within Karen refugee camps, giving credence to at least the potential for arms and ammunition being stored and transported through the camps.[64] Although the Karen community, and specifically the KNU and KNLA, have disavowed any relationship to illicit drug (namely heroin) and human trafficking, it is possible that some funds were also acquired via taxation or other fees related to black marketing or other activities. Weapons and ammunition were recovered from the battlefield or acquired on the black market, and anything that could be salvaged or reused was put to use. Challenges to sustainment were mitigated with the deactivation of some brigades and battalions, meaning that there were fewer forces to supply.[65]

METHODS OF OBTAINING LEGITIMACY

The KNU/KNLA have been fighting since 1949 and represent the only major ethno-nationalist movement that has not agreed to a cease-fire, giving it tremendous credibility within Burma.[66] The KNU/ KNLA have established legitimacy by solidifying relationships with the Karen diaspora community; maintaining both a standing military and political apparatus as well as an information operations campaign for both Karen and external audiences; and by sheer survival for more than 60 years of fighting. Moreover, the KNU/KNLA have a long-standing culture, complete with flags, songs, an educational

---

[62] Ibid.

[63] A. South, "Conflict and Displacement in Burma/Myanmar," in *Myanmar: The State, Community, and the Environment*, eds. M. Skidmore and T. Wilson (Canberra, Australia: Asia Pacific Press, 2007).

[64] A. Rajah, "Burma: Protracted Conflict, Governance and Non-Traditional Security Issues" (working paper, Institute of Defence and Strategic Studies, 2001), 6, http://www.rsis.edu.sg/publications/workingpapers/wp14.pdf.

[65] Core, "Burma/Myanmar," 95–105.

[66] South, "Conflict and Displacement in Burma/Myanmar."

system, and a village-based governing system, that serves as a de facto alternative government in the area that they control.

Their legitimacy has been challenged as a result of accusations that some KNLA military leaders have personally profited during the conflict by establishing relationships with illicit businesses along the Burma–Thai border. Alleged activities have included taxation of black-market activities, such as distribution of gems and illegal teak logging.[67]

EXTERNAL SUPPORT

Nongovernmental organizations (NGOs) have played a role in reporting and publicizing allegations of human rights abuses against the Karen people, lending tacit public affairs support to the greater KNU/KNLA cause. For several years, organizations such as the Karen Human Rights Group have reported on forced labor, SPDC military activities against Karen noncombatants, and the care and treatment of internally displaced persons within Burma. These NGOs, generally working without the consent of the SPDC, have provided valuable external verification and reporting of military and nonmilitary activities undertaken against the Karen. The Karen diaspora and international sympathizers have played significant roles in raising international awareness by using all forms of media, raising funds, and building a community of interest.

The KNLA has maintained a strong support base along the Thai border for decades. This support base includes means of funding as well as providing supplies (including arms), safe haven, access to medical care, and communications and electronic equipment, among other things. After fighting for more than 60 years, the KNLA has a mature network of international support, focused on the Thai border. This support does not necessarily translate into operational capability, as KNLA leadership has often referred to their efforts as a "lonely struggle."[68] Moreover, the government of Thailand tactically supported the mutually beneficial trade along the border; however, as the relationship between Bangkok and Rangoon has improved since 1988, this activity has diminished.[69]

---

[67] MacLean, "Spaces of Extraction."
[68] "Rebel in Myanmar Promises a Long War," *The New York Times*, March 1, 2005.
[69] Smith, *Burma: Insurgence and the Politics of Ethnicity*, 432.

COUNTERMEASURES TAKEN BY THE GOVERNMENT

The Burmese military (Tatmadaw) has made little effort to engage in population-centric counterinsurgency. Instead, they have waged concerted campaigns against insurgent forces and the populace through fear, intimidation, and extreme violence. The Tatmadaw has developed an active defense strategy based on guerrilla warfare with limited conventional military capabilities, designed to cope with low-intensity conflicts from external and internal foes that threaten the security of the state. This strategy is founded on a system of a total people's defense, where the armed forces provide the first line of defense and the training and leadership of the nation. It is designed to deter potential aggressors by publicizing the message that defeat of Tatmadaw's regular forces in conventional warfare would result in persistent guerrilla warfare following the Maoist doctrine of strategic defensive, strategic stalemate, and strategic offensive.[70]

The first phase of postcolonial Tatmadaw doctrine was developed in the early 1950s when threats to the security of the state were perceived to be external. After 1962, the focus turned inward, and the second phase focused on counterinsurgency warfare. During this phase, a foreign policy based on isolation minimized any link between external threats and internal problems. Principles of anti-guerrilla warfare were outlined and counterinsurgency courses were emphasized at the training schools. The doctrine was based on "three totalities: the population, time, and space; and 'four strengths': manpower, material, time, and morale."[71] Patterned after the British "new village" operations in Malaysia and the US "strategic hamlet" operations in Vietnam, Ne Win implemented a policy called "Four-Cuts." The term "Four-Cuts" is believed to be a derivation of the Japanese army's "three all" policy ("kill all; burn all, destroy all") in China.[72]

In April 1968, Tatmadaw introduced special warfare training programs at various regional commands, with special emphasis placed on ambush and counterambush operations, counterinsurgency weapons and tactics, individual battle initiative for tactical independence, commando tactics, and reconnaissance. The Burma Socialist Programme Party established directives for the "complete annihilation of the insurgents as one of the tasks for national defense and state security" and called for "liquidation of insurgents through the strength of the working people as the immediate objective."[73]

---

[70] MacLean, "Spaces of Extraction," 40.
[71] Oo and Min, *Assessing Burma's Ceasefire Accords*.
[72] MacLean, "Spaces of Extraction," 40.
[73] Oo and Min, *Assessing Burma's Ceasefire Accords*, 40.

However, Karen villages have predominately been the areas of greatest interest and focus for Tatmadaw activity, not for insurgent military forces. The Tatmadaw denied access to land, enforced curfews, restricted freedom of movement, burned and looted homes, forcibly conscripted villagers into military units or labor camps, stole crops, and relocated entire villages.[74]

From the early 1980s into the 1990s, the Burmese military government began massive offensives on Karen strongholds along the Thai border. The third phase of doctrinal development of Tatmadaw came after the military coup and formation of the SLORC in September 1988. Military leadership was concerned that foreign powers might arm the insurgents on the border (specifically the Karen) to exploit the political situation and tensions in the country. This new potential threat, previously insignificant under the nation's isolationist foreign policy, led Tatmadaw leaders to review their defense capability. This third phase entailed confronting external threats with an approach of strategic denial under the Total People's Defense concept.[75] The Tatmadaw, which expanded from 1988 to 2000 to become the second-largest standing army in Southeast Asia, has implemented a major expansion plan, establishing military bases in rural areas. The presence of bases represents government power and control while also imposing restrictions on activities of the local populace. Once established, military bases may impose forced labor of local villagers, demand food and supplies, and attempt to supplant the authority of village elders—all tools of intimidation intended to control the populace by negative reinforcement.[76] The Tatmadaw also changed their organizational structure with their expansion plan, creating more than twelve new divisions and limiting the control of regional commanders, who had traditionally ruled their areas of responsibility as de facto fiefdoms.[77]

Major offensives to dislodge the KNU from their headquarters at Manerplaw commenced in early 1992. However, it was not until the fallout between the 400-strong dissident group DKBO/A and the KNLA in late 1994 and early 1995 that the headquarters fell to the Burmese government troops. On January 27, 1995, Burmese forces, at least 10,000 strong, marched into Manerplaw, led by several hundred KNLA troops that had defected to their side nearly two months before.

---

[74] Malseed, "Networks of Noncompliance."

[75] B. Lintner, "Recent Developments on the Thai-Burma Border," *IBRU Boundary and Security Bulletin* 3, no. 1 (1995): 72–76.

[76] Malseed, "Networks of Noncompliance."

[77] W. Min, "Looking Inside the Burmese Military," *Asian Survey* 48, no. 6 (2008): 1018–1037.

The KNU headquarters had already been torched by the defenders and not a single building was left standing. The last remaining troops blew up the command post and the nearby Saw Ba U Gyi statue so that the revolution would survive.[78]

Since approximately 1990, forced relocation has been a principal means of targeting KNLA supporters and sympathizers; however, Karen villagers have been relocated as far back as the occupation of Burma by Japanese soldiers during World War II. The Tatmadaw have not generally allocated resources and personnel for affixing the displaced villagers to the relocation sites and, thus, some displaced Karen eventually returned to their original lands. In other instances villages have been reestablished in relocation camps, particularly if the new sites offered benefits in education, health, or jobs. The Tatmadaw placed mines in former village sites as a means of denying or degrading access to returning villagers. The Tatmadaw has also been accused of forcing conscripted Karen soldiers to lead assaults against KNLA forces, but this activity has not been confirmed.[79]

SHORT- AND LONG-TERM EFFECTS

CHANGES IN GOVERNMENT

By the 1970s, there were more than twenty ethnic-based armies fighting against the Burmese military government. In 1973, the government abandoned its policy of the "Burmese Way to Socialism," resulting in economic revisions and requests for international aid to bolster internal development. By the 1980s, military forces launched several offensive campaigns against ethno-nationalist groups, including the KNLA. The first Karen refugee camps were established along the border with Thailand and were filled with civilians and insurgent fighters who were fleeing contested areas. In 1988, there was a prodemocracy uprising, supported by numerous armed ethnic groups including the KNLA.

The SPDC did not make sweeping, broad changes in government; the military regime serves in the role as "guardian of the nation" and has been in power since 1962.[80] Since that time, the government has focused on its survival by isolation and confrontation of internal threats, including the myriad ethno-nationalist movements. The goal of the state is consolidation of power, elimination of threats, and

---

[78] Khin, "Fifty Years of Struggle."
[79] MacLean, "Spaces of Extraction."
[80] Rajah, "Burma: Protracted Conflict," 13.

predominance of the military above all. The regime has demonstrated resilience, even when faced with coordinated internal threats and numerous tides of external political pressure. Policy guidelines have remained constant to "prevent the disintegration of the nation, to unify the multi-ethnic nation, and to preserve national sovereignty."[81]

In March 1988, student-led demonstrations broke out in Rangoon in response to the worsening economic situation; these demonstrations evolved into a call for regime change. Despite repeated violent crackdowns by the military and police, the demonstrations increased in size, and many in the general public joined the students. During mass demonstrations on August 8, 1988, military forces killed more than 1,000 demonstrators. At a rally after this massacre, Aung San Suu Kyi, the daughter of General Aung San, made her first political speech and assumed the role of opposition leader. In September 1988, a group of generals deposed Ne Win's Burmese Socialist Program Party, suspended the constitution, and established a new ruling junta, the SLORC. In an effort to restore order, the SLORC deployed the army into the streets to suppress the ongoing public demonstrations. It is estimated that an additional 3,000 people were killed, and more than 10,000 fled into the hills and border areas; many of those who fled left Burma altogether.[82]

During 1989, most of the ethnic-based armies agreed to a cease-fire with the government. The KNLA was the one major exception. The SLORC ruled by martial law until national parliamentary elections were held in May 1990, which resulted in a victory for Aung San Suu Kyi's National League for Democracy party. The SLORC refused to honor the results or call the parliament into session, instead imprisoning many political activists and maintaining its grip on power. Although Aung San Suu Kyi received international attention for her political actions, the SLORC generally considered the KNLA to be a greater threat to the government.[83]

The ruling junta changed its name to the State Peace and Development Council (SPDC) in 1997, but it did not change its policy of autocratic control and repression of the democratic opposition. In 2000, the SPDC began talks with the political opposition led by Aung San Suu Kyi (who remains under house arrest).[84]

In October 2004, members of the SPDC senior leadership consolidated their power by removing Prime Minister General Khin

---

[81] Ibid., 12–13.

[82] US Department of State, "Background Note: Burma," accessed September 9, 2010, http://www.state.gov/r/pa/ei/bgn/35910.htm.

[83] Ibid.

[84] Ibid.

Nyunt from control of the government and military intelligence apparatus. In late November 2004, the regime announced it would release approximately 9,000 prisoners it claimed had been jailed by the National Intelligence Bureau. Over the years, the government has continued to release significant numbers of prisoners, although only a small fraction of those released have been political prisoners. On August 28, 2007, as popular dissatisfaction spread, Buddhist monks began leading peaceful marches. On September 5, 2007, security forces in the town of Pakokku violently broke up demonstrations by monks, resulting in injuries and triggering calls for a nationwide response and a government apology for the incident.[85]

In October 2007, the SPDC appointed fifty-four proregime persons to sit on a constitution-drafting committee. The government declared the completion of the committee's work in February 2008 and announced that it would hold a national referendum on the constitution in May 2008, with multiparty elections planned for 2010. Although the referendum law provided for a secret ballot, free debate was not permitted, and activities considered to be "interfering with the referendum" carried a three-year prison sentence. The government carried out the referendum on May 10 and May 24 in an atmosphere of fear and intimidation.[86]

Possibly the greatest change in the Burmese government since the inception of the Karen insurgency has been the expansion of skills that military officers are expected to master. With the presumption that government power and superiority will continue to be exercised by military officers, not civilian bureaucrats, these officers are not only expected to understand the art of war, but they are also expected to understand economics, engineering, or other technical fields. The military holds a special class in Burmese society, with opportunity that is unparalleled in the civilian or private sectors.[87]

CHANGES IN POLICY

During 1989, seventeen separate ethnic groups signed cease-fire agreements with the government. Over the following years, six additional insurgent groups signed cease-fire agreements, further isolating the KNLA, which has continued to refuse any cease-fire accord with the government. The agreements included a no-contact policy, meaning that the military could have access to ethnic sites, but

---

[85] Ibid.
[86] Ibid.
[87] Rajah, "Burma: Protracted Conflict."

no other ethnic groups would have access to the same location. The government had effectively isolated adversarial parties while removing a major internal threat to its existence. They had also ceased any potential of uniting various ethnic armies against the state.[88] Adoption of these cease-fires led to two types of engagement with the military forces: cease-fire forces and non-cease-fire forces.

The first kind of engagement involves the Tatmadaw in offensive deployments and counterinsurgency operations. Depending on the regional operation command area, counterinsurgency warfare may include the co-optation of cease-fire forces, acting as surrogate militias against non-cease-fire forces. Tatmadaw forces have continued to employ the ruthless Four-Cuts strategy. This strategy aims to cut off the insurgents' access to food, funds, intelligence, and recruits and often entails forced relocations of entire communities into "strategic villages," confiscation of food that is then reissued as rations, destruction of crops, "taxes," and a shoot-on-sight policy after curfew hours.[89]

During 2003, the SPDC adopted a seven-step road map:

1. Reconvene the National Convention
2. Implement the process for a "disciplined democratic system"
3. Draft a new constitution
4. Adopt the constitution through national referendum
5. Hold fair and free elections
6. Convene assemblies in accordance with the new constitution
7. Build a modern, developed, and democratic nation.[90]

In 2005, the United Nations (UN) Security Council approved a mechanism to monitor abuses, including murder, rape, and the use of child soldiers, and considered sanctions against offenders. Under the plan, UN-led task forces were established in eleven conflict zones to monitor the conduct of all parties and send regular reports to a central task force based at UN headquarters. The reports were then used as a basis for targeted action against offenders. Both the KNLA and Burmese military were among the fifty-four offending parties to be monitored.[91]

---

[88] Heaney, "Burma: Assessing Options for U.S. Engagement."

[89] Rajah, "Burma: Protracted Conflict," 6.

[90] South, *Conflict and Displacement in Burma/Myanmar*, 54.

[91] AFP, "Briefly: Plan Seeks to Protect Children in War Zones," *The New York Times*, July 27, 2005, http://www.nytimes.com/2005/07/26/world/americas/26iht-briefs.html?_r=1&scp=4&sq=%22Karen%20National%20Liberation%20Army%22&st=cse.

CHANGES IN THE REVOLUTIONARY MOVEMENT

In September 1974, the Ninth KNU Congress saw KNU policy substantially shift toward the political right; this shift continued in force into the late 1990s. Over the years, most members of rebel/ ethnic minority groups in Burma had redefined their political goals, generally toning down separatist language, but for the first time, the demand for the right of secession by all National Democratic Front members, including the KNU, was explicitly dropped and the political goals were rewritten in terms designed to win support from the Burmese majority.[92] In 1994, there was an internal struggle within the ethnic Karen community, leading to a splinter of power, as the DKBO/A was formed. That same year, the KNU headquarters at Manerplaw was overrun by government forces. Since 1995, the KNLA has become an all-volunteer force, with operations radically scaled back.[93]

The weaknesses of the KNU/KNLA include lack of freedom of movement, limited access to arms, lack of external support, and a decentralized command and control network, all of which have been exacerbated by only informal ties that frequently exist from one village to the next. From the beginning, the Karen people's political objectives were subsumed by their animosity toward the Burmans.[94] The KNU's loss of its Manerplaw headquarters in 1995, and subsequently its stronghold Kawmoorah in early 1997, was a significant blow to the command and control of the KNLA. This loss limited military action to primarily defensive operations, not only because of the diminished physical infrastructure but also to retain the option of negations with the regime.

By 2007, the KNLA was in decline. They had become isolated, as almost all other political and military ethno-nationalist movements (numbering more than one hundred at one time) had entered into cease-fire agreements with the government. Other organizations had simply ceased to exist. In January 2007, the KNLA entered talks, through the KNU, with the government. Upon conclusion of the talks, the Burmese army agreed to withdraw from designated border areas. The informal agreement caused a stir within Karen leadership; some wanted to fight on, noting that the government had actually agreed to very little, while others hoped to negotiate a binding cease-fire.

---

[92] Khin, "Fifty Years of Struggle."
[93] South, "Conflict and Displacement in Burma/Myanmar," 54.
[94] Khin, "Fifty Years of Struggle."

Disunity followed, and soldiers from the KNLA 7th Brigade deserted, resettled in Thailand, or even joined the SPDC army.[95]

OTHER EFFECTS

Since 2004, Chinese investment in Burma, including loans in support of infrastructure development and offers of trade expansion, has increased substantially. China has also taken a position of sympathy toward Burma at the UN, including at Security Council sessions, where they have opposed economic sanctions proposed by the greater international community and have represented Burma's greater interest on the floor of the Council. China's relations with Burma, however, have not always been smooth. During 2006, Burma raised concerns of illegal logging by Chinese traders. The environmental group Global Witness reported that Burmese soldiers shot and killed Chinese loggers.[96] In addition, China has at times delayed payment of loans to Burma as a means of exerting influence, particularly related to political or economic goals and activities.[97] However, trade with China increased 46% in the first three quarters of 2007, and Burma's exports to China increased 5.2% from 2006. China has publicly stated that it supports the Burmese implementation of the seven-step road map. Chinese Assistant Foreign Minister He Yafei has publicly said, "We cannot permit Myanmar to fall into chaos, we cannot permit Myanmar to become another Iraq. No matter what ideas other countries have, China's stance on this is staunch."[98]

Drug trafficking within the border region of Laos, Thailand, and Burma, known as the Golden Triangle, continues. Production of methamphetamines, not heroin or other illicit drugs, is now the key business trade in this area controlled by the Shan State, and Burmese army officers have been alleged to be profiting directly from the drug trade. China, India, and Thailand have all been pressuring Burma to be more proactive in disrupting drug trafficking.[99]

For decades the United States has supported sanctions against Burma to compel the military regime to reform its human rights record. After 1988 and the prodemocracy uprising, the United States severed financial assistance and arms sales and downgraded diplomatic representation. Both the Clinton and G.W. Bush administrations

---

[95] R. H. Taylor, "Myanmar in 2007: Growing Pressure for Change But the Regime Remains Obdurate," *Southeast Asian Affairs 2008* (2008): 247–273.

[96] MacLean, "Spaces of Extraction," 87.

[97] Ibid.

[98] Taylor, "Myanmar in 2007," 247–273.

[99] Heaney, "Burma: Assessing Options for U.S. Engagement."

imposed economic sanctions (1997 and 2003), but the results have not been positive or forthcoming.[100]

Because of the challenges of drug trafficking and other illegal trading activities being conducted between the Thai and Burma border, Thailand has increased its military presence along the border provinces since approximately 1997. Military confrontations between Burmese and Thai forces have been reported on numerous occasions. The Royal Thai Third Army has the mission of maintaining border sovereignty along the Burmese border. Altercations between Thai and Burmese forces remain possible.[101]

In September 2007, the military regime conducted a crackdown on a popular uprising in an event known as the Saffron Revolution. Beginning on September 18, Buddhist monks resumed peaceful protests in several cities throughout the country. These marches quickly grew to include ordinary citizens, culminating in a large gathering of protesters in Rangoon on September 24. On September 26 and 27, the regime renewed its violent crackdown, shooting, beating, and arbitrarily detaining thousands of monks, prodemocracy activists, and onlookers. The event was unique in that protesters used the Internet: from Internet cafes in the Thai border regions, they flooded websites with photos and videos visually depicting images of the Tatmadaw using methods of violence against unarmed civilians, even the killing of monks. The SPDC government was unable to effectively restrict dissemination of news on the events, even though there was a dramatic increase in the number of sites that were blocked.

The Saffron Revolution was considered significant because citizen protesters were able to use transnational networks to distribute their message without using journalists as intermediaries and effectively (for a time) bypassing government restrictions and filters.[102] The regime confirmed the deaths of only ten protesters. However, some NGOs estimated the number of casualties to be much higher, and in their December 2007 report to the UN General Assembly, the Special Rapporteur on the Situation of Human Rights in Myanmar stated that there were more than thirty fatalities associated with the protests in Rangoon. In retribution for leading protest marches, monks were beaten and arrested, and many were disrobed. Additionally,

---

[100] Ibid.

[101] D. Ball, "Security Developments in the Thailand-Burma Borderlands" (working paper no. 9, Australian Mekong Resource Centre, University of Sydney, 2003), http://www.usyd.edu.au/mekong/documents/wp9.pdf.

[102] M. Chowdhury, "The Role of the Internet in Burma's Saffron Revolution," Berkman Center Research Publication no. 2008-08 (Harvard University, Berkman Center for Internet & Society, Cambridge, MA, 2008).

several monasteries were raided, ransacked, and closed. In addition to the more than 1,100 political prisoners whose arrests predate the crackdown, another thousand or more were detained because of their participation.[103]

BIBLIOGRAPHY

AFP. "Briefly: Plan Seeks to Protect Children in War Zones." *The New York Times*, July 27, 2005. http://www.nytimes.com/2005/07/26/world/americas/26iht-briefs.html?_r=1&scp=4&sq=%22Karen%20National%20Liberation%20Army%22&st=cse.

Ball, D. "Security Developments in the Thailand-Burma Borderlands" Working paper no. 9, Australian Mekong Resource Centre, University of Sydney, 2003, http://www.usyd.edu.au/mekong/documents/wp9.pdf.

Bandow, Doug. "Forgotten War in a Forgotten Country." The Cato Institute Web. Accessed September 9, 2010. http://www.cato.org/pub_display.php?pub_id=4622.

Becker, Joe. "Children at War." *The New York Times*, October 16, 2002.

Brees, I. "Burmese Refugee Transnationalism: What is the Effect?" *Journal of Current Southeast Asian Affairs* 28, no. 2 (2009): 23–46.

Briscoe, C. H. D. "Kachin Rangers: Allied Guerrillas in WWII Burma." HtoiGinTwang.Over-Blog.com (blog), http://htoigintawng.over-blog.com/article-kachin-rangers-allied-guerrillas-in-ww-ii-burma-51989324.html.

Central Intelligence Agency. "Burma." *The World Factbook*. https://www.cia.gov/library/publications/the-world-factbook/geos/bm.html.

Chowdhury, M. "The Role of the Internet in Burma's Saffron Revolution." *Harvard University, Berkman Center for Internet & Society, Research Publication* (2008).

Core, P. "Burma/Myanmar: Challenges of a Ceasefire Accord in Karen State." *Journal of Current Southeast Asian Affairs* 28, no. 3 (2009): 95–105.

Heaney, D. S. "Burma: Assessing Options for U.S. Engagement." Master's thesis, Naval Postgraduate School, 2009.

Jane's Intelligence Review. "Mahn Sha, General Secretary, Karen National Union (KNU), Interview." Accessed October 10, 2010. http://www.janes.com/articles/Janes-Intelligence-Review-2006/Interview--Mahn-Sha-General-secretary-Karen-National-Union-KNU.html.

---

[103] US Department of State, "Background Note: Burma," http://www.state.gov/r/pa/ei/bgn/35910.htm.

Khin, Ba S. "Fifty Years of Struggle: A Review of the Fight for Karen People's Autonomy." Life In Picture. Accessed October 10, 2010. http://www.kwekalu.net/photojournal1/soldier/story6.htm#hist.

Lintner, B. "Recent Developments on the Thai-Burma Border.." *IBRU Boundary and Security Bulletin* 3, no. 1 (1995): 72–76.

MacLachlan, H. "The Don Dance: An Expression of Karen Nationalism." *Voices: The Journal of New York Folklore* 32 (2007): 26–34.

MacLean, K. "Spaces of Extraction." In *Myanmar: The State, Community, and the Environment.* Edited by M. Skidmore and T. Wilson. Canberra, Australia: Asian Pacific Press, 2007.

Malseed, K. "Networks of Noncompliance: Grassroots Resistance and Sovereignty in Militarised Burma." Presentation, Yale Agrarian Studies Colloquium, April 25, 2008. Accessed October 10, 2010. http://www.khrg.org/khrg2008/khrg08w3.pdf.

Marshall, H. I. *The Karen People of Burma: A Study in Anthropology and Ethnology.* Columbus, OH: Dodo Press, 1922.

Min, W. "Looking Inside the Burmese Military." *Asian Survey* 48, no. 6 (2008): 1018–1037.

Oo, Z., and W. Min. *Assessing Burma's Ceasefire Accords.* Washington DC: East-West Center, 2007.

Pauker, Guy J. *Government Responses to Armed Insurgency in Southeast Asia: A Comparative Examination of Failures and Successes and their Likely Implications for the Future.* Santa Monica, CA: RAND Corporation, 1985.

Rajah, A. "A 'Nation of Intent' in Burma: Karen Ethno-Nationalism, Nationalism and Narrations of Nation." *The Pacific Review* 15, no. 4 (2002): 517–537.

———. "Burma: Protracted Conflict, Governance and Non-Traditional Security Issues." Working Paper, Institute of Defence and Strategic Studies, 2001.

"Rebel in Myanmar Promises a Long War." *The New York Times*, March 1, 2005.

Rogers, E. W. "Burma on the Brink: Complications for U.S. Policy in Burma." Master's thesis, Naval Postgraduate School, 1991.

Smith, M. *Burma: Insurgence and the Politics of Ethnicity.* Dhaka, Bangladesh: The University Press, 1999.

South, A. "Conflict and Displacement in Burma/Myanmar." In *Myanmar: The State, Community, and the Environment.* Edited by M. Skidmore and T. Wilson. Canberra, Australia: Asia Pacific Press, 2007.

Tan, A. T. H., ed. *A Handbook of Terrorism and Insurgency in Southeast Asia.* New York: Edward Elgar Publishing, 2007.

Taylor, R. H. "Myanmar in 2007: Growing Pressure for Change Bur the Regime Remains Obdurate." *Southeast Asian Affairs 2008* (2008): 247–273.

Thawnghmung, A. T. *The Karen Revolution in Burma: Diverse Voices, Uncertain Ends.* Washington DC: East-West Center, 2008.

SECTION II

······················

REVOLUTION BASED ON
IDENTITY OR ETHNIC ISSUES

GENERAL DISCUSSION

This section looks at revolutions in which conflicts are driven by strong undercurrents of (or in some cases, completely overt) identity or ethnic issues. Although the desired end-result is to overthrow the government, the motivating spirit of the revolution as well as its methods, operations, and support structures are heavily dependent on specific identity groups or ethnic divisions within the country. The cases within this section are indicative of this broad type of revolution, which almost invariably occurs when artificial national boundaries are created in the aftermath of colonialism or war, or when two very distinct ethnic groups are collocated without a democratic governance structure that accommodates the two groups equally.

Clashes between ethnic groups or races, or some other form of identity distinction, will always be a source of war and conflict. Of interest in this volume are those conflicts that directly impact the choice of governmental system. Such conflicts are most often precipitated by a government system that is biased, usually intentionally but sometimes through happenstance, toward one identity-based group to the disadvantage of another group. The lack of equal representation of the Catholics in Northern Ireland or the lack of a proportional political voice of the Tamil in Sri Lanka are examples. The ability of the government system and the surrounding cultural context to accept dissent and accommodate change as a result of the disadvantage is a strong factor in whether the situation escalates into warfare. For example, the twentieth-century civil rights marches in the United States were successful due to an eventual adoption of legal changes and cultural maturation that allowed for equal representation without a necessary turn to open warfare. Conversely, similar civil rights marches turned violent in Northern Ireland, where Catholics and Protestants clashed early and often and marches escalated into riots and a call for the British army to intervene and occupy the country.

It is self-evident that such revolutions will heavily depend on the resources and networks within the resisting ethnic/identity groups. To establish and perpetuate these support systems, two major factors seem to be critical. First is the establishment of a motivational or objective narrative that draws on the identity issues. To motivate the group, the revolutionaries must convince them of their "oneness," usually by finding or using a narrative that revolves around the distinctness of their race, ethnicity, religion, etc., and by incorporating that narrative into a tale of oppression, victimhood, or status lower than that of the ruling group. The ethnic and linguistic distinctiveness of the Tamil people and their lack of voice within the Sri Lankan government combined into

230

a powerful recruiting and motivation tool for the Liberation Tigers of Tamil Eelam (LTTE). The Provisional Irish Republican Army (PIRA) was not only nationalist but also Catholic, separated by neighborhood, church, and traditions that were evident in murals, press releases, and manifestos.

The second factor in many of these cases is an escalating tit-for-tat retribution cycle that occurs during the revolution. This escalation cycle provides a natural recruiting mechanism, for as the violence becomes more widespread and intense, more of the population can be persuaded that they should take sides. Psychologically, the gradual increase in violence can abet the radicalization process by resetting the norms of acceptable behavior and providing rationalization mechanisms for increased levels of violence and a broader target set.[1]

In conflicts of this sort, there is a delicate balance between the legitimacy of the claims made by the oppressed group, its sanctioned operations, and the counteractions of the government. If either the government or the revolutionary group goes beyond the publicly held "acceptability" in bombings or assassinations (see, for, example, Bloody Sunday in the PIRA case), the support can quickly turn to the other side as a reactive display of regret or disapproval. Often, the other group then overreaches within its own operations (as Bloody Friday would attest), bringing about a rebalance of the conflict.

The escalation of violence can also be perpetuated between competing insurgent groups as well. Often, when a splinter group is breaking off from the original insurgency or a new group is vying for relevance, the heightening of action is one way in which the new group tries to distinguish itself from the other. Other times, the violence is directed at the other group as internecine warfare takes over in a struggle for dominance. The PIRA fought against the original (official) IRA for many years as the Troubles started, with each side trying to gain allegiance and weaken the other side.

Revolutions with strong undercurrents of ethnic tension are incredibly enduring. Political solutions are unable to address many of the underlying identity divisions, which may stem from centuries-old prejudices. Identity hatred keeps the battle spirit going in times when operational success is fleeting. History is also full of examples when political solutions are averted solely by a small but well-targeted operation that reignites the underlying tension (e.g., the Palestinian/Israel situation).

---

[1] For a description of the psychological and sociological mechanisms of the radicalization process, see Chuck Crossett and Jason Spitaletta, "Radicalization: Relevant Psychological and Sociological Concepts" (Ft. Meade, MD: US Army Asymmetric Warfare Group, 2010).

LIBERATION TIGERS OF TAMIL EELAM (LTTE)

Maegen Nix and Shana Marshall

SYNOPSIS

Since 1976, the Liberation Tigers of Tamil Eelam (LTTE) has waged a separatist campaign against the government of Sri Lanka in pursuit of an *eelam* (homeland) for ethnic Tamils. The historical roots for this ethnic conflict date back to the period of colonial rule when British authorities favored the minority Tamils in matters of employment and the allocation of official resources, creating conditions ripe for ethnic strife after decolonization in 1948.[1] Once the majority Sinhalese gained control of the government, they instituted a range of discriminatory policies that marginalized the Tamil population and led to a cycle of violence and retribution. Although numerous Tamil nationalist groups would emerge in the struggle against the Sinhalese-led government, the LTTE became the dominant nationalist group by the mid to late 1980s and went on to achieve a stunning array of victories against the much larger Sri Lankan army. The LTTE was especially noteworthy for its innovations in suicide tactics, incorporation of women into fighting units, vast global network of fund-raising and smuggling operations, incorporation of conventional and guerrilla tactics, and a singularly successful public relations campaign aimed at garnering international sympathy for the Tamil cause. During three decades of fighting, the LTTE and the Sri Lankan government were both accused of employing heavy-handed military tactics in pursuit of their objectives at an estimated cost of 80,000–100,000 lives.[2] Despite achieving relative military parity with the Sri Lankan army by the mid-1990s and controlling some 15% of the island's territory, the LTTE was significantly weakened by the decision of several Western governments to designate the group as a terrorist organization and choke off its financing. This was followed by LTTE's loss of the strategically important eastern provinces in the summer of 2007 to an alliance of LTTE defectors and the Sri Lankan government. In

---

[1] Because the British did not anticipate the eruption of significant ethnic conflict in Sri Lanka, they rejected Tamil requests to insert principles of minority protection into the constitution. By contrast, the British colonial government in Malaysia considered the country an ethnic powder keg and encouraged structures of political compromise that promoted representation of the country's minority ethnic groups.

[2] For details on the Sri Lankan government's harsh stance, see International Crisis Group, "Sri Lanka's Return to War: Limiting the Damage," Asia Report no. 146, February 20, 2008, http://www.unhcr.org/refworld/country,,ICG,,LKA,4562d8cf2,47bc2e5c2,0.html.

May 2009, the Sri Lankan army conducted a northern offensive that resulted in a declaration of final victory by the Sri Lankan government and a concession of defeat by the LTTE.

TIMELINE

1948	Ceylon independence from British rule established by cooperation of Sinhalese and Tamil elites.
1956	Sri Lanka Freedom Party (SLFP) wins national election on the basis of "Sinhalese Only" platform.
	Sinhala Only Act of 1956 sparked the first anti-Tamil riots.
1958	Riots and protests against proposals of Tamil self-rule.
1961	Sri Lankan army stationed in northeast Sri Lanka to suppress peaceful Tamil protests against discrimination.
1972	Anti-Tamil policies formally incorporated into the constitution.
	The Tamil New Tigers (TNT) established in 1972.
January 1974	Police attacked the Fourth International Tamil Conference in Jaffna, killing eleven Tamils.
1976	LTTE formed from the TNT under Velupillai Prabhakaran.
1981	Burning of the Jaffna Library, which housed 90,000 Tamil books and manuscripts.
July 1983	LTTE ambush of a Sri Lankan army convoy that kills thirteen soldiers and sparks riots that kill 2,500 Tamils.
1987	LTTE employs first noted suicide bombing of a Sri Lankan army camp followed by conventional tactics.
July 1987	India and the Sri Lankan government sign the Indo-Lankan Peace Accord; India deploys military forces to Sri Lanka.
Mar 1990	India withdraws forces from Sri Lanka.
May 1991	LTTE employs a suicide bomber to assassinate Indian Prime Minister Rajiv Gandhi.
May 1993	LTTE employs a female Black Tiger to assassinate Sri Lankan President Ranasinghe Premadasa.
October 1997	LTTE is placed on US State Department list of foreign terrorist organizations.
2002	Norway brokers cease-fire agreement between LTTE and Sri Lankan government.

2004	Tsunami hits Sri Lanka and causes 40,000 deaths.
March 2004	Colonel Karuna splits LTTE Eastern command away from Prabhakaran-led Northern command.
2005	The Sri Lankan government incorporates national military draft system that substantially increases the size of the Sri Lankan army.
May 2005	LTTE assassinates Sri Lankan government Foreign Minister Lakshman Kadirgamar.
November 2005	Anti-LTTE hard-liner Mahinda Rajapaksa wins national elections.
2006	Colonel Karuna founds the *Tamil Makkal Viduthalai Pulikal* (TMVP) in opposition to the LTTE.
	The Sri Lankan government begins military campaign against LTTE and Tamil population with support of Tamil opposition.
2007	LTTE Air Tiger attack against the Colombo airport.
July 2007	Sri Lankan army controls Eastern Sri Lanka.
December 2007	US government suspends military aid to Sri Lanka because of Sri Lankan government human rights violations.
January 2008	The Sri Lankan government formally withdraws from cease-fire and Norwegian monitors depart Sri Lanka.
2009	250,000 civilians displaced because of fighting in northern Sri Lanka.
May 2009	Sri Lankan government claims victory over LTTE after large military operation.

THE ENVIRONMENT OF THE REVOLUTION

PHYSICAL ENVIRONMENT

Sri Lanka is a pear-shaped island the size of Ireland with a population slightly smaller than that of Australia. Sri Lanka is separated from the southern coast of India by the twenty-mile-wide Palk Strait. Composed of mostly flat terrain with mountains in the south-central region, Sri Lanka has a tropical climate, but temperatures are moderated by ocean winds from the island's 800 miles of coastline. The northeast and the southwest regions of the country are arid and relatively unsuitable for agriculture, although the rest of the island receives ample rain from monsoons. Much of the fighting during the LTTE insurgency occurred directly across from India on the northern Jaffna peninsula, as well as down the country's eastern border, both areas where the Tamil population is concentrated. Sri Lanka's location astride key Indian Ocean shipping routes amplified the regional significance of the insurgency because of the potential disruption of global transport.[3]

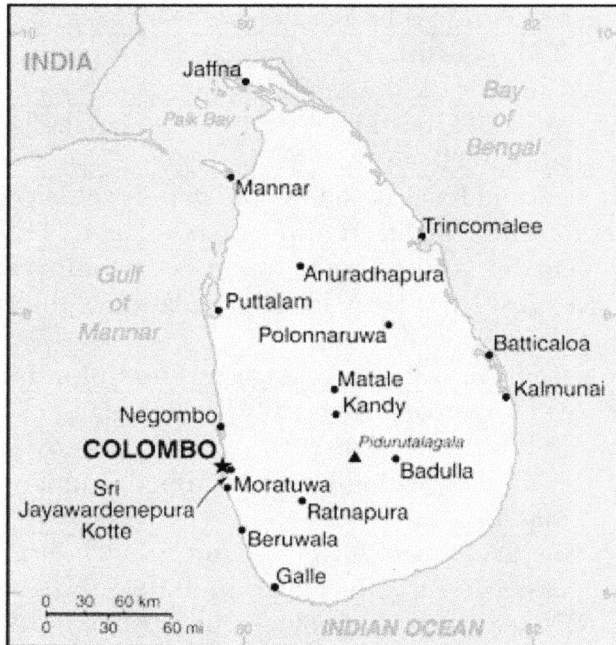

Figure 1. Map of Sri Lanka.[4]

CULTURAL AND DEMOGRAPHIC ENVIRONMENT

Sri Lanka's population of roughly twenty million is, by almost all accounts, well over two-thirds Sinhalese. According to the official 2001 Sri Lankan census, the population distribution is approximately 82% Sinhalese, 9.4% Tamil, and 8% Muslim.[5] Population statistics are highly politicized and unreliable, however, with alternative sources suggesting an accurate estimate for the Tamil population as closer to 18%.[6] The Tamil concentration in the north and east makes them

---

[3] China's growing trade and investment infrastructure necessitates secure access to these routes, which some analysts cite as a primary motivation in China's recent military and political support for the Sri Lankan government.

[4] Central Intelligence Agency, "Sri Lanka," *The World Factbook*, accessed August 9, 2010, https://www.cia.gov/library/publications/the-world-factbook/geos/ce.html.

[5] The census did not include LTTE-dominated areas (Jaffna, Kilinochchi, and Mulativu districts), although the Sri Lankan government reports that these final data cover 94% of the country's population.

[6] The colonial-era estimate of the Tamil population was 11%.

the majority in some districts, although less so in the east where large portions of the Muslim minority are also concentrated.[7] The Sinhala language has classical Indian roots, but today is a distinct tongue spoken only in Sri Lanka.[8]

The vast majority of Tamils are Hindu, and the Sinhalese are overwhelmingly Buddhist. Although Buddhism originated in India, it was driven to the peripheries of the subcontinent by centuries of Hindu expansion. Sinhalese, therefore, view themselves and their island home as the last line of defense for a besieged religion. Despite this religious divide, the dynamics of the conflict have largely revolved around issues of resource distribution and political representation, not religious differences. The presence of sixty million Tamils across the narrow Palk Strait in the Indian state of *Tamil Nadu* feeds into historical Sinhalese fears of renewed subjugation by Hindu invaders—a narrative that is frequently exploited by politicians and religious figures.[9] Moreover, a 2002 NASA satellite image of the strait initiated a predictable rise in tensions when it revealed a submerged chain of islands linking the two countries. The geological formation was seized upon by many Hindu nationalists in India and Sri Lanka as the ancient bridge built by the Hindu King Rama to rescue his kidnapped wife. Myths, legends, ancient religious texts, and archaeology have all been routinely exploited by Tamil and Sinhalese nationalists attempting to write the island's history to suit their own purposes.

SOCIOECONOMIC ENVIRONMENT

Although relatively poor, Sri Lanka ranks above most other South Asian nations in important development indicators, notably the United Nation's Human Development Index (HDI). Among the favorable factors contributing to Sri Lanka's HDI ranking are a long life expectancy (74 years), high adult literacy (91%), and a relatively high gross domestic product (GDP) per capita.[10] However, income inequality in Sri Lanka greatly exceeds that of neighboring

---

[7] For alternative population figures, see Stephen Hopgood, "Tamil Tigers, 1987–2002," in *Making Sense of Suicide Missions*, ed. Diego Gambetta (Oxford: Oxford University Press, 2005), 43.

[8] K. M. De Silva, *A History of Sri Lanka* (Berkeley, CA: University of California Press, 1981).

[9] Robert D. Kaplan, "The Buddha's Savage Peace," *The Atlantic* (September 2009), and Stephen Hopgood, "Tamil Tigers, 1987–2002" both give a figure of 18% for the Tamil minority. The figure of sixty million Indian Tamils is given in "A World of Exiles," *The Economist* 366 (January 2, 2003): 41.

[10] "Sri Lanka," in Human Development Report 2009 (New York: United Nations Development Programme).

India, Pakistan, and Bangladesh and is more in line with trends seen in Sub-Saharan Africa.[11] Many observers attribute Sri Lanka's history of youth radicalization to a combination of high literacy rates and an underdeveloped economy that is unable to absorb skilled labor.[12] Moreover, before the conflict resulted in massive migration, the Tamils were better educated and more affluent than the majority of their Sinhalese counterparts because of British colonial favoritism and the concentration of American missionary schools in Tamil population centers.[13]

Agriculture accounts for roughly 30% of the island's employment and is primarily centered on rice production and tea cultivation. Sri Lanka was previously a leading rice exporter for much of Southeast Asia, but changing weather patterns and the escalating conflict turned the country into a net importer. Sri Lanka is the world's largest exporter of tea, which remains a staple crop for the country and accounts for roughly 15% of its GDP.[14] The tea plantations of the central highlands are home to the Indian Tamils—relatively recent migrants of Tamil ethnicity who were brought by British colonists to labor on the plantations.[15]

Economic underdevelopment in the postcolonial era greatly exacerbated the ethnic conflict between the Tamil minority and Sinhalese majority. During the LTTE's formative period in the 1970s, Marxist thought dominated the rhetoric of most antiestablishment movements, including Tamil separatism. However, Marxism was also a major ideological theme for Sinhala nationalists who drew on the Tamil's historically privileged status to argue their own case for fighting against Tamil elitism.

---

[11] Sri Lanka's Gini coefficient, a measure of income inequality, was 41 in 2002, the last year for which data are available from the World Bank. This is more similar to the measurements for Ghana and Nigeria (both 43), than India (37), Pakistan (31), or Bangladesh (31).

[12] Gamini Samaranayake, "Patterns of Political Violence and Responses of the Government of Sri Lanka, 1971–1996," *Terrorism and Political Violence* 11, no. 1 (1999): 113.

[13] No adequate statistics exist to judge current rates of inequality between the Sinhala and Tamil populations.

[14] "Sri Lanka Moves to Protect Tea Industry," *BBC News*, February 19, 2003, accessed August 9, 2010, http://news.bbc.co.uk/2/hi/business/2779267.stm.

[15] Indian Tamils have historically been a more isolated and poor population, working mostly on tea plantations in central Sri Lanka. They are generally less active politically than their indigenous co-ethnics. Sinnappah Arasaratnam, "Nationalism," in *India and Ceylon: Unity and Diversity*, ed. Philip Mason (London: Oxford University Press, 1967), 262.

HISTORICAL FACTORS

The historical record of interactions between Tamils and Sinhalese is central to understanding the LTTE insurgency and is also highly contentious. Before the Dutch, Portuguese, and British came to Sri Lanka, Hindu Tamils from India invaded the prosperous Buddhist city-state of Anuradhapura and, over the course of several centuries, established a Tamil dynasty.[16] This introduction was followed by centuries of intense intermingling of Buddhism and Hinduism, as demonstrated by the island's many temples that house gods of both faiths as well as the historical accounts of systematic intermarriages within the ruling dynasties. When the British overthrew these dynasties, they eliminated many of the traditional institutions (such as elite intermarriage) that previously engineered close ties between the two groups. Once established, the colonial government insisted that political and civil associations organized around ethnicity were divisive and backward and that the unitary nation-state should serve as the organizational and political foundation of the people.[17] This British policy of alienating existing organizations coincided with the establishment of democratic "majority-rule" principles—without minority protections—and greatly weakened the Tamil voice in politics. Ultimately, British efforts to modernize what was then Ceylon by recreating Western institutions of statehood in a multiethnic society would destroy the very mechanisms that had evolved over centuries to mediate intergroup conflict.

Because Sri Lanka was under full British control, arriving American missionaries were relegated to the agriculturally inhospitable area of Jaffna—the center of Sri Lanka's Tamil population. These missionaries proved adept at building and operating English-language schools, which created a reserve of well-trained, English-speaking Tamils. Because of their language skills and a high level of migration from Jaffna's nonarable territory, Tamils were disproportionately represented in both the British colonial administration and in commercial industry.[18] Predictably, this led to perceptions, both real and imagined, of Tamil

---

[16] The Dutch ceded their coastal claims in Sri Lanka in 1802, whereby it became Britain's first crown colony.

[17] This is in contrast to British policy in India, which accommodated religious, ethnic and linguistic differences by establishing a constitution based on a federation of states. No accommodation was made for similar Sri Lankan cleavages. Michael Roberts, "Ethnic Conflict in Sri Lanka and Sinhalese Perspectives: Barriers to Accommodation," *Modern Asian Studies* 12, no. 3 (1978): 361.

[18] Donald L. Horowitz, *Ethnic Groups in Conflict* (Berkeley, CA: University of California Press, 1986): 156.

cooperation with the British colonial government.[19] In a development common to many cases of ethnic rivalry, demands from the Sinhalese for equality with the Tamils eventually gave way to demands for preferential treatment.[20]

Aided by the introduction of the printing press and rising literacy rates, Sinhala scholarship and social commentary on ancient Tamil invasions led to a climate wherein all Tamils, even those who could trace their Sri Lankan heritage back for centuries, were considered *kallathoni*—or illegal Indian immigrants.[21] In the discourse of Sinhala nationalism, Tamils were merely the first of many waves of invaders that included European colonists and other foreign capitalists. In addition, a particularly active and militant Buddhist clergy created myths and legends to explain away generations of mixing between Sinhalese and Tamils, thus further exacerbating religious tensions.[22]

Overall, this legacy of socioeconomic privilege granted to the Tamils by the British colonial government led to decades of reactionary Sinhalese nationalist policies that effectively barred Tamils access to civil service employment, the university system, electoral politics, and social services; recognized the Sinhala language and religion (Buddhism) as those of the state; and resettled Sinhala peasants on Tamil land. In addition, the Sinhala-dominated military launched indiscriminate campaigns in Tamil territory in response to attacks by Tamil separatists. Consequently, academics consider Tamil separatism as a unique case of a relatively more affluent minority trying to secede from a comparatively "backward society" as opposed to the more common case of traditional or fundamentalist movements attempting to withdraw from modern, liberal states.[23]

---

[19] Gallege Punyawardana, secretary of the Federation of Buddhist Organizations, evidenced this common impression when he blamed the British for an LTTE attack on a Buddhist shrine, commenting that the Tamils "fought along with their English masters against the Sri Lankans. They are the originators of our problem." See Reuters, "British Blamed for Tamil Attack," *The Independent* (January 27, 1998). The Kingdom of Kandy, the center of Sinhala/Buddhist power and culture in Sri Lanka, was also the last province to fall to British tutelage, further contributing to this perception.

[20] Donald L. Horowitz, *Ethnic Groups in Conflict*, 197.

[21] Ceylon Daily News (Colombo), May 15, 1970, cited in Donald L. Horowitz, *Ethnic Groups in Conflict*, 210. Early Tamil migration, notably the one that helped establish the thirteenth-century Tamil dynasty in the north, took on added significance during the Sinhalese struggle against British imperialism.

[22] One such legend concerns the adoption of Tamil (Hindu) gods by Sinhalese (Buddhist) worshippers—a common feature in religiously diverse societies. In the Sinhalese narrative, the god Kandeswami abandons his Tamil worshippers in favor of the Sinhalese when the latter agree to carry him across a river after this responsibility is shirked by a passing band of Tamils. Donald L. Horowitz, *Ethnic Groups in Conflict*, 141.

[23] Donald L. Horowitz, *Ethnic Groups in Conflict*, 155–156.

GOVERNING ENVIRONMENT

Sri Lanka's post-independence governing environment has been characterized by cycles of accommodation and obstruction of Tamil rights, all within an atmosphere of increasing ethnic polarization. Upon gaining independence in 1948, the ruling United National Party (UNP) portrayed itself as a multiethnic party—albeit one with Sinhalese leadership—and made electoral appeals to all ethnic groups. The UNP owed its early success to relatively high levels of integration of the Sinhalese and Tamil elite, with many attending the same schools and having struggled side by side in the independence movement.[24] However, as the memory of British imperialism receded, so too did this brief experiment in non-identity-based politics. In 1956, the Sri Lanka Freedom Party (SLFP) swept to victory on a tide of ultraexclusionary Sinhalese nationalism. Since then, political success in Sri Lanka has rested largely on the commitment of leaders to the parochial interests of their co-ethnic constituents, amounting to a situation where interethnic clashes are frequently precipitated by intraethnic political competition in which the most hard-line ethnic politicians garner the support of their constituents.[25]

In the period immediately after independence, Sri Lanka's parliamentary system awarded seats on the basis of election by plurality in largely homogenous, ethnic regions of the country. These conditions led to the election of candidates who appealed to the most extreme elements of their ethnic constituencies as the candidates had no reason to accommodate appeals from other ethnic groups whose members were not present in any electorally meaningful numbers within their districts. Likewise, candidates running for national office had much more to gain from appealing to extreme Sinhalese voters than moderate Tamil voters because of the overall majority numbers of the Sinhalese. The Parliament was eventually dominated by two main Sinhalese parties, although it also included some smaller, third parties that represented other groups, including Tamils, Muslims, and Up-Country (or Indian) Tamils.[26]

---

[24] This partially explains the British failure to predict the coming tide of ethnic conflict. Because the very uppermost echelons of the elite were relatively well integrated, the British mistook this condition as representative of the entire population. Indeed, most of the attempts at accommodation have occurred under the UNP, whose leadership has been drawn primarily from among these colonial-era elites and their descendents.

[25] Donald L. Horowitz, "Incentives and Behaviour in the Ethnic Politics of Sri Lanka and Malaysia," *Third World Quarterly* 11, no. 4 (October 1989): 18–35.

[26] Indian Tamils are also known as "up-country" Tamils because of their residence on the tea plantations of the central highlands where their ancestors were brought by the British to work.

The sheer numerical superiority of the Sinhalese electorate meant that, even if a party aspired to gain ruling status, it need not appeal to any of the state's ethnic minorities. This was clearly demonstrated on numerous occasions when Sinhala-led governments attempted to implement policies to accommodate Tamil grievances but which invariably mobilized such powerful opposition from Sinhala nationalists that the issue of Tamil autonomy singularly determined the political fortunes of the major Sinhalese parties. If one party supported the proposal, the opposition party would oppose it and leech supporters away from the more "accomodationist" politicians.[27] In 1957, the Sri Lankan prime minister agreed to a plan that would have granted significant powers of self-rule to local Tamil authorities in the north and east, but this plan was met with sustained opposition from Sinhala groups who toured the country with maps that displayed large black footprints over the areas that would be "ceded" to the Tamils.[28] These activities prompted a series of riots and protests in 1958 during which Tamils were attacked by civilian mobs. In 1957, and again in 1965, pacts were signed between Tamil and Sinhalese leaders to accommodate Tamil grievances, and Tamil politicians were briefly included in a coalition government from 1965–1968, but these agreements were again abrogated in the face of virulent Sinhalese opposition.[29] Finally in 1978, the electoral system was altered to include a separately elected executive and a party-list system that included proportional representation in multimember districts. By this time, however, most Tamil politicians had already boycotted the Parliament and armed insurrection had taken hold.[30] The strength of Sinhala nationalism and the failure of mainstream Tamil politicians to secure any noticeable gains, therefore, encouraged the eventual growth and sustainment of radical Tamil separatist movements.

---

[27] Donald L. Horowitz, *Ethnic Groups in Conflict,* 133.

[28] The most significant effects of this devolution, known as the Bandaranaike-Chelvanayakam Pact, would have been linguistic autonomy and the return of Tamil civil servants, company clerks, and traders to Tamil-dominated areas. This would have reversed the flow of migration of educated and wealthy Tamils back to Jaffna and away from the capital Colombo, a move that would have strengthened Sinhalese economic prospects. Nonetheless, its symbolism prevented its adoption. Robert N. Kearney, *Communalism and Language in the Politics of Ceylon* (Durham: Duke University, 1967), 117–118.

[29] The Tamil Federal Party members of Parliament supported Sinhalese leaders at various points when those leaders made pacts with Tamil leaders or agreed to implement certain legislation. However, one accomodationist Sinhalese prime minister, Bandaranaike, was assassinated by an extremist Buddhist monk for making a pact with a Tamil leader, after which most of the Sinhalese–Tamil agreements were abrogated, the Federal Party resumed protest against the government, and many Tamil members of Parliament were imprisoned.

[30] Horowitz, "Incentives and Behaviour."

WEAKNESSES OF THE PREREVOLUTIONARY ENVIRONMENT AND CATALYSTS

Although the two decades after independence would see an oscillation of intercommunal violence and formal attempts at reconciliation, the events of the 1970s eventually set the two rival ethnic groups on a collision course. In 1972 the Sinhalese-led government formalized a number of preexisting, anti-Tamil policies by incorporating them into the constitution. These amendments made Sinhala the sole language of the court system and government administration, accorded special status to the Buddhist religion, and codified various policies that gave the Sinhalese preference in civil service employment, university admissions, and professional exams.[31] The adoption of the new constitution, combined with a militant ethos promoted by some of the Tamil student movements, prompted a proliferation of separatist organizations, including as many as thirty-six explicitly militant groups.[32] The Tamil Youth League's open advocation of violent acts against the state led to the arrest of not only its leaders but also the leaders of many other Tamil organizations.[33] There were also less extremist organizations that were actively engaged in the political system, including the Tamil United Front (TUF) from which the more militant Tamil United Liberation Front (TULF) eventually broke away. The TULF, although also active in politics, included a clandestine military wing. Communal violence continued to escalate in the 1970s and into the 1980s, marked by the burning of the Jaffna Library (which housed 90,000 Tamil books and manuscripts) in 1981[34] and culminating in the anti-Tamil riots that killed an estimated 2,500 Tamils in 1983.[35] Sparked by an LTTE attack on a Sri Lankan army convoy, which killed thirteen soldiers, the 1983 riots would reinforce the LTTE message that nothing short

---

[31] In 1971, legislation was passed that led to higher university exam requirements for Tamil speakers. The introduction of a district quota system meant that many well-educated Tamils from Jaffna and Colombo (the historic education centers of the country) could not gain entrance to universities because spaces were reserved for lower-scoring students from less-affluent rural districts.

[32] Rohan Gunaratna, *War and Peace in Sri Lanka* (Sri Lanka: Institute of Fundamental Studies, 1987), 27.

[33] Bruce Hoffman, *Inside Terrorism* (New York: Colombia University Press, 2006), 138.

[34] The LTTE is blamed for the assassination of UNP presidential candidate Gamini Dissanayake in 1994, presumably for his role in the library burning. Stephen Hopgood, "Tamil Tigers."

[35] Jayshree Bajoria, "Backgrounder: The Sri Lankan Conflict" (Council on Foreign Relations, May 18, 2009), accessed August 9, 2010, http://www.cfr.org/publication/11407/. Official Sri Lankan government statistics claimed that only 350 Tamils were killed in the rioting.

of a separate Tamil state would provide the community with security.[36] No government response to the riots was forthcoming other than the issuance of an edict outlawing separatism. The anti-Tamil violence spurred mass migration, resulting in further geographic polarization along ethnic lines and a large supply of displaced persons who were quickly absorbed by the recruitment arms of the militant groups.[37]

In addition to this escalation in ethnic violence, resettlement and state irrigation projects sponsored by the Sri Lankan government in the 1980s further exacerbated tensions by relocating tens of thousands of landless Sinhalese from the south and west into the Tamil-majority lands of the east. These projects brought Sinhala and Tamil farmers into direct conflict over water resources that were increasingly being diverted to support the sugarcane production practiced primarily by resettled Sinhalese. The projects were also framed as a return to the days of the ancient Sinhala kingdoms that dominated the east, thus further enhancing both Tamil and Muslim fears of being overrun. These government-designed settlement programs exposed Sinhala settlers to acts of vengeance from Tamil groups who considered the Sinhala settlers legitimate military targets, resulting in an anti-Tamil mind-set among settler families whose sons were increasingly recruited into either the regular army or paramilitary organizations.[38] In a country where 85% of the population was rural, the resettlement schemes had a major impact on demographic composition in many provinces, shifting the majority–minority distribution between ethnic groups and further exacerbating ethnic tensions.[39]

---

[36] The Sri Lankan army publicly displayed the corpses of the slain soldiers, and anti-Tamil riots broke out the very next day. Narayan Swamy, *Tigers of Sri Lanka* (Delhi: Konark Publishers Pvt Ltd., 1994), reports that Prabhakaran himself took part in this attack, which is credited with sparking the onset of major hostilities.

[37] Rohan Gunaratna, "Sri Lanka: Feeding the Tamil Tigers," in *The Political Economy of Armed Conflict: Beyond Greed and Grievance*, ed. K. Ballentine and J. Sherman (Boulder, CO: L. Rienner Publishers, 2003), 199.

[38] International Crisis Group, "Sri Lanka's Eastern Province: Land, Development, Conflict," Asia Report no. 159, October 15, 2008, 5, http://www.crisisgroup.org/en/regions/asia/south-asia/sri-lanka/159-sri-lankas-eastern-province-land-development-conflict.aspx.

[39] See "Sri Lanka's Eastern Province: Land, Development, Conflict," for population figures.

FORM AND CHARACTERISTICS OF THE REVOLUTION

OBJECTIVES AND GOALS

The terrorist attacks of September 11, 2001, significantly impacted both popular and academic narratives about the LTTE movement. The group's use of suicide tactics and its expansive transnational support network were attributes it shared in common with Al Qaeda, but attempts by analysts to categorize the LTTE as motivated primarily by ideology or religion were overstated. Although the LTTE did target religious sites, such as the "Temple of the Tooth" in Kandy, the majority of their targets were overtly political.[40] As such, the LTTE was primarily an ethno-nationalist movement seeking secession, not a Hindu or Marxist-Leninist movement—although both of these belief systems did provide material for recruiting and maintaining support.[41]

As the LTTE's power shifted vis-à-vis the Sri Lankan army and as alternatively accomodationist and hard-line Sinhalese governments came into power, the demands and objectives of the LTTE also shifted. Overall, the group espoused very specific, concrete political goals. Official statements by the LTTE's political wing cited the group's "clearly defined political programme" which aimed to secure "self-determination" for the Tamil people. LTTE literature underscored those characteristics of the Tamil people that classified them as a nation meriting the right to self-determination, including a "well-defined, contiguous territory," "distinct language and culture," "unique economic life," and "lengthy history."[42] This call for self-determination manifested itself in efforts to secure either an independent Tamil state (a demand first propagated by the TULF in 1976) or at least substantial autonomy in the form of devolution of central authority to local rulers or the establishment of an ethnic confederation. In 2002, a cease-fire was based on a deal that included a de facto state partition. During these peace negotiations, the LTTE declared that if substantial autonomy for the Tamil areas was

---

[40] The "Temple of the Tooth" is Sri Lanka's holiest Buddhist shrine, built around a sacred relic believed to be the canine tooth of the Buddha smuggled to Sri Lanka as Indian Buddhists fled Hindu invaders in 313 AD.

[41] For LTTE writings that make reference to Marxist principles, see Anton Balasingham, "Liberation Tigers of Tamil Eelam: The Birth of the Tiger Movement," 1983, accessed August 9, 2010, http://www.tamilnation.org/ltte/83birth.htm.

[42] Letter from the international secretariat of the LTTE to the US Court of Appeal, District of Columbia Circuit (November 6, 1997). Available online at the website of the Association of Tamils of Sri Lanka in the USA, http://www.sangam.org/NEWSEXTRA/ltte.htm.

not forthcoming, it would revert to its demand for an independent homeland. The scaling back of LTTE demands (from independence to autonomy) took many parties by surprise, but because autonomy was never forthcoming from the Sri Lankan government, the LTTE never had to make good on this promise. Despite the relative clarity of LTTE's goals, many Tamil intellectuals consistently criticized not only the LTTE but also the entire Tamil nationalist movement for making unreasonable demands of Sinhala leaders, including early Tamil calls for overrepresentation in official government institutions.[43]

Figure 2. Emblem of the LTTE.[44]

In pursuit of its goal to achieve an independent Tamil state, the LTTE first sought to achieve dominance as the sole legitimate representative of the Tamil people. This goal meant eliminating rival Tamil groups, something the LTTE did to great effect through political maneuvering and through violence. Many of the LTTE's operations—such as political assassinations and suicide missions—contributed to the image of an elite, professional force that distinguished it from rival Tamil organizations.[45] This distinction was reinforced on numerous occasions when other Tamil groups joined with the Sri Lankan

---

[43] These include requests for 50/50 ethnic representation in Parliament—50% for Sinhalese, 50% for other minorities—made by prominent Tamil politicians, as well as demands for a separate state.

[44] "File:Ltte emblem.jpg," *Wikipedia*, accessed August 9, 2010, http://en.wikipedia.org/wiki/File:Ltte_emblem.jpg.

[45] Bruce Hoffman, *Inside Terrorism*, 137.

government to support government-sponsored devolution schemes that were inevitably watered down or thrown out.

Although Tamil politicians were elected to Parliament through provincial elections beginning in 1977, militant activity continued. In fact, the elections themselves became a major vehicle for political violence, with each round eliciting increasingly organized and systematic attacks by both parties.[46] Then, in 1983, the LTTE called for a boycott of the provincial elections and later ambushed a Sri Lankan army convoy. This ambush sparked a wave of anti-Tamil riots that killed thousands of Tamils and destroyed tens of thousands of their homes and businesses. Considered a watershed point by many analysts, the 1983 riots were largely ignored by the government, which passed legislation outlawing secessionism but made no serious effort to prosecute the anti-Tamil offenders. Recruitment among armed groups exploded and Tamil politicians forfeited their parliamentary seats in protest. Subsequent government responses to the armed violence, including the imposition of martial law and the blockade of the Jaffna peninsula, only served to increase the ethnic violence.

The Indo-Lankan Peace Accords of 1987 brought Indian Peacekeeping Forces (IPKF) into Sri Lanka and had the unintended consequence of radicalizing both sides of the conflict. Sinhala nationalists resented and distrusted the largely Hindu troops whose government had armed, trained, and otherwise supported the LTTE for years in an effort to bolster its popularity among its own Tamil population, only to reverse course over fears that the conflict could spill over into southern India. The Sinhala nationalists who opposed the Indian presence were subjected to a brutal campaign of repression by the Sri Lankan government. For its part, the LTTE targeted both the peacekeepers and the Tamil groups that cooperated with the peacekeepers. Eventually, the LTTE and Sri Lankan government joined forces to expel the peacekeepers in an informal alliance that dissolved quickly after the withdrawal of the Indian troops in March 1990.

In addition to the larger goals and objectives of the LTTE movement, groups within Sri Lanka have also exploited the conflict to further their own narrow interests and political aspirations, as President Premadasa and the UNP did when they falsely accused the left-wing Sinhala *Janatha Vimukhthi Peramuna*, or People's Liberation Front (JVP), of participating in the 1983 anti-Tamil pogroms as a pretense to

---

[46] Gamini Samaranayake, "Patterns of Political Violence and Responses of the Government of Sri Lanka."

banning the group from participating in national elections.[47] Radical Buddhist clergy have also capitalized on the conflict in support of their efforts to increase their religious status. Reactionary xenophobia has thus become a common feature of Sri Lankan politics and has led many observers to characterize Sinhala society as a paranoid "majority community with a minority complex," that sees itself squeezed between a powerful northern neighbor, the international community's sympathy for the Tamil cause, and a historical trajectory leading to the demise of its religion and culture.[48] The impact of this dynamic on the governing environment has been clearly visible throughout the conflict.

External mediation efforts—including numerous cease-fires and peace processes—have been highly contentious for a number of reasons and have often served to heighten already strong nationalist sentiments. Many rounds of negotiation included economic development components that were viewed by many concerned Sinhalese nationalists as a way to channel international capital and foreign-aid dollars away from the south and toward the LTTE-dominated north and east. The focus on preventing conflict between the major parties also meant that human rights violations within the groups received little attention. These oversights served to intensify Sinhala fears and suspicion of peace efforts while simultaneously allowing for the widespread abuse of Tamil civilians by their LTTE "protectors" as well as the abuse of Sinhalese citizens by the central government.

---

[47] International Crisis Group, "Sri Lanka: Sinhala Nationalism and the Elusive Southern Consensus," Asia Report no. 141, November 7, 2007, 11, http://www.crisisgroup.org/en/regions/asia/south-asia/sri-lanka/141-sri-lanka-sinhala-nationalism-and-the-elusive-southern-consensus.aspx. The JVP, which emerged in the mid-1960s from within the Maoist wing of the Communist Party, is perhaps the most well-known militant actor within the Sinhala nationalist movement. It launched numerous insurrections against the government including an uprising in 1987 that left nearly 60,000 dead on both sides. Each JVP offensive was met with even more brutal counter-campaigns by the government until eventually the group's leadership was eliminated. What remained of the JVP slowly evolved into a political party and now has several members of Parliament in the government. Many experts actually credit UNP activists (under the leadership of Sirimavo Bandaranaike, one of the UNP's most virulently nationalist party leaders) with organizing the anti-Tamil riots—suggesting that the UNP's persecution of the JVP was in fact an effort to divert attention from its own responsibility. Stanley Jeyaraja Tambiah, *Sri Lanka: Ethnic Fratricide and the Dismantling of Democracy* (Chicago: University of Chicago Press, 1986), 32.

[48] Michael Roberts, "Ethnic Conflict in Sri Lanka and Sinhalese Perspectives," 368. See also Robert D. Kaplan, "The Buddha's Savage Peace."

LEADERSHIP AND ORGANIZATIONAL STRUCTURE

The LTTE mirrored the organizational structure of many separatist and revolutionary groups in that it was composed of two primary wings, one military and one political, with the latter largely subservient to the leadership of the former. The Central Governing Committee oversaw both wings as well as several military subunits, including a naval unit (the Sea Tigers), an air unit (the Air Tigers), a conventional unit (the Charles Anthony Regiment),[49] an elite unit often tasked with suicide missions (the Black Tigers), an intelligence unit, and a political office.[50] An international secretariat oversaw the LTTE's global network, including publicity and propaganda, arms procurement and shipping, and fund-raising.[51]

Figure 3. Velupillai Prabhakaran.[52]

Strong leadership was a critical aspect of the Tamil struggle and the LTTE was nearly synonymous with its founder, Velupillai Prabhakaran. His veneration by LTTE members led some observers to liken the LTTE to a religious cult rather than an ethno-nationalist movement,[53] although Prabhakaran himself was believed to be

---

[49] This unit, composed mostly of Northern (as opposed to Eastern) Tamils, was first trained in guerrilla tactics, slowly evolving into a conventional unit.

[50] Peter Chalk, "Liberation Tigers of Tamil Eelam's (LTTE) International Organization and Operations–A Preliminary Analysis," *Commentary* no. 77 (Winter 1999).

[51] Ibid.

[52] "File:Velupillai Prabhakaran.jpg," *Wikipedia,* accessed August 9, 2010, http://en.wikipedia.org/wiki/File:Velupillai_Prabhakaran.jpg.

[53] Bruce Hoffman, *Inside Terrorism,* 132.

an atheist.[54] There is disagreement as to his background—some analysts claim he was the son of a smuggler, while others report that his father was a tax commissioner and his grandfather a postman.[55] This latter account would indicate that his family had access to more educational resources than the average rural Tamil family. It is known that Prabhakaran was recruited into the TULF's military wing in the early 1970s. However, specific information on LTTE founding dates, group leadership, and chronologies of defections and the formation of splinter Tamil groups diverges wildly.[56] It is unclear whether Prabhakaran founded an earlier group called Tamil New Tigers or whether he just assumed the leadership position after the original founder's arrest and subsequently renamed the group as the LTTE. Either way, the LTTE emerged in 1976 with Prabhakaran as its leader, and his ruthless consolidation of power targeted both moderate and extremist rivals.

LTTE membership was overwhelmingly Hindu, although there were some Christians, and except for a number of long-surviving members most members were very young. The leadership tended to be drawn from the upper-class "warrior-fisherman" caste, which was composed of university-educated English speakers with significant international linkages, but the rank and file were of the lower caste.[57] Geographically, members were drawn from the Tamil population centers in the north and east, and those who came from other regions, such as Tamils of more recent Indian heritage living in the central highlands, were reportedly treated as second-class members.[58] This corresponded to Prabhakaran's reported vision of the LTTE in its early years as a very small group of professional and well-disciplined fighters.[59] Especially gifted fighters were selected for membership in the Black Tigers, believed to date back to 1987. This special unit was tasked with the most difficult missions, including suicide bombings

---

[54] Stephen Hopgood, "Tamil Tigers, 1987–2002," 45. Hopgood also sites Narayan Swamy *(Tigers of Sri Lanka)*, who characterized Prabhakaran as "quietly pious" and "disinterest[ed] in Marxist politics and ideology."

[55] Bruce Hoffman, *Inside Terrorism*, 138.

[56] Perhaps because the LTTE eventually succeeded in displacing all rival groups, Prabhakaran's role in some of the earlier organizations is exaggerated. Some analysts claim he founded the Tamil New Tigers (TNT), which he later renamed the LTTE, as an offshoot of some of the larger militant student movements; others report that he was only a member of TNT, which was founded by Chetti Thanabalasingam as a breakaway group of the TULF, designed specifically to target Tamil collaborators. For an account that depicts Prabhakaran as a less central figure in the early days, see Bruce Hoffman, *Inside Terrorism*.

[57] R. A. Hudson, *Who Becomes a Terrorist and Why: The 1999 Government Report on Profiling Terrorists* (Guilford, CT: The Lyons Press, 2000).

[58] Ibid.

[59] Gunaratna, *War and Peace in Sri Lanka*.

and assassinations.[60] Although some analysts have theorized that groups are more likely to utilize suicide missions during periods when insurgent tactics are deemed to be less effective, the LTTE's use of the Black Tigers did not appear to conform to this model.[61] Instead, the Black Tigers were employed as part of the overall insurgency strategy once the LTTE became the dominant Tamil opposition force.[62] Indeed, their first suicide attack was utilized as a breaching action for a much larger operation with a truck bomb, allowing LTTE regular forces to storm the target site (a Sri Lankan army camp) in order to halt an impending army offensive.[63]

Estimates of the overall size of the LTTE vary considerably, from several hundred in the early 1980s to Sri Lankan government reports of 22,000 rebels killed in the final battle of 2009.[64] A 2003 US Department of State estimate put the size of the LTTE at around 10,000. However, even the largest estimates for the LTTE were mere fractions of the number of government fighters, which continued to increase until reaching a high of 200,000 in 2009.

COMMUNICATIONS

Coordinating and equipping the LTTE required a vast communication network. Many analysts cite the LTTE's superior intelligence-gathering capability as a major factor in many of its tactical victories over the Sri Lankan army,[65] with some analysts also claiming that LTTE intelligence actors infiltrated the Sri Lankan state security establishment. The LTTE's early intelligence training probably came from the Research and Analysis Wing (RAW) in India during the 1970s when Tamil separatist groups were trained in camps in Tamil Nadu. Early in the movement, Prabhakaran instructed operatives to

---

[60] Hopgood, "Tamil Tigers, 1987–2002," 43.

[61] Ibid., 46.

[62] Ibid.

[63] Ibid., 50. Hopgood provides evidence that this attack might not have been one of intentional suicide and indeed that the incorporation of suicide attacks and the existence of the elite "Black Tigers" might have been constructed after the attack to enhance the LTTE's reputation. Another LTTE fighter launched a similar attack shortly after this initial bombing but was able to flee the scene before the bomb exploded.

[64] Hopgood, "Tamil Tigers, 1987–2002," provides the low-end estimates.

[65] Some regional observers claim that the Air Tiger attack on the Sri Lankan army's Katunayake air force base in March 2007 was only possible because most of the base's radar systems were undergoing maintenance—something the Tigers must have known in advance. If they had been operational, the base's air defenses would have shot down the Tiger aircraft before the Tigers could carry out their mission. N. Manoharan, "Air Tigers' Maiden Attack: Motives and Implications," Issue Brief no. 45, Institute of Peace and Conflict Studies (April 2007), http://www.ipcs.org/pdf_file/issue/1734776809IPCS-IssueBrief-No45.pdf.

collect the intelligence manuals of other countries—including the United States, United Kingdom, and Israel—and translate them into Tamil. One Canadian intelligence report contends that the LTTE had communication hubs in Singapore and Hong Kong to facilitate its weapons procurement activities, with secondary cells in Thailand, Pakistan, and Myanmar and front companies in Europe and Africa. From these locales, LTTE operatives coordinated purchases and shipments from Asia, Eastern Europe, the Middle East, and Africa.[66]

Communication with the Tamil expatriate community was also crucial to LTTE operations. Numerous newspapers reported on homeland developments as well as developments within major Tamil communities outside Sri Lanka, while Tamil websites were linked to well-respected humanitarian and development agencies that reported on the conflict.[67] Most LTTE appeals to the international community were channeled through sympathetic pressure groups and media outlets, with activities coordinated by umbrella organizations located in major expatriate sites.[68] These public relations campaigns were crucial in establishing LTTE's dominance over rival groups with less effective communications and were also considered far superior to those of the Sri Lankan government. Among other activities, the LTTE sent daily faxes via satellite phone links to diplomatic missions detailing battlefield reports. The LTTE also put out videos, pamphlets, and calendars detailing the results of government strikes against LTTE strongholds[69] as well as footage of Black Sea Tiger suicide attacks against Sri Lankan naval ships and dramatizations of successful operations.[70] Before its presence on the web, the group utilized telephone networks to dispense local news in major expatriate centers.[71] The sense of community and regularity of communication enabled by these forms of media greatly increased the LTTE's fund-raising capabilities.

---

[66] Chalk, "Liberation Tigers of Tamil Eelam's (LTTE)."

[67] Prominent examples include the Canadian Relief Organization for Peace in Sri Lanka, the International Educational Development Inc. (IED), the World Council of Churches, the Australian Human Rights Foundation, the International Human Rights Group, the International Federation of Journalists, Pax Romana, the International Peace Bureau, the International Human Rights Law Group, and the Robert F. Kennedy Memorial Center for Human Rights. Chalk, "Liberation Tigers of Tamil Eelam's (LTTE)."

[68] These groups include the Australasian Federation of Tamil Associations, the Swiss Federation of Tamil Associations, the French Federation of Tamil Associations, the Federation of Associations of Canadian Tamils, the Ilankai Tamil Sangam in the United States, the Tamil Coordinating Committee in Norway, and the International Federation of Tamils in the United Kingdom. Chalk, "Liberation Tigers of Tamil Eelam's (LTTE)."

[69] Ibid.

[70] Hopgood, "Tamil Tigers, 1987–2002," fn40.

[71] Rohan Gunaratna, *Sri Lanka's Ethnic Conflict and National Security* (Colombo: South Asian Network on Conflict Research, 1998), 4.

METHODS OF ACTION AND VIOLENCE

LTTE operations can be roughly separated into three categories: conventional operations, guerrilla warfare, and targeted bombings/assassinations,[72] most of which took place in the Tamil population's centers in the north and east. The overwhelming majority of LTTE assassinations targeted officials and police, while bomb attacks focused primarily on strategic assets of the Sri Lankan army and, to a lesser extent, large infrastructure such as oil depots. Despite this state-centric targeting, many civilians died in attacks blamed on the LTTE.[73]

Tactically, the LTTE relied heavily on assassinations of rival Tamil separatists and politicians as well as police informants. In fact, twenty-four of the thirty-seven high-ranking politicians assassinated by the LTTE were Tamils, and the rest were Sinhalese (nine), Muslim (three), and Indian (one).[74] The first assassination for which Prabhakaran claimed responsibility was that of Alfred Duraiappah, the Tamil mayor of Jaffna and SLFP member, killed in 1975.[75] Few large operations were carried out by the LTTE between 1977 and 1983, during which time the LTTE killed eleven Tamil politicians, thirteen informants, and sixteen civilians.[76] After the riots of 1983, however, operations and assassinations expanded. Many assassinations attributed to the LTTE were impossible to verify because its practice was to neither claim nor deny involvement in specific activities, instead pointing to continued paramilitary and police violence against Tamil civilians as contributing to specific assassinations. Some of the suspected LTTE assassinations include the death of TULF Vice President Neelan Thiruchelvam; the attempt on then-President Kumaratunga in 1999; the death of Rajiv Gandhi, whom it was feared would redeploy the highly unpopular IPKF if elected again as prime minister of India, and in whose death

---

[72] Hopgood, "Tamil Tigers, 1987–2002," 44.

[73] These attacks include a 1984 attack on Sinhala prisoners that had been resettled by the government on former Tamil land (believed to be the first LTTE attack on civilians), a 1985 attack that killed one hundred Buddhist pilgrims in Anuradhapura, the 1986 bombing of Air Lanka Flight 512 that killed twenty-one civilians, the 1996 attack on the Central Bank that killed eighty, and a 2006 claymore mine attack on a bus twenty kilometers outside Colombo that killed sixty.

[74] Gamini Samaranayake, "Political Terrorism of the Liberation Tigers of Tamil Eelam (LTTE) in Sri Lanka," *Journal of South Asian Studies* 30, no. 1 (April 2007): 177.

[75] This assassination probably took place when the LTTE was still operating as the TNT, an organization that Prabhakaran took over after its founder Chetti Thanabalasingam was arrested. The TNT under Thanabalasingam primarily targeted Tamil collaborators, and this characteristic stuck with the LTTE in subsequent years.

[76] W. I. Siriweera, "Recent Development in Sinhala-Tamil Relations," *Asian Survey* 20, no. 9 (September 1980): 903–913.

Prabhakaran took the unusual step of denying LTTE involvement;[77] President Premadasa;[78] Lieutenant General Sarath Fonseka; and Lakshman Kadirgamar, a two-time foreign minister and ethnic Tamil. Kadirgamar was a harsh LTTE critic and central to the Sri Lankan government's successful efforts to have the LTTE listed as a terrorist organization in the United States and United Kingdom, a legal step that severely hampered the organization's ability to raise funds.[79] LTTE leaders denied involvement in Kadirgamar's death, as they often did in cases of high-level assassinations, and investigations produced little credible evidence.[80]

Figure 4. A stone mosaic stands at the spot of assassination of the late Indian Prime Minister Rajiv Gandhi.[81]

[77] Dagmar Hellmann-Rajanayagam, *The Tamil Tigers: Armed Struggle for Identity* (Stuttgart: Franz Steiner Verlag, 1994), points out, intuitively, that launching an attack inside Tamil Nadu, where the LTTE enjoyed broad public support, would have been foreseen by the LTTE leadership as a public relations disaster.

[78] The LTTE leadership explicitly denied involvement in the murder of Premadasa, who was also very unpopular with the right-wing Sinhala parties and was widely suspected of having ordered the assassination of a Sinhala opposition figure just days before his own assassination. Hopgood, "Tamil Tigers, 1987–2002," 56. These statistics make the LTTE the only armed group to have assassinated three sitting or former heads of state.

[79] "Senior Sri Lankan Minister Killed," *BBC News* (August 13, 2005), http://news.bbc.co.uk/2/hi/south_asia/4147196.stm.

[80] "Tigers Deny Killing Minister," *BBC News* (August 13, 2005), accessed August 9, 2010, http://www.bbc.co.uk/sinhala/news/story/2005/08/050813_kadir_ltte.shtml.

[81] "File:Rajiv Gandhi Memorial bombsite.jpg," *Wikipedia*, posted by user PlaneMad, accessed March 11, 2011, http://en.wikipedia.org/wiki/File:Rajiv_Gandhi_Memorial_bombsite.jpg.

Suicide missions were also a key component of LTTE tactics.[82] How the organization assimilated the ideas and tactics associated with suicide bombing is of some dispute, but the first attack occurred in 1987, four years after the commencement of major hostilities.[83] Strategically, suicide operations appear to have been initiated in response to the Sri Lankan government's economic blockade that cut off the flow of raw materials to LTTE-controlled areas, rendering the group's weapon caches insufficient to launch conventional attacks against the large concentrations of Sri Lankan army soldiers stationed in cities in the north and east as part of a stepped-up government offensive.[84] The first suicide attack, on a former Tamil high school in Jaffna that had been turned into a makeshift army camp, involved a vehicle laden with explosives—similar to the attack by Hizbollah against the US Marine barracks in Lebanon four years earlier.[85] However, unlike the attack in Lebanon, LTTE regulars were stationed nearby and rushed the camp after the explosion. Some forty soldiers were killed and a planned Sri Lankan offensive was scrapped as a result of this preemptive LTTE strike.[86] Analysts disagree as to whether this first mission was intended to be suicidal or whether the LTTE hierarchy constructed the story of a suicide mission after the operative failed to make it clear of the bomb.[87] Regardless, the group's commitment to self-immolation was made clear in 1991 when a female Black Tiger (or Freedom Bird, as they were often referred) detonated the first concealed suicide vest to assassinate former Indian Prime Minister Rajiv Gandhi.

In the LTTE case, the utilization of suicide missions appears to have depended more on the missions' efficiency—they were estimated to have achieved their instrumental aim 80% of the time—than on any possible symbolism.[88] Nonetheless Prabhakaran portrayed suicide

---

[82] "Suicide Terrorism: A Global Threat," *Jane's Information Group* (October 20, 2000) puts the number at 168, while Ehud Sprinzak, "Rational Fanatics," *Foreign Policy* 120 (September/October 2000): 66–73, gives 171. Robert Pape, *Dying to Win* (New York: Random House, 2005) gives a figure of 137 between the first suicide attack by truck bomb in 1987 and the last suicide attack in May 2009.

[83] Hopgood, "Tamil Tigers, 1987–2002," 51, suggests that this time lapse indicates reservations about the efficacy and appropriateness of suicide bombing within the LTTE's leadership.

[84] Swamy, *Tigers of Sri Lanka*, gives the following list of LTTE equipment for this period: AK-47 rifles, self-loading rifles, light machine guns, heavy machine guns, Singapore assault rifles, M-16 rifles, Mausers, hand grenades, rocket-propelled grenades, Browning automatic pistols, 25-inch mortars, and land and claymore mines.

[85] Ibid., 97–101, claims that the LTTE sent sixteen fighters to be trained by the Palestine Liberation Organization (PLO) in 1983, implying that they certainly would have been familiar with the truck bombing in Lebanon.

[86] Hoffman, *Inside Terrorism*, 142.

[87] Hopgood, "Tamil Tigers, 1987–2002," 50–51.

[88] Chalk, "Liberation Tigers of Tamil Eelam's (LTTE)."

missions as a way of achieving a Tamil homeland more quickly and thereby reducing the suffering inflicted on the Tamil population in the long run.[89] For example, dual suicide attacks launched against civilian and military airports in Colombo had the express purpose of countering an intensified aerial bombardment of Tamil population centers. Some analysts also believe that in conventional LTTE battles, suicide bomb units would be deployed initially to take out enemy fortifications and breach the lines ahead of an assault by regular forces.[90] Suicide missions against the Sri Lankan army were highly effective, but when India sent in peacekeepers in 1987, their superior numbers and weaponry forced the LTTE back into guerrilla-style tactics.[91] Moreover, from the mid to late 1990s until the 2002 cease-fire, the Sri Lankan army became increasingly well equipped (benefiting from a market flush with excess Cold War weaponry) and elicited higher casualty rates from the LTTE, who themselves began to make more extensive use of Black Tiger suicide missions.[92]

The LTTE also incorporated suicide tactics into naval and air operations. Sea Tiger attacks often included use of fiberglass boats for increased speed and maneuverability; the boats were outfitted with penetration rods to puncture the outer hulls of target ships and amplify the shock waves caused by the explosion. Forty such assaults were carried out between 1990 and 2004, frequently by injured LTTE fighters unable to carry out other operations. The success of these tactics dramatically reduced recruitment into the Sri Lankan Navy. The Sri Lankan government also reported the discovery of seven LTTE air strips—two that were in frequent use and five that appeared to be emergency landing strips—in the months after the LTTE's 2009 defeat. Interviews with recruits demonstrate that the narratives offered by the LTTE leadership regarding the principles of self-sacrifice were effectively internalized by members of the Black Tigers who carried laminated cards during missions instructing those who would interfere (in English and Sinhala) that, "I am filled with explosives. If my journey is blocked I will explode it. Let me go."[93] Overall, the LTTE's innovation of the suicide vest; use of female fighters; execution of suicide missions on land, sea, and air; and use

---

[89] Hoffman, *Inside Terrorism*, 141.
[90] Hopgood, "Tamil Tigers, 1987–2002," 55.
[91] Ibid., 52.
[92] Ibid., 54.
[93] Hoffman, *Inside Terrorism*, 142.

of cyanide pills by captured LTTE fighters contributed greatly to the "elite" image the group cultivated.[94]

The LTTE's extensive financing network also made possible the group's acquisition of increasingly heavier equipment. In addition to its vessels, including submarines, the LTTE had a small contingent of aircraft that enabled its use of conventional tactics. LTTE losses increased as the fighting became increasingly conventional, but the naval operations were key to maintaining LTTE supply lines, as well as intercepting arms shipments bound for the Sri Lankan army, all of which relied on sea access.

Figure 5. A Black Sea Tiger fast-attack fiberglass suicide boat passing a Sri Lankan freighter sunken by the Sea Tigers just north of the village of Mulativu, northeastern Sri Lanka.[95]

METHODS OF RECRUITMENT

Many of the LTTE's recruitment activities were essentially passive, with the group's ranks swelled by the indiscriminate bombing

[94] Cyanide capsules were carried in glass vials around the neck of fighters, who would bite the glass, thereby lacerating the gums and allowing the poison to quickly enter the bloodstream. The distribution of the vials became a highly ritualized aspect of the recruitment process, and high-ranking members always displayed the vials prominently for photos. Interestingly, Michael Roberts, "Suicidal Political Action II: Ponnudurai Sivakumaran," *The Sacrificial Devotion and Virulent Politics Research Network*, accessed December 8, 2009, http://sacrificialdevotionnetwork.wordpress.com/2009/04/08/suicidal_political_action_2/, traces the first use of cyanide in the Sri Lankan conflict to the student leader of a small cell who swallowed the poison after a failed assassination attempt in 1974. Hoffman, *Inside Terrorism*, 141, on the other hand, traces this tactic to an announcement made by Prabhakaran much later in 1983.

[95] "File:LTTE Sea Tigers attack vessel by sunken SL freighter.JPG," *Wikipedia*, photograph taken by Isak Berntsen in Mulativu, Sri Lanka, 2003, accessed March 11, 2011, http://en.wikipedia.org/wiki/File:LTTE_Sea_Tigers_attack_vessel_by_sunken_SL_freighter.JPG.

campaigns of the Sri Lankan army.[96] The island's large number of well-educated, yet unemployed or underemployed, youth provided a reserve of radicalized opposition,[97] as did the families who were targeted for persecution by local Sinhalese security forces, on whom the LTTE often focused its recruitment efforts.[98] In addition to this latent source of fighters, the LTTE did engage in active recruitment, with varying degrees of volunteerism and conscription. Prabhakaran was rumored to have demanded one son of every family living in LTTE-controlled areas as a LTTE recruit—a policy probably linked to the group's 1999 attempt to establish a Universal People's Militia that would impose military training on anyone over the age of 15 living in LTTE-controlled territory.[99]

The incorporation of women into the LTTE was initially in response to a shortage of male recruits, although female Tigers proved to be as eager and as lethal as their male counterparts.[100] Nearly 860 male fighters were lost in the five years leading up to the first female casualty in 1987, with almost 500 of those losses occurring in 1987 alone. If estimates are correct, they indicate that the LTTE lost about 8% of its fighters during these years, suggesting that the number of members and fighting strength were indeed pressing concerns.[101] The LTTE also lagged behind other Tamil separatist groups that incorporated women into their operations much earlier, owing in part to these groups' explicit Marxist orientation and the tenets of gender equality within the Marxist dogma.[102]

The issue of the use of child soldiers by the LTTE is highly politicized, and it is unclear to what extent children were incorporated into LTTE

---

[96] Hopgood, "Tamil Tigers, 1987–2002," 45.

[97] Sri Lanka has some of the highest literacy rates in South Asia and a relatively well-functioning educational system.

[98] Hoffman, *Inside Terrorism*, 140, cites interviews with local nongovernmental organization (NGO) workers and militants, all of whom identify abuse by the Sinhalese security forces as the primary motivating factor behind joining the LTTE.

[99] Hudson, *Who Becomes a Terrorist and Why*.

[100] Numerous studies have been conducted on the role of women in the LTTE; these studies were partly driven by the fact that Rajiv Gandhi's assassin was an eighteen-year-old female LTTE member. Y. Schweitzer, "Suicide Terrorism: Development and Main Characteristics," in *Countering Suicide Terrorism* (New York: Anti-Defamation League, 2002), reports that 40% of the suicide missions in 2000 were carried out by female LTTE members.

[101] Alisa Stack-O'Connor, "Lions, Tigers, and Freedom Birds: How and Why the Liberation Tigers of Tamil Eelam Employs Women," *Terrorism and Political Violence* 19, no. 1 (Spring 2007): 47.

[102] Gender equality was much more a feature of Marxist–Leninist groups, such as the People's Liberation Organization of Tamil Eelam (PLOTE) and Eelam People's Revolutionary Liberation Front (EPRLF), than it was of the LTTE, with its more ethno-nationalist platform. Rajan Hoole, Daya Somasundaram, K. Sritharan, and Rajani Thiranagama, *The Broken Palmyra* (Claremont, CA: The Sri Lankan Studies Institute, 1988), 78.

operations. Studies that cite official Sri Lankan government statistics report incredibly high numbers of child soldiers, with recruitment and fatality rates of fighters under eighteen years old as high as 60%.[103] Other sources identify child soldiers as an elite unit within the LTTE, citing a 1997 battle between the Leopard Brigade (LTTE orphans) and a commando unit from the Sri Lankan army that ended with 200 Sri Lankan army casualties.[104]

Although by no means a religious movement, the LTTE did actively propagate a "cult of martyrdom" to support recruitment.[105] This approach incorporated the use of religious symbolism, but it was seldom of an explicitly Hindu variant. Much of the symbolism utilized Judeo-Christian terminology including the idea of a "Zion" for the Tamil people and a sort of death and resurrection narrative that constructed the physical *Eelam* from the bodies of martyred fighters. Elements of mysticism and ceremony were present in the ritual dissemination of cyanide capsules to fighters as well as in a "planting ceremony" that symbolically transferred a martyr's impending death into an act of fertilizing the soil of the *Eelam*.[106] LTTE leaders also substituted the word *thadkodai* (to give one's self) for *thadkolai* (committing suicide) when discussing suicide missions, and they emphasized the role of such operations in hastening the establishment of a homeland, which would spare Tamil civilians from the hardship of a more lengthy struggle.[107] Some observers believe the success of the LTTE's suicide missions contributed to a culture that internalized this tactic as part and parcel of the Tamil nationalist movement, suggesting that suicide missions would continue regardless of their efficacy or the environment of political accommodation.[108]

METHODS OF SUSTAINMENT

The LTTE had numerous methods by which it mobilized resources to sustain operations. These methods included various criminal activities, such as bank robberies,[109] extortion, and the smuggling of drugs and other contraband, but also more traditional fund-raising

---

[103] Cecile Van de Voorde, "Assessing and Responding to the Liberation Tigers of Tamil Eelam (LTTE)," *Police Practice and Research* 6, no. 2 (May 2005): 186.

[104] Peter Warren Singer, *Children at War* (Berkeley, CA: University of California Press, 2006), 87.

[105] Peter Schalk, "The Revival of Martyr Cults among Ilavar," *Temenos* 33 (1997): 151–190.

[106] Schalk, "The Revival of Martyr Cults among Ilavar."

[107] Hoffman, *Inside Terrorism*, 141.

[108] Chalk, "Liberation Tigers of Tamil Eelam's (LTTE)."

[109] Hopgood, "Tamil Tigers, 1987–2002," 48.

activities, which themselves incorporated varying levels of coercion. The vast majority of the group's financing came from the large Tamil expatriate community, especially those contingents in Western countries (Canada, the United Kingdom, Australia, the United States, and Scandinavia) but also those living in the Indian province of Tamil Nadu (Land of the Tamil). Indeed, many analysts identify the overseas Tamil communities as the single most important actor enabling the insurgency.[110] A 2001 United Nations estimate put the Tamil expatriate community at 817,000,[111] whereas another source estimated that one-third (or about one million) of the population of Sri Lankan Tamils were living overseas as of 2002.[112] The early waves of migration were dominated by more economically mobile, often English-speaking, Tamils fleeing the Sri Lankan government's discriminatory regime. The later waves of expatriates, especially those that came after the 1983 riots, arrived in their new homelands having experienced much greater violence. This 1983 experience, as well as the difficulty they encountered in assimilating to their new countries, contributed to their willingness to support the separatist movement.

Expatriate support included voluntary contributions from individuals and Tamil-owned businesses, as well as extortion.[113] Collection methods evolved over time, from poorly coordinated, often violent acts of coercion to scheduled collections facilitated by computerized databases, allowing overseas collectors to avoid paying visits to individuals who supported rival Tamil groups or who were already regular contributors.[114] The collection schedule was similarly regimented on a monthly or annual basis; additional collections were made according to special dates commemorating specific battles or individual "martyrs."[115] Information was also collected on extended families residing elsewhere in order to lend credibility to threats in the event that a donation was not forthcoming. A 2009 Canadian intelligence report revealed that the country was one of the top contributors to the LTTE, with donations of approximately $12 million per year. After the LTTE lost control of the Jaffna Peninsula in the mid-1990s, this source of financing became increasingly important,

---

[110] C. Christine Fair, "Diaspora Involvement in Insurgencies: Insights from the Khalistan and Tamil Eelam Movements," *Nationalism and Ethnic Politics* 11, no. 1 (Spring 2005): 125–156.

[111] "A World of Exiles," *Economist* 366 (January 2, 2003), 41.

[112] Gunaratna, "Sri Lanka: Feeding the Tamil Tigers," 201.

[113] Jo Becker, "Funding the 'Final War': LTTE Intimidation and Extortion in the Tamil Diaspora," Human Rights Watch 18, no. 1(C) (2006), http://www.hrw.org/en/reports/2006/03/14/funding-final-war-2.

[114] Gunaratna, "Sri Lanka: Feeding the Tamil Tigers," 212.

[115] Ibid., 213.

by some accounts providing up to 90% of the group's operating funds.[116] However, the classification of the LTTE as a terrorist group by most Western countries, the result of an intense lobbying effort by anti-LTTE forces, put a serious strain on the ability to raise and transfer expatriate funds and was probably a major contributor to LTTE's defeat in 2009.[117] LTTE also raised funds domestically by levying taxes on the population, especially in the early stages of the conflict. Those who could not pay were often incarcerated, but those families that had sons or daughters serving in the cadres were exempted.[118] Proof of payment of this tax served as a pass for traveling through LTTE-controlled territory and for serving in administrative positions. Extracting payment was relatively inexpensive; once the LTTE established its reputation for ruthlessness, few families had to be reprimanded.

The LTTE also developed extensive links with international arms smugglers. In one particularly infamous case, an Israeli subcontractor agreed to divert into LTTE-designated ports a large weapons shipment destined for the Sri Lankan army. The arms were crucial to the LTTE, which was engaged in a battle for control over the highway linking the capital Colombo with the Jaffna peninsula.[119] In addition to smuggling, the LTTE also armed itself through indigenous production and by capturing weapons from the Sri Lankan army and the IPKF. The LTTE also proved adept at exploiting dual-use technologies to manufacture at least four types of maritime attack-craft from material purchased overseas,[120] as well as its own Jony- and Claymore-type mines.[121] The group made extensive use of improvised explosive devices (often explosive-laden petrol cans equipped with tripwires) and was recognized internationally as experts in the use of these devices.[122] The group also had its own unit that specialized in weapons procurement, facilitated by an extensive maritime

---

[116] Chalk, "Liberation Tigers of Tamil Eelam's (LTTE)"; Gunaratna, *Sri Lanka's Ethnic Conflict and National Security*, gives a contradictory figure of only 60%.

[117] "A World of Exiles," 41.

[118] On the other hand, families with children abroad had to pay an additional tax, as did those who owned businesses or commercial property. Gunaratna, "Sri Lanka: Feeding the Tamil Tigers," 210.

[119] Chalk, "Liberation Tigers of Tamil Eelam's (LTTE)."

[120] Ibid.

[121] The Jony mine is a small wooden box filled with explosives that detonates upon applied pressure; the Claymore is a directional anti-personnel mine that can be command-detonated with a range of about one hundred meters.

[122] International Campaign to Ban Landmines, "Landmine Monitor Report 2007: Toward a Mine-Free World" (Ottawa: Mines Action Canada, 2007), 534, http://www.the-monitor.org/index.php/publications/display?url=lm/2007/sri_lanka.html.

network[123] that included a fleet of merchant ships, a large number of fishing trawlers, high-speed motor launches, professionally trained crew,[124] submersibles, mini-submarines, and possibly vessels capable of carrying one to two shipping containers.[125] Weapons of Chinese origin, including automatic rifles, anti-tank weapons, and grenade launchers, were believed to have reached the LTTE through these channels, most likely via Myanmar.[126] Weapons acquisitions facilitated by this maritime network also included large amounts of explosive materials and surface-to-air missiles (including SA-7s and possibly US-made Stinger missiles).[127] LTTE access to these extensive international procurement networks also contributed to the belief that drug trafficking provided the LTTE with significant resources.[128]

In addition to exploiting various resource flows, the LTTE had a demonstrated capability of adapting its strategy to changing political and military realities. The LTTE's ability to operate as both a conventional military and a guerrilla insurgency enabled it to react to structural changes in the conflict environment, including the introduction of the IPKF, large-scale Sri Lankan army offenses, and changes in the materiel capabilities of the Sri Lankan army.

METHODS OF OBTAINING LEGITIMACY

The LTTE derived considerable legitimacy from its military achievements, its provision of social services to Tamil populations, and the merits of its appeals for Tamil autonomy. The LTTE's string of successes against the numerically superior Sri Lankan army, as well as its consolidation of power after years of intra-Tamil violence, contributed significantly to its popularity locally and within

---

[123] Chalk, "Liberation Tigers of Tamil Eelam's (LTTE)."

[124] Chalk, "Liberation Tigers of Tamil Eelam's (LTTE)," reports that LTTE fighters have received glider, micro-light, and speedboat training in Europe and Southeast Asia.

[125] The LTTE is presumed to have developed these capacities in the wake of the assassination of Rajiv Gandhi, after which India withdrew most of its logistical support including shipping assistance. Vijay Sakhuja, *The Dynamics of LTTE's Commercial Maritime Infrastructure* (New Delhi: Observer Research Foundation, 2006).

[126] Ibid.

[127] Chalk, "Liberation Tigers of Tamil Eelam's (LTTE)."

[128] The majority of evidence for this claim comes from a report by the Mackenzie Institute, a Canadian think tank. The Institute has been widely criticized for its right-wing orientation and engagement in fear tactics. See John Thompson, "Terrorism and Transnational Crime: The Case of the LTTE," in *After 9/11: Terrorism and Crime in a Globalised World*, eds. David A. Charters and Graham F. Walker (Canada: Centre for Conflict Studies/Centre for Foreign Policy Studies, 2004), 223–230. Sakhuja, *The Dynamics of LTTE's Commercial Maritime Infrastructure*, points out that no LTTE owned ship has ever been discovered carrying narcotics.

expatriate populations.[129] The group's extensive networking with other revolutionary and secessionist movements not only provided the group with training in new techniques but also ensured that the Tamil struggle was included in popular transnational narratives of self-determination and minority rights.[130] The perceived group commitment and the dedication of Tamils more generally, and LTTE fighters specifically, was highly coveted by Sinhala commentators, who blamed their own continued military failure on a lack of Sinhala cohesiveness.

The LTTE's well-established system of disseminating information to sympathetic individuals and organizations abroad was incredibly successful in raising awareness of Tamil grievances and gaining significant support for Tamil autonomy from human rights organizations and policy makers across the globe. Tamil academics provided a wealth of convincing historical and cultural evidence for the classification of Tamils as a distinct "nation," and the LTTE incorporated this academic analysis into its own discourse. LTTE-affiliated NGOs were also widely perceived as effective and responsible in their provision of assistance to Tamil populations. Evaluations of post-tsunami relief efforts were nearly universal in their praise for the LTTE-affiliated Tamils Rehabilitation Organization (TRO) and almost as universal in their condemnation of the bungled response of the Sri Lankan government to the disaster.[131] This was largely in line with the perception, whether real or imagined, that the LTTE was also efficient and honest in the utilization of donations made to its military wing.[132]

The group's state-building activities in the Tamil population centers of Sri Lanka were perhaps their greatest source of legitimacy. These activities not only provided credibility, but they also demonstrated the feasibility of an independent Tamil homeland. The scope and formality of these activities grew in parallel with the organization itself. The provision of justice, for instance, began with the establishment of village mediation boards in the 1980s and expanded to include

---

[129] Stack-O'Connor, "Lions, Tigers, and Freedom Birds."

[130] Gunaratna, *Sri Lanka's Ethnic Conflict and National Security*, 3.

[131] Georg Frerks and Bart Klem, "Tsunami Response in Sri Lanka: Report on a Field Visit from 6-20 February 2005," Disaster Studies, Wageningen University, and Conflict Research Unit, Clingendael Institute (March 14, 2005), www.clingendael.nl/publications/2005/20050300_cru_other_frerks.pdf. This report includes fifty-seven individual and group interviews with intergovernmental organization and NGO organizers as well as indigenous civil and religious authorities.

[132] Gunaratna, *Sri Lanka's Ethnic Conflict and National Security*, 4.

a Tamil judiciary, legal code, and college of law by the mid-1990s.[133] The new legal code focused on modernizing the traditional system by providing for the rights of women and the lower castes, both causes that were championed by the LTTE in official discourse. The court system, combined with the LTTE-coordinated police forces, maintained a high degree of rule of law in their jurisdictions, although critics claim this was a result of intimidation more than community policing. In addition to legal and penal institutions, LTTE administration also included revenue collection, public service provision in health and education, and economic development initiatives.[134] These quasi-state capacities may also have contributed to the LTTE's ability to secure equal status with the Sri Lankan government in various mediated talks, reinforcing LTTE's claim as the sole legitimate representative of the Tamil independence movement.[135]

Finally, the organization's use of female fighters also provided a source of legitimacy, especially given the intensity with which patriarchal and caste-based structures restricted women's freedom in Sri Lanka. Driven partly by demographic imperatives, notably the shortage of males eligible for recruitment, the LTTE's employment of female fighters was also a tool in siphoning off support from rival Tamil groups that did not employ women and a direct response to women's demands to be incorporated into the struggle.[136]

EXTERNAL SUPPORT

Second to the Tamil expatriate community, the Indian government was probably the most influential external actor in the LTTE insurgency. The Indian government intervened both directly and indirectly in the conflict, reversing policy trajectories as implications for India's security shifted. In the 1980s, Indian politicians of Tamil ethnicity in the southern Indian state of Tamil Nadu provided safe haven, training, and support to Indian groups agitating for an independent Tamil state within India, and later to Sri Lankan Tamil separatist groups including the LTTE.[137] It was in Tamil Nadu where

---

[133] Kristian Stokke, "Building the Tamil Eelam State: Emerging State Institutions and Forms of Governance in LTTE-Controlled Areas in Sri Lanka," *Third World Quarterly* 27, no. 6 (September 2006): 1027. Stokke reports that the Tamil courts are considered effective and professional and preferred by many over the state courts.

[134] Ibid., 1022.

[135] International Crisis Group, "Sri Lanka: Sinhala Nationalism and the Elusive Southern Consensus."

[136] Stack-O'Connor, "Lions, Tigers, and Freedom Birds," 43.

[137] For a detailed description of the specific training camps in Tamil Nadu, see Sakhuja, *The Dynamics of LTTE's Commercial Maritime Infrastructure.*

many of the Sri Lankan separatist groups fought their first battles for sole ownership of the Tamil cause.[138] As Indian fears mounted over the potential for the conflict in Sri Lanka to spread, they turned to mediating. In 1987, India and the Sri Lankan government signed the Indo-Lankan Peace Accord, granting greater autonomy to the Sri Lankan Tamils and deploying 100,000 members of the IPKF to Sri Lanka; these IPKF members were charged with monitoring the cease-fire.[139] Although the LTTE initially welcomed the Indian presence, LTTE support quickly eroded. Within three months the LTTE declared war on the IPKF and the latter proceeded to attempt to disarm the LTTE. The IPKF presence grew increasingly unpopular with all parties, and in 1988, the newly elected Sri Lankan president (Ranasinghe Premadasa) allied himself with the LTTE to drive out the IPKF.[140] Soon after, the Indians halted operations against the LTTE and ultimately left Sri Lanka at the request of the Sri Lankan government in 1990. Only after these hostilities, which included a formal LTTE declaration of war against the IPKF, did India begin to crack down on Tamil training camps and support networks within its borders. However, by this time the LTTE had secured routes for smuggling weapons and other resources, and these routes did not require Indian cooperation.[141] The shared resentment toward the Indian peacekeepers led to a brief rapprochement between the LTTE and the Sri Lankan government, but this was followed by the resumption of hostilities that escalated in intensity until a cease-fire in 2002.

Although assistance to Tamil separatists from India is a matter of record, claims of assistance in the form of training, equipment, or other support from other state and non-state entities are both numerous and highly contested. Among these most likely external entities are contacts with militant Palestinian groups in the late 1970s and early 1980s, including the PLO and Hizbollah,[142] as well as numerous anti-Indian insurgent groups and possible contacts with the Pakistani state.[143]

---

[138] Hoole et al., *The Broken Palmyra*, 85.

[139] Gunaratna, "Sri Lanka: Feeding the Tamil Tigers," 199.

[140] Hopgood, "Tamil Tigers, 1987–2002," 49.

[141] Gunaratna, "Sri Lanka: Feeding the Tamil Tigers," 199.

[142] See Pape, *Dying to Win*, on Hizbollah, and Swamy, *Tigers of Sri Lanka*, on the PLO.

[143] Gunaratna, "Sri Lanka: Feeding the Tamil Tigers," 199.

COUNTERMEASURES TAKEN BY THE GOVERNMENT

Government countermeasures, both coercive and noncoercive, have been dictated by a number of conditions that varied over time, including the political will of the Sinhala leadership, the capabilities of the Sri Lankan army, external support to the government, and the contemporaneous character of LTTE activity.[144] Although the state authorities had already begun to target Tamil rebel groups in the 1970s, state response was much more coordinated after the violence, mass migration, and subsequent Tamil radicalization following the 1983 riots. This response included the institution of a national military draft in 1985, as well as economic blockages that cut off from the LTTE both foodstuffs and raw materials for making weapons.[145] The Sri Lankan government also employed martial law several times throughout the conflict. The first instance was in 1979, when the army occupied Jaffna, leading Prabhakaran and many other militants to flee to Tamil Nadu in southern India. The government instituted martial law again in 2005 after the murder of Foreign Minister Lakshman Kadirgamar.

Kadirgamar's assassination prompted repeated cease-fire violations by both sides until, in 2006, the Sri Lankan government formally withdrew from the cease-fire and eventually stepped up its military campaign against the LTTE. A later offensive within the broader campaign, which some observers claim was prompted by the LTTE's air attack against military-controlled areas of the Colombo airport in 2007, allowed the Sri Lankan army to gain control of the eastern provinces by the summer of 2007. Many analysts attribute the army's ultimate victory to its assimilation of guerrilla tactics, targeting militants as opposed to holding territory, because the LTTE's many conventional capabilities were useless against these tactics.[146] It is widely believed that the Sri Lankan army received this training from US, UK, and Israeli Special Forces. Although the United States and India cut off supplies of military assistance and hardware, respectively, over concern about tactics of the Sri Lankan army, other countries stepped in to fill the supply vacuum.[147] China provided significant

---

[144] Samaranayake, "Patterns of Political Violence and Responses," 117.

[145] Hopgood, "Tamil Tigers, 1987–2002," suggests that the establishment of the Black Tigers may have been partially in response to the supply restrictions caused by the blockade.

[146] Feizal Samath, "Foreign Forces Look to Sri Lanka for Plan," *The National* (UAE), August 27, 2009, accessed August 9, 2010, http://www.thenational.ae/apps/pbcs.dll/article?AID=/20090828/FOREIGN/708279920.

[147] This came after China's earlier refusals to provide material support to the Sri Lankan government because of Indian sensitivities. Sakhuja, *The Dynamics of LTTE's Commercial Maritime Infrastructure*. Sakhuja also suggests that China was successful in encouraging Pakistan to support the Sri Lankan government.

amounts of heavy equipment designed to target the LTTE's crucial capabilities, including its airstrips, command posts, underground bunkers, ammunition caches, and shipyards.[148] Also crucial was China's success in keeping the conflict off the agenda of the United Nation's Security Council, allowing the Sri Lankan government greater freedom to pursue the LTTE.[149]

Finally, the government's success in co-opting Karuna Amman (full name Vinayagamoorthy Muralitharan), then the eastern commander of the LTTE and a former bodyguard for Prabhakaran, also represented a successful government countermeasure. Both Karuna's group, the TMVP (*Tamil Makkal Viduthalai Pulikal* or Tamil People's Liberation Tigers) and the LTTE requested assistance from the Sri Lankan government in fighting the other, the LTTE as part of its 2006 peace negotiations and the TMVP in exchange for consolidating support from eastern Tamils.[150] Although the TMVP initially lost control of the east to the LTTE, they regained it later with the support of the Sri Lankan army. Although some observers claim that the TMVP owed its success in the east to electoral fraud and intimidation perpetrated with government support,[151] others cite a perceived LTTE bias against eastern Tamils that weakened the LTTE's hold on that area.[152] Regardless, the TMVP was instrumental in the eventual defeat of the LTTE, with TMVP's operational and intelligence-related cooperation with the Sri Lankan government helping to oust the LTTE from its last stronghold in the east and restricting the group to its operating bases in the north. Amman's defection also deprived the LTTE of one of its most valuable fighters; observers report he was often dispatched to the frontline of the LTTE's toughest battles.[153] In addition to fighting alongside the TMVP in the eastern province, the Sri Lankan government hailed the party as a sign of democratic progress in the east and supported the party's electoral activities in order to sideline the LTTE.

---

[148] Ibid.

[149] Ibid.

[150] When the Sri Lankan government refused to disband the TMVP, the LTTE pulled out of the peace talks.

[151] International Crisis Group, "Sri Lanka's Eastern Province."

[152] This latter perception is at least partially confirmed by the disproportionate attention that LTTE NGOs are said to have paid to the north after the 2004 tsunami, although this could have merely reflected the fact that there was no Eastern organizational base upon which LTTE-affiliated organizations could rely to distribute aid and supplies. Frerks and Klem, "Tsunami Response in Sri Lanka," 17.

[153] IPS Correspondents, "Deep Plot Seen in Former Tiger Turning MP," Inter Press Services, India, October 11, 2008, accessed August 9, 2010, http://ipsnews.net/news.asp?idnews=44224.

SHORT- AND LONG-TERM EFFECTS

CHANGES IN THE ENVIRONMENT

The decades of direct conflict in Sri Lanka resulted in significant ethnic polarization, both geographically, in terms of communal intermixing, and culturally, in terms of the exclusivity of ethnic politics. Nearly all Sri Lankan government efforts at reconciliation with the Tamils, notably cease-fires and peace processes but also foreign- and domestic-led development projects in the north and east, precipitated violent responses from Sinhalese and Muslims.[154] Muslims had grievances against both groups—they had grievances against the government for population transfers that resettled Sinhalese peasants on Muslim land, and they had grievances against the LTTE for confiscation of property and harassment.[155] This fact, combined with the inability of Muslim members of Parliament to deliver any meaningful services to their constituents, as brought into sharp focus by the aftermath of the 2004 tsunami, led to significant disenchantment with the established system.[156] In addition to increased ethnic polarization, religious fundamentalism continued to gain currency in the country's electoral politics. The *Jathika Hela Urumaya* (JHU), or National Heritage Party, led by Buddhist Monks who advocated criminalizing "unethical conversions" and shutting down stores that sold meat and liquor on Buddhist holidays, earned an astounding 6% of the national vote in their first contested election in 2004, making it the fourth largest party in Sri Lanka.[157]

Overall, the conflict has impacted Tamil society in a variety of ways. Restrictive gender and caste systems have been moderated, but in many instances these systems have been replaced by militarized hierarchies of command and obedience dependent on association with the Tamil resistance. In terms of international impacts, many observers agree that the 2009 Sri Lankan government's victory over the LTTE will be

---

[154] International Crisis Group, "Sri Lanka: Sinhala Nationalism and the Elusive Southern Consensus," 10.

[155] International Crisis Group, "Sri Lanka's Eastern Province."

[156] Ibid. Although the Muslims are also Tamil-speakers, they do not share much more than linguistic affinity with the Tamil population.

[157] These principles are presented as part of efforts to realize the *dharmarajya* (righteous state), believed to be the basis for policies pursued by the ancient dynasties during the golden age of Buddhism. In addition to the group's religious credentials, the JHU's exhortations on corruption in the ruling Sinhala parties and the untimely death of a popular activist Monk in the months leading up to campaign season also contributed to its electoral success. Mahinda Deegalle, "Politics of the Jathika Hela Urumaya Monks: Buddhism and Ethnicity in Contemporary Sri Lanka," *Contemporary Buddhism* 5 (2004): 83–103.

treated as a case study by states confronting secessionist and insurgent movements, perhaps increasing the levels of state-sponsored violence tolerated by the international community. The practices employed in 2007–2009 may likely be incorporated by many besieged governments and their strategists; practices include indiscriminate attacks on Tamil population centers, receiving advanced training for the Sri Lanka Army in counterinsurgency tactics from developed-country militaries, co-optation of LTTE-defectors equally guilty of violating humanitarian law, and the collective punishment of LTTE supporters. In addition, President Rajapaksa's visit to Myanmar in the immediate aftermath of the conflict was viewed by some analysts as an information-sharing mission designed to advise the Burmese leadership on ending conflicts with minority ethnic groups within its own territory.

CHANGES IN GOVERNMENT

Coercive countermeasures taken by the government generally increased in brutality alongside the increasing radicalization of mainstream politics. Beginning in the period immediately following independence when the ruling UNP adopted increasingly exclusionary politics in response to the electoral challenge represented by the Sri Lankan Freedom Party, political competition created a self-perpetuating cycle of increasingly radical nationalism. Similarly, the Tamil militant groups were formed in response to the impotence of their elected leaders.

The repeated failures of both political parties and governments to effectively address the conflict led to a substantial transfer of power from civilian institutions to the military. Sri Lankan army numbers grew from 12,000 in 1963 to 100,000 in 1999 and 200,000 in 2009. Defense budgets similarly increased. The country's security institutions took on many responsibilities generally reserved for civilian leaders, including issuing identification cards to residents in the northeast, determining the movement of people (including resettlement of returning refugees), and dictating NGO operations.

CHANGES IN POLICY

Many conflict resolution efforts centered on redesigning state institutions, such as the electoral system, to ameliorate ethnic divisions. Among the most significant changes was the establishment of provincial council elections, although many critics condemned these elections for further exacerbating ethnic cleavages, claiming they were an official effort to engineer an electoral victory for Karuna, the LTTE defector, in the eastern province. In addition to splitting the Tamil vote, the government also removed many of the administrative structures that linked the northern and eastern parts of the island (making a contiguous Tamil homeland nearly impossible to realize);[158] appointed a Sinhalese-dominated administration for the provinces, and continued to resettle Sinhalese in Tamil-dominated areas.[159]

In addition to the restriction of LTTE financing, the post-9/11 environment provided the Sri Lankan government with the necessary political space to crack down on the LTTE. Although the US State Department issued a license for Military Professional Resources Inc. (MPRI) to train the Sri Lankan army in 1996, the US government ultimately decided against the partnership, presumably because of fear of negative political fall-out.[160] However, it is widely believed that US, UK, and Israeli Special Forces later provided training to the Sri Lankan army, and this training was integral to the 2009 defeat of the LTTE.[161]

CHANGES IN THE REVOLUTIONARY MOVEMENT

Although the LTTE was able to retain the majority of its senior leadership and avoid defection until the last stages of the conflict, it often employed brutal methods to maintain cohesion. There is evidence that early in the group's evolution it was sheltered and supported by the TULF, a relatively moderate political party that brought in other militant groups to be absorbed into the LTTE movement.[162] However, this early period of cooperation belies the LTTE's subsequent consolidation of power, which targeted not only rival militant groups but also Tamil politicians, including members of the TULF. The LTTE is estimated to have killed more than 300 members of competing movements, some of whom were pursued as far as India, France, Germany, the United Kingdom, and Canada.[163] Once established, the LTTE was able to gain control over, and

---

[158] The more accomodationist Sinhala UNP voted to allow the north and east to remain as one administrative unit as a concession to Tamil nationalist sentiment, but this was rejected by the ruling SLFP.

[159] International Crisis Group, "Sri Lanka's Eastern Province," ii. See also International Crisis Group Asia Report no. 134.

[160] Ken Silverstein and Daniel Burton Rose, *Private Warriors* (New York: Verso, 2000), 171.

[161] Samath, "Foreign Forces Look to Sri Lanka for Plan."

[162] Gunaratna, *Sri Lanka's Ethnic Conflict and National Security*, 2–3.

[163] Samaranayake, "Political Terrorism of the Liberation Tigers of Tamil Eelam (LTTE) in Sri Lanka," 117.

administer services to, much of the Tamil-majority territory. They were also able to mobilize significant human and material resources from a wide variety of sources, including the expatriate population and the international community at large.

Since the Sri Lankan government declared victory over the LTTE in 2009, fears have surfaced concerning the emergence of new radical elements, notably from among second-generation youth living abroad. The promise made by some Tamil student organizers to take up arms to pursue the LTTE cause is exacerbated by the perception that the Sri Lankan government is neither seriously committed to granting rights to the Tamil minority nor interested in resettling the hundreds of thousands of civilians still living in government camps. However, a more radicalized expatriate community is a common feature of many enduring ethnic conflicts, and there is some indication that the Tamil population in Sri Lanka is responding positively to government overtures, among them increasing Tamil representation in the state's police forces.[164]

After Kadirgamar's assassination, the LTTE's political opponents gained traction in national politics and defeated the relatively more accomodationist United National Party. The victorious coalition consisted of the anti-LTTE hard-line Sri Lankan Freedom Party, the Marxist *Janatha Vimukhthi Peramuna*, the nationalist *Jathika Hela Urumaya*, and supportive Muslim parliamentarians.[165] The dominant political families of the UNP, the Senanayake and Bandaranaike, represented the centrist Colombo elite: urbanized and educated. The Rajapaksa family, which dominated the new SLFP-led coalition, represented the more rural, less well-educated component of Sinhalese society that tended to also be more xenophobic.

BIBLIOGRAPHY

Arasaratnam, S. "Nationalism." In *India and Ceylon: Unity and Diversity*, edited by Philip Mason. London: Oxford University Press, 1967.

Bajoria, Jayshree. "Backgrounder: The Sri Lanka Conflict." Council on Foreign Relations. May 18, 2009. http://www.cfr.org/publication/11407/sri_lankan_conflict.html#p1.

Becker, Jo. "Funding the 'Final War': LTTE Intimidation and Extortion in the Tamil Diaspora." *Human Rights Watch* 18, no. 1(C) (2006).

---

[164] N. Parameswaran, "Sri Lanka Recruits Tamil Police in Former War Zone," *Reuters*, September 28, 2009, accessed August 9, 2010, http://in.reuters.com/article/idINCOL46456420090928.

[165] Jayshree Bajoria, "Backgrounder: The Sri Lanka Conflict."

http://www.hrw.org/en/reports/2006/03/14/funding-final-war-2.

Central Intelligence Agency. "Sri Lanka." *The World Factbook*. https://www.cia.gov/library/publications/the-world-factbook/geos/ce.html.

Chalk, Peter. "Liberation Tigers of Tamil Eelam's (LTTE) International Organization and Operations–A Preliminary Analysis." Commentary no. 77, *Canadian Security Intelligence Service*. March 17, 2000. http://www.fas.org/irp/world/para/docs/com77e.htm.

Deegalle, Mahinda. "Politics of the Jathika Hela Urumaya Monks: Buddhism and Ethnicity in Contemporary Sri Lanka." *Contemporary Buddhism* 5 (2004): 83–103.

De Silva, K. M. *A History of Sri Lanka*. Berkeley, CA: University of California Press, 1981.

Fair, C. Christine. "Diaspora Involvement in Insurgencies: Insights from the Khalistan and Tamil Eelam Movements." *Nationalism and Ethnic Politics* 11, no. 1 (Spring 2005): 125–156.

Frerks, Georg, and Bart Klem. "Tsunami Response in Sri Lanka: Report on a Field Visit from 6-20 February 2005." Disaster Studies, Wageningen University, and Conflict Research Unit, Clingendael Institute, March 14, 2005.

Gunaratna, Rohan. "Sri Lanka: Feeding the Tamil Tigers." In *The Political Economy of Armed Conflict: Beyond Greed and Grievance*, edited by K. Ballentine and J. Sherman. Boulder, CO: L. Rienner Publishers, 2003.

———. *War and Peace in Sri Lanka*. Sri Lanka: Institute of Fundamental Studies, 1987.

———. *Sri Lanka's Ethnic Conflict and National Security*. Colombo: South Asian Network on Conflict Research, 1998.

Hellmann-Rajanayagam, Dagmar. *The Tamil Tigers: Armed Struggle for Identity*. Stuttgart: Franz Steiner Verlag, 1994.

Hoffman, Bruce. *Inside Terrorism*. New York: Colombia University Press, 2006.

Hoole, Rajan, Daya Somasundaram, K. Sritharan, and Rajani Thiranagama. *The Broken Palmyra*. Claremont, CA: The Sri Lankan Studies Institute, 1988.

Hopgood, Stephen. "Tamil Tigers, 1987–2002." In *Making Sense of Suicide Missions*, edited by Diego Gambetta. Oxford: Oxford University Press, 2005.

Horowitz, Donald L. *Ethnic Groups in Conflict*. Berkeley, CA: University of California Press, 1986.

———. "Incentives and Behaviour in the Ethnic Politics of Sri Lanka and Malaysia." *Third World Quarterly* 11, no. 4 (October 1989): 18–35.

Hudson, R. A. *Who Becomes a Terrorist and Why: The 1999 Government Report on Profiling Terrorists.* Guilford, CT: The Lyons Press, 2000.

International Campaign to Ban Landmines. "Landmine Monitor Report 2007: Toward a Mine-Free World." Ottawa: Mines Action Canada, 2007, 534, http://www.the-monitor.org/index.php/publications/display?url=lm/2007/sri_lanka.html.

International Crisis Group. "Sri Lanka: Sinhala Nationalism and the Elusive Southern Consensus." Asia Report no. 141. November 7, 2007. http://www.crisisgroup.org/en/regions/asia/south-asia/sri-lanka/141-sri-lanka-sinhala-nationalism-and-the-elusive-southern-consensus.aspx.

International Crisis Group. "Sri Lanka's Return to War: Limiting the Damage." Asia Report no. 146. February 20, 2008. http://www.unhcr.org/refworld/country,,ICG,,LKA,4562d8cf2,47bc2e5c2,0.html.

IPS Correspondents. "Deep Plot Seen in Former Tiger Turning MP." Inter Press Services, India. October 11, 2008. http://ipsnews.net/news.asp?idnews=44224.

Kaplan, Robert D. "The Buddha's Savage Peace." *The Atlantic* (September 2009).

Kearney, Robert N. *Communalism and Language in the Politics of Ceylon.* Durham: Duke University, 1967.

Manoharan, N. "Air Tigers' Maiden Attack: Motives and Implications." Institute of Peace and Conflict Studies, Issue Brief no. 45. April 2007. http://www.ipcs.org/pdf_file/issue/1734776809IPCS-IssueBrief-No45.pdf.

Pape, Robert. *Dying to Win* (New York: Random House, 2005).

Parameswaran, N. "Sri Lanka Recruits Tamil Police in Former War Zone." *Reuters.* September 28, 2009. http://in.reuters.com/article/idINCOL46456420090928.

Reuters. "British Blamed for Tamil Attack." *The Independent* (January 27, 1998).

Roberts, Michael. "Ethnic Conflict in Sri Lanka and Sinhalese Perspectives: Barriers to Accommodation." *Modern Asian Studies* 12, no. 3 (1978): 353–376.

———. "Suicidal Political Action II: Ponnudurai Sivakumaran." *The Sacrificial Devotion and Virulent Politics Research Network.* December 8, 2009. http://sacrificialdevotionnetwork.wordpress.com/2009/04/08/suicidal_political_action_2/.

Sakhuja, Vijay. *The Dynamics of LTTE's Commercial Maritime Infrastructure.* New Delhi: Observer Research Foundation, 2006.

Samaranayake, Gamini. "Patterns of Political Violence and Responses of the Government of Sri Lanka, 1971–1996." *Terrorism and Political Violence* 11, no. 1 (1999).

———. "Political Terrorism of the Liberation Tigers of Tamil Eelam (LTTE) in Sri Lanka." *Journal of South Asian Studies* 30, no. 1 (April 2007).

Samath, Feizal. "Foreign Forces Look to Sri Lanka for Plan." *The National* (UAE). August 27, 2009. http://www.thenational.ae/apps/pbcs.dll/article?AID=/20090828/FOREIGN/708279920.

Schalk, Peter. "The Revival of Martyr Cults among Ilavar." *Temenos* 33 (1997): 151–190.

Schweitzer, Y. "Suicide terrorism: Development and Main Characteristics." In *Countering Suicide Terrorism.* New York: Anti-Defamation League, 2002.

"Senior Sri Lankan Minister Killed." *BBC News.* August 13, 2005. http://news.bbc.co.uk/2/hi/south_asia/4147196.stm.

Silverstein, Ken, and Daniel Burton Rose. *Private Warriors.* New York: Verso, 2000, 171.

Singer, Peter Warren. *Children at War.* Berkeley: University of California Press, 2006.

Siriweera, W. I. "Recent Development in Sinhala-Tamil Relations." *Asian Survey* 20, no. 9 (September 1980): 903–913.

Sprinzak, Ehud. "Rational Fanatics." *Foreign Policy* 120 (September/October 2000): 66–73.

"Sri Lanka." In *Human Development Report* 2009. New York: United Nations Development Programme.

International Crisis Group. "Sri Lanka's Eastern Province: Land, Development, Conflict." Asia Report no. 159. October 15, 2008. http://www.crisisgroup.org/en/regions/asia/south-asia/sri-lanka/159-sri-lankas-eastern-province-land-development-conflict.aspx.

"Sri Lanka Moves to Protect Tea Industry." *BBC News.* February 19, 2003. http://news.bbc.co.uk/2/hi/business/2779267.stm.

Stack-O'Connor, Alisa. "Lions, Tigers, and Freedom Birds: How and Why the Liberation Tigers of Tamil Eelam Employs Women." *Terrorism and Political Violence* 19, no. 1 (Spring 2007).

Stokke, Kristian. "Building the Tamil Eelam State: Emerging State Institutions and Forms of Governance in LTTE-Controlled Areas in Sri Lanka." *Third World Quarterly* 27, no. 6 (September 2006): 1021–1040.

"Suicide Terrorism: A Global Threat," *Jane's Information Group*. October 20, 2000.

Swamy, Narayan. *Tigers of Sri Lanka*. Delhi: Konark Publishers Pvt Ltd., 1994.

Thompson, John. "Terrorism and Transnational Crime: The Case of the LTTE." In *After 9/11: Terrorism and Crime in a Globalised World*, edited by David A. Charters and Graham F. Walker. Canada: Centre for Conflict Studies/Centre for Foreign Policy Studies, 2004, 223–230.

Van de Voorde, Cecile. "Assessing and Responding to the Liberation Tigers of Tamil Eelam (LTTE)." *Police Practice and Research* 6, no. 2 (May 2005).

"A World of Exiles." *Economist* 366. January 2, 2003.

PALESTINE LIBERATION ORGANIZATION (PLO): 1964–2009

Sanaz Mirzaei

SYNOPSIS

Understanding the Palestinian–Israeli conflict is central to understanding politics of the Middle East. The inability to resolve this struggle led to the rise of the Palestine Liberation Organization (PLO), represented chiefly by Fatah. Although throughout the last thirty years the insurgency's main objective has been to liberate Palestine from Israeli control and to establish Palestinian sovereignty, the group has often been engaged in intergroup conflict as well as conflict against its Israeli occupiers. The group's varying methods of action and sources of sustainment, both external and domestic, have led to waves of relative success and failure. However, the PLO's ultimate goal of a sovereign Palestinian state is yet to be achieved as its struggle continues.

TIMELINE

1917	British issue Balfour Declaration, which promised a national home for Jews in Palestine.
1936–1939	The Arab Revolt—5,000 Arabs killed by the British and hundreds of Jews killed by Arabs.
1948	Israel declared a Jewish State. British leave Palestine.
1949	Armistice between Israel and Arab state, increasing Israeli territory.
1964	PLO founded.
June 5–10, 1967	Six-Day War.
September 1970	Black September—PLO and Jordan engage in confrontation. Marks the beginning of international terrorist operations for the PLO.
1971	PLO expelled from Jordan and moves its headquarters to Lebanon.
October 6, 1973	Yom Kippur War—Egypt and Syria take back Sinai and Golan Heights (respectively) in a surprise attack.

1973	Arab League recognizes PLO as the official representatives of the Palestinians.
1974	The United Nations General Assembly recognizes Palestine's sovereignty and national independence.
1982	Israel invades Lebanon to fight the PLO.
1983	PLO splits.
December 8, 1987	First *intifada*.
1988	Arafat declares Palestinian independence.
	Arafat accepts Israel's right to exist and renounces terrorism.
1993	Oslo Declaration.
1995	Oslo Interim Agreement; Palestinian Authority (PA) is established.
1996	Arafat is elected president of the PA in the first Palestinian general elections.
September 28, 2000	Second *intifada*.
2003	Mahmoud Abbas becomes first Palestinian prime minister.
2004	PA President Yasser Arafat dies.
2005	Mahmoud Abbas elected president of the PA.
	Ariel Sharon and Mahmoud Abbas agree to a cease-fire.
January 2006	Hamas gains parliament.

THE ENVIRONMENT OF THE REVOLUTION

PHYSICAL ENVIRONMENT

Israel is a relatively small country bordering the Mediterranean Sea and located centrally in the Middle East between Egypt, Jordan, Lebanon, and Syria. With a total of 22,072 square kilometers, it is roughly the size of New Jersey. After World War II when the British gave up their mandate of Palestine, the United Nations (UN) reorganized the area into Arab and Israeli states. This reorganization lies at the crux of the Israeli–Palestinian debate. Israel occupies and controls two Palestinian territories—the West Bank and Gaza Strip—within its borders. The West Bank, a rugged and mostly barren land slightly smaller than the size of Delaware, is located in the east, shares a border with Jordan, and flanks the Dead Sea. The Gaza Strip, roughly twice the size of Washington, DC, lies in the southern part of Israel, borders the Mediterranean Sea, and shares a border with Egypt.

CULTURAL AND DEMOGRAPHIC ENVIRONMENT

There are roughly four million Palestinians living in the Palestinian Authority (PA), with 1.5 million living in the Gaza Strip and 2.5 million in the West Bank. The population of Palestinians in the PA is young, with the median age of about seventeen years in Gaza and twenty in the West Bank, and 98% of the population is younger than sixty-five in Gaza and 97% is younger than sixty-five in the West Bank. The fertility rate of about five children born per woman in the Gaza Strip and 3.22 children born per woman in the West Bank is one reason for the young population. This is a largely urban territory, with 72% of the population living in urban centers. In the Gaza Strip, approximately 99.3% of the population is Muslim and primarily Sunni, leaving only 0.7% Christian. In the West Bank, however, there are several Jewish settlements, which is why only about 75% of the population there are Muslim, 17% are Jewish, and 8% are Christian and other.[1] All Palestinians speak Arabic, and most also speak Hebrew and understand English.[2]

[1] Central Intelligence Agency, "West Bank," *The World Factbook*, https://www.cia.gov/library/publications/the-world-factbook/geos/we.html.

[2] Minorities at Risk, "Assessment for Palestinians in Israel," http://www.cidcm.umd.edu/mar/assessment.asp?groupId=66603; Central Intelligence Agency, "Gaza Strip," *The World Factbook*, https://www.cia.gov/library/publications/the-world-factbook/geos/gz.html.

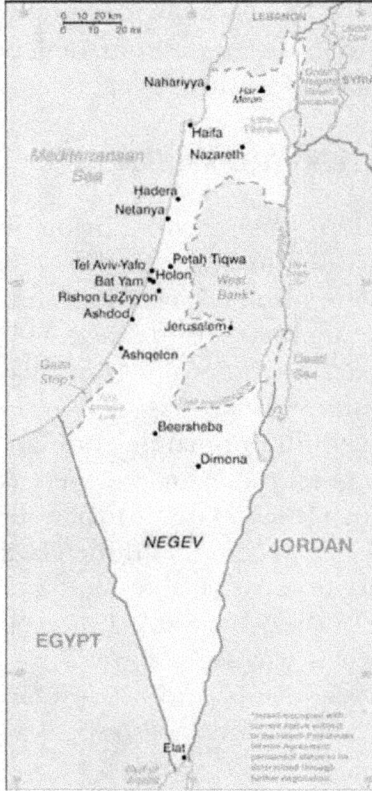

Figure 1. Israel 2010.[3]   Figure 2. West Bank 2010.[4]

Figure 3. Gaza Strip 2010.[5]

---

[3]   Central Intelligence Agency, "Israel," *The World Factbook*, https://www.cia.gov/library/publications/the-world-factbook/geos/is.html.

[4]   Central Intelligence Agency, "West Bank," *The World Factbook*.

[5]   Central Intelligence Agency, "Gaza Strip," *The World Factbook*.

SOCIOECONOMIC ENVIRONMENT

Although Israel has one of the Middle East's strongest economies, the economic situation of the Palestinian territories is one of the worst in the region, with high unemployment, rising birth rates, poor public health conditions, limited access to land and sea, high population density, and strict restriction on the movement of people and goods.[6] Even though the population in both the Gaza Strip and the West Bank is relatively young, meaning that a good portion of the population is of working age, the people of Gaza suffer from 40% unemployment and 70% of the population live below the poverty line. Although the situation in the West Bank is slightly better, there is still a 19% unemployment rate and 46% of the population live below poverty line. In 2009, the gross domestic product (GDP) of the West Bank was estimated to be $12.79 billion.[7]

Israeli policy toward the Gaza Strip continues to exacerbate the already tense political situation in the country. With very strict security controls and restrictions on movement, the Palestinian economy remains very limited. These restrictions have become even stricter in response to increased violence by Hamas—especially after its violent capture of the territories in June 2007. Continued fighting between Hamas and Israel led to the collapse of the majority of the private sector in the winter of 2009, resulting in citizens turning to a Hamas-controlled black market for goods.[8]

Unlike in the Gaza Strip, the economic situation in the West Bank demonstrated some growth in 2009 due to donor assistance, improvement in economic reform and security, and more lenient restrictions on movement. However, the standard of living in the West Bank is still subpar and worse than before the second *intifada* in 2000, primarily due to Israeli-imposed restrictions on movement that hurt labor flows as well as both external and internal manufacturing and commerce. However, because of some successful economic institutional changes, the Palestinian government in the West Bank has been able to attract foreign aid, which has marginally helped its economy.[9]

[6] Minorities at Risk, "Assessment for Palestinians in Israel."

[7] Central Intelligence Agency, "West Bank" and "Gaza Strip," *The World Factbook.*

[8] Central Intelligence Agency, "Gaza Strip," *The World Factbook.*

[9] Central Intelligence Agency, "West Bank," *The World Factbook.*

HISTORICAL FACTORS

Palestine came under British control in December 1917 as a result of its war against the Ottoman Empire.[10] At this time, the British, advocating a Zionist mission, called on Jews to immigrate to Palestine, and the Balfour Declaration of 1917 promised them a national home. During the interwar period, Palestine saw a large influx of Jewish people seeking to take advantage of Britain's promise. The combination of thousands of Jewish immigrants and land purchases led to greater Arab–Palestinian disenfranchisement, and several violent riots between the Palestinians and Zionists ensued, with parties on both sides killed. In the 1920s, Palestinian peasants and urban poor rioted against the Jewish settlers; however, active violent resistance had not yet been aimed at the British authorities. In 1920, a Palestinian Arab Congress convened to unite the Palestinians and to push for a change in the British policy to advocate, or at the very least permit, Jewish settlement of Palestine.[11] However, despite some offers by the British to establish agencies for the Palestinian Arabs similar to those of the Jewish people, the Palestinians turned down the British offers, claiming that they were not fair and analogous measures.

The continuing British support for Jewish Zionism led to a large violent outbreak, known as the Arab Revolt, from 1936 to 1939. What began as civil disobedience quickly turned into armed resistance, leaving many dead and injured.[12] In 1939, the British government issued a white paper that reversed its previous policy and "capped Jewish immigration at seventy-five thousand over five years, restricted land transfers to limited areas, and proposed to make Palestine independent within ten years, if Arab-Jewish relations improved."[13] As Britain began to engage in the Second World War and in the aftermath of a harsh suppression of the Palestinians during the Arab Revolt, Palestinian political activity was destroyed and Palestinian political parties were made illegal in 1939.[14] With many Palestinian activists and political leaders in jail, Palestinian political activity was somewhat stifled as the Israelis became empowered. On November 29, 1947, the British government withdrew from Palestine and left it under control of a UN mandate that partitioned the land into Arab and Israeli states.

---

[10] Samih K. Farsoun and Naseer H. Aruri, *Palestine and the Palestinians: A Social and Political History* (Boulder, CO: Westview, 2006), 60.

[11] Ibid., 85.

[12] Ibid., 91.

[13] Ibid., 92–93.

[14] Ibid., 93.

The Palestinian Arabs, refusing to accept the UN's partition plan, waged a guerrilla war against Israel in May 1948 with help from four Arab-state armies—Iraq, Syria, Egypt, and Jordan—and some assistance from Lebanon and Saudi Arabia.[15] The Palestinians anticipated an Israeli defeat but were left disheartened and humiliated when Israel defeated the Arab armies. This led to *al-Nakbah* day, or "catastrophe," leaving 630,000–800,000 Palestinian refugees, the settlement of urban neighborhoods by Jews, and Jewish occupation of the once-Palestinian state.[16] The British promise of a national home for the Jews in Palestine materialized on May 15, 1948, when Israel was declared a Jewish state.[17] After *al-Nakbah,* Palestinian Arabs were divided into three segments: 100,000–180,000 Palestinians stayed in their homes and lands in what became Israel; approximately half a million Palestinians stayed behind Arab military lines in east-central Palestine and the Gaza Strip; and, of the total Palestinian population of 1.4 million in 1948, more than 750,000 became refugees in the Gaza Strip, the West Bank, and nearby Arab countries.[18]

As the Palestinian political situation continued to deteriorate and Israel was aiming to demonstrate its new strength to its Arab neighbors, the surrounding Arab countries took an interest in the Palestinian cause. One of the most powerful and influential of these countries was Israel's southern neighbor, Egypt. At that time, Egyptian President Gamal Abdel Nasser promoted a pan-Arab nationalist and socialist agenda, which was attractive to the Palestinians. It was at this time that Yasser Arafat, leader of Fatah and later the PLO, was recruited by Nasser to be trained to attack Israel.[19] Committed to spreading Arab nationalism, in 1964, Nasser, along with other Arab states, established the PLO, a Palestinian army, to assist Palestinians in fighting for the land lost to Israel.[20] Although Arafat and his colleagues were drawn to Nasser's pan-Arabism, they were also attracted to several other political views—most notably Marxism and Islamism. The PLO was initially led by Ahmad Shuqayri; however, Yasser Arafat, being the leader of Fatah, the most powerful and influential of the Palestinian parties represented in the PLO umbrella, remained a prominent figure in the Palestinian struggle. Unlike Shuqayri, Arafat was greatly influenced by his 1962 visit to Algeria, where he was convinced that

---

[15] Barry Rubin, *Revolution Until Victory?: The Politics and History of the PLO* (Cambridge: Harvard University Press, 1994), 4.

[16] Ibid.; Farsoun and Aruri, *Palestine and the Palestinians: A Social and Political History,* 105.

[17] Ibid., 64.

[18] Ibid., 105.

[19] Rubin, *Revolution Until Victory?,* 6.

[20] Ibid., 8.

guerrilla warfare and violent resistance was the best way to liberate Palestine.[21] In pursuit of this goal, Arafat's Fatah established military training bases in neighboring Jordan and Syria from 1963 to 1964 to prepare for their operations in January 1965.[22]

GOVERNING ENVIRONMENT

After World War II when the British withdrew from Palestine, the UN redistributed that land into Jewish (Israeli) and Arab (Palestinian) states. Although there was much contention regarding the division of the land, in 1967 Israel waged a successful war against its Arab counterparts, capturing several disputed territories and gaining effective control of the Palestinian land. In an initiative to resolve the territorial conflict, Yasser Arafat, the PLO leader, organized an emergency meeting of the Palestine National Council (PNC) in 1988. During this meeting, he recognized the UN General Assembly Resolution 181, which had been passed in 1947 and called for the division of the area into two states. As a result and almost overnight, the PLO and Fatah[23] were treated as though they constituted a makeshift government.[24]

After almost twenty years of conservative rule in Israel, the left-wing Labor party won the Knesset in 1992,[25] marking a greater commitment to the peace process. As a result, Prime Minister Yitzhak Rabin placed a freeze on new housing settlements built in the territories and also engaged in dialogue with the PLO on the side in order to establish a self-rule agreement for the Palestinians.[26] After September 1993 when Israel and the PLO recognized each other, they signed a basic plan outlining the steps toward Palestinian self-rule at the Oslo Accords.[27] In the period after the Oslo Accords, between May 1994 and September 1999, Israel gradually transferred some security and civilian control to the PA (West Bank and Gaza Strip). However, to most Palestinians, this interim government was not enough, and demands for a permanent solution to the Palestinian issue led to a second *intifada* in September 2000.[28] In 2003, in the aftermath of the second *intifada*, the United

---

[21] Ibid., 6–8.

[22] Ibid., 6–7.

[23] Fatah is the leading PLO faction led by Arafat. Its inception predates the PLO, and Fatah remains today as one of the main PLO groups.

[24] Jonathan Schanzer, "The Iranian Gambit in Gaza," *Commentary* 127, no. 2 (2009): 29–32.

[25] The Knesset is the Israeli legislative body.

[26] Minorities at Risk, "Assessment for Palestinians in Israel."

[27] Central Intelligence Agency, "Israel," *The World Factbook.*

[28] Central Intelligence Agency, "West Bank," *The World Factbook.*

States, European Union, UN, and Russia formed a group named "the Quartet" to help negotiate a final settlement to the conflict; this settlement centered on a two-state solution for Israel and a democratic Palestine by 2005.

Palestinian leader Yasser Arafat led the PA until his death in 2004. Afterward, Mahmoud Abbas was elected to succeed him in January 2005. Soon thereafter, Israel and the new Palestinian government signed the Sharm el-Sheikh Commitments in a continuing effort toward a peaceful solution to the Palestinian issue. By September of that year, Israel "withdrew all settlers and soldiers and dismantled its military facilities in the Gaza Strip and withdrew settlers and redeployed soldiers from four small northern West Bank settlements."[29] Although Israel has withdrawn from Gaza and the West Bank, it still controls access to those territories by air, sea, and a restricted border zone.

Although from their inception the PA and the Palestinian Legislative Council were largely under control of various PLO factions, Hamas won control of the Palestinian Legislative Council in January 2006 and the PA by March 2006. Until the 2006 elections, Hamas had chosen not to run for any positions, as it was fundamentally opposed to Israel's right to exist, which was the basis for the initial Oslo Accords from which the PA was established. There was much international debate whether to recognize Hamas as the official Palestinian government because it is widely recognized as a terrorist group, even though the organization came to power via democratic elections. As a result of Hamas's election, tensions between Palestine and Israel continue to this day.

WEAKNESSES OF THE PREREVOLUTIONARY ENVIRONMENT AND CATALYSTS

Since the British invasion, the Palestinians have struggled for the ability to rule over their land. However, beginning with the Balfour Declaration in 1917, followed by the mass immigration of Jews after the Second World War combined with a renewed sensitivity and commitment to the idea of having a Jewish national home, the Palestinians slowly continued to lose this ability. The creation of the Israeli state in 1948 was the main catalyst for the Palestinians to focus their quest through an armed, violent resistance. The Palestinians felt that they had been pushed out of their home and that they had lost the right to exist in their own state. Furthermore, when the Israeli state took over the Palestinian land, thousands of Palestinians were

---

[29] Central Intelligence Agency, "Gaza Strip," *The World Factbook.*

forced to live in refugee camps under very poor circumstances. They were discriminated against in the workplace and felt humiliated at their loss. This extreme loss of power influenced the PLO to unite in consolidating a "Palestinian" identity, separate from their Arab or Muslim identities, and further pushed them to seek their goals through violent resistance.[30]

FORM AND CHARACTERISTICS OF THE REVOLUTION

OBJECTIVES AND GOALS

Although the Palestinian issue is best represented by the need to "liberate" the Palestinian land from the Israeli foreign occupier, the goals of the Palestinian groups are much more complex. A persisting debate between the PLO and other Palestinian groups, particularly Hamas,[31] is whether their main goals and objectives are nationalist or Islamist. At the time of its inception, the PLO was greatly influenced by Gamal Abdel Nasser's socialist-nationalist pan-Arab movement. However, as time passed, Yasser Arafat, leader of Fatah, pushed for a separate Palestinian identity, rather than a pan-Arab or Islamic state.[32] This focus on a Palestinian identity is a departure from the original statement in the PLO Charter that it is "an indivisible part of the Arab homeland," and the Palestinians are "an integral part of the Arab nation."[33] Instead of Arab unity leading to Palestinian liberation, Fatah reversed this statement.[34] Although the initial PLO Charter pledged to regain all of Palestine and to destroy Israel, about thirty years later the PLO tried to balance its goal to once again rule all of Palestine and be open to an interim solution, i.e., the two-state solution.[35]

---

[30] Helga Baumgarten, "The Three Faces/Phases of Palestinian Nationalism, 1948–2005," *Journal of Palestine Studies* 34, no. 4 (Summer 2005): 25–48.

[31] Hamas, founded in 1987 as an extension of the Egyptian Muslim Brotherhood, defines Palestinian nationalism through Islam. This is one major point of departure from the PLO. Ziad Abu-Amr, "Hamas: A Historical and Political Background," *Journal of Palestine Studies* 22, no. 4 (1993): 5–19.

[32] Rubin, *Revolution Until Victory?*, 9.

[33] Ibid., 20.

[34] Baumgarten, "The Three Faces/Phases of Palestnian Nationalism, 1948–2005," 25–48.

[35] Khaled Hroub, *Hamas: Political Thought and Practice* (Washington, DC: Institute for Palestine Studies, 2000), 7.

LEADERSHIP AND ORGANIZATIONAL STRUCTURE

In 1964 the PLO was set up as an umbrella organization for the main Palestinian parties in a meeting led by Egypt's then president, Gamal Abdel Nasser, and the Arab League. Accordingly, participants at the meeting organized the PLO to meet Egypt's needs, including a parliament, the PNC, an army, and a treasury.[36] The PLO's charter, which acted as a constitution, established these institutions. However, the real decision-making power was with Ahmad Shuqayri, the PLO's first leader, and his handpicked PLO Executive Committee.[37] Shuqayri ran the PLO until 1967 when the Palestinian lawyer Yahya Hammuda replaced him after Shuqayri was discredited by Fatah as lacking strategy and vision.[38] Soon thereafter, in 1969, Yasser Arafat and his allies took over the PLO, making several significant organizational changes to the group.[39]

Given that the PLO was an umbrella organization with several parties that were not always in agreement, it lacked a strong central identity and leadership. In 1971, the PLO consisted of an Executive Committee with representatives from Fatah, the pro-Syria al-Sa'iqa, the Democratic Front for the Liberation of Palestine (DFLP), the Popular Front for the Liberation of Palestine (PFLP), the Palestine Salvation Front, and the Palestine Liberation Army.[40] This weak confederation suffered from often-conflicting views, including views on ideology (Marxist vs. Nationalist) and operational strategy (whether attacks should take place domestically or internationally). This caused confusion within the organization as well as between the organization and its followers. Several of the PLO factions, especially the PFLP, adopted a more revolutionary Marxist rhetoric focused on class struggle, an idea that the more nationalist PLO groups, such as Fatah, feared would work against the PLO's main goal of creating a united Palestine among the other Arab states.[41] Arafat was especially frustrated that, as the leader of the PLO, he was unable to control the disparate groups. He unsuccessfully tried to establish a Permanent Office for Commando Action in 1968 and a Unified Command of Palestine

---

[36] Jonathan Schanzer, *Hamas vs. Fatah: The Struggle for Palestine* (New York: Palgrave, 2008), 5.

[37] Rubin, *Revolution Until Victory?*, 2.

[38] Ibid., 15.

[39] Neil C. Livingstone and David Halevy, *Inside the PLO: Covert Units, Secret Funds, and the War Against Israel and the United States* (New York: William Morrow and Company, 1990), 164.

[40] Rubin, *Revolution Until Victory?*, 36.

[41] Ibid., 32.

Resistance in 1970.[42] However, the PLO member organizations were able to maintain their own operational and ideological autonomy.

Palestinian politics was transformed, beginning in 1982, by three patterns of internal change: (1) "changing role and status of the formal political organizations—the guerrilla groups—that compose the PLO"; (2) confirming "Arafat as the single most important national symbol and arbiter of Palestinian politics"; and (3) relocating "focus of the Palestinian national struggle from the diaspora into the Occupied Territories of the West Bank and Gaza Strip."[43] During this time in the early 1980s, there was effectually "three PLOs: one headed by Arafat; the Syrian-controlled Palestine National Alliance, consisting of the Fatah rebels and three Syrian-controlled groups; and the Democratic Alliance, including the DFLP and PFLP, along with two Iraq-sponsored (and, hence, anti-Syria) groups."[44] Tensions continued between the PLO subgroups—especially between those that were for Palestinian independence but against Arafat. However, these subgroups were unable to consolidate popular support for their cause, and Arafat remained central to the PLO. During the 1980s, he pursued a strategy of pushing Resolution 242, which sought to establish greater Palestinian authority but was controversial within the Palestinians because it made no mention of the pre-1948 boundaries.[45] After the 1988 PNC session in Algiers, which resulted in the declaration of an independent Palestinian state as well as US authorization to begin dialogue with the PLO, the PLO entered a new phase in its movement toward establishing statehood.[46]

After the Oslo Declaration of Principles on September 13, 1993, the Israeli Prime Minister Yitzhak Rabin and Yasser Arafat agreed to establish Palestinian autonomy in the West Bank and Gaza Strip, which led to the creation of the PA in 1994.[47] The PA established an interim administrative organization to govern parts of the West Bank and the Gaza Strip.[48] The PLO, unlike Hamas, was receptive to this new autonomous role.

---

[42] Ibid., 32.

[43] Yezid Sayigh, "Struggle within, Struggle without: The Transformation of PLO Politics since 1982," *International Affairs (Royal Institute of International Affairs 1944-)* 65, no. 2 (Spring 1989): 249.

[44] Rubin, *Revolution Until Victory?*, 66.

[45] Sayigh, *Struggle Within, Struggle Without*, 260.

[46] Ibid., 247.

[47] Schanzer, *Hamas vs. Fatah: The Struggle for Palestine*, 40.

[48] Ibid., 40.

COMMUNICATIONS

The PLO used communiqués and leaflets as vehicles for communicating with the Palestinian public. These communications were central to their strategy, especially during the initial stages of their insurgency but also well after the Oslo process. In 1994, the PLO established the Negotiations Affairs Department (NAD) in Gaza to serve as its communications wing during the Oslo Process and to communicate steps taken toward the implementation of the Interim Agreement it signed with Israel.[49] The NAD was headed by Mahmoud Abbas until 2003, when he shifted his attention to heading the PLO as Arafat weakened. The NAD now has two offices—one in Gaza, in charge of reporting on Israeli relations with the PLO, and one in Ramallah, which reports on the Interim Agreement's implementation status.

METHODS OF ACTION AND VIOLENCE

When the Palestinians fell under Israeli occupation in 1948, some, especially those in Gaza, were able to make some economic progress because they were able to cross borders and work in Israel. However, these workers often faced discrimination in their new jobs. The Palestinians began to feel indignation and were frustrated that they had lost their land and, along with it, their pride. As a result, Palestinian nationalism began to play an important role in this minority group.[50] Nationalist and Islamic leaders not only discouraged Palestinians from mixing with the Israelis but also encouraged resistance.[51]

The PLO engaged in both violent and nonviolent action from its inception in 1964 until the present. Motivated by socialist and nationalist struggles as embodied by the Algerian resistance and guerrillas in China, Cuba, and Vietnam, the PLO's use of terrorism was a rational choice for the group.[52] In the minds of the PLO leaders, terrorist "operations induced a sense of achievement among Palestinians and PLO activists, mobilized Palestinian and Arab support for the PLO, raised the Palestine issue's international priority, coerced Arab states or other Palestinians into rejecting negotiations with Israel, and made many European sates eager to appease the PLO."[53] Although

---

[49] "About Us," PLO Negotiations Affairs Department website, accessed October 15, 2010, http://www.nad-plo.org/etemplate.php?id=182.

[50] Azzim Tamimi, *Hamas: A History from Within* (Northampton, MA: Olive Branch Press, 2007), 11–12.

[51] Ibid., 12

[52] Rubin, *Revolution Until Victory?*, 26.

[53] Ibid., 24–25.

attempts were not very successful at first, the PLO placed several small explosives that targeted Israelis. The PLO prepared to begin with a series of "hit-and-run cross-border raids in rural areas, gradually building to a higher level of combat: bigger units, the transfer of bases to the enemy's territory, the seizure and holding of 'liberated zones,' and, finally, a march on the cities and the enemy regime's collapse."[54] Although the PLO's followers supported its violent attacks, they were not yet ready to be involved in a long and involved guerrilla war.[55]

However, this lack of initial support did not deter the group's operations. In its first military operation on January 3, 1965, the PLO set off a small explosive in the water system, but it never detonated. The PLO's first few attacks were similarly unsuccessful, with bombs that did not detonate and rifles that misfired. Nonetheless, the PLO showed its dedication to the use of violence and "carried out ten raids against Israel in the first three months of 1965."[56] Through these attacks, the PLO aimed to encourage Jewish emigration and to prevent further Jewish immigration and settlement in the Palestinian land.[57] Arafat suggested that the best way to achieve these goals was to destroy tourism and to weaken the Israeli economy.[58] By making it more difficult for the Israelis to live in Palestine, the PLO hoped that the Israelis would realize that living in Israel was not ideal.[59] However, the PLO was far from realizing these goals, and with each failed mission, the PLO lost legitimacy as an effective organization in destabilizing Israel.[60] Instead, Israel illustrated its strength and determination by quickly repressing the Palestinian attacks, making it more difficult for the Palestinians—not the Israelis—to live in their land.[61] After facing embarrassing defeat by Israel in the Six-Day War in 1967, the PLO sought a new home base. However, it struggled with the top two contenders—Lebanon and Jordan. Neither country was open to the initially Marxist rhetoric of the PLO. Although the PLO was quite successful in confronting Lebanon and taking advantage of Lebanon's unorganized army, Jordan was much stronger and more unified, making things difficult for the PLO, especially given that one of its largest bases was in Jordan. After a series of failed missions and as a result of growing Palestinian frustration with the movement, in 1968, Arafat took control of the PLO for the next thirty-six years

---

[54] Ibid., 26.
[55] Ibid., 26.
[56] Schanzer, *Hamas vs. Fatah: The Struggle for Palestine,* 17.
[57] Rubin, *Revolution Until Victory?,* 29.
[58] Ibid., 29.
[59] Ibid., 29.
[60] Ibid., 17.
[61] Ibid., 17.

and positioned the organization under the leadership of his own Fatah[62] organization.

A turning point for the Fatah faction was the Battle of Karameh in February 1968. As the Israeli army sought to push the Palestinians out of their main headquarters to the Jordan Valley, instead of running away from the offensive, Arafat, in a bold but successful move, ordered his Fatah forces to attack the Israelis. Fatah gained considerable legitimacy after the attack as the "first Arab force to put up a fight against the Israeli enemy and force it to withdraw with material and human losses."[63] However, this success also led the PLO to believe that armed resistance was the only way to reach its goals. After the Battle of Karameh, Arafat endeavored to form an armed resistance in the West Bank and East Jerusalem, and this became the dominant strategy for the PLO until 1988.[64] Although diplomacy was sometimes attempted, it was often done in secret.

In response to escalating tensions, the PFLP, one of the more extremist PLO factions, hijacked three international passenger planes on September 6, 1970, landing two of them in Jordan and blowing up the third in Cairo.[65] Three days later, they hijacked a British plane, adding it to the collection from days before; they blew up the planes and kept fifty-four of the passengers, including about twenty American Jews, as prisoners.[66] In response to this event, on September 15, the Jordanian army attacked the Palestinian forces, resulting in full-scale fighting for ten days in Amman. As the army surrounded the capital so that the PLO could not call on reinforcements from its Lebanese bases, the Jordanian army blasted the PLO in refugee camps. As a result, PLO leaders were captured, leaving their forces defeated and cut off from food and supplies.[67] Syria came to the aid of the PLO, but a US- and Israeli-backed Jordan was able to stop them. Eventually Nasser stepped in to broker a deal between the PLO and Jordan, exchanging prisoners and hostages to end the confrontation. By 1971, the PLO was expelled from Jordan and turned its attention to creating a mini-state inside Lebanon.[68] September 1970 became known as "Black September."

---

[62] Fatah is the reverse acronym for *Harakat al-Tahrir al-Filastiniya* (The Palestine Liberation Movement). Fatah means "conquest," "victory," or "triumph."

[63] Baumgarten, "The Three Faces/Phases of Palestinian Nationalism, 1948–2005," 25–48.

[64] Ibid.

[65] Barry M. Rubin, *Revolutionaries and Reformers: Contemporary Islamist Movements in the Middle East* (Albany, NY: State University of New York Press, 2003), 35.

[66] Ibid., 35.

[67] Ibid., 35.

[68] Schanzer, *Hamas vs. Fatah: The Struggle for Palestine*, 8.

Subsequent to losing its main bases and suffering an embarrassing defeat, Fatah embraced international terrorism,[69] even naming one group Black September. The Palestinians then engaged in several worldwide international terrorist attacks, including the "1972 massacre of Israeli athletes at the Munich Olympic games; the 1973 attack on the Saudi embassy in Khartoum, Sudan, that led to the murder of the US embassy's chief of mission; and the 1985 attack on the cruise ship *Achille Lauro*, in which a wheelchair-bound American Jew was shot dead and dumped into the water."[70] One common trait of the PLO's terrorist attacks was that they all targeted Israel, or Israelis or Jews, whether they took place in Israel or not. The PLO forces also failed to make a distinction between soldiers and civilians, as Arafat claimed in 1972, "they're equally guilty of wanting to destroy our people."[71] It is estimated that between 1969 and 1985, the PLO "committed over 8,000 terrorist acts—primarily in Israel, but at least 435 abroad—and killed more than 650 Israelis, over three-quarters of them civilians, and hundreds of people from other countries."[72] In May of 1970, members of the PLO forces came in from the Lebanese border and fired a rocket into an Israeli school bus, killing nine children and three teachers and injuring nineteen children.[73] Nevertheless, the PLO's foreign-based attacks were not especially effective in hurting the Israelis, as only thirty-two Israelis were killed and twenty-four wounded outside the country's borders between 1968 and 1973.[74] These attacks helped Israel gain international sympathy.

On the Jewish holy day of Yom Kippur in 1973, as Syria and Egypt staged a surprise reprisal to the 1967 Six-Day War, the PLO helped its Arab allies by trying to distract the Israeli Defense Forces (IDF) through opening a third front against Israel from the south of Lebanon. Aiming to divert IDF attention from the Golan[75] front to

---

[69] Rubin, *Revolutionaries and Reformers: Contemporary Islamist Movements in the Middle East*, 37.

[70] Schanzer, *Hamas vs. Fatah: The Struggle for Palestine*, 8.

[71] Rubin, *Revolutionaries and Reformers*, 24–25.

[72] Ibid., 24–25.

[73] Ibid., 24–25.

[74] Rubin, *Revolution Until Victory*, 39.

[75] The Golan Heights is a contested mountainous area between Israel and Syria, which provides a geopolitical strategic advantage to the state owning it. It has gone back and forth between Israeli and Syrian ownership. Israel captured the Golan in 1967 in the Six-Day War. Syria occupied the land in the 1973 Yom Kippur War but signed a disengagement agreement allowing Israeli control of the land. In 1981, Israel applied Israeli law to the area and still maintains control of the Golan. However, Syria continues to demand Israeli withdrawal of the area and wants to return to the pre-1967 borders, when Syria controlled the territory.

Lebanon, the PLO launched guerrilla attacks on northern Israel.[76] Nonetheless, Israel was able to once again hold its own ground, gaining even more territory than it had after the 1967 War. By this time, the PLO had established a strong base in Lebanon. Although some of the PLO factions, particularly the PFLP and DFLP, were starting to get involved with Lebanon's ensuing civil war, Arafat's Fatah tried to keep its distance from that conflict.

In the next year, Arafat made a speech at the UN General Assembly claiming that Zionism was equal to racism. To commemorate the anniversary of this speech, Fatah exploded a twenty-three-pound bomb in front of a coffee house in Jerusalem, killing seven and injuring forty people.[77] It is estimated that between 1971 and 1982 "Palestinian attacks within Israel killed 250 civilians and wounded 1,628."[78] By 1974, Fatah preferred to attack Israel directly from Lebanon and decreased its international terror attacks.[79] Consequently, Fatah decided to discontinue the Black September international terrorist group, causing some Fatah members to walk out in protest. At this point, internal cleavages within Fatah and between more moderate Palestinian factions in the PLO and more revolutionary factions caused the group to reconsider its organizational structure, with some smaller groups splintering. Later that year Abu Nidal formed his own Fatah group called the Fatah Revolutionary Council (FRC), which continued both international terrorism and even attacked the PLO.[80]

In the period between 1980 and 1982, although the PLO was based primarily in Lebanon and was caught in the beginnings of the Lebanese civil war, the disconnect between the PLO factions' rhetoric and action became increasingly apparent. The PLO was let down primarily by the Arab countries that had promised to support it, namely Egypt, which had made a peace deal with Israel; Iraq, which was entangled in a war with Iran; Jordan, which had removed the PLO from the country in the early 1970s and formed a friendship with Israel; Lebanon, which was caught in its own civil war; and Syria, which had caused a divide within the PLO. In 1982, Arafat left Lebanon and moved his operations to Tunisia.

In the 1980s, the PLO was being pulled in different directions, with Syria, Jordan, Iraq, and Libya seeking to control the group. While the United States was hoping to "circumvent [the PLO] by

---

[76] Yezid Sayigh, *Armed Struggle and the Search for the State: The Palestinian National Movement 1949–1993* (Oxford, UK: Oxford University Press, 2000), 331.

[77] Rubin, *Revolutionaries and Reformers*, 48.

[78] Ibid., 48.

[79] Ibid., 39–40.

[80] Ibid., 39–40.

negotiating with a joint delegation of Jordanians and non-PLO Palestinians, Jordan and Egypt wanted to moderate it and push it into peace talks as their clients."[81] However, Arafat tried to maintain good relations with all of these different parties. In 1985, King Hussein of Jordan tried to influence the PLO in the aftermath of its splits and defeats in Lebanon, offering Arafat a land-for-peace deal in exchange for accepting the conditions in the UN resolutions calling for Israel's withdrawal from the West Bank and Gaza.[82] However, the agreement was somewhat vague and, in some points, even contradictory, causing the PLO to question whether they should accept this deal from Jordan. While Arafat did not want to completely dismiss the offer, at the same time, Syria was pressuring the PLO to reject the negotiations with Jordan. From its beginning, Syria had attempted to influence the umbrella organization; however, the PLO tried to distance itself from Syria. Nonetheless, several PLO factions were quite loyal to Syria. One of these groups was the newly formed Abu Nidal FRC, which murdered a PLO moderate in 1983. As a result, Nidal's group moved its headquarters to Damascus. Syria, dissatisfied with the PLO's more moderate stance, launched an attack on a Palestinian refugee camp in Beirut, Lebanon, which "killed 600 Palestinian civilians and guerrillas, while 1,500 others were missing."[83]

The Palestinian resistance was emboldened in 1987 when an almost spontaneous event led to a large armed resistance. Arafat decided to welcome negotiations, shifting his position from decades of launching terrorist attacks after the first *intifada* in 1987. At the time, Arafat was based in Tunisia, to which he had been exiled after being ousted from his perch in Lebanon in 1982. Trying to manage the PLO's response to the incident from afar proved too difficult for Arafat. In the midst of this chaos, the Muslim Brotherhood formed Hamas, which took advantage of Arafat's distance from the group to spread leaflets challenging Arafat's authority.[84] However, at this point in the insurgency, the PLO preferred to demonstrate to the world that it was capable of ruling itself and decided to give up its "traditional role as a guerrilla fighting squad."[85] This role was taken up by Hamas, which was committed to Palestinian liberation by any means necessary—particularly violence. As the PLO withdrew from

---

[81] Ibid., 84.
[82] Ibid., 72.
[83] Ibid., 73.
[84] Schanzer, "The Iranian Gambit in Gaza," 29–32.
[85] Schanzer, *Hamas vs. Fatah* 28.

the uprising, Hamas continued to fill the vacancy and came to control the *intifada*.[86]

A year later, on November 12, 1988, Arafat called an emergency meeting of the PNC to announce Palestine's independence and invited Israel to a "peace conference based on UN Resolutions 242 and 338, which called for Israel to withdraw from territories it had conquered during the Six-Day War."[87] This was a historic turn of events for the revolutionary movement, as in the next two weeks fifty-five other countries recognized the PLO's declaration of independence and set up the PLO to run the transitional government.[88] By the next month, the PLO had accepted Israel's right to exist and renounced terrorism—a move that enabled Hamas to gain support.

As Hamas continued its violence not only toward Israel but also toward Fatah, Arafat sought to secure his role as the sole and legitimate leader of the Palestinian people. Therefore, he decided to join the Jordanian delegation in the 1991 Madrid Conference for peace. As a result, Arafat was able to earn support from many countries worldwide and was lauded for his attempts at resolving this tense issue after half a century of violence.[89] On September 13, 1993, Israel and the PLO agreed to mutual recognition and limited Palestinian self-rule in the West Bank and Gaza in what became known as the Oslo process. Given that the PLO was open to settling for an interim solution to the Palestinian problem by recognizing Israel's right to exist, Israel also modified its 1967 platform that rejected the possibility of an independent Palestinian state with limited sovereignty. However, Israel did not necessarily welcome a completely independent Palestinian state.[90] Israel and the PLO continued on this road to peace when, in 2000, President Bill Clinton invited Yasser Arafat and the Israeli Prime Minister Ehud Barak to settle several outstanding issues standing in the way of peace between the two countries. However, Arafat rejected the Israeli land-for-peace deal.

---

[86] Ibid., 28.

[87] Ibid., 28.

[88] Ibid., 28.

[89] Ibid., 38.

[90] Jason Hicks, "Reflections in a Mirror: Hamas and the Israeli Politik," *Washington Report Middle East Affairs* 28, no. 1 (2009): 20–21.

Figure 4. Yitzhak Rabin, Bill Clinton, and Yasser Arafat at the Oslo Accords signing ceremony on September 13, 1993.[91]

Shortly thereafter, when Ariel Sharon visited the Haram Sharif in Jerusalem, shooting ensued in what became known as "*al-Aqsa intifada*," or the second *intifada*, against Israel.[92] This return to violence suggested that the PLO was frustrated at the failure of the second Camp David talks, the lack of Palestinian independence after the 1993 accords, and a worry that Arafat had lost connection to the Palestinian people who were now supporting the openly violent Hamas group. Three Palestinian groups, a minority of Arafat's PA, a group of middle-command Fatah and Tanzim leaders, and Hamas were responsible for starting the second *intifada* and, for the most part, acted independently.[93] However, none of these groups had planned for the second *intifada* to be as involved or to last so long. In the first year of the *intifada*, the insurgents killed 164 Israelis. This number increased to 694 by the end of 2002 and to 1,000 by September 2004.[94] However, the Palestinians lost nearly three times that number. Tensions remained high between the Palestinians and the Israelis as attacks between the two groups continued after the second *intifada* until Mahmoud Abbas was elected the new president of the PA in 2005 and agreed to a cease-fire. Nonetheless, this cease-fire did not last long

---

[91] "File:Bill Clinton, Yitzhak Rabin, Yasser Arafat at the White House 1993-09-13. jpg," *Wikipedia Commons*, accessed March 15, 2011, http://commons.wikimedia.org/wiki/File:Rabin_at_peace_talks.jpg.

[92] Schanzer, "The Iranian Gambit in Gaza," 29–32. For a detailed timeline of the second *intifada*, see BBC News, "Al Aqsa Intifada Ttimeline," http://news.bbc.co.uk/2/hi/middle_east/3677206.stm.

[93] Amal Jamal, *The Palestinian National Movement: Politics of Contention 1967–2005* (Bloomington, IN: Indiana University Press, 2005), 156.

[94] Ibid., 157.

as the election of Hamas to the Palestinian Legislative Council caused further tensions in the region and attacks resumed from both parties.

METHODS OF RECRUITMENT

The PLO used a multifaceted approach to recruitment, including staging spectacular attacks, issuing leaflets and communiqués, offering financial incentives, and integrating women in the movement. Sensationalization of PLO attacks in the media gave legitimacy to the group and helped to attract new followers.[95] Furthermore, the communiqués and leaflets that the PLO distributed also helped to garner support for their missions by reinforcing the Palestinian nationalist agenda on which the group ran. By inciting the right to a Palestinian state, the PLO also gathered recruits because this message resonated deeply with the public.

In addition to the information strategy to attracting recruits, the PLO also offered financial incentives. It is estimated that a male adult recruit to Fatah was paid between 700 to 1,000 Lebanese pounds per month.[96] The PLO also paid money to the Palestinians in the camps who supported their cause. Moreover, in addition to these more traditional methods of recruitment, the PLO utilized the General Union of Palestinian Women, which was made up of the various PLO factions. This group was used to help families in need, perpetuate the memory of martyrs, recruit, teach the illiterate, and participate in propaganda.

METHODS OF SUSTAINMENT

The PNC established the Palestinian National Fund (PNF) in its first meeting from May 28 to June 2, 1964, to finance all PLO activities.[97] To raise revenue for the organization, in the late 1960s, the PNC started collecting a 5–7% income tax from Palestinian workers living in Arab countries.[98] The Arab countries were supposed to deduct this tax from the wages of all Palestinian workers. However, some countries were more dedicated to the tax than others. This tax was later expanded to all Palestinian workers in the Islamic world and was named the Palestinian Liberation Tax Fund (PLTF). In addition to collecting the income tax of Palestinians, the PNF also accepted

---

[95] Rubin, *Revolution Until Victory*, 31–32.

[96] Jillian Becker, *The PLO: The Rise and Fall of the Palestine Liberation Organization* (London: Widenfield & Nicholson, 1984), 148.

[97] Livingstone and Halevy, *Inside the PLO*, 164–166.

[98] Ibid., 164–166.

donations from private and foreign state sponsors and benefited from income from its own investments.[99] It is estimated that the PLTF generated between $25 million to $30 million a year for the PLO.[100] This increased to about $50 million in 1987 during the first *intifada*. After the PLO was expelled from Lebanon, annual donations to the PNF jumped to more than $100 million. However, by the 1990s, the PNF received only about $50–$65 million annually.[101] In addition to establishing the PNF, Arafat created an alternative fund called the Chairman's Secret Fund. Furthermore, each faction had their own sources of funding—mostly from other Arab countries seeking to influence the powerful Palestinian organization.[102] Since 1979, six Arab countries have made annual contributions to the PNF ranging from $250 to $300 million.[103]

METHODS OF OBTAINING LEGITIMACY

The PLO relied on spectacular acts of terrorism and military operations to garner the legitimacy of the public.[104] In this way the group illustrated determination to reach its goals in any way necessary and also demonstrated steadfast defiance of Israel for the Palestinian cause. With the launch of the PLO's terror campaign against Israel in the late 1960s/early 1970s, the PLO became known by the Arab League as the "sole legitimate representative of the Palestinian people."[105] For every attack that the Israelis countered, the Palestinian organizations gained more legitimacy because the events showcased Israel's violent occupation of Palestinian land.[106] The harsher the PLO was against the Israelis, the more support they were able to garner from the public.[107] When Israel declared Hamas an illegal organization on September 28, 1989, this designation helped to legitimize the organization in the view of the Palestinian public.

However, the PLO faced several challenges to obtaining legitimacy. Because it was made up of so many different factions and lacked strong centralization within the group, the PLO could not consolidate its identity and goals to be consistent over time, as was evident in its 1967

---

[99] Ibid., 164–166.
[100] Ibid.
[101] Ibid.
[102] Ibid.
[103] Ibid.
[104] Rubin, *Revolutionaries and Reformers*, 242.
[105] Schanzer, *Hamas vs. Fatah*, 19.
[106] Ibid., 18.
[107] Ibid., 33.

loss.[108] Because of its inability to unify its goals with its actions, the PLO lost some legitimacy in the eyes of the public, especially in the eyes of those who found Hamas's commitment to its stated goals to be unwavering and clear.[109] The PLO's rise to legitimacy reached its peak in 1991 with the Madrid Conference. The group's "inability to end Israel's post-1967 occupation via an endless series of negotiations came to erode its political and national capital."[110] In addition, the PLO also lost legitimacy as a result of charges of corruption and personal aggrandizement in the organization. By the 1990s, there were allegations that Arafat and his colleagues were pocketing billions of dollars in international aid, while the Palestinian people continued to suffer.[111]

Furthermore, the PLO wrestled with Hamas over the public's "hearts and minds." Additionally, the PLO's view of Hamas changed over time. Although at first the PLO denied that Hamas even existed, it later suggested that Hamas was "operating outside the sphere of legitimate Palestinian action."[112] Then later, the PLO recognized Hamas and even invited it to join the PLO and PNC.[113] When Hamas rejected this offer, the PLO tried to promote divisions within Hamas or, in several cases, physically fought the group.

However, when the PLO leader, Arafat, tried to suppress Hamas, it weakened his own legitimacy and popularity. Hamas gained legitimacy when it stood up to the PLO regarding the Oslo process in the early 1990s, at which time the PLO lost support from some Palestinians. The PLO was accused of negotiating with Israel and settling for an interim solution that seemed (to the public) to serve the PLO's own goals rather than promote the greater good of the Palestinians.[114] In response to his loss of popularity, in July 1994, Arafat staged a motorcade to escort him back from Egypt in a grand display reinforcing his desired image as president.[115] However, once he returned, Hamas continued to challenge the PLO's authority.[116]

---

[108] Sayigh, *Armed Struggle and the Search for the State*, 174.

[109] Rubin, *Revolutionaries and Reformers*, 231.

[110] Khaled Hroub, "Hamas After the Gaza War," *openDemocracy*, accessed August 1, 2010, http://www.opendemocracy.net/article/hamas-after-the-gaza-war.

[111] Rubin, *Revolutionaries and Reformers*, 231; Hroub, *Hamas: Political Thought and Practice*, 235.

[112] Ibid., 92.

[113] Ibid.

[114] Schanzer, "The Iranian Gambit in Gaza," 29–32.

[115] Schanzer, *Hamas vs. Fatah: The Struggle for Palestine*, 42.

[116] Ibid., 43.

EXTERNAL SUPPORT

From the very beginning of the insurgency, the Palestinians were supported by their Arab neighbors. Early in the movement, Syria allowed Fatah to broadcast its military communiqués from its state radio and base its operations from Syria.[117] However, Fatah operated from all states bordering Israel.[118] Financially, the PLO was heavily subsidized by other Arab countries. Saudi Arabia contributed close to $40 million a year.[119] The United Arab Emirates (UAE) promised to send the PLO protection money if it refrained from being active in the small Gulf state. In 1979, at a summit in Baghdad, seven Arab countries pledged the following to the PNF: Saudi Arabia, $85.7 million; UAE, $34.3 million; Algeria, $21.4 million; Iraq, $44.6 million; Qatar, $19.8 million; Kuwait, $47.1 million; and Libya, $47.1 million (for a total of $300 million).[120] In addition to pledging this money, these seven Arab countries promised $50 million annually to help the Palestinians living under Israeli occupation, and these funds were administered by the Joint Jordanian-PLO Committee for the Occupied Territories.[121] The joint task force "agreed to match the contribution of the Arab governments with another $50 million of its own."[122] In addition to the financial support, other Arab countries offered weapons and munitions to the PLO, and the Libyan government provided munitions, a radio station, vehicles, and guns. Furthermore, Iraq and Saudi Arabia offered arms, "including M-16 automatic weapons sold to it by the United States."[123]

The PLO also received considerable support from the Soviet Union, Eastern Europe, and Eastern Asian countries.[124] These countries provided the PLO arms, training, and a safe haven for its terrorists.[125] In 1972, Arafat met with Romanian leader Nicolae Ceausescu, and during this meeting Arafat "exchange[d] his organization's experience in conducting kidnappings and assassinations for Romanian advice on running disinformation and influence-buying operations to improve the PLO's political standing in the West."[126] In addition to this informational exchange, Romania provided PLO operatives with

---

117 Ibid., 17.
118 Ibid., 17.
119 Rubin, *Revolution Until Victory*, 52–53.
120 Livingstone and Halevy, *Inside the PLO*, 168.
121 Ibid., 168.
122 Ibid.
123 Rubin, *Revolutionaries and Reformers*, 52–53.
124 Including China, Vietnam, North Korea, Pakistan, and India.
125 Ibid., 52–53.
126 Ibid., 39.

forged passports and training on its military bases.[127] The PLO also welcomed external support from East Germany in a 1979 agreement in which the PLO offered intelligence on the United States and Israel in return for "safe haven for terrorists, training facilities, electronic equipment, and explosives."[128] Other Eastern European countries that supported the PLO included Czechoslovakia, which gave explosives to the group, and Poland, which "helped build factories to manufacture antitank guns and shells at Palestinian refugee camps."[129] The PLO also appeared to benefit from illegal trade with Eastern Europe, engaging in drug smuggling from Lebanon through Bulgaria.[130]

The PLO also established a relationship with Iran. In exchange for Arafat's support in training the Iranians in guerrilla military tactics and providing weapons, Ayatollah Khomeini "closed the Israeli embassy in Tehran, handed the keys over to Arafat, and flew a Palestinian flag overhead," converting the building into an "official PLO entity, complete with an ambassador."[131] However, this relationship soured after the Oslo process as the Arafat-led PLO negotiated with Israel, causing Iran to shift its support to Hamas and to members of the PLO in exile who were against Arafat.[132] Arafat noted that Iran provided some $30 million to Hamas in the early 1990s.[133] However, in 2000, as Arafat's PLO adopted a more Islamist stance after the second round of Camp David talks, Iran resumed its support to the group in the form of training and weapons.[134]

COUNTERMEASURES TAKEN BY THE GOVERNMENT

Israel was largely effective in repressing and countering the various PLO factions, especially at the onset of the movement. The 1967 Six-Day War gave a considerable boost in legitimacy to the Israeli state, which was able to defeat its Arab neighbors in such a short time. Seeking to stand up to its new image, Israel was not prepared to allow a small resistance group to weaken this position. As a result, Israel took a two-pronged approach to dealing with the Palestinians. First it would capture and imprison activists and deport others who would

---

[127] Ibid.

[128] Ibid.

[129] Ibid., 52–53.

[130] Ibid.

[131] Schanzer, "The Iranian Gambit in Gaza," 29–32.

[132] Ibid.

[133] Ibid.

[134] Ibid.

try to organize rebellion. Second, it tried to allow Palestinians who abstained from the armed resistance to lead normal lives.[135]

Israeli repression of Palestinians did not end in Palestine, however. The IDF frequently engaged in counterattacks on Palestinians operating from Lebanon, Jordan, and Syria. On November 13, 1966, in response to sabotage operations ordered by Arafat, Israel attacked a Jordanian village in the West Bank, resulting in the death of more than seventy people.[136] After the Yom Kippur War, only three of the PLO factions—Fatah, PFLP, and the DFLP—were still active in the occupied territories.[137] However, constant Israeli pressure made it very difficult for the PLO to operate in the 1970s.[138] In the 1980s Ariel Sharon ordered the "red berets," a group of Israeli paratroopers, to intimidate and humiliate Arabs who were suspected of helping the resistance.[139] This policy effectively turned Gaza into a prison as travel restrictions on the Palestinians increased.[140]

As Hamas entered the scene in 1987 with harsher terrorist tactics, Israel responded even harder, and this affected the PLO as well. Within a few months of the first *intifada*, strict Israeli repression made it difficult for Arafat to leave his house, as he was surrounded by Israeli tanks and "the rest of the Palestinian Authority infrastructure was reduced to rubble."[141] During the first year of the *intifada* it is estimated that between 390 and 500 Palestinians were killed, close to 30,000 were wounded, and about 21,000 were detained. In the aftermath of the first *intifada*, the Palestinians began to become more self-reliant and to focus on ways to survive even when faced with very strict countermeasures by the Israeli government. Although the situation in the West Bank was a little different than that in Gaza because of the relative economic advantages in the West Bank, people living in both areas struggled.[142]

In efforts to reduce Israeli repression and establish the PA's legitimacy, in its first year, the PA adopted a strict policy against the Palestinian resistance by which it confiscated their arms and tracked down Palestinian resistance military cells.[143] The Palestinian police also assassinated some military wing leaders, arrested and interrogated

---

[135] Rubin, *Revolution Until Victory*, 17.
[136] Schanzer, *Hamas vs. Fatah* 18.
[137] Sayigh, *Armed Struggle and the Search for the State*, 347.
[138] Ibid., 350.
[139] Tamimi, *Hamas: A History from Within*, 13.
[140] Ibid., 13.
[141] Schanzer, "The Iranian Gambit in Gaza," 29–32.
[142] Sayigh, *Struggle Within, Struggle Without*, 266.
[143] Hroub, *Hamas: Political Thought and Practice*, 107.

scores of Hamas members and shaved their beards in detention to humiliate them, and mounted raids on Hamas mosques, agencies, and the Islamic University.[144]

SHORT- AND LONG-TERM EFFECTS

CHANGES IN THE ENVIRONMENT

As a result of years of Palestinian-Israel fighting, many Palestinians remain stateless, seeking refuge in neighboring countries or abroad. The Palestinians who did continue to live within Palestine endured strict travel restrictions and poor living and working conditions. Over the past sixty years, the boundaries of the Palestinian state have changed several times as a result of numerous conflicts between Israel and its Arab neighbors.

CHANGES IN GOVERNMENT

Arguably the most considerable change in the Palestinian government since its establishment in 1994 came in January 2006 when Hamas gained control of the Palestinian Legislative Council and several months later also gained control of the PA.[145] As a result of Hamas's elections, the Western world, with the United States leading the way, imposed strict sanctions on the new Palestinian government as an extension of the US antiterrorism policy. Although PA President Mahmoud Abbas attempted to negotiate with Hamas to moderate its stance in order to alleviate these strict sanctions, he was not successful. As a result, from 2006 to 2007, the Gaza Strip endured a violent outbreak between Fatah and Hamas supporters.[146] In February 2007, President Abbas and the Hamas chief, Khalid Mishal, signed an agreement forming the Palestinian National Unity Government, led by Hamas representative Ismail Haniyeh, who became the prime minister in 2006.[147] This new alliance did not, however, end the fighting in Gaza. In June of 2007, Hamas won in a "violent takeover of all military and governmental institutions in the Gaza Strip."[148] Fighting continued among the Palestinians and between the Palestinians and Israelis as the power struggle remained unsettled.

---

[144] Ibid., 108.
[145] Central Intelligence Agency, "Gaza Strip," *The World Factbook*.
[146] Ibid.
[147] Ibid.
[148] Ibid.

CHANGES IN POLICY

In the midst of the rise of Hamas as an alternative representative to Palestinian issues, the PLO forged ahead with its policy of pursuing dialogue in establishing a Palestinian state. In 2010, after fourteen months of communications breakdown between the organization and Israel, the PLO agreed to US-mediated talks with Israel.[149] These talks suggested that the Palestinian leadership was softening after stating that it would not engage in dialogue with Israel until Israel stopped constructing settlements in the West Bank and Jerusalem.[150] This event demonstrated that efforts to seek autonomy and independence for the Palestinians continued, and that the PLO remained the primary representative for Palestinian issues in the international realm.

CHANGES IN THE REVOLUTIONARY MOVEMENT

After Arafat's death in 2004 and considerable losses in the second *intifada*, the PLO, now led by Mahmoud Abbas, represented sixty years of failed negotiations and military operations, as they had yet to achieve their primary goal—Palestinian liberation. Dealing with losing the face of the PLO for the last thirty years, Mahmoud Abbas failed to consolidate the Palestinians as the new leader of the PA. Many tired PLO supporters switched allegiance to Hamas in 2005.

BIBLIOGRAPHY

Abu-Amr, Ziad. "Hamas: A Historical and Political Background." *Journal of Palestine Studies* 22, no. 4 (1993): 5–19.

Associated Press. "Palestinians, Skeptical, Agree to Talks with Israel." *The New York Times.* March 8, 2010.

Baumgarten, Helga. "The Three Faces/Phases of Palestinian Nationalism, 1948–2005." *Journal of Palestine Studies* 34, no. 4 (Summer 2005): 25–48.

BBC News. "Al Aqsa Intifada Timeline." http://news.bbc.co.uk/2/hi/middle_east/3677206.stm

Becker, Jillian. *The PLO: The Rise and Fall of the Palestine Liberation Organization.* London, UK: Widenfield & Nicholson, 1984.

---

[149] Associated Press, "Palestinians, Skeptical, Agree to Talks with Israel," *The New York Times*, March 8, 2010.

[150] Ibid.

Central Intelligence Agency. "Gaza Strip." *The World Factbook.* https://www.cia.gov/library/publications/the-world-factbook/geos/gz.html

———. "Israel." *The World Factbook.* https://www.cia.gov/library/publications/the-world-factbook/geos/is.html.

———. "West Bank." *The World Factbook.* https://www.cia.gov/library/publications/the-world-factbook/geos/we.html.

Farsoun, Samih K., and Naseer H. Aruri. *Palestine and the Palestinians: A Social and Political History.* Boulder, CO: Westview, 2006.

Hicks, Jason. "Reflections in a Mirror: Hamas and the Israeli Politik." *Washington Report Middle East Affairs* 28, no. 1 (2009): 20–21.

Hroub, Khaled. "Hamas After the Gaza War." *openDemocracy.* Accessed August 1, 2010. http://www.opendemocracy.net/article/hamas-after-the-gaza-war.

———. *Hamas: Political Thought and Practice.* Washington, DC: Institute for Palestine Studies, 2000.

Jamal, Amal. *The Palestinian National Movement: Politics of Contention 1967–2005.* Bloomington, IN: Indiana University Press, 2005.

Livingstone, Neil C., and David Halevy. *Inside the PLO: Covert Units, Secret Funds, and the War Against Israel and the United States.* New York, NY: William Morrow & Company, 1990.

Minorities at Risk, "Assessment for Palestinians in Israel," http://www.cidcm.umd.edu/mar/assessment.asp?groupId=66603

PLO Negotiations Affairs Department. "About Us." Accessed October 15, 2010. http://www.nad-plo.org/etemplate.php?id=182.

Rubin, Barry. *Revolution Until Victory?: The Politics and History of the PLO.* Cambridge, MA: Harvard University Press, 1994.

Rubin, Barry M. *Revolutionaries and Reformers: Contemporary Islamist Movements in the Middle East.* Albany, NY: State University of New York Press, 2003.

Sayigh, Yezid. *Armed Struggle and the Search for the State: The Palestinian National Movement 1949–1993.* Oxford, UK: Oxford University Press, 2000.

Sayigh, Yezid. "Struggle Within, Struggle Without: The Transformation of PLO Politics Since 1982." *International Affairs (Royal Institute of International Affairs 1944-)* 65, no. 2 (Spring 1989): 247–271.

Schanzer, Jonathan. *Hamas vs. Fatah: The Struggle for Palestine.* New York, NY: Palgrave, 2008.

———. "The Iranian Gambit in Gaza." *Commentary* 127, no. 2 (2009): 29–32.

Tamimi, Azzim. *Hamas: A History from Within.* Northampton, MA: Olive Branch Press, 2007.

HUTU–TUTSI GENOCIDES

Bryan Gervais

SYNOPSIS

The Tutsi and Hutu are two ethnic groups from the central African nations of Burundi and Rwanda. After political independence from Belgian colonial rule in 1962, both Burundi and Rwanda became embroiled in ethnic clashes between the majority Hutu and the minority Tutsi that sporadically continued for the next four decades. Although there is much debate as to whether the Hutu and Tutsi composed one single ethnic group before European colonization,[1] most scholars agree that European favoritism toward the Tutsi led to extensive social divisions and tension between the Hutu and Tutsi. Despite these similarities, substantial differences exist between the Burundian and Rwandan conflicts, and, thus, they should be treated as two distinct processes.

Violence between the Hutu and Tutsi has been common in both countries in the postcolonial era. In Rwanda, ethnic conflict occurred during the *inyenzi* attacks in the 1960s, the Rwandan Civil War, and following genocide in the early 1990s. In Burundi, conflict between the Tutsi and Hutu broke out from time to time between 1965 and 1993.[2] Hutu–Tutsi violence continues to agitate peace processes in both Rwanda and Burundi.[3] Although incidents in both countries are discussed in this chapter, special attention is paid to the events surrounding the 1994 Rwandan genocide that produced an estimated 800,000 casualties.

---

[1] To what extent social divisions existed—if they were two distinct groups—is also debated.

[2] Noted clashes occurred in 1965, 1969, 1972, 1988, and 1993, with the 1972 and 1993 events regarded as the most severe.

[3] Hutu–Tutsi violence also affects the Democratic Republic of the Congo.

TIMELINE

1916	During World War I, the Belgians forced the Germans out of the Rwanda and Burundi (then Ruanda-Urundi) territory and continued the German policy of favoring the Tutsi over the Hutu.
1957	The *Hutu Manifesto* increased Hutu group consciousness and the idea that the Tutsi were invaders of Hutu land.
1959	The Belgians ceased to support the Tutsis in favor of the Hutu as Rwanda and Burundi prepared for independence.
1960	The "Hutu Revolution" occurred, during which the Hutu won control of the Rwandan government and implemented anti-Tutsi policies. A mass exodus of Tutsi to neighboring countries (particularly Uganda) occurred, and the first *inyenzi* attacks began.
September 1961	The Tutsi monarchy was abolished and Ruanda-Urundi again became the two separate states of Rwanda and Burundi.
May 1965	After King Mwambutsa's refusal to appoint a Hutu prime minister (PM), a large-scale Hutu revolt occurred in Burundi, which the government brutally suppressed. A Hutu PM was put in place but was assassinated later in the year, and the assassination initiated more violence.
November 1966	In a year that saw multiple political coups in Burundi, Michel Micombero began his period in office as president.
May–July 1972	The Tutsi-dominated Burundian government slaughtered an estimated 100,000–300,000 (mostly educated) Hutu over the course of two months. The genocide, the first to occur in the Great Lakes region, was in response to an attempted coup by the Hutu that claimed the lives of 10,000 Tutsi.
July 1973	President Grégoire Kayibanda of Rwanda was overthrown by General Juvénal Habyarimana, who established a military dictatorship.
1988	Government troops in Burundi killed an estimated 20,000 Hutu, and more than 60,000 fled the country after armed Hutu killed hundreds of Tutsi.

October 1990	The Rwandan Patriotic Front (RPF) launched its first attacks from Uganda, beginning the Rwandan Civil War.
July 1992	The Arusha Accords commenced with the aim of ending the war in Rwanda.
October 1993	Violence occurred in Burundi after the assassination of President Ndadaye and other Hutu leaders. An estimated 50,000–100,000 Burundians were killed via attacks perpetrated by both Hutu and Tutsi.
April–July 1994	Habyarimana was assassinated. An organized genocide of mostly Rwandan Tutsi began almost immediately afterward. An estimated 800,000 were killed before the RPF took control of Kigali and the country in July.
November 1996–May 1997	The First Congo War began when Rwanda, Uganda, and Angola invaded eastern Zaire with hopes of destroying Hutu military camps.
August 1998–July 2003	The Second Congo War occurred and included major combat roles for Hutu and Tutsi military groups. The war claimed 5.4 million lives.

THE ENVIRONMENT OF THE REVOLUTION

PHYSICAL ENVIRONMENT

Rwanda and Burundi are both landlocked countries in the east-central Africa's Great Lakes region. Rwanda shares borders with Uganda, Burundi, the Democratic Republic of the Congo (DRC), and Tanzania, with Lake Kivu on its western boundary. Besides the border with Rwanda to the north, Burundi is encircled by the DRC and Tanzania, with the Great Lake of Tanganyika in the central and southwest region of the country.

With an area of roughly 10,000 square miles (more than 26,000 kilometers), Rwanda is slightly smaller than the US states of Massachusetts and Maryland. Known as the "Land of a Thousand Hills," Rwanda is mostly made up of grassy uplands and hills, with a mountainous region extending from the southeast to the northwest, increasing in altitude heading northwest toward the country's border with the DRC. Despite its geographic position slightly below the equator, Rwanda has a temperate climate because of its altitude, with two rainy seasons lasting roughly from February to May and November to January.

Excluding its territory in Lake Tanganyika, Burundi is incrementally smaller than Rwanda. Burundi is significantly more mountainous than Rwanda with fewer grasslands and a plateau in the east. Lying farther south than Rwanda, Burundi is more tropical and equatorial in climate, but its rainy seasons occur at roughly the same time as those in Rwanda.

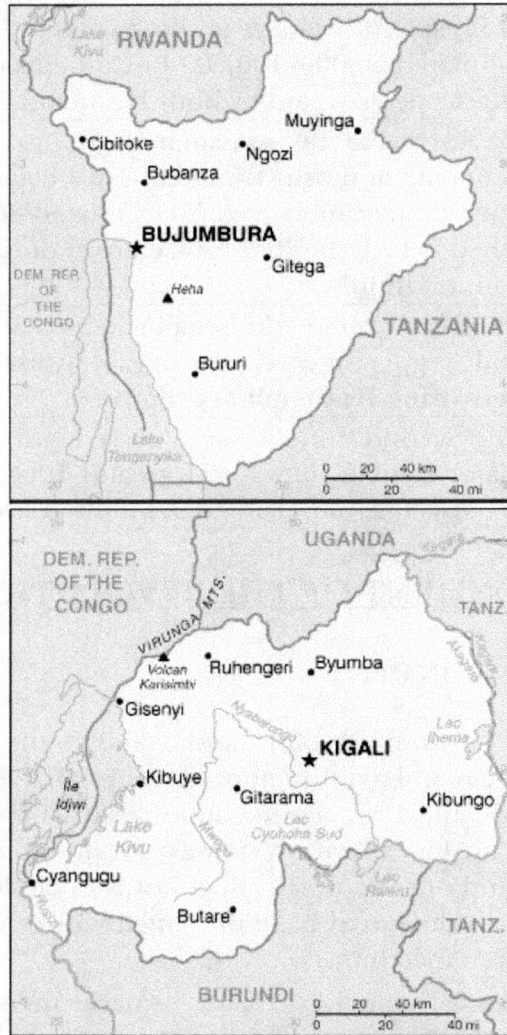

Figure 1. Maps of Africa: Burundi and Rwanda.[4]

    ⁴ (Top) Central Intelligence Agency, "Burundi," *The World Factbook*, accessed December 10, 2010, https://www.cia.gov/library/publications/the-world-factbook/maps/maptemplate_by.html; (bottom) Central Intelligence Agency, "Rwanda," *The World Factbook*, https://www.cia.gov/library/publications/the-world-factbook/maps/maptemplate_rw.html.

CULTURAL AND DEMOGRAPHIC ENVIRONMENT

The ethnic composition of both Burundi and Rwanda is roughly 85% Hutu, 14% Tutsi, and 1% Twa—the pygmy people who have inhabited the Great Lakes region longer than any other ethnic group. The Bantu language of Kirundi is the main language in Burundi and is mutually intelligible with Kinyarwanda, the main language of Rwanda. Swahili is also spoken, and French was an official language in both countries for decades[5] and primarily used in government documents and by the educated. Christianity—in particular, Roman Catholicism (62% of population in Burundi and 57% in Rwanda)—is the majority religion in both Burundi (67% of the population) and Rwanda (82.5%).[6]

Besides sharing language, traditions, social taboos, and religion, there is little evidence of any anthropological differences between Hutu and Tutsi. Intermarriage has been common, and it is possible to become a member of the other group through marriage or even a change in financial standing. For example, an increase in livestock—specifically cattle—could allow a Hutu to become a Tutsi.[7]

SOCIOECONOMIC ENVIRONMENT

Rwanda's economy is predominately agricultural. The country has the highest population density in Africa, with a population of nearly 10.5 million. Burundi's economy is also largely agrarian, with close to 90% of Burundians relying on subsistence farming. Its population is slightly below 9 million, and it has the second highest population density in Africa after Rwanda. The median age is 16.7 years in Burundi, and in 2009, ranked the ninth worst in the world on the United Nations Development Programme (UNDP) Human Development Index.[8] With approximately 338,000 Burundians living

---

[5] Rwanda replaced French with English in 2008 in an effort to increase participation in the East African Community.

[6] Central Intelligence Agency, "Rwanda," *The World Factbook*; Central Intelligence Agency, "Burundi," *The World Factbook*; Peter Uvin, "Ethnicity and Power in Burundi and Rwanda: Different Paths to Mass Violence," *Comparative Politics* 31, no. 3 (1999); Thomas Patrick Melady, "Burundi and Rwanda: A Tragic Past, A Cloudy Future," accessed September 3, 2010, http://www.americanambassadors.org/index.cfm?fuseaction=publications. article&articleid=37; Ruth Ann Hudson, "Breaking the Cycle of Violence: Cohesion Across Ethnic Barriers in Burundi," *Elections Today* 11 (2003): 6.

[7] Mahmood Mamdani, *When Victims Become Killers: Colonialism, Nativism, and the Genocide in Rwanda* (Princeton, NJ: Princeton University Press, 2001); Uvin, "Ethnicity and Power in Burundi and Rwanda"; Hudson, "Breaking the Cycle of Violence," 6.

[8] This is calculated using gross domestic product per capita, life expectancy after birth, adult literacy rates, and school enrollment.

in camps, Burundi has the largest internally displaced population among the African Great Lakes nations. The Burundian refugee population is somewhere near 640,000. The median age in Rwanda is 18.7 years, and the country ranked sixteenth worst on the UNDP Human Development Index in 2009. In 2007, 2.8% of the Rwandan population and 2.0% of the Burundian population had HIV/AIDS, the twenty-fifth and thirty-second worst rates in the world, respectively. Infrastructure is limited in both countries.[9]

HISTORICAL FACTORS

It has been theorized that the Hutu arrived in the Great Lakes region sometime around the eleventh century from elsewhere in central Africa and joined the pygmy Twa. The Tutsi, fleeing famine and drought, arrived in Burundi and Rwanda sometime during the fifteenth and sixteenth centuries. The three ethnic groups became extensively integrated, sharing religion, language, and other cultural elements. Before European colonization, both Burundi and Rwanda were hierarchically organized kingdoms with Tutsi on the top and middle levels and the Hutu and Twa on the lower levels. However, the sociopolitical division was much less fluid in the Rwandan hierarchy than the Burundian. Under the Tutsi *mwami* (or "king") Kigeri Rwabugiri's reign (1860–1895), power in Rwanda became considerably centralized, and only the upper echelons of the society were considered "Tutsi."[10]

A roughly seventy-five-year history exists of European colonizers exploiting the Hutu–Tutsi distinction for economic and political gain. The Europeans introduced the nineteenth century theory of the Hamitic race[11]—the idea that a nomadic pastoralist subgroup of the Caucasian race had migrated into sub-Saharan Africa, bringing with it technology and civilization. The myth, popularized by British explorer John Hanning Speke, articulated that "Hamites" were vastly superior to the "Negroids" who had inhabited the land well before the Hamities arrived. To the Europeans, it was clear that the Tutsi fit the role of Hamites and that the Hutu were the inferior Negroids.

---

[9] Central Intelligence Agency, "Rwanda," *The World Factbook*; Central Intelligence Agency, "Burundi," *The World Factbook*; UNDP 2009; Uvin, "Ethnicity and Power in Burundi and Rwanda."

[10] Mamdani, *When Victims Become Killers*; Isaac A. Kamola, "The Global Coffee Economy and the Production of Genocide in Rwanda," *Third World Quarterly* 28, no. 3 (2007): 571–592; Linda Kirschke, "Broadcasting Genocide: Censorship, Propaganda and State-sponsored Violence in Rwanda 1990-1994," *Article 19* (1996): 3; Uvin, "Ethnicity and Power in Burundi and Rwanda."

[11] This theory is now referred to as the Hamitic myth.

Nevertheless, the Hamitic myth was incessant, both during the colonial era and afterward. The Hutu in Rwanda, for example, rallied behind this idea as the colonial era came to a close, arguing that the Tutsi were invaders of Hutu land. The Hamitic myth claimed that the Tutsi, like the Ethiopians, Berbers, Egyptians, Bahuma, and Masai, were descendants of Caucasians and a separate and superior race to the Hutu. These Hamites were said to have civilized Africa before European exploration. This idea would dictate the way Europeans governed Rwanda and Burundi and how both the Tutsi and Hutu regarded themselves before and after independence.[12]

After the 1885 Berlin Conference, the African continent was divided between the European powers, and Rwanda and Burundi were placed under German control as provinces of German East Africa. The Germans were drawn to the region because of its favorable agricultural conditions, as well as the fact that they indirectly governed the two provinces, keeping the Tutsi monarchical and hierarchical systems in place. The Germans observed that the Tutsi tended to be tall and raised cattle, which led them to believe that the Tutsi were superior to the shorter, agrarian-focused Hutu. The Germans provided the Tutsi with privileges (including education and elite positions) not offered to the Hutu, provoking increased ethnic tensions between the two groups. Although the period of German rule (as well as subsequent Belgian rule) was marked by relatively few ethnic clashes, a Hutu uprising occurred in northern Rwanda in 1911 because of Hutu resentment toward the Germans and the Tutsi hierarchy, as well as the Catholic Church, which had recently set up missions around Rwanda. Although the German–Tutsi alliance easily crushed the Hutu uprising, the event is indicative of the increased ethnic tension between Hutu and Tutsi during the colonial era.[13]

In 1916, during World War I, the Belgians forced the Germans out of Rwanda and Burundi. After Germany's defeat in the war, a 1924 League of Nations mandate officially placed Rwanda and Burundi under Belgian control and combined the two into the single province of Ruanda-Urundi. The Belgians continued the German policy of indirect rule and increased Tutsi dominance. Height restrictions were put on those seeking higher education in an effort to keep Hutu out

---

[12] Mamdani, *When Victims Become Killers*; Philip Gourevitch, *We Wish to Inform You that Tomorrow We Will be Killed with Our Families: Stories from Rwanda* (New York: Picador, 1999); Alison Liebhafsky Des Forges, *Leave None to Tell the Story: Genocide in Rwanda* (New York: Human Rights Watch, 1999); Linda Melvern, *A People Betrayed: The Role of the West in Rwanda's Genocide* (London: Zed Books, 2000); Kamola, "The Global Coffee Economy."

[13] Kirschke, "Broadcasting Genocide."

of universities,[14] and the traditional custom of only Tutsi owning cattle was turned into law. In order to maximize their economic gain in the territory, the Belgians instilled a system of Tutsi administrators and Hutu workers. Tutsi domination extended down to local governance where the Tutsi controlled the local chiefdoms. By many accounts, an emphasis on the Hamitic myth and a focus on ethnic differences between the Hutu and the Tutsi were the cornerstones of Belgian policy.[15]

In 1945, as World War II came to a close, the newly formed United Nations (UN) made Ruanda-Urundi a "trustee territory" of Belgium. A critical change in the Hutu–Tutsi conflict occurred soon after when, in the 1950s, the Belgians began to reform the Ruanda-Urundi sociopolitical structure. The ethnic and racist disaster of World War II, including the Holocaust, led the Belgians (and Catholic Church) to rethink the caste system in place. As the West rallied ideas of democracy, the elitist Tutsis began to be antithetical to democratic governance and reform and came to represent pan-Africanism, which was viewed as a threat to the West and Europe. Thus, the Hutu, for the first time in the colonial era, became the favored ethnic group. Signs of Hutu opposition to Tutsi oppression began to arise in the mid to late 1950s as the UN pushed forward plans to end colonial rule. In 1961, the Tutsi monarchy was abolished and Ruanda-Urundi again became two separate states: the Republic of Rwanda and the Kingdom of Burundi.[16]

GOVERNING ENVIRONMENT

The Rwandan Tutsi, no longer aided by Belgian support, lost control of the Rwandan government in the years immediately before independence to a faction of Catholic-educated Hutu. With the centuries-old Tutsi monarchy overthrown, the *mwami* (or "king") deposed and exiled, and the Rwandan Hutu in control of the newly formed Rwandan republic,[17] Tutsi began fleeing Rwanda for neighboring countries—particularly Uganda and Burundi. After Rwandan independence, the refugee Tutsi, referring to themselves as

---

[14] This was despite the fact that height often did little to distinguish between the two groups, and thus Hutu and Tutsi were given identity cards listing their ethnicity so there would be no mistake.

[15] Gourevitch, *We Wish to Inform You.*

[16] Ibid., 3.

[17] In particular, the anti-Tutsi Parti du Mouvement de l'Emancipation Hutu (or *Parmehutu*) had vast electoral success.

inyenzi (or "cockroaches"), began launching attacks in Rwanda with the aim of overthrowing the Hutu nationalist government.[18]

Because of the success of the Kayibanda regime in excluding Tutsi from politics and the establishing Hutu dominance at all levels of government, ethnic disputes faded and regional disputes between Hutu began to arise. Believing the central region of Rwanda, home to the president, was getting favorable treatment, Hutu from the north made plans for a coup d'etat. In 1973, President Grégoire Kayibanda was overthrown and murdered by General Juvénal Habyarimana, a Hutu from northern Rwanda who was then serving as defense minister. Habyarimana established a military dictatorship and replaced *Parmehutu* with his *Mouvement Revolutionnaire National pour le Développement* (MRND) party. Despite accumulating much power for himself and his northwestern power base, Habyarimana reintroduced the parliament, renaming it the *Conseil National de Développement* (National Council for Development, or CND).[19]

By the late 1980s, the RPF[20] had formed from a Tutsi refugee organization based in Uganda, the Rwanda Alliance for National Unity (RANU), and a Ugandan rebel group, the National Resistance Army (NRA). Despite a last-ditch attempt by Habyarimana to repatriate the Tutsi refugees in order to undermine invasion plans, the RPF invaded Rwanda in the fall of 1990. Led by Fred Rwigyema, along with fellow RPF leader Paul Kagame, a long-time deputy of the Ugandan president, and former NRA leader Yoweri Museveni, the invasion went poorly, and Rwigyema was killed on the second day of fighting. French, Belgian, and Zairian forces aided the FAR (*Forces Armées Rwandaises* or Rwandan Armed Forces) in holding back the RPF. With the deaths of more RPF commanders, Kagame took over leadership of the organization. After fighting the FAR with limited success, in 1992 the RPF succeeded in capturing territory in the province of Byumba in northern Rwanda. This development, as well as international pressure, led Habyarimana to initiate multipartism[21] in 1991 and to arrange

---

[18] Uvin, "Ethnicity and Power in Burundi and Rwanda"; Alan J. Kuperman, "Explaining the Ultimate Escalation in Rwanda: How and Why Tutsi Rebels Provoked a Retaliatory Genocide" (paper presented at the annual meeting of the American Political Science Association, Philadelphia, PA, August 28–31, 2003).

[19] Uvin, "Ethnicity and Power in Burundi and Rwanda."

[20] Note: In many sources, a distinction is made between the RPF, a political organization, and its military arm, the Rwandan Patriotic Army (RPA). For simplicity's sake, both the RPF and RPA are addressed as one unitary body in this chapter and are solely referred to as the RPF.

[21] Nearly all the opposition parties were made up of Hutu extremists and fringe members of the MRND; the most influential of these parties, *Coalition pour la Défense de la République* (CDR, or Coalition for the Defence of the Republic), was a mere front for the Akazu. Gourevitch, *We Wish to Inform You.*

peace negotiations with the RPF in the Tanzanian city of Arusha in the summer of 1992. To enforce the peace negotiations made at Arusha, international actors deployed the United Nations Assistance Mission for Rwanda (UNAMIR), a contingent of peacekeeping troops under the command of Canadian Lieutenant-General Roméo Dallaire.[22]

The Arusha Accords would produce a peace treaty between the Habyarimana regime and the RPF and planned to integrate the RPF into the FAR. The treaty, which would reduce Habyarimana's authority to ceremonial powers, infuriated other "Hutu Power" leaders. In an effort to solidify power and wipe out political opposition, Hutu Power leaders (also known as the *Akazu*, the northwesterners who were the source of Habyarimana's power), had formed the *Interahamwe*,[23] a paramilitary organization consisting of armed youth gangs. The *Interahamwe* began its first massacres of Tutsi and Hutu moderates in 1992 and increased its activity in 1993. All of this activity went on with impunity. In early 1992, the *Akazu* massacred Tutsi in the Bugesera region, after which they launched the *Radio Télévision Libre des Mille Collines* (RTLM), a radio station dedicated to promoting genocide, as well attacking the Arusha Accords, a mere four days after the accords were signed.[24]

On April 6, 1994, while on a returning flight from Dar es Salaam, Burundi, Habyarimana's plane was shot down and crashed near his presidential palace, killing Habyarimana and Cyprien Ntaryamira, the president of Burundi, who was also on board. The *Akazu* seized power immediately after Habyarimana's assassination. Under the direction of Colonel Théoneste Bagosora, who had been serving the Secretary-General of the Ministry of Defense in the Habyarimana regime, FAR forces assassinated PM Madame Agathe Uwilingiyimana and other moderate Hutu leaders. Ten Belgian UNAMIR troops, who had been protecting Uwilingiyimana, were tortured and killed. Bagosora and the *Akazu* negotiated with the military and the UN representative to install a pro-genocide extremist government disguised as a temporary transitional government. This government, composed completely of extremist Hutu leaders, was created within forty-eight hours of Habyarimana's assassination and quickly organized local leaders and the militia gangs in preparation for the slaughtering of Tutsi. As the *Interahamwe* and other gangs killed Tutsi, older Hutu joined in on the

---

[22] Uvin, "Ethnicity and Power in Burundi and Rwanda"; Jeffrey H. Powell, "Amnesty, Reintegration, and Reconciliation in Rwanda," *Military Review* (September-October 2008): 84; Kuperman, "Explaining the Ultimate Escalation in Rwanda."

[23] *Interahamwe* means "Those Who Stand Together" or "Those Who Attack Together."

[24] Gourevitch, *We Wish to Inform You*; Forges, *Leave None to Tell the Story*.

slaughter. Paul Kagame became the head of the RPF and resumed the war effort with hopes of capturing Kigali and ending the genocide.[25]

In post-independence Burundi, the Tutsi monarchical government survived. The constitutional monarchy, headed by Tutsi King Mwambutsa IV, was designed to give equal representation to both Hutu and Tutsi. As PM, Prince Louis Rwagasore headed UPRONA (*Union pour le Progrès National,* or Union for National Progress), a party that incorporated both Tutsi and Hutu. Violence nonetheless occurred, Rwagasore was assassinated, leading to the collapse of UPRONA, and, in an effort to maintain power, the Tutsi executed hundreds of Hutu leaders over the next several years. The Hutu managed to win control of the parliament elections of 1965, but Mwambusta refused to appoint a Hutu PM. In response, the Hutu attempted an unsuccessful coup—the first large-scale revolt in post-independence Burundi—that was brutally suppressed by the government. Mwambusta eventually put a Hutu PM in place, Pierre Ngendandumwe, but he was soon assassinated by a Tutsi gunman. In response to riots in the wake of the assassination, the Tutsi regime increased repression of civil liberties, and Burundi became increasingly unstable.[26]

The following year, in 1966, Ntare V, another son of King Mwambutsa, killed his father and crowned himself king. Four months later, Tutsi military commander Michel Micombero led a coup d'etat that formed a military regime disguised as a republic and killed Ntare. With Micombero as president, the army effectively controlled Burundi. The Tutsi-Hima, an ethnic group of which Micombero was a member, controlled the military and had a stranglehold on power in Burundi. As a result, they banished other Tutsi and Hutu groups from the government. In 1969, a collection of Hutu leaders were accused of plotting to overthrow the government and were subsequently executed. Consequently, the absence of Hutu elites led to infighting among the Tutsi.[27]

WEAKNESSES OF THE PREREVOLUTIONARY ENVIRONMENT AND CATALYSTS

As in many post-independence African nations, the state alone provides avenues to riches and success. Thus, in both Rwanda and

---

[25] Powell, "Amnesty, Reintegration, and Reconciliation in Rwanda"; Gourevitch, *We Wish to Inform You*; Forges, *Leave None to Tell the Story.*

[26] Melady, "Burundi and Rwanda"; Uvin, "Ethnicity and Power in Burundi and Rwanda"; Warren Weinstein, "Conflict and Confrontation in Central Africa: The Revolt in Burundi," *Africa Today* 19, no. 4 (1972): 17–37.

[27] Ibid.; Weinstein, "Conflict and Confrontation in Central Africa."

Burundi, control of the state was critical for individual achievements, and out-of-power elites exploited ethnic divides to usurp power from the ruling class.[28] The source of ethnic tension between the Hutu and Tutsi originated with colonial exploitation and institutionalization of the Tutsi hierarchy in both Burundi and Rwanda. The Germans, Belgians, and the French all adhered to a "divide and rule" policy, understanding that it was easier to maintain control of the lands if the population was divided among itself. Belgian control of Ruanda-Urundi was particularly repressive of Hutu, considerably increasing Tutsi dominance. Although Hutu–Tutsi transethnic identities most likely existed before the colonial period, the Belgians in particular succeeded in constructing the notion that Hutu and Tutsi were two separate races and that the Tutsi, as a Hamitic branch, were foreigners to the land.[29]

The Belgians had guaranteed a Tutsi monopoly on power, going beyond instituting a "divide and rule" strategy and implementing policies reflective of the nineteenth and twentieth century racist ethnic tensions between the groups. Hutu resentment of Tutsi had reached a high in 1959 when the Belgians ceased to support the Tutsis in favor of the Hutu and as Rwanda and Burundi prepared for independence. The Hutu were gradually given more pastoral lands and cattle, the traditional measures of wealth and social standing. The practice of Hutu indentured servitude ended, giving the Hutu a sense that full liberation from Tutsi dominance was possible and within reach.[30]

It was during this period that the Hutu rallied behind concepts associated with the Hamitic myth, particularly the idea that the Tutsi had migrated to the Great Lakes region well after the Hutu. The 1957 *Hutu Manifesto*, written by a group of Rwandan Hutu academics, increased group consciousness around the idea that the Tutsi were effectively invaders of Hutu land. Thus, Hutu identity and an "us-versus-them" mentality became cornerstones of Hutu empowerment in the region.[31]

In Rwanda, the Hutu began pushing for reform and change, but Tutsi elites resisted. In response, the Belgians replaced many Tutsi chiefs and subchiefs with Hutu. The Hutu were not satisfied, and the Tutsi were not willing to give up more influence. In 1960, in what would be referred to as the "Hutu Revolution," the Hutu took control of the Rwandan government in Belgian-run elections, increasing a

---

[28] Ibid.

[29] Mamdani, *When Victims Become Killers*; Forges, *Leave None to Tell the Story*; Gourevitch, *We Wish to Inform You.*

[30] Ibid.

[31] Ibid.

sense of urgency among the Tutsi that they were losing control of the country. The Hutu government began policies of denying Tutsi citizens an equitable distribution of resources through quotas in reprisal for the years of inequitable treatment the Hutu suffered under Tutsi–colonial rule. Two of these policies were the *Paysannat*, a program that uprooted and resettled Tutsi families in underdeveloped areas of the country, and the reinstitution of ethnic identity cards to maximize Tutsi repression. These events triggered the first examples of Tutsi resistance. After a Tutsi assassination attempt on Grégoire Kayibanda, *Parmehutu* founder and coauthor of the *Hutu Manifesto*, Hutu–Tutsi tension plunged into a period of violence beginning in 1960, resulting in the deaths of 10,000 Tutsi between 1960 and 1962, and the first waves of Rwandan Tutsi refugees arrived in the surrounding countries. As the Tutsi fled Rwanda, the *inyenzi* militant movement formed and began launching attacks in Rwanda to combat the anti-Tutsi policies of the Hutu government.[32]

In response to the *inyenzi* attacks, President Kayibanda's Hutu-led Rwandan government murdered masses of Tutsi civilians who remained in Rwanda. The most successful of the *inyenzi* attacks, a 1963 invasion launched from Burundi, brought the Tutsi refugees to within ten miles of the Rwandan capital of Kigali. Although the Tutsi invaders were soundly defeated, in retaliation, the FAR massacred thousands of Rwandan Tutsi[33] in a few days. These reprisal killings escalated the Rwandan Tutsi diaspora, and by the mid-1960s, the Tutsi population in Rwanda had dropped from 17 to 9%, with at least 20,000 having been murdered at the hands of the FAR and up to a quarter million Tutsi having fled the country.[34]

In Burundi, after the 1962 revolution and the systematic exclusion of a generation of Hutu from various resources, Hutu elites used education and elite positions to rally other Hutu to engage in violence against the Tutsi. The violence seemed to have an additive effect. The events of the 1972 genocide, for example, crystallized Hutu and Tutsi identities for decades, allowing future conflicts to unfold. President Ndadaye's assassination, which triggered the 1993 violence in Burundi, was seen among Hutu as the beginning of a second coming of the 1972 violence afflicted on Hutu by Tutsi. Thus, the Burundian Hutu launched a preemptive strike against the Tutsi only to have the Tutsi-controlled military respond in kind. The 1972 genocide in Burundi

---

[32] Mamdani, *When Victims Become Killers*; Gourevitch, *We Wish to Inform You*; Kuperman, "Explaining the Ultimate Escalation in Rwanda."

[33] One report suggested more than 14,000.

[34] Gourevitch, *We Wish to Inform You*; Uvin, "Ethnicity and Power in Burundi and Rwanda"; Kuperman, "Explaining the Ultimate Escalation in Rwanda."

also had an effect on Hutu–Tutsi relations in Rwanda. Tutsi children were expelled from schools; civil services were denied to Tutsi; and hundreds of Tutsi were killed, all in reaction to the slaughter of Hutu in Burundi.[35]

The 1994 genocide in Rwanda resulted from a set of complex social and economic relationships. For one, Habyarimana and MRND failed to put policies in place to promote political liberalization and multipartism soon enough. For example, the 1986 formal ban on the return of Tutsi refugees in Uganda to Rwanda galvanized the refugees and led to the eventual creation of the RPF. The situation in Uganda was a key variable in the development of the RPF. Uganda instituted very harsh refugee laws, especially under presidents Idi Amin and Milton Obote, where refugee status could transfer from parent to child, even if the child was born in Uganda. These Tutsi children, however, qualified for UN aid, which led to much resentment from the rest of the Ugandan population. *Inyenzi* attacks launched from Uganda, as well as Tutsi alliances with the Ugandan Hima tribe— which was viewed unfavorably by other Ugandans as an elitist group not all too different from the Tutsi in Rwanda and Burundi—further turned Ugandans against the Tutsi refugees. Once it became obvious to the Tutsi population in Uganda that they would always be viewed as outsiders and not accepted into the population at large, the refugee population got behind the idea of returning to Rwanda by force.[36]

Many Tutsi refugees in Uganda got their teeth cut in warfare as part of Yoweri Museveni's overthrow of Obote in 1986 and were subsequently compensated with positions in the government, military, and business. This resulted in increased backlash from Ugandan nationals, resentful of the munificent treatment that refugees in their own country were receiving. Violence against the Tutsi refugees followed. In response, Tutsi leaders in Uganda formed the RANU with the explicit goal of planning a return of the Tutsi refugees to Rwanda. RANU was headed by Fred Rwigyema, who had served in Museveni's NRA, as well as Paul Kagame. RANU, a Marxist group, was initially a pacifist organization dedicated to peacefully returning the Tutsi refugees. However, with a change in ideology in 1987, RANU

---

[35] Uvin, "Ethnicity and Power in Burundi and Rwanda"; René Lemarchand, "Genocide in the Great Lakes: Which Genocide? Whose Genocide?" *African Studies Review* 41, no. 1 (1998): 3–16.

[36] Forges, *Leave None to Tell the Story*; René Lemarchand, "Case Study: The Burundi Killings of 1972," *Online Encyclopedia of Mass Violence*, accessed September 3, 2010, http://www.massviolence.org/The-Burundi-Killings-of-1972.

became the RPF and became open to using force against the Hutu-led Rwandan government to push the Tutsi homecoming.[37]

Throughout the Habyarimana regime, coffee growth and exports became an increasingly crucial component of the Rwandan economy. Little of the profits were reinvested in Rwanda; rather, they were claimed and spent by the *Akazu* and other Hutu elites. In June 1989, Rwanda found itself entering an economic crisis brought on by the collapse of international coffee prices[38] and severe drought. Facing bankruptcy, state-run health and education agencies severely cut back services, resulting in a huge increase in cases of malaria, especially among children. As coffee prices fell and the economic base of Rwanda weakened, tension between Hutu elites arose, threatening the Habyarimana regime's monopoly on power and worrying the *Akazu*.[39]

In 1990, Habyarimana was losing support because of the economic crisis when the RPF launched its first attacks from Uganda. French pressure finally convinced Habyarimana to make some concessions at the Arusha Accords, which allowed for the Tutsi to reenter Rwandan politics and possibly take control of the country. These developments further weakened Habyarimana and infuriated the *Akazu*. Realizing the opportunities his administration had of regaining support, the Habyarimana administration (prompted by the *Akazu*) exaggerated the RPF threat and began suggesting that all Rwandan Tutsi were part of a conspiracy to launch war in Rwanda. Using the memories of Tutsi domination and the Hutu revolution of 1959, the Habyarimana administration began to successfully amplify resentment toward Tutsi and Hutu moderates. Preaching the concept of Hutu Power, the *Akazu* formed anti-Tutsi media such as the *Kangura* newspaper and the radio station RTLM, which reached a wide audience. The unpunished massacres carried out by the *Interahamwe* had also fostered a sense of "normalcy" among the population, in terms of violence.[40]

It is now commonly believed that these Hutu extremists were behind the Habyarimana assassination, although some dispute this.[41] Habyarimana had angered many Hutu hardliners by making concessions to the Arusha Accords. His assassination sparked

---

[37] Kuperman, "Explaining the Ultimate Escalation in Rwanda"; Forges, *Leave None to Tell the Story*.

[38] The International Coffee Agreement (ICA) broke down in part because of pressure from United State coffee traders. Melvern, *A People Betrayed*; Kamola, "The Global Coffee Economy."

[39] Ibid.

[40] Ibid. Forges, *Leave None to Tell the Story*, 789; Gourevitch, *We Wish to Inform You.*

[41] A 2004 French report by counterterrorism magistrate Jean-Louis Bruguière concluded that the RPF and Paul Kagame (and possibly even the CIA) were responsible for Habyarimana's assassination.

widespread anger from Hutu, who believed the assassination to have been the work of the RPF, and this was enough to galvanize the masses into a genocidal rage. Thus, in all likelihood, Habyarimana was sacrificed by his own backers in order to push through their plans of genocide, thus stalling the Arusha Accords and keeping power of Rwanda in their hands.[42]

FORM AND CHARACTERISTICS OF THE REVOLUTION

OBJECTIVES AND GOALS

The resentment the Tutsi and Hutu held toward each other had continuously been taken advantage of by elites in both groups to claim, solidify, or protect political influence. The 1994 Rwandan genocide was Hutu extremists' attempt to secure power for themselves. Habyarimana's northwestern power bloc, the *Akazu*, threatened by the Arusha Accords' promise of power sharing, determined that violence against Tutsi civilians would undermine any compromises emerging from Arusha and create ethnic solidarity—eclipsing any regional disputes among Rwandan Hutu and protecting their oligarchic control of the country. The RPF's political agenda included not only returning the hundreds of thousands of Tutsi refugees to Rwanda, but also removing Habyarimana from power and ensuring significant political power for Rwandan Tutsi. This is evidenced by the fact that the RPF continued its military offensive even after the Habyarimana regime began to address democratic aspirations.[43]

LEADERSHIP AND ORGANIZATIONAL STRUCTURE

The RPF formed from descendents of Tutsi refugees in Uganda who had left in mass exodus during the violence of the Hutu revolution in Rwanda from 1959 to 1961. The RPF was very institutionalized and democratic, often making decisions through debate, working for consensus, and allowing members of the RPF to vote on matters when consensus could not be reached. The RPF was small but well trained and well equipped and used guerrilla-style warfare.[44]

---

[42] René Lemarchand, "Rwanda: The Rationality of Genocide," *Issue: A Journal of Opinion* 23, no. 2 (1995): 8; Forges, *Leave None to Tell the Story*; Gourevitch, *We Wish to Inform You.*

[43] Lemarchand, "Rwanda"; Forges, *Leave None to Tell the Story*; Kuperman, "Explaining the Ultimate Escalation in Rwanda."

[44] Ibid.

The source of Habyarimana's power and the creator and sponsor of the *Interahamwe* was the *Akazu* from the northwestern region of Rwanda. The *Akazu* has been referred to as the "court within the court," a tightly knit oligarchic inner circle of political, economic, and military leaders who both pushed and benefited from a patronage system. The *Akazu* was also referred to as Hutu Power, and the ridding of both Tutsi and Hutu opposition in Rwanda was their idea, viewing it as a way to increase their power and derail any attempts at democratic reforms. Publicly decrying reforms would not work. However, the *Akazu* believed that they could exterminate their enemies by instilling fear of a Tutsi "threat" in the population. Madame Agathe Habyarimana, the wife of President Habyarimana, was from an influential northern family responsible for Habyarimana's rise,[45] and she was the central figure of the *Akazu*. Thus, both Habyarimana and the MRND dared not cross the *Akazu*. These oligarchs from the north profiteered from foreign aid and treated Rwanda like a personal business, attempting to extract as much profit as possible.[46]

The *Interahamwe* was a nation-wide network of gangs composed of uneducated and unemployed young men. The *Akazu* funded and supervised the *Interahamwe* and other genocidaire groups. These groups were further supported by the FAR. Centrally organized, the *Interahamwe*, as well as the *Impuzamugambi* (meaning "those with a single purpose"), the *Interahamwe's* counterpart to the *Akazu* front party CDR, had national leaders, vice presidents, supervisors, and regional and neighborhood commanders.[47]

COMMUNICATIONS

Radio and print publications played a significant role in the Hutu–Tutsi conflict, particularly in Rwanda. Even before independence, Hutu publications spread anti-Tutsi rhetoric and ideas in Rwanda. The 1957 *Hutu Manifesto* embraced the Hamitic myth, suggesting that the Tutsi were invaders of Hutu land. The *Manifesto* urged Hutu in both Burundi and Rwanda to acknowledge their position as the majority ethnic group and to prepare for future ethnic struggles.[48]

---

[45] Habyarimana did not come from an influential family, and it has been rumored that he was in fact born in Uganda. Without the support of his wife's family, Habyarimana would most likely not have been able to climb the political ladder and eventually oust Kayibanda as president.

[46] Melvern, *A People Betrayed*; Gourevitch, *We Wish to Inform You.*

[47] Ibid. Kirschke, "Broadcasting Genocide"; Forges, *Leave None to Tell the Story.*

[48] Mamdani, *When Victims Become Killers*; Gourevitch, *We Wish to Inform You*; Melvern, *A People Betrayed.*

In the late 1980s, an anti-Habyarimana newspaper called *Kanguka* (or "Wake Up") began circulating in Rwanda. In response, the MRND sponsored a rival newspaper called *Kangura* (or "Wake It Up"), which, serving as the *Akazu's* and CDR's mouthpiece after it was formed, fomented hatred toward Tutsi and the RPF and began to stoke genocide among the Hutu. Widely read by the literate Hutu population, *Kangura's* content was also read at public meetings and rallies in order to inform the large portion of Hutu who could not read. Its stories, primarily sensationalist in nature, warned of an impending war launched by the Tutsi and of Tutsi plans to reinstitute their monarch while leaving the Hutu enslaved. These stories were widely spread through word of mouth. A December 1990 edition first published the "Hutu Ten Commandments," which promoted the idea that Hutu should not interact with Tutsi at all and that any who did were traitors. The newspaper began suggesting that the only way to stave off the Tutsi and their plans was to exterminate them. The *Interahamwe* read *Kangura* during their rallies in order to bolster support for genocide.[49]

Administrative authorities spread misinformation and rumors to provoke local residents to attack their Tutsi neighbors even before the RPF launched its first attacks in October 1990. Leaflets were often distributed and warned of attacks by the Tutsi, encouraging preemptive strikes.[50] The RPF and the Tutsi in general were blamed for the violence that took place, with the government reporting that Tutsi extremists incited attacks against the population. There are reports that local administrators concocted fake RPF attacks, which were broadcast by the media in order to drum up support for violence against Tutsi.[51]

It was radio, however, that played a crucial role in fomenting the 1994 genocide. High illiteracy rates in Rwanda guaranteed large radio audiences. Rwanda's three main radio stations began broadcasting hate-filled messages against Tutsis, opposition parties, and politicians. Radio broadcasts, like the *Kangura*, accentuated the idea that the RPF and the Tutsi were bloodthirsty and that preemptive action was necessary to protect the Hutu. Radio Rwanda, the state-owned radio under the control of the Habyarimana regime, began broadcasting false reports about the RPF, often citing that the atrocities were committed by the Tutsi-led organization shortly after the RPF first launched its invasion. Radio Rwanda falsely reported the slaughter of Tutsi in the

---

[49] Ibid. Gourevitch, *We Wish to Inform You*; Kirschke, "Broadcasting Genocide."

[50] Hassan Ngeze, editor of the *Kangura*, was responsible for creating and distributing some of these leaflets.

[51] Gourevitch, *We Wish to Inform You*; Kirschke, "Broadcasting Genocide."

eastern Bugesera region of Rwanda in March 1992 as a RPF-initiated massacre. After Bugesera, Hutu, FAR forces, and the *Interahamwe* preemptively slaughtered hundreds of Tutsi, Radio Rwanda claimed the violence was initiated and carried out by the RPF. In response to the Bugesera incident, the RPF created Radio Muhabura, which broadcast pro-Tutsi and anti-Habyarimana messages from Uganda. Radio Rwanda told its listeners that all information broadcast by Radio Muhabura was false and designed to pull the country apart.[52]

The *Akazu* formed the RTLM, a nominally private radio station, in response to the Arusha Accords in 1993. Radio Rwanda had encouraged the Hutu to partake in attacks against Tutsi for some time; however, it was RTLM in particular that played a crucial role in dehumanizing the Tutsi and fomenting genocide. Modeling itself after Western-style talk radio shows, the crude RTLM was considered much more entertaining than the other more traditional and formal radio stations in Rwanda and quickly developed a large following in the country—even among RPF soldiers. Its coverage, infused with pop music, consisted of commentaries and interviews, instead of objective reporting, and was the first radio station in Rwanda to allow listeners to call into the station and be on air.[53]

RTLM unremittingly preached Hutu solidarity and described the Tutsi as foreign invaders of Hutu land. Building on fear of the RPF, RTLM suggested that all Tutsi were a part of a united clan that would bring Rwanda to an era of Hutu feudal subjugation and likely carry out a genocide of Hutu people. By describing the Tutsi as both "alien" and "clever," RTLM simultaneously excluded the Tutsi from the national community and reinforced the idea that they would be an unrelenting threat. It was the station of choice for members of the Habyarimana regime, FAR, and militia groups, all to which RTLM had strong connections.[54]

In the months and weeks leading up to the genocide in 1994, RTLM warned listeners of a coming bloodbath, but it did not elaborate on the details. RTLM was the first media source to announce the death of Habyarimana on April 6 and immediately recognized the transitional government formed two days later by Bagosora. In return,

---

[52] Forges, *Leave None to Tell the Story*; Kirschke, "Broadcasting Genocide"; Gourevitch, *We Wish to Inform You*.

[53] Kirschke, "Broadcasting Genocide"; Gourevitch, *We Wish to Inform You*; Forges, *Leave None to Tell the Story*.

[54] Lemarchand, "Genocide in the Great Lakes"; Lemarchand, "Rwanda: The Rationality of Genocide"; Gourevitch, *We Wish to Inform You*; Melvern, *A People Betrayed*; Kirschke, "Broadcasting Genocide."

RTLM journalists were protected from the violence that silenced independent journalists throughout the genocide.[55]

Broadcasting twenty-four hours a day as the genocide began, RTLM sent out incessant reminders of the RPF threat, now suggesting the rebels were "evil incarnate," and adamantly proscribed exterminating the RPF and all its "accomplices" as the only solution. It pushed public participation in this "final battle," stating the FAR would not be able to defeat the RPF alone and overtly claiming that "real men" would defend themselves against the threat "dressed as civilians and unarmed." It provided specific instructions to listeners on how to man the roadblocks and what weapons should be used. RTML would also congratulate participants on their "heroism," such as when the death toll reached 20,000 a week after the genocide began. The names of the Tutsi and their locations, as well as the license plates of those fleeing in cars and places where Tutsi were likely seeking refuge, were read on air to help the genocidaires hunt down Tutsi with efficiency.[56]

METHODS OF ACTION AND VIOLENCE

In 1972, the first act of genocide in the Hutu–Tutsi conflict occurred on Burundian soil. Inspired by the Hutu revolution of 1959 in Rwanda and the establishment of the Hutu-led Rwandan Republic in 1962, Hutu leaders in Burundi organized a rebellion in one of Burundi's southern provinces. It had become obvious to many Hutu leaders that the Tutsi hegemony would only solidify and that insurgency was their only hope of gaining political power in Burundi. Hutu refugees from surrounding nations carried out an unsuccessful coup—killing more than 10,000 Tutsi in the process. The Tutsi-led Burundian government responded with an organized two-month-long slaughter of Hutu by the Tutsi-dominated army, focusing on the educated. The death count is estimated to be anywhere from 100,000 to 300,000. Martial law was imposed, and government repression reached a high.[57]

The infighting among various Tutsi cliques continued, and Micombero was overthrown in 1976 in a coup led by Deputy Chief of Staff Jean-Baptiste Bagaza, a distant cousin of Micombero and also a Tutsi-Hima. Bagaza's regime continued the repressive practices of

---

[55] Forges, *Leave None to Tell the Story*; Gourevitch, *We Wish to Inform You*; Kirschke, "Broadcasting Genocide."

[56] Ibid.

[57] Lemarchand, "Genocide in the Great Lakes"; Uvin, "Ethnicity and Power in Burundi and Rwanda"; Lemarchand, "Case Study"; Weinstein, "Conflict and Confrontation in Central Africa."

Micombero, and many Hutu who attempted to gain any significant political or economic influence were targeted and assassinated. In 1987, Pierre Buyoya, who made promises of liberalization and improved relations between Hutu and Tutsi, killed Bagaza.[58]

In 1988, a group of armed Hutu killed hundreds of Tutsi living in one of Burundi's Northern provinces. In response, the Buyoya regime slaughtered an estimated 20,000 Hutu and sent upwards of 60,000 Hutu fleeing the country, the vast majority of whom made their way into Rwanda. These events, which triggered international pressure for reform from a post-Cold War West, led Buyoya to initiate democratic reforms in Burundi. In 1992, a new Burundian constitution went into effect, limiting ethnic-based parties and mandating popular and provincial checks on partisan activities. These reforms resulted in the June 1993 election of Melchior Ndadaye as the first democratically elected president and the first Hutu to lead from UPRONA. The Tutsi Buyoya ostensibly accepted his defeat and vacated the presidency.[59] However, democratic reform supporters did not have much time to take pleasure in these apparent successes in Burundi. In October 1993, a mere four months after Ndadaye's election and three months after he took office, Tutsi soldiers assassinated Ndadaye and a contingent of Hutu leaders, including the president and vice president of the General Assembly, who, per the new constitution, would have been next in line for the presidency. In reaction, Hutu across the country began slaughtering Tutsi, which, in turn, resulted in the Tutsi-led army massacre of Hutu. In the end, an estimated 50,000 to 100,000 Burundians were killed.[60]

A new Hutu president, Cyprien Ntaryamira, was selected by the Parliament in January 1994, but he died alongside Habyarimana in the April 1994 plane crash that launched the Rwandan genocide. An October 1994 convention resulted in a largely unstable government comprising UPRONA and Front for Democracy in Burundi (*Front pour la Démocratie au Burundi*, or FRODEBU) officials. Failing to govern well, the government was overthrown in a 1996 coup that returned Buyoya to power as president. In response to the coup, neighboring countries imposed an economic embargo on Burundi.[61] Violence was prevalent between the Tutsi-led army and the Hutu militias. One such militia was the *Forces pour la Defense de la Democratie* (FDD), a militarized

---

[58] Uvin, "Ethnicity and Power in Burundi and Rwanda"; Janvier D. Nkurunziza and Floribert Ngaruko, *Explaining Growth in Burundi: 1960-2000* (Oxford, United Kingdom: Centre for the Study of African Economies, 2002).

[59] Uvin, "Ethnicity and Power in Burundi and Rwanda."

[60] Ibid.

[61] Ibid. Nkurunziza and Ngaruko, *Explaining Growth in Burundi*.

group that split off from the FRODEBU and randomly committed brutal acts of violence against Tutsi. Recently, peace agreements have slowed the violence, but outbreaks have been frequent. It is estimated that at least 500,000 Burundians have died from Hutu–Tutsi conflict violence since Burundi gained its independence, and hundreds of thousands more have been uprooted and displaced.[62]

The Hutu–Tutsi conflict was marked by sporadic violence between the two groups. Both groups had the attitude that they were victims of unjust treatment from the other group, the perpetrators. Thus, vast differences existed between the two groups' interpretations of the conflict. This resulted in both groups experiencing "selective collective memory," each maintaining that they had each been severely wronged and were in the moral right in their attempts to correct these wrongs. Having the mind-set of victims, members of one group would strike out of vengeance or fear of attacks orchestrated by the other group. Because members of the other group were regarded as "socially dead" individuals, the decision to murder them did not carry the same weight as would the decision to murder members of an individual's own group. The other group responded with the same mind-set. The pattern continued successively, with members of each group viewing themselves as victims.[63] This pattern seemed even more severe in Rwanda.

Throughout the Hutu–Tutsi conflict, insurgency groups amassed weapons and increased the numbers within their ranks in the attempt to topple the current regime and subjugate the other group. Guerrilla warfare was common, and the most drawn-out and obvious example was the RPF's tactics during the Rwandan civil war. After the RPF was held back by FAR forces supported by the French, Belgians, and Zairians during the RPF's initial assault, the Rwandan civil war declined into a guerrilla conflict, and the RPF was able to successfully take control of an area within Rwanda. Hutu refugees based in the DRC camps used guerrilla tactics to strike settlements in western Rwanda.[64]

Following the evacuation of foreign troops from Rwanda, the UN peacekeeping force and all other foreigners exited the country, and the targeted population was left at the mercy of Bagosora's plans. Throughout the next four months, FAR, militia gangs, and groups

---

[62] Ibid. Uvin, "Ethnicity and Power in Burundi and Rwanda."

[63] Lemarchand, "Genocide in the Great Lakes"; Uvin, "Ethnicity and Power in Burundi and Rwanda."

[64] Forges, *Leave None to Tell the Story*; William G. Thom, "Congo-Zaire's 1996-97 Civil War in the Context of Evolving Patterns of Military Conflict in Africa in the Era of Independence," *Journal of Conflict Studies* 19, no. 2 (1999): 93–123.

of Hutu slaughtered an estimated 800,000 Rwandans.[65] Although most of the casualties were Tutsi, some Hutu from the south-central region of the country were also targeted because they opposed the Habyarimana regime's plans. When the RPF took Kigali in July, the genocide was effectively brought to a close and an estimated 2 million Hutu—most of whom were associated in some way with the slaughter—fled to Zaire.[66]

The 1994 genocide in Rwanda is notable for the large percentage of the population that partook in it. Many Hutu civilians, armed with machetes and other types of non-firearm weapons, helped massacre their Tutsi neighbors in communities across the country. The *Akazu* wished to distribute firearms to every member of the *Interahamwe*; however, some, such as Colonel Théoneste Bagosora, argued that this would cost far too much. Thus, with the help of businessmen with close ties to Habyarimana, massive numbers of machetes were imported into Rwanda and distributed to the youth militias. Although the genocidaires primarily used clubs, masus,[67] grenades, hammers, and tear gas to commit the massacre, some *Interahamwe* were given AK-47 assault rifles. Nearly all the weapons were purchased and imported from France, Egypt, and South Africa.[68]

The genocide was very organized and structured. The militia groups were organized hierarchically. Former soldiers and policemen were recruited to train the *Interahamwe*, direct them during the attacks, and participate in the massacres themselves. The commands and tactical knowledge provided by members of the National Police and soldiers allowed the killings in the first few days to occur quickly and efficiently, with machine guns, grenades, and mortars being used to commit the murders. This leadership was referred to as the "civilian self-defense" program and was headed by Bagosora.

After the initial slaughter, completed by the military, Presidential Guard, National Police forces, and the *Interahamwe*, Hutu civilians joined in, sometimes by compulsion from the military and sometimes of their own will. In places like Kigali, all Tutsi were forced to register, and lists of Tutsi and anti-Habyarimana Hutu who were to be targeted were made up in advance. Genocide "administrators" were tasked with rounding up Tutsi from their homes and transporting them to "slaughter" gatherings. Road blocks, security checkpoints, and search patrols were organized to catch Tutsi trying to flee, and administrators

---

[65] Conservative estimates are no lower than 500,000.

[66] Forges, *Leave None to Tell the Story*; Uvin, "Ethnicity and Power in Burundi and Rwanda"; Gourevitch, *We Wish to Inform You.*

[67] Clubs with nails.

[68] Ibid. Forges, *Leave None to Tell the Story*; Kamola, "The Global Coffee Economy."

directed the removal of dead bodies and supervised the looting. Political leaders arranged for *Interahamwe* and other militia to be dispatched to different areas of the country when needed.[69]

The *Akazu* and Hutu Power leaders framed participation in the genocide as something close to "community service." From 1990 to 1993, the *Interahamwe* had been involved in massacres of Tutsi, with the effect of "routinizing" violence and the slaughter of Tutsi among the militias and the general public. With leaders at all levels of government and society encouraging attacks, and with RTLM broadcasting nonstop messages of across-the-board extermination of Tutsi and moderate Hutu, Hutu of all sorts murdered their Tutsi friends, neighbors, and colleagues, often in the victims' homes. RTLM pushed listeners to take no pity on women and children, and women were often raped and tortured before they were slaughtered. The chance to rape young girls was bartered for looted goods. Looting, of both Tutsi homes and businesses, was commonplace, and the *Interahamwe* were said to have conducted slaughter while intoxicated with drugs from the pharmacies they raided. The heads of Tutsi were sold and bartered, according to some reports.[70]

METHODS OF RECRUITMENT

The RPF membership grew out of Rwandan Tutsi refugees in Uganda. Although many Tutsi were rewarded with posts and jobs via the UN after service in Museveni's army, this treatment led Ugandan nationals to engage in violence against the refugees. In response, the Tutsi organized the Rwanda Refugee Welfare Foundation, later renamed RANU.[71]

The *Akazu* formed soccer clubs to initially draw in young men for the militia gangs that would eventually turn into the *Interahamwe* and other death groups, such as "Network Zero." Because of the economic conditions of the early 1990s in Rwanda, where many young men had no hope of getting any sort of employment, many were drawn to Hutu Power, which offered them food, shelter, clothing, and beer. To keep their training secret, recruits were brought to camps far from Kigali where they were treated to barbecues and festivals to keep them content.[72] Then, selling genocide as something of a street party, they preached ethnic solidarity and trained the men using military

---

[69] Forges, *Leave None to Tell the Story*; Gourevitch, *We Wish to Inform You.*

[70] Ibid. Forges, *Leave None to Tell the Story*; Melady, "Burundi and Rwanda."

[71] Kuperman, "Explaining the Ultimate Escalation in Rwanda."

[72] Rewarding militia with beer and tobacco for killing continued throughout the genocide.

drills. The *Interahamwe* practiced wielding machetes against dummies. Former soldiers and members of the National Police trained and supervised the men and directed them during the genocide.[73]

The overall preparations for the genocide in Rwanda were slow-moving and organized. Once Bagsora's regime was in charge after Habyarimana's assassination, Bagosora's group began recruiting political leaders and administrators for the genocide campaign. Preaching Hutu Power, they were able to convince leaders of opposition parties to join in on the plans against the vilified Tutsi. Local leaders and military chiefs opposed to the genocide plans were coerced into joining through meetings in their localities and through threats and ridicule broadcast on RTLM.[74]

It was thought by some members of the MRDN and *Akazu* that a civilian "self-defense" program would be more cost effective than a militia. Thus, well before 1994, plans were laid out as to how to organize, train, and arm civilians for participation in the genocide. When the genocide began, some civilians willingly joined in because they believed in the cause, while others were encouraged and often threatened into joining in on the slaughter. RTLM, incessantly broadcasting about the threat the Tutsi represented, not only convinced many Hutu to partake in the slaughter but also instructed listeners to not trust Hutu who did not help in the killing and to view them as enemies as well. Participation was further incentivized through the provision of foods, drinks, various intoxicants, and sometimes even small payments by genocide administrators. Pillaged Tutsi goods, such as land, homes, and livestock, were also offered in exchange for participation in the massacre.[75]

METHODS OF SUSTAINMENT

The Hutu–Tutsi conflict's four-decades-long existence was sustained through a combination of developments. Domestically, a rejection of the possibility of power sharing by both Hutu and Tutsi in Rwanda and Burundi kept the conflict—which has ceased to end in Burundi even in 2010—going. Furthermore, members of both groups in the two countries have failed to see the role that their groups have played in protracting the violence. Both groups viewed

---

[73] Gourevitch, *We Wish to Inform You*; Kirschke, "Broadcasting Genocide"; Forges, *Leave None to Tell the Story*.

[74] Kirschke, "Broadcasting Genocide"; Gourevitch, *We Wish to Inform You*.

[75] Forges, *Leave None to Tell the Story*; Kirschke, "Broadcasting Genocide"; Gourevitch, *We Wish to Inform You*.

their own members as victims and members of the other groups as the perpetrators of the problem.[76]

Foreign influence sustained the conflict in a number of ways. The Rwandan Tutsi refugees in Uganda would most likely never have created such an organized and professional union as the RPF without having been involved in the Ugandan political struggle. Museveni's and his army's training and subsequent appointments of many of these refugees, including Paul Kagame, were critical in the Tutsi refugees' ability to launch an effective military campaign in Rwanda. Foreign aid to both the RPF and the Habyarimana regime allowed both establishments to subsist and continue with their respective goals.[77]

METHODS OF OBTAINING LEGITIMACY

Throughout the Hutu–Tutsi conflict, perpetrators of violence claimed that they were forced to act when they did in order to preemptively thwart atrocities that the other group was planning. In their view, these actions were not reprehensible because they were done out of self-defense. The Rwandan civilian genocidaires deflected culpability for the 800,000 lives lost in 1994 by way of a "masquerade of legitimacy." They argued that the interim government and local officials ordered them to kill throughout the genocide to maintain the moral authority of the state. Furthermore, they cited RPF/Tutsi war plans as a legitimate threat at the time that necessitated a preemptive attack against all Tutsi.[78]

The Hutu's rise to power and subsequent fight to maintain or expand influence in Burundi and Rwanda has sometimes been seen as legitimate and necessary—regardless of the brutality—because of the centuries of oppression under Tutsi and colonial regimes, connections to the Hamitic myth that the Tutsis invaded the Hutu land, and the Hutu's status as the majority ethnic group. As is evidenced by the Belgians' and Catholic Church's change in support post-World War II, the Hutu's fight was seen as good and democratic, and this appealed to the postwar West. Following the fall of Kigali, ex-FAR and *Interahamwe* forces were granted a form of legitimacy when they were asked to join in on the Democratic Republic of Congo's fight against Tutsi-led Rwanda.[79]

---

[76] Lemarchand, "Genocide in the Great Lakes."

[77] Kuperman, "Explaining the Ultimate Escalation in Rwanda."

[78] Lemarchand, "Genocide in the Great Lakes"; Forges, *Leave None to Tell the Story.*

[79] Jeffrey H. Powell, "Amnesty, Reintegration, and Reconciliation in Rwanda"; Lemarchand, "Rwanda: The Rationality of Genocide"; Helen M. Hintjens, "Explaining the 1994 Genocide in Rwanda," *The Journal of Modern African Studies* 37, no. 2 (1999): 241–286.

Rwanda's apparent accomplishments under the Habyarimana regime in the 1970s and 1980s—improved roads, lowered crime rates, limited corruption, and decent church attendance—demonstrated some economic and social progress. This made Rwanda appear to the West as something of an African success story, and, thus, the country was considered a great place in which to invest foreign aid. Some in the West also viewed Rwanda as truly belonging to the Hutu after the group ended centuries of brutal subjugation under Tutsi rule. Thus, despite the fact that power continued to be concentrated into the hands of Habyarimana and the *Akazu*, the relative tranquility within Rwanda's borders and relative respect toward human rights during this time resulted in the view that the Habyarimana regime was a triumph in the Great Lakes region—especially when compared with some of Rwanda's neighbors in sub-Saharan Africa.[80]

EXTERNAL SUPPORT

As the colonial period came to an end, the newly formed UN, sympathetic to the plight of the Hutu after the Holocaust, began pushing for democratization in Ruanda-Urundi. A change in allegiance from the Tutsi to the Hutu by the Belgians and the Catholic Church, due to UN pressure and a change in their views after World War II, played a substantial role in the ascension of the Hutu to political power in Rwanda. Both the colonial authorities and the Church began expanding economic and political opportunities for the Hutu. Furthermore, from 1959 to 1962, the Catholic Church and the *tutelle* authorities supplied the Hutu with logistical support as well as moral and political arguments, which allowed the Hutu to continue the violence against Tutsi, leading to the mass exodus of Tutsi across Rwanda's borders to neighboring lands.[81]

French, Belgian, and Zairian military support enabled President Juvénal Habyarimana to thwart (at least in the short term) the RPF's 1990 invasion of Rwanda, which launched the Rwandan Civil War. Belgian forces withdrew after a couple of months, and the Zairian forces were asked to withdraw after reports of widespread looting by Zairian troops became known to the Habyarimana regime. The French, however, aided the FAR for the next three years, throughout the entire civil war. From the French perspective, the RPF, which was launching attacks from English-speaking Uganda, was a threat to what

---

[80] Uvin, "Ethnicity and Power in Burundi and Rwanda"; Gourevitch, *We Wish to Inform You*; Melvern, *A People Betrayed*.

[81] Gourevitch, *We Wish to Inform You*; Kamola, "The Global Coffee Economy"; Lemarchand, "Rwanda: The Rationality of Genocide."

they referred to as *chez nous* ("our house"), a semi-francophone country that was an extended part of the French empire. The RPF invasion renewed French bitterness toward the perceived Anglo-Saxon threat to French culture. Furthermore, Rwanda was a particularly esteemed case among Paris's African allies in that it had never been a French colony but had successfully gained independence from Belgium, a French rival. Defeating the RPF would help to humiliate Anglo-Saxon forces, including the United States, which the Mitterrand regime believed had "hegemonic aims" in the Great Lakes region.[82]

Personal ties also existed between then-French President Francois Mitterrand and the Habyarimana regime. Mitterrand's son, Jean-Christophe Mitterrand, a supposed arms dealer, was an admirer of Habyarimana as was the French president himself. Habyarimana's apparent personal assimilation of French culture impressed both the president and his son. Paris not only continued to supply arms and ammunition to the FAR in its effort to hold back the RPF, but it also trained Rwandan forces—including the elite Presidential Guard—and provided French soldiers and commanders for the military effort against the RPF. It is possible that French intervention prevented the well-trained RPF from defeating the FAR before the genocide and deposing Habyarimana. In June 1994, in the midst of the genocide, the French launched "Operation Turquoise," ostensibly a humanitarian mission to halt the slaughter of Tutsi and guard against reprisal killings of Hutu. In effect, however, the French forces' presence between the invading RPF and the section of Rwanda still under FAR control allowed for many genocidaires (FAR and *Interahamwe* forces) to flee into Zaire. This, along with the fact that the French did not capture any perpetrators of the violence, left the true motives of the Mitterrand regime in launching the operation up for debate.[83]

More broadly, international aid given to promote development in Rwanda helped to lay the foundation for genocide by contributing to a system of inequality, racism, and oppression. As discussed above, the international community was willing to give aid because of apparent successes made during the Habyarimana regime. Habyarimana, boasting of "development," was said to have been very talented at impressing American and European foreign aid bureaucrats. It is speculated that World Bank aid was used by the Hutu elite to buy up land and livestock of Tutsi fleeing their homes and land. Both the

---

[82] Kuperman, "Explaining the Ultimate Escalation in Rwanda"; Lemarchand, "Rwanda: The Rationality of Genocide"; Gourevitch, *We Wish to Inform You.*

[83] Forges, *Leave None to Tell the Story*; Kuperman, "Explaining the Ultimate Escalation in Rwanda"; Lemarchand, "Rwanda: The Rationality of Genocide"; Gourevitch, *We Wish to Inform You*; Thom, "Congo-Zaire's 1996-97 Civil War."

International Monetary Fund (IMF) and World Bank increased their aid to Rwanda in the early 1990s, as the genocide was being planned, in hopes of helping to stabilize its post-coffee-bust faltering economy. Instead, more than $112 million of foreign aid was used to purchase weapons—mostly machetes—from France, South Africa, and Egypt. Foreign aid also allowed the FAR to grow from 5,000 soldiers to 40,000 in the weeks after RPF first launched its attack. Although the IMF and World Bank suspended new loans in 1993, as it became obvious the money was not being used to recover the economy, other aid continued to come in, and no assets in foreign banks were ever frozen.[84]

The RPF also benefited from foreign aid because both the American and Ugandan governments[85] backed their initial invasion of northern Rwanda. The Tutsi diaspora, primarily those in America and Europe, donated $1 million annually to the RPF cause throughout the first couple years of the Rwandan civil war.[86]

Western forces did not attempt to prevent the 1994 genocide, despite warnings from UNAMIR commander Major General Dallaire. Dallaire became aware that weapons were being illegally stockpiled in Kigali—a weapons-free zone per the Arusha Accords—and relayed this information to then-chief of peacekeeping at the UN, Kofi Annan. Annan's office rejected Dallaire's suggestion of raiding the arms stockpiles. Despite the murder of Belgian UNAMIR soldiers as the genocide began and UNAMIR troops witnessing of the slaughter, the peacekeeping mission did little to prevent or diminish the massacre of Tutsi. It has been suggested that the embarrassment the United States experienced after the failure of its Operation Gothic Serpent in Somalia in late 1993 played a role in the Western powers' unwillingness to send more troops into Rwanda or to let the peacekeeping troops already deployed engage, allowing the genocide to continue.[87]

COUNTERMEASURES TAKEN BY THE GOVERNMENT

The Hutu–Tutsi conflict has been marked by frequent brutal government repression in response to attacks by the current out-group, whether the Hutu or Tutsi. Perhaps the two greatest tragedies

---

[84] Gourevitch, *We Wish to Inform You*; Kamola, "The Global Coffee Economy."

[85] How much the Museveni regime in Uganda indeed backed the RPF's invasion is a matter of debate. It has been hypothesized that although Museveni wanted the Tutsi refugees in his country to return to Rwanda, he did not want this to happen militarily because he feared that his country would lose foreign aid if it appeared it had fostered an insurgency group. Kuperman, "Explaining the Ultimate Escalation in Rwanda."

[86] Ibid. Kamola, "The Global Coffee Economy."

[87] Gourevitch, *We Wish to Inform You*; Forges, *Leave None to Tell the Story*.

of the conflict, the 1972 genocide in Burundi and 1994 genocide in Rwanda, both exemplified a government regime responding to an apparent threat with brutality. The 1972 genocide, however, focused on the execution of Hutu intellectuals and leaders in hopes of ending the movement to include Hutu in the governance of the country. On the contrary, the Hutu regime behind the 1994 genocide was much less selective in its victims and targeted all Tutsi in hopes of wiping out the entire population.[88]

The extent to which the governments of Burundi and Rwanda are creators or instigators of insurgent attacks varies by case. The 1994 Rwandan genocide, for example, represents a case where the government was sponsoring the militia, partly in response to the threat posed by the invading RPF army. However, the degree to which the Burundian government was involved in slaughtering Tutsi in 1993 following the assassination of Ndadaye is less clear and is a matter of debate.[89]

SHORT- AND LONG-TERM EFFECTS

CHANGES IN THE ENVIRONMENT

Following the 1994 genocide, Rwanda was left with a devastated economy and the near destruction of its infrastructure, civil services, and basic industry. Despite the RPF's victory, the new government faced complex internal and external difficulties. For one, the killing did not completely halt with the fall of Kigali. Violence was rampant in Hutu refugee sites in both Rwanda[90] and Zaire. FAR leaders and *Interahamwe* soldiers fled into Zaire (now the Democratic Republic of Congo), bringing the Hutu–Tutsi conflict (as well as their weapons and vehicles) with them and undermining regional security. The conflicts and wars that followed were commonly referred to as the "Great Lakes Refugee Crisis." Two million Hutu refugees, representing more than a third of the Rwandan population, escaped into Zaire from 1994 to 1996.[91]

---

[88] Lemarchand, "Genocide in the Great Lakes."

[89] Ibid.

[90] A notable example is the Kibeho incident; the southern Rwandan town of Kibeho was a Hutu IDP (internally displaced person) site in 1995 when the RPF opened fire on the Hutu refugees, killing 5,000.

[91] Forges, *Leave None to Tell the Story*; Lemarchand, "Genocide in the Great Lakes"; Hintjens, "Explaining the 1994 Genocide in Rwanda," 241–286; Gérard Prunier, *The Rwanda Crisis: History of a Genocide* (New York: Columbia University Press, 1997).

The ex-FAR forces, allowed by Zairian officials to keep most of their small arms, established military camps and began launching guerrilla attacks into the western region of the RPF-controlled Rwanda, targeting Tutsi. This continued for two years after the fall of Kigali, before Rwanda, Uganda, and Angola jointly invaded eastern Zaire in 1996, forming the Alliance of Democratic Forces for the Liberation of Congo-Zaire (ADFL) with Zairian rebel leader Laurent-Désiré Kabila. This conflict, later known as the First Congo War, resulted in Zairian President Mobutu Sésé Seko, who had aided Habyarimana and the FAR during the Rwandan Civil War, being overthrown. Afterward, Kabila installed himself as president and renamed Zaire as the DRC.[92]

Tensions between the newly formed Kabila regime and Kagame's Rwanda and Museveni's Uganda rose after Rwandan and Ugandan forces failed to leave the DRC. The foreign armies allegedly raided homes, seized possessions, and killed many natives in anti-Hutu refugee operations. Thus, public opinion toward Rwanda and the Tutsis in general grew sour in the DRC. In 1998, Rwandan Tutsi living in the DRC, known as the Banyamulenge, formed a rebel militia Rally for Congolese Democracy (RCD), which Rwanda and Uganda quickly armed and backed. The Banyamulenge had been targeted in the DRC by Hutu refugees and native Congolese, who perceived the Tutsi as having hegemonic ambitions. Following Rwandan and Ugandan invasion into the northeastern DRC and successful advances by the RCD, Kabila sought and received the aid of Hutu refugee *Interahamwe* and ex-FAR forces in the DRC.[93]

Thus began the Second Congo War, or what has been referred to as the "First African World War."[94] The Hutu–Tutsi crisis had expanded not only into the wider Great Lakes region but also into greater sub-Saharan Africa. Many African nations, particularly Zimbabwe and Angola, rushed in to help the DRC fight against what they felt was unnecessary aggression by Rwanda in particular. Although the 1999 Lusaka Ceasefire Agreement was agreed to by all the belligerent countries, the Tutsi RCD continued to fight against Congolese forces. The UN intervened, but fighting, often involving Rwandan forces, continued for several more years. In 2002, Rwanda and the

---

[92] Powell, "Amnesty, Reintegration, and Reconciliation in Rwanda"; Prunier, *The Rwanda Crisis*; Hintjens, "Explaining the 1994 Genocide in Rwanda."

[93] Ibid.

[94] This designation was made by then-US Assistant Secretary of State for African Affairs and current (as of 2010) United States Ambassador to the UN Susan Rice. The war would be the deadliest conflict since World War II, with an estimated 5.4 million casualties from combat and displacement after the war. Hintjens, "Explaining the 1994 Genocide in Rwanda," 241–286; Joe Bavier, "Congo War-Driven Crisis Kills 45,000 a Month: Study," Reuters, January 22, 2008, http://www.reuters.com/article/idUSL2280201220080122.

DRC reached a peace deal where Rwanda agreed to withdraw all military forces and the DRC would round up and arrest all remaining *Interahamwe* and ex-FAR forces in the DRC; however, conflict between Banyamulenge and the DRC government continued even until 2009 when the *Interahamwe* and ex-FAR forces resurfaced in eastern DRC. After the recruitment of Congolese Hutu, the forces were collectively referred to as the Democratic Forces for the Liberation of Rwanda (*Forces démocratiques de libération du Rwanda*, or FDLR).[95]

In Burundi, after Buyoya returned to power in 1996, violence was prevalent between the Tutsi-led Army and the Hutu militias, such as the FDD, a militarized group that split off from the FRODEBU and that randomly committed brutal acts of violence against the Tutsi. Despite this, Buyoya was successful in carrying on long peace talks, resulting in the 2001 Arusha Peace and Reconciliation Accords (APRA). In 2001, Hutu militants carried out an attempted coup. In 2002, the government and the FDD, the main Hutu militant group, agreed to a cease-fire. After more violence by Hutu militants, another cease-fire was agreed upon in 2003, and FDD leader Pierre Nkurunziza and other FDD officials were given posts in the government, ushering in a new era of power sharing. One smaller Hutu militant group, the Forces for National Liberation (FNL), continued its violence— especially toward Tutsi—until finally a peace agreement was signed in 2006. After promising changes, including the UN' decision to refocus its aims from peacekeeping in Burundi to reconstruction and a new economic agreement between the DRC, Rwanda, and Burundi (known as the Great Lakes Countries Economic Community, or CEPGL), the FNL renewed its campaign of violence. However, in 2009, the FNL again agreed to a cease-fire and transformed into a political party under African Union supervision.[96]

CHANGES IN GOVERNMENT

After the fall of Kigali, Pasteur Bizimungu, a moderate Hutu, became president of Rwanda, and Paul Kagame served as vice president and defense minister. Kagame, however, remained head of the RPF with Bizimungu serving as deputy. The two clashed over issues, especially over the issue of the Hutu in the DRC. In 2000, Bizimungu resigned, allowing the Tutsi Kagame to become president. The Kagame regime presided over a semiauthoritarian government. The 2003 legislative elections, which featured a constitutional referendum, created what

---

[95] Hintjens, "Explaining the 1994 Genocide in Rwanda"; Prunier, *The Rwanda Crisis.*
[96] Hudson, "Breaking the Cycle of Violence."

has been described as a constitutional dictatorship, with Kagame serving as head of state, president of the RPF, and military commander. Multipluralism was thus elusive; the two elections that took place after the 1994 genocide (in 2003 and 2008) resulted in huge victories for the RPF. Nevertheless, these elections represented only the second and third examples of multiparty legislative elections in the history of Rwanda.[97]

In 1998, Burundian president Pierre Buyoya and the Burundian parliament approved a new transitional constitution and a transitional government, with Buyoya still at the head. As per power negotiations agreed upon in 2001, a Hutu, Domitien Ndayizeye, succeeded Buyoya as president in 2003, and leaders of the FDD joined the government later that year. After a national vote to approve a power-sharing constitution in 2005, Hutu and FDD leader Pierre Nkurunziza was elected as president.[98]

CHANGES IN POLICY

Although Rwanda began the reconciliation process between Tutsi and Hutu, the Kagame regime failed to grant amnesty to many Hutu in Rwanda and has also yet to acknowledge some illegal actions executed by RPF forces in the refugee camps. Furthermore, power sharing between Hutu, Tutsi, and other political groups was severely limited; Hutu, for the most part, were not included in the new postgenocide government. Two policies for reconciliation and reintegration—the *gacaca* (grassroots) courts, which allowed local communities to try *Interahamwe* and parcel out punishments, and the *ingando* camps, which have been used to indoctrinate all Rwandans with the idea that they form a single people—have had limited success. The Multi-Country Demobilization and Reintegration Program (MDRP), funded by the World Bank, led to some demobilization of genocide-era combatants (from the RPF and Hutu militias), repatriated some Hutu refugees, and reallocated some government funds from defense programs

---

[97] The other being the 1961 "Hutu Revolution" elections that were supervised by the Belgians. Alexander Stroh, "The Effects of Electoral Institutions in Rwanda: Why Proportional Representation Supports the Authoritarian Regime," Working Paper No. 105, GIGA Research Programme: Legitimacy and Efficiency of Political Systems, 2009; Powell, "Amnesty, Reintegration, and Reconciliation in Rwanda"; Thom, "Congo-Zaire's 1996-97 Civil War."

[98] BBC News, "Timeline: Burundi," November 21, 2011, http://news.bbc.co.uk/2/hi/africa/1068991.stm.

to programs designed to improve social and economic conditions in Rwanda.[99]

For a long time, both Hutu and Tutsi in Burundi rejected the idea of power sharing; however, political agreements in the mid-2000s began to change this. As evidence of the downtick in violence and growing political stability, a curfew first imposed in the 1970s because of Hutu–Tutsi violence was eliminated in 2006. By comparison, Burundi did a much better job at reconciling and including all members of its society into the political process than Rwanda, as well as the DRC, which has had minimal success in working out power-sharing agreements with the Tutsi-led RCD rebel group.[100]

CHANGES IN THE REVOLUTIONARY MOVEMENT

The *Interahamwe*, ex-FAR forces, and other Hutu Power remnants in the DRC formed the *Armée pour la Libération du Rwanda*, or Army for the Liberation of Rwanda (ALiR) in 1997. The ALiR was active in guerrilla attacks before and after the Second Congo War against Ugandan, Rwandan, and Congolese forces, as well as civilian populations. In 2001, the ALiR joined with another rebel group composed of Congolese Hutu; the forces are now collectively referred to as the Democratic Forces for the Liberation of Rwanda (FDLR). By 2010, the FDLR was carrying out violence in the Great Lakes region. Rwanda, Burundi, and the DRC have committed to disbanding the organization.[101]

OTHER EFFECTS

Sexual violence against women conducted during (and after) the Rwandan genocide resulted in a surge of unplanned pregnancies and births, as well as a spike in sexually transmitted diseases, including HIV/AIDS. Inaction of many Western powers to intervene in the Rwandan genocide reopened debate about the Convention on the Prevention and Punishment of the Crime of Genocide, first passed in 1948, and its effectiveness.[102]

---

[99] Powell, "Amnesty, Reintegration, and Reconciliation in Rwanda," 84; Lemarchand, "Genocide in the Great Lakes"; Lemarchand, "Case Study."

[100] Hudson, "Breaking the Cycle of Violence," 6; Lemarchand, "Genocide in the Great Lakes"; Lemarchand, "Case Study."

[101] Prunier, *The Rwanda Crisis.*

[102] Melvern, *A People Betrayed*; Anne-Marie de Brouwer, *Supranational Criminal Prosecution of Sexual Violence* (Antwerpen, Oxford: Intersentia, 2005).

BIBLIOGRAPHY

Bavier, Joe. "Congo War-Driven Crisis Kills 45,000 a Month: Study." Reuters, January 22, 2008. http://www.reuters.com/article/idUSL2280201220080122.

BBC News. "Timeline: Burundi." November 21, 2011. http://news.bbc.co.uk/2/hi/africa/1068991.stm.

de Brouwer, Anne-Marie. *Supranational Criminal Prosecution of Sexual Violence.* Antwerpen, Oxford: Intersentia, 2005.

Central Intelligence Agency. "Burundi." *The World Factbook.* Accessed October 12, 2009. https://www.cia.gov/library/publications/the-world-factbook/geos/by.html

————. "Rwanda." *The World Factbook.* Accessed October 12, 2009. https://www.cia.gov/library/publications/the-world-factbook/geos/rw.html.

Forges, Alison Liebhafsky Des. *Leave None to Tell the Story: Genocide in Rwanda.* New York: Human Rights Watch, 1999.

Gourevitch, Philip. *We Wish to Inform You that Tomorrow We Will be Killed with Our Families: Stories from Rwanda.* New York: Picador, 1999.

Hintjens, Helen M. "Explaining the 1994 Genocide in Rwanda." *The Journal of Modern African Studies* 37, no. 2 (1999): 241–286.

Hudson, Ruth Ann. "Breaking the Cycle of Violence: Cohesion Across Ethnic Barriers in Burundi." *Elections Today* 11 (2003): 6.

Kamola, Isaac A. "The Global Coffee Economy and the Production of Genocide in Rwanda." *Third World Quarterly* 28, no. 3 (2007): 571–592.

Kirschke, Linda. "Broadcasting Genocide: Censorship, Propaganda & State-sponsored Violence in Rwanda 1990-1994." *Article 19* (1996): 3.

Kuperman, Alan J. "Explaining the Ultimate Escalation in Rwanda: How and Why Tutsi Rebels Provoked a Retaliatory Genocide." Paper presented at the annual meeting of the American Political Science Association, Philadelphia, PA, August 28–31, 2003.

Lemarchand, René. "Case Study: The Burundi Killings of 1972." *Online Encyclopedia of Mass Violence.* Accessed September 3, 2010. http://www.massviolence.org/The-Burundi-Killings-of-1972.

————. "Genocide in the Great Lakes: Which Genocide? Whose Genocide?" *African Studies Review* 41, no. 1 (1998): 3–16.

————. "Rwanda: The Rationality of Genocide." *Issue: A Journal of Opinion* 23, no. 2 (1995): 8.

Mamdani, Mahmood. *When Victims Become Killers: Colonialism, Nativism, and the Genocide in Rwanda.* Princeton, NJ: Princeton University Press, 2001.

Melady, Thomas Patrick. "Burundi and Rwanda: A Tragic Past, A Cloudy Future." Accessed September 3, 2010. http://www.americanambassadors.org/index.cfm?fuseaction=publications.article&articleid=37.

Melvern, Linda. *A People Betrayed: The Role of the West in Rwanda's Genocide.* London: Zed Books, 2000.

Nkurunziza, Janvier D., and Floribert Ngaruko. *Explaining Growth in Burundi: 1960-2000.* Working Paper 162. The Centre for the Study of African Economies, Oxford, United Kingdom, 2002.

Powell, Jeffrey H. "Amnesty, Reintegration, and Reconciliation in Rwanda." *Military Review* (September-October 2008): 84.

Prunier, Gérard. *The Rwanda Crisis: History of a Genocide.* New York: Columbia University Press, 1997.

Stroh, Alexander. "The Effects of Electoral Institutions in Rwanda: Why Proportional Representation Supports the Authoritarian Regime," Working Paper No. 105, GIGA Research Programme: Legitimacy and Efficiency of Political Systems, 2009.

Thom, William G. "Congo-Zaire's 1996-97 Civil War in the Context of Evolving Patterns of Military Conflict in Africa in the Era of Independence." *Journal of Conflict Studies* 19, no. 2 (1999): 93–123.

Uvin, Peter. "Ethnicity and Power in Burundi and Rwanda: Different Paths to Mass Violence." *Comparative Politics* 31, no. 3 (1999): 253.

Weinstein, Warren. "Conflict and Confrontation in Central Africa: The Revolt in Burundi," *Africa Today* 19, no. 4 (1972): 17–37.

KOSOVO LIBERATION ARMY (KLA): 1996–1999

Maegen Nix and Dru Daubon

SYNOPSIS

The Kosovo Liberation Army (KLA) conducted an armed insurgency against Serbian police and military forces in the late 1990s with the goal of attaining Kosovo's independence from Serbia. With an ethnic population that was primarily Kosovo Albanian ("Kosovar"), the region of Kosovo became a domestic symbol of Serbian nationalism and an international symbol of Serbian aggression and human rights violations. The KLA made extensive use of the expatriate community and the media to present a one-sided story of the Kosovar–Serbian struggle, and it also benefited from the recent negative experiences that Western governments had during the Bosnian crisis. The 1997 collapse of the Albanian government created a significant opportunity for the KLA to acquire arms and training opportunities provided by Germany, the United States, and other governments, enabling the KLA to develop basic military capabilities and discipline. By relying on small-unit ambush tactics and targeting Serbian officials and Albanian collaborators, the KLA achieved significant local successes. In addition, the use of heavy-handed reprisal attacks by Serbian police and military forces validated the KLA message that armed resistance was the only option. The eventual involvement of the international community in negotiations and the spring 1999 North Atlantic Treaty Organization (NATO) air campaign led to the withdrawal from Kosovo of Serbian police and military forces, the introduction of NATO-led peacekeeping forces, and the disbanding of the KLA in June 1999. The passage of United Nations (UN) Resolution 1244 at the same time established Kosovo's autonomy within Serbia until a final political status for Kosovo could be determined. In February 2008, Kosovo unilaterally declared its independence, and the International Court of Justice ruled this declaration lawful in July 2010.

TIMELINE

1987	Slobodan Milosevic utilizes Kosovo in the Serbian nationalist dialogue.
1991	Breakup of the Federal Republic of Yugoslavia.
1993	Popular League for the Republic of Kosovo (LPRK) splits into two factions: Popular Movement for Kosova (LPK), the political wing; and the national Movement for the Liberation of Kosovo, the armed wing. Kosovo Liberation Army (KLA) founded; first KLA communiqués issued.
1996	KLA member Shaban Shala travels to Albania to meet with purported intelligence services of United States, United Kingdom, and Switzerland. Initial KLA leadership (LPK's "group of four") consisting of Kadri Veseli (chief of security service), Hashim Thaçi (head), Xhavit Haliti, and Abaz Xhuka begin building network of secret cells in Kosovo.
March 1997	Albanian government implodes, and army and police dissolve; first uniformed KLA men appear in Kosovo to establish "liberated zones"; KLA strength estimated to be about 150 members.
October–November 1997	KLA suffers its first casualties; funerals attract thousands of protesters; KLA begins to overshadow existing visions of passive resistance.
March 1998	Massacre by Serbian police in the village of Donji Prekaz ignites the call for war among Kosovar Albanians; Adem Jashari (the Kazak) becomes martyr.
March 1999	Hashim Thaçi (political leader of the KLA) declares that KLA's end strength is 30,000; US estimates varied between 15,000 and 17,000 plus 5,000 in Albania.
March 24, 1999	NATO air campaign against Serbian forces commences and lasts for seventy-eight days; 850,000 civilians driven from Kosovo.
June 1999	Milosevic agrees to peace accord. UN Security Council passes Security Council Resolution 1244 in which it establishes a UN interim administration and invites NATO troops to police the peace. *Vojska Jugoslavije* and Interior Ministry Police depart Kosovo, and KLA disbands.

THE ENVIRONMENT OF THE REVOLUTION

PHYSICAL ENVIRONMENT

Figure 1. Map of Kosovo.[1]

Slightly larger than Delaware, Kosovo is a small land-locked country[2] within the Western Balkans that is covered by numerous mountain ranges, plains, and rivers that crisscross the topography and feed its lakes, gorges, and falls. Kosovo is surrounded by the former Yugoslavian republics of Serbia to the north and northeast, Montenegro to the west, and Macedonia to the southeast.[3] Albania is to the southwest and across the Šar Mountains, which contain Kosovo's highest peak at 8,000 feet. The country's natural resources include chrome, lead, nickel, bauxite, zinc, and magnesium.[4]

---

[1] Central Intelligence Agency, "Kosovo," *The World Factbook,* accessed August 12, 2010, https://www.cia.gov/library/publications/the-world-factbook/geos/kv.html.

[2] On July 22, 2010, the International Court of Justice ruled that Kosovo's 2008 declaration of independence was legal. At the time of this ruling, sixty-nine of the 192 UN member countries—including the United States—recognized Kosovo as an independent state. See "Kosovo Independence Move Not Illeal, Says UN Court," *BBC News,* July 22, 2010, http://www.bbc.co.uk/news/world-europe-10730573.

[3] The other provinces of the former Yugoslavia include Bosnia, Croatia, Herzegovina, and Slovenia.

[4] Central Intelligence Agency, "Kosovo," *The World Factbook.*

CULTURAL AND DEMOGRAPHIC ENVIRONMENT

Kosovo has a population of approximately two million people, of which 88% are ethnic Albanians, 7% are Serbs, and the remaining 5% comprise Bosniak, Gorani, Roma, Turk, Ashkali, and Egyptian ethnic groups. The term "Kosovar" is commonly used to describe ethnic Albanians, and the term "Kosovac" is used to describe ethnic Serbians living in Kosovo.[5] The Albanian people who reside within Kosovo and the state of Albania trace their presence within the Balkan Peninsula back to the Illyrians of the fourth century BC who remain part of the historic Albanian narrative. Although a large majority of Albanians are Muslim, Albanian identity is defined less by religion and race than it is by language, culture, and history.[6] The observance of Islam by the Kosovars tends to be very relaxed, as demonstrated by one Gallup poll that reported only 5.8% of Kosovo Albanian Muslims attend religious services daily.[7] In addition, unorthodox Islamic sects, such as the Sufi Bektashi, are present in certain regions of Kosovo; these sects allow unveiled women to lead religious rituals and their members openly drink alcohol.[8] Overall, Kosovar and Albanian nationalism is framed by a history of geographic isolation, the central role of extended families, the unique attributes of their folk music, their marriage and harvest traditions, and the governing influence of their clan-based law.[9] These attributes contrast with other ethnic groups of the former Yugoslavia who closely link their identity with religion. The Serbs, for example, are aligned with their own Orthodox Church, the Bosnians with Islam, and the Croats with Catholicism. But for Albanians, their identity is well expressed in a nineteenth-century quotation by the Albanian patriot Pashko Vasa who stated, "The religion of the Albanians is Albanianism."[10]

Albanian populations can be found in a number of different geographic regions. Although current reports estimate a population of roughly two million Albanians within Kosovo, the Republic of Albania holds approximately 3.5 million. Additionally, half a million

---

[5] Ibid. The term "Kosovan" can also be used as a neutral term to describe anyone from Kosovo.

[6] H. H. Perritt, *Kosovo Liberation Army: The Inside Story of an Insurgency* (Champaign, IL: University of Illinois Press, 2008), 5; Tim Judah, *What Everyone Needs to Know* (New York: Oxford University Press, 2008), 9. With respect to the high attachment to dialect, the Albanian language itself is quite distinct from the Slavic languages used by the Croats, Serbs, and most Macedonians.

[7] International Crisis Group, "Religion in Kosovo," Balkans Report No. 105, January 31, 2001, 2.

[8] Ibid., 4.

[9] H. H. Perritt, *Kosovo Liberation Army*, 20.

[10] International Crisis Group, "Religion in Kosovo," 3.

Albanians likely reside in northern and western Macedonia, and another 100,000 live in Montenegro.[11] Diaspora locations include Switzerland, Italy, Germany, Austria, Scandinavia, Greece, Britain, and the United States. Although many Albanians have left or fled Kosovo throughout history, these Kosovars have increased their presence demographically within the country since World War II. According to available census data and population estimates, between 1948 and 1981, the Serbian and Montenegrin demographic within Kosovo decreased from 27.5% to 14.9%, while the Kosovar Albanian population increased from 68.5% to 77.4%.[12] By 1991, census figures revealed an even larger shift, with the Albanian presence in Kosovo growing to 90% of the resident population.

SOCIOECONOMIC ENVIRONMENT

The traditional Kosovar Albanian society before World War II centered on family-run agriculture and cattle businesses but also included local artisan work and simple manufacturing.[13] After World War II, however, Yugoslavia's socialist system emphasized state production and economic centralization over private-sector assets and family-based business enterprises. During subsequent years, Kosovar Albanians remained the poorest ethnic group of the former Yugoslavia and were frequently discriminated against by the Serbian-led government for employment opportunities within the state. Between 1971 and 1984, unemployment grew from 18.6% of the working-age population to 29.1%, where the majority of unemployed were from the younger, yet increasingly literate, Albanian population.[14]

When Slobodan Milosevic gained control of the Serbian government in 1989, the socioeconomic situation within Kosovo rapidly disintegrated for the politically marginalized Albanian population. More than 80,000 Albanian workers lost their jobs, and many evicted residents became dependent on humanitarian aid. Additionally, although official figures indicated that 67,000 people registered for

---

[11] Ibid., 5.

[12] Depending on the source, some figures vary by a percentage point.

[13] These businesses were often dominated by large patriarchal families that numbered between twenty and sixty members, with each family member having a clearly defined role and responsibilities. World Bank Group, "Kosovo Economic History in a Nutshell" (Report based on Essay by Professor Mustafa and News Agency Reports, 2001), *Beyond Transition Newsletter*, http://web.archive.org/web/20110509045917/http://www.worldbank.org/html/prddr/trans/marapr98/boxpg19.htm.

[14] Dick Leurdijk and Dick Zandee, *Kosovo: From Crisis to Crisis* (Burlington, VT: Ashgate Publishing, 2001),17. High birth rates also "led to a continued growth of the Albanian population during the 1980s, with 2.3 percent a year." By 2010, more than 50% of the population in Kosovo was under the age of nineteen.

unemployment, observers estimated the actual unemployment figure to be roughly 250,000.

In response to political and economic repression by the Serbians, the Kosovar leadership established a parallel system of government, education, social and health services, information flows, tax collection, and cultural and scientific activities.[15] The result was underground school programs that reached approximately 500,000 students and even the establishment of private clinics and hospitals by Albanian physicians and nurses who were dismissed from official Serbian medical programs.[16] The Kosovar Albanian shadow state was also supported by the 600,000–700,000 Kosovo Albanians abroad who sent money back home to Kosovo.[17] In this regard, two separate societies arose: the official Serbian state that repressed the socioeconomic opportunities of the Albanian majority (roughly 90% of Kosovo's total population) and an underground "shadow state" that was tolerated by the Serbs as long as the Albanians refrained from taking up arms.[18]

HISTORICAL FACTORS

Kosovo's history played an important role in both the Albanian and Serbian nationalist narratives that surrounded the KLA insurgency. Albanians argued that they were descendents of the Illyrian and Dardanian tribes that inhabited the Kosovo region more than 2,000 years ago, and that the Slavic population from which the Serbs descended did not arrive until the sixth century. Serbians, on the other hand, stated that despite this early presence of Albanians in Kosovo dating back to the Middle Ages, the population of the region became predominately Serbian with the establishment of the Nemanjic Dynasty in the twelfth century. This was followed by the founding of the Serbian Orthodox Church in neighboring Montenegro and the subsequent moving of this Serbian religious center to Kosovo in the thirteenth century. It was only after the arrival of the Ottomans, according to the Serbian narrative, that large numbers of Albanians migrated into what is now Kosovo and the ethnic balance of the region shifted to

---

[15] Katariina Simonen, "Operation Allied Force: A Case of Humanitarian Intervention?" (Athena Papers Series, Partnership for Peace, Consortium of Defense Academies and Security Studies Institutes, 2004), 6; World Bank Group, "Kosovo Economic History in a Nutshell."

[16] Leurdijk and Zandee, *Kosovo*, 21; World Bank Group, "Kosovo Economic History in a Nutshell."

[17] World Bank Group, "Kosovo Economic History in a Nutshell."

[18] Leurdijk and Zandee, *Kosovo*, 21; World Bank Group, "Kosovo Economic History in a Nutshell."

majority Albanian.[19] Therefore, whereas Albanians see their claim to Kosovo as originating before that of the Serbs, the Serbian narrative disregards the significance of this chronological status and instead focuses on the importance of the Kosovo region to the emergence of Serbian political power and religious identity.

In 1389, Serbian Prince Lazar made a final stand against the Ottomans in the Battle of Kosovo, with Albanians and Serbs fighting together on both sides of the battle, some in support of Prince Lazar, and some in support of the Ottoman sultan.[20] By 1459, however, the Kosovo region was dominated by the Ottoman Empire. Although Serbian unity was fostered and maintained by the Orthodox Church, "the Serbian and Orthodox population gradually shifted northward, to Hungary, to what is today Vojvodina, and to Bosnia, Dalmatia, and Croatia."[21] Over the next four centuries, Serbia did not exist as a formal geopolitical entity within the Ottoman Empire, and Albanian clans within the region were given increasing autonomy by Constantinople. By the end of the nineteenth century, Albanian nationalist aspirations began to emerge when Albanian clan leaders established the League of Prizren in 1878, with a view toward setting up an Ottoman Albanian province with administrative autonomy.

As the Albanians began to work for autonomy within the Ottoman system, Serbia and Montenegro waged war to establish their independence from the waning Ottoman Empire and were recognized as states by the international community in 1878. Aware that Serbia and Montenegro remained dissatisfied with their territorial limitations, the Albanian creation of the League of Prizren also served to deter the Serbs' expansion into Kosovo. By this time, the Serbs represented an existential threat to Kosovar Albanians because of their expulsion of Muslims from the newly independent Serbian state, a practice the Serbs would likely continue if they took control of Kosovo. As in many parts of the Balkans and Europe, Serbian identity was romanticized, and this ethnic nationalism increased hatred and discrimination between Serbs and other ethnic groups. The creation of a Serbian state in 1878 added a political dimension to this ethnic tension and coincided with the emergence of the medieval battle of Kosovo within Serbian ideology as "some sort of nationally-defining historical and spiritual event."[22] When the Serbian and Montenegrin armies finally gained control of the Kosovo region in 1912 during the

---

[19]   Judah, *What Everyone Needs to Know*, 19.

[20]   Noel Malcolm, *Kosovo: A Short History* (New York: New York University Press, 1998), xxix.

[21]   Judah, *What Everyone Needs to Know*, 32.

[22]   Malcolm, *Kosovo: A Short History*, xxx.

First Balkan War, much of the local population fled en masse because of the feared and actual violence that they experienced at the hands of the arriving forces.[23] The London Conference in May 1913 that ended the war, therefore, saw a resulting shift in Kosovar Albanian demands, with independence now being sought rather than autonomy.[24] At that time, however, independence was not granted and most of Kosovo was annexed by Serbia, and the Metohija region of Kosovo was annexed by Montenegro.

During World War I, Kosovo changed hands between the Serbs, the Austro-Hungarians, and the Bulgarians. In 1918, Serbian troops reclaimed Kosovo with the support of the French and Italians, "liberating" the minority Serbian population (30–40%).[25] The new kingdom of Serbs, Croats, and Slovenes was declared and unofficially called "The Kingdom of Yugoslavia." This was soon followed by a deliberate government campaign to "colonize" the Kosovo region with Serbs, many of whom were veterans from the First World War. These Serbian colonists were given the land and often the homes of Albanian residents.[26] Albanian resistance to this Serbian colonization soon began, with guerrilla attacks by poorly armed rebels. At the same time an underground school system emerged to educate Albanian youth.

In April 1939, Mussolini occupied Albania and, soon after, the Germans, Bulgarians, and the Italians divided the Balkan region. Because the Serbs sided with the Allied forces during World War I, the collapse of Yugoslavia at the beginning of World War II led to reprisal attacks on Kosovo Serbian villages by the invading Axis forces and produced a great Serbian migration out of Kosovo. Many Serbs were also sent to concentration camps or to labor camps in local mines. The Nazis also fostered Albanian autonomy, access to schools, and self-government during this time. Germany, however, declined the Kosovar Albanian request to be united with greater Albania.

Once the Germans retreated from the Kosovo region in late November 1944, Serbians began to return, and their influence again began to increase.[27] The end of World War II did not bring about an end to hostilities in the Balkans, however, and fighting

---

[23] Katariina Simonen, "Operation Allied Force," 93, fn 78.

[24] Ibid., 4. In October of 1912, Bulgaria, Greece, Montenegro, and Serbia attacked the Ottoman Empire and initiated the first of two Balkan Wars.

[25] Judah, *What Everyone Needs to Know*, 39.

[26] Ibid., 45. The Institute of History in Pristina, which provides an account of this colonization from the Albanian perspective, estimates that 13,938 Serbian families were settled into Kosovo during this campaign. See Kosovo Information Center, "Expulsion of Albanians and Colonisation of Kosova," http://www.kosova.com/arkivi1997/expuls/chap2.htm#n6.

[27] Ibid., 49.

continued in the region as different religious, ethnic, and political factions struggled for power. Eventually, communist forces under Josip Broz ("Tito") and Enver Hoxha established control, and Tito consolidated his territories into a state called "Yugoslavia" (the union of the southern Slavs) while Hoxha assumed leadership of the state of Albania.[28] Despite Tito's promise that Kosovo citizens would decide by referendum whether to remain part of Yugoslavia or join Albania, he divided Yugoslavia into six republics—Bosnia, Croatia, Macedonia, Montenegro, Slovenia, and Serbia—and incorporated Kosovo into Serbia as an autonomous region.

During the early communist period, Serbians held all of the prominent government and Communist Party positions in Kosovo. From 1947 to 1966, the Serbian ministry of interior under the leadership of Aleksandar Rankovic enacted heavy-handed security measures against Kosovar Albanians. The display of Albanian flags and other nationalist symbols was prohibited, Albanian weapons were confiscated, the teaching of Albanian history and literature was viewed as anticommunist, and many ethnic Albanians were forced to emigrate from Kosovo. In addition, the emergence of a strategic relationship between Hoxha's Albania and Stalin's Soviet Union, focused against Tito and Yugoslavia, further increased Serbian suspicion and mistrust of Kosovar Albanians.[29]

Figure 2. Marshall Tito.[30]

[28] Perritt, *Kosovo Liberation Army*, 6.

[29] Perritt, *Kosovo Liberation Army*, 7.

[30] "File:Marsal Tito.jpg," *Wikipedia*, accessed March 15, 2011, http://en.wikipedia.org/wiki/File:Marsal_Tito.jpg.

In the late 1960s and early 1970s, however, some of the repressive measures against ethnic Albanians in Kosovo were reduced. The removal of Rankovic in 1966 as the head of the Ministry of the Interior and the 1968 offer by Tito to increase opportunities for Kosovars signaled a lessening of official repression. These new freedoms, however, enabled the resurgence of Albanian nationalism and led to riots in 1968 related to Kosovar demands for the integration of Kosovo into Albania or the joining of Kosovo with Albanian regions of Macedonia to form a new ethnic-Albanian republic in Yugoslavia. As a compromise, Tito created an Albanian-language university in Pristina in 1969, and in the 1974 revision of the constitution he established Kosovo, as well as Vojvodina, as autonomous provinces within Yugoslavia with certain self-governing authorities.[31] Similar to a republic, these provinces could now provide for their own "banking, police, legal, and parliamentary system."[32] Additionally, although Kosovo technically remained part of Serbia, this heightened status enabled Kosovar Albanians to control their own assembly and to send representatives to sit in the greater Serbian and Yugoslav federal parliaments.[33]

Beginning with the death of Tito in May 1980, the turmoil and ethnic tensions within Kosovo began a decade-long downturn that included the eventual breakup of Yugoslavia and the outbreak of organized violence between Serbian and Albanian groups in Kosovo. In March 1981, Kosovar Albanian students began to openly protest the overcrowded and underfunded conditions of their education system, especially at Pristina University.[34] These protests turned violent and not only resulted in attacks by Serbian security forces on students but also attacks by Kosovar students on local Serbian and Montenegrin businesses and homes. A state of emergency was declared, with tanks and riot police deployed to control the student protests. Despite claims by Serbian authorities within Yugoslavia that the students were promoting Albanian nationalism and seeking stronger links between Kosovo and Albania, many foreign observers noted that there was little enthusiasm among the protestors for the policies of Enver Hoxha in Albania. Rather, these initial protests were a genuine representation of student frustrations with the quality of their segregated education system.[35]

---

[31] Perritt, *Kosovo Liberation Army*, 7–22. In the fifteen-year period from 1952 to 1967, approximately 175,000 Muslims immigrated to Turkey. See Judah, *What Everyone Needs to Know*, 52.

[32] Simonen, "Operation Allied Force," 4.

[33] Judah, *What Everyone Needs to Know*, 57.

[34] Simonen, "Operation Allied Force," 4.

[35] Malcolm, *Kosovo: A Short History*, 337.

Figure 3. The New Serbian Provinces of Kosovo and Vojvodina in 1974.[36]

As a result of the spring 1981 demonstrations, more than 2,000 people were arrested. Many youth were sentenced to jail for a range of minor offenses, including possession of tape-cassette recordings of radio broadcasts about the protests, as well as for writing anti-Yugoslavian slogans in chalk on walls and sidewalks.[37] These arrests also set the stage for more than eight years of aggressive imprisonment policies against Kosovar Albanians that resulted in more than half a million people being arrested or questioned by police before the decade was over. As a consequence of these anti-Kosovar policies, a high percentage of Albanian families had members directly affected by the imprisonment policies. Moreover, the high number of imprisonments created a shared, unifying experience that served to radicalize many Kosovars against Yugoslavia and Serbia and also produced a breeding

---

[36] "File:Serbian Provinces.jpg" *Wikimedia Commons*, http://commons.wikimedia.org/wiki/File:Serbian_Provinces.jpg.

[37] Leurdijk and Zandee, *Kosovo*, 18.

ground for future leaders of the KLA, with many of them serving jail time in 1981 or soon after.[38]

However, the 1981 spring protests also created fear among Serbians living in Kosovo who witnessed the effects of student violence against fellow Serbians, their businesses, and their homes. Serbian political rhetoric and propaganda induced Serbian citizens to believe that they were being pushed out of Kosovo because of increased harassment, discrimination, and hostility by ethnic Albanians. In addition, Serbian nationalists highlighted the risk of rising Serbian emigration from the Kosovo region, which shifted the political power structure back in favor of ethnic Albanians. This Serbian sentiment was underscored by a September 1986 memorandum published by the Serbian Academy of Arts and Sciences that warned Kosovo Serbs of an impending genocide against them unless government policies were put in place to promote the permanent return of exiled Serbs.[39] The memorandum also advocated the "de-Albanianisation" of Kosovo and the "immediate limitation of Kosovo's autonomy"—two tenets that soon became central to the political platform of a Communist party leader named Slobodan Milosevic.[40]

GOVERNING ENVIRONMENT

In the 1980s, Slobodan Milosevic, the head of one of Yugoslavia's largest banks, began his rise through the ranks of the Communist parties in Belgrade and then Serbia to become the president of Serbia on May 8, 1989.[41] During this ascent to power, Milosevic utilized Serb-Albanian tensions in Kosovo to garner political support from Serbs across the country; he made repeated promises to protect Kosovo Serbs from Albanian violence and to reduce and then eliminate

---

[38] Judah, *What Everyone Needs to Know*, 58. Upon release, many influential leaders went into exile, with some of them establishing a tiny radical party, the Levizja Popullore e Kosoves (the Popular Movement for Kosovo, or LPK) in 1982. The party argued that Kosovo would achieve freedom only through an armed uprising. This clandestine organization, active within Switzerland, Germany and Albania by 1985, worked to intensify the sense of Albanian nationalism within Kosovo. See Tim Judah, "The Growing Pains of the Kosovo Liberation Army," in *The Politics of Delusion*, eds. Michael Waller, Kyril Drezov, and Bulent Gokay (Portland, OR: Frank Cass Publishers, 2001), 21. "Planners in Exile" were initially called the LPRK (Popular League for the Republic of Kosovo) and later became the LPK (Popular League for Kosovo). See Perritt, *Kosovo Liberation Army*, 7.

[39] Simonen, "Operation Allied Force," 4. According to census data, in 1948 there were roughly 27.5% Serbs and Montenegrins in Kosovo, with 68.5% Albanians. In 1991, even though Kosovar Albanians abstained from the census, projections estimated the population to be roughly 77.4% Albanian, with only 14.9% Serb and Montenegrin.

[40] Leurdijk and Zandee, *Kosovo: From Crisis to Crisis*, 18.

[41] In 1991, the formal name of Serbia changed from the Socialist Republic of Serbia to the Republic of Serbia.

Kosovo autonomy. The speech that perhaps sealed Milosevic's rise to power occurred on April 24, 1987, when he was a leader within the Serbian League of Communists and he was sent to Kosovo Polje to address thousands of Serb protestors who were upset by what they perceived as repressive policies by Kosovo's Albanian leadership. As Milosevic heard accounts of police officers beating the Serb protestors, he famously proclaimed: "No one should dare to beat you again!"[42] Milosevic's placement of Kosovo at the center of the Serbian political and nationalist dialogue significantly elevated the stakes for both Albanians and Serbs living in Kosovo, as their status and relations were now the foundation of the president's political credibility and power.

After the election of Milosevic, legal changes to the Serbian constitution were implemented in 1990 to abolish the autonomous status of Serbian provinces and to also abolish the autonomous status of the provincial governments, thus making Kosovo completely subordinate to Serbian national authority.[43] A by-product of these actions was a surge in Kosovar Albanian nationalism and growing membership in both moderate and radical groups seeking full Kosovo independence from Serbia.[44] By July 1990, a referendum was held in Kosovo in which 114 of the 123 Albanian members of Kosovo's parliament voted in favor of establishing a Republic of Kosovo that remained part of Yugoslavia yet was independent of Serbia.[45] In response to this vote, the Serbian government dissolved Albanian participation in the Kosovo government and assumed full control over the administration of the province. Conversely, the Kosovar Albanian elite formed the Democratic League of Kosovo (the LDK) and established a government in exile in Germany.[46]

---

[42] Adam Lebor, *Milosevic: A Biography* (New Haven, CT: Yale University Press, 2002), 79–84.

[43] Simonen, "Operation Allied Force," 5.

[44] Alberto Coll, "Kosovo and the Moral Burdens of Powers," in *War over Kosovo: Politics and Strategy in a Global Age*, eds. Andrew Bacevich and Eliot Cohen (New York: Columbia University Press, 2001), 131.

[45] Leurdijk and Zandee, *Kosovo: From Crisis to Crisis*, 19.

[46] Perritt, *Kosovo Liberation Army*, 8.

Figure 4. Slobodan Milosevic (seated, third from left) preparing to sign the Dayton Peace Accords.[47]

Coinciding with Serbian efforts to control the governing status of Kosovo, the other province in Serbia (Vojvodina) and other non-Serbian republics in Yugoslavia (Slovenia, Croatia, Macedonia, and Bosnia-Herzegovina) sought their independence. Although the international community recognized the independence of Slovenia, Croatia, Macedonia, and Bosnia-Herzegovina in the latter half of 1991, international legal restrictions mandated the exclusion of Kosovo and Vojvodina because they were technically provinces of Serbia and not equal Republics within the Yugoslav federation.

In 1992, Kosovar Albanians held their own presidential elections (declared illegal by Belgrade) and brought Ibrahim Rugova into office as the president of the unofficial Republic of Kosovo. In addition, the Kosovars established a parallel system of government that included tax collection, schools, and medical clinics.[48] However, Kosovo's economy continued to decline, with high unemployment rates, resulting in many Albanians being evicted from their homes, dependent on humanitarian aid, or both. In 1993 alone, a quarter of a million Kosovar Albanians were dependent on the food supplies provided by an organization that was affiliated with Mother Teresa.[49] Despite these hardships, most Albanians continued to follow Rugova's lead and participated in passive resistance to Serbian control. Certain

---

[47] "File:DaytonAgreement.jpg," *Wikimedia Commons*, photo by US Air Force/Staff Sgt. Brian Schlumbohm, accessed March 15, 2011, http://commons.wikimedia.org/wiki/File:DaytonAgreement.jpg.

[48] Katariina Simonen, "Operation Allied Force," 6.

[49] Ibid., 20.

radical leaders who disagreed with Rugova, however, organized training camps for guerrilla fighters, and small armed groups began to conduct hit-and-run attacks on Serbian police officers and official targets.[50] As the number of Albanians who supported the use of force began to grow in 1993 and 1994, a division between moderates and hard-liners emerged, as did open criticism of Rugova.[51] As this momentum in support of armed resistance increased in late 1995, the US-backed Dayton Peace Accords were signed. These accords ended the secessionist conflict between Serbia, Bosnia, and Herzegovina and simultaneously stated that sanctions against Serbia would remain in place until it engaged in focused negotiations with Kosovar Albanians and stopped violating human rights in Kosovo. This singular reference to Kosovo within the Accords signaled to the Kosovars that the international community was going to treat Kosovo as an internal issue for Serbia and that the nonviolent means promoted by Rugova in order to achieve independence had fallen well short of their goal.[52] Moreover, for the more hard-line Albanian Kosovars, the Bosnian example supported their contention that independence could come only through mobilization and the intervention of the international community in response to armed conflict.

WEAKNESSES OF THE PREREVOLUTIONARY ENVIRONMENT AND CATALYSTS

From the perspective of Kosovar Albanians, three key elements energized the prerevolutionary environment within Kosovo during the 1990s.[53] Firstly, the emerging Serbian nationalistic narrative, which Milosevic eventually pushed to a tipping point, set the two rival ethnic groups on a collision course. Included in this nationalist rhetoric was the belief that, for centuries, Albanians had collaborated with the Ottomans in suppressing Serbs. Additionally, the media joined in this rhetorical campaign and published false stories of Serbian degradation and injustice at the hands of Kosovar Albanians, and scholarly Serbian journal articles published claims of Kosovar-led

---

[50] This initial guerrilla training program was infiltrated by the Serbian secret police in 1993 and resulted in the arrest of many participants, with the remaining going into temporary exile. See Perritt, *Kosovo Liberation Army*, 8.

[51] Leurdijk and Zandee, *Kosovo: From Crisis to Crisis*, 21.

[52] William Hayden, "The Kosovo Conflict and Forced Migration: The Strategic Use of Displacement and the Obstacles to International Protection," *Journal of Humanitarian Assistance*, February 14, 1999, http://jha.ac/articles/a039.htm.

[53] This section focuses on those factors that triggered an armed insurgency by the Kosovar Albanians. As such, the factors are slanted towards the Kosovar perspective and therefore somewhat biased.

genocide in the early 1980s.[54] And for his part, Milosevic saw the clear benefits of manipulating these tensions for political gain and ethnic dominance.[55] In general, his rise to power within Serbia and Yugoslavia fractured the country and brought him extreme power and popularity among the Serbs, but it also instilled fear in the surrounding ethnic populations and fed their own nationalistic aspirations.[56]

The second formative prerevolutionary catalyst was linked to the Kosovar Albanian expatriate movement. Clandestine organizations in exile learned about guerrilla warfare and insurgency by studying cases from Algeria, the Basque region, Ireland, and Vietnam. Central to these studies was the desire to understand how these insurgencies acquired weapons and how they designed and managed fund-raising networks.[57] Through their fund-raising activities and the recruitment of fighters from the diaspora, these expatriate groups provided leadership and momentum for Albanians in Kosovo and helped spur them toward armed resistance.[58] The expatriate community in Britain and Germany also played a critical role in drawing international attention to Serbian human rights violations in Kosovo by leveraging the existing media attention on Bosnia and broadening that media coverage to include Kosovo.[59]

Finally, although small hit-and-run skirmishes between Kosovar insurgents and Serbian forces began in 1996, large-scale fighting in Kosovo did not commence until two key events occurred. The first event was the collapse of the Albanian state, just south of Kosovo, during the spring of 1997. At that time, the absence of security surrounding weapons stockpiles of former communists in Albania enabled massive arms transfers to support Kosovar militarization. Second, the deaths of Adem Jashari, a hero within the Kosovar liberation movement, and his family at the hands of a Serbian paramilitary force in Drenica Valley in 1998 created a national icon for Kosovo Albanians and helped spark the 1999 Kosovo War.[60]

---

[54] Malcolm, *Kosovo: A Short History*, 337.

[55] Leurdijk and Zandee, *Kosovo: From Crisis to Crisis*, 8.

[56] Judah, *What Everyone Needs to Know*, 65.

[57] Perritt, *Kosovo Liberation Army*, 8.

[58] Ibid., 3.

[59] Ibid.

[60] Adem Jashari led a clan that controlled the city of Prekaz and its surroundings, which were a stronghold for the KLA. In late February and early March 1998, Serbian forces killed approximately eighty Albanians in the region of Prekaz, including Jashari and his family. See Judah, *What Everyone Needs to Know*, 27.

FORM AND CHARACTERISTICS OF THE REVOLUTION

OBJECTIVES AND GOALS

The main Kosovar combatant force during the insurgent movement in the late 1990s was the KLA, and the overall strategic objective of the KLA was straightforward—to obtain independence from Serbia for the Province of Kosovo and Metohija.[61] To achieve this specific, yet challenging, objective, the leadership of the KLA understood that the insurgency had to be fought as much on the battlefields as on the diplomatic and political fronts. Harmonizing these two efforts proved to be extremely challenging for the KLA because it was a young organization with leadership and stakeholders dispersed domestically and abroad. Within the sphere of armed resistance, the KLA based its operations around a key set of principles and tasks: the direct targeting of key members of the Serbian military, police, and security establishments; the elimination of Albanian collaborators; the defense of Albanian civilians in Kosovo; the maintenance of supply routes from Albania; and the interdiction of internal Serbian supply routes, especially those through the Llap region.[62] In support of its political and diplomatic efforts, the KLA understood that it had to address domestic and international audiences. For the domestic audience, the KLA emphasized that resistance was possible against the Serbs. For the international community, the KLA tried to discredit the Milosevic regime by presenting it as a foreign occupying force in Kosovo that violated human rights; the KLA also presented itself as a defensive army and not a terrorist organization.[63]

The KLA's early realization of the need to integrate combat actions with diplomatic and political outreach helped to form objectives that were geared toward manipulating information outlets; this realization also served to consolidate the resolve of their fighters and support base. Specifically, the ability to simultaneously shape the nationalist sentiment of the guerrillas and of the external supporters became a valuable weapon of the insurgency. By selecting concise tactical and operational objectives and relying on the use of limited force, the KLA was often able to not only contain the physical impact of its armed resistance but also control the informational impact of its operations. Stated differently, the KLA adopted objectives and

---

[61] Metohija is the western part of Kosovo.

[62] Perritt, *Kosovo Liberation Army*, 62. The Llap region to the north of Kosovo included control of the roads that connected Mitrovica and Pristina.

[63] Ibid., 62.

operational methods that enabled it to present itself to the world—through a sympathetic media—as a defender against Serbian aggression.[64] Finally, the KLA carefully chose its words and actions to underscore that its goals were purely nationalistic and not tied to religious ideology, thus avoiding the potential appearance of the Kosovar struggle as a religion-based conflict that could impact the willingness of some Western governments to become involved.

LEADERSHIP AND ORGANIZATIONAL STRUCTURE

The Kosovar insurgency benefited from the concurrent availability of able-bodied young supporters inside Kosovo, as well as older, revolutionary-minded leaders within the expatriate community. Together, these two groups formed the core of the organizational structure and leadership of the KLA movement. In addition, this integration of an older, Marxist–Leninist generation with a younger, well-networked generation living inside Kosovo marked a unique aspect of the KLA's ability to place differences aside and organize for armed resistance. Before 1998, several attempts were made by the more radical elements of the Kosovar political wing to form armed groups in preparation for the looming insurgency. Individuals were assessed on the basis of their paramilitary or police skills and were then trained to fight in small groups.[65] These efforts produced very limited results, however, because many of the groups were infiltrated by members of the Serbian secret police and disbanded. In one significant case, hundreds of Kosovar Albanians were arrested and placed on a very public trial by Serbian authorities.[66]

Overall, it is difficult to identify any singular leader who was the primary force behind the organization of the KLA, but the group's 1993 emergence can be traced to the National Movement for the Liberation of Kosovo, which was the armed wing of the politically based Popular Movement for Kosova (LPK). During this early period of the insurgency, several influential Kosovars living overseas emerged

---

[64] This is not to imply that KLA forces were not also guilty of atrocities during the Kosovo War but rather that they were much more successful in controlling what the media did—and did not—cover about the conflict. For a detailed account, see "Under Orders: War Crimes in Kosovo," Human Rights Watch (2001), http://www.hrw.org/reports/2001/kosovo/.

[65] Ibid., 118.

[66] Ibid., 88. Similar trials occurred in 1993 for former Yugoslav army officers who were members of the "Kosovo reserve territorial defense forces" and were accused of participating in militant activities as part of the "Popular Movement for the Republic of Kosovo." Among the expatriate organizers who had significant connections abroad were Mentor Kaqi (a goldsmith), who founded the "National Front of Albanians," and Pleurat Sejdiu (a doctor), who in 1998 was a spokesman in London for the KLA.

as focal points for the armed movement after giving carefully crafted speeches concerning the plight of Kosovo Albanians.[67] Although many Kosovars claimed to have served in different KLA leadership positions, four individuals stood out during late 1996 for their contributions. This "group of four" consisted of Hashim Thaçi (KLA leader), Kadri Veseli (KLA chief of security service), Xhavit Haliti, and Abaz Xhuka. They were tasked with building a network of secret cells in Kosovo[68] and began arming the cells with the sudden availability of Albanian weapons that became prevalent during the Albanian financial meltdown in early 1997.[69] One other significant leader to emerge during the initial buildup phase of the KLA was Rexhep Selimi (also known as "The Sultan") who operated in the Drenica Valley, a traditional stronghold of Kosovar Albanians.[70] Because of the decentralized structure of the KLA organization, however, many influential Kosovars took it upon themselves to fill the local leadership vacuums created by the rapid stand-up of the armed resistance.[71]

In the early months of 1999, after a temporary cease-fire in the insurgency from December 1998 to March 1999, the structure of the KLA became more defined and two primary figures emerged as military and political leaders within the KLA: General Agim Ceku, who represented the KLA as military chief of staff, and Hashim Thaçi, who continued his service as the KLA political leader. Both Ceku and Thaçi served on the KLA's eighteen-member general staff, which was the organization's main decision-making body. Second in command in the military structure was Ramush Haradinaj.[72] At its height in the spring of 1999, it was estimated that the KLA membership ranged anywhere from 20,000 to 30,000 strong.[73] The KLA was now organized into seven districts that encompassed the operational regions of

---

[67] Judah, *Kosovo War and Revenge,* 103 and 115.

[68] Ibid., 115.

[69] Ibid., 127–128. Judah reports that after the Albanian government collapsed in 1997, the army dissolved and police abandoned their posts, resulting in Kalashnikov rifles being sold for $5 to $16.

[70] Ibid., 138.

[71] Judah, *Kosovo War and Revenge,* 168. Reporters noted that the decentralized nature of the KLA also made it difficult for Western journalists to identify KLA leaders for conducting interviews.

[72] W. Finnegan, "Letter from Kosovo: The Countdown," *The New Yorker,* October 15, 2007, 1. After the Kosovo conflict, Haradinaj would stand accused at The Hague for rape, torture, and indiscriminate killings.

[73] Tim Judah, "The Kosovo Liberation Army," 20–25. Hashim Thaçi (political leader of the KLA) declared that the KLA end strength was 30,000. US estimates, however, varied between 15,000 and 17,000, plus 5,000 in Albania.

Drenica, Pastrik, Dukagjin, Shalja, Llap, Nerodimlje, and Karadak.[74] Within the districts, operational units were structured as battalions, companies, and platoons, with the platoons containing approximately thirty personnel with a dedicated medic or doctor.[75] Local unit leaders were appointed in an ad hoc fashion and possibly vetted by the general staff of the KLA.

Figure 5. KLA members.[76]

As the insurgency matured in the spring of 1999, so did the organization's ability to develop sophisticated plans and coordinate efforts between districts. The northern areas were primarily focused on disrupting traffic with Belgrade, the central areas performed low-level attacks, and the southern zones concentrated on keeping the supply lines with Albania wide open. However, a major obstacle emerged in the planning and organization of the insurgency between the KLA and the Bukoshi's Forces of the Army of the Republic of Kosovo (FARK). Despite advanced coordination and a common cause, personality clashes, as well as power jockeying, tended to undermine any joint planning done by the two groups.[77] In one serious event that almost broke the back of the KLA, more than 2,000 FARK soldiers

---

[74] BIA/Security Information Agency, "Albanian Terrorism and Organized Crime in Kosovo and Metohija," September 2003, http://www.kosovo.net/albterrorism.html. Within these districts, the command structures were loosely affiliated and rather decentralized.

[75] Perritt, *Kosovo Liberation Army,* 71. Fighters who were not armed would conduct logistical support and prepare defensive barriers and trenches.

[76] "File:Kla members.jpg," *Wikipedia,* photo by Sgt. Craig J. Shell, US Marine Corps (US Department of Defense, 990630-M-5696S-002), accessed March 14, 2011, http://en.wikipedia.org/wiki/File:Kla_members.jpg.

[77] Ibid., 85. Comments from a confidential US observer who worked closely with the KLA.

deserted the KLA and retreated to Albania after brief skirmishes with the Serbs.[78]

COMMUNICATIONS

Modern communications and the emerging interconnectivity of the Internet played critical roles in almost every aspect of the formation and functioning of the KLA. Telecommunications and the increased speed of data transmission made it possible for certain strategic aspects of the organization to be controlled or supported from external and friendly areas. During the early stages of the insurgency, the use of nonsecure cell phones and personal mobile radios provided a primary means of tactical and operational communication for intraunit and interunit coordination. With the onset of major conflict in 1998, however, the primary methods for military coordination shifted to the discussions using unsecure commercial cell phone, faxes, and face-to-face meetings. By this point in the insurgency, KLA leaders expected that their cell phones were being monitored by Serbian intelligence services so all their discussions were kept to an absolute minimum and strict adherence to operational security was followed.

Mass communication in the early stages of the insurgency was achieved through underground printed newspapers that also doubled as an information source to spread the word on significant military movements. These newspapers became essential elements of the underground, especially as the Serbian government clamped down and oppressed common methods of information dissemination among Kosovars. KLA messages were also spread through a publication named *Bujku,* a farmer's almanac dedicated to local agricultural issues.[79] In addition, printed forms of news existed for the overseas expatriate community and for the more fringe elements of the armed revolt. The news source for the latter group, the *Clirimi,* was an illegal newspaper, and possession of it was grounds for arrest.[80]

Finally, proper management of media outlets, through close personal contact with reporters, was an important element of the KLA communication strategy. Through coordinated efforts, the KLA was able to reach out to expatriate communities in London,

---

[78] Perritt, *Kosovo Liberation Army,* 87. FARK leadership did not immediately concur with the operational planning presented by the KLA, which they viewed as a disorganized, unprofessional, leaderless group.

[79] Judah, *Kosovo War and Revenge,* 92. Many Kosovar Albanians owned satellite dishes so they could tune in to the Albanian channel, which had dedicated Kosovo programming. This was another avenue by which Kosovars kept attuned to world events and internal happenings.

[80] Ibid., 116.

Switzerland, and Germany to develop sympathy and build a media image of Kosovars as underdogs and overmatched peasants fighting for their inherited rights. By creating a close relationship with the international media, the KLA ensured that powerful images reflected only the atrocities committed by the Serbian faction.[81] Although some external media organizations did cover the Serbian perspective of the conflict, or at least a more balanced perspective, the sudden rise of the KLA and a well-organized, biased, and integrated Western media were too difficult for the Serbs to counter in the brief time period of the conflict. This was especially true given the brutal operations that were captured on film during the run-up and commencement of full hostilities. The Serbian use of armored vehicles against Kosovars who were perceived to be lightly armed provided an opportune and direct linkage to the recent tragedies of the Bosnian war and drew upon the international community's fear of a continuation of human rights violations in the Balkans.

METHODS OF ACTION AND VIOLENCE

The KLA primarily used small-unit tactics and maneuver to perform "off-axis" attacks on a superior Serbian force. The KLA would fight with small teams of insurgents, avoiding direct confrontations against the main Serbian forces while still crippling the main forces by attacking their supply lines or communication hubs or targeting their leadership.[82] The KLA's violent methods were defined by ambushes on logistics routes, abductions, hit-and-run attacks against major communication hubs, and attacks on poorly manned police checkpoints and outposts. Even though many of these attacks were unconditionally conducted against Serbian military and police forces, many civilians who were implicated in complicity with the Serbian forces—wrongly or not—were also targeted by the KLA.[83]

Despite having achieved only a few major battlefield successes, the KLA was able to dominate most of Kosovo after six short months of fighting because of the group's successful employment of small-unit tactics, as well as the assistance of NATO airpower. The initial KLA

---

[81] S. Taylor, *Regardless of the Consequences: Images of Serbia & the Kosovo Conflict* (Ottawa: Esprit de Corps Books, 2000), 34. During the initial period of NATO bombing, the Serbian authorities rounded up and expelled most Western reporters from Belgrade. This decision proved disastrous for the Serbian image during the escalation of the NATO bombing campaign in the spring of 1999.

[82] There was only one case in which the KLA attempted to engage the Serbian military in a direct, force-on-force major battle. This occurred in Orahovac in July 1998 and the offensive failed after the Serbian forces retook the town in two days. "Under Orders," 103.

[83] BIA/Security Information Agency, "Albanian Terrorism and Organized Crime."

attacks were very violent and occurred in isolated locations. Usually small teams of insurgents would conduct attacks on isolated police checkpoints during which any pursuit or counterattack by the police would be drawn into an ambush. Even though these initial small-scale attacks were sporadic, they served as a catalyst and unifying campaign for the Kosovar Albanians and elevated the reputation of the KLA as the primary defenders of the Kosovars.[84] Unfortunately, the extreme nature of the KLA led to numerous abuses, including the murder of suspected informants and of some Kosovar leaders who were opposed to the KLA's views.[85]

The KLA was somewhat conservative in its tactics concerning the use of explosives. The KLA never developed a mature capability to produce improvised explosive devices (IEDs), nor did they employ complex IEDs during the insurgency. In one reported case, the rector of Pristina University was assassinated by a car bomb for which the KLA assumed credit, but other reports highlighted different motives for the murder and pointed to possible Serbian criminal entities rather than the KLA as the culprits.[86] During the insurgency, the KLA did utilize some anti-personnel mines and conventional land mines that were booby trapped, but that was the main extent of their use of explosives. The land mines were of particular use along the main highways that interconnected Serbia and Montenegro to Kosovo and Albania because the mines limited the mobility of the Serbian forces. The KLA did not employ suicide bombers as a weapon.

Before the escalation in hostilities in 1999, the KLA was able to improve tactics of its units and individuals by establishing training camps across the border in Albania. These camps enabled groups of traditional hunters with average shooting skills and superior knowledge of the local environment to become more disciplined and coordinated groups. These new members of the KLA had a greater understanding of the surrounding geography and used it extremely effectively against Serbian forces. However, most KLA members possessed little or no real military skills because their training occurred over very short periods of time. This lack of real training forced the insurgents to quickly disperse and "melt away" if the opposition forces were too great, but these small groups were able to quickly regroup and hit back when the Serbs least expected them. The rapid dispersion also forced the Serbian forces to spread thin while in pursuit of the KLA, which aided the KLA if it decided to counterattack.

---

[84] Perritt, *Kosovo Liberation Army*, 27.

[85] Judah, *Kosovo War and Revenge*, 130.

[86] Ibid., 131.

In one of the most successful strategic employments of these small teams, the KLA was able to establish control of the long highways that interconnected Serbia and Montenegro to Kosovo and Albania. With the inclusion of land mines to slow or stop Serbian convoys along these main thoroughfares, the Kosovars were able to deploy sniper teams to target the Serb soldiers and vehicles. Using this technique, the KLA created such havoc on the main Pristina–Belgrade thoroughfare that Serbians would not travel at night because they feared attack.[87]

As the size and complexity of the KLA small-unit attacks grew, so too did the Serbian counterattacks whose aim was to recapture territory that was lost to the KLA. The KLA leadership soon abandoned the premise of disparate guerrilla groups and issued military uniforms to its fighters, creating the image of an organized army and fostering a sense of nationalism for Albanians and, for Westerners, creating the perception that the KLA was a rightful movement. As the conflict ensued, the KLA fighters achieved greater success with the arrival of NATO airpower. Although direct, real-time coordination between the KLA and the overhead NATO planes was limited, the KLA forces often achieved their military objectives by intentionally drawing the superior Serbian forces out into the open where they could be seen and engaged by the NATO aircraft.[88]

METHODS OF RECRUITMENT

The success of the KLA in its armed guerrilla attacks elevated its prestige and significantly facilitated its ability to recruit new members. Moreover, the KLA was perceived as an all-embracing, diverse, and multifaceted solution for all oppressed Kosovars. KLA soldiers were recruited from all walks of life and across all social scales, with women and men fighting side by side.[89] In addition, expatriates also returned home to Kosovo to join the KLA. Before 1998, recruitment was accomplished via underground methods, such as in overseas locations or on a small, localized scale within Kosovo. The underground methods included the use of unofficial newspapers that circulated quite freely in overseas *Gastarbeiter* (or foreign guest worker) communities and within closely guarded social circles in Kosovo.[90] Several websites also

---

[87] Taylor, *Regardless of the Consequences*, 99.

[88] Finnegan, "The Countdown," 1–2.

[89] Judah, *Kosovo War and Revenge*, 176. Many women joined the ranks of the KLA and fought alongside the men.

[90] Ibid., 69 and 174. These Kosovar guest worker communities were formed in the early 1960s in Germany, Switzerland, Austria, and Scandinavia.

existed in support of the KLA, but it is uncertain how early in the conflict they came online.

However, the pace of KLA recruitment picked up significantly in March 1998 after the perceived martyrdom of Adem Jashari.[91] The violent death of Jashari and his family at the hands of the Serbian police provided an immediate symbol of Serbian aggression and Kosovar oppression and served as a call to arms for many individuals who had not initially join the insurgency.[92] Jashari posters soon appeared everywhere. Within weeks, the KLA was overwhelmed with new recruits from within Kosovo, the expatriate community, and Albania.

Figure 6. Poster of Adem Jashari at the Palace of Youth and Sport of Pristina.[93]

Despite its apparent success, the KLA recruiting process lacked an overarching central plan that integrated the numerous recruiting efforts underway across Kosovo and abroad. Nonetheless, there are some indications that former Yugoslav soldiers were specifically recruited into the Kosovar liberation movement in the early 1990s by extreme factions of the LDK that understood the importance of

---

[91] Jashari was regarded as one of the original leaders of the armed revolution, and he was considered to symbolize the "Kazak" revolutionary spirit.

[92] Taylor, *Regardless of the Consequences*, 29.

[93] "File:Palace of Youth and Adem Jashari.JPG," *Wikimedia Commons*, photo by Ferran Cornellà, accessed March 14, 2011, http://commons.wikimedia.org/wiki/File:Palace_of_Youth_and_Adem_Jashari.JPG.

having professional soldiers within the movement.[94] General Agim Ceku was the most prominent example; he was a former Croatian army general, and his leadership was vital in organizing the KLA and its military efforts. Among the factors that enticed former soldiers to join the KLA were personal ties to the Kosovo Muslim cause, lack of other employment opportunities, or the desire to fulfill vendettas against Serbs. Although the KLA was known to welcome almost all willing fighters, it did apparently exclude groups of Islamic jihadists from joining the insurgency because the KLA did not consider the fight to be a religious battle, and it did not consider Islamic extremism to represent the views of the majority of Kosovar Albanians.[95]

METHODS OF SUSTAINMENT

The Kosovar insurgency was sustained through multiple internal and external sources. Financially, one of the most effective support systems was the expatriate "Homeward Calling Fund." Through fund-raising associations in Western Europe, the United States, and Albania, sympathizers who could not fight—or who did not want direct complicity in the violent methods—could contribute funds that directly upheld the KLA movement.[96] The expatriate community in the United States did, however, appear to provide sniper rifles to the KLA by using a loophole in federal laws that allowed these weapons to be shipped to hunting clubs, and, therefore, a hunting club in Albania was utilized as the shipping address for the KLA.[97] Reports also indicate that the KLA generated substantial funding through illicit activity and criminal entities.[98] These illicit activities were associated with or facilitated by established criminal organizations operating in and about the region and produced funds for general operating costs and enabled the purchase of military hardware as well. KLA involvement in drug smuggling was also suspected and, in 2000, was confirmed by INTERPOL.[99] Although several attempts to provide

---

[94] Judah, *Kosovo War and Revenge*, 174.

[95] Ibid. One KLA leader is quoted as saying, "They came to offer their help, but were declined . . . Some of us are Catholics and this is not a religious struggle."

[96] In all, the estimated contributions in support of the KLA ranged from $75 million to $100 million. Perritt, *Kosovo Liberation Army*, 88.

[97] Tom Walker and Aidan Laverty, "CIA Aided Kosovo Guerrilla Army," *The Sunday Times*, London, March 12, 2000.

[98] BIA/Security Information Agency, "Albanian Terrorism and Organized Crime," 16. According to the author, "almost DM 900 million that reached Kosovo and Metohija in between 1996 and 1999, nearly a half was the money earned in drug trafficking, which made Interpol experts conclude (in December 2000) that the activities of fund raising for Kosmet and KLA were used for laundering of illegally acquired money."

[99] Ibid., 17. This included money laundering and smuggling activities.

funding also emanated from charities linked to Al Qaeda, it appears that these donations were turned down by the KLA because of the risk of creating a negative perception with the West.[100] Finally, training sites for most KLA activities were funded and organized outside of Tropoja, just over the southern border of Kosovo in Albania proper.[101]

METHODS OF OBTAINING LEGITIMACY

To emphasize the legitimacy of its struggle, the KLA employed several tools in its interactions with its domestic and international audiences. Domestically, the KLA highlighted the history of Kosovar oppression, as well as the ongoing heavy-handed measures being employed by the Serbian government. The recent fighting in Bosnia provided additional justification because it gave an example of the scope of Serbian aggression, while the subsequent Dayton Accords underscored that Kosovo could not count on the international community to fight for Kosovo's independence.[102] Coupled with numerous high-profile Serbian atrocities within Kosovo, such as the murder of Adem Jashari and his family, the KLA had an easy message to present and a receptive audience that was listening. To the domestic and expatriate communities, therefore, the KLA message was that the cause was just and that the KLA provided the only viable and genuine solutions.[103]

---

[100] Finnegan, "The Countdown," 1–3. Even though foreign Islamic fighters were suspected of participating in military operations in Bosnia years earlier, by 1998 it was assumed that accepting funds from an enemy of the West would have created greater issues for the KLA cause.

[101] Judah, *Kosovo War and Revenge*, 172.

[102] Ibid., 125.

[103] Ibid., 63. S. Taylor, *Regardless of the Consequences*. The resultant effects of the 1991–1996 civil war that fragmented Croatia, Slovenia, and Bosnia from the Federation of Yugoslav Republics created the ethnic cleansing of an estimated 750,000 Serbs from those regions. The perception that Serbia was intentionally manipulating the demographics in Kosovo by relocating thousands of Serbs had an immense influence on legitimizing the KLA cause.

Figure 7. Brigadier General Bantz J. Craddock (right), Commander, US Kosovo Forces/Joint Task Force Falcon, exchanges documents with Colonel Ahmed Isusi, Commander, Karadak Zone of the Kosovo Liberation Army, near Cernica, Kosovo, during the summer of 1999. The document signing was part of a local agreement establishing a phased demobilization and demilitarization of KLA Forces in the US Sector in support of NATO Operation Joint Guardian.[104]

Internationally, the KLA made heavy use of the media to present a sympathetic image of the Kosovar insurgency as a defensive struggle and an issue of human rights. The recent involvement of Western governments in the Bosnian crisis also helped because of these other governments' own experiences with and perception of Serbian aggression. One excellent example of media exploitation by the KLA was attributed to an American diplomatic misstep when the US presidential envoy was photographed sitting next to an armed KLA fighter. This picture circulated around the world and provided instant legitimacy for the KLA cause.[105] The KLA was also extremely effective at controlling media access to sites where the KLA may have been the perpetrator of human rights violations against Serbian victims. Overall, the KLA was very successful in convincing the world that theirs was a human rights fight, not a basic land grab and not a religious conflict.

EXTERNAL SUPPORT

During the KLA's buildup to and execution of its armed insurgency, it was able to garner the critical support of several foreign governments,

---

[104] "File:Bantz J. Craddock 1999.jpg," *Wikimedia Commons*, photo by Marcus D. McAllister, accessed March 14, 2011, http://commons.wikimedia.org/wiki/File:Bantz_J._Craddock_1999.jpg.

[105] Judah, *Kosovo War and Revenge*, 156.

as well as some international criminal networks, in order to achieve its financial and operational requirements. Germany in particular was a significant political ally of the Kosovar independence movement, and it also served as a major home for the Kosovar expatriate community, with an estimated 400,000 Kosovars living there by September 1998.[106] In February 1995, Germany and Albania signed a declaration to find a "solution to the Kosovo question" by advocating for self-determination for Kosovar Albanians.[107] This joint declaration was viewed as a continuation of Germany's foreign policy that began in the early 1990s to deliberately fracture Yugoslavia, undermine the authority of Milosevic, and increase German influence in the Balkans.[108] As the KLA organization began to formalize in 1996, the German intelligence service (the *Bundesnachrichtendienst*, or BND) reportedly took an active role in the recruitment of KLA leaders from among the 500,000 Kosovars living in Albania. Moreover, although several countries would provide training and equipment to the KLA, Germany appears to be the only Western government to directly provide arms to the KLA, which caused a rift between Germany and other countries, such as the United States and the United Kingdom, that were less eager to push for an armed confrontation in Kosovo.[109] Despite its reluctance to directly arm the KLA, the United States was actively involved in the training and education of KLA fighters, even though the KLA was labeled as a "terrorist organization" by the US special envoy to Kosovo in early 1998.[110] Specifically, the Central Intelligence Agency (CIA) provided the KLA with military training manuals and operational planning guidance when the KLA was fighting the Serbian police forces and Yugoslav army. During the 1998–1999 cease-fire, CIA officers served undercover as cease-fire monitors as part of the

---

[106] Roger Faligot, "How Germany Backed the KLA," *The European*, September 21, 1998.

[107] Matthias Küntzel, "Germany and the Kosovo: How Germany's Independent Line Paved the Way to the Kosovo War," contribution to the 2nd International Hearing of the European Tribunal Concerning NATO's War Against Yugoslavia, Hamburg: April 16, 2000, www.matthiaskuentzel.de.

[108] In 1991, Germany recognized Croatia's right to self-determination. Faligot, "How Germany Backed the KLA" and Kuntzel, "Germany and Kosovo."

[109] Faligot, "How Germany Backed the KLA." A central concern for Washington and London was that fighting in Kosovo could stir up additional areas of violence in the Balkans.

[110] In February 1998, Robert Gelbard, who was at the time the US special envoy for Kosovo, stated that "We condemn very strongly terrorist actions in Kosovo. The UCK [the Albanian initials for the KLA] is, without any questions, a terrorist group." A month later, however, Gelbard appeared before the House Committee on International Relations and carefully rephrased his assessment of the KLA to state that although the KLA has committed "terrorist acts," it has "not been classified legally by the U.S. Government as a terrorist organization." See "The Kosovo Liberation Army: Does Clinton Policy Support Group with Terror, Drug Ties? From 'Terrorists' to 'Partners,'" United States Senate Republican Policy Committee, March 31, 1999, http://rpc.senate.gov/releases/1999/fr033199.htm.

Kosovo Verification Mission for the Organization for Security and Co-operation in Europe (OSCE). When they left Kosovo just before the commencement of NATO airstrikes in 1999, these CIA operatives left their satellite phones and global positioning systems with the KLA in order to facilitate the KLA's continued coordination with Washington and NATO.[111]

COUNTERMEASURES TAKEN BY THE GOVERNMENT

The primary Serbian forces that battled the KLA consisted of special police and regular and reserve members of the Yugoslav Army.[112] Initially, the suppression of the KLA was carried out by the existing police forces in Kosovo known as the Interior Ministry Police (MUP). When these local police assets were overwhelmed by the summer of 1998, the regular Army was deployed.[113] The primary operational goal of these police and military forces was to "decapitate" the KLA leadership. The goal failed, however, because of the KLA's reliance on close familial lines of trust, excellent operational security measures, and the compartmentalization of valuable leaders and their families. By preventing Serbian intelligence penetration, the KLA effectively pushed Serbs to adopt more visible and cruel methods of counterinsurgency that increased the anti-Serbian sentiment of the Kosovar population. In May 1999 at the farming village of Qyshk, Serbian paramilitary forces entered the village forcefully and publicly tortured, killed, and burned the father of Agim Ceku (KLA chief of staff) and other family members.[114] Scenes like these were repeated over and over and only served to induce more Kosovars to take up arms against the Serbs. The most damning strategic approach utilized by the Serbian regime, however, was the mass expulsions of ethnic Albanians from villages suspected of cooperating with the KLA. The images of thousands of poor civilians being expatriated by force were immediately made public and became an enormous media catastrophe for the Serbian leadership in Belgrade. Serbian forces also attacked civilians directly or used them as human shields when NATO forces entered the conflict. The most effective strategy employed by the Serbs was their attempt to cut off all KLA supply lines from Albania through the use of booby traps, shelling, and the emplacement of land mines.

---

[111] Walker and Laverty, "CIA Aided Kosovo Guerrilla Army."

[112] The Yugoslav Army (Vojska Jugoslavije, VJ) was considered a modern military force.

[113] Perritt, *Kosovo Liberation Army*, 48. By the summer of 1998, MUP numbered 11,000 and VJ numbered 12,000. These numbers grew by a factor of 50%—plus thousands of armored pieces and paramilitaries—before the commencement of NATO bombing in March 1999.

[114] Finnegan, "The Countdown," 1.

It is assumed that this strategy could have eventually choked off the KLA had NATO not commenced aerial support.

SHORT- AND LONG-TERM EFFECTS

CHANGES IN THE ENVIRONMENT

After the extensive NATO bombing campaign that began on March 24, 1999, Milosevic agreed to a peace plan on June 3, 1999. On June 10, 1999, UN Security Council Resolution 1244 was passed and it placed Kosovo under transitional administration of the UN (the UN Interim Administration Mission in Kosovo, UNMIK) while also authorizing the deployment of a NATO-led peacekeeping force into Kosovo (KFOR).[115] Although the exact number of casualties from the Kosovo war remains uncertain, a detailed listing of victims published by the Humanitarian Law Center in 2008 showed 13,472 people killed in Kosovo during the period of January 1998 to December 2000. Of these victims, 9,260 were Albanian, 2,488 were Serbian, and the remaining 1,254 were of undetermined ethnicity.[116] With the departure of Serbian military and security forces in June 1999 and the arrival of NATO-led peacekeeping forces, a major demographic shift occurred again in Kosovo; approximately 200,000–280,000 Serbs reportedly left Kosovo and many of the 850,000 Kosovars who fled Kosovo in early 1999 returned.[117]

CHANGES IN GOVERNMENT

Under UN Security Council Resolution 1244, Kosovo was granted autonomy as a province within Serbia until its final status could be determined. UNMIK established four main "pillars" for its operations in Kosovo: interim civil administration (led by the UN), humanitarian affairs (led by the UN High Commissioner for Refugees), reconstruction (led by the European Union), and institution building (led by the OSCE).[118] Milosevic remained the president of Yugoslavia until the presidential elections of 2000, by which time he had been charged with crimes against humanity, genocide, embezzlement, and

---

[115] Central Intelligence Agency, "Kosovo," *The World Factbook.*

[116] "Kosovo Memory Book," Human Law Center, 2008, http://www.hlc-rdc.org/FHPKosovo/KOSOVO-KNJIGA-PAMCENJA/index.1.en.html.

[117] Alvaro Gil-Robles, "Kosovo: The Human Rights Situation and the Fate of Persons Displaced From Their Homes," Council of Europe, Office of the Commission for Human Rights (Strasburg, Germany: October 16, 2002).

[118] Gil-Robles, "Kosovo: The Human Rights Situation," 7.

other war crimes. He spent five years on trial at The Hague before dying there of a heart attack before jury deliberations. Elections were held in November 2001 for the 120 seats of the provisional parliament of Kosovo, and in March 2002, the Provisional Institutions for Democratic and Autonomous Self-Government (PISG) took effect in Kosovo, establishing ministerial positions and structures for Kosovo until the determination of its final status.[119] In late 2005, the UN-led process to determine Kosovo's future began and continued through 2006 and 2007, but the negotiations did not result in an agreement between the parties in Pristina and Belgrade. Finally, on February 17, 2008, the Kosovo Assembly formally declared that Kosovo was an independent state. Although Serbia rejected that declaration of independence, over the next two years approximately sixty countries recognized Kosovo as an independent country, and in July 2010, the International Court of Justice ruled that Kosovo's declaration did not violate international law.[120]

Figure 8. Ethnic composition of Kosovo in 2005.[121]

---

[119] Ibid., 7–8.

[120] Central Intelligence Agency, "Kosovo," *The World Factbook*.

[121] "File:Kosovo ethnic 2005.png," *Wikimedia Commons*, accessed March 14, 2011, http://commons.wikimedia.org/wiki/File:Kosovo_ethnic_2005.png.

CHANGES IN POLICY

As an independent state, Kosovo maintains a republic form of government. Within its national assembly, one hundred seats are directly elected, ten seats are guaranteed for ethnic Serbs, and another ten seats are reserved for other ethnic minorities. However, Kosovo Serbs have protested the unilateral declaration of Kosovo independence and have undertaken a campaign to take control of key municipal institutions in regions of Kosovo where they are in the majority. This has resulted in Serbian control of local police, customs stations, judicial systems, and municipalities in the region north of the Ibar River.[122]

CHANGES IN THE REVOLUTIONARY MOVEMENT

With the passage of UN Security Council Resolution 1244 and the entrance of NATO-led peacekeeping forces in June 1999, the KLA movement was disbanded but reorganized into the Kosovo Protection Corps (KPC), which operated under KFOR and had the stated purpose of providing internal defense and supporting emergency relief within Kosovo. Composed primarily of former KLA combatants and organized along military lines, to include special forces, the KPC was accused of numerous cases of human rights violations during the exodus of Kosovo Serbs in late 1999.[123]

Although the formal structure of the KLA ceased to exist in 1999, its legacy and political influence remained strong in Kosovo more than a decade later. General Agim Ceku (former KLA military chief) became the prime minister of Kosovo, followed by Hashim Thaçi (former KLA political leader) and Ramush Haradinaj (former KLA military commander). Hashim Thaçi would become the leader of the Democratic Party of Kosovo (PDK) and in 2008 served as prime minister of Kosovo for a second time. Despite their political fortunes, all three former KLA commanders were considered for prosecution in The Hague for war crimes, but only Haradinaj would be indicted—and eventually acquitted—by the tribunal.[124] In addition, Thaçi, Ceku, and other former KLA leaders continued to face accusations of

---

[122] International Crisis Group, "Kosovo's Fragile Transition," Europe Report no. 196 (September 25, 2008).

[123] "Under Orders," 559, fn54.

[124] "Kosovo ex-PM War Charges Revealed," *BBC News*, March 10, 2005, http://news.bbc.co.uk/2/hi/europe/4337085.stm.

continued direct involvement in organized crime even after the KLA movement was over.[125]

BIBLIOGRAPHY

BIA/Security Information Agency. "Albanian Terrorism and Organized Crime in Kosovo and Metohija." September 2003. http://www.kosovo.net/albterrorism.html.

Chossudovsky, Michel. "Kosovo: The US and the EU Support a Political Process Linked to Organized Crime." Center for Research on Globalization, February 12, 2008. http://www.globalresearch.ca/index.php?context=va&aid=8055.

Coll, Alberto. "Kosovo and the Moral Burdens of Powers." In *War over Kosovo: Politics and Strategy in a Global Age*, edited by Andrew Bacevich and Eliot Cohen. New York: Columbia University Press, 2001.

Faligot, Roger. "How Germany Backed the KLA." *The European*, September 21, 1998.

Finnegan, W. "Letter from Kosovo: The Countdown." The New Yorker, October 15, 2007.

"The Kosovo Liberation Army: Does Clinton Policy Support Group with Terror, Drug Ties? From 'Terrorists' to 'Partners.'" United States Senate Republican Policy Committee, March 31, 1999. http://rpc.senate.gov/releases/1999/fr033199.htm

Gil-Robles, Alvaro. "Kosovo: The Human Rights Situation and the Fate of Persons Displaced From Their Homes." Council of Europe, Office of the Commission for Human Rights, Strasburg, Germany. October 16, 2002.

Hayden, William. "The Kosovo Conflict and Forced Migration: The Strategic Use of Displacement and the Obstacles to International Protection." *Journal of Humanitarian Assistance*, February 14, 1999. http://jha.ac/articles/a039.htm.

International Crisis Group. "Kosovo's Fragile Transition." Europe Report no. 196. September 25, 2008.

International Crisis Group. "Religion in Kosovo." Balkans Report no. 105. January 31, 2001.

---

[125] For a detailed listing of media reports on this subject, see "Kosovo Liberation Army: Freedom Fighters or . . . Truth in Facts and Testimonies" at http://www.kosovo.net/kla3.html, and Michel Chossudovsky, "Kosovo: The US and the EU Support a Political Process Linked to Organized Crime," Center for Research on Globalization, February 12, 2008, http://www.globalresearch.ca/index.php?context=va&aid=8055.

Judah, Tim. "The Growing Pains of the Kosovo Liberation Army." In *The Politics of Delusion*, edited by Michael Waller, Kyril Drezov, and Bulent Gokay. Portland, OR: Frank Cass Publishers, 2001.

Judah, Tim. *What Everyone Needs to Know*. New York: Oxford University Press, 2008.

"Kosovo Independence Move Not Illegal, Says UN Court." *BBC News*, July 22, 2010. http://www.bbc.co.uk/news/world-europe-10730573.

"Kosovo Liberation Army: Freedom Fighters or . . . Truth in Facts and Testimonies," http://www.kosovo.net/kla3.html.

"Kosovo Memory Book." Human Law Center, 2008. http://www.hlc-rdc.org/FHPKosovo/KOSOVO-KNJIGA-PAMCENJA/index.1.en.html.

Küntzel, Matthias. "Germany and the Kosovo: How Germany's Independent Line Paved the Way to the Kosovo War." Contribution to the 2nd International Hearing of the European Tribunal Concerning NATO's War Against Yugoslavia, Hamburg, Germany. April 16, 2000. www.matthiaskuentzel.de.

Lebor, Adam. *Milosevic: A Biography*. New Haven, CT: Yale University Press, 2002.

Leurdijk, Dick, and Dick Zandee. *Kosovo: From Crisis to Crisis*. Burlington, VT: Ashgate Publishing, 2001.

Malcolm, Noel. *Kosovo: A Short History*. New York: New York University Press, 1998.

Perritt, H. H. *Kosovo Liberation Army: The Inside Story of an Insurgency*. Champaign, Illinois: University of Illinois Press, 2008.

Simonen, Katariina. "Operation Allied Force: A Case of Humanitarian Intervention?" Athena Papers Series, Partnership for Peace, Consortium of Defense Academies and Security Studies Institutes. 2004.

Taylor, S. *Regardless of the Consequences: Images of Serbia & the Kosovo Conflict*. Ottawa, Canada: Esprit de Corps Books, 2000.

"Under Orders: War Crimes in Kosovo." Human Rights Watch. 2001. http://www.hrw.org/reports/2001/kosovo/.

Walker, Tom, and Aidan Laverty. "CIA Aided Kosovo Guerrilla Army." *The Sunday Times*, March 12, 2000.

World Bank Group. "Kosovo Economic History in a Nutshell" (Report based on Essay by Professor Mustafa and News Agency Reports, 2001). *Beyond Transition Newsletter*. http://web.archive.org/web/20110509045917/http://www.worldbank.org/html/prddr/trans/marapr98/boxpg19.htm.

THE PROVISIONAL IRISH REPUBLICAN ARMY (PIRA): 1969–2001

Chuck Crossett and Summer Newton

SYNOPSIS

Waves of violence broke out in Northern Ireland in 1969 as the minority Catholic population marched in civil rights parades, calling for more equitable rights from the Protestant-dominated government. Catholic and Protestant mobs attacked each other across Belfast and Derry neighborhoods, and the British Army was finally called in to regain control. As the violence continued to grow, a new offshoot of the Irish Republican Army (IRA) stood up to protect the Catholics from Protestant attacks and to drive the British back to England. The new Provisional IRA (PIRA), just like the IRA from which it separated, intended to fight until Northern Ireland was allowed to merge with the independent Republic of Ireland. The Provisionals undertook a three-decade-long campaign of sniping, assassinations, and bombings across Northern Ireland and England that included waves of indiscriminate attacks, cease-fires, leadership changes, and shifting tactics. After the Provisionals modified their strategy to place an increased emphasis on political diplomacy, the two sides agreed to a power-sharing arrangement, self-determination, and disarmament.

TIMELINE

1912–1921	The IRA conducts campaign for Home Rule.
1921	Britain establishes the Irish Free State (Republic of Ireland).
1956–1962	The IRA conducts the Border Campaign.
October 5, 1968	The beginning of "the Troubles." A civil rights protest in Derry ends with Protestant and Royal Ulster Constabulary (RUC) attacks on the marchers.
January 1969	Belfast-to-Derry march ends with an attack at Burntollet Bridge outside Derry. Thirteen marchers go to the hospital. "Free Derry" is created.
August 12–14, 1969	Battle of the Bogside in Derry after a march by the Apprentice Boys.

August 14, 1969	British troops are deployed on the streets of Northern Ireland.
December 1969	The PIRA splits from the "Official" IRA.
August 9, 1971	Operation Demetrius begins "internment without trial."
January 30, 1972	Bloody Sunday in Derry. Fourteen civilians are killed by British troops.
March 4, 1972	PIRA bombing in Belfast. No warning; bombing kills two and injures 130.
July 1972	Secret talks take place between PIRA and the Secretary of State.
July 21, 1972	"Bloody Friday"—PIRA explodes twenty-two bombs across Belfast in 75 minutes. Nine are killed and 130 are injured.
July 1972	Operation Motorman is begun by the British Army and results in clearing "no-go zones" in Derry and Belfast.
March 1973	Interdiction of the *Claudia* carrying weapons from Libya.
February 1975	PIRA declares cease-fire.
September 1976	Blanket protests begin at Maze prison.
May 5, 1981	Bobby Sands is the first of ten hunger strikers to die.
October 12, 1984	Bombing of the Grand Hotel, located in Brighton, targets Prime Minister (PM) Margaret Thatcher.
June 15, 1996	Manchester bombing injures 200. Largest bombing in the United Kingdom since World War II.
April 10, 1998	Good Friday Agreement is signed.

THE ENVIRONMENT OF THE REVOLUTION

PHYSICAL ENVIRONMENT

Northern Ireland is located in the uppermost right corner of the island of Ireland. The total area of the country is 5,456 square miles, and the country has 232 miles of coastline. The country is divided into six counties: Antrim, Armagh, Derry, Down, Fermanagh, and Tyrone. Belfast is the capital and the largest city. Four other locations have been designated cities: Armagh, Derry, Lisburn, and Newry. The primary rivers in the country are the River Foyle and the Upper and Lower Bann. Northern Ireland also has several mountain ranges of significance—the Sperrin Mountains in the northwest, the Antrim Plateau in the northeast, and the Mourne Mountains in the southeast. The country's climate is influenced by the warm waters of the Gulf Stream, permitting a mild, temperate climate in the region.

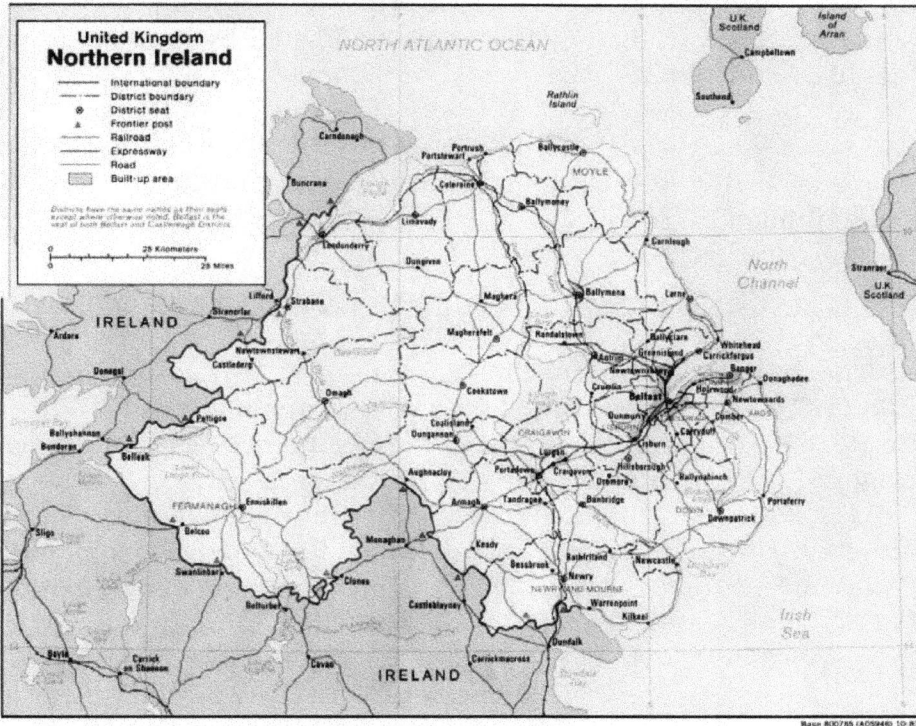

Figure 1. Map of Northern Ireland.[1]

The conflict in Northern Ireland was first and foremost an urban one. The majority of conflict fatalities occurred in urban areas, partially because of the high (65%) urban population.[2] Belfast experienced the brunt of the conflict, with other urban areas, especially Derry, the second largest city in Northern Ireland after Belfast, seeing significant

---

[1] Perry-Castañeda Library Map Collection, "Northern Ireland (U.K) (Political) 1987 Map," The University of Texas at Austin, accessed March 13, 2011, http://www.lib.utexas. edu/maps/europe/northern_ireland_pol87.jpg.

[2] Northern Ireland Statistics Research Agency, *Theme Tables. 1. Demography: People, Family, and Households* (Belfast, Northern Ireland: Her Majesty's Stationary Office, 2001); Northern Ireland Statistics and Research Agency, *Northern Ireland Census for 2001: Key Statistics for Settlements* (Belfast, Northern Ireland: Her Majesty's Stationary Office, 2005).

action as well.[3] The Belfast Urban Area (BUA), the eleventh most populated urban area in the United Kingdom, has a population of 483,418 living in a 62.42-square-mile area, resulting in a population density of 7,745 people per square mile.[4] At the time, the population of Belfast accounted for 29.4% of the overall population of Northern Ireland. Between 1968 and 1993, 54.5% of the fatalities resulting from the insurgency occurred in Belfast, primarily in the northern and western sections of the city that were dominated by Protestant and Catholic populations, respectively.[5] Conflict-related fatalities occurred at a rate of 2.95 per person in the city of Belfast but only at a rate of 1.02 per person throughout the rest of the country, with most of the deaths occurring in the highly urbanized areas and the central city and its immediate suburbs.[6]

CULTURAL AND DEMOGRAPHIC ENVIRONMENT

The 2001 census places the population of Northern Ireland at 1,685,267.[7] The largest city, Belfast, has a population of 277,391. Derry, the second largest city, has a population of 83,562.[8] Catholics comprise 44% of the overall population of Northern Ireland at 737,412, and Protestants account for 53% of the population at 895,277. Although Northern Ireland was partitioned in order to maintain a Protestant majority, throughout the 25 years of the conflict, neither the Catholics nor the Protestants comprised a significant majority of the population. At the time of the 1971 census, several years after

---

[3] According to Malcolm Sutton's *An Index of Deaths from the Conflict in Ireland, 1969–1993* (2001), 217 people were killed in Derry as a result of the conflict with an additional 127 killed in the county of Derry. See the database on the CAIN Web Service, accessed October 27, 2009, http://cain.ulst.ac.uk/sutton/index.html.

[4] Graham Pointer, "The UK's Major Urban Areas," in *Focus on People and Migration 2005*, ed. Office of National Statistics, United Kingdom (London: Office of National Statistics, 2005).

[5] Sutton's *An Index of Deaths from the Conflict in Ireland, 1969–1993* (Belfast: Beyond the Pale Publications, 2001) lists different figures. Of the 3,526 killed, 1,541 are from Belfast, or 44%. The breakdown is as follows: Belfast East, 128; Belfast North, 577; Belfast South, 213; Belfast West, 623. See the database on the CAIN Web Service, accessed October 27, 2009, http://cain.ulst.ac.uk/sutton/index.html.

[6] Paul Doherty and Michael A. Poole, "Ethnic Residential Segregation in Belfast, Northern Ireland, 1971-1991," *Geographical Review* 87, no. 4 (1997): 520–536.

[7] The Northern Ireland Statistics and Research Agency (NISRA) conducts a census every ten years.

[8] Northern Ireland Statistics and Research Agency, *Northern Ireland Census for 2001: Key Statistics* (Belfast, Northern Ireland: Her Majesty's Stationary Office, 2005). When the surrounding suburbs are included, the population of the entire Belfast region (the Belfast Urban Area) is 483,418.

the civil rights disturbances of the late 1960s, Catholics accounted for 36.8% of the population.[9]

Northern Ireland, along with the Republic of Ireland, has consistently been among the most religious European countries. When most European nations were becoming increasingly secular, religious commitment and the authority of the Church increased in Ireland, especially in working-class Catholic communities at the onset of the Troubles in the 1970s. Many of the Provisionals were themselves reasonably devout and some notably so. One Provisional is noted for having refused to allow contraceptives into Northern Ireland even though they were useful for constructing explosives. His social conservatism led him to remark that he would rather be caught with a gun in his pocket than contraceptives.[10] The authority of the Church began to diminish only in the 1980s during the H-Block prison protests and the hunger strike. Secularizing forces, long having passed over Northern Ireland, increased in the 1990s as religious practice in the country dwindled. Compared with the rest of Europe, however, Northern Ireland remained among the most Christian and most supportive of institutionalized religious activities.[11]

SOCIOECONOMIC ENVIRONMENT

During the post-World War II period, after experiencing an initial economic boom, Northern Ireland slid into economic recession as the staple industries in the country withered. The linen and shipbuilding industries, as well as agriculture, provided the majority of employment opportunities for the workforce.[12] However, because of increased global competition, modernization, and downshifting in wartime production, the powerhouse industries of Northern Ireland downsized significantly, leading to widespread unemployment.[13]

---

[9] Paul A. Compton, *Northern Ireland: A Census Atlas* (Dublin, Ireland: Gill and Macmillan, 1978), 6, 169.

[10] Richard English, *Armed Struggle: The History of the IRA* (New York: Oxford University Press, 2003), 486.

[11] Tony Fahey, Bernadette Hayes, and Richard Sinnott, *Conflict and Consensus: A Study of Values and Attitudes in the Republic of Ireland and Northern Ireland* (Leiden, The Netherlands: Brill, 2006).

[12] Of the total workforce employed in the manufacturing sector, 30% were employed in textiles and linen while 20% were employed in the shipbuilding industry. Approximately one-sixth of the entire workforce was employed in agriculture. Paul Bew, Peter Gibbon, and Henry Patterson, *The State in Northern Ireland, 1921-72: Political Forces and Social Classes* (New York: St. Martin's Press, 1979), 231.

[13] Ibid., 134–135. Between 1950 and 1960, agricultural employment dropped by a third. Employment in the shipbuilding industry declined 22% from its post-war peak in the early 1960s. Similarly, employment in the linen industry fell by 40%.

In response, Stormont—the parliament of Northern Ireland—
implemented welfare programs similar to those in Great Britain.
Significant increases in health, education, and social service
expenditures expanded the public sector, thereby offsetting many
of the employment losses in the private sectors. Stormont also took
successful measures to attract foreign investment and reinvigorate
industrial production in Northern Ireland. Catholics, however, did
not benefit as much from these new employment opportunities in
the private sector, although they did benefit from increased spending
on education, which improved the quality and availability of Catholic
education.[14] These and other measures taken by Stormont helped to
mitigate, but not eliminate, disparities between the two communities.
Moreover, Stormont's decisions to close railway and ferry services to
Derry, which had a Catholic majority; move a proposed university from
Derry to Coleraine, a Protestant stronghold; and increase industrial
expansion in Protestant areas all combined to amplify Catholic
perceptions of discrimination.

In the 1960s, inspired in part by the civil rights movement in
the United States and elsewhere, Catholics, and occasionally liberal
Protestants, initiated civil rights campaigns seeking equal treatment.
Previous Catholic mobilization was primarily confined to the IRA,
which sought to dismantle the state rather than seek reforms within
it. Those involved in the civil rights movement, however, increasingly
saw their future in the context of a Northern Irish state rather than
in the context of a united Ireland as espoused in the IRA ideology.
Local committees agitating for reform in housing and employment
opportunities were prevalent, especially in Derry. The Cameron
Report, an investigation launched by British authorities after the civil
rights disturbances of 1968 and 1969, cited employment and housing
inequalities, rather than political or constitutional matters, as the most
pressing concerns for those involved in the civil rights campaign.[15]

Disparities in employment rates among Catholics and Protestants
manifested in several ways. At the onset of the Troubles, Catholics
were far more likely to be unemployed than Protestants, and for those

---

[14] From the onset of partition, Catholics preferred to send their children to voluntary
schools administered by Catholics rather than have them attend the state Protestant-
dominated schooling system. As a result, Northern Ireland had a largely self-segregated
primary and secondary schooling system until the 1980s. The state also did not provide
full funding for Catholic schools until 1992. See John Darby and Minority Rights Group,
Scorpions in a Bottle: Conflicting Cultures in Northern Ireland (London: Minority Rights
Publications, 1997), 83–85.

[15] Cameron Report, *Disturbances in Northern Ireland: Report of the Commission Appointed
by the Governor of Northern Ireland* (Belfast, Northern Ireland: Her Majesty's Stationary Office,
1969).

Catholics who were employed, social mobility was limited.[16] According to the 1971 census data, the unemployment rate for Catholics was 13.9% compared with 5.6% for Protestants.[17] Not surprisingly, the Campaign for Social Justice, doing a great deal of the research that supported activists' claims, dedicated a significant portion of its comprehensive pamphlet, *The Plain Truth*, to the issue of employment discrimination. The Unionist Regime, according to the pamphlet, was making a concerted attempt to keep Catholics as "second-class citizens," making them the "white negroes of Ulster."[18] As a result, Catholics were more likely to be unemployed, or "on the dole," than unionists, which produced the net effect of Catholic emigration from Northern Ireland. This, according to *The Plain Truth*, would lead to a balancing of the existing population growth disparity that was caused by the higher Catholic birth rate.

HISTORICAL FACTORS

After a sustained campaign by the IRA in 1912–1921 for Home Rule and following negotiations with Sinn Féin,[19] the IRA's political wing, Britain established the Irish Free State (later to become the Republic of Ireland), which Northern Ireland opted out of in 1921.[20] The decision to limit the North to six counties, and the choice of the counties themselves, was an explicit attempt by the statesmen of Northern Ireland and Britain to ensure a unionist[21] majority in the region. Ulster, one of the four provinces of Ireland, traditionally comprised nine counties, which at the time of partition held 900,000 Protestants, most of whom supported the British connection, and 700,000 Catholics, most of whom wished to separate from Britain. These figures are in comparison with those of the six counties that were to become Northern Ireland; those counties were home to

---

[16] John Whyte, "How Much Discrimination Was There Under the Unionist Regime, 1921 - 1968?" in *Contemporary Irish Studies*, ed. Tom Gallagher and James O'Connell (Manchester, UK: Manchester University Press, 1983).

[17] Northern Ireland Population Census, 1971.

[18] Campaign for Social Justice, *Northern Ireland: The Plain Truth*, 2nd ed. (Dungannon, Northern Ireland: Campaign for Social Justice, 1969).

[19] After the Treaty, Sinn Féin overtook the IPP as the most electorally successful party in Ireland.

[20] The two-state solution was proposed by the Government of Ireland Act in 1920.

[21] "Unionists" were supporters of the Union with Britain. They stem from the opponents of the Home Rule movement beginning in the nineteenth century and subsequently settled for the partitioned Northern Irish state. The Ulster Unionist Party (UUP) is the most electorally successful, and moderate, of the unionist political parties. The UUP formed all governments from 1921 to 1972. The Democratic Unionist Party (DUP) is the more populist and virulently antinationalist of the unionist parties and was founded by the controversial Rev. Ian Paisley.

820,000 Protestants and 430,000 Catholics. Many in the Catholic community therefore refused to engage in any activity that lent support or legitimacy to the new state. In some instances, Catholic teachers refused salaries and even sent children to schools in Dublin for their examinations.[22]

The IRA was active in the interim period between the Treaty and the 1969 split with the Provisionals, but their support among nationalists began to wane considerably in the post-World War II period. Britain implemented extensive welfare state reforms both at home and in Northern Ireland after the cessation of the war, which precipitated a drop in popular support for militant republicanism. In 1956, the IRA launched its Border Campaign but, because of a pronounced lack of public support, halted the operation while also maintaining that it still had the military capacity to continue the operation indefinitely if necessary.[23] This disconnect between the IRA's emphasis on armed resistance and the population's need for economic and social relief highlighted a major weakness of the IRA. Before 1962, the IRA's strategy focused on military action to achieve an "independent, united, democratic, Irish republic," but the strategy failed to address any widespread economic or social reforms.

Following the failed border campaign of 1956–1962, however, Cathal Goulding, a charismatic radical nationalist with an impressive republican pedigree, assumed leadership of the IRA. He inherited an organization that was ideologically and logistically weakened. Funding, arms supplies, and recruits to replace those imprisoned and killed were all in short order. Some significant "re-thinking" about the IRA's strategy and operations was therefore in order.[24] Goulding shifted the IRA's emphasis from a military strategy to one based on reform through the political process and nonviolent opposition, a remarkable shift considering the republicans' traditional antipathy toward the state. The shift was not just rhetorical—the IRA sold a goodly portion of its arms to a Welsh nationalist group.[25] Those arms that remained were few and nearly obsolete.

The bridge between the IRA and the civil rights movement was realized through the creation of the Wolfe Tone societies, which were formed and manned in large part by the IRA in an effort to reeducate the populace in the aftermath of the failed Border Campaign. The Wolfe Tone intellectuals melded socialist theory with republican objectives, hoping to improve the quality of life for

---

[22] Darby and Minority Rights Group, *Scorpions in a Bottle*, 28.
[23] English, *Armed Struggle*, 486.
[24] Ibid.
[25] Ibid.

Catholics by creating a national liberation front that united working-class Protestants and Catholics while simultaneously undermining Irish capitalism and partition.

The IRA did draw up several military plans during the 1960s, and certainly Goulding himself was not opposed to the use of force as he "continued to believe in its appropriateness." But Goulding and the IRA recognized that the times required transformation of the IRA, and they were convinced that the left would "sweep the world." In 1966, the IRA Army Council met and developed plans for a renewed northern onslaught. Between 1962 and 1969, membership increased, especially in Belfast, from 650 to 1,000. These military rumblings of the IRA during the 1960s were, in part, strategic moves by Goulding to prevent alienating members more comfortable with traditional notions of republican force. Nonetheless, the IRA training increasingly focused on leftist republican struggles and was noticeably slim on military training.[26] This transformation of the IRA under Goulding, combined with the violent responses to the civil rights movement seen from police forces and loyalist supporters, played a prominent role in the split of the Provisionals from the IRA in December 1969.

GOVERNING ENVIRONMENT

Though intended to resolve the communal differences between the north and south of Ireland, the establishment of Northern Ireland actually exacerbated existing divisions among the majority Protestant and minority Catholic communities.[27] After the partition of Northern Ireland in 1921, the state was largely autonomous from Britain but did send its own members of parliament (MPs) to Westminster. In Belfast, Northern Ireland maintained its own bi-cameral parliament, which was often referred to as "Stormont" because it was located in the Stormont area of the city. Stormont had discretionary authority over policing, education, local government, and social services, with local government divided into numerous local councils. Britain retained ultimate executive authority, however, with the power to suspend the legislative and executive powers of the Northern Irish government, which it would do in 1972. It was not until the signing of the Good Friday Agreement in 1998 that the slow process of devolving government to Northern Ireland commenced. Until that time, Britain administered Northern Ireland through a secretary of state who handled all

---

26 Ibid.

27 Darby and Minority Rights Group, *Scorpions in a Bottle*, 242.

executive and legislative affairs. Political, constitutional, and security issues were administered through a Northern Ireland Office.[28]

The boundaries of Northern Ireland were selected during partition to ensure a Protestant and unionist majority. After partition, Protestant power became entrenched in nearly all of the institutions of the state, including the judiciary, security, and upper levels of the civil service,[29] in part because of the Catholics' refusal to participate in—and thereby legitimize—the state. In addition, through discriminatory franchises and electoral gerrymandering, the unionists were able to dominate local governing authorities despite Catholic majorities in some areas. Moreover, legislation that provided the state with broad discretionary powers to maintain "peace and order," as well as a Protestant-heavy security force, contributed significantly to increased sectarian tensions and the emergence of the PIRA.

However, movement toward a decrease in republican violence and sectarian tension seemed imminent at the onset of the 1960s. PM Lord Brookeborough,[30] thought by many Catholics to be the quintessential example of unionist control, retired in 1963 and was replaced by the more benign Captain Terence O'Neill. Although the extent to which O'Neill sought to combat sectarianism is debated,[31] on several occasions, he appeared to support an end to discrimination of Catholics and to "build bridges between the two traditions."[32] O'Neill, breaking with antipartitionist tradition, also re-established friendly relations with the south, meeting with Ireland's PM Seán LeMass on several occasions. The meetings also precipitated the agreement of the Nationalist Party of Northern Ireland, for the first time, to cease their policy of absentionism and join Stormont as the official opposition party.

PM O'Neill's hesitant steps at reform met with increased demands from the Catholic community and the defense of the status quo by unionists. Whereas moderate Protestants (or "Orangemen") at least hoped the reforms might ease sectarian tensions, their more hard-line counterparts feared a sellout by O'Neill and viewed the dealings

---

[28] English, *Armed Struggle*, 486.

[29] David John Smith and Gerald Chambers, *Inequality in Northern Ireland* (New York: Clarendon, 1991), 369.

[30] In 1956, PM Brookeborough issued a white paper detailing his staunch defense of the partition and maintained that the partition resulted from the will of the majority of Northern Irish people and its maintenance was necessary because of the "ideological gulf between the two peoples," referring to the Irish and the Northern Irish. See The Northern Ireland Government, *Why the Border Must Be: The Northern Irish Case in Brief* (Belfast, Northern Ireland: The Northern Ireland Government, 1956).

[31] English, *Armed Struggle*, 486.

[32] As quoted in Darby and Minority Rights Group, *Scorpions in a Bottle*, 31.

with "Papists"[33] as "tantamount to treason." Among the most vocal hard-liner was Rev. Ian Paisley, founder of the Democratic Unionist Party (DUP), who reportedly "castigated O'Neill in the language of a seventeenth-century Anabaptist."[34] More ominously, a newly reformed Ulster Volunteer Force emerged that was ready to defend Ulster against the "inroads of liberalism" and that was responsible for the subsequent shooting of two men in Belfast for the crime of being Catholic.[35] After riots in Belfast and Derry that left seven post offices set on fire and a hundred casualties, respectively, O'Neill resigned, having lost credibility and the support of his Unionist Party.[36] He was replaced by Major James Chichester-Clark in April 1969. Although mostly in line with O'Neill's, PM Chichester-Clark's policies were formed without the burden of "old grudges and past feuds."[37] Despite the promises of reform and moderation, it became apparent that no Unionist government could acquiesce to the demands of the civil rights activists quickly enough and certainly not without raising the hackles of militant Orangemen.[38] Therefore, there existed little space for compromise in the tense Northern Irish atmosphere at the onset of the 1970s.

After extensive civil rights disturbances in 1968 and 1969, the Northern Irish government suggested a number of reforms. The Cameron Report called for a "comprehensive reform and modernization of the local government structure," the abolishment of the limited franchise, and review of the finances and structures of local governing bodies.[39] Already under discussion as early as 1966, substantial local government reforms were implemented in 1972. The Local Government (Northern Ireland) Act of 1972 abolished the existing local authorities and divided the country into twenty-six local government districts. In one of largest concessions of Stormont to the civil rights disturbances, Londonderry Corporation, noted for its egregious discriminatory practices that led to severe civil unrest in Derry, was abolished and replaced with the Londonderry Development Commission in 1969.[40] However, when this new Derry

---

[33] Papist, or Papalist, is a term for a Roman Catholic or something associated with the Roman Catholic Church.

[34] J. Bowyer Bell, *The Secret Army: The IRA* (New Brunswick, NJ: Transaction Publishers, 1997), 348.

[35] Ibid., 348.

[36] In the election that preceded his resignation, O'Neill received 7,745 votes in his own Bannside district to 6,331 votes for Paisley. A few years later, Paisley would win the elected seat previously held by O'Neill. Ibid.

[37] Ibid.

[38] Ibid.

[39] Cameron Report, *Disturbances in Northern Ireland.*

[40] Niall O'Dochartaigh, *From Civil Rights to Armalites: Derry and the Birth of the Irish Troubles* (Cork, Ireland: Cork University Press, 1997), 104.

council was finally elected in 1973, it had been stripped of many of its powers, suggesting that the "greatest change was perhaps symbolic."[41]

WEAKNESSES OF THE PREREVOLUTIONARY ENVIRONMENT AND CATALYSTS

Despite a nationwide poll in December 1966 that showed that 43% of the population supported the legal abolishment of discrimination, the Northern Ireland Civil Rights Association (NICRA) had difficulty tapping into that sentiment. It was not until 1968, when a Catholic family was denied housing by the local council in Caledon in favor of an unmarried Protestant woman, that the NICRA had its first opportunity for a large-scale march, which passed without incident. Later that year, a march was planned to protest the discriminatory housing, unemployment, and electoral policies of the Londonderry Corporation. Although banned by the Home Affairs Minister William Craig,[42] the march through Derry went ahead and was intended to end in a Protestant stronghold. Estimates differ as to the number of participants, but approximately 400 people lined up to march and another 250 spectators watched from the sidelines.[43] Among the marchers were three British Labour MPs, Gerry Fitt, the Republican Labour MP, several Stormont MPs, and members of the media. Before the protest was well underway, the marchers were ambushed by loyalist supporters as well as the local constables who batoned the crowd, including MP Fitt. This attack served to further mobilize civil rights supporters in Northern Ireland and abroad. In the weeks after the march, Derry and Belfast saw thousands of Catholics, and a number of unionist-weary Protestants, take to the streets in support of the much-beleaguered Catholic cause. Several weeks after the October 5 march, a commission was appointed by PM O'Neill to investigate the grievances of the Catholic community and the events that transpired at the march.

Overtures by the Unionist government after the October 1968 march led to a brief respite, but this was broken a few months later by the Burntollet march, which was organized by the radical student's movement. The march, a four-day journey from Belfast to Derry, took place in early January 1969. The RUC did little to prevent the attacks or to protect the protestors.[44] The most egregious violence occurred just outside of Derry at the Burntollet Bridge, where the

---

[41] Ibid., 101.

[42] Bell, *The Secret Army*, 702.

[43] See the chronology of events surrounding the Derry march at the CAIN Web Service, accessed November 9, 2009, http://cain.ulst.ac.uk/events/derry/chron.htm.

[44] Cameron Report, *Disturbances in Northern Ireland*.

marchers were ambushed by loyalist mobs, approximately 200 strong, including several off-duty members of the reserve security squadron, the B-Specials; the attacks resulted in thirteen marchers going to the hospital.[45] Members of the RUC also rampaged though the Catholic Bogside, smashing windows and intimidating people.[46] The riots led to the first "Free Derry," or the barricading of sections of Catholic Derry, as a measure to stop the RUC and create "no-go" areas.

Increasingly isolated within his own party, having never achieved any significant reforms, failing to crush the demonstrations, and not reassuring to the conservatives, O'Neill resigned in April 1969.[47] Hard-liner Rev. Ian Paisley subsequently won O'Neill's seat in Parliament, and Chichester-Clark became the new PM. Although Chichester-Clark's policies were not significantly different than O'Neill's, he was not haunted or burdened by the past as was O'Neill's administration. Before leaving office, O'Neill proposed a measure, the Public Order Bill, that prohibited or restricted the tactics that could be used by the civil rights movement, such as sit-ins and demonstrations. The bill passed in Parliament but with strenuous objections from the new activist MPs John Hume and Ivan Cooper. In protest against the bill, as well as the failure to initiate a "one man, one vote" franchise, the activists took to the streets again. Riots ensued when a march re-traced the October 5 route. The unrest was so severe that some moderates, including Cooper, sought to put an end to the marches because they were exacerbating already heated sectarian tensions. More marches, and more riots, in both Derry and Belfast followed and ended in what some reference as the first death of the Troubles. The RUC, chasing rioters in the Bogside, entered the home of Sam Devenney, an innocent man with no activist connections. The RUC officers beat Devenney severely enough that several months later, he died of his injuries.

Many anticipated an upswing in violence when the Apprentice Boys of Derry,[48] 15,000 strong, were scheduled to march through the streets of restive Derry on August 12, 1969. The day after Devenney's funeral in July, Derry republicans established the Derry Citizens' Defence Association (DCDA) to defend Catholic neighborhoods from marauding security forces and loyalist mobs during the August

---

[45] See the Summary of the People's Democracy March on the CAIN Web Service, accessed on November 16, 2009, http://cain.ulst.ac.uk/events/pdmarch/sum.htm.

[46] Bell, *The Secret Army*, 702.

[47] Ibid.

[48] The Apprentice Boys of Derry is a Protestant society that holds annual gatherings to remember the 1689 siege of Derry by King James II of England (a Catholic). The society has worldwide membership.

parade.[49] In a meeting between the DCDA and Cathal Goulding, Chief of Staff for the IRA, the DCDA requested IRA protection for the Bogside neighborhood. Goulding replied that he had neither the guns nor the men necessary for the job.[50] The DCDA therefore prepared for the worst, setting up barricades in Derry the night before the march, and "tens of thousands" of empty milk bottles were gathered for making petrol bombs.[51]

As anticipated, violence erupted during the march of the Apprentice Boys of Derry. After three days of rioting in and around the Bogside, British troops had to march into Derry to restore order. Because the RUC could not hold out on their own, the B-Specials of the RUC were activated. When the B-Specials descended on the Fountain neighborhood, they joined the Protestant crowd, some seeking to restrain the rioters but others joining in the attacks on the Bogsiders. As the Protestant crowd from the Fountain approached the Bogside, the Catholics anticipated another concerted assault by security forces and the Protestant crowd. Instead, the British Army arrived, replaced the RUC, and eventually dispersed the approaching crowds. Within 24 hours, the streets of Derry calmed considerably.[52] It would take weeks of careful negotiations, however, to get the barricades taken down in Derry and Belfast.[53]

Figure 2. Entering Free Derry.[54]

---

[49] O'Dochartaigh, *From Civil Rights to Armalites.*

[50] Ibid. After the IRA was unable to provide protection, the DCDA made the decision to not use guns in defense of Catholic neighborhoods, in part because they had access to only a few guns. Instead, the DCDA relied on sticks, stones, and petrol bombs. Many of the older republicans, remobilizing after an extended respite, bitterly complained about the IRA's failure to militarily protect the community.

[51] Ibid.

[52] Ibid.

[53] Bell, *The Secret Army,* 702.

[54] "File:Derry mural.jpg," *Wikimedia Commons,* accessed March 13, 2011, http://en.wikipedia.org/wiki/File:Derry_mural.jpg.

FORM AND CHARACTERISTICS OF THE REVOLUTION

OBJECTIVES AND GOALS

The differences within the Irish republican movement became stark as the protests in Derry, Belfast, and Burntollet turned violent. The energy and conflict unleashed during the civil rights marches uncovered two important facts about the IRA in 1969. First, at least some of the IRA members had grown dissatisfied with the leftist turn of the organization and felt that the vigor and traditional ways of the IRA were being ignored. They felt Goulding had unwisely involved the organization in a political arena where it could be outmaneuvered by the O'Neill government and where the dream of a united Ireland could be squashed. The second realization during the violence of late 1969 was that the IRA could not protect Catholic citizens. Dissatisfied IRA members were irritated that the leadership refused to arm its members or, even worse, was unable to arm its members. Belfast units had only a few dozen small arms, and Dublin had nothing to send north.[55]

The split within the organization began in Belfast in August at a meeting to discuss replacing the local IRA leadership. Many of the key figures of the as-yet-unformed Provisional movement were present, and the meeting led to an armed confrontation with the Belfast unit commanders, which resulted in a compromise. The new faction gained some leadership responsibility but not full command.[56] Finally, in October 1969, the split became official during an Army Council meeting. The Council passed two motions that were desired by the Goulding-controlled Dublin leadership but opposed by the future Provisionals. The IRA was to drop its customary practice of parliamentary abstentionism and would align itself with the radical leftist political parties. Because these motions were expected to win, the dissidents had arranged to meet quickly to form a new organization that they believed reflected traditional republican ideals. Army Council member Seán MacStiofáin was made its first Chief of Staff, and thirteen other IRA Convention members joined thirteen supporters to become the PIRA.[57] Their first public statement reaffirmed the new group's

---

[55] Bell, *The Secret Army.*

[56] English, *Armed Struggle*, 105.

[57] Bowyer Bell states the name "Provisional" comes from the recognition that a full Army Convention would have to formally create the group, but the name was not changed at the first PIRA convention. J. Bowyer Bell, *The IRA, 1968-2000: Analysis of a Secret Army* (Portland, OR: Frank Cass, 2000), 126–127.

allegiance with the ideal of a unified Ireland "suppressed to this day by the existing British-imposed . . . partition states."[58] The political arm of the IRA, the Sinn Féin, had its own split in January 1970. The remaining elements of Goulding's organization soon became known as the "Official" IRA, to distinguish it from the new "PIRA."

The PIRA emerged as a nationalist group that, when coupled with Sinn Féin (more accurately, the Provisional Sinn Féin), constituted a movement that aimed to support the establishment of a unitary Irish republic. The PIRA stated that it was the legal authority of a government denied to its people, and, therefore, the protection of the sovereignty of that republican government rested with it. With the long-term objective of reunification of the entire Ireland under a single Irish-run government, the short-term goals for the PIRA were to protect the people from the injustices of their government and to, simply put, "get the Brits out." The British had imposed the two-state partition, and when British troops arrived to impose peace, the British Army and government became full-fledged combatants in the eyes of the IRA. The Northern Irish government had requested the Army's deployment when it was clear that the police could no longer contain the violence, and on August 14, 1969, the Army's deployment began. Withdrawal of British troops and the elimination of support to the Loyalist Northern Ireland government, therefore, became prime motivating factors for the PIRA's campaign and operations.

The move of the "Official IRA" in 1969 to end the practice of abstentionism, which had been so vital to the struggle, and to turn toward political engagement and accommodation with the British system undermined the basic military role of the IRA. The PIRA was determined to reinstitute the disciplined, trained army structure and operations of its ancestor organization and to finish that organization's quest. However, it was the PIRA's decision almost three decades later to engage in politics and to allow concessions that allowed an end to the violence in the late 1990s and enabled some autonomy for the Northern Irish from Britain. The initial use of violence by the PIRA therefore changed the dynamics of the government and the situation in the north, but the eventual acceptance of compromise was the only thing that allowed peace.

LEADERSHIP AND ORGANIZATIONAL STRUCTURE

The best estimate of the PIRA as an organization was that it contained only a few hundred official Army members after it reached

---

[58] English, *Armed Struggle*, 106.

a steady-state campaign in the mid-1970s. After the Troubles and the formation of the Provisional movement, volunteers were very easy to come by, and membership peaked during 1972. As the British stepped up their countermeasures and narrowed the opportunities for action and movement, the PIRA implemented a smaller cellular structure for its active units and diminished the number of volunteers needed. The support network for the Army, however, stretched into the thousands in both Northern Ireland and the Republic.[59]

Active Service Units contained the bulk of the membership of the PIRA and carried out its military operations. These units included both part-time volunteers and full-time members of the Army. Some men and women held normal jobs in the communities and participated in Army operations on the weekends or after work hours. Far fewer members were paid by the PIRA with a weekly allowance to enable their full-time support to the Army. Members that had such a stipend often received additional support from the local community in the form of purchased clothes, donations, and food.[60]

Active Service Units usually had four volunteers and one operations commander. If needed, an intelligence officer or an education officer could be made available for operational support or training, respectively. Each ASU was believed to be responsible for the bulk of their operational expenses, their local safe houses, and any required transportation. For security purposes, each operations commander most likely knew the identity of only one higher commander, the Brigade Adjutant. In turn, the Brigade Adjutant took orders from the higher geographical and/or central command.[61]

The PIRA was divided into two commands, the Northern and Southern, with the Northern Command covering the entirety of Northern Ireland plus the five Republic border counties of Louth, Cavan, Monaghan, Donegal, and Leitrim. This command was created in 1976 and was the predominate area of operations for the PIRA. Its leadership was tightly coupled with the commanders above it. The creation of the Northern Command coincided with a marked shift in the PIRA leadership from being dominated by members from the Republic to being dominated by members from Belfast and Northern Ireland. Southern Command, created slightly later, encompassed the rest of the Republic and was far smaller in membership and importance. Southern Command functioned mainly to provide logistical support,

---

[59] Bell, *The IRA, 1968-2000*, 84–85; John Horgan and Max Taylor, "The Provisional Irish Republican Army: Command and Functional Structure," *Terrorism and Political Violence* 9, no. 3 (2007): 3.

[60] Ibid., 18–19.

[61] Ibid., 20.

including training, funding, safe houses, and the storage and movement of arms, to the operations in the Northern Command.

The upper authority echelons of PIRA consisted of the General Army Convention, the Army Executive, the Army Council, and General Headquarters (GHQ). GHQ regulated and executed the daily tasks of the Army, ensuring the maintenance of the Commands, and centralized the functions of the Army, including Finance, Security, Quartermaster, Operations, Foreign Operations, Training, Engineering, Intelligence, Education, and Publicity. GHQ consisted of approximately fifty to sixty people who were responsible for the overall conduct of the PIRA. GHQ was officially located in Dublin, Ireland, although many of its staff were based elsewhere.

GHQ was headed by the Army Chief of Staff, who was selected by vote of the Army Council and was the key decision maker within the Army. The first Chief of Staff was Seán MacStiofáin, one of the original troika that formed the Provisionals. Traditionally, the Chief of Staff and other major positions were held by the trusted core members, with little turnover unless one was arrested. Members were required to give up their positions upon capture, and they were banned from partaking in decisions for a period of time after their release. There were nine Chiefs of Staff from the PIRA's inception to the peace accord in 2002.[62]

The Army Council was a seven-member panel, usually including the GHQ Chief of Staff, the Adjutant General, the Quartermaster, the head of Intelligence, the Head of Publicity, and the Head of Finance. These men approved all major actions and established the majority of the strategy and policies of the PIRA, therefore serving as the main authority of the PIRA. The Council met at least once a month to review operations and to vote on major items, such as cease-fires or the expansion of operations to Britain.[63]

The Army Executive was a board of twelve very senior and experienced PIRA veterans who met every six months or so to review the activities of the Army Council and to serve as the voting body that elected the seven members of the Army Council. Members were barred from sitting on both the Executive and the Council.[64]

The Constitution of the PIRA placed the supreme authority of the movement within the General Army Convention. The Convention was a large body of delegates from the structures of the entire PIRA and designed to meet to approve and vote on only the most important

---

[62] Ed Moloney, *A Secret History of the IRA* (New York: Penguin Books, 2007), 513.
[63] Horgan and Taylor, *The Provisional Irish Republican Army*, 5–7.
[64] Ibid., 5

of issues, such as declarations of peace and the election of the Army Executive. The size of the Convention ranged from one hundred to two hundred members and included active volunteers, representatives for the imprisoned, staff from local brigades, the Commands, and GHQ, and usually all members of the Army Council. The constitution allowed for meetings to be held every two years unless security deemed it prudent to delay such a large group of members being collected. The first convention instituted the new PIRA as an entity separate from the Official IRA, and others may have been held to debate major cease-fires, the end of abstentionism, and the final accords.[65]

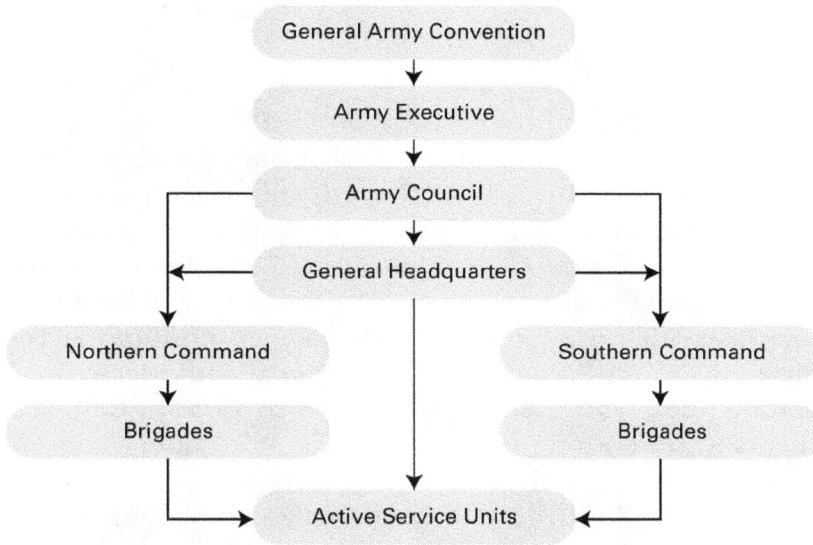

Figure 3. IRA organization.[66]

The relationship between the PIRA and its political arm, Sinn Féin, was often viewed as complementary and equal, but for the majority of the period of violence, the Army was the central organization and decision-making apparatus, while Sinn Féin was peripheral. The break of the Provisionals from the Official IRA in 1969 was due to the old army being seen as having given up the armed fight and becoming a political organization. Thus, the PIRA existed primarily as a fighting Army; it saw political solutions as undesirable and not worth pursuing

[65] Ibid., 4–5; Moloney, *A Secret History of the IRA*, 445, 475, 481.
[66] Data from Horgan and Taylor, "The Provisional Irish Republican Army," 26.

until it had beaten the British into submission. Sinn Féin, therefore, operated mostly to explain the military operations to the public in the early days. It would take a generational change from within the Provisionals for it to be possible for a political solution to take place in the 1990s. With the ascendancy of Gerry Adams to the top post of the political party in 1983, Sinn Féin became a powerhouse to the republican cause, putting candidates up for office and acting as the proxy for the PIRA in the peace process. However, when the PIRA engaged in military actions, the British government would refuse Sinn Féin a place at the negotiating table until a cease-fire was called.

COMMUNICATIONS

In February 1970, the first issue of *An Phoblacht* was published in Dublin with the approval of the Army Council. Seán Ó Brádaigh served as the editor of this PIRA news organ, which communicated official Army political statements, interviews with members, and details about British and government atrocities and major skirmishes mostly to the faithful every Thursday. The first issue sold 20,000 copies, and the newspaper was printed weekly through most of the 1970s.[67] The newspaper's staff acted as an internal forum for ideas, somewhat like a think tank, and often provided a work location for transitory volunteers who were reintegrating into the PIRA after prison or for other reasons.[68] The north also had its own newspaper, the *Republican News*, which started being published in Belfast at the same time as *An Phoblacht*. The two papers officially merged in 1979. The merger gave control of the public voice of the PIRA to those in Belfast, particularly the new leadership that was emerging under Gerry Adams.[69]

METHODS OF ACTION AND VIOLENCE

The initial strategy of the PIRA during the violence of 1969 was to defend and protect the Catholic population and to re-establish the strength and purpose of the organization. The Provisionals quickly started to gather materiel and financing and began training so that any Loyalist or government action could be countered with a capable PIRA unit in the neighborhood. By early 1970, there were fourteen new units in Belfast. Arms were arranged through Dublin contacts, imported into Ireland, and smuggled into the north. Hundreds of

---

[67] Bell, *The Secret Army*, 368.
[68] Bell, *The IRA, 1968-2000*.
[69] Moloney, *A Secret History of the IRA*, 179–180.

light ArmaLite rifles went through Ireland in 1970, and in 1971, the Northern Ireland security forces captured 700 weapons and more than 150,000 rounds of ammunition, which came primarily from the United States.[70] In May 1970, the Republic of Ireland had to fire two senior ministers because of their alleged involvement in an arms importation ring, although both were eventually exonerated because their involvement appeared to be state sanctioned. The early support of the Republic, both under-the-table government support and citizen support, was crucial to the formative days of the PIRA.

As conflicts between the Catholic and Protestant populations continued into 1970, the Provisionals started defensive operations. The PIRA also took responsibility to stamp out drug sales and burglary in the Catholic parts of Belfast and Derry. Miscreants could expect a beating or a knee-capping, punishments far harsher than those that they would face from the police, and the Catholic population seemed quite supportive of these actions.[71] Sectarian attacks became commonplace, and tit-for-tat killings between Catholics and Protestants were constant. In the absence of the British Army or a police presence, the PIRA Belfast units took up the defense of St. Matthew's Church from petrol-bomb-wielding mobs. The resulting five-hour gun battle ended with the successful defense of the Catholic building and became the legendary start of the PIRA's defense of the oppressed people. The fight also led to a British-enforced curfew in the Catholic neighborhoods. Gerry Adams noted that this curfew "made popular opposition to the British Army absolute in Belfast . . . After that, recruitment to the IRA was massive."[72] Through 1970, however, operations were not directed toward the British Army itself.

As the PIRA expanded and trained enough people to safeguard Catholic neighborhoods, its strategy turned to a more offensive tone designed to expand the area of operations to include the British Army. The Army Council felt that the Catholics were sufficiently alienated from the British Army to support operations against soldiers. Volunteers started to appear in "uniform" during operations, wearing army surplus combat jackets, berets, or helmets. Direct harassment operations against British Army outposts began. Skirmishes soon escalated between the two armies, with assassinations and hit-and-run gunfights solidifying into hatred on both sides.

The bombing campaign also expanded with this new offensive. Incendiary bombs were placed in other sections of the city of Belfast

---

[70] English, *Armed Struggle*, 116–117.

[71] Patrick Bishop and Eamonn Mallie, *The Provisional IRA* (London: Corgi Books, 1987), 133.

[72] Quoted in English, *Armed Struggle*, 136.

so that British attention was turned away from the Catholic ghettos. Patrols and bomb squads soon had to operate all over the city, with 153 bombings taking place in 1970. This bombing campaign quickly escalated because the bombs were simple to make and because they generated widespread television and newspaper coverage in both Northern Ireland and Britain. In 1971, 1,022 bombs were placed. The British were fairly successful in finding them, and more than one-third of those devices were defused.[73]

At first, the PIRA built simple incendiary bombs and stole gelignite from local limestone quarries. Bombs were placed with simple fuses or timers. Anti-handling devices were introduced around 1971 to counter the defusing success, making it impossible to move the bombs after they were armed without causing detonation. As the supply of gelignite dried up, the PIRA turned to fertilizer-based mixtures. By March 1972, they found that using cars or trucks as delivery devices allowed them to create larger bombs; limit the risk of being seen placing the bomb; and reduce the danger associated with arming the bomb.[74] The success of the car bomb led to a vast expansion of traffic checkpoints around Belfast, which both absorbed more British Army resources and increased the irritation of and inconvenience to the population. By 1974, using parts from remote-controlled airplanes, the PIRA experimented with and perfected the remote-controlled bomb, which became the staple of their operations.[75]

The beginning of the PIRA offensive in 1971 led to further calls from the Northern Irish government to London for additional action. In August, Operation Demetrius commenced and involved a broad sweep to arrest and detain known PIRA members. The intelligence, upon which the British were operating, however, was very dated, and of the 342 people arrested in the first twenty-four hours, less than one hundred were members of either the Provisional or Official IRA. Immediately, 116 suspects were released. Many members had already gone into hiding upon hearing that the British had reinstated the policy of detention and internment without trial. The operation, which was undoubtedly aimed solely at the trouble makers on the Catholic-Republican side of the conflict, galvanized the view that the British Army was taking sides with the Loyalists, who were not arrested or interned until much later in the conflict.

Internment without trial became a political disaster as well, spurring more violence and agitation, and led to an increase in

---

[73] Bishop and Mallie, *The Provisional IRA*, 131.
[74] Ibid., 171; Moloney, *A Secret History of the IRA*, 115.
[75] Bishop and Mallie, *The Provisional IRA*, 158.

PIRA membership as opposed to stemming it. Despite being used effectively twice before—in the 1920s and 1950s—the policy this time triggered a much stronger reaction within the Catholic community and undermined security efforts that were disproportionate to the gain of a few significant arrests. It also broadened the rejection of the Northern Irish government as a legitimate body. Exacerbating the situation were accounts of how the interned were treated to the "five techniques"—a hood over the head; standing spread-eagled against a wall for long periods; sleep deprivation; irregular delivery of food and water; and subjection to white noise. Internment without trial quickly became a public relations disaster.[76]

The overall situation in the conflict was becoming bloodier as well. In 1971, nine PIRA members and thirty-three Catholic civilians were killed by security forces and fifty-six security force members were killed by the PIRA. It became a self-perpetuating war, and it turned dramatically worse in the beginning of 1972. An anti-internment march was planned in Derry for January 30, and because of the recent violence in the city, the Army deployed the First Battalion of the Parachute Regiment to contain the march away from the city center. The plan was to allow the march to take a particular route and then surround and arrest any stragglers and rioters that followed. The British were ready for snipers and bombs and expected some form of PIRA action. On that Sunday, more than 10,000 marchers walked the prescribed route and then mostly dispersed. As the soldiers stepped in to arrest the remaining marchers who had not peacefully left, they were pelted with rocks and bricks.

What happened next is subject to debate and continued inquiry but there is no doubt of the resulting consequences and their overall impact on how the conflict would proceed. The British Army claimed that they came under gunfire, a claim that has always been fiercely disputed by the marchers and observers. The PIRA had promised the march organizers that it would not undertake any violence that day and insisted that it did not. The reaction of the Parachute Regiment, whether instigated or not, had deadly consequences—thirteen civilians were killed and thirteen more were wounded.[77] All of the victims were Catholic, seven were under nineteen years old, and none were armed or even wanted by the security forces. "Bloody Sunday" brought strong reactions from both the Republic of Ireland and the

---

[76] English, *Armed Struggle*, 140–141.

[77] One later died of his wounds, bringing the total killed to fourteen.

Catholic community in the north. The British embassy was burned in Dublin, and fighting escalated across the north.[78]

The massacre also had political ramifications. The event was followed by a decision from the British PM to suspend Stormont, the Northern Irish parliament, and take direct control of the government. The suspension was a huge defeat for the Unionist government and seen as a major victory for the Republican movement. The PIRA sensed victory in early 1972, and the leadership felt it was entering the final phase of the struggle. However, a series of missteps by both the Officials and the PIRA took away their perceived advantage and hastened calls for a cease-fire and talks. One misstep occurred on February 22, 1972, when the Official IRA bombed the headquarters of the 16th Parachute Brigade in Aldershot, England, in retaliation for Bloody Sunday but only managed to kill five female kitchen workers, an elderly gardener, and a Catholic army chaplain.[79] Then, in March, a bomb went off without warning in a busy restaurant in Belfast, killing two women and injuring seventy.[80] Although the PIRA did not claim responsibility for the attack and had previously avoided civilian causalities by providing some warning and time for evacuation, the attack was viewed as the work of the Provisionals. By June, the PIRA and the Officials had both declared a cease-fire to let conditions cool a bit, and the British Army followed suit.

The Provisionals openly called for talks in June with the British Secretary of State for Northern Ireland, but these calls were publicly rebuffed. However, secret talks did commence in London on July 7, 1972, with the Secretary of State and the PIRA leadership. MacStiofáin laid out the PIRA demands: the Irish people should be allowed to choose the future of Ireland as a unified country; the British should declare its intention to withdraw by January 1, 1975; and internment must end with a broad amnesty in effect. The British considered these demands completely unrealistic, and the high-level meeting ended without any accord.[81] Two days later, the cease-fire ended, but the PIRA appeared to be positioned in a new and significant role for future political agreements. The moment was quickly lost, however, because of two decisions the PIRA made immediately after the talks failed.

The Army Council thought it could gain some benefit from the failed talks by announcing they had talked with the British government, unwittingly embarrassing the Secretary of State for Northern Ireland.

---

[78] Bell, *The Secret Army*, 384–385; English, *Armed Struggle*, 148–156; Moloney, *A Secret History of the IRA*, 110–111.

[79] "Northern Ireland: Now, Bloody Tuesday," *Time*, March 6, 1972.

[80] Bell, *The Secret Army*, 393.

[81] English, *Armed Struggle*, 157–158.

This effectively ended any chance for another bargaining session in the near term. In addition, bombing operations were to be redoubled, underlining the Council's belief that politics could only be driven by violence. Ironically, the return to all-out operations started a dramatic reversal for the PIRA over the next two years in terms of its power and viability.[82]

On Friday, July 21, just two weeks after the secret meeting with the Secretary of State, more than twenty car bombs were delivered to Belfast and detonated over a period of an hour and fifteen minutes. Nine people were killed and 130 were injured. One bomb at a bus station killed six and scattered body parts over the streets. The chaos across Belfast was televised, with screaming shoppers and firemen shoveling human remains into plastic bags. "Bloody Friday" quickly matched "Bloody Sunday" in importance and in the images that came out of the events.

The operation had been planned by Seamus Twomey and the Belfast Brigade[83] prior to the secret talks and was meant as a way of driving the British to negotiate. The operation was approved after the failed talks to "demonstrate the IRA was still in business."[84] The PIRA insisted that it had provided multiple warnings about each bomb's position and did not intend to kill any civilians. The number of bombs and the scale of the operation across the city, however, proved to be too much for the security forces to handle. "We put it down to the Brits allowing bombs to go off, but the real reason was it was too much for the Brits to cope with, the bombs went off too close together, the town was too small, people were being shepherded from one bomb to another," conceded one PIRA member.[85]

The reaction was immediate and not good for the PIRA. The population, both Catholic and Protestant, was horrified at the carnage, and the British Army felt the momentum shift in their favor. Within two weeks, the Army mounted a major expansion of their efforts, including Operation Motorman, which invaded and cleared the Catholic areas in Derry and Belfast that had been "no-go" zones. The PIRA was warned and escaped, but it lost a region where it could freely operate and that it could claim as a legitimate area of jurisdiction and protection. The Army also enhanced its intelligence gathering

---

[82] Moloney, *A Secret History of the IRA*, 116; Bishop and Mallie, *The Provisional IRA*, 180.

[83] Twomey was convinced to step down and move to Dublin as a result of the debacle, and Gerry Adams replaced him as Belfast commander. Adams had been imprisoned in Long Kesh when he was released as a precondition of the cease-fire and secret summit with the British Secretary of State. The PIRA had demanded his participation in the talks.

[84] Moloney, *A Secret History of the IRA*, 117.

[85] Quoted in Moloney, *A Secret History of the IRA*, 117.

and successfully captured middle- and upper-level leaders of the PIRA in the next few months. MacStiofáin was arrested in November and never again took a role in the PIRA. He was replaced by Joe Cahill, who took over as Chief of Staff.[86]

Protestant paramilitary groups that had been minor annoyances up to this point also dramatically increased their presence and violence against the Catholic population and the PIRA. The Ulster Defense Association (UDA), founded in 1971, began a series of tit-for-tat assassinations and violence with the PIRA. By the end of the year, the UDA and other loyalist groups had killed more than 120 people.[87]

On the basis of its experiences in Northern Ireland, the British Army adjusted its training for its soldiers, teaching them better methods for arresting suspects in order to avoid potential release on technicalities, and the laws were also modified to shift more of the burden of proof on to the defendant to prove their innocence.[88] In addition, the British increased their use of informants and undercover agents within the PIRA. For their part, the PIRA mounted operations against the British intelligence network and reconnaissance force, raiding a massage parlor, ice cream shop, and laundromat that were run by the British operation. Although the PIRA was able to disrupt this particular network, it began to understand how infiltrated it had become. It learned about the faux businesses only through the interrogation of a low-level volunteer.[89] By the end of 1972, the PIRA was weakened and found its recruitment suffering. Its operational effectiveness fell steadily as well. The PIRA killed 103 soldiers in 1972, fifty-eight in 1973, twenty-four in 1974, and merely fourteen in 1975. The consequences of Bloody Friday were lowered temptations to use car bombs in Belfast and the return to sniping and small engagements by the PIRA.[90]

Throughout these early years of the PIRA, pressure mounted on the Republic of Ireland to not support or even acknowledge the PIRA. In March 1973, the British government tipped off the Irish Navy about the arrival of a potential arms shipment. The Navy boarded the *Claudia* off the Irish coast and found five tons of weapons from Libya, as well as Joe Cahill, who was tried and sentenced to three years in prison for importing arms and for his membership in the IRA.[91]

---

[86]   Ibid., 118.
[87]   English, *Armed Struggle*, 160.
[88]   Bishop and Mallie, *The Provisional IRA*, 188.
[89]   Moloney, *A Secret History of the IRA*, 119–120.
[90]   Bishop and Mallie, *The Provisional IRA*, 193.
[91]   English, *Armed Struggle*, 161; Bell, *The Secret Army*, 395.

In the midst of these crackdowns, the PIRA started to consider extending its operations to mainland Britain. In early 1973, the Army Council formally approved action in Britain, to include strikes on economic, military, political, and judicial targets. These actions were designed to re-engage the attention of the British public, generate a weariness of the problem in Northern Ireland, and increase pressure on their political parties to pull out. It would also allow the PIRA to conduct operations without risking harm to Irish civilians while simultaneously taking pressure off of the Belfast and Derry units. The first mainland operation involved four car bombs in the heart of London—two were defused before exploding and the remaining two detonated and injured 180 people.[92] The discovery and diffusion of the two bombs tipped off police to the bombing plot and allowed them to capture most of the PIRA members before they could depart on their planned flights out of Heathrow. This capture led the Army Council to prefer using "in-place" teams for future operations over trying to make a quick escape. The next team of operatives planted nine bombs in mid-England and a series of small incendiary and booby-trap bombs in well-known shops around London. Team members integrated themselves into the Irish communities around London and other big cities, and women couriers carried money and orders from GHQ.[93]

Innocent civilians were supposed to be avoided in these mainland bombing attacks, but in reality, the bombings were often done indiscriminately and without any worry over accompanying civilian deaths. Pubs were occasionally targeted, especially those frequented by British soldiers. More often, though, "establishment" targets were chosen that affected the ruling and influential people of British government and business. Residences of the MPs, high-end hotels, clubs for the wealthy, and similar locations were favorite targets. The highest-profile bombing was perhaps the attempt on PM Margaret Thatcher's life in 1984 at the Brighton Grand Hotel. The PIRA planted a large Semtex bomb twenty-four days before its detonation using a sophisticated long-delay timer. The cabinet and Thatcher were attending a conference at the hotel, and the blast came very close to Thatcher's suite. She was unharmed, although the Trade Secretary was badly hurt and several senior Tories were among those killed. Five people died in the blast, and significant visible damage inside and outside the hotel showed the reach and power of the PIRA.[94]

---

[92] Bishop and Mallie, *The Provisional IRA*, 198–199.

[93] Ibid., 200.

[94] English, *Armed Struggle*, 248.

Figure 4. Exterior view of the Brighton Grand Hotel after the bombing attempt on PM Thatcher.[95]

The crackdown by the British Army starting in 1972 began to deprive the PIRA of both leadership and recruits. The Army Council agreed to talks with the British, hoping that a period of respite would allow them to regroup and train recruits. The cease-fire agreement of 1975 was unpopular with the PIRA members from the north, however, and led to a gradual change of the organization's leadership from being dominated by men from the "south" to being dominated by those from Belfast and the North. The reorganization of the Army into two geographical commands, the Northern Command and the Southern Command, was part of this process even though the Northern Command was led by a "southerner" at first.[96]

The cease-fire allowed the British to also build up while the PIRA was increasingly involved in fighting with the Official IRA and the loyalist paramilitary groups. Morale started to wane and questions about the leadership emerged from within. Some volunteers began to see that the cease-fire was purely to the advantage of the British. Younger members, especially from the north, began to move to positions of authority, starting with the editorship of the *Republican News*. The paper helped give a voice to the new, younger PIRA

---

[95] "File:Grand-Hotel-Following-Bomb-Attack-1984-10-12.jpg," *Wikimedia Commons*, by D4444n at the English language Wikipedia, accessed March 14, 2011, http://en.wikipedia.org/wiki/File:Grand-Hotel-Following-Bomb-Attack-1984-10-12.jpg.

[96] Moloney, *A Secret History of the IRA*, 157–159.

volunteers, including Gerry Adams,[97] who wrote under the pseudonym "Brownie." The peace didn't last through the end of 1975, though, as the Northern Brigades increasingly took action on their own without notifying the upper echelons.

That same year, the Secretary of State for Northern Ireland announced a change in the status for prisoners. Since 1972, as a result of Billy McKee's hunger strike, politically motivated prisoners had been granted special-category status within the legal system. As of March 1, 1976, however, all new prisoners would henceforth be treated as ordinary criminals by the legal system. This downgrade was an attempt to delegitimize the republicans' struggle and provoked a strong reaction by the PIRA. Kieran Nugent was the first Provisional subjected to the new status, and in protest, he refused to put on his prison uniform. He was placed in a cell without any clothes at all, only a single blanket. Because prison rules required inmates to wear clothes when outside their cell, he was confined to his cell twenty-four hours a day. This protest led the prison to remove access to television, radio, reading material, or any remission of sentences. The protest escalated into 1977 and 1978 in reaction to what the republicans felt was harassment, brutality, and inhumane treatment. The prisoners stopped washing themselves, and then chamber pots were thrown at the warders or their contents spread across the walls. The "blanket protest" had turned into the "dirty protest." By late 1980, there were 800 republican prisoners at the Long Kesh prison, of whom 300 were "dirty" protesters.

A traditional tactic of Irish republicans was the hunger strike, which was used during the 1920s revolt. When Seán MacStiofáin was arrested in the post-Bloody Friday crackdown, he immediately announced that he was not only on a hunger strike but also on a thirst strike at the Mountjoy prison. By the tenth day, a republican priest convinced MacStiofáin to agree to water, although he maintained his refusal of food.[98] In 1980, the situation in Long Kesh prison became so bad that the PIRA prisoners felt that another hunger strike was warranted. Seven prisoners started refusing food on October 27, three women joined on December 1, and another thirty joined by mid-month. After

---

[97] Adams denies ever being a member of the PIRA. "Adams Denies IRA Book Allegations," *BBC News*, September 30, 2002, http://news.bbc.co.uk/2/hi/uk_news/northern_ireland/2288775.stm.

[98] MacStiofáin ended up lasting fifty-seven days, primarily because his water contained glucose, and he stopped the strike under orders from the PIRA. His inability to carry the strike through (until death or capitulation to his demands) greatly diminished his stature within the Provisionals. He also refused to abide by the internal jail PIRA hierarchy. Upon his release the following April, he never regained status within the Army, was reprimanded, and was generally ignored in all future PIRA publications. Bishop and Mallie, *The Provisional IRA*, 191–192.

fifty-three days, with one of the strikers near death and a deal from the government on the table, the internal PIRA hierarchy ordered an end to the strike.

The deal that ended the strike was quickly disputed. The PIRA thought its demands had been met, whereas the government claimed that it had only offered additional amenities and privileges within the jail. By the following March, the conditions had reached another stand-off, and another strike began. Bobby Sands initiated this hunger strike, which led to nine republicans dying without food within the jail. In addition, the strike generated very high levels of public attention because four days into the protest, Bobby Sands was nominated to be an MP from Northern Ireland.[99] On April 9, he narrowly defeated the challenger from his jail cell, and his plight reached new heights of attention as everyone wondered if the government would allow one of their own MPs to die. On May 5, 1981, his sixty-sixth day without food, Bobby Sands died and massive international publicity followed. The republicans took full advantage of the situation, smuggling out messages from the prisoners and strikers, and each subsequent death put additional pressure on the British government to concede. Yet the government did not, even in the face of international protests and condemnation. It was Bobby Sands's death, however, that revitalized the PIRA's motivation and started to give it a more overtly political dimension than it had previously had. Although the PIRA was suspicious of politics and equated participation in politics with compromise, some of the leadership began to see its value.[100]

Figure 5. Bobby Sands in 1973.[101]

---

[99] Frank Maquire was an MP for Fermanagh/South Tyrone and had been an IRA internee himself as well as a staunch supporter of gaining political status for the prisoners.

[100] English, *Armed Struggle*, 187–205.

[101] Image courtesy of Bobby Sands Trust, http://www.bobbysandstrust.com/.

The PIRA's refusal to officially participate in any form of government would change slightly in 1981. Though still abstaining from the London and Dublin governments, the PIRA decided to pursue and occupy seats in the Northern Ireland government. Although the Sinn Féin candidates for these political offices were often defeated, the potential for a violence-advocating political party to potentially win seats did cause both London and the unionists to worry. In 1985, the United Kingdom and the Republic of Ireland struck an accord that stated that the status of Northern Ireland would not change without the consent of its population and that the Republic would work with the United Kingdom on issues of security, human rights, and reconciliation. This agreement angered the unionists in Northern Ireland more than it angered the PIRA. The PIRA continued its dual operations in both violence and now politics, with Sinn Féin now becoming as important as the PIRA military wing. Gerry Adams took charge of the political party in 1983. By 1986, the Provisionals had removed their ban on pursuing seats in the Dublin parliament and had also removed the official taboo on volunteers openly discussing the end of abstentionism.[102]

Bombings and attacks continued in parallel with the political process, as tit-for-tat killings continued with the unionist paramilitaries and anyone working with or for the British security forces was considered a target. In November 1987, a bomb brought down a community hall in Enniskillen during a remembrance service, killing eleven Protestants. The following day, the PIRA issued a statement expressing regret that the bomb went off during the service because the timer was supposed to have detonated the bomb during preceremony activities by the security forces. The embarrassing event was soon followed by an escalated series of killings. It would be one of a number of bombings that mistakenly killed civilians during the late 1980s and early 1990s. The PIRA soon admitted that their support was again dwindling.[103]

The combined use of violence and politics continued up to 1994 when the PIRA took up a new strategy called TUAS, the Tactical Use of Armed Struggle,[104] which did not alter the goal of a united Ireland but did argue that a broader nationalist alliance might move the struggle forward. The PIRA unilaterally declared a three-day cease-fire in April 1994, and by August, it had announced a complete stop to its military operations. "We are therefore entering into a new situation in a spirit of determination and confidence: determined that the injustices which

---

[102] English, *Armed Struggle*, 227.
[103] Ibid., 260.
[104] Many believed the acronym stood for Totally Unarmed Strategy.

created the conflict will be removed and confident in the strength and justice of our struggle to achieve this."[105] Talks between the British government and Sinn Féin were explored, but the British government required the PIRA to disarm before allowing negotiations, and Sinn Féin rejected this requirement. Former US Senator George Mitchell brokered an agreement to allow disarmament to take place in parallel with talks. But the PIRA had already grown frustrated with the pace of the process. In January 1996, the Army Council voted unanimously to end the cease-fire. On February 7, a huge truck bomb was left in the underground parking garage of a six-story office building in South London. The blast killed two and injured many others, causing more than £85 million in damage to one of London's most expensive commercial areas. In June, another bombing caused major damage in central Manchester and injured 200.[106]

A change of governing party in Britain from Conservative to Labour brought hope back to the peace process, and the PIRA Army Executive held off a challenge from volunteers who rejected the idea of holding any talks with the British. In August 1997, the PIRA announced a resumption of the cease-fire, and this decision resulted in a split within the movement. The "Real IRA" emerged from the faction that wished to continue the violence, and this new group pursued armed resistance despite the ongoing peace process.[107]

METHODS OF RECRUITMENT

When the Troubles started in August 1969, approximately one hundred PIRA active members were in Belfast, with another 300–500 supporters providing cover, hiding, or communications. Through 1971 and 1972, the Provisionals had smaller membership numbers than the Official IRA, but the perception that the Provisionals were protecting the Catholic population against the brutality of the British Army and Protestant mobs gained them a following and support that began to drain the popularity and support from the older, politically oriented Official IRA group.

Because the Provisionals broke off from the Official membership, the early leaders understood what kind of volunteers were needed and how to train them to operate as an underground army. The breakaway members desired action and to fight, and they were disappointed by the loss of that emphasis within the original Army. They therefore

---

[105] IRA statement, August 31, 1994, quoted in English, *Armed Struggle*, 285.

[106] Moloney, *A Secret History of the IRA*, 441–443.

[107] English, *Armed Struggle*, 293–296.

instituted a strict and disciplined training and recruitment policy. The new PIRA leaders also understood the full scope of capabilities and functions needed, to include a political wing, a women's support movement, scouts, and active units.

The Provisionals did not have a problem finding volunteers during the Troubles of 1969 through at least 1972, and all were welcomed. The sectarian violence and actions of the British Army were sufficient catalysts to drive young men to the group. Many volunteers had already experienced a violent confrontation with mobs or the British Army, and retribution was more of a motivating factor for men seeking out PIRA membership than politics. But the leadership carefully weeded out those who were motivated purely by revenge and violence; those who remained were sometimes disappointed by the lack of violence in their daily activities. New recruits sat through long and tedious lectures and had reading assignments about republicanism. The purpose was to fully convert the person to the cause of a united Ireland and discourage those who were looking solely for prestige or violence.[108] The volunteers who made it through the process were often in their late teens or earlier twenties, were mostly from the middle class, and had usually received an adequate education. They also had had a heavy exposure to Irish history and cultural pride because these were a focus of the Catholic school curricula in Northern Ireland.[109] To some, the PIRA offered a "vocation" even though few became full-time members who could collect a wage. Those who did, however, received a small salary and relied on the kindness and support of the community for their drinks at the pub, extra needs, holiday presents, and the like. Most PIRA members had day jobs and took part in operations during the weekends or evenings. Their regular jobs often helped in some way with the support of the PIRA. If they worked in a government office, then collecting official paperwork, forms, or intelligence was part of their PIRA duty.

Those volunteers who had training or experience in politics or in organizational planning were put to work in Sinn Féin. The volunteers who wished to serve without exposure to violence or who lacked total commitment were asked to be part of the support network, which provided safe houses, transportation, donations, and communications. The support and cover that the PIRA received from this network and from the broader Catholic community allowed them to operate fairly freely, hide caches, or quickly disappear after an ambush. This deep connection to the community allowed the PIRA

---

[108] Bell, *The IRA, 1968-2000*, 84–85.
[109] Bishop and Mallie, *The Provisional IRA*, 118.

tremendous operational movement and advantage but also put them at the mercy of popular opinion and whim. When the promise of a political solution to the crisis came about, or a particularly egregious PIRA bombing upset the sensitivities of the community, this support base withered and could vastly curtail the available options and movement of the PIRA.

During the Troubles, because of the lack of active members, new recruits were needed immediately to serve in protection units and safeguard neighborhoods. But as the Army grew, it focused on gaining members who were motivated by a desire to serve, were committed, and were somewhat disinterested in the daily horrors around them. The PIRA was primarily looking for sound, stable members, not academics or thrill-seekers, not the well connected or the rich. They were recruiting a professional army.

By 1972, the PIRA had a lower rate of new members because of at least three factors. One factor was the clampdown by the British Army and the start of internment without trial, which led to fewer willing volunteers. In response, the PIRA decided that a smaller, cell-like organization was required in order to avoid large round-up raids and to protect the leadership, therefore leading to a requirement for fewer new members. Third, actions by the PIRA, such as Bloody Friday, contributed to lower recruitment rates because the accelerated bombing campaign and the willingness to target all non-Catholics, Irish or not, were controversial and made the PIRA less desirable to those who disapproved. Overall, the protection and defense of the population was an easier sell to recruits than a violent offensive campaign that destroyed property, assassinated politicians, and involved large-scale bombings.

METHODS OF SUSTAINMENT

The Quartermaster General, a member of the Army Council, had the duty of procuring, transporting, and storing armaments. The Quartermaster was often supported by regional Quartermasters. Weapons were strictly controlled; they were stored in bunkers and issued solely for operations. However, in some areas, small numbers of weapons were held by the local service units for guarding or conducting limited operations, but on the whole, the disciplined control of arms was maintained.[110]

The lack of arms to defend against Protestant mobs was one of the major dissatisfactions with the original IRA leadership in 1969 that led

---

[110] Horgan and Taylor, *The Provisional Irish Republican Army*, 10.

to the breakaway Provisional movement. Arms were therefore in short supply during the early years, and the leaders concentrated much of their initial efforts on pushing the republican-leaning movement for cash and arms. Firearms were restricted in Northern Ireland, so the Provisionals knew they had to establish a major flow into the country from outside sources. The flow of arms into the operational areas of the Northern Command was often conducted via Southern Command routes, sometimes on open roads in daylight. Theft and extortion were other options for collecting what was not obtainable within the country. Members were sent abroad to the United States to collect money and any arms they could get. By 1972, shipments of RPG-7 rocket launchers, military machine guns, and explosives were arriving from Libya.[111]

Early in the campaign, mines and explosives were in short supply. Gelignite was stolen from local mines to make the first explosives, and then fertilizers were used until the arrival of Semtex from Libya. The standard procedure was to warn the local police or British Army of the presence of the bomb and to give them enough time to evacuate civilians. But the success of the British defusing capability led the group to move from simple timers to more complex detonation devices that included anti-handling techniques. They also introduced the car bomb, which allowed a larger explosive charge, a more discreet delivery, and safer emplacement.[112] The PIRA also learned to make remote-control detonators from small consumer electronics.[113]

The implicit and explicit support of the Catholic population ebbed and flowed throughout the PIRA campaign. There was always support for the cause, but the violence against civilians or the periodic hope for a political solution would often erode that support. Catholics regarded the PIRA members as members of their community and had strong familial and community ties to the men that they knew (or guessed) were members. The Catholics could turn a blind eye or deaf ear when they saw or heard of an operation. Vehicles might be loaned or intelligence passed on. Catholics in government administrative jobs also provided rich information to the Provisionals, such as the home addresses of policemen or loyalist paramilitary members. Those who provided information to the security forces about the PIRA were intimidated or executed. The killing of civilians, however, often placed the PIRA on the defensive, and marches and organizations were formed by some Catholics to express counter-balances to the PIRA's policy of violence. One group, the Peace People, even gained a great deal of

---

[111] Bell, *The IRA, 1968-2000*, 170–172.

[112] Ibid., 171; Moloney, *A Secret History of the IRA*, 115.

[113] Bishop and Mallie, *The Provisional IRA*, 158.

notoriety when their founders were awarded the Nobel Peace Prize in 1976. Their reputations soon faded, however, when they began to advocate for passing information about the Provisionals to the security forces. Asking the population to turn in their brothers and neighbors was not successful or popular.[114]

METHODS OF OBTAINING LEGITIMACY

The deeply entrenched social networks within the Catholic community combined with the long-standing narrative of the republican movement allowed the PIRA to immediately have an implied legitimacy across the entire Irish Catholic population. By invoking the name and mission of the IRA and calling for a return to its violent struggle for independence, the Provisionals began their campaign with an established cause that was seen as being historically legitimate. As violence escalated and the PIRA took action to protect the neighborhoods against Protestant mobs and RUC aggression, the PIRA continued to highlight the need for their armed resistance.

The PIRA also built upon the traditions of the earlier IRA through its institution of a nominal uniform; the conduct of operations within the community; and the insistence upon abstentionism—all of which had been abandoned by the Goulding IRA. It quickly adopted maintenance and supply routes that harkened back to the tactics and gun running of the 1920s. Even the use of the hunger strike was a traditional tool meant to show dedication and commitment.

Despite the early swell of Catholic support for the PIRA, it had difficulty maintaining its place as the best alternative for the broader community once its operations focused on the British Army and its bombing campaign expanded. The PIRA was originally seen as a protector of the Catholics against the RUC and Protestant mobs, but as it turned its attention to the Brits, it provoked stronger retaliation that spilled into the populace. The bombing campaign proved more difficult, however, as inevitable civilian deaths turned public opinion away from tacit or open support. At first the PIRA tried to mitigate the loss of civilian life by carefully selecting its targets or warning the police in time to clear the area, but these efforts failed too often.

The evolution of the Provisional Sinn Féin was also an outgrowth of the PIRA's campaign to create and maintain support. The group initially operated as a propaganda machine, facilitating the public announcement of PIRA positions, demands, or intentions. As the PIRA began to see the need for or allure of engagement in the political

---

[114] Ibid., 227–231.

realm, the Provisional Sinn Féin grew into an organization that could maintain communications with the public, negotiate with the enemy, and allow the population to engage in the struggle through nonviolent means by electing their members.

EXTERNAL SUPPORT

As soon as the Provisionals started their own organization, they knew they would need outside assistance in the form of money and armaments. Their reliance on their Irish peers across the border was vital during the entire campaign, although less so as the Northern arm of the PIRA took control in the late 1970s. Arms and money were primarily routed through the Republic of Ireland no matter the origin, and safe houses, staff offices, caches, storage, and transportation were all available in the south.

As soon as the PIRA was sufficiently organized to send agents abroad, it turned to the United States for money and arms, with the Irish communities in Boston and New York proving especially supportive. In 1969, the United States had five times as many Irish as there were in Ireland. The Irish Northern Aid Committee was set up in New York City in 1970 to provide a steady stream of money to the PIRA, mostly for the purchase of weapons. Republican sympathizers cooperated at many points along the weapons shipping routes to Dublin. For example, furniture might be filled with weapons at a warehouse and customs in both New York and Dublin could be taken care of with a phone call. These supply lines were often set up with Irish émigrés in the United States. One network alone shipped hundreds of collapsible (and concealable) ArmaLite AR-18 rifles during the 1970s. The security forces confiscated more than 700 weapons, two tons of explosives, and more than 150,000 rounds of ammunition in 1971 alone, most of which came from the United States.[115]

---

[115] English, *Armed Struggle*, 117; Bishop and Mallie, *The Provisional IRA*, 233–235.

Figure 6: The ArmaLite AR-18 (upper) and the RPG-7 anti-tank rocket launcher (lower).[116]

The fortunes of the later PIRA turned on the relationship they established with Colonel Muammar Gaddafi in 1972. The head of Libya saw himself as an enabler of revolutionary movements around the world and agreed to meet with an IRA representative. Shipments began to flow to Dublin in 1973, often without success. The *Claudia* was boarded by Irish authorities to reveal that the PIRA was willing to accept large amounts of money and weapons from the state sponsor. The relationship was a long one, further reinforced with the later leadership of the PIRA under Adams.[117] On November 1, 1987, French authorities captured the *Eksund*, which was carrying more than 150 tons of armaments: 1,000 AK-47s, one million rounds of ammunition, 430 grenades, twelve rocket-propelled grenade launchers, twelve DHSK machine guns, more than fifty SA-7 surface-to-air missiles, 2,000 electric detonators, 4,700 fuses, 106-mm cannons, anti-tank missiles, and two tons of Semtex.[118]

COUNTERMEASURES TAKEN BY THE GOVERNMENT

After the outbreak of the violence of 1969, the British parliament and the O'Neill administration enacted a few reforms intended to

---

[116] (Upper) "File:AR-18.jpg," *Wikipedia*, accessed March 14, 2011, http://upload.wikimedia.org/wikipedia/commons/2/22/AR-18.jpg; (lower) "File:Rpg-7.jpg," *Wikimedia Commons*, accessed March 14, 2011, http://upload.wikimedia.org/wikipedia/commons/8/8d/Rpg-7.jpg.

[117] Bell, *The IRA, 1968-2000,* 184–185.

[118] Moloney, *A Secret History of the IRA*, 5–6, 242.

ameliorate the situation, including the establishment of a Community Relations Commission, legislation prohibiting incitement to racial hatred, and centralization of the local authorities and housing functions within the government. However, by the time the commission and central authorities were enacted and running, they were ineffective because of the entrenchment of the protests, as well as growing sectarian violence. The law prohibiting incitement based on race was little used because of the difficulty in securing prosecutors as well as the challenge in proving intent of incitement against an individual as opposed to against a general group, as stipulated in the law.[119]

As the British Army grew more entrenched in Northern Ireland and their counter-operations increased, the British decided to change the legal means by which they could arrest and incarcerate PIRA members to include internment without trial, the relaxation of evidentiary standards, and more aggressive interrogation tactics within the prisons. The move came, however, with a concession to grant the prisoners special rights with regard to prison amenities and status. Although this did not avoid the strong public reaction to the use of internment and harsh treatment, the special category status mattered a great deal to the PIRA, as evidenced by their dirty protests and hunger strikes when these rights and special status were revoked. The Maze/Long Kesh prison, built specifically to house the paramilitary members, was a means to allow military protection and to lessen the possibility of escape, which had become a problem.

Politically, the British government initially coerced the Irish Republic to cease any support of or obvious sympathy with the PIRA and soon convinced them to take measures against gun running, importation of arms, and open financial solicitation. As it became more apparent that only politics could solve the situation, agreements between the governments allowed the Republic to advise and participate in talks and also allowed for the removal of claims on the territory and the removal of mandatory reunification as a goal.

SHORT- AND LONG-TERM EFFECTS

CHANGES IN THE ENVIRONMENT

Reactions to the Good Friday Agreement demonstrated that the population of Northern Ireland was tired of the violence and the off-again /on-again peace process. "There are no big celebrations and

---

[119] Michael J. Cunningham, *British Government Policy in Northern Ireland, 1969-2000* (Manchester, UK: Manchester University Press, 2001), 8–9.

there are no big arguments either. I think everybody just realizes that it has to work this time," said a bartender and firm republican supporter. The reaction from much of the republican-supporting population was a resigned wait-and-see. The unionists felt more betrayed, seeing their leaders agree to a power-sharing arrangement with those they regarded as terrorists. The leaders from both sides engaged in a massive publicity campaign with their respective constituents and pushed for a "yes" vote on the referendum. The desire to try for peace again, even without much hope for success, won out over the disagreements over the compromises that were reached, and the referendum passed.

CHANGES IN GOVERNMENT

The Good Friday Agreement was signed on April 10, 1998. The talks were chaired by former US Senator George Mitchell, Canadian General John de Chastelain, and Finnish ex-PM Harri Holkeri and produced a sixty five-page document establishing the way to peace that had to be voted on in a general election that year in both Northern Ireland and the Republic. The referendum passed in Northern Ireland with 71% of the vote and in Ireland with more than 94% of the vote.

The agreement created a power-sharing Assembly to replace the Stormont and stipulated that popular vote should decide any change to the constitutional status of Northern Ireland, such as unification with the Republic. It called for the release of political prisoners and the decommissioning of all paramilitary weapons within two years, as well as the institution of reforms for human rights and for the police force.

The new Assembly was suspended when the DUP, headed by Rev. Ian Paisley, refused to abide by the agreement, and members of Sinn Féin were arrested on charges of collecting intelligence for the PIRA in 2002.[120] The PIRA men were acquitted in 2005, and negotiations continued between the parties until 2007 when a new Transitional Assembly was created with new power-sharing agreements. The unionist DUP and Sinn Féin co-shared the powers of government, with the DUP acting as First Minister and the republicans as Deputy First Minister.

---

[120] One of those arrested was Denis Donaldson, a veteran Sinn Féin member who confirmed that he was a British MI5 spy for more than twenty years. He also claimed that the intelligence collection for which he had been arrested was a scam created by the British special branch. Donaldson was killed less than five months after this admission. "Mystery of Sinn Féin Man Who Spied for British," accessed February 22, 2010, http://www.guardian.co.uk/uk/2005/dec/17/northernireland.northernireland, and "Sinn Fein British Agent Shot Dead," accessed February 22, 2010, http://news.bbc.co.uk/2/hi/uk_news/northern_ireland/4877516.stm.

CHANGES IN POLICY

The peace agreements since 1998 have stipulated that a majority vote of the Northern Irish population is required for the country to join the Republic of Ireland, and as part of process, the Republic has renounced all territorial claims to the northern counties. A number of Northern-Republic joint councils and British-Republic joint councils were established to negotiate and administer policies. The Irish Republic also has a consulting role concerning certain matters of Northern Irish policy.

CHANGES IN THE REVOLUTIONARY MOVEMENT

In September, 2005, the man in charge of the PIRA's decommissioning, General John de Chastelain, said that the group's weapons had been destroyed "beyond any shadow of doubt." The final event was observed by the General and two men, a Catholic priest and a Methodist minister, but without any representative of the Unionist parties being present or any photographs being taken.[121]

Sinn Féin emerged from the Troubles as one of the largest political parties in Northern Ireland and as one of four parties[122] holding executive positions in the Northern Ireland Assembly. Under the 2007 power-sharing agreement, the First Minister and Deputy First Minister positions were held by the DUP and Sinn Féin, respectively.

The original organization, the Official IRA, continued its socialist positions and officially ended its armed struggle in 2010, decommissioning its weapons in February under the same auspices as the PIRA.[123] When the PIRA discussed the end of abstentionism in 1986, a few disillusioned members, including a few original PIRA members, broke and founded the Republican Sinn Féin, which soon created its own military wing, the Continuity IRA. This group maintained a low level of armed activity after the 2010 decommissioning of weapons by the Official IRA. The Real IRA was created by PIRA members upset with a cease-fire and peace talks by the PIRA. In 1997, the members tried to gain control of the General Army Convention but were outmaneuvered by Gerry Adams's supporters. They promptly started the Real IRA and began a bombing campaign. The bloodiest event of

---

[121] "IRA 'Has Destroyed All Its Arms,'" accessed February 22, 2010, http://news.bbc.co.uk/2/hi/uk_news/northern_ireland/4283444.stm.

[122] Sinn Féin, DUP, Social Democratic and Labour Party, and UUP hold the executive positions in the 2010 Assembly.

[123] "Official IRA Gets Rid of Weapons," *BBC News*, accessed February 22, 2010, http://news.bbc.co.uk/2/hi/uk_news/northern_ireland/northern_ireland_politics/8504374.stm.

the entire thirty-year insurgency was a Real IRA car bomb on August 15, 1998, in Omagh, which killed 29 people, wounded 220 wounded, and prompted a short cease-fire. The Real IRA, however, continued its low-level campaign against the new Northern Ireland government on the basis of its belief that violence, not peace, was the proper means for achieving unification.[124]

BIBLIOGRAPHY

Bell, J. Bowyer. *The IRA, 1968-2000: Analysis of a Secret Army*. Cass Series on Political Violence. Portland, OR: Frank Cass, 2000.

———. *The Secret Army: The IRA*. New Brunswick, NJ: Transaction Publishers, 1997.

Bew, Paul, Peter Gibbon, and Henry Patterson. *The State in Northern Ireland, 1921-72: Political Forces and Social Classes*. New York: St. Martin's Press, 1979.

Bishop, Patrick, and Eamonn Mallie. *The Provisional IRA*. London: Corgi Books, 1987.

Cameron Report. *Disturbances in Northern Ireland: Report of the Commission Appointed by the Governor of Northern Ireland*. Belfast, Northern Ireland: Her Majesty's Stationary Office, 1969.

Campaign for Social Justice. *Northern Ireland: The Plain Truth*. 2nd ed. Dungannon, Northern Ireland: Campaign for Social Justice, 1969.

Compton, Paul A. *Northern Ireland: A Census Atlas*. Dublin, Ireland: Gill and Macmillan, 1978.

Cunningham, Michael J. *British Government Policy in Northern Ireland, 1969-2000*. Manchester, UK: Manchester University Press, 2001.

Darby, John, and Minority Rights Group. *Scorpions in a Bottle: Conflicting Cultures in Northern Ireland*. London: Minority Rights Group, 1997.

Doherty, Paul, and Michael A. Poole. "Ethnic Residential Segregation in Belfast, Northern Ireland, 1971-1991." *Geographical Review* 87, no. 4 (1997): 520–536.

English, Richard. *Armed Struggle: The History of the IRA*. New York, NY: Oxford University Press, 2003.

Fahey, Tony, Bernadette Hayes, and Richard Sinnott. *Conflict and Consensus: A Study of Values and Attitudes in the Republic of Ireland and Northern Ireland*. Leiden, The Netherlands: Brill, 2006.

Horgan, John, and Max Taylor. "The Provisional Irish Republican Army: Command and Functional Structure." *Terrorism and Political Violence* 9, no. 3 (1997): 1–32.

---

[124] English, *Armed Struggle*, 251–252, 318.

Moloney, Ed. *A Secret History of the IRA.* New York: Penguin Books, 2007.

The Northern Ireland Government, *Why the Border Must Be: The Northern Irish Case in Brief.* Belfast, Northern Ireland: The Northern Ireland Government, 1956.

Northern Ireland Statistics and Research Agency. *Northern Ireland Census for 2001: Key Statistics.* Belfast, Northern Ireland: Her Majesty's Stationary Office, 2005.

———. *Northern Ireland Census for 2001: Key Statistics for Settlements.* Belfast, Northern Ireland: Her Majesty's Stationary Office, 2005.

———. *Theme Tables. 1. Demography: People, Family, and Households.* Belfast, Northern Ireland: Her Majesty's Stationary Office, 2001.

O'Dochartaigh, Niall. *From Civil Rights to Armalites: Derry and the Birth of the Irish Troubles.* Cork, Ireland: Cork University Press, 1997.

Pointer, Graham. "The UK's Major Urban Areas." In *Focus on People and Migration 2005*, edited by the Office of National Statistics, United Kingdom. London: Office of National Statistics, 2005.

Sutton, Malcolm. *An Index of Deaths from the Conflict in Ireland, 1969–1993.* Belfast: Beyond the Pale Publications, 2001

Smith, David John, and Gerald Chambers. *Inequality in Northern Ireland.* New York: Clarendon, 1991.

The Northern Ireland Government. *Why the Border Must Be: The Northern Irish Case in Brief.* Belfast, Northern Ireland: The Northern Ireland Government, 1956.

Whyte, John. "How Much Discrimination was there Under the Unionist Regime, 1921 - 1968?" In *Contemporary Irish Studies.* Edited by Tom Gallagher and James O'Connell. Manchester, UK: Manchester University Press, 1983.

SECTION III

·················

REVOLUTION TO DRIVE OUT
A FOREIGN POWER

GENERAL DISCUSSION

The presence of a foreign military or the dependence of a regime on overt and pervasive foreign military support often sparks opposition groups that are able to utilize this fact as a unifying narrative. We have chosen cases for this section on the basis of the overriding use of this motivational technique, as well as aspects of the movements' operations that are primarily based against the foreign military, rather than the national government. There is, of course, a fine line to these distinctions, and we readily admit that others would have categorized some of these cases differently.

Like the use of ethnic or identity divisions to fuel an underlying need for change in the previous section, the revolutions' use of the foreign power as the "other" is a powerful technique in establishing grievances and providing motivations. The foreign army provides an easy, identifiable group toward which active and psychological operations can be directed more easily than if the military were made up of a similar ethnic and/or national composition as the revolutionary group. The classification of an "out-group" enables individuals or groups to exploit preexisting biases toward "in-group" cohesion while assigning a negative identity to a clearly defined adversary.

Operationally, the distinctness of the foreign military provides advantages to the insurgent groups as well. The population can be easily trained to fear the foreign power (often at its own behest), as well as to fear the violence that takes place when the foreign military is nearby, due to the conflict with the revolutionary group. The separation of the foreign military from the population denies them any popular legitimacy and can isolate the foreign power for easier targeting. Another advantage of the visual distinctness of the foreign power is that the revolutionary group can utilize their "sameness" with the local population to blend in before and after operations. By avoiding any distinguishing visual characteristics, the foreign military has to resort to pervasive searches and checks to provide security, often at an exorbitant cost.

Another characteristic found with revolutions with foreign military presence is what may be called a "jujitsu" strategy. We can infer that regimes that use (or are forced to use) a foreign military presence within their borders do so because of the strength that said military provides. This strong military presence may vastly outweigh the numbers, firepower, and training of the insurgent groups (such as the Afghan Mujahidin). In such cases, the small groups try to use the strength and reach of the foreign military against themselves. Operations of the insurgent group may be designed to expand the foreign military's geographic reach, stretching them as far as possible. Terror and guerilla tactics are preferred so as to limit losses and are designed to harass and provoke more than attrit.

These provocations are also meant to draw the military into harsh overreactions. The moral victories attained when a force uses overly aggressive tactics to suppress or counter smaller operations by the opposition can be far greater than the materiel or personnel losses incurred. Using the strength of the military to anger the populace, produce moral outrage, and weaken the resolve of both the military and its population in its home country have proven very successful tactics over the past forty years, from the US experience in Vietnam to the Soviet's in Afghanistan.

AFGHAN MUJAHIDIN: 1979–1989

Sanaz Mirazei

SYNOPSIS

The Afghan Mujahidin waged a successful insurgency against its Soviet-backed communist government. Beginning in 1979 with the Soviet invasion of troops, the Mujahidin were able to effectively organize an armed resistance largely from Pakistan where the party's headquarters and leaders were based. The Mujahidin were largely organized into two categories: traditionalists and fundamentalists. Although both of these groups' main goal was to remove the Soviet-backed government from power, the fundamentalists also wished to create an Islamic government for the country. These two dimensions of the Mujahidin were then divided into several factions: seven predominately Sunni groups and several Shi'a factions. Waging a multidimensional war and strategically using the geography, the Mujahidin, who seemed at first to be at a disadvantage, were able to expel the Soviets from their homeland by acquiring significant foreign backing from the United States, Saudi Arabia, and other countries. As a result of their insurgency, the Mujahidin established and led an Islamic government in Afghanistan.

TIMELINE[1]

1953	General Mohammed Daoud becomes prime minister. Soviet Union gives economic and military assistance.
1963	Forced resignation of Mohammed Daoud.
1973	Republic declared after Mohammed Daoud seizes power in a coup.
1978	General Daoud is overthrown and killed in a leftist coup by the People's Democratic Party of Afghanistan (PDPA). Conservative Islamic leaders begin armed revolt in countryside.

---

[1] Adapted from BBC News, "Timeline: Afghanistan," March 18, 2012, http://news.bbc.co.uk/2/hi/1162108.stm.

1979	Leftist leader Hafizullah Amin wins power struggle. Revolts in countryside continue, and Afghan army faces collapse. Soviet Union sends in troops to help remove Amin, who is executed.
1980	Babrak Karmal, leader of Parcham, is installed as ruler and backed by Soviet troops. Mujahidin resistance fights Soviet forces. The United States, Pakistan, China, Iran, and Saudi Arabia supply money and arms.
1985	Mujahidin come together in Pakistan to form an alliance against Soviet forces. Gorbachev says he will withdraw troops from Afghanistan.
1986	Najibullah replaces Babrak Karmal as head of the Soviet-backed regime.
1989	Last Soviet troops leave. Civil war continues as Mujahidin try to overthrow Najibullah.
1991	The United States and Soviet Union end military aid to both sides.
1992	Mujahidin resistance captures Kabul, and Najibullah falls from power. Rival militias fight for influence.
1993	Mujahidin factions agree to form a government with an ethnic Tajik, Burhanuddin Rabbani, as president.
1994	Factional contests continue, and the Pashtun-dominated Taliban emerge as a major challenge to the Rabbani government.
1996	Taliban seize control of Kabul and introduce a hard-line version of Islam. Rabbani flees to join anti-Taliban northern alliance.

THE ENVIRONMENT OF THE REVOLUTION

PHYSICAL ENVIRONMENT

Figure 1. Map of Afghanistan.[2]

Afghanistan lies east of Iran, north and west of Pakistan, south of the Central Asian states of Uzbekistan, Turkmenistan, and Tajikistan, and southwest of China, encompassing approximately 652,290 square kilometers, roughly the area of Texas. Afghanistan features a rugged, mountainous terrain with a major mountain range, the Hindu Kush, which runs northeast to southwest and divides the northern provinces from the rest of the country. Flanking the Hindu Kush mountains are fertile but isolated valleys and the deserts and river valleys. A landlocked country, Afghanistan primarily relies on water from surrounding rivers. Having limited natural fresh water resources and inadequate supplies of potable water are sources of environmental concern for the country. The climate is arid to semi-arid.

Afghanistan is divided into eleven geographic zones. The first six, including the Wakhan Corridor-Pamir Knot, Badakhshan, the Central Mountains, the Eastern Mountains, the Northern Mountains and Foothills, and the Southern Mountains and Foothills, are located in the mountainous region of the Hindu Kush. The remaining five are

---

[2] Central Intelligence Agency, "Afghanistan," accessed March 15, 2011, https://www.cia.gov/library/publications/the-world-factbook/maps/maptemplate_af.html.

mostly plains and dessert plains. They are the Turkistan Plains, Herat-Farah Lowlands, Sistan Basin-Helmand Valley, Western Stony Desert, and Southwestern Sandy Desert.

CULTURAL AND DEMOGRAPHIC ENVIRONMENT

Afghanistan has a population of approximately 17,000,000, with almost 1,000,000 having died in recent wars through 2001 and a refugee population of close to 5,000,000.[3] Afghanistan represents a culturally diverse country with several, often clashing, ethnic and linguistic groups. Organized by tribal and kin lineages that mobilize the people both politically and economically, Afghanistan's deep ethnic divisions have been the source of several domestic conflicts.[4] Afghanistan's patrilineal society consists of approximately 40–50% Pashtuns, 25% Tajiks, 9% Uzbeks, and 12–15% Hazara and minor ethnic groups, including the Chahar Aimaks, Turkmen, Baloch, and others.[5] Corresponding with the different ethnic groups, there are also several spoken languages in Afghanistan. Approximately 50% speak Pashtu, 35% speak Dari, 11% speak Turkic languages (primarily the Uzbek and Turkmen), and about 4% speak any of thirty minor languages (primarily Balochi and Pashai).[6]

The majority of Afghans—approximately 99% of the population—are Muslim.[7] There are a few tens of thousands Hindus and Sikhs, plus small numbers of Armenian Christians and Jews in the major cities.[8] Roughly 85% of the Muslims are Sunnis of the Hanafi school. The rest are Shi'a, and most are Twelver Shi'a, like those of Iran.[9]

With a long history of patrilineal organization, this male-dominated society is rooted in the Islamic tradition, which dominates Afghanistan. The male gender controls both the private and public spheres and few rights are given to women, who are subjected to forced covering of their bodies by head-to-toe burqas, arranged marriages, and other Islamic practices. Women are precluded from political life and are deprived of the opportunity to get an education.[10]

---

[3] Nasreen Ghufran, "The Taliban and the Civil War Entanglement in Afghanistan," *Asian Survey* 41, no. 3 (May, 2001), 462–487.

[4] Barnett Rubin, *The Fragmentation of Afghanistan: State Formation and Collapse in the International System* (New Haven, CT: Yale University Press, 2002).

[5] Ibid.

[6] Ibid.

[7] Ibid.

[8] Ibid.

[9] Ibid.

[10] Ahmed Rashid, *Taliban: Militant Islam, Oil, and Fundamentalism in Central Asia* (New Haven, CT: Yale University Press, 2000).

Instead, they are usually married at very young ages and expected to stay at home, raise families, and maintain the honor of the family, tribe, and kin group. These practices were especially present after the Taliban rule. The Taliban sought not only to combat the corruption rampant in Afghanistan but also to reinforce Islam in the everyday lives of Afghans. As such, they "issued decrees in which they required men to wear turbans, beards, short hair, and shalwar kameez."[11] The Taliban also banned "music, games, any representation of the human or animal form, and entertainment including television, chess, kites, cards, etc."[12]

SOCIOECONOMIC ENVIRONMENT

Before 1978, Afghanistan was a rigidly feudal, agrarian country.[13] With the exception of Kabul, the country had practically no electricity, no railroads, and very few highways.[14] Estimates from the UN and other sources suggest that Afghanistan was among the poorest and least developed countries in the world. In 1975, the country's per capita income was only $160, and two-thirds of the national income came from agriculture. More than 85% of the population was engaged in subsistence rural cultivation, and an additional 9% was engaged in nomadic pastoralism.[15]

Beginning in the 1970s, revenue from trucking between Afghanistan and Pakistan was a growing economic source and led to the development of a smuggling economy. Yet another illicit economy in Afghanistan was poppy cultivation for opium. Poppy cultivation in Afghanistan increased in the mid-1970s when opium crops in the Southeast Asian countries declined. By the late 1970s, poppies were grown and cultivated in more than half of Afghanistan's provinces.[16]

HISTORICAL FACTORS

In 1973, with help from the Parcham faction, Mohammad Daoud, the former prime minister who had been out of power for ten years, staged a military coup and established a republic that ruled until 1978,

---

[11] Ibid.

[12] Ibid.

[13] M. Nazif Shahrani and Robert L. Canfield, eds., *Revolutions and Rebellions in Afghanistan: Anthropological Perspectives* (Berkeley: Institute of International Studies, University of California, 1984), 10.

[14] Ibid.

[15] Ibid.; Rubin, *The Fragmentation of Afghanistan*, 32.

[16] Shahrani and Canfield, *Revolutions and Rebellions in Afghanistan*.

ending almost 150 years of monarchic rule. During this time, Daoud heavily suppressed the opposition and sought to promote Soviet-influenced communist elements in his government. While the leftist parties were battling internal power struggles, a group of students at Kabul University named the Muslim Youth Organization were studying and discussing political Islam and critiquing Marxism. With links to Sayyid Qutb's Muslim Brotherhood in Egypt, this organization started to become a direct threat to the Daoud and subsequent Democratic Republic of Afghanistan (DRA) regime. In response, Daoud launched a violent attack on the Muslim Youth Organization, imprisoning and killing several of its leaders. In response, the Muslim Youth retaliated with armed attacks against the regime in the summer of 1975.[17] Although at first the Muslim Youth appeared successful, eventually the government squelched their efforts, forcing the group underground or into exile. It is argued that for these students, this uprising represented an attempt at Che Guevara-style *foco* insurgency. In this model, "a guerrilla band enters a rural area where it has never operated before with the hope of serving as the 'insurrectional focus.'"[18] These events led to the beginning of the Islamic Mujahidin movement, as well as to internal factions within this Islamist movement.

Islam offered a concrete framework with which the Afghans were familiar, understood, and had long practiced. However, the idea that Islam could be political was new and was readily embraced, especially by those who felt that the communism imposed by the new government was a foreign, imported concept.[19] As political Islam gathered a more domestic following, communist counterparts with support from the Daoud regime also drew significant attention. In 1977, the two major communist groups Khalq and Parcham reunited after ten years of separation. However, this coalition was rather short lived. When the Saur Revolution took place in 1978, the Khalq–Parcham party faced very little resistance because the Daoud regime expunged all of the opposition and made political parties and organization illegal. However, power struggles between the two factions in the summer of 1978, as a result of the inability of the regime to effectively deal with the rising domestic unrest, Islamic *jihad,* and tense relations with the Soviets, caused the party to again split, with the Khalq faction winning.[20]

---

[17] Shahrani and Canfield, *Revolutions and Rebellions in Afghanistan,* 42.

[18] Rubin, *The Fragmentation of Afghanistan,* 104.

[19] Grant M. Farr and John Merriam, eds., *Afghan Resistance: The Politics of Survival* (Boulder, CO: Westview, 1987), 64.

[20] Shahrani and Canfield, *Revolutions and Rebellions in Afghanistan,* 44, 60.

GOVERNING ENVIRONMENT

Afghanistan has been ruled by several different governments over the last one hundred years. Traditionally, Afghanistan has been organized by two power structures. The first is the local government administration, which is directed from the central government. The second is the tribal or village structures within each region.[21] Mohammad Daoud, a politician who overthrew the previous monarchy and became president in 1973, attempted to maintain these two separate power structures, allowing local systems to maintain self-governance, usually through tribal institutions.[22] After the Daoud regime, these two power structures came into conflict as the old regime's elites, which held most of the official government power, and the local village leaders, or khans, clashed.[23] This tension resulted in an attempt to consolidate the two power structures into one communist government.

From 1979 to 1989, Afghanistan was led by a Soviet-style communist government until the Soviets withdrew in 1989, after which the Mujahidin led the new government. On April 27, 1978, in what is known as the Saur Revolution,[24] the PDPA overthrew the Daoud government in a bloody coup. The PDPA then installed a new communist government, referred to as the Democratic Republic of Afghanistan. Subsequent to killing Daoud and most of his family, Nur Muhammad Taraki, leader of the PDPA, became the new president and prime minister of the new Afghani government. This fledgling government faced almost immediate opposition when, during the first eighteen months of its rule, the PDPA forced a communist reform program that directly challenged many deep-rooted Afghan customs and traditions. Because of how unpopular these new reforms were and how strong the opposition was to these policies, the new government took severe measures—including imprisonment, torture, and executions of the traditional elders, the religious establishment, and intellectual leaders—to maintain its position and authority.[25]

A few months later, a revolt in eastern Afghanistan gave way to a larger countrywide insurgency. Facing increasing domestic pressure from the insurgency, the new government in Afghanistan signed a bilateral treaty of friendship and cooperation with the Soviet Union in December. As a result of this treaty, Soviet military assistance greatly

---

[21] Ibid., 170–171.

[22] Rubin, *The Fragmentation of Afghanistan*, 151.

[23] Ibid., 184.

[24] Saur is April in the Afghan language.

[25] Bureau of South and Central Asian Affairs, "Background Note: Afghanistan," US Department of State, accessed September 16, 2010, http://www.state.gov/r/pa/ei/bgn/5380.htm.

increased, causing the fledgling regime to become more dependent on this foreign assistance to combat the spreading insurgency as the Afghan army began to collapse.

In addition to external conflict between the new government and the people, the PDPA itself was in disarray. It split into the two factions from which it was created: the Khalq and Parcham. In 1978, Khalq had exiled and expelled leading Parchami members and others were arrested, tortured, and even executed. By October 1979, Afghan–Soviet relations began to deteriorate because the new Afghanistan administration, led by Hafizullah Amin, did not agree with Soviet recommendations on how to deal with the domestic instability. As a result, in December 1979, the Soviet Union invaded Kabul, killed Amin, and installed Babrak Karmal, leader of the Parcham faction, as prime minister.[26] At first, the traditional elites, especially in rural Afghanistan, stood idly by as they observed the fallout from the Soviet invasion. Even though they were in a better position to organize an opposition to the regime, it was the religious leaders—who pointed to the un-Islamic and, therefore, un-Afghan nature of the invasion and new regime—who were able to gather support for resistance. In this regard, they were able to consolidate support for a *jihadi* war.[27]

The new Karmal government terminated several PDPA reform programs that were coercively enforced and were a cause of major revolt in Afghanistan. However, given that Karmal was installed by a foreign power and not domestically, the revolts quickly turned into a nationwide uprising led by Mujahidin forces.[28] Moreover, the new Karmal regime was unable to establish authority outside of Kabul. It is estimated that as much as 80% of the countryside evaded the new government.[29] Organized by the Mujahidin, the new regime was unable to maintain effective government control outside of urban centers. Seeking to establish its legitimacy, the new government banned all opposing political parties, leaving the Afghan people with no way to legally organize the struggle for the human rights of which they were deprived during this time.[30] In addition, the new government increased surveillance of the people through the armed forces and the secret police.[31]

As the quest for maintaining its legitimacy continued both internally and externally by the Soviets, by 1984, the Afghan government seemed to lack definitive leadership and ability to effectively contain resistance

---

[26] Ibid.

[27] Farr and Merriam, *Afghan Resistance*, 64.

[28] Rubin, *The Fragmentation of Afghanistan*, 122.

[29] Bureau of South and Central Asian Affairs, *Background Note: Afghanistan*.

[30] Shahrani and Canfield, *Revolutions and Rebellions in Afghanistan*, 10.

[31] Rubin, *The Fragmentation of Afghanistan*, 130.

by the Mujahidin. As a result, the top three defense generals under the minister were replaced with Khalqis dominating the new positions.[32]

WEAKNESSES OF THE PREREVOLUTIONARY ENVIRONMENT AND CATALYSTS

In response to the Soviet occupation, the Mujahidin resistance was aroused by three main catalysts.[33] The first was a response to the political ideology of the new government—chiefly Soviet-style communism. Second, the Mujahidin were dismayed with the Soviet-led political domination of Afghanistan. This political domination was embodied by several changes to Afghan legislation, reorganizing the community and challenging long-held cultural and religious beliefs, such as the role of women in society and new land laws. Finally, as a consequence of this Soviet domination, the new communist Afghan government severely oppressed the resistance, engaging in torture, killings, and other human rights violations. Some would argue that the resistance movement, however, started much earlier, during the period when Daoud was prime minster (1953–1963) and again when he was president (1973–1978).[34]

One of the main factors attributed to the formation of the resistance movement Mujahidin spearheaded was the imposition of the Soviet-style government embodied in the social reform programs. After only two weeks in power, the ruling PDPA party announced the new social "democratic" reforms in May 1978 in a radio broadcast. This thirty-point program included several different items aimed at creating a more democratic government. However, many of these programs, especially Revolutionary Decrees Nos. 6, 7, and 8, really touched a nerve with the largely traditional, Islamic public, causing widespread dissent and dissatisfaction with the newly established government. These three programs in particular addressed land mortgage and indebtedness; the democratic rights of women (including a limitation on brideprice and allowing women freedom of choice in marriage); and land reform through confiscation and redistribution.[35] Although the new government sought to instill these reform programs in an effort to modernize and liberalize Afghanistan, they directly challenged the social fabric of the country, causing widespread agitation with the extent to which the government was interfering with the everyday life of the average citizen.

---

[32]   Ibid., 131.

[33]   Farr and Merriam, *Afghan Resistance*, 13.

[34]   Ibid., 21.

[35]   Shahrani and Canfield, *Revolutions and Rebellions in Afghanistan*, 12.

Figure 2. Afghan Mujahidin return to a village after its destruction by Soviet forces.[36]

The reform programs, especially with regard to economic aspects of land reform and land mortgage, exacerbated the already poor position of the peasants and middle class. Peasants were unable to meet government taxes and other costs that were socially expected of them, and therefore, would take out loans at very high rates, often using their land as security. Furthermore, given Afghanistan's rough terrain, lack of water, and erratic weather combined with the people's low level of agricultural technology, a deficit of knowledge in how to maximize returns from the little arable land in the country was more important to the people than merely the equal distribution of that land. However, the Khalq-Parcham party did not share these beliefs, and, in the end, this disconnect alienated the people from their new government and caused an intensification of the armed resistance. In addition, the PDPA programs failed because the people did not understand the communist ideology behind the reforms; the mechanisms by which the PDPA attempted to install the reforms were inappropriate and ill-equipped for any positive action; and finally, at a local level, tribal and religious organizations were more efficient than the central government.[37]

A second motivation for the Mujahidin movement was that the reform policies were directly associated with the foreign concept of

[36] DefenseImagery.mil, DD-ST-86-06668, accessed March 15, 2011, http://www.defenseimagery.mil/imagery.html#a=search&s=DD-ST-86-06668.

[37] Shahrani and Canfield, *Revolutions and Rebellions in Afghanistan*, 180.

Soviet-style communism. The Afghan people did not like the fact that their central government was acting on behalf of a foreign superpower—in this case, the Soviets. The ruling party eventually became totally dependent on Soviet command, especially after 1979. The people's worries about communism became manifest when the Soviets invaded, increasing coercive measures against suspected enemies and implementing largely unpopular new reforms.[38] In response, the people turned to a domestic, viable alternative—Islam. The choice of an Islamic *jihad* as a solution to the foreign communist problem was welcomed by many Afghans.

Finally, the third catalyst was the coercive tactics displayed by the institutionally weak and ideologically detached new government. The PDPA's reliance on force, especially at the local level where it was unable to gain respect from the people, increased the intensity of the armed resistance.[39] In a quest to establish its legitimacy, the PDPA was forced to use coercion in order to prove its strength and position. Many were killed, tortured, imprisoned, and sent into exile in Pakistan and other areas. This led to underground efforts by some of the resistance and forced others to gather arms to defend their way of life. Islam became an important counterbalance to the foreign ideology that dominated the new government. Leadership by charismatic religious figures and ties to groups that were exiled and others that shared their political understanding, such as the Muslim Brotherhood, enabled these activists to set off a national revolt.

FORM AND CHARACTERISTICS OF THE REVOLUTION

OBJECTIVES AND GOALS

The main objective of the Mujahidin resistance was clear: to reclaim their land from the foreign communist power of the Soviet Union. According to which faction of the Mujahidin individuals belonged, secondary objectives changed. The traditionalists sought only to rid the country of the foreign-imposed communist ruling. In addition to regaining independence for Afghanistan, the fundamentalists also sought to defend Islam, which was under attack by the atheist, communist reforms, through an Islamic *jihad,* or religious war. They were weary of the way in which Western capitalist imperialism and communism had affected Islam in their country and wanted a

---

[38] Ibid., 86–84; Rubin, *The Fragmentation of Afghanistan*, 118.
[39] Shahrani and Canfield, *Revolutions and Rebellions in Afghanistan*, 23–24.

government that respected these traditional roots and a country where Islam was a revered part of their everyday life. Thus, the goal for these fundamentalists was not only to defend Islam, but to also wage an armed struggle to establish an Islamic social and political order in the country.

LEADERSHIP AND ORGANIZATIONAL STRUCTURE

From the onset of the Saur Revolution until Najibullah's overthrow, the Mujahidin faced two main problems. The first was that there was a "political vacuum and social fragmentation left by the old regime" and the second was "the necessity of forming that national leadership on foreign soil."[40] Before the 1979 Soviet invasion, there were some local resistance groups led by local elites; however, the Mujahidin lacked stable domestic groups because party leaders of most of the opposition had been either exiled to Pakistan or killed. Therefore, establishing regional or national leadership remained a serious challenge, especially given ethnic divisions between the groups. After the Soviet invasion and with increased foreign support, the Mujahidin groups became more organized. Most of the Mujahidin groups had the following simple command structure: "a commander and a small group of men linked to each other by some local social network."[41] While commanders did not usually align themselves with specific parties, the Mujahidin followed their commanders on the basis of local social networks.[42]

At the outset of the *jihad*, various strands of Mujahidin emerged. These different groups can be categorized into two alignments: traditionalists and fundamentalists.[43] There were seven different Mujahidin groups that all had external bases in Pakistan. There were three traditionalist organizations: *Jabhai Milli Nijat* (National Liberation Front, or NLF), led by Sibghatullah Mujadidi; (2) *Mahazi Milli Islami* (Islamic National Front) led by Sayyid Ahamad Gailani; and (3) Harakati Inqilab-I Islami (Islamic Revolutionary Movement), led by Mawlawi Muhammad Nabi Muhammadi. These traditionalists groups "include leading figures from the former regime, tribal chiefs and traditionalist religious leaders trained in nongovernmental religious institutions."[44] The fundamentalists organizations were: (1) *Jamiati Islami Afghanistan* (Islamic Society of Afghanistan, or JIA),

---

[40] Rubin, *The Fragmentation of Afghanistan*, 192.

[41] Ibid., 188.

[42] Ibid., 202.

[43] Farr and Merriam, *Afghan Resistance*, 64.

[44] Shahrani and Canfield, *Revolutions and Rebellions in Afghanistan*, 45–46.

led by Burhanuddin Rabbani; (2) *Hizbi Islami* (Islamic Party), led by Gulbudin Hikmatyar; (3) *Hizbi Islami* (Islamic Party), led by Mawlawi M. Yunus Khalis; and (4) *Itihadi Islami Baray Azadyi Afghanistan* (Islamic Alliance for the Liberation of Afghanistan), led by 'Abdur Rabbur Rasul Sayyaf.[45] The fundamentalists groups were made up primarily of Islamic activists from the rural and urban youth and middle to lower class.

The NLF was a traditionalist Islamic group led by Sibghatullah Mujadidi. This party formed a patrimonial structure around Mujadidi's family. They were typically moderate Islamists and strongly anticommunism. Although the different Mujahidin groups did cooperate at times, given that the character of the groups were so determined by their leaders and that the leaders did not always agree with each other, there was some intergroup antagonism. The NLF was especially antagonistic towards Hikmatyar, the leader of Islamic Party or *Hizbi Islami*. The latter was founded in 1978, and its principal goals were "to defend our national traditions" and "the establishment of an Islamic society in which all the political, economic, and social affairs [will] be founded on the teachings of Islam."[46]

Mahazi Milli Islami was the most nationalist of the three traditionalists.[47] Led by Sayyid Ahamad Gailani, the party maintained close ties with the old royalist regime. With a patrimonial organization, after the leader and his family, a series of elite from the old regime made up most of the party's second-tier leadership. Although the commanders did include *ulema*, or religious leaders, the majority of these positions were filled by khans who favored a moderate nationalist government. Gailani and his commanders led his troops in "small, traditionally organized units."[48]

Harakati Inqilab-I Islami, or the Islamic Revolutionary Movement, is another traditionalist–nationalist party. Led by Mawlawi Muhammad Nabi Muhammadi, this party, although it attracted a lot of members, was one of the worst organized and most corrupt.[49] Like the NLF, this party was also patrimonially organized. The party drew almost 90% of its commanders from *ulama*, or religious leaders.[50] Because of this party structure, this group also "received a significant number of weapons from the [Pakistan Inter-Services Intelligence] (ISI)."[51]

---

[45] Ibid., 46.
[46] Rubin, *The Fragmentation of Afghanistan*, 210–211.
[47] Ibid., 203.
[48] Ibid., 205.
[49] Ibid., 212.
[50] Ibid.
[51] Ibid., 213.

However, the organization did not gain the support of Arab Islamists who heavily financed the other parties.

Of the fundamentalist parties, JIA was more moderate, with a mostly Tajik base. Led by Burhanuddin Rabbani, commanders were given freedom in using their own strategies and organizational structures in their own groups because they were transferred property rights of the weapons.[52] This autonomy at the lower level enabled JIA to be one of the more effective resistance fronts. Another factor that helped JIA was its distance from the conflict-ridden Pashtun areas on the Pakistan border. In terms of foreign assistance, JIA had warm relations with the Arab Islamists. However, in 1986, because Rabbani met with President Ronald Reagan, the Arabs temporarily severed assistance. This did not damage JIA's ability to perform because it continued to have excellent relations with Pakistan.

The Islamic Party led by Gulbuddin Hikmatyar was the "most revolutionary and most disciplined of the Islamist parties."[53] With a strong Islamist core, the Islamic Party was heavily influenced by the Muslim Youth movement, of which Hikmatyar was a member. This helped to gain favor and funding with the Arab Islamists and Pakistanis. The party was also one of the best-organized groups, holding internal elections, giving membership cards to its followers, and having a strong central organization in which all weapons belonged to the party—not individual commanders like most of the other groups.[54] Recruitment for this party was based more on skill and ideology than merely being a member of a particular social group. The Islamic Party split in 1979, forming a second party led by Mawlawi M. Yunus Khalis. Islamic Party-Khalis was mostly led by fundamentalist, tribal *ulama*. Because of Khalis's Pashtun lineage, party members followed a strong tribal, Pushtun puritanism in which they sought to combine their tribal traditions with Islam.

The Islamic Alliance for the Liberation of Afghanistan, led by 'Abdur Rabbur Rasul Sayyaf, was another fundamentalist party that Pakistan recognized. This group practiced Salafi, or Wahhabi, Islam, making it a prime candidate for Arab funding, and it was favored particularly by Saudi Arabia. However, it did not draw as much support internally from the Afghan people because they did not identify with its ideology as much.

There were also several Shi'a parties that were active members of the Mujahidin resistance. Although they were not recognized

---

[52] Ibid., 220.

[53] Ibid., 213.

[54] Ibid., 214.

by Pakistan, they did receive support from Iran. There were four dominant Shi'a parties: the Shura, a traditionalist, rural mostly Hazara party; the *Harakati Islami* (Islamic Movement), a moderate Islamist party led by Ayatollah Asif Muhsini; *Subhi Danish* (Dawn of Knowledge), a group revived from the 1960s; and *Sazmani Nasri Islami Afghanistan* (The Islamic Victory Organization of Afghanistan), which was a Shi'a, Afghan youth organization.

There were several attempts to create coalitions between the seven Sunni groups even though they were led by very different leaders. In May 1985, the Islamic Unity of Afghan Mujahidin was formed with limited success. The new alliance represented both the traditionalist and fundamentalist groups. Islamic Unity had a rotating spokesman— one leader from each group would have a chance to be the alliance spokesman for three months. This rotating leadership enabled each group to have a chance to lead the coalition. However, each group maintained control over its own organization and finances. Although this coalition was meant to be a unifying voice for the opposition, the guerrilla operations did not come under unified command, which led to a considerable amount of waste in terms of money and ammunition, as well as manpower.[55] As a result, the three traditionalist parties withdrew from the Unity alliance in February and formed their own moderate alliance, which did not receive any extra financial support from the Arabs or Pakistan. With pressure from Arab funding, the Shi'a parties also formed coalitions. After the Soviet withdrawal, the Iranians combined the Shi'a parties into *Hizbi Wahdat* (Unity Party), which received significant Arab funding.[56]

COMMUNICATIONS

Given that the different Mujahidin parties were located in various parts of Afghanistan, communications became important for four reasons. First, it was needed to coordinate activities of each local area with their command teams. Second, it was used to broadcast the parties' activities to solicit further funding from foreign donors. Third, it was used to continue recruiting. And fourth, it helped the parties connect their activities with other Islamist movements, namely the Muslim Brotherhood in Egypt and the Pakistani Jam'ati Islami, from which they drew much of their beginning ideology.[57] In order to achieve these four goals, all of the Mujahidin groups made good

---

[55] Farr and Merriam, *Afghan Resistance*, 5–6.
[56] Rubin, *The Fragmentation of Afghanistan*, 223.
[57] Ibid., 84.

use of radio, newspapers, and telephones. Additionally, some of the fundamentalist groups, such as Islamic Party-Khalis, led Islamic sermons over the radio, wrote several Islamic publications, translated books, and taught in institutions and madrassas to spread their word and attract new recruits.[58]

METHODS OF ACTION AND VIOLENCE

Outside of a few uncoordinated attacks led from Pakistan by the Islamic Party, most of the other Mujahidin groups did not act from the Saur Revolution until the late summer of 1978. This waiting period could be attributed to the fact that the coup was not predicted and that some of the groups were willing to give the new government a chance.[59] However, the situation changed for these Mujahidin groups as the government kept imposing its communist–foreign agenda on the everyday lives of the people that were heavily exacerbated by the Soviet troops' invasion.

However, there were some smaller-scale attacks that were started by indigenous Afghan Mujahidin for the narrow self-interests of certain tribes. In October 1978, several Nuristani tribal groups attacked a central government post in Kamdesh. In a three-day-long battle, these Nuristanis captured the government post and acquired some weapons. After their success, they phoned the news to a nearby village in Bragimatal, which sparked these neighboring villages to follow a similar pattern of action to capture more weapons. In the next six months, these tribal groups in eastern Nuristan began the first successful insurgency against the new communist government and were effectually no longer controlled by the central government.[60]

The rebellion continued when a JIA group of *ulama* and peasants started an armed attack in June 1979. The group followed a pattern of action similar to that of the Nuristanis—it commandeered a central government post, killed the sub-district officer, and captured weapons and ammunition from the station. The JIA group, led by military leader and strategist Ahmad Shah Massoud, chose this area in Kuran for tactical purposes; the area had no drivable roads. After its success, the group broadcast its victory to groups in Peshawar who then sent more troops to Kuran for a sustained attack. Within ten days, the newly reinforced Mujahidin in Kuran spread to nearby Jurm, where they

---

[58] Ibid., 216.
[59] Shahrani and Canfield, *Revolutions and Rebellions in Afghanistan*, 66.
[60] Ibid., 77.

continued gather more weapons and ammunition as they captured that town.[61]

After the 1979 Soviet invasion, the Mujahidin began with sporadic attacks led by loose bands of guerrilla forces or *lashkars*, defined as "temporary aggregations of tribesmen and tribal segments, each intent upon outdoing its opposite numbers and, in the process, defeating the common enemy."[62] These *lashkars* were based on Afghan kin units. Each unit was organized according to intergroup lineages where they would traditionally compete against each other unless there was an outside force that threatened their inter-group lineage. In such a case, the tribes would unite to resist and defeat the invader.[63] When there were outside threats, in this case the Soviets and the Soviet-backed central government, these *lashkars* would bombard the adversaries and then retreat to the mountains, go home, and come back the next day.[64] Most fighters lived at home and were connected to their commander through traditional personal networks.[65] The commander often led the operations from their guest houses, and the fighters tended not to receive systematic training.[66] However this sort of *lashkar* fighting was expensive and labor intensive. Although the mountainous terrain favored this type of guerrilla activity, it was difficult for them to continue fighting in this way. In the beginning, Mujahidin used weapons appropriate for hand-to-hand fighting, such as flint rifles, breechloaders, Martini-Henris, Lee-Enfields, Nagants, and some old Soviet weapons.[67]

By 1980, the Mujahidin were able to stifle the central government's movements because the Mujahidin controlled most of the roadways with strategic strongholds, making the government's ability to control issues outside of major urban centers almost impossible.[68] This strategy contributed to the continued loss of the central government's legitimacy because the central government was unable to collect taxes and make payments to its workers.[69] As the Mujahidin weakened the new domestic government, the Soviet attacks in 1980–1982 became more oppressive as the Soviet troops launched massive offenses that

---

[61]   Ibid., 162.

[62]   Farr and Merriam, *Afghan Resistance*, 44.

[63]   Shahrani and Canfield, *Revolutions and Rebellions in Afghanistan*, 71.

[64]   Farr and Merriam, *Afghan Resistance*, 38.

[65]   Rubin, *The Fragmentation of Afghanistan*, 190.

[66]   Ibid., 190.

[67]   Olivier Roy, *Islam and Resistance in Afghanistan* (Cambridge, MA: Cambridge University Press, 1990), 184.

[68]   Bureau of South and Central Asian Affairs, *Background Note: Afghanistan.*

[69]   Rubin, *The Fragmentation of Afghanistan*; Roy, *Islam and Resistance in Afghanistan*, 189.

destroyed and depopulated many areas.[70] However, given that the Soviets were accustomed to fighting in flat terrains using a slew of tanks, the Mujahidin were able to have both a strategic and tactical advantage over the slow, highly visible Soviets by using the mountainous terrain to their advantage and using several small, mobile bands of offensive attacks. Moreover, the severe nature of the Soviet attacks led to some positive developments in the Mujahidin strategy. Because the villages of the guerrilla leaders had been destroyed and many of their families had been killed or had moved out of the area, the fighters had no worries other than meeting their goal of expelling the Soviet powers and restoring a representative Islamic government.[71]

Although the Mujahidin were relatively poorly armed in the beginning stage of the resistance and faced large setbacks when the Soviets arrived, not being able to compete with their sophisticated weapons, after 1984, when the Mujahidin received a sizeable amount of foreign funding for weapons and training from the United States,[72] they were able to make considerable progress. However, as the Mujahidin received more funding, better weapons, and training from foreign powers, the Soviets also improved their fighting capabilities, leading to more losses for the Mujahidin in 1986. Furthermore, given that the Mujahidin were not coordinated and suffered from inadequate tactics and strategies and poor leadership, the Soviets continued to win the upper hand.[73] In order to revive the movement and to equip the Mujahidin with the appropriate weapons to deal with heliborne assaults, the foreign powers increased funding and training initiatives. At this point in the resistance, the Mujahidin used AK-47 assault rifles,[74] as well as shoulder-held, laser-guided Stinger antiaircraft missiles that were supplied to them by the United States in September 1986.[75] By 1986, in addition to Stingers, the Mujahidin had a whole range of machine guns, including RPKs, RPDs, and Goryunovs; antitank weapons, such as RPG7s and RPG2s; DShK machine guns; grenades, including F1s, RDG-5s, RG-42s, and RKG-3Ms; Makarov and Takarev revolvers; KPV/ZPU antiaircraft machine guns; and portable antiaircraft missiles such as the SAM-7.[76]

The Mujahidin continued ambush tactics, which they preferred to set battles. In these ambushes, one or two groups provided cover

[70] Rubin, *The Fragmentation of Afghanistan*, 180.

[71] Shahrani and Canfield, *Revolutions and Rebellions in Afghanistan*, 72.

[72] Bureau of South and Central Asian Affairs, *Background Note: Afghanistan*.

[73] Farr and Merriam, *Afghan Resistance*, 4.

[74] Ibid., 14.

[75] Rubin, *The Fragmentation of Afghanistan*, 181.

[76] Shahrani and Canfield, *Revolutions and Rebellions in Afghanistan*, 185–186.

while one or two other group members used RPG7 antitank rocket launchers to move forward. When attacking government posts, the resistance used this ambush tactic and then sent one of their commandos in as close as possible to the post to open fire on possible targets. Then, once the area was secure, the rest of the group came in to mark the victory and capture any weapons and ammunition available. In addition, the Mujahidin had a few trained commandos to infiltrate enemy lines where they set explosives, mostly rockets, on some strategic targets, like electricity power stations. The Mujahidin relied heavily on mountains and caves as a natural cover; however, after 1984, they also began to dig trenches and build air raid shelters.[77]

Because of the lack of coordination among the Mujahidin groups, there were some attempts to create regional coalitions to have a greater impact against the Soviets and central government.[78] However, these attempts had mixed outcomes depending on their geographic location and the ethnic composition of the groups involved. In the Tajik-dominated northeast, Jamiat and Islamic Party-Hikmatyar worked alongside one another to meet their common goals of regaining their autonomy from the central government. Being strategically far from the more ethnically tense regions closer to Pakistan, they were able to make some progress in this region in establishing several institutions and managing their own affairs. In addition, Ahmad Shah Massoud, the region's commander, was able to draw recruits from local villages and create a stationary militia.[79] His efforts were successful in negotiating with the Soviets, and a one-year truce was reached in 1983. In response to new attacks from the Soviets in 1984, Massoud met with the elders to restrategize and develop a new council to execute the new plan. In this new initiative, the Mujahidin under Massoud's command were divided into two different groups: locals and mobile groups. The locals were to stay at the base and act as defensive units, some of whom were also part of the better armed shock troops. The mobile groups were self-sufficient professional soldiers who were to carry out air raids yet travel light to escape potential capture. These uniformed mobile groups, usually consisting of thirty-three men, had no territorial bases and "fought together in mixed units without regard to tribe, locality, or party."[80]

The northern parts of Afghanistan housed a majority of the government forces and mixed representation from the Mujahidin. However, there was no concerted effort by the Mujahidin in this

---

[77] Ibid., 183.
[78] Ibid.
[79] Ibid., 235.
[80] Ibid., 236–237. supra note 65, 181

region. Iran influenced the western region of Afghanistan. While some conflict between different Mujahidin factions in this region existed, and given that the geographic landscape of the region did not lend well to guerrilla style fighting, the Mujahidin in this area formed a more extensive military organization.

Another region in which the Mujahidin were not able to consolidate and coordinate efforts was in the central provinces around the capital of Kabul. Although the fighting in this area was intense, ethnic and ideological differences made cooperation difficult for the Mujahidin. Moreover, this region was close to Pakistan, enabling the foreign government to hinder the ability of the Afghan Mujahidin to gain autonomy.[81] The south also faced problems trying to coordinate efforts for reasons similar to those facing the central region.

The Soviets developed new strategies, including night ambushes, against the Mujahidin, which, by 1986–1987, had received considerable external funding, weapons, and training. The Mujahidin had successfully set strongholds on the strategic path from Kabul to the east and were able to move more than half of their weapons and ammunition to their destination. The Soviets were only successful in destroying one-third of the resistance's weapons, and the remainder were typically stolen by other Mujahidin groups. By 1987, the Mujahidin had been so successful that the majority of Soviet attacks were defensive.[82]

After the Soviet retreat, Najibullah governed Afghanistan as the Mujahidin formed a rival Afghan Interim Government with American and Pakistani advisement and created an alternative capital in Jalalabad. Although there was not a clear-cut Mujahidin victory, they were able to gain effective rule by 1992. As the Najibullah government tried to maintain its authority in the absence of Soviet backing and with the rising Mujahidin power, it sought to control Kabul, Mazar-i-Sharif, Kandahar, Herat, Jalalabad, and other smaller cities. However, the Mujahidin continued to attack government positions in these areas and launch rocket attacks at the capital of Kabul. The government was able to survive these attacks and attempted to maintain its political stronghold for two main reasons: first, it still had a substantial amount of resources from the Soviets; and second, the Mujahidin forces suffered from political, ethnic, and ideological differences.[83]

---

[81] Ibid., 241.

[82] Roy, *Islam and Resistance in Afghanistan*, 209.

[83] Peter Marsden, *The Taliban: War and Religion in Afghanistan* (London: Zed Books, 2002), 34–35.

Figure 3. An Afghan Mujahidin demonstrates positioning of a handheld surface-to-air missile.[84]

METHODS OF RECRUITMENT

In their efforts to mobilize support and attract recruits, the Mujahidin focused on several elements: a strong commitment to Islam; a recognized enemy—the Soviet power and the new communist government; the charisma of the group leaders; financial incentives; security benefits; and weapons.[85] Some groups, like the Islamic Party led by Hikmatyar, based their recruitment more on skill and ideology than social status, whereas others focused mostly on existing tribal structures. All of the groups also recruited fighters from areas "where tribal structures have broken down or which have a mixture of groups originating from different tribe."[86] An additional recruiting strategy of the Mujahidin was sharing security and financial benefits with the new members. One area of particular interest was the refugee community. The better-financed Mujahidin groups built schools to attract refugee youth to join their party. Finally, the parties used the foreign aid they received from donors, newspapers, and demonstrations as a way to recruit more followers.[87]

---

[84] DefenseImagery.mil, DD-ST-88-09407, accessed March 15, 2011, http://www.defenseimagery.mil/imagery.html#a=search&s=DD-ST-88-09407.

[85] Shahrani and Canfield, *Revolutions and Rebellions in Afghanistan*, 49.

[86] Rubin, *The Fragmentation of Afghanistan*, 215.

[87] Ibid., 210, 215.

METHODS OF SUSTAINMENT

Sustainment of the Mujahidin forces came primarily from foreign funding. However, there were some indigenous sources of income for the resistance. Until about 1982, *ushr* and *zakat*, Islamic alms giving, provided much of the first year's sustainment for the Mujahidin. Most of the groups also received donations, the amount of which is unknown. The Mujahidin also indirectly taxed some Afghan internal production, such as carpets, salt, trucks and transport, and emeralds and lapis lazuli.[88] As the Mujahidin received more foreign funding, they became less dependent on sustainment from the local population.

METHODS OF OBTAINING LEGITIMACY

The Mujahidin insurgency also took into consideration its relationship with society in general. Often the group leaders would consult with tribal elders and *ulama* to help consolidate the group's legitimacy. Some of the groups, like Islamic Party-Hikmatyar, did not consult with elders at all and only spoke to the *ulama*. Others took more intermediate positions. In addition, each group sought to reinforce its devotion to *jihad* and, as a result, promised security benefits for the local community.[89] Another way in which the Mujahidin forces secured their legitimacy was by bypassing the local community and establishing schools for fighters.

Because political parties were outlawed in Daoud's regime, none of the parties could claim that they "[represented] a national constituency through past electoral results or through any other form of activity."[90] Because the Soviets were using significant propaganda to delegitimize the Mujahidin, the resistance relied on the tried and true method of emphasizing its shared Islamic values to unite the opposition. Two allegations made by the Soviet propaganda machine were first that the Mujahidin were linked to the Central Intelligence Agency (CIA) and, therefore, another mechanism for Western imperialism, and second that the fundamentalist nature of the *jihad* was against progressive change and reform and that the Mujahidin wished to return to a feudalistic society.[91]

---

[88] Ibid., 180–189.
[89] Ibid., 232.
[90] Ibid., 192.
[91] Shahrani and Canfield, *Revolutions and Rebellions in Afghanistan*, 54.

EXTERNAL SUPPORT

The Mujahidin were fortunate to receive a significant amount of funding from foreign powers. At the time of the conflict, which took place during the Cold War era, the United States was willing to do almost anything to contain the growing Soviet power, and the United States took advantage of this strategic opportunity not only by supplying a majority of the funding, but also by lobbying for Arab funding, mainly from Saudi Arabia. In addition, it was also in the best interest of the United States and Sunni Arab countries, which have traditionally had rivalries with the Shi'a country, to have a strong, independent Afghanistan to act as a balance to the rising Iranian theocracy. Moreover, because much of the motivation for the Mujahidin and rhetoric on political Islam and Islamic *jihad* came from Egypt's Muslim Brotherhood, and because some groups were aligned with Saudi Wahhabism, funding the cause became more attractive to Arab groups. As a result, in 1984, the Mujahidin began to receive substantial aid, weapons, and training from the United States and other powers.[92] Other prominent foreign funders were Pakistan, Iran (who mostly supported the Shi'a Mujahidin), China, and some Islamic and anticommunist movements.[93]

The Mujahidin began to receive US assistance from the Carter administration, receiving $30 million for the program in 1980 and about $50 million in 1981. Although these numbers are substantial, under the Reagan administration, funding almost quadrupled to $120 million by 1984 and increased to $250 million by 1985. US assistance was not only in the form of money but also in the form of weapons, training, food, clothes, and other supplies. The arms pipeline was a cooperation between the CIA, ISI, and the Mujahidin resistance. First, the CIA would use money from the Saudis and the US administration to buy weapons from China, Egypt and Israel. Because the CIA did not want to be directly connected with the Mujahidin in the initial stages, it did not supply any weapons to the movement. However in 1986, President Reagan authorized the sending of several stingers to the Mujahidin.[94]

After 1985, US aid increased substantially and included giving more money and supplying better weapons, food, clothing, and other supplies to the movement. Not only did the Mujahidin receive financial and military support from the US, but in April 1985, President Reagan also gave them political support with the National Security Directive

---

[92] Bureau of South and Central Asian Affairs, *Background Note: Afghanistan*.
[93] Rubin, *The Fragmentation of Afghanistan*, 179.
[94] Ibid., 197.

166, which authorized a new policy of expelling the Soviets from Afghanistan "by all means possible."[95] At this point, all US funding was matched by Saudi Arabia, bringing funds to "$470 million in 1986 and $630 million in 1987."[96] In September 1986, the United States provided "shoulder-held, laser-guided Stinger antiaircraft missiles to the Mujahidin, the first time this ultra-sophisticated weapon had been distributed outside NATO."[97] In addition to continuing military aid, the US also offered humanitarian assistance through the United States Agency for International Development (USAID), which supplied "$60.6 million on health, $30.2 on education, and $60 million on agriculture" from 1986 to 1990.[98]

After the Soviet withdrawal, the United States began to provide support to exiled Mujahidin parties. Encouraged by the Mujahidin's success, in 1989, the United States and Saudi Arabia gave "$600 million each to the Mujahidin; . . . [and] an additional $100 million from the United States brought the total to $1.3 billion. The weapons included Stingers, heavy artillery, and other arms considered appropriate for a shift from guerrilla to conventional warfare."[99] However, the funding did not stop but rather continued to increase to about $715 million for fiscal year 1990.[100]

Pakistan was also a major contributor to the Mujahidin. First and foremost, its political support in allowing the Mujahidin parties to operate from its territory was an important factor in the Mujahidin strategy. The Pakistani government recognized the seven Sunni Mujahidin parties; however, it did not want a unified resistance party in its territory and played favorites among the groups. In addition to housing the Mujahidin bases, Pakistan also took in several thousands of refugees.[101] Pakistan also provided humanitarian assistance in ordering Jalaluddin to build a base along the Durand Line to house "electric generators, flush toilets, and [provide] large storage areas for weapons and other supplies."[102]

---

[95]   Ibid., 181.

[96]   Ibid.

[97]   Ibid.

[98]   Ibid.

[99]   Ibid., 182.

[100]   Ibid.

[101]   Ibid., 142.

[102]   Ibid., 217.

Figure 4. President Ronald Reagan meets with members of the Afghan Mujahidin.[103]

In the beginning of the insurgency, Pakistan supplied the Mujahidin with some small weapons. The Pakistani ISI was also responsible for distributing US and Saudi aid, giving Pakistan an important political role in the insurgency and allowing it to influence the Mujahidin forces to gather more support for Pakistani goals.[104] Pakistan also provided training for Mujahidin between 1983 and 1987.[105] In 1984, the Mujahidin received support from a new organization called *Maktab Al Khidamat,* or Services Bureau, which raised funds and recruited Arabs and other foreign Mujahidin, usually of the fundamentalist *jihadi* persuasion, to help the Afghans fight the Soviets. This organization, founded by Abdullah Azzam and Osama bin Laden, was a precursor to Al Qaeda.[106]

COUNTERMEASURES TAKEN BY THE GOVERNMENT

In response to the Mujahidin uprisings, the Khalq regime appointed a domestic "Commander of the Revolutionary Defense

---

[103] Ronald Reagan Library, accessed March 15, 2011, http://www.reagan.utexas.edu/archives/photographs/large/c12820-32.jpg.

[104] Ibid., 216.

[105] Ibid., 199.

[106] David Bukay, *From Muhammad to Bin Laden: Religious and Ideological Sources of the Homicide Bomber Phenomenon* (New Brunswick, NJ: Transaction Publishers, 2008).

Forces"[107] led by Hashimi in the northern areas of Afghanistan. As part of this initial attempt to defend the new government against the armed resistance, Hashimi organized three different defense units: "1) Sazman (Organization)—largely middle school and high school students serving as police and intelligence units; 2) Watan Parast (Patriots)—recruited from among the illiterate masses by means of draft or levy on villages; given large quantities of small arms and some basic training to defend their own areas; and 3) Defa-I Inqilab (Defense of the Revolution)—mainly teachers who coordinated the activities of the other two units."[108] The Afghan police continued their efforts by forming village militias in which the government provided arms and ammunition to the villagers; that the villagers willingly joined these militias came as a shock to many of the resistance commanders, who had assumed that most of the villagers would be antigovernment and support the Mujahidin.[109] The northern parts of Afghanistan became dominated by government forces and progovernment village militias.

After a few months, in the summer of 1978, the new Afghan government benefited from a Soviet-equipped and trained army.[110] With these new weapons, the Khalqi government started bombing and napalming villages, causing much destruction. By the fall and winter of 1978–1979, the resistance had spread nationwide. The tribesmen who normally began their agricultural cycles in the spring did not do so and, instead, stayed on to fight in the civil conflict to overthrow the Khalqi regime. As a response, the DRA requested and received more sophisticated equipment and Soviet military advisers.[111] The DRA continued military operations to secure the urban centers and break up Mujahidin offenses, with thousands dying in the process on both sides.[112]

Government countermeasures intensified with the Soviet invasion in 1979. From 1980 to 1982, Soviet military forces "launched massive, indiscriminate offensives that depopulated certain areas, took back several provincial and district centers, and established posts along major communication arteries."[113] In addition, the Soviets used Mi-24s to implement their harsh strategy against the Afghans, including "rubbleization of the countryside and migratory genocide" against

---

[107] Shahrani and Canfield, *Revolutions and Rebellions in Afghanistan*, 160–161.
[108] Ibid
[109] Farr and Merriam, *Afghan Resistance*, 6–7.
[110] Shahrani and Canfield, *Revolutions and Rebellions in Afghanistan*, 68.
[111] Ibid., 68–69.
[112] Rubin, *The Fragmentation of Afghanistan*, 172.
[113] Ibid., 180.

which the Mujahidin did not have any defense.[114] The Soviets commonly used a combined land–air attack on the Mujahidin in which they would send in a large number of troops by helicopter and then send in reinforcements via slower armored vehicles to trap the Mujahidin.[115] By 1986, there was an increase in fighting, especially because the Mujahidin had received an influx of better weapons and training. This included new tactics by the Soviets, such as night ambushes targeted mostly at Paktya, Herat, and Kandahar.[116]

However, the Mujahidin were successful in using the mountainous terrain against the Soviets and the DRA. By the fall of 1988, the Mujahidin had won effective control over eastern Afghanistan. At that point, the Soviets had withdrawn their air attacks, which led to the cessation of blindly bombing the rural areas.[117] Instead, the DRA tried to draft Mujahidin commanders and (unsuccessfully) incorporate them into their efforts. On April 7, 1988, the Soviet Union announced withdrawal of troops from Afghanistan. By February 14, 1989, all Soviet soldiers had left Afghanistan.

SHORT- AND LONG-TERM EFFECTS

CHANGES IN THE ENVIRONMENT

Much of Afghanistan was left devastated after this war because Soviet attacks had indiscriminately bombed and destroyed villages all over the country. Crop production had decreased because tribesmen had fought instead of planted. As a result, the influx of foreign aid, especially from the US government and USAID, focused on rebuilding the Afghan economy and on development and humanitarian issues. Furthermore, the lack of an effective political authority led to an increase in criminal activities, such as heroin production and trafficking, that began to create alternate economies for the local Mujahidin commanders.

CHANGES IN GOVERNMENT

In May 1986, Babrak Karmal was replaced by Najibullah, who launched the "national reconciliation" policy to draw more support from traditionalist Mujahidin. However, the Najibullah government

---

[114] Shahrani and Canfield, *Revolutions and Rebellions in Afghanistan*, 72.

[115] Roy, *Islam and Resistance in Afghanistan*, 195.

[116] Ibid., 208.

[117] Rubin, *The Fragmentation of Afghanistan*, 172.

began to weaken between 1989 and 1992 when it had lost support from the northern militia. By 1990, the Mujahidin had established an Islamic government in Afghanistan. The new constitution proclaimed that Islam was the national religion and that *shari'a*[118] law would rule the country.[119] The 1987 and 1990 versions of the new constitution allowed for "institutionalized representation at the local level and for some devolution of control over administration to these local bodies."[120] Local Mujahidin groups were incorporated into the new government, with much of the eastern part of Afghanistan remaining under Mujahidin rule. The Mujahidin factions maintained relative autonomy in their respective areas.

As Najibullah was planning to allow Afghanistan to be ruled by the Mujahidin in April 1992, the initial peaceful transfer of power turned violent when Hikmatyar resorted to arms and a violent overthrow. Although the Mujahidin factions attempted to form an alliance, largely to attract more funds from foreign powers, political differences and grievances regarding power sharing led to a new arrangement. A new power-sharing deal was formed, keeping Burhanuddin Rabbani, the Tajik leader of JIA, as the new interim president and appointing Hikmatyar as the prime minister. However, shortly thereafter, Rabbani was shot, causing concern for Hikmatyar's safety. As the fighting in Kabul continued in the wake of the dissolution of Najibullah's regime, Massoud, the lead commander of JIA, came to have a significant role in the new government.[121]

In 1992, leaders of the exiled Mujahidin in Pakistan created the Islamic Jihad Council to assume power in Kabul, appointing Sibghatullah Mojaddedi as an interim chair for two months until the Mujahidin decided on who should be represented in the ten-member leadership council.[122] From 1992 to 1996, Afghanistan, outside of Kabul, was essentially ruled as separate fiefdoms, where the local Mujahidin commanders served as warlords controlling those areas. The Mujahidin government of the Islamic State of Afghanistan, as it was called, was merely an extension of the Afghan Interim Government.[123]

As the Mujahidin led the new Afghan government, the ethnic composition of the different groups became more important, and this new council was led by Rabbani for the next four months. However,

---

[118] *Shari'a* is the main document on which Islamic law is based and is sometimes more generally referred to as the Islamic legal system.

[119] Ibid.

[120] Ibid., 174.

[121] Marsden, *The Taliban*, 37–38.

[122] Bureau of South and Central Asian Affairs, *Background Note: Afghanistan*.

[123] Marsden, *The Taliban*, 39–40.

the new government led by Rabbani and his leading commander, Massoud, both ethnic Tajiks, did not bode well with the Pashtuns, who made up most of Afghanistan's ethnic composition, because they felt they had lost political sovereignty over their people.[124] This development was also attributed to increased support for the Taliban.

CHANGES IN POLICY

After the Soviet retreat, a significant change in policy was an effort by Najibullah to distance the government from some of the previous communist rhetoric through his "national reconciliation policy." The government changed its name to the Republic of Afghanistan, dropping the word democratic.[125] In addition, Islam became the official state religion. Once the Mujahidin took power in 1992, Islam became even more prevalent in the government, transforming the communist-style government into an Islamic system in which Afghanistan was no longer governed by state law but instead by Islamic law or *shari'a.*

However, this Islamic government quickly became more fundamentalist, and this had direct consequences for women in Afghanistan. Although women traditionally had had a secondary role in Afghan society, the imposition of *shari'a* law reduced women's rights even more by imposing stricter dress codes and taking several measures to keep women indoors. The Islamic government revoked the Soviet-style reforms, which had acted as a catalyst for resistance against the Soviet-backed government. Women, who had gained some rights under the PDPA, which had removed the brideprice and increased the age of marriage, once again lost these rights. As the Taliban ascended to power, women's rights became even more limited—many girls' schools were closed, and women were effectually restricted from appearing in public.[126]

CHANGES IN THE REVOLUTIONARY MOVEMENT

The lack of a formal political authority, the rise of local commanders who were taking on new roles as warlords, and the rising dissatisfaction of ethnic Pashtuns with their new mostly Tajik government exacerbated the original divides among the Mujahidin. The traditionalists and fundamentalists became wearier of their ethnic

---

[124] Rasul Bakhsh Rais, *Recovering the Frontier State: War, Ethnicity, and the State in Afghanistan* (Lanham, MD: Lexington Books, 2008), 44.

[125] Roy, *Islam and Resistance in Afghanistan*, 213.

[126] Bukay, *From Muhammad to Bin Laden*, 109.

and ideological differences, and they were more and more incapable of coming to a power-sharing agreement for the new government. All of these factors led the Pashtuns to lend support to the Taliban, a largely Pashtun Islamic fundamentalist organization championed by Mullah Omar and other *ulama* and heavily influenced by the Egyptian Islamic Jihad and other *jihadi* movements. The Taliban, seen as an organization with pure devotion to the principles of Islam and the *shari'a* rule of law, was appealing to Pashtuns tired of the chaos and warlordism that was rampant in the country at that time.[127] The Taliban championed Wahhabi Islamic fundamentalism. which began to take over Islamism, and *jihad*, which motivated the Mujahidin. This Wahhabi movement was buttressed by the *Maktab Al Khidamat*, which brought in foreign Muslim volunteers for the Mujahidin.[128]

The minority ethnic Mujahidin groups did not buy into the Taliban's claim to represent Islamic unity above ethnicity and were very much aware of the Pashtun composition of the group. These minority groups, especially the Tajiks, Uzbeks, and Hazaras, were ready to defend their territory, and claims for regional autonomy by any means necessary started yet another civil war in the country as the Taliban ascended to power in 1996.[129]

OTHER EFFECTS

Once the flow of millions of dollars of foreign aid had ended, and in the wake of the political and ethnic fragmentation had occurred in Afghanistan, local Mujahidin commanders turned to criminal activities, including heroin production, trafficking, extortion, and kidnapping for ransom using the weapons from the insurgency that were left in their possession. Because of the devastation from the war and the increasing spread of political Islam, and because it did not ethnically represent the majority of the population, the Afghan government was not able to establish the legitimacy it needed to construct and maintain effective control of its people. This lack of political authority led these commanders to evolve into warlords with their own local gendarmes, effectively splitting Afghanistan into several different fiefdoms.[130]

---

[127] Ibid., 44–45.

[128] Roy, *Islam and Resistance in Afghanistan*, 218.

[129] Ibid., 44.

[130] Rais, *Recovering the Frontier State*.

BIBLIOGRAPHY

Bukay, David. *From Muhammad to Bin Laden: Religious and Ideological Sources of the Homicide Bomber Phenomenon.* New Brunswick, NJ: Transaction Publishers, 2008.

Bureau of South and Central Asian Affairs. "Background Note: Afghanistan." US Department of State. Accessed September 16, 2010. http://www.state.gov/r/pa/ei/bgn/5380.htm.

Central Intelligence Agency. "Afghanistan." *The World Factbook.* Accessed December 5, 2009. https://www.cia.gov/library/publications/the-world-factbook/geos/af.html.

Farr, Grant M., and John Merriam, eds. *Afghan Resistance: The Politics of Survival.* Boulder, CO: Westview, 1987.

Ghufran, Nasreen. "The Taliban and the Civil War Entanglement in Afghanistan." *Asian Survey* 41, no. 3 (May 2001): 462–487.

Marsden, Peter. *The Taliban: War and Religion in Afghanistan.* London: Zed Books, 2002.

Rais, Rasul Bakhsh. *Recovering the Frontier State: War, Ethnicity, and the State in Afghanistan.* Lanham, MD: Lexington Books, 2008.

Rashid, Ahmed. *Taliban: Militant Islam, Oil, and Fundamentalism in Central Asia.* New Haven, CT: Yale University Press, 2000.

Roy, Olivier. *Islam and Resistance in Afghanistan.* Cambridge, MA: Cambridge University Press, 1990.

Rubin, Barnett. *The Fragmentation of Afghanistan: State Formation and Collapse in the International System.* New Haven, CT: Yale University Press, 2002.

Shahrani, Nazif M., and Robert L. Canfield. *Revolutions and Rebellions in Afghanistan: Anthropological Perspectives.* Berkeley: Institute of International Studies, University of California, 1984.

Shahrani, Nazif M. "War, Factionalism, and the State in Afghanistan." *American Anthropologist* 104, no. 3 (2002): 715–722.

VIET CONG: 1954–1976

Bryan Gervais

SYNOPSIS

The National Liberation Front for South Vietnam (NLF), also known as the Viet Cong, was a political and military organization based in South Vietnam (and Cambodia) that fought the government of the Republic of Vietnam (or South Vietnam) and its ally, the United States, in the Vietnam War[1] from 1959 to 1975. The Viet Cong insurgency fought South Vietnamese and American forces on two fronts: an armed conflict and a political battle for the hearts and minds of villagers in rural South Vietnam. Initially, the insurgency used subversive and terrorist tactics to destabilize the South Vietnamese government, but they began to use guerrilla warfare and eventually large conventional military units after increased intervention by the United States.

After the end of French–Indochina War in 1954, the Communist Viet Minh left affiliates in South Vietnam that would develop into a political-paramilitary organization commanded by the North Vietnamese. This organization, which the South Vietnamese government dubbed the Viet Cong, attempted to subvert the US-backed South Vietnamese government by turning the rural peasants of the South Vietnamese countryside against the government. Threatened by the insurgency and the North, Ngo Dinh Diem, the president of South Vietnam, created increasingly repressive policies to control the population. Diem's alienation of the rural peasantry and North Vietnamese sponsorship led to the creation of the NLF in 1960.

The Viet Cong was commanded and supported by the Communist government of North Vietnam, led by Ho Chi Minh. The primary goal of the North Vietnamese and Viet Cong was to reunite both North and South Vietnam under a single Communist regime. The United States, hoping to contain the spread of communism, and the Communist blocs of both the USSR and China, provided substantial support to both the South Vietnamese and North Vietnamese causes, respectively.

Using guerrilla-style warfare initially, and later more conventional warfare tactics, the Viet Cong combated the United States and South Vietnamese forces alongside the People's Army of Vietnam (the North Vietnamese military) throughout the Vietnam War. After the acclimation of significant casualties during the Tet Offensive of 1968,

---

[1] The Vietnam War is also known as the Second Indochina War.

the Viet Cong's ranks were increasingly supplemented with regulars from the People's Army of Vietnam (PAVN). Upon the official reunification of North and South Vietnam under the Communist regime in 1976, the Viet Cong organization was formally disbanded.

TIMELINE

January 1930	Ho Chi Minh organizes the Indochinese Communist Party (ICP).
May 1941	After the occupation of French Indochina by the Japanese during World War II, the Vietnamese Communists organize and form the Viet Minh to force out both the Japanese and the French.
September 1945	Ho Chi Minh announces the creation of the Democratic Republic of Vietnam, with Ho serving as president; this claim is disputed by the USSR, United States, and United Kingdom.
December 1946–August 1954	The French–Indochina War takes place. The Viet Minh begin construction of a tunnel system in the late 1940s to battle the French.
April–July 1954	The Geneva Accords end the Indochina War and grant independence to a divided Indochina. Ho establishes a Communist government of North Vietnam, while a US-backed regime is created in South Vietnam. A number of Viet Minh units are left behind in South Vietnam, from which the Viet Cong network will form.
October 1955	Ngo Dinh Diem wins a (likely rigged) election to become president of the Republic of Vietnam (South Vietnam).
April 1957	The Viet Cong begins an assassination campaign in South Vietnam.
March 1959	Ho Chi Minh announces the start of an armed revolution, referred to as "the People's War," against the South Vietnamese to reunify the North and South under a single Communist regime.
May 1959	The Central Office of South Vietnam (COSVN) is established by the North Vietnamese to oversee the Viet Cong guerrilla units in South Vietnam and work begins on "Ho Chi Minh Trail."

January 1963	The Viet Cong win their first battle against the South Vietnamese Army (ARVN) and US Special Forces at Ap Bac.
December 1963	After the Buddhist Monk Crisis, Diem is successfully thrown out of office in a coup, after a couple of previous failed attempts.
January–March 1968	The Tet Offensive takes place. Heavy Viet Cong casualties effectively end the insurgency's involvement in the war.
September 1969	Ho Chi Minh dies, and Le Duan succeeds him as president.
January 1973	Direct US involvement in the war is suspended.
April 1975	The Fall of Saigon occurs.
July 1976	North and South Vietnam reunite as the Socialist Republic of Vietnam.

THE ENVIRONMENT OF THE REVOLUTION

PHYSICAL ENVIRONMENT

Vietnam is a country in Southeast Asia and is located on the Indochinese peninsula. The country forms an S-shaped curve and is only thirty miles (fifty kilometers) wide at its narrowest position. Being part of a peninsula, the country also has an extensive coastline (2,025 miles or 3,444 kilometers). It is bordered by the Gulf of Tonkin to the east, the South China Sea to the southeast, the Gulf of Thailand on the southwest, Laos and Cambodia to the west, and China to the north. At 128,527 square miles (331,210 square kilometers), it is slightly larger than the American state of New Mexico.

The south is part of the Mekong River Delta and is flat, as are parts the northern half of the country. The northwest and far north areas are mountainous, and the central region is made up of highlands. The climate of the southern half of the country (formerly South Vietnam) has a tropical climate, whereas the north is monsoonal.

Vietnam's capital and second-largest city, Hanoi, is located in northern Vietnam and also served as the capital of North Vietnam. The largest city in the country is Ho Chi Minh City, which was formerly known as Saigon before the reunification of the North and South and served as the capital of South Vietnam.

Figure 1. Maps of Vietnam.[2]

CULTURAL AND DEMOGRAPHIC ENVIRONMENT

Vietnamese is the official language of Vietnam. Because of the country's history as a French colony, French is widely spoken among the older members of the population, but English has become increasingly popular as a second language among the younger

[2] The two maps on the left are from Central Intelligence Agency, "Vietnam," *The World Factbook*, accessed January 12, 2010, https://www.cia.gov/library/publications/the-world-factbook/geos/vm.html; the map on the right is from US Department of State, "Questions and Answers: Viet-Nam: The Struggle for Freedom," Department of State Publication 7724, Far Eastern Series 127 (Washington, DC: Government Printing Office, August 1964).

generation. More than 86% of the population is represented by the Kinh (or Viet) ethnicity. Tay and Thai are the next largest ethnic groups, but each make up less than 2% of the population. A total of fifty-four ethnic groups reside in Vietnam. The population was slightly less than eight-six million as of 2009.[3]

More than 80% of Vietnamese claim to have no religious affiliation, with Buddhism (around 9%) and Catholicism (slightly under 7%) representing the largest two religious groups. A majority of the population claims to identify with Buddhism in some way, suggesting that although many Vietnamese do not practice on a regular basis, they hold and accept many Buddhist beliefs.[4] Two major religious movements, which originally pushed for independence from France, were prominent in the period after World War II: Cao Dai and the militant Buddhist group, the Hoa Hao. The two groups are the next-largest religious groups in Vietnam after mainline Buddhism and Catholicism, but their ranks are comparatively small, with each only accounting for slightly more than 1% of the population. At the time of the Vietnam War, 85% of the population of South Vietnam was concentrated in the delta and northern coastal regions.[5]

SOCIOECONOMIC ENVIRONMENT

In the early 1950s, the Viet Minh implemented a set of arbitrary taxation policies modeled after the taxation policies implemented in Communist China. These policies meant to produce "economic leveling" as part of a more general transition to Communist doctrine by targeting businessmen and more affluent peasants. The French limited the effectiveness of these policies to some degree.[6]

After the 1954 Geneva Accords divided Vietnam into the North and South, the South maintained a free-market economy with ties to the West, while the North became a Communist state, aligned with both the USSR and the People's Republic of China. Among the significant policies enacted by the North Vietnamese government during this time were land reform programs, which involved seizing property from large landowners and redistributing the land throughout the populace. As North Vietnam had a comparatively small industrial base

---

[3] US Department of State, "Background Note: Vietnam," accessed February 1, 2010, http://www.state.gov/r/pa/ei/bgn/4130.htm.

[4] Central Intelligence Agency, "Vietnam," *The World Factbook.*

[5] Jeffery M. Paige, "Inequality and Insurgency in Vietnam: A Re-Analysis," *World Politics* 23, no. 1 (October 1970): 24–37.

[6] George A. Carver Jr., "The Faceless Viet Cong," *Foreign Affairs* 44, no. 3 (April 1966): 347–372.

at this time, the support provided by the USSR and China was all the more imperative to the subsistence of the regime.

The United States funded the majority of the South Vietnamese military expenditures through military and economic aid. In addition to this external support, South Vietnam enjoyed food surpluses, particularly rice, in the post-French period. However, unemployment was an issue in the country, which led to growing dissatisfaction with the Diem government. Government corruption was claimed to be a source of widespread monopolies of resources and industry that benefited a few private interests and slowed economic development.[7]

The perceived growing threat of the Viet Cong in South Vietnam throughout the 1950s led the South Vietnamese government to implement more repressive policies, including reeducation and relocation programs, military tribunals, and arbitrary arrests for anyone believed to have had any association with the Viet Cong. The Diem regime did attempt to implement social and economic reforms, including building schools, hospitals, and roads; however, these attempts did little to assuage resentment among the South Vietnamese populace toward the Diem government (see *Governing Environment*).

The illiteracy rate was quite high at the time the Viet Cong first formed; it had been as high as 90% circa 1945. Illiteracy, as well as a lack of overall education, was problematic for the Viet Cong when training recruits from South Vietnamese villages. Programs, including political indoctrination, were used by the Viet Cong to educate recruits.

HISTORICAL FACTORS

Vietnam has a long history of military operations and struggles, with more than a thousand years of invasions, occupations, rebellions, and revolutions. Vietnam was a part of China for more than a millennium, from 111 BCE until 939 CE, and remained independent until the nineteenth century. A host of dynasties ruled the land for the next 800 years, including the last ruling family, the Nguyen Dynasty. Interaction with European nations began as far back as 200 CE but began to pick up in the seventeenth and eighteenth centuries, primarily based on trade and Catholic missions. The French increasingly became involved in the region (then known as Indochina) in the nineteenth century and

---

[7] US Department of Defense, *The Pentagon Papers: The Defense Department History of United States Decisionmaking on Vietnam*, Senator Gravel Ed. (Boston: Beacon Press, 1971).

began setting up Catholic missions, which members of Nguyen family viewed as a threat. Thus, persecution of Catholic missionaries began.[8]

The purported persecution of missionaries, real or not, provided the French with the opportunity and justification to intervene in Indochina. In 1857 during the French Second Empire, Napoleon III ordered French naval forces to Indochina to protect the Catholic missionaries. The French also hoped to develop a port in the country that could be used to engage in trade in Hong Kong. The following year, France initiated a conquest of the area, choosing to invade the southern half of the country first, because of the belief that the inhabitants of the southern half were less connected to the Vietnamese culture. The initial invasion, led by Rigault de Genouilly (with Spanish assistance[9]), was only mildly successful, but the French did manage to establish a foothold in southern Vietnam.[10]

In the 1860s after years of skirmishes, Emperor Tu Duc, based in the north, signed a treaty with the French that recognized French sovereignty over a group of possessions in the south. Although guerrilla activity against the French commenced, it did little to curb French territorial expansion, which led to substantial gains in the northern Tonkin region in the 1870s. Concerned about rapid French expansion in the region, the Chinese engaged the French in what would be known as the Tonkin Wars.[11]

War between France and China ended in 1885 with the signing of the Treaty of Tientsin, recognizing the French claim to Indochina. The territory, which in 1887 the French named *Indochine française*, or French Indochina, was run by a governor-general who reported directly to Paris. Vietnamese nationalists continued guerrilla strikes, led by Phan Dinh Phung, against the French.

Ho Chi Minh, the Indochinese Communist Party (ICP), was organized in January 1930. The ICP slowly gained influence in Vietnam over the next decade, taking advantage of anti-French, nationalist sentiment in the country while subverting or annexing all competing nationalist parties. The Viet Nam Doc Lap Dong Minh Hoi (which means the League for Vietnamese Independence, shortened to the

---

[8] Oscar Chapuis, *A History of Vietnam: From Hong Bang to Tu Duc* (Westport, CT: Greenwood Press, 1995), 216; Spencer C. Tucker, *Vietnam* (Lexington, KY: The University Press of Kentucky, 1999); Gordon M. Wells, "No More Vietnams: CORDS as a Model for Counterinsurgency Campaign Design: A Monograph" (School of Advanced Military Science, US Army Command and General Staff College, 1991).

[9] Spanish missionaries had also been victims of persecutions, and thus they also hoped to push the Nguyens for redresses.

[10] Richard Shultz, "Limits of Terrorism in Insurgency Warfare: The Case of the Viet Cong," *Polity* 11, no. 1 (1978).

[11] Tucker, *Vietnam*. The Tonkin Wars are also known as the Black Flag Wars.

Viet Minh) formed in 1941 to push for Vietnamese independence. The Vietnamese Communists joined with the Viet Minh that same year. Japan occupied Vietnam during World War II, at which time Ho Chi Minh and the Viet Minh cooperated with United States intelligence[12] to combat Japanese forces. In 1945, the Japanese installed Bao Dai as emperor of Vietnam. After the Japanese surrender later that year, Bao abdicated, and the ICP formally disbanded in an effort to increase the appeal of the transformed Viet Minh to more Vietnamese. The purpose of the Communists' takeover of the Viet Minh, like previous takeovers of other political organizations, was to use the group to gain more power and influence in Hanoi.[13]

On September 2, 1945, after the end of World War II, Ho Chi Minh proclaimed the creation of the Democratic Republic of Vietnam with Ho serving as its president. The USSR, United States, and United Kingdom all agreed that French Indochina was to remain a French colony, despite Ho Chi Minh and Viet Minh's claim that an independent Democratic Republic of Vietnam (or North Vietnam) had been formed.

The Viet Minh won elections across Vietnam in 1946.[14] The same year Ho, hoping to expand the base of communist nationalist support, announced the establishment of a new Popular National Front (Lien-Hiep Quoc Dan Viet Nam, or the Lien Viet) for the purposes of gaining independence and establishing democracy for Vietnam. The Lien Viet ultimately absorbed the Viet Minh, but the term Viet Minh continued to be used to describe the movement, individuals, and organization working to end French rule. The Lien Viet would later take the name Lao Dong, or Worker's Party.

From 1946 to 1954, the Viet Minh nationalist movement combated the French during the French–Indochina War.[15] The nationalist forces were commanded by General Vo Nguyen Giap, who defeated French forces in the climatic battle of Dien Bien Phu to conclude the Vietnamese victory in the war. Subsequently, Vietnam was divided into two distinct countries: North Vietnam and South Vietnam (see *Governing Environment*).

During the war in 1950, Communist China (the People's Republic of China) recognized North Vietnam, while non-Communist states, particularly American allies, recognized the French-backed State of Vietnam led by Emperor Bao Dai in Saigon. The United States

---

[12] Specifically, the Office of Strategic Services.

[13] Carver Jr., *The Faceless Viet Cong*, 347–372.

[14] Jonathan Neale, *The American War: Vietnam 1960–1975* (Chicago: Bookmarks, 2001).

[15] The French–Indochina War is also known as the First Indochina War.

hoped to have a non-Communist ally in the Southeast Asia region and beginning with the Eisenhower administration began to provide military and economic aid to the South Vietnamese state.

In 1954, the United States became significantly more involved in Vietnam, when a military advisory commission (originally called the Military Assistance Advisory Group and later known as the US Military Assistance Command, Vietnam) was sent to South Vietnam to train the Army of the Republic of Vietnam (ARVN) and provide equipment for combating the Viet Cong and the Vietnam People's Army (PAVN). American participation at this time also included the deployment of helicopter units and airlift support for ARVN.

GOVERNING ENVIRONMENT

The 1954 Geneva Accords between France and the Viet Minh ended the Indochina War and granted independence to the divided Indochina. Ho Chi Minh established the Communist government of North Vietnam (known as the Democratic Republic of Vietnam, or DRV), while a US-backed government was created in South Vietnam. A military demarcation line was established between the north and south, with more than three miles (around five kilometers) of demilitarized zones on both sides, known as the Seventeenth Parallel. The split into two countries, however, was meant to be temporary until a political solution for reunification could be worked out.[16] Both the Soviet Union and the People's Republic of China pushed the North Vietnamese delegates to the Geneva Conference to accept a short-term partition of Vietnam at the seventeenth parallel. This was likely in an effort to help France save face, as both the USSR and People's Republic of China hoped to avoid provoking the United States or western Europe so soon after the Korean War had ended.

In 1955 South Vietnam held a referendum to form a new government. Emperor Bao Dai tried to reinstate the monarchy but lost the election to Ngo Dinh Diem, who became the first president of South Vietnam, now the Republic of Vietnam. The new government of the Republic of Vietnam and its unwavering, anti-Communist president instantly received substantial American economic, military, and political aid. Diem's government was a centralized, single-party oligarchy, and nearly all high-level government posts were given to his friends and family members, including his brother, Ngo Dinh Nhu, who served as his chief adviser.

---

[16] George K. Tanham, *Communist Revolutionary Warfare: From the Vietminh to the Viet Cong* (Westport, CT: Praeger Security International, 2006).

Although Diem was a Roman Catholic, Vietnam at the time was primarily a Buddhist country. Nevertheless, Diem encouraged Catholics living in the North to flee to South Vietnam, resulting in a migration of nearly one million people. This coincided with 90,000 Southern Communist sympathizers heading north. North Vietnam, however, instructed 10,000 Viet Minh veterans from the war with the French to remain in the south; these Viet Minh would be known as the "left-behinds" and would launch the Viet Cong[17] insurgency.[18]

Diem was a detached and autocratic leader who was unable to rule his land, which was at the time very chaotic, without authoritarian use of force. After the French Indochina War, the civil service in South Vietnam was decimated and the government was severely limited. Diem blamed North Vietnam and its Communist allies for most of these troubles and began a vicious counterattack that included severely repressive policies for dealing with suspected Communists. Diem came to believe that any liberalization would aid the Viet Minh forces within the South Vietnamese borders and North Vietnam. Thus, the rural peasantry was kept out of the political process while South Vietnam sought to deal with the Viet Minh, causing it to deteriorate into a quasi-police state. For instance, Diem abolished municipal elections in 1956 out of fear that the Viet Minh would gain seats.[19]

While the South Vietnamese situation saw a period of improvement from 1956 to 1957, the Diem regime failed to gain the favor of the rural peasantry and could not convince them to support the South's resistance to the communist North. Besides the failure of the Diem regime to politicize the peasantry, the peasantry's aversion to the Diem regime also stemmed from their dislike of South Vietnamese officials who were posted to the villages, many of whom were corrupt and nearly all of whom were not from the village. Vietnamese villagers were known for holding strong provincial sentiments, so the combination of the government officials' "foreignness" and corruption led to extensive distrust of the Diem regime overall.[20] After a failed coup in 1960, Diem managed to arrest and jail his most significant political opponent, Dr. Phan Quang Dan, as well as members of the Caravelle Group, a group of eighteen prominent South Vietnamese politicians and intellectuals who urged Diem to liberalize and reform his corrupt

---

[17] Viet Cong is an abbreviation of "Viet Nam Cong-San," which means Vietnamese Communist.

[18] Ibid.

[19] Robert K. Brigham, *Guerrilla Diplomacy: The NLF's Foreign Relations and the Viet Nam War* (Ithaca, NY: Cornell University Press, 1999); US Department of Defense, *The Pentagon Papers*.

[20] Carver Jr., *The Faceless Viet Cong*; US Department of Defense, *The Pentagon Papers*.

and undemocratic government. In response to the coup attempt, a harsh crackdown was orchestrated by Diem's brother Nhu, and thousands of innocent individuals were tortured and executed during this crackdown.[21]

In 1963 the Buddhist Monk Crisis occurred, during which Diem was accused of provoking demonstrations among the South Vietnamese populace. The unpopularity of Diem led to a loss of American support, and the Kennedy administration made clear to the ARVN that the United States would not intervene should the military attempt a coup. Tensions between Diem and the ARVN were consistent throughout Diem's tenure, and the military forces were involved in unsuccessful coup attempts in both 1960 and 1962, with the former being more significant.[22] On December 1, 1963, a successful coup was completed, and Diem was assassinated. Designed and implemented by generals of the ARVN, the coup led to a political vacuum and a period of coups and countercoups. Widespread political instability ensued, and many government institutions and functions ceased to operate. This instability aided the Viet Cong, which capitalized on the disappearing presence of South Vietnam. Saigon did not see anything resembling political stability until 1965 when General Nguyen Van Thieu was named the chief of state and Air Vice Marshal Nguyen Cao Ky was installed as the new premier. The two were inaugurated as President Thieu and Vice President Ky in 1967.[23]

WEAKNESSES OF THE PREREVOLUTIONARY ENVIRONMENT AND CATALYSTS

The failure of the 1954 Geneva Accords to bind the South Vietnamese to its stipulations is suggested to be one of the primary catalysts for the Viet Cong insurgency. Both Diem and the United States refused to sign the Geneva Accords, permitting them to excuse themselves from its obligations. The United States, as explained by then-Secretary of State John Foster Dulles, did not support the Geneva Accords because of the fear that too much power was given to the Vietnam Communists.[24]

The Geneva Accords stipulated that elections were to be held in 1956 to reunify the North and South. However, Diem refused to hold elections, arguing that because South Vietnam had not signed the

---

[21] Ibid.
[22] Ibid.
[23] Ibid.
[24] Brigham, *Guerrilla Diplomacy*.

treaty, it was not bound to the election requirement. Diem, fearful of the Communists and North Vietnam—and fearful that Ho Chi Minh would win the elections, should they be held—named all Viet Minh Communists and defended the cancellation of elections by claiming that the 1956 elections would not be free because of North Vietnam's intervention.

The Diem regime focused on security, with little expenditure on social services such as schools and hospitals, in the rural countryside. The South Vietnamese government used arbitrary arrests, torture, and relocation and reeducation programs to combat the perceived Communist threat in the rural countryside. These repressive tactics were used to increase security. The distrust of the Diem regime gave the Communists the opportunity to make inroads with the villagers and provide services.[25]

The Viet Minh "left-behinds"—those who been commanded to remain in South Vietnam after the French–Indochina War—were extensively disconcerted with both Diem's refusal to hold elections and the Diem regime's increasingly brutal anti-Communist policies. The Viet Minh ranks in South Vietnam included many non-Communists, who *ex facie* may have supported the Diem regime had they not been clustered into the "Communist" category by Diem. Brutal tactics by South Vietnamese agents, including mass arrests, torture, and execution, were employed to eliminate the "left-behinds."[26]

The South Vietnamese crackdown that followed the failed 1960 coup sent thousands fleeing to North Vietnam; however, North Vietnam ultimately sent many of these refugees (or *regroupees*) back to the South to join Viet Minh "left-behinds" in the South to form the People's Liberation Armed Forces. Diem (pejoratively) dubbed the group "Viet Cong," hoping to cast them all as Communist Vietnamese. Members of the Viet Cong, intermingling with rural villagers, were more or less indistinguishable from other South Vietnamese and began to carry out missions meant to weaken the Diem regime.

The Diem regime managed to successfully suppress a number of the armed militias that had made their presence known in South Vietnam, including the Hoa Hao and Binh Xuyen between 1955 and 1956. After this accomplishment, Diem and South Vietnam began to focus on eradicating the Viet Cong, which had begun its political agitation strategy meant to undermine South Vietnam and turn the populace against Diem.

[25] US Department of Defense, *The Pentagon Papers.*
[26] Ibid.

In 1959 Ho Chi Minh announced the beginning of an armed revolution against the South Vietnamese to reunify the North and South under a single Communist regime; he referred to this revolution as "the People's War." It is during this time that the *regroupees* were sent back to South Vietnam and the Central Office for South Vietnam (COSVN) was formed (see *Leadership and Organizational Structure*). In the following year North Vietnam established a political arm for the South Vietnamese Viet Cong, named the NLF, designed to coordinate Viet Cong activity and membership.

FORM AND CHARACTERISTICS OF THE REVOLUTION

OBJECTIVES AND GOALS

The Viet Cong's primary political objective was the same as that of its sponsor and supervisor, the North Vietnamese government: to reunify North and South Vietnam under a single Communist government. Although the primary enemy was the Saigon-based South Vietnam, the United States' role, which increased throughout the 1960s, shifted the Viet Cong's focus from devastating the social structure of the South to defeating the imperialist alliance of the United States and South Vietnam. American forces increasingly became the primary targets of the Viet Cong in what was termed "a just war waged by the people over the unjust war by aggressive imperialism."[27]

To meet the reunification objective, the Viet Cong attempted to restructure and recruit the South Vietnamese rural villages to their cause, utilizing political terrorism in the process. Taking control of the rural population of South Vietnam was a central aspect of the Viet Cong insurgency—the objective was to gain control of not just territory but also the people who inhabited that territory. Terrorist tactics were used to force the South Vietnamese out of the rural villages, and once a governmental void was created, the NLF could substitute their own infrastructure and ideology.[28]

The nationalistic goal of reunifying the North and South grew out of the objectives of the group's predecessors. The Viet Minh had ostensibly been committed to bringing independence and democracy

---

[27] B. L. Horton, "A Content Analysis of Viet Cong Leaflets as Propaganda, 1963–68" (master's thesis, Texas Tech University, 2008).

[28] Shultz, *The Limits of Terrorism in Insurgency Warfare;* Douglas Pike, *Viet Cong: The Organization and Techniques of the National Liberation Front of South Vietnam* (Cambridge: Massachusetts Institute of Technology, 1966); Carver, *The Faceless Viet Cong.*

to Vietnam, with the declared objective of throwing out the French during the first years of the French–Indochina War. However, in the early 1950s, the Party conformed to orthodox Communist dogma and began to push for further societal and economic change consistent with Marxist ideology.

Their strategy for achieving the ultimate goal of reunification changed throughout the 1950s and 1960s. In the mid to late 1950s, the Viet Minh looked to the French to push the reunification process via diplomacy. This was followed by the strategy of creating chaos and anarchy throughout rural South Vietnam to reduce confidence in the Diem regime and South Vietnamese government. However, beginning in 1965, the Viet Cong began to engage in a big-unit war against the allied South Vietnamese and American forces.

As the United States prepared to disengage in the 1970s, and the "Vietnamization" process was to take place,[29] the Viet Cong leadership developed the following set of objectives that they hoped would lead to victory in South Vietnam: (1) demonstrate that the ARVN could not have military success without American support, (2) prove the South Vietnamese government in Saigon is an American puppet, (3) destroy any faith that villagers still had in the South Vietnamese government by sabotaging its Pacification Program, and (4) increase American casualties.[30]

LEADERSHIP AND ORGANIZATIONAL STRUCTURE

The Viet Cong formed from the Viet Minh units left behind in the South after the end of the French–Indochina War. At the time, American and South Vietnamese estimates suggested there were about 5,000–10,000 soldiers left behind, but it is now believed to have been an extensively larger number. Further masses of *regroupees*, untrained men originally from the South who had received significant training in the North, returned to the South to join the "left-behinds" and this swelled the Viet Cong ranks. The newly trained *regroupees* also provided extensive leadership and recruited and organized South Vietnamese villagers into the Viet Cong ranks.

The post-1960 Viet Cong can best be understood as a multifaceted system in which many independent organizations and elements were coordinated and infused to form a united political and military

---

[29] "Vietnamization" was an American exit strategy developed by the Nixon administration to increase, train, and equip the ARVN so that they could replace US forces in combat, allowing the Americans to withdraw.

[30] Pike, *Viet Cong.*

organization. Ho and the North Vietnamese government implemented a "superstructure" of sorts for the inchoate insurgency, known as the NLF. The creation was announced over *Radio Hanoi* and a ten-point manifesto was read, calling for the "overthrow [of] the disguised colonial regime of the imperialists and the dictatorial administration, and to form a national and democratic coalition administration."[31] The Viet Cong and North Vietnam hoped that the NLF would appeal to non-Communists, and thus, its independence from Hanoi was stressed.[32] The NLF was exceptionally organized in rural South Vietnam, and this organization greatly aided the group's military and political activities and arranged the Viet Cong into two components: a military component referred to as the Liberation Army and a political component referred to as the People's Revolutionary Party.[33]

The Liberation Army was a combination of guerrilla units, large local force units, and main force units, all of which interacted with and supported one another. The Central Office of South Vietnam (COSVN), a Cambodian-based command structure, was formed by the North Vietnamese Army (NVA) in 1959 to supervise the guerrilla, local force, and main force units in the southern regions of South Vietnam and also served as the headquarters of the PAVN. The civil administrative elements of the NLF were also based at COSVN. COSVN had been preceded by another command structure in the Mekong Delta, formed the previous year, where thirty-seven armed companies had been arranged. Although membership in the Liberation Army consisted of all NLF members (including women and children), its tasks were usually carried out by trained military units.[34] Le Duan, the chief policy maker of North Vietnam, played a large role in directing the Viet Minh in the 1950s and served as secretary of COSVN.[35] Later, however, he shared direction of the inchoate Viet Cong with Le Duc Tho. General Giap, the commander-in-chief of the PAVN throughout the Vietnam War, also had significant control over the Viet Cong.

The Viet Cong leadership comprised a group of unknown South Vietnamese (with the exception of a few, such as Tran Van Tra, who served as commander of the Liberation Army). Although lacking conventional military experience and know-how, the Viet Cong became organized through a central committee of allied groups

---

[31] Cheng Guan Ang, *The Vietnam War from the Other Side* (New York: Routledge, 2002).

[32] Also, the North Vietnamese were wary of being accused of violating the Geneva Accords.

[33] Note: The NLF itself has sometimes been identified as a separate, third component.

[34] Carver Jr., *The Faceless Viet Cong.*

[35] Le Duan later became the leader of North Vietnam after Ho Chi Minh's death in 1969.

and villages throughout South Vietnam. Its membership grew exponentially throughout the years of 1959 to 1962, with an estimated force of 300,000. Members usually belonged to multiple "liberation associations" designed to carry out various administrative functions for the NLF.

Three "action" programs were utilized by the Viet Cong; these programs collectively made up the *dau tranh* and were used to promote and sustain the Viet Cong's cause. The programs included *dich van*, a propaganda effort (see *Methods of Obtaining Legitimacy*); *binh van*, an effort designed to promote desertion and defection from the South Vietnam (see *Methods of Recruitment*); and *don van*, meaning "action among the people," which involved administration of areas controlled by the Viet Cong, including tax collection and recruitment. The *don van* program organized Viet Cong-controlled villages down to individual families, which were arranged into cells and supervised by Viet Cong officials. The Viet Cong provided social services and implemented policies and programs that aided the villagers, but to enforce control of these policies and programs, the Viet Cong furnished punishments, including admonition, public humiliation, reeducation, and death.

Every member of the village was mandated to participate in labor projects, and the populace was subjected to high taxation, with taxes collected in both money and rice. The villages were administered by a "Village Party Chapter"; the secretary of this chapter was normally the most powerful person in the village and controlled the local guerrilla unit. *Don van* provided safe havens for both the NVA and Viet Cong forces.[36]

The principal political party within the NLF was The People's Revolutionary Party, or the "Marxist–Leninist Party of South Vietnam." Although the party denied that it was officially tied to the North Vietnamese Communist party, the United States believed the People's Revolutionary Party to be the southern division of the North Vietnamese party, through which Hanoi could control and sustain the revolt against the Diem regime. The People's Revolutionary Party was launched in 1962, but Communist influence within and on the NLF likely began much earlier.[37]

The PAVN High Command, which supervised the regular North Vietnamese forces, also commanded some Viet Cong regiments positioned in the northern B-3 Front and the two northernmost

---

[36] W. P. Davison, *Some Observations on Viet Cong Operations in the Villages* (Santa Monica, CA: RAND Corporation, 1967).

[37] US Department of State, *The Pentagon Papers*.

military regions (see Figure 1). PAVN forces took on more supervisory and command roles in the Viet Cong forces as the North Vietnamese units moved farther into South Vietnamese territory. The Military Affairs Committees were the liaisons between the Viet Cong and the Central Reunification Department, the North Vietnamese organization tasked with coordinating the reunification cause. The Vietnamese Politburo[38] had a supervisory role over the fronts, and each front also had an independent military command.

COMMUNICATIONS

The true "mass medium" used by the Viet Cong was the propaganda leaflet. Leaflets were dispersed throughout South Vietnam and included slogans, allegations of American atrocities and desertions, promotion of reunification, appeals to ARVN troops to desert, and urgings for the South Vietnamese to rise up against the US-backed Saigon government. Some of the leaflets were aimed at American soldiers and included messages asking African American soldiers to ponder why they were helping the US government as well as messages praising American activists in the United States who were against the war.[39]

Radio also played an important role in the Viet Cong propaganda effort. A clandestine radio station began to push for armed attacks against the Diem regime in October 1957.[40] *Radio Liberation*, run by the Viet Cong, was used for propaganda purposes and was aimed at both the South Vietnamese population and the American troops. Other clandestine propaganda broadcasts were made by *Radio Pathet Lao* and *Radio of the Patriotic Neutralist Forces.*

The North Vietnamese-controlled *Radio Hanoi*, including the "Hanoi Hannah" broadcasts, also contributed to the propaganda effort. The Manifesto of the NLF, which declared the creation of the revolutionary organization bent on removing the Diem regime, was read over *Radio Hanoi* in December 1960, and in September 1967, *Radio Hanoi* broadcast a message by General Vo Nguyen Giap called "The Big Victory, The Great Task," which, unbeknownst to American forces, was a description of the Tet Offensive that would launch five months later.

---

[38] A Politburo is the executive committee for a Communist party and is the primary policy-making body. The Vietnamese Politboro was based in Hanoi and its membership (at the time of the Vietnam War) included Ho Chi Minh, Le Duan, and Vo Nguyen Giap. The Politburo was organized through regional Party organizations.

[39] B. L. Horton, "A Content Analysis of Viet Cong Leaflets as Propaganda, 1963–68."

[40] Ibid.

METHODS OF ACTION AND VIOLENCE

The Viet Cong can be classified as a Maoist-style insurgency, as it was modeled after Mao Tse-Tung's modern revolutionary guerrilla warfare techniques. As Mao outlined, the first step in this type of warfare is to utilize irregular warfare techniques to increase the strength of the insurgency, in terms of both political and military force, at the expense of the enemy. This is followed by a "strategic stalemate," where the insurgency takes control of an area and then uses both irregular and conventional tactics to maintain and expand control. The final stage relies more on conventional warfare techniques in an effort to defeat the enemy entirely.[41]

The particular type of Maoist-style insurgency practiced by the Viet Cong was known as *dau tranh*, or "the struggle." *Dau tranh* incorporated political, economic, and military elements into a single synergistic strategy and was devised by Ho Chi Minh, General Vo Nguyen Giap, and members of the Hanoi Politburo. The combination of agitation and propaganda (or the "agit-prop" tactic) was used extensively by the Viet Cong to weaken the South Vietnamese government and to gain the loyalty of the South Vietnamese peasantry.

As advocated by the Maoist-style insurgency strategy, the Viet Cong developed a symbiotic relationship with the rural villagers. The villagers provided the Viet Cong with economic assistance and information. The Viet Cong utilized a form of "agitational terror" in the villages of South Vietnam, with the primary goal of disrupting, highlighting the weaknesses of, and ultimately eliminating the government of Vietnam. From 1957 on, terrorism began to rise as a part of this effort to weaken the South Vietnamese government and turn public opinion against it. By 1960, *Time Magazine* reported that 250–300 South Vietnamese officials were being killed per month.[42]

Particular terrorist tactics included assassinations of government officials and attacks on ARVN forces and representatives of an "organized political society" such as school teachers, health workers, etc., to create political instability. The Viet Cong believed that the elimination these groups and individuals, particularly popular village leaders, was necessary for the NLF's sociopolitical program to be initiated and effectual. The NLF's aim was to purge the areas of these targeted individuals by either physical means (executions) or psychological means, whereby the costs of remaining in the area would

---

[41] Wells, "No More Vietnams."

[42] "Rural and Violence South Vietnamese Counters," in US Department of State, *The Pentagon Papers*.

appear to be so steep that the targeted individuals would choose to vacate the area, stay neutral, or become part of the NLF.

Methods used to eliminate South Vietnamese officials included kidnappings, assassinations, and trials. Kidnappings were utilized to a greater extent than assassinations; from 1958 to 1965, 9,700 assassinations took place in South Vietnam, compared to 36,800 kidnappings. The NLF focused on and succeeded in wiping out village leaders who were well respected and wielded much influence in the villages; the NLF did this largely to strengthen its own influence in the villages and to limit the possibility of political opposition. Also, by taking out corrupt officials of the Diem regime, the NLF gained further favor with the villagers. The NLF capitalized on the villagers' growing dislike of some targeted officials, creating public shows of the sentencing and punishments of these officials. By contrast, officials who remained popular with the general populace were executed clandestinely. Viet Cong forces also targeted members of other paramilitary organizations (such as the Hoa Hao), Catholics, relatives of persons found to have been guilty of working against the movement, and individuals with "suspicious" backgrounds. Overall, there were fifteen categories of Viet Cong targets, and an estimated three million Vietnamese qualified for one or more of these categories.[43]

However, terrorism was only one of several successful tactics utilized by the NLF in the South Vietnamese countryside. More severe approaches were used when the NLF first arrived at a targeted village. After a successful conquer, coercive tactics tended to decrease in severity, and social methods, such as propaganda and reeducation, were used to a greater extent. Various forms of repression were used, including a range of punitive measures, from warnings to in-house reform, to extended confinement and hard labor in thought-reform camps.

As the Vietnam War began, and as American forces became more present in combat, the Viet Cong utilized guerrilla-warfare tactics to combat the Americans and the South Vietnamese, taking advantage of the jungle terrain. Crucial to this tactic was the use of tunnel networks (including the Cu Chi tunnels, an underground fortress based in and around Saigon), which allowed the Viet Cong to navigate underground throughout South Vietnam. Camouflaged on top, trapdoors allowed the Viet Cong to launch surprise attacks before retreating back into the tunnel systems. The tunnels were often booby-trapped with punji

---

[43] Pike, *Viet Cong.*

stakes and included false tunnels, making American efforts to deal with the tunnel system largely unsuccessful.[44]

The tunnel systems, also known as the Iron Triangle, began in the late 1940s as a strategy to battle the French; the systems included amenities such as kitchens, conference rooms, hospitals, sleeping areas, and storerooms.[45] Using the tunnel systems, the Viet Cong would often strike at night when visibility was low, with forces spending their days inside the tunnel systems. Complete with poisonous insects, disease (malaria was particularly rampant), and cramped living spaces, life in the tunnel systems was very difficult.

Beginning in 1959, sizable Viet Cong military units began confronting and engaging ARVN forces, and the following year saw the first battalion-size Viet Cong units engaged in combat. Larger forces would become more frequent throughout the early 1960s. However, beginning in 1965, Hanoi pushed for the Viet Cong to return to more guerrilla-style combat involving ambushes and hit-and-run tactics. The Viet Cong began to steer clear of pitched battles with American forces, except when the odds were overwhelmingly in the insurgents' favor. As the number of American troops in Vietnam increased, the Viet Cong countered with more recruitment efforts.

On January 31, 1968, the first day of the Tet holiday, the Viet Cong launched the Tet Offensive, issuing attacks on military and civilian command posts in more than a hundred cities and towns throughout South Vietnam. Utilizing the tunnel systems, the strategy for the attacks involved Viet Cong sapper-commandos (special forces) delivering shock attacks followed by waves of Viet Cong regulars continuing the onslaught.

---

[44] Tom Mangold and John Penycate, *The Tunnels of Cu Chi* (New York: Berkley, 1986).

[45] The systems even included theaters, where politically motivated plays were performed.

Figure 2. Command center in the Cu Chi tunnels.[46]

The offensive, which lasted until the end of March 1968, resulted in 37,000 Viet Cong causalities, including many of the Viet Cong's elite soldiers and officers, with many more troops captured and injured. Although the offensive hurt American morale and negatively affected opinion of the American war effort back in the United States, the offensive also seriously weakened the Viet Cong infrastructure, and 1968 would continue to be a high-casualty year for the Viet Cong (with phases II and III of the offensive, or the "mini-Tets," occurring in May and August, respectively). The result was the replacement of more than a third of Viet Cong troops with PAVN forces. Having lost more than half of their fighting force by some estimates, the Viet Cong ceased being an effective military organization for the rest of the Vietnam War. Although the Tet Offensive significantly reduced the Viet Cong's numbers and influence, North Vietnam continued to have military success, ultimately winning the war and reuniting North and South Vietnam under the new Socialist Republic of Vietnam.

METHODS OF RECRUITMENT

From 1954 to about 1960, North Vietnam managed a subsidiary branch in the South. However, it is believed that although Viet Minh members were left behind in the South, they did not initially have

---

[46] "File:VietnamCuChiTunnelsCommand.jpg," *Wikimedia Commons*, photo by Kevyn Jacobs, accessed March 11, 2011, http://commons.wikimedia.org/wiki/File:VietnamCuChiTunnelsCommand.jpg.

orders to create an insurgency; rather, the period served as a time of reformation and recruitment.[47] The Viet Cong used revolutionary insurgency techniques for recruitment. Party cadres in South Vietnam were usually set up in villages with which leaders were familiar after the French–Indochina War, and members of these cadres led austere, sober lives. This posed a stark contrast to the foreign governmental officers in the Diem regime, who were often corrupt. Communist doctrine was not normally stressed and instead was downplayed, and the cadres would typically study the political and social structures of the villages before attempting to recruit.

Viet Cong's recruitment of villagers was strategic, with recruiters often collecting information on individuals before actively recruiting them. This information was used to start casual conversations and friendships with villagers targeted for recruitment. The Viet Cong took pains to stress its independence from the North Vietnamese government and the Hanoi-based Communist Party. One method for achieving this image of independence was creating or infiltrating front groups (e.g., a farming cooperative); once such groups were created or infiltrated, Viet Cong members would portray the Viet Cong as populist and concerned with local issues. Traditional festivals were also infiltrated and used for mobilization and recruitment efforts. Members of the groups were slowly influenced into adopting the Viet Cong cause as their own. The Viet Cong also took advantage of antiforeign sentiment (resulting from both French and American interference) and addressed local issues to recruit villagers and develop their trust. Drafts were also used in Viet Cong-controlled villages to fill military units with young men.[48] Viet Cong recruits normally received training at locations outside of their villages, where they also were subjected to extensive political indoctrination. The training was usually conventional (i.e., weapons use, marching, small-unit tactics, etc.). Recruits who displayed the most potential received additional advanced training, as well as indoctrination, and were assigned to the main force units. The rest of the recruits were used for guerrilla combat missions and other part-time assignments.[49]

Binh van (meaning "action among the military"), one of the three action programs the Viet Cong implemented under *dau tranh*, was designed to convince South Vietnamese civil service and military personnel to either defect to the Viet Cong cause or desert their posts. Tactics included the offering of rewards, the use of undercover agents to spread dissension, intimidation, and pressure from friends and

---

[47] US Department of State, *The Pentagon Papers*.

[48] Pike, *Viet Cong*, 490; Davison, *Some Observations on Viet Cong Operations*.

[49] Ibid.; Douglas Pike, *PAVN: Peoples Army of Vietnam* (New York: Presidio, 1996).

family members. The extent to which the program was successful is not known, but it is likely to have had a positive impact on the overall Viet Cong cause.[50]

METHODS OF SUSTAINMENT

The provision of supplies to the Viet Cong in the field was critical to the Communist victory. Chief among the means of provision was the Ho Chi Minh trail,[51] a road system used to provide logistical support for both NVA and Viet Cong forces. The 9,940-mile-long route (16,000 kilometers) consisted of a multitude of roads, waterways, trails, and tracks through the Vietnamese jungles. Construction first began in 1959 by a special PAVN unit (Group 559), and with the consent of Cambodia's Prince Sihanouk, the route was established on the Vietnamese–Cambodian border and stretched from the western coast of North Vietnam through Laos and Cambodia into South Vietnam. The trail provided a supply route to furnish the Viet Cong with ammunition and equipment and also gave tens of thousands of soldiers (Viet Cong and PAVN) a way to move through South Vietnam.

The NLF set up base camps (known as *Binh Trams*) and medical stations along the trail, and most of these camps and stations were underground. The *Binh Trams* were usually staffed with guides, specialists, and infantry, as well as anti-aircraft weaponry.[52]

Vast improvements to the road allowed Viet Cong operatives to travel its length in six weeks—down from six months before the upgrades. The NLF also used Soviet military vehicles to move supplies quickly down the trail. When blocking the trail was deemed impossible (partly because of Viet Cong ambushes and partly because of the official neutrality of Laos and Cambodia), American forces began an aerial bombing campaign; although this increased Viet Cong casualties, supplies were still effectively delivered to the insurgents.[53]

The three programs of the *dau tranh* were significant in sustaining the Viet Cong effort. *Dich van* in particular raised revenues for the Viet Cong via taxes, helped portray NLF rule as one offering societal stability, kept villages in Viet Cong command, and provided havens for NVA and Viet Cong forces. Mandatory labor policies had villagers building fortifications for the Viet Cong and also delivering supplies

---

[50] Wells, "No More Vietnams."

[51] The Ho Chi Minh trail is known as Duong Truong Son, or Truong Son Road, in Vietnam.

[52] John Prados, *The Blood Road: The Ho Chi Minh Trail and the Vietnam War* (New York: John Wiley & Sons, 2000)

[53] Ibid.

to and from the insurgents by using specialized bicycles with widened handlebars. The Viet Cong wielded extensive economic power at the local level through *dau tranh*.

The villagers provided critical economic, intelligence, and other logistical aid to the Viet Cong. Homemade weapons—such as booby traps, mines, and various explosives—were produced in Viet Cong-controlled villages. American bombs that failed to explode were also collected, reworked, and "recommissioned" into the war effort.[54]

METHODS OF OBTAINING LEGITIMACY

The Viet Cong was largely successful in its attempts to gain the trust and acceptance of the rural peasantry. The group gained trust and acceptance by providing civil services that the constrained and feeble Diem regime could not provide and by appealing to the villagers who became increasingly alienated by the corrupt, oligarchic, and repressive South Vietnamese government.

Dich van (meaning "action among the enemy") was a propaganda effort to increase the legitimacy of North Vietnam (and the Viet Cong cause) at the expense of South Vietnam, which was portrayed as a puppet government of the United States. Aimed at the South Vietnamese people, many methods and media were utilized, including plays, protests, rallies, leaflets, meetings, and rumor campaigns. The purpose of the *dich van* action propaganda program was to convince the South Vietnamese people that the South Vietnamese government was merely a puppet government of the United States and that the North Vietnamese government was the legitimate government for all of Vietnam.

In an effort to use public opinion against the American government, with the *dich van* effort the Viet Cong tried to convince the American public that an American victory was not possible. Through diplomatic channels and media (such as *Radio Hanoi*, which was aimed at American troops), extensive propaganda was used. It was hoped that the American people would recognize the legitimacy of Hanoi's claim to all of Vietnam and that they would increase political pressure for an American exit from the war.[55]

---

[54] Wells, "No More Vietnams."
[55] Ibid.

EXTERNAL SUPPORT

Aid and support by North Vietnam and PAVN was extensive because the North Vietnamese government was the NLF's ultimate authority. The greater Communist bloc also significantly supported the Viet Cong effort. The victory of the Chinese communists in 1949 and the Viet Minh subsequently gaining a border with China in 1950 during the French–Indochina War was a crucial development for the Viet Cong cause.

North Vietnam's communist allies, China and the Soviet Union, provided critical support for the war effort in South Vietnam. The Soviet Union provided the most aid in the form of fuel, munitions, and heavy equipment, including sophisticated air defense systems. The Chinese provided medical assistance, training facilities, and infantry arms. Despite support from both the Chinese and the Soviets, there was considerable tension between Peking and Moscow over the Vietnam situation. The Chinese were more adamant about participating in the armed conflict, as Mao considered the United States to be a threat to his own Communist state.[56] The strained relationship between the Soviet Union and China led the Soviets to reduce aid to the Viet Cong and North Vietnam because of North Vietnam's relationship with the Chinese; however, military success would lead to an increase in Soviet aid the following year.

The influx of modern and more technologically advanced equipment allowed General Giap to transform the Viet Cong from a primarily guerrilla insurgency into a more conventional military including five light infantry divisions and a heavy division. The armaments and equipment provided by the Chinese were ironically made in America, having been used by the Chinese Nationalists before their defeat by Mao in the 1949 revolution. By the mid-1960s, most Viet Cong regular forces were armed with a multitude of Soviet and Chinese machine guns (light, medium, and heavy), including AK-47 submachine guns (the Chinese models), which were particularly effective in combating American helicopter forces.[57]

COUNTERMEASURES TAKEN BY THE GOVERNMENT

The South Vietnamese government attempted to counter the emergent Viet Cong insurgency by augmenting its regular military

---

[56] Ang, *The Vietnam War from the Other Side;* Qiang Zhai, *China and the Vietnam Wars, 1950–1975* (Chapel Hill, NC: University of North Carolina Press, 2000).

[57] "The Vietnam War," *The History Place*, accessed August 26, 2010, http://www.historyplace.com/unitedstates/vietnam/.

and paramilitary forces and developing its Pacification Programs. "Pacification Programs" or "Revolutionary Development" are umbrella terms used to describe South Vietnam's efforts to retain or regain control in contested villages. A number of social, political, and economic programs were launched; however, the Viet Cong attempted to nullify or destroy any progress the pacification policies had in improving village life or gaining the trust of rural villagers.[58] South Vietnam also launched the National Reconciliation Program, its own attempt at reunifying the North and South.

Other pacification programs included the Civil Operations and Revolutionary Development Support (CORDS) program, launched by the United States in South Vietnam. CORDS involved providing security for the South Vietnamese peasantry, followed by subsequent attempts at weakening the NLF infrastructure and developing programs that would lead the peasants to trust and esteem South Vietnam. CORDS was believed to be effective in limiting the Viet Cong's ability to gain support from the peasantry, but the program was regarded as being "too little, too late."[59]

To combat the guerrilla-style warfare used by the Viet Cong, the United States adopted a strategy referred to as "Search and Destroy"; with this strategy, American forces would enter enemy territory, seeking out the insurgents to engage them in what was ultimately a war of attrition. Part of the Search and Destroy strategy was the tunnel-rat tactic used to deal with the Viet Cong's tunnel attacks, where soldiers were sent down into uncovered tunnels to plant explosives and kill any remaining Viet Cong in the complex.

In 1961, the United States and South Vietnam launched a joint campaign to combat the Viet Cong insurgency through a combination of social, economic, political, military, and psychological tactics. Through what was known as the Strategic Hamlet Program, which involved purging an area of Viet Cong and subsequently providing security and civil services for the area, it was hoped that denizens of these areas would begin to trust and support South Vietnam. By 1962, the US–South Vietnamese united effort appeared to have changed the course of the war. With the dramatic political tumult of the following year, however, when Diem was ultimately removed from office, South Vietnamese forces began to struggle. As the South Vietnamese paramilitary units began to disperse and disappear, the

---

[58] Pike, *Viet Cong.*
[59] Robert W. Komer made this statement.

insurgents gained important and strategic villages and increased the confiscation of South Vietnamese arms and weaponry.[60]

Other American–South Vietnamese counterinsurgency efforts included the Phoenix Program, which was devised by the Central Intelligence Agency (CIA) and aimed to substantially weaken the Viet Cong infrastructure. The Phoenix Program involved gathering intelligence on the NLF and subsequently "neutralizing" NLF members. The CIA was also involved in training tens of thousands Lao Hmong people for special missions against Communist Pathet Lao forces, which supported North Vietnam.

SHORT- AND LONG-TERM EFFECTS

CHANGES IN THE ENVIRONMENT

The Viet Cong insurgency played a crucial part in the withdrawal of American forces and the eventual defeat of the South Vietnamese. Although the Tet Offensive was a devastating blow to the Viet Cong infrastructure, it nevertheless has been cited as a decisive moment in turning American public opinion against the war.

Peace within Vietnam came about after the war, but conservative policies and extensive persecution of individuals led to a mass exodus of many innovative people as well as to limited economic growth and isolation by the international community. The government enacted several policies, including reeducation camps for anyone who had been associated with the South Vietnamese government and a collectivization campaign of private properties. Economic modernization and liberalization would come about in the mid-1980s (see *Changes in Policy*), leading to extensive economic growth. Vietnam was also involved in two wars with Cambodia against the Khmer Rouge and China (which caused a mass exodus of ethnic Chinese) in the late 1970s. The Vietnamese government continues to be cited for human rights abuses, including limits on freedom of speech and press.

CHANGES IN GOVERNMENT

After the fall of Saigon and the North Vietnamese victory, the Provisional Revolutionary Government of the Republic of South Vietnam was set up to govern South Vietnam, before the North and

---

[60] Bernard W. Rogers, *Cedar Falls Junction City: A Turning Point*, Vietnam Studies Series no. 2, foreword by Verne L. Bowers (Washington, DC: Government Reprints Press, 2001).

South were finally reunified under the Socialist Republic of Vietnam in 1976. As of 2010, Vietnam remains a single-party state, with the Communist Party of Vietnam dominating all of politics and society.

CHANGES IN POLICY

After its reunification, Vietnam went through a period of economic stagnation from 1975 to 1985. Extensive economic reforms (referred to as *Doi Moi,* or renovation) were introduced in 1986, including market reforms that opened up Vietnam to foreign investment and considerably improved the business environment. The Vietnam economy grew rapidly—one of the fastest-growing in the world—with an average annual gross domestic product growth of about 8% from 1990 to 1997.

CHANGES IN THE REVOLUTIONARY MOVEMENT

Although it were successful in setting up a shadow government during the Diem regime, the Viet Cong had hoped to set up a permanent governmental structure, which they termed the Revolutionary Administration, in South Vietnam. However, after the Tet Offensive, during which the Viet Cong incurred significant casualties, the insurgents lost control of many of their villages, and their governmental apparatus fell apart. The Viet Cong eventually retreated to Cambodia, even as North Vietnam moved closer to victory, losing all influence in South Vietnam before its eventual abolition after the end of the war.[61]

In 1970, the North Vietnamese-friendly Prince Sihanouk was ousted from power in Cambodia, and a new Cambodian government allowed the United States to attack Viet Cong bases within its borders. However, the Viet Cong spent the rest of the war as an accessory to PAVN forces. The Viet Cong was permanently disbanded after reunification, and the last remnant of the insurgency, the NLF, was merged with the North Vietnam-based Vietnamese Fatherland Front in 1977.

---

[61] "Failure of the Viet Cong to Establish Liberation Committees," Central Intelligence Agency Collection, The Vietnam Center and Archive, Texas Tech University, http://www.vietnam.ttu.edu/virtualarchive/items.php?item=0410691002.

BIBLIOGRAPHY

Ang, Cheng Guan. *The Vietnam War from the Other Side.* New York, NY: Routledge, 2002.

Brigham, Robert K. *Guerrilla Diplomacy: The NLF's Foreign Relations and the Viet Nam War.* Ithaca, NY: Cornell University Press, 1999.

Carver Jr., George A. "The Faceless Viet Cong." *Foreign Affairs* 44, no. 3 (April 1966): 347–372.

Central Intelligence Agency. "Vietnam." *The World Factbook.* Accessed January 12, 2010. https://www.cia.gov/library/publications/the-world-factbook/geos/vm.html.

Chapuis, Oscar. *A History of Vietnam: From Hong Bang to Tu Duc.* Westport, CT.: Greenwood Press, 1995.

Davison, W. P. *Some Observations on Viet Cong Operations in the Villages.* Santa Monica, CA: RAND Corporation, 1967.

"The Vietnam War." *The History Place.* Accessed August 26, 2010. http://www.historyplace.com/unitedstates/vietnam/.

Horton, B. L. "A Content Analysis of Viet Cong Leaflets as Propaganda, 1963–68." Master's thesis, Texas Tech University, 2008.

Mangold, Tom, and John Penycate. *The Tunnels of Cu Chi.* New York: Berkley, 1986.

Neale, Jonathan. *The American War: Vietnam 1960–1975.* Chicago: Bookmarks, 2001.

Paige, Jeffery M. "Inequality and Insurgency in Vietnam: A Re-Analysis." *World Politics* 23, no. 1 (October 1970): 24–37.

Pike, Douglas. *PAVN: Peoples Army of Vietnam.* Presidio. 1996.

———, *Viet Cong: The Organization and Techniques of the National Liberation Front of South Vietnam.* Cambridge: Massachusetts Institute of Technology, 1966.

Prados, John. *The Blood Road: The Ho Chi Minh Trail and the Vietnam War.* New York: John Wiley & Sons, 2000.

Rogers, Bernard W. *Cedar Falls Junction City: A Turning Point.* Vietnam Studies Series no. 2. With a foreword by Verne L. Bowers. Washington DC: Government Reprints Press, 2001.

Shultz, Richard. "The Limits of Terrorism in Insurgency Warfare: The Case of the Viet Cong." *Polity* 11, no. 1 (1978).

Tanham, George K. *Communist Revolutionary Warfare: From the Vietminh to the Viet Cong.* Westport, CT: Praeger Security International, 2006.

Tucker, Spencer C. *Vietnam.* Lexington, KY: The University Press of Kentucky, 1999.

US Department of Defense. *The Pentagon Papers: The Defense Department History of United States Decisionmaking on Vietnam.* Senator Gravel Ed. Boston: Beacon Press, 1971.

US Department of State. "Background Note: Vietnam." http://www.state.gov/r/pa/ei/bgn/4130.htm.

US Department of State. "Questions and Answers: Viet-Nam: The Struggle for Freedom." Department of State Publication 7724, Far Eastern Series 127. Washington, DC: Government Printing Office, August 1964.

Wells, Gordon M. "No More Vietnams: CORDS as a Model for Counterinsurgency Campaign Design: A Monograph." School of Advanced Military Science, US Army Command and General Staff College, 1991.

Zhai, Qiang. *China and the Vietnam Wars, 1950–1975.* Chapel Hill, NC: University of North Carolina Press, 2000.

CHECHEN REVOLUTION: 1991–2002

Maegen Nix and Shana Marshall

SYNOPSIS

After the collapse of the Soviet Union in 1991, the Republic of Chechnya in southern Russia declared its independence from the Russian Federation. Because of the turmoil in Moscow surrounding the disintegration of the USSR, the Chechen people were initially able to achieve self-rule with minimal interference from Moscow. In December 1994, Russian troops entered Grozny, the capital of Chechnya, to assert Moscow's control over the breakaway republic. Over the next twenty months, approximately 100,000 people, mostly civilians, died during the Russian siege on Grozny and the resulting Chechen insurgency to drive the Russians out. The Chechen rebels were well armed and better trained than the Russians expected, and the Russian forces were initially ineffective and ill disciplined in their operations. As the skill and effectiveness of the Russian forces improved in 1995, the Chechen rebels increased their use of ambushes, sniper attacks, and bombings. More radical Islamic elements within the Chechen movement assumed leadership and advisory positions and the Chechen rebels expanded their scope of attacks to include the large-scale taking of civilian hostages in towns in neighboring Dagestan. The First Chechen War ended with a cease-fire agreement signed in August 1996. Three years later, the Second Chechen War commenced in response to Chechen rebel attacks against a perceived Russian-supported puppet government in Grozny and purported rebel attacks against military and civilian apartments in Dagestan. By the end of 2002, the city of Grozny was destroyed, more than 120 civilians had died during a siege at a Moscow theater, and the simmering insurgency had developed deeper connections to Islamic extremism.

TIMELINE

1991	Collapse of the Soviet Union.
1992	Chechnya declares independence and adopts a constitution.
December 1994	Russian troops enter Chechnya. First Chechen War begins.
June 1995	Chechen rebels led by Shamil Basayev seize hundreds of hostages at a hospital in Budyonnovsk (Dagestan). More than one hundred killed.
April 1996	Dzhokhar Dudayev killed in Russian missile attack.
August 1996	Chechen attack on Grozny. Cease-fire signed on August 22.
January 1997	Maskhadov elected president of Chechnya.
March 1997	Yeltsin and Maskhadov sign treaty on peace and bilateral relations.
June 1998	Maskhadov issues a state of emergency in response to growing lawlessness in Chechnya.
March 1999	Moscow's top envoy to Chechnya is kidnapped and murdered.
September 1999	Bomb attacks on military housing and apartments in Dagestan.
October 1999	Russian forces advance through Chechnya.
February 2000	Grozny is captured by Russian forces.
May 2000	Putin declares direct rule from Moscow.
September 2001	Chechen rebel offensive against the town of Gudermes.
October 2002	Moscow theater siege. More than forty rebels and 126 hostages killed.
December 2002	Suicide bomb attack on Russian-backed Chechen government headquarters in Grozny kills more than eighty.

THE ENVIRONMENT OF THE REVOLUTION

PHYSICAL ENVIRONMENT

Figure 1. Chechnya and surrounding area.[1]

Chechnya is a Russian republic that sits along the Russian–Georgian border, west of the Caspian Sea and landlocked between the Russian republics of Dagestan, Ingushetia, and North Ossetia. Spread across a region roughly the size of New Jersey, Chechen terrain includes the Argun, Terek, and Sunzha rivers; the plains north of the Terek; arable rolling hills and low mountain ridges in the middle of the republic; and densely forested mountains along the southern and southwestern borders. The chain of mountains that runs along the base of the Caucasus stretches 600 miles from the Caspian Sea to the Black Sea.[2]

Chechnya's proximity to northwestern Iran and northeastern Turkey makes it a conduit between Russia and the Middle East. Access routes for oil pipelines and natural gas make the area important to the Russian economy. Although Chechnya's internal oil fields are nearly depleted, Russia would lose economically and geopolitically if the pipeline that runs through the region were to fall outside its control.[3]

---

[1] "File:Chechnya and Caucasus.png," *Wikipedia*, Image by Kbh3rd, accessed March 15, 2011, http://en.wikipedia.org/wiki/File:Chechnya_and_Caucasus.png.

[2] Chechens are adept, and have been so historically, at mountain warfare.

[3] Rajan Menon and Graham Fuller, "Russia's Ruinous Chechen War," *Foreign Affairs* 79, no. 2 (2000): 39.

CULTURAL AND DEMOGRAPHIC ENVIRONMENT

Composed of a predominantly Muslim population since the eleventh century, the Caucasus region is home to more than seventy-five different ethnicities.[4] Chechens make up the area's most homogenous society and are also its most numerous people. Approximately 70% of the republic's inhabitants are Chechen and only 0.2% of the population considers Russian as their national language.[5] The neighboring Republic of Dagestan, in contrast, hosts dozens of ethnic groups, which interact in more than thirty languages.

Although it is difficult to obtain an accurate population count, there were well over one million people living in Chechnya before 1994. During the first period of violent conflict with Russia, between December 1994 and August 1996, 100,000 Chechens were estimated to have died.[6] In 2010, speculation placed the number of inhabitants in Chechnya at 700,000, with tens of thousands living outside its borders as refugees.[7] United Nations' figures indicated that in 2002, there were "160,000 displaced persons in Chechnya and 150,000 in Ingushetia, some 50% of them children."[8]

SOCIOECONOMIC ENVIRONMENT

In accordance with the region's geography, traditional Chechen society centered on a horizontal clan structure of highlanders (Lamanroi) and lowlanders (Checkharniakh), similar to the traditional clan relationships of Scotland. Although farming the arable land of the plains provided for clan subsistence, the raising of livestock provided the basis of the clan's wealth.[9] The Chechen economy remained largely agrarian throughout the twentieth century, although the decade-long war with Russia left many farms contaminated with

---

[4] The Soviet government frequently manipulated the number of ethnic groups in the Caucasus by combining multiple groups together under one rubric in censuses. Thus, arguably, this number is significantly higher.

[5] Gail W. Lapidus, "Contested Sovereignty: The Tragedy of Chechnya," *International Security* 23, no. 1 (Summer 1998): 10.

[6] Jean-Herve Bradol, "MSF Presentation to Council of Europe Regarding Chechnya and Ingushetia," Medécins Sans Frontières report, January 23, 2002, http://js.static. reliefweb.int/node/94650.

[7] Mark Kramer, "Guerrilla Warfare, Counterinsurgency and Terrorism in the North Caucasus: The Military Dimension of the Russian-Chechen Conflict," *Europe-Asia Studies* 57, no. 2 (2005): 210.

[8] Office of the Special Representative of the Secretary-General for Children and Armed Conflict (June 24, 2002), accessed January 26, 2010, http://www.un.org/children/conflict/pr/2002-06-2447.html.

[9] Moshe Gammer, *The Lone Wolf and the Bear, Three Centuries of Chechen Defiance of Russian Rule* (Pittsburgh: University of Pittsburgh Press, 2006), 4.

land mines and toxic ordnance. Grozny, the Republic's only major city and locus of industry, was the primary target of an intense Russian bombing campaign in 1994 that decimated its central infrastructure, including the sewage, water, and electricity systems.

Except for manufacturing within Grozny, Chechnya remained an area of low industrialization even after World War II when other areas of the Soviet Union began to increase their production capacity. The Soviet government processed fuel and lubricants for aviation within the city, also producing machinery and parts for the oil industry. However, the Soviet government discriminated against native Chechens in matters of labor, employing primarily ethnic Russian workers, and 95% of profits from resource extraction and industry went to the central Soviet budget.[10] The indigenous population remained dependent on agriculture for subsistence, with rural areas rearing livestock and focusing on small-scale poultry production to provide food and to generate income.[11] Once the Soviet Union disintegrated and war with Russia began in earnest, routine means of subsistence became impossible and standards of living dropped. Much of this resulted from deliberate Russian measures during the war that were meant to eliminate the Chechens' ability to resist. For example, when Chechens claimed independence in 1991, Yeltsin severed financial and economic ties with the new government and imposed a blockade. Chechens rejected political participation in the Russian federated system and launched their own currency, which was rejected by Russian banks. The limited rudimentary political and economic organization that did exist deteriorated quickly, and subsequently, "formal institutions were replaced by informal family and clan ties, the only instrument available to ensure socially enforceable commitments, safety and the sharing of resources."[12] Ultimately, a shadow economy developed in Chechnya based on informal and criminal networks often sponsored by elites within Russia who benefited from the unregulated economy.

Although economic data on the Republic of Chechnya during the conflict are not available, the region experienced high unemployment

---

[10] Valery Tishkov, *Chechnya, Life in a War-Torn Society* (Berkeley, CA: University of California Press, 2004), 41.

[11] Technical Cooperation Department, The Food and Agriculture Organization of the United Nations (FAO), "The FAO Component of the Inter-agency Consolidated Appeals 2005" (Rome: Emergency Operations and Rehabilitation Division, 2004), http://www.fao.org/docrep/007/y5805e/y5805e06.htm#TopOfPage.

[12] Svetlana Glinkina and Dorothy Rosenberg, "The Socioeconomic Roots of Conflict in the Caucasus," *Journal of International Development* 15, no. 4 (2003): 519.

and pervasive poverty even before the onset of violence.[13] Agriculture could not employ a growing population of educated Chechen youth, and estimates indicate that "the labor surplus reached perhaps 100,000 to 200,000, or 20–30% of the able-bodied population."[14] After Chechnya claimed independence, the expansion of the black market put legal governance and the development of the rule of law in direct competition with the interests of local power brokers who divided control over political authority and resource flows.[15] Business development within the unregulated area, based on personal relationships and supported by external Russian elites, saw the growth of illegal entrepreneurship and included "racketeering, money laundering, smuggling, criminal privatization, intentional bankruptcy, fraudulent securities, counterfeiting, unfair competition, illegal trade, tax crimes, etc."[16] In order for these illicit activities to thrive, resident criminal groups required partnerships with elements outside of Chechnya, both in Russia and abroad.

HISTORICAL FACTORS

Chechen resistance to Russia dates back to the sixteenth century when Russians first entered Chechnya in 1552 under the reign of Ivan the Terrible. Catherine the Great also attempted to subdue the region during the late eighteenth century. Records dating back to 1816 indicate that Russian expansion came "by means of punitive raids on mountain villages, collective punishment, razing of houses and crops, deforestation, forced mass deportation, and settlement of Cossacks on lands vacated by the Chechens."[17] When tsarist Russia intensified its colonial expansion into Chechnya, the Chechen resistance unified under the leadership of Sufi Naqshbandi Islamic preachers. "Islamic discipline was seen as the best way of securing unity against the Russians . . . and also set in place a differentiated tax

---

[13] Yuka Takeda, "Economic Growth and its Effect on Poverty Reduction in Russia," Global COE Hi-Stat Discussion Paper Series 075, Institute of Economic Research, Hitotsubashi University (July 2009), http://gcoe.ier.hit-u.ac.jp/research/discussion/2008/pdf/gd09-075.pdf.

[14] Tishkov, *Chechnya, Life in a War-Torn Society*, 41. Chechen families felt compelled to regain their numbers after deportation but could not support the growing population. Many of these Chechens became involved in fighting the Russians.

[15] Glinkina and Rosenberg, "The Socioeconomic Roots of Conflict in the Caucasus," 517–519.

[16] Ibid., 520.

[17] Tony Wood, *Chechnya, The Case for Independence* (New York: New Left Books/Verso, 2007), 21. Also see Moshe Gammer, *Muslim Resistance to the Tsar: Shamil and the Conquest of Chechnya and Daghestan* (London: Frank Cass & Co. Ltd., 1994).

system, pensions for widows and invalids, and military hospitals."[18] The most successful Islamic leader was Shaykh Shamil, who united North Caucasian Muslims from Chechnya and Dagestan under an Imamate that lasted for roughly thirty years, 1830–1859, and combined *shari'a* law, *ghazavat* (holy war), and jihad within the Sufi mystical tradition of *tariqa*.[19] Although the Naqshbandi were excluded from official Russian administrative and legal institutions upon their defeat, "the Sufi brotherhoods never ceased to wield influence and formed an alternative system of administration. This system permeated all levels of social, religious, and political life in Chechnya and Dagestan and, based as it was on a clandestine network of Murid (students of Sufism) organizations, remained largely outside Russian reach."[20] Shamil's efforts and the resulting Russian policy promoted Islam from a cultural link into a way of life, and it became an integral part of North Caucasian self-identity.

When Lenin visited Chechnya in December of 1917, he recruited many Chechens to fight, declaring that the Chechen people were free and inviolable. At the time, the region had proclaimed itself to be a separate state called the Mountainous Republic of the Northern Caucasus and included a 27,000-square-mile footprint that covered areas that now comprise Chechnya, Ingushetia, North Ossetia-Alania, Kabardino-Balkaria, and Dagestan. In 1919, an agreement was signed between Communists and mullahs that guaranteed "Chechen autonomy and religious practices within a Soviet system."[21] However, once the Communists assumed power, the Red Army moved into the area and began to behave as occupiers. Chechen resistance against the Russians therefore continued.

Much as Lenin had done, Stalin in 1921 "pledged full autonomy for the rechristened Soviet Mountain Republic and acceptance of local customary law; then in 1922 the Mountain Republic's various components were sliced away and incorporated, one by one, as regions of Russia."[22] Chechen resistance continued with uprisings in 1922, 1924, 1925, and into the 1930s.[23] Although roughly 20,000 Chechens joined the Soviet Army during World War II, Stalin deported the entire population to Central Asia in 1944. Soldiers and officers of Caucasian origin, many of them highly decorated, were recalled from the German

---

[18] Chechens converted to Sufi Islam during the eleventh century. Tony Wood, Chechnya, *The Case for Independence*, 24.

[19] Anna Zelkina, "Jihad in the Name of God: Shaykh Shamil as the Religious Leader of the Caucasus," *Central Asian Survey* 21, no. 3 (2002): 249–264.

[20] Ibid., 260.

[21] Wood, *Chechnya, The Case for Independence*, 31.

[22] Ibid., 31.

[23] Lapidus, "Contested Sovereignty," 8.

front without explanation and sent to work in the Siberian Gulags.[24] The Chechen people as a whole were rehabilitated only in 1957. This genocidal deportation of the population in 1944, in which 30% of the Chechens perished, and their subsequent marginalization upon return to Chechnya, solidified an anti-Russian sentiment among the Chechen people and continues to play a pivotal role in the Chechen national consciousness.[25]

Although the ethnic Chechen presence within the Caucasus dates back thousands of years, and their historic narrative is peppered with repudiations of outside rule, including struggles against the Arabs, the Persians, and the Ottomans, it is the struggle with Russia that has defined the concept of Chechen nationhood. At the time of their formal push toward independence in 1991, one in three Chechens was a survivor of the 1944 deportations.[26] This included President Dzhokhar Dudayev and President Aslan Maskhadov. Although deportees were allowed to return to Chechnya under Khrushchev in 1957, they returned to find that settlers sent by the Soviet government occupied their homes and lands.

GOVERNING ENVIRONMENT

The Soviet Union was created by binding together numerous nationalities within a system of ethno-territorial federalism—establishing a hierarchy of ethnicities that claimed different rights and benefits on the basis of their status within the system. Thus, when the Soviet Union broke apart in 1991, certain nations were allowed to secede with little fanfare while others were not.[27] However, additional complexities emerged from the post-World War II political landscape. First, the federal system provided privileges to certain ethnic elites for public and military service—thereby creating in each ethnic enclave networks of powerful individuals whose aspirations were nonetheless

---

[24] Many Russian commanders misrepresented the identities of their Chechen soldiers, whose courage was widely revered. One unit, commanded by the famous Chechen Movlit Visaitov, was among the first to break through the Berlin line and meet up with the American troops on the other side. "Remembering Stalin's Deportations," *BBC News*, February 23, 2004, http://news.bbc.co.uk/2/hi/3509933.stm.

[25] Wood, *Chechnya, The Case for Independence*, 2; and Lapidus, "Contested Sovereignty," 9.

[26] Wood, *Chechnya, The Case for Independence*, 38. "Even by the most conservative estimate, the Chechen and Ingush people lost, across the years of exile, over a third of their total number." See Tishkov, *Chechnya, Life in a War-Torn Society*, 27.

[27] "In the former USSR the borderline between the status of a 'Union' republic, which had the right 'to leave' the Union and that of an 'autonomous' republic, which had no such right, was never impassable. Kazakhstan was at first an autonomy within the Russian Federation; Moldova was once also an autonomy within Ukraine." Wood, *Chechnya, The Case for Independence*, 187.

constrained by the central pull of the Soviet government.[28] Second, policy discrepancies regarding Chechnya were caused by infighting in the Soviet/Russian leadership—notably between reformist Soviet President Mikhail Gorbachev and Boris Yeltsin, the president of the new Russian Republic, each of whom attempted to gain the allegiance of various ethnic elites.[29] When Dzhokhar Dudayev, a retired Soviet Air Force general newly returned from service in the Baltics, was invited to lead the Executive Committee of the All-National Congress of the Chechen People (ANCCP), he was supported by Yeltsin, who hoped to outmaneuver the reigning local Soviet elites.

Dudayev spearheaded the Chechen national movement and dissolved the Supreme Soviet leadership within Chechnya—again, with the initial support of Yeltsin against Gorbachev's Communist party. However, after Dudayev's 1991 election as president of the Chechen Republic of Ichkeria (CRI), his separatist rhetoric and subsequent declaration of Chechen independence alienated Yeltsin.[30] Dudayev's declaration occurred less than two months before the breakup of the Soviet Union, at which point the Soviet constitution fell apart and the region entered a period of tense ambiguity during which a number of republics rejected participation in the nascent Commonwealth of Independent States.

The regions of the USSR that declared independence included Lithuania (March 1990); Armenia (August 1990); Georgia (April 1991); Estonia, Latvia, Ukraine, Belarus, Moldova, Azerbaijan, Kyrgyzstan, and Uzbekistan (August 1991); Tajikistan (September 1991); Turkmenistan (October 1991); and Kazakhstan (December 1991).[31] Consequently, the Russians turned their attention first to the Baltic States and Tatarstan to prevent their withdrawal from the federation, believing these areas held a greater security risk should they be lost.[32] Russia's preoccupation with these regions, in addition to serious domestic challenges, meant that Chechnya served as its own de facto state between 1991 and 1994. During this time, Yeltsin supported a growing anti-Dudayev movement that "demanded a

---

[28] Lapidus, "Contested Sovereignty," 9.

[29] Gorbachev was the president of the Soviet Union while Yeltsin was the president of the Russian Republic within the Soviet Union.

[30] Julie Wilhelmsen, "Between a Rock and a Hard Place: The Islamisation of the Chechen Separatist Movement," *Europe-Asia Studies* 57, no. 1 (January 2005): 36.

[31] Wood, *Chechnya, The Case for Independence,* 51–52.

[32] It should be noted that Lithuania declared independence from the USSR even before the coup. And Tatarstan, a Republic within Russia, was the only region threatening to secede from the Russian Federation itself. Finally, many Russian leaders did not take Chechen claims seriously, believing that the republic could not support itself as an independent state.

referendum on independence, accusing Dudayev of having usurped power and of violating the Chechen constitution."[33] When the Constitutional Court ruled that his actions were illegal and criminal, he had the court dissolved.

Although Yeltsin adopted a hard-line stance on Chechnya, the new Russian Federation legislature[34] made several conciliatory overtures in an attempt to bring the Chechens to the negotiating table, including the *de jure* recognition of Chechen independence. On March 14, 1992, after negotiating on a range of legal, economic, and security issues, Chechen and Russian representatives signed protocols explicitly referring to the "political independence and state sovereignty of the Chechen Republic," language that was also included in documents signed on May 28 and September 25 of the same year.[35] These conciliatory signals, which were at odds with Yeltsin's intentions in the region, continued until the large-scale Russian invasion of Chechnya commenced in December 1994.

It should be noted that during the Russian legislature's legal and diplomatic efforts, Yeltsin sent forces to expel the Chechen government on at least five occasions. In fact, one hypothesis for the Russia invasion in 1994 was that Yeltsin waged war in order to "forestall any parliamentary investigation of the previous five failed coups undertaken by the government in Chechnya."[36] Each coup was directly traceable to Yeltsin's office and the Intelligence Service, which used Russian regular troops to carry out the mission. The final infraction against Dudayev's government occurred on November 26, 1994, when a small contingent attempted to capture Grozny. Often compared to the failure of the Bay of Pigs invasion, the debacle "presented the Kremlin with a difficult choice between an ignominious retreat and a decisive military intervention by Russian federal forces."[37] Three days later, Russian air assets initiated a bombing campaign in Grozny, followed by the introduction of ground troops on December 11.

---

[33] Dmitri Trenin and Aleksei Malashenko, *Russia's Restless Frontier, the Chechnya Factor in Post-Soviet Russia* (Washington, DC: The Brookings Institution Press, 2004): 20.

[34] From 1990 to 1993, the Russian legislature was the Congress of People's Deputies of the Russian Federation. This was replaced by the Federal Assembly after the new 1993 constitution.

[35] Wood, *Chechnya, The Case for Independence*, 52.

[36] Stephen J. Blank and Earl H. Tilford Jr., "Russia's Invasion of Chechnya: A Preliminary Assessment," Strategic Studies Institute Special Report, US Army War College, 1995, 7.

[37] Trenin and Malashenko, *Russia's Restless Frontier*, 21.

WEAKNESSES OF THE PREREVOLUTIONARY ENVIRONMENT AND CATALYSTS

The abusive historic relationship between the Russians and the Chechens strongly influenced the prerevolutionary environment. The effects of deportation and forced relocation, the Soviet suppression of Islamic religious practice, high unemployment rates resulting from systematic discrimination, and the unmet promises of autonomy produced cultural and economic conditions that were easily exploited by a national secessionist narrative. At the same time, the Soviet practice of ethno-territorial federalism, designed to develop a cadre of pro-Soviet ethnic elites within the government and the military, actually fostered the very leaders who eventually led the Chechen nationalist movement.[38] This included Chechnya's first president Dzhokhar Dudayev, who was born in Chechnya in 1944 and spent the first thirteen years of his life living with his family in exile in Kazakhstan.[39] Dudayev achieved the rank of general in the Soviet Air Force, and as president of Chechnya he successfully countered the 1994–1996 Russian offensives. He was killed by the Russians in 1996 by a simultaneous bomb and rocket attack while using a satellite phone.[40] Shamil Basayev, groomed by Russian special operations and intelligence alongside other Chechen fighters, deployed under the Soviet army and gained critical combat experience in Russian-sponsored conflicts against Georgia and Moldova. Basayev later participated in the new Chechen government and served as a Chechen field commander until he broke away independently in 1998 to form the Wahhabi terrorist organization, the Islamic International Peacekeeping Brigade (IIPB).[41]

Amidst this prerevolutionary context, a primary catalyst for war was fostered by the collapse of the Soviet Union during 1990 and 1991. Political liberalization and the decline of communism opened up opportunities for ethno-political groups to mobilize.[42] In addition, the dissolution of the Soviet Union left the federal system itself open to renegotiation.[43] Moreover, the competition between Gorbachev

---

[38] The very existence of these "tactical nation-states" fostered, however unintentionally, the development of national elites and cultures while constraining their economic and political expression. Lapidus, "Contested Sovereignty," 9.

[39] Andrew Higgins, "Profile: Dzhokhar Dudayev: Lone Wolf of Grozny," *The Independent* January 22, 1995, http://www.independent.co.uk/opinion/profile-dzhokhar-dudayev-lone-wolf-of-grozny-1569145.html.

[40] Tom de Waal, "Dual Attack Killed President," *BBC News,* April 21, 1999, http://news.bbc.co.uk/2/hi/europe/325347.stm.

[41] Menon and Fuller, "Russia's Ruinous Chechen War," 38.

[42] Lapidus, "Contested Sovereignty," 9.

[43] Ibid., 11.

and Yeltsin yielded support to unknown political elements who served the short-term interests of the feuding leaders; in this case, Yeltsin backed Dudayev and the ANCCP in order to displace the resident Communist Party leadership that supported Gorbachev.

An additional catalyst emerged under Dudayev's initial leadership as the new government looked at options to solidify support, counter internal opposition, and unify the Chechen people. In this regard, 1991 proved to be a watershed point. The dissolution of the Soviet Union brought issues of self-determination and democratic governance to the fore, creating the potential for a democratic constitution based on the liberal rule of law. Unfortunately, although the first Constitution did in fact propose this type of construction, the Chechen leadership lacked experienced political elites who could mobilize a constituency for reform. For the Chechens, as with many of the former Soviet republics, the liberal democratic moment slipped away. As state services disappeared and the Chechen economy crumbled, Dudayev disbanded the state's formal political opposition, empowered new corrupt elites who appropriated the republic's natural resources for personal gain, and turned to narrow religious appeals in order to garner domestic support and attract financial resources from abroad.

Although religion was not a decisive factor in the prerevolutionary environment, it became the ideological locus of the conflict—for both the Russian state and the Chechen resistance—after the 1996 cease-fire. Historically, Chechen religious practice was based on a form of Islam that merged pre-Islamic cultural traditions with Sufism, the most mystical form of Islam. Although the first Chechen constitution did not include religious language, political leaders soon turned toward Islam to rally support. Dudayev, for example, made direct appeals to Islamic institutions to support his government and in 1993 invited religious representatives to serve in his government. Two Sufi leaders who declined to support his administration were labeled traitors and infidels.

Chechen appeals for religious support were soon answered by organizations outside of the republic. Consequently, political and military leaders connected with a "network of Arab financiers and facilitators" that emerged in the Caucasus between 1988 and 1994 after the war between Armenia and Azerbaijan. "Initial alliances between emerging indigenous Salafists and their Middle Eastern counterparts at this critical historical juncture was one of the key enablers that opened up the region to foreign fighters."[44] Although large numbers

---

[44] Cerwyn Moore and Paul Tumelty, "Foreign Fighters and the Case of Chechnya: A Critical Assessment," *Studies in Conflict and Terrorism* 31, no. 5 (2008): 416.

of foreign fighters did not incorporate into the resistance, those who did held enabling positions within the movement.

FORM AND CHARACTERISTICS OF THE REVOLUTION

OBJECTIVES AND GOALS

The objectives and goals of the Chechen resistance oscillate between three positions that were related to political autonomy. The first position was held by multiple Chechen leaders and involved the belief that Chechnya should have political autonomy while remaining in federation with Russia. In May 1990, the ANCCP advocated sovereignty for Chechnya as a separate republic of the Soviet Union. This concept was explored in the 1996–1998 cease-fire during communications between the Russian government and Maskhadov, Chechnya's second elected president (1997) and a retired Soviet Army colonel. Although the legitimacy of this referendum is unclear, during a March 2003 open vote, the majority of Chechens accepted political autonomy under Russia.

A second position held at other times by some of these same Chechen leaders called for the complete independence of Chechnya as a self-governing state free from Russia. As president, Dudayev, with the ANCCP, declared independence from the Soviet Union in 1991. In January 1997, while running unsuccessfully for president against Maskhadov, Basayev told journalists that "he envisaged Chechnya as a moderate Islamic state within the Commonwealth of Independent States (CIS) that could serve as an intermediary between Russia and the Muslim world."[45] Maskhadov reaffirmed Chechen independence when his peace plan was rejected by Russia in 1999 and Russia invaded Chechnya for the second time.

Last, a third position advocated by some extremist Islamic leaders demanded the establishment of a Northern Caucasus caliphate to unify regional Islamic populations under fundamental Salafi and Wahhabi principles. Geographically, this included Chechnya and Dagestan within Russia, as well as the Islamic areas of Georgia in Abkhazia. In 1990, the Islamic Renaissance Party (IRP), the first Sunni political party, emerged within the North Caucasus. Ideologically aligned with the Muslim Brotherhood and the Pakistani Jama'at-i Islami, the IRP established connections with Islamist groups in the Middle East.

---

[45] Liz Fuller, "Chechnya: Shamil Basayev's Life of War and Terror," *Radio Free Europe/Radio Liberty*, July 10, 2006, http://www.rferl.org/content/article/1069740.html.

Although it fragmented when the Soviet Union collapsed, the IRP's goals represented the first modern political call to reestablish the North Caucasian Imamate.[46] This political call reemerged in 1998 and 1999 with the establishment of a number of Islamic fundamentalist terrorist organizations that frequently worked in coordination. The IIPB and the Islamic Special Purpose Regiment (SPIR), led by Basayev, transformed the Chechen national struggle into a regional religious jihad. Together, these groups had "countless, deep-seated personal and organizational linkages, ostensibly sharing fighters, weapons, and materiel in their ethno-nationalist struggle for an independent homeland free from what they [saw] as Russian subjugation."[47] These groups would later claim to form a North Caucasian Imamate along Salafi jihadist lines.

LEADERSHIP AND ORGANIZATIONAL STRUCTURE

The leadership and organizational structure of the Chechen separatist movement changed course after its initial establishment in 1991. Although Dudayev's national campaign began within the Soviet political framework, Chechnya soon divided into areas of command on the basis of clan structure.[48] Chechen army units were often clan-based groups that had been given an official status because they were likely to exist anyway. Field commanders, who held both political and military authority, took advantage of the ensuing power vacuum to profit from the black market and other illicit activities. In addition, the dissolution of legislative and judicial institutions confounded the establishment of a functioning central government, strengthening the emerging trend of warlordism.

Aslan Maskhadov served as Dudayev's deputy and also governed the Chechen defense forces during the 1994–1996 conflict. He assumed leadership of Chechnya after Dudayev's death in 1996 and then beat out thirteen other candidates, including Shamil Basayev, to be elected president in 1997. Widely viewed as a moderate who enjoyed a high degree of popular support, Maskhadov came to power during the cease-fire with Russia, when Chechens perceived their "victory" over Russia to be "a kind of miracle."[49] However, the prevailing social

---

[46] Moore and Tumelty, "Foreign Fighters and the Case of Chechnya," 416.

[47] Armond Caglar, "In the Spotlight: Islamic International Peacekeeping Brigade," Center for Defense Information, May 30, 2003, http://www.cdi.org/program/issue/document.cfm?DocumentID=1116&IssueID=56&StartRow=1&ListRows=10&appendURL=&Orderby=DateLastUpdated&ProgramID=39&issueID=56.

[48] This does not include the republic's northern regions, which were briefly controlled by a unified Chechen opposition.

[49] Tishkov, *Chechnya, Life in a War-Torn Society*, 181.

and military context within Chechnya prevented Maskhadov from exerting control over his more radical field commanders, precluding any chance of accommodation with Russia.[50] Even though hostilities ceased in 1996, little support for reconstruction came from Russia, and the nascent republic's civil institutions were all but destroyed. Ultimately, daily Chechen life descended into anarchy, while opportunistic field commanders undermined Maskhadov's authority and divided control over Chechen territory between themselves.[51]

Figure 2. Chechen Leaders Dzhokhar Dudayev (left), Aslan Maskhadov (center), and Shamil Basayev (right).[52]

Several influential Sunni leaders played key roles, both to Russia and to Maskhadov's leadership, in the development of the Chechen resistance. Much of their influence can be traced to former President Dudayev, who invited a long-standing member of the Muslim Brotherhood, Khabib Ali Fathi, to be the principal Chechen religious adviser during the first period of conflict.[53] By 1995, approximately fifty hand-selected Arab fighters with Wahhabi allegiances were paired

---

[50]   Elisabeth Smick, "Backgrounder: The Chechen Separatist Movement" (Council on Foreign Relations, July 18, 2006), accessed January 26, 2010, http://www.cfr.org/publication/11121/chechen_separatist_movement.html.

[51]   The following field commanders/warlords underwent a process of radicalization/Islamization during the first war: Shamil Basayev, Salman Raduev, Arbi and Movsar Barayev, Movladi Udugov, and Zelimkhan Yandarbiev. Wilhelmsen, "Between a Rock and a Hard Place," 37.

[52]   (Left) Dzhokhar Dudayev Facebook page, accessed March 14, 2011, http://www.facebook.com/pages/Dzhokhar-Dudayev/49360364744?sk=photos; (middle) "File:Aslan Maskhadov.jpg," *Wikimedia Commons*, photo by Natalia Medvedeva, accessed March 14, 2011, http://commons.wikimedia.org/wiki/File:Aslan_Maskhadov.jpg; (right) "File:Shamil Basaev.jpg," *Wikimedia Commons*, photo by Natalia Medvedeva, accessed March 14, 2011, http://commons.wikimedia.org/wiki/File:Shamil_Basaev.jpg.

[53]   Recall that Sufi orders refrained from supporting Dudayev's policies.

with Chechen field commanders under the leadership of Omar Ibn al-Khattab, who in 1997 was promoted to lieutenant colonel and given two Chechen commendations.[54]

Basayev and al-Khattab established a strong relationship and eventually formed the terrorist organization, the IIPB, in 1998.[55] Other groups were soon to follow. Overall, the period between 1996 and 1999 saw increasing ties between the new Chechen warlords and Sunni extremists, both regional and foreign. These partnerships yielded "further militarization, the aggression of armed groups toward adjacent regions, massive levels of hostage taking, "and the republic's failure to comply with stipulations of the 1996 cease fire agreement."[56] Although many of the native Chechen fighters did undergo a process of religious radicalization, others resisted the "internationalization" of the Chechen struggle, preferring to focus on the achievement of political independence or, more ambitiously, on the founding of a Pan-Caucasus Islamic state. Basayev, for example, appeared to reject the concept of the global Islamic fight against the "far enemy," instead focusing directly on the Caucasus region and its relationship with Russia.[57]

The operational differences between Maskhadov and the Sunni extremists eventually came to a head, and Maskhadov attempted to counter the activities of the growing Wahhabi organizations, condemning their actions and methods. Maskhadov submitted to Wahhabi pressure, however, and established *shari'a* law and a system of Islamic regional courts within the republic in 1999. Once the Russians invaded later that year, Maskhadov became the political figurehead of the secessionist movement and commanded his own militia. His forces exacted a heavy toll on the Russian forces within Grozny by using the city's sewer system to conceal their movement. "A parallel command existed from the war's outset in the form of the Supreme Military Majlis ul-Shura around Basayev and al-Khattab standing against a small coterie of respected Maskhadov deputies in the State Defence Committee. This led to an uncoordinated and disparate command structure and allowed some commanders to depart from traditional

---

[54] Trenin and Malashenko, *Russia's Restless Frontier*, 96.

[55] The IIPB is also known as the Islamic International Brigade, the Islamic International Battalion, the Islamic Peacekeeping International Brigade, the Peacekeeping Battalion, the International Brigade, the Islamic Peacekeeping Army, and the Islamic Peacekeeping Brigade. Armond Caglar, "In the Spotlight: Islamic International Peacekeeping Brigade."

[56] Tishkov, *Chechnya, Life in a War-Torn Society*, 183.

[57] It is interesting to note that ideological variations between the Chechen groups did not prevent or impede their ability to work together.

methods."[58] Although Maskhadov repeatedly called for cease-fires and negotiations, video evidence illustrates his involvement in Chechen terrorist attacks within Russia after 2000.

In July 2000, once the Chechen resistance was pushed out of Grozny, Putin appointed Akhmad Kadyrov as the acting head of the new Russian-supported Chechen government. Kadyrov was a Sufi mufti who supported Dudayev during the first war. He opposed the growing Wahhabi movement during the interwar period and defected from the Chechen resistance in 1999. After 2000, Moscow effectively clamped down on the Chechen republic and slowly began to pick off key leaders within the Wahhabi organizations, which were the foundation of the Chechen resistance. The Russian-supported government in Grozny began to claim more control of the Chechen territory, and the resistance retreated to bases in ungoverned territories within the Caucasus region. As a result, the Chechen resistance movement began to operate as local and regional cells rather than a national structure and it used informal networks linked to Salafi groups outside of Chechnya proper. "Thus a network of linked cells provided support for the planning and execution of terror attacks, often drawing on regional affiliates."[59] Many attacks, such as the attack on the Mozdok Army Hospital in August of 2003, comprised a joint operation by the Stavropol and Ingush Wahhabi *jamaats*, as well as the Kabardino-Balkaria *jamaat* alliance. Attacks against Russian soft targets were perpetrated by multiethnic and indigenous terrorist forces. More recent groups were often led by non-Chechen commanders, including Turks, North Africans, and Gulf state leaders. Although it was likely that there were still several thousand participants dispersed across several areas of the North Caucasus region in 2010, it was difficult to estimate the number of armed Islamists. "Different sources put the number of potential and active Salafite 'soldiers' at between 2,000–10,000."[60]

COMMUNICATIONS

Communications in support of the Chechen resistance began with a political campaign that mobilized the open press, including print

[58] Cerwyn Moore and Paul Tumelty, "Assessing Unholy Alliances in Chechnya: From Communism and Nationalism to Islamism and Salafism," *Journal of Communist Studies and Transition Politics* 25, no. 1 (2009): 87. "In mid-2002 Maskhadov yielded to Salafi pressure and a State Defence Committee-Majlis ul-Shura was formed to coordinate the resistance better and to bring the nationalist Sufi commanders together with the Salafis. Despite this, the rebel movement lacked cohesion until Maskhadov's death in 2005."

[59] Ibid., 87.

[60] Trenin and Malashenko, *Russia's Restless Frontier*, 85.

media, radio, television, and the Internet. For example, the Azzam website was established in the United Kingdom largely to highlight al-Khattab's exploits. Throughout the first period of conflict, Yeltsin allowed Russian and international news organizations to report on the course of the war, and reports were used by sympathetic organizations to raise funding and foster awareness for Chechen support efforts. The Chechen resistance also appointed a director of information operations, Movladi Udugov. After Russian forces destroyed Chechnya's infrastructure, only high-level separatist leaders and politicians were able to gain access to modern communication equipment.[61] Even so, the Chechen communications and outreach efforts were considered to pioneer the use of the web to disseminate information and video throughout the global jihadist world.

Military leaders, such as Basayev, frequently videotaped resistance activities, posted them on Islamic websites, and sent footage to Arabic-language media organizations such as Al Jazeera. "Shamil Basayev more than anyone else recognized the importance of extensive media coverage. He demanded coverage during terrorist activities and otherwise courted the media openly. He issued personally signed safe passage documents for some correspondents, gave interviews from his command post or living quarters very frequently, and on occasion had correspondents as guests in his home."[62] During the second war, in addition to the release of operational tapes, planning videos were also distributed. For example, before infrastructure attacks were conducted in and around Moscow in 2004, Chechen rebels released a planning video of leaders, such as Basayev and Maskhadov, reviewing the Moscow subway, regional water-heating power stations, and gas pipelines.

The Islamic radicalization of the Chechen resistance movement was reflected in the evolution of its separatist communications. Udugov's messages exhibited a dramatic change in both content and tone. By the end of the first war, "articles posted on his website Kavkaz-Centre portray not only Russia as the enemy of Chechnya but the whole of Western civilization as a threat to the Islamic world."[63] Zelimkhan Yandarbiev, the vice president under Dudayev and originally a poet and author of children's literature, made little mention of radical Islamic rhetoric in his 1996 book, *Chechnya—the Fight for Freedom*. "After the first war, however, Yandarbiev promoted the establishment

---

[61] Tishkov, *Chechnya, Life in a War-Torn Society*, 189.

[62] Dianne L. Sumner, "Success of Terrorism in War: The Case of Chechnya" (master's thesis, Naval Postgraduate School, 1999), 5.

[63] Wilhelmsen, "Between a Rock and a Hard Place," 37. Other websites include the Arabic Voice of the Caucasus and Jihad in Chechnya run by Azzam Publications.

of an Islamic state in Chechnya, and eventually, he represented the violent fight as a Muslim duty."[64]

METHODS OF ACTION AND VIOLENCE

As early as 1992, Chechen forces began to capture former Soviet stockpiles in order to gain weapons and materiel. They also established relationships with Russian service members who were open to bribes and payment for weapons. In 1994, when Russian troops first entered Chechnya, mass demonstrations by civilians effectively delayed the advance of the tanks on their way to Grozny. A handful of units and Russian commanders also refused to participate in military action against the civilians. This gave components of the Chechen force additional time to organize and make defensive preparations as the Chechen government took responsibility for units centered about Grozny and field commanders took responsibility for areas within their traditional territories. Russian troops arrived in the city unprepared, believing that resistance would be minimal and failing to conduct intelligence preparation before moving troops.[65] "Instead of light resistance from a few small bands, the 6,000-man Russian force that attempted to penetrate the city on New Year's Eve found itself fighting an enemy far better prepared for battle and much larger than expected (estimates vary widely, from a low of about 1,000 to a high of ten times that amount)."[66]

Initially, Chechen forces relied on conventional tactics and focused their efforts on traditional military targets. For example, using the collection of former Soviet weaponry, Chechens shot down Russian helicopters and aircraft, successfully using surface to air missiles against platforms, such as the Su-25 Frogfoot and Su-24 Fencer.[67] Additionally, because Russian troops remained confined to a limited number of roads and rails for ingress and egress, ambushes along well-known routes increased. Night raids against Russian camps and targets also became frequent; although they were lacking in night-vision capabilities, Chechen snipers would often wait for soldiers "to

---

[64] Ibid., 38. Yandarbiev briefly became president of Chechnya after Dudayev's death and before the 1997 election of Maskhadov. During his time as president, Yandarbiev established *shari'a* courts, introduced Sudan's criminal law code, supported Wahhabi schools, and formed Islamic security regiments.

[65] By the 1980s, urban warfare was not included in Russian military training (with the exception of training provided to some special forces units). Olga Oliker, *Russia's Chechen Wars 1994-2000: Lessons from Urban Combat* (Santa Monica, CA: RAND Corporation, 2001), 8.

[66] Oliker, *Russia's Chechen Wars*, 13.

[67] Kramer, "Guerrilla Warfare, Counterinsurgency and Terrorism in the North Caucasus," 232.

take out pocket lighters or matches to light their cigarettes, offering a conspicuous target."[68] The consistent use of seemingly random sniper attacks took a psychological toll on Russian soldiers and disrupted the work of specialized units. In addition, because Russian forces transmitted open communications as a result of limitations in their equipment and training, the Chechens were able to release false reports over Russian radio channels in order to draw Russian forces into certain areas for attack.[69]

In February, 1995, Russian forces grew to 30,000 and included more-experienced, better-trained, and better-equipped reinforcements. Forces became more efficient, were reorganized into smaller assault groups, and implemented lessons learned. By late April, Chechen fighters were pushed into the southern mountains and Russian forces began to "attack the last remaining Chechen strongholds."[70] At this point, Chechen methods were transformed, and terrorist tactics became an element of the movement. On June 14, 150–200 Chechen troops led by Basayev moved into the Russian border town of Budennovsk and seized two bank buildings, an administrative center, and the local hospital. Taking more than 1,500 civilian hostages and booby-trapping the area, "the rebels promised that the hostages would be released if the Russians agreed to cease hostilities in Chechnya and withdraw their forces from the region."[71] Although the Chechens did not receive their demand for Russian withdrawal, they successfully entered negotiations with the federal government and "return[ed] to Chechnya unimpeded, leaving behind 150 dead civilians."[72] On June 18–19, Russian Prime Minister Viktor Chernomyrdin actually negotiated with Basayev on live television and agreed to halt military operations in Chechnya and to guarantee Basayev and his men safe passage back to Chechnya.[73] By the end of the month, the Chechens achieved a cease-fire agreement with the Russian government and militants quickly began to return to Chechen towns and villages.

In August 1995, the Chechens disavowed the post-Budennovsk cease-fire and began a larger effort to retake Chechen cities and towns. For example, Chechen militants succeeded in achieving a stalemate with Russian troops in Gudermes, Chechnya's second-largest city, late in 1995. Soon after, in January 1996, those same Chechen forces took hostages in Kizlyar, another Russian border town, for ten days, forcing

---

[68] Ibid., 240.
[69] Oliker, *Russia's Chechen Wars*, 18.
[70] Sumner, "Success of Terrorism in War," 76.
[71] Oliker, *Russia's Chechen Wars*, 28.
[72] Ibid., 29.
[73] Sumner, "Success of Terrorism in War," 76.

a standoff with Russian troops. Although Basayev's men sent an initial expedition to reengage Grozny in March 1996, a full Chechen offensive was not launched until August. At this point, seasoned Russian troops had rotated out of the area and the new troops that were in place did not know how to respond efficiently to the attacks. After two weeks of fighting, total Russian casualties for the battle included 500 dead and 1,400 missing and wounded. When the battle finally ended, it was not with a military victory but with a cease-fire agreement finalized on August 22 by negotiators Aleksandr Lebed and Aslan Maskhadov.[74]

The interwar period for the Chechens proved internally divisive, and, eventually, the rift between Maskhadov and the Chechen field commanders came to a tipping point. During the summer of 1999, al-Khattab and Basayev led their forces across the Chechen border into the Russian province of Dagestan in an attempt to support three villages that had declared *shari'a* law. "However, their actions failed dramatically as members of a number of Dagestani ethnic groups took up arms against them."[75] The failure of the Dagestanis to rise in support of the Chechens was likely linked to the January 1996 raid by Chechens militants into Kizlyar. This cross-border raid into Dagestan in the summer of 1999 prompted the Russian government to retake Chechnya later that year.

Chechen forces utilized classic guerrilla tactics, detonating roadside bombs and employing heavy gunfire during the enemy's confusion. "In rural terrain they camouflaged cave entrances with rocks, cobblestones, and anything else that came to hand to create shelters from artillery and air strikes. In towns and villages they used lower floors and basements of buildings as fighting positions."[76] During the conflict, particularly when the Chechen resistance was pushed out of Grozny, they began to rely on targeted assassinations and frequent suicide bombings.[77] During the long fight for Grozny, however, the operational key to resistance included a network of underground passages and sewers that enabled the Chechens to move freely and undetected for purposes of ambush, resupply, and evacuating the wounded.

Resistance forces frequently employed land mines to inflict casualties on the Russians. "Terrorists are now organized in their preparations, in their accumulation of stockpiles of high-explosive munitions, in their development of a network of clandestine

---

[74] Oliker, *Russia's Chechen Wars*, 31.

[75] Moore and Tumelty, "Assessing Unholy Alliances in Chechnya."

[76] Oliker, *Russia's Chechen Wars*, 41.

[77] Kramer, "Guerrilla Warfare, Counterinsurgency and Terrorism in the North Caucasus," 240.

laboratories to construct improvised explosive devices and radio-controlled detonators."[78] When the force ratio between groups could not support direct Chechen confrontation with Russian troops, insurgents also enlisted children into their ranks, providing financial incentives for them to plant land mines and explosives. Eventually, there would be approximately 500,000 mines planted in Chechnya, making it one of the most heavily mined areas in the world.[79]

When the Second Chechen War began in September 1999, Basayev augmented the Chechen forces with hundreds of youths from al-Khattab's training camps. He also expanded the underground support network for the resistance to include the broader North Caucasus region and made use of alliances he had created while serving in the Confederation of Mountain Peoples and fighting in Abkhazia. These new relations and networks would continue to thrive after the end of the Second Chechen War in 2002.[80]

METHODS OF RECRUITMENT

Initially, the Chechen political movement inspired mobilization of the population. Local residents, for example, independently "attacked various military installations" in order to secure weapons.[81] Mobilization was also spurred by the behavior of Russian troops after 1994. This included indiscriminate air and ground strikes on villages, targeting civilian infrastructure such as residences and hospitals, not allowing civilians to leave, shooting at fleeing civilians, establishing filtration centers, and taking hostages. Often Russian troops were intoxicated and they frequently committed robbery, assault, and rape of the local population.

Recruitment opportunities also originated within the Chechen expatriate community. As a result of the historical context of the Russian–Chechen relationship, during the nineteenth century thousands of Chechens had migrated out of the region and into present-day Turkey, Syria, Iraq, and Jordan. Although Turkey, Syria, and Iraq largely assimilated their ethnic populations, "there still exists a unique community of around 8,000 Jordanian-Chechens who have

---

[78] Ibid., 226.

[79] Office of the Special Representative of the Secretary-General for Children and Armed Conflict, "Special Representative for Children and Armed Conflict Concludes Russian Federation Trip; Welcomes Assurances on Voluntary Return of Displaced Chechen Populations," United Nations press release, June 24, 2002, http://www.un.org/children/conflict/english/pr/2002-06-2447.html.

[80] Moore and Tumelty, "Assessing Unholy Alliances in Chechnya."

[81] Tishkov, *Chechnya, Life in a War-Torn Society.*

preserved their language and cultural traditions through time."[82] Many individuals within this population participated in Arab struggles against Russia as members of local mujahideen, including those in Afghanistan, Tajikistan, Bosnia, and Kosovo.[83] When the Soviet Union opened travel restrictions, these expatriates took the opportunity to return to Chechnya.

In 1993, one such traveler was Khabib Ali Fathi, a former Muslim Brother with experience in Afghanistan. "In conjunction with local Islamists, [he] established a Salafi Islamic jamaat known in Islamist circles as al-Jama'at al-Islamiyya. Capitalizing on his Chechen ancestry, Fathi organized his group and began *da'wa* (literally "the call," but more accurately proselytizing) among the Chechen population in alliance with a small number of Jordanian-Chechens, quickly creating a following numbering around ninety."[84] Fathi was asked to be Dudayev's religious adviser, providing added benefits that enabled recruitment from the Middle East and also providing funding channels for the resistance. Fathi's principal recruit, a Saudi named Omar Ibn al-Khattab, arrived in Chechnya in 1995. Al-Khattab, who had fought in the Tajik Civil War (1992) and maintained some loose connection with Osama bin Laden, incorporated Fathi's followers into a fighting corps and established the military links between his foreign fighters and the independent Chechen field commanders.[85] Both Dudayev and Maskhadov used al-Khattab to train Chechen forces in guerrilla warfare.

A systematic incorporation of women into the resistance began at the end of the interwar period. The first female suicide bombers, or "Black Widows," conducted a successful operation against a Russian headquarters unit in June of 2000. The two Chechen women used a van filled with explosives to kill at least two soldiers and wound numerous others. By using women who lost husbands and family members in the Chechen war to recruit others in a similar circumstance, Basayev brought female recruits to train in local terrorist camps. "Many Chechen widows have been convinced by separatists that they have

---

[82] Moore and Tumelty, "Foreign Fighters and the Case of Chechnya," 416.

[83] Ibid., 414. In Afghanistan, Osama bin Laden eulogized the victories of Shamil's nineteenth-century resistance to Russia in the Caucasus, even naming one unit after him.

[84] Ibid., 416.

[85] Ibid., 417. "Khattab organized half a dozen experienced Arab commanders into small sub units that were under his general command, in turn subordinating them to the Chechen Armed Forces." While in command, al-Khattab screened any incoming Arab fighters for experience and capabilities, keeping the number of non-Chechen forces low. This was in accordance with Chechen wishes but was also the result of challenges posed by logistics, terrain, custom, and language.

become burdens and that the loss of their husband was a punishment for their sins, leaving suicide bombing as their last resort."[86]

METHODS OF SUSTAINMENT

At the beginning of the 1990s, kidnapping and hostage taking for ransom were primarily methods of settling debts and conducting business. Eventually, however, kidnapping became a common practice for all sides of the conflict to generate revenue, conduct hostilities, promote renewed violence, and prevent positive relations between parties. For example, "high ranking officials or foreigners were sometimes abducted to influence political and economic decisions, such as preventing Maskhadov from holding talks with Russia's federal government, or Chechnya's participation in constructing a pipeline for Caspian Sea oil."[87] Overall, Chechen forces, criminal elements, the internal Chechen opposition, and numerous Russian organizations kidnapped and held for ransom at least 2,000 victims. In accordance with popular sentiment, Maskhadov's unsuccessful offensive against the Chechen warlords was launched in the fall of 1998 under "a slogan of combating the hostage-takers."[88]

Of the numerous forms of illicit activity within the area, drug production in the highland and foothill districts of Chechnya became extremely lucrative for local warlords associated with criminal Islamic elements in central Asia. Field commanders supported plantations and facilities to produce drugs in their own controlled areas, such as poppy crops that were grown to generate and facilitate the sale of narcotics.[89] Illicit activity also included the illegal pillaging of oil resources, or "bunkering," that began in 1998. This illicit activity included small-scale, private oil production, which was extremely harmful to Chechnya's local environment. "In one settlement, there were over 200 clandestine oil refineries."[90]

An additional form of funding for sustainment of the Chechen resistance came from external Islamic organizations and Middle

---

[86] "Terrorist Organization Profile: Black Widows," National Consortium for The Study of Terrorism and the Response to Terrorism (START), University of Maryland, accessed January 26, 2010, http://www.start.umd.edu/start/data_collections/tops/terrorist_organization_profile.asp?id=3971. There is some mention that methods of brainwashing and coercion, possibly even drugging, were used on female recruits.

[87] Tishkov, *Chechnya, Life in a War-Torn Society*, 124.

[88] Ibid., 193.

[89] Trenin and Malashenko, *Russia's Restless Frontier*, 77.

[90] Tishkov, *Chechnya, Life in a War-Torn Society*, 188.

Eastern *hawala* networks.[91] Although certain Chechen warlords fully adopted the Wahhabi ideology, others professed statements of faith in order to maintain positive relations. For example, donations came from the following organizations based in Riyadh: the World Islamic League, the International Islamic Rescue Organization al-Igasa, and the World High Council for Mosques. Other prominent centers and foundations include al Haramain, the Al Qaeda-linked Benevolence International Foundation, "the International Charity Association Taiba, the Ibrahim al-Ibrahim Foundation, the International Association of Islamic Appeal, and the Sudan-based World Islamic Appeal League. These provide assistance by training Muslim clerics, by financing new religious schools and universities, and by sponsoring various scientific and religious seminars."[92] Because there were numerous sources of funds, it was also hard to determine how much money was being received by the resistance. "Foreign funds have been flowing to Chechnya through various channels: bogus firms or intermediaries (both in Russia and abroad); foreign emissaries bringing cash directly to field commanders; Chechen diaspora communities in the Middle East that collect money; and Chechen politicians on fundraising missions. What is not known is how much money is being transferred."[93] Some of this money was used for recruitment purposes and also to establish regional training camps. For example, al-Khattab and Basayev, who visited Afghanistan training camps before 1994, opened such a camp in the Serzhen'-Yurt village. By 1999 another base was established for "rebels and religious fundamentalists" around the village of Karamakhi.[94]

METHODS OF OBTAINING LEGITIMACY

At the outset, the Chechen resistance gained immediate popular legitimacy because of the history of the Russian–Chechen relationship and the continued discriminatory policies practiced by the Soviets. In this light, participation could be seen as retribution or "compensation for diminished social status and second-rate treatment."[95] Once warfare commenced, Chechen operational successes and the inadequacy of the Russian military highlighted the falsity of statements made by the

---

[91] *Hawala* networks use an informal system of money collection through couriers and acquaintances.

[92] Trenin and Malashenko, *Russia's Restless Frontier*, 92.

[93] Ibid., 93.

[94] Sharon LaFraniere, "How Jihad Made Its Way to Chechnya, Secular Separatist Movement Transformed by Militant Vanguard," *The Washington Post*, April 26, 2003, A01.

[95] Trenin and Malashenko, *Russia's Restless Frontier*, 80.

Yeltsin administration about Russian prospects for victory. The very nature of success, therefore, provided the resistance with legitimacy.

Over time, the Islamic narrative became integral to the concepts of legitimacy. Indigenous organizations competing for power within Chechnya adopted the use of Chechen history and parallels to the independent Imamate founded by Shamil during the nineteenth century. The secular nationalists, the traditional Sufi authorities, and radical Wahhabi Islamists in both Chechnya and Dagestan referred to the time of the Imamate as a model time period and derived legitimacy from an appeal to its legacy.[96] Although the level of religion practiced by most of the Chechen population did not support the drive toward a *shari'a*-based state, field commanders targeted local young men who were alienated by the effects of the war. Attempts to overhaul the entire Sufi and secular population within Chechnya "would be difficult without a civil war, an option rejected by most Muslims in the North Caucasus. Most Muslims dislike the religious rigidity of Salafiyya,[97] its rejection of a so-called people's Islam, and its call for radical actions against local authorities."[98] However, many within the younger generation, who grew up in an atmosphere of fear and war, found the tenets and benefits of radical Islam to be an avenue for justice and retaliation.

The Chechen warlords succeeded in obtaining legitimacy for external support from the Middle East, sending representatives to meetings and councils abroad. In order to obtain political legitimacy at home, gaining a say in the political process, warlords gained power through coercion and threats on their own government. For example, Basayev and al-Khattab threatened Maskhadov at gunpoint in order to secure their places on "a State Defense Council as the highest organ of the state." In the council the radical warlords were given seats together with Maskhadov and decisions were taken in a "collegial manner."[99] Maskhadov was ultimately caught between two untenable positions. If he were supported by the Russian government, field commanders could claim that he was only Moscow's puppet, but without Russian support, Maskhadov was dependent on the resources of the warlords.

One of Maskhadov's primary challenges during his administration was gaining legitimacy from the perspective of the population, the

---

[96] Zelkina, "Jihad in the Name of God," 261.

[97] Salafiyya is similar to Wahhabism, in which the strict writings of the Prophet are followed. Wahhabism is often referred to as the Saudi version of Salafiyya. Febe Armanios, "The Islamic Traditions of Wahhabism and Salafiyya," Congressional Research Service Report for Congress, December 22, 2003, http://www.fas.org/irp/crs/RS21695.pdf.

[98] Trenin and Malashenko, *Russia's Restless Frontier*, 85.

[99] Wilhelmsen, "Between a Rock and a Hard Place," 49.

radical warlords and their associated terrorist organizations, and also the Russian government. Successful in gaining popular support, Maskhadov was unable to establish legitimacy in the eyes of either the independent Chechen commanders or Russia.[100] Often, he vacillated between cracking down on the warlords and including them in decision making.[101] For example, Maskhadov condemned al-Khattab and Basayev's terrorist behavior, including the 1999 raid into Dagestan and the continued abductions of Russian ministry. On the very day that Maskhadov offered to hand Basayev and al-Khattab over to the Russian government, Putin publicly proclaimed the illegitimacy of Maskhadov's presidency.[102] When Russian troops began to advance toward Chechnya, Maskhadov fell back in his position and joined Chechen army troops with the forces of the field commanders in defense of the region.

EXTERNAL SUPPORT

Support from the international community for the Chechen cause was much greater during the first war. International organizations were likewise more involved with open access to the region. The events of 9/11 changed the West's perception of the Chechen conflict. During the 2000 presidential campaign and the early months of the George W. Bush presidency, National Security Advisor Condoleezza Rice urged the Russian government to recognize Chechnya's legitimate aspirations for a political solution; after 9/11, the Russians were praised by Secretary of State Colin Powell and others for their fight against Chechen terrorists. One argument, for example, purports that al-Khattab was sent to Chechnya in order to "carry out a special mission assigned to him by Osama Ben [sic] Laden to organize training camps for international terrorists"; three schools in Chechnya, one in Ingushetia, and one in Dagestan were to train converted "Europeans, Russians, Ukrainians, Cossacks, and Ossetians" in kidnappings and terrorist activities.[103] During the first war effort, although the military influence of foreign fighters was minor, foreign "militant ideas and religious influence began to percolate through war-torn Chechen society after August 1996, in part, hastening the divisions in Chechen

---

[100] Miriam Lanskoy, "Daghestan and Chechnya: The Wahhabi Challenge to the State," *SAIS Review* XXII, no. 2 (Summer-Fall 2002): 183–194.

[101] Wilhelmsen, "Between a Rock and a Hard Place," 49. The warlords retaliated against Maskhadov with assassination attempts and an offensive campaign.

[102] It is difficult to determine the degree of complicity between Maskhadov and the Chechen warlords.

[103] This redacted DIA assessment is available at Judicial Watch, accessed January 26, 2010, http://www.judicialwatch.org/cases/102/dia.pdf.

society and ultimately inspiring some of the events that led to the resurgence of the Russo-Chechen war in 1999."[104]

Local support for the Chechen movement from other Muslim communities was also mixed. For example, Maskhadov hoped to obtain the support of Russian Tatarstan. "In 1999, the State Council of Tatarstan adopted a resolution in order to" suspend conscription as a way "to prevent Tatarstan-born Muslims from confronting their fellow Muslims in Chechnya."[105] Tatarstan's president, Shaymiyev, however, made no further motions to support Chechen efforts, eschewing the turn toward Islamic fundamentalism. The neighboring region of Dagestan, on the other hand, provided a number of supportive Wahhabi communities that felt persecuted. "During the 1990s, Dagestan adopted increasingly repressive legislation against the adherents of the minority Salafi (or Wahhabi) strain of Islam. In 1998, due to unrest in Dagestan, several hundred Wahhabis left for Chechnya where they combined forces with rogue Chechen"[106] In 1999, when Basayev and al-Khattab raided Dagestan, it was in support of these same communities.

"The radical warlords and politicians would probably never have managed to gain the upper hand over the more moderate actors in Chechnya had it not been for the attempts by international Islamist actors to co-opt the Chechen conflict."[107] Many Sunni extremists were able to draw on the military and financial connections they had made during their participation in previous conflicts—in Afghanistan, Tajikistan, Azerbaijan, Georgia, and other locations—to recruit fighters and elicit funding. Eventually, the Pankisi Gorge region within northern Georgia became a holding place for Islamic militants. During al-Khattab's leadership of the foreign fighters in Chechnya, these militants were screened and very few were allowed into Chechnya proper. "According to Georgian officials, in early 2002, some sixty Arab computer, communications, and financial specialists, military trainers, chemists, and bomb-makers settled in the gorge. The group used sophisticated satellite and encrypted communications to support both Ibn al-Khattab's operations in Chechnya and terrorists planning attacks against Western targets."[108] After al-Khattab's death in 2002, his successor Abu Walid al-Ghamidi continued al-Khattab's practices.

---

[104] Moore and Tumelty, "Foreign Fighters and the Case of Chechnya," 418.

[105] Trenin and Malashenko, *Russia's Restless Frontier*.

[106] Lanskoy, "Daghestan and Chechnya," 183–184.

[107] Wilhelmsen, "Between a Rock and a Hard Place," 52

[108] Lorenzo Vidino, "How Chechnya Became a Breeding Ground for Terror," *Middle East Quarterly* XII, no. 3 (Summer 2005): 57–66. Note: In April 2002, the US Special Forces supported the "Georgia Train and Equip Program" to enhance the counterterrorism capabilities of Georgian troops in the Pankisi Gorge.

COUNTERMEASURES TAKEN BY THE GOVERNMENT

The initial government response to Chechnya's nationalist movement centered on political and economic measures aimed at bringing the region back into the Russian federation. This was followed by Yeltsin's support of local political opposition and small-scale attempts at regime change. In 1994, Yeltsin initiated large-scale military operations that decimated the Chechen physical infrastructure, beginning first in Grozny and then moving out to smaller villages. Participating units were composed of organizations from the Internal and Defense forces, including special forces, airborne forces, anti-riot police, reconnaissance and logistical personnel, intelligence assets, attack and transport personnel, armor and infantry personnel, aviation support, communications specialists, search and rescue squads, transport regiments, and emergency management personnel.[109]

Despite superior numbers, Russian troops struggled with low morale, corruption, and substance abuse; arrived ill prepared and often unpaid; and were frequently misled about the nature, length, and locations of their missions. Even during the second military conflict, "Russian troops in Chechnya have been hindered by deficient training, outdated equipment, poor nutrition, abysmal health care and the physical and psychological tribulations of violent bullying."[110] Illicit activity within Chechnya was also frequently enabled by Russian personnel who were open to Chechen bribes for targeting information, weaponry, and explosives.[111] In some cases, Russian forces were noted to have refused assignments, such as mountain warfare, preferring instead to remain close to regional bases.

Throughout both conflicts, Russian operations were often undermined by behavior and tactics of their own troops. "Russian units in Chechnya have been plagued by rampant corruption and have been linked with narcotics trafficking, prostitution rings, illegal arms dealing and kidnappings for ransom. In many cases when Chechen guerrillas have bribed Russian conscripts or officers, they have been able to gain access to sensitive facilities or have been allowed to drive explosive-laden vehicles near government buildings."[112] Common items for sale by Russian troops included "shoulder-held missiles, anti-tank guns,

---

[109] Kramer, "Guerrilla Warfare, Counterinsurgency and Terrorism," 218–219.

[110] Ibid., 220.

[111] Ibid., 221. Although he describes events that occurred beyond the end date of 2002 for this case study, Kramer indicates that "the Russian government has acknowledged that corrupt MVD officers were paid off by Chechen terrorists who seized hostages at the Dubrovka theatre in October 2002 and at Middle School No. 1 in Beslan in September 2004."

[112] Ibid., 221–222.

mortars, artillery shells, rocket-propelled grenades, automatic rifles and other firearms."[113] Additionally, terrorism countermeasures by Russian officials backfired, as was the case in October, 2002, when more than forty terrorists took roughly 850 hostages prisoner at a Moscow theater. When the terrorists demanded an end to the Russian presence in Chechnya, threatening to execute the hostages in the event of unmet conditions, Russian forces pumped an incapacitating gas into the theater ventilation system, which resulted in the deaths of 129 hostages. Most of the terrorists died from either the gas or the subsequent shoot-out with Russian special forces.[114] A high percentage of the 700 surviving hostages were poisoned and seriously debilitated. Russian forces did not provide chemical antidotes to the gas, nor did they release the type of chemical agent used or properly plan for the medical treatment of those rescued.

During the second conflict, once the Russian-backed Sufi government was installed within Grozny, "both Russian forces and pro-Kremlin Chechen groups have sought to isolate and kill (foreign) fighters and also the foreign financiers and ideologues who played a prominent role in the Chechen resistance movement in the second conflict."[115] Targeted assassinations were also successful against Dudayev, Maskhadov, al-Khattab, and his replacement, Abu Walid al-Ghamidi.[116] Al-Khattab, for example, was killed in 2002 by a poisoned letter sent from Russia's Federal Security Service.[117]

SHORT- AND LONG-TERM EFFECTS

CHANGES IN THE ENVIRONMENT

The physical damage to the Chechen habitat was immense. "The economic collapse of state resources that had begun under Dudayev was exacerbated by the destruction of a large part of Grozny and more than twenty large villages, the damage caused to roads and bridges, the destruction of forests, the spoiling of nearly a third of the republic's arable fields by military vehicles and landmines, and the contamination

---

[113] Ibid., 222.

[114] Many terrorists were shot in the head once the police raided the theater.

[115] Moore and Tumelty, "Assessing Unholy Alliances in Chechnya," 89–90.

[116] Basayev was killed in July 2006 in Ingushetia when a truck full of explosives he was escorting blew up. Russian special services claimed responsibility for remotely setting off the explosion, although it is still disputed whether the explosives were intentionally detonated or they exploded when mishandled.

[117] "Obituary: Chechen Rebel Khattab," *BBC News,* April 26, 2002, http://news.bbc.co.uk/2/hi/europe/1952053.stm.

of water sources."[118] In the decade following the end of the conflict, little was done to rebuild the Chechen infrastructure. According to certain local inhabitants, the Russians tried to create a facade of improvements; for example, they created apartment complexes, but the apartments did not have electricity or running water.

The cultural environment of the region also changed dramatically. "The internal impetus toward Islamization of the Chechen separatist movement did not come from the Chechen population in general but rather from a group of warlords and politicians who acquired prominent positions in Chechnya because of the war."[119] Extremist Islamic organizations attempted to influence the outcome in Chechnya through radical propaganda, financing, and support for recruitment and training.[120] "Across the North Caucasus young people ages fourteen to seventeen are increasingly attracted to Islam. Further, their Islam is not the traditional 'Islam of their forefathers,' but a politicized Islam, one dominated by the ideas of sharia, the Islamic state, jihad, and even Islamic revolution."[121]

CHANGES IN GOVERNMENT

The beginning of the Chechen campaign saw Gorbachev still at the lead of the Soviet Union. Yeltsin's rise to the presidency and the disintegration of the former Soviet empire yielded a renegotiation of regional relationships and enabled the temporary Chechen withdrawal from the federation. Under Yeltsin's leadership, Russian forces were unable to overcome the Chechen resistance, leading to an embarrassing cease-fire in 1996. Before Yeltsin resigned as president at the end of 1999, he named Vladimir Putin as his successor and the status of relations with Chechnya changed quickly. As al-Khattab and Basayev's men raided Dagestan in the name of a North Caucasus Imamate, Putin made the decision to renew bombings and raids against Chechnya through large-scale military operations that September. He rejected any consideration of negotiations with Chechen representatives and focused the Russian discourse on terrorism.

A key Russian weakness centered on the government's limited capacity for command and control of federal forces. Without a

---

[118] Tishkov, *Chechnya, Life in a War-Torn Society*, 184.

[119] Wilhelmsen, "Between a Rock and a Hard Place," 37.

[120] Trenin and Malashenko, *Russia's Restless Frontier*, 77.

[121] Ibid., 81–87. At the same time, Trenin and Malashenko argue that "it is difficult to assess objectively the influence that radical Islam has had in the North Caucasus. Any assessment must rely on indirect evidence and often intuition rather than hard facts and statistical data (which are frequently doctored)."

unified command structure, numerous ministries within the central government operated independently, with little to no coordination, within Chechnya. "To mitigate that problem in the latest war, the Russian government created a "Unified Grouping of Federal Forces (OVG), which exercises jurisdiction over all military and security troops in Chechnya."[122] Even so, the tension between the Defense Ministry and the Ministry of Internal Affairs (MVD) forces continued and delineation of control remained ambiguous.

CHANGES IN POLICY

During the Russian elections in March 2000, a number of presidential candidates called for negotiations with the Chechen government. The public, however, conveyed their approval of President Putin's hard-line tactics against the rebels and elected him into office with 53% of the vote.[123] Negotiating with the Chechen government or any representatives of the Chechen resistance was taken off the table. Whereas Yeltsin's drive into Chechnya met with eventual negotiations, Putin had no intention of withdrawing the military from the region. "Moscow has convinced itself that Muslim extremists are the essence, not a part, of the problem. As a result, Russia has no viable strategy to govern an increasingly turbulent area."[124]

Another change within Russian policy occurred when Putin began to dominate his opposition within the Russian government, eventually controlling the legislature and any potentially adversarial political organizations—including the media. In fact, during the first 1994–1996 confrontation between Chechnya and Russia, the press severely impacted Russia's political and military responses by undermining the credibility of the government's reporting and its statements about the war. After 1999, with much greater control over the press, Putin successfully prevented the bloodshed in Chechnya from reaching the political and public agenda, unless it served the state's interests. It was even suggested that the Russian government played a role in some of the most egregious terrorist actions against Russian civilians in order to garner public support.[125]

---

[122] Kramer, "Guerrilla Warfare, Counterinsurgency and Terrorism in the North Caucasus," 217.

[123] Ibid., 213.

[124] Menon and Fuller. "Russia's Ruinous Chechen War," 38.

[125] Kramer, "Guerrilla Warfare, Counterinsurgency and Terrorism in the North Caucasus," 258.

CHANGES IN THE REVOLUTIONARY MOVEMENT

The Chechen nationalist movement originated within the framework of the former Soviet government. Chechen leaders, however, were unable to constitute a viable state. Accordingly, and over time, the popularly elected Chechen government lost control of field commanders who generated income on the black market and the government turned to outside communities for financial support. Chechen warlords developed relationships with Wahhabi radicals from the Middle East, established terrorist organizations, and adopted political goals to foster a North Caucasus Imamate based on *shari'a* law. After 1994, the younger Chechen generation that turned to Salafi practices and rewards increasingly challenged the traditional society of elders and *adat* (customary law). "While their message was yet to be heard amid the mass mobilization of the nationalist-separatist Chechen movement in the first half of the 1990s, their power has increased exponentially and by 2008 become the dominant influence over the ideology of the Chechen-led North Caucasus resistance."[126]

Historically, Sufi brotherhoods adapted Islamic practice to local traditions and customs. "Chechen society and the Sufi brotherhoods have long struggled to reconcile aspirations for Sharia law with local customary law. Previous attempts to impose the Sharia by Chechen and Dagestani leaders of the *gazowat*, or holy war, against the Russians have failed, although they still attained status as heroic, national figureheads."[127] Thus, there seem to be contemporary parallels with the Wahhabi movement. Although with the Chechen war there was an attempt to spread Islamic radicalism across the region, "the failure to create an Islamic state in Chechnya set limits to that radicalism. Chechens rejected the idea of an Islamist state, and attempts to impose Islamic rule by force have further discredited the Islamists."[128] However, the Salafi jihadists continued to exist in the region and continued to expand their influence, even if the bulk of the Chechen people rejected them.

Although the Salafi status and influence increased briefly after the second Russian invasion, because fewer Chechens remained actively involved in the resistance and field commanders relied on their own militias and terrorist organizations, Chechen resistance tactics and targets were dramatically different than they were during the first war.[129] For example, tactics became more extreme against

---

[126] Moore and Tumelty, "Assessing Unholy Alliances in Chechnya," 73–94.
[127] Moore and Tumelty, "Foreign Fighters and the Case of Chechnya."
[128] Trenin and Malashenko, *Russia's Restless Frontier*, 88.
[129] Moore and Tumelty, "Assessing Unholy Alliances in Chechnya," 87.

the civilian Russian population, often involving hundreds of civilians. Regionalization of the movement enabled the resistance to continue but also altered the initial Chechen focus for national autonomy. In May 2005, regional *jamaats* pledged support to Shaykh Abdul Khalim Sadulaev, Maskhadov's successor. These *jamaats* included Ingushetia, North Ossetia, Kabardino-Balkaria, Karachaevo-Cherkessia, Stavropol, Adygea, Krasnodar, and Dagestan.[130] "Chechnya has become, within certain limits, an exporter of radical Islam to Muslim regions of Russia and the Commonwealth of Independent States."[131] This change in focus caused deep cleavages within the movement. In October 2007, Dokka Umarov, Sadulaev's successor, declared an Islamic North Caucasus emirate. "Umarov's declaration caused a furor within the wider Chechen resistance movement, effectively signaling the end of the independence project under the banner of the Chechen Republick of Ichkeria."[132]

BIBLIOGRAPHY

Armanios, Febe. "The Islamic Traditions of Wahhabism and Salafiyya." Congressional Research Service Report for Congress. December 22, 2003. http://www.fas.org/irp/crs/RS21695.pdf.

Blank, Stephen J., and Earl H. Tilford Jr. "Russia's Invasion of Chechnya: A Preliminary Assessment." Strategic Studies Institute Special Report, US Army War College, 1995.

Bradol, Jean-Herve. "MSF Presentation to Council of Europe Regarding Chechnya and Ingushetia." Medécins Sans Frontières report. January 23, 2002. http://js.static.reliefweb.int/node/94650.

Caglar, Armond. "In the Spotlight: Islamic International Peacekeeping Brigade," Center for Defense Information. May 30, 2003. http://www.cdi.org/program/issue/document.cfm?DocumentID=1116 &IssueID=56&StartRow=1&ListRows=10&appendURL=&Orderby =DateLastUpdated&ProgramID=39&issueID=56.

de Waal, Tom. "Dual Attack Killed President." BBC News, April 21, 1999, http://news.bbc.co.uk/2/hi/europe/325347.stm.

Fuller, Liz. "Chechnya: Shamil Basayev's Life of War and Terror." *Radio Free Europe/Radio Liberty,* July 10, 2006, http://www.rferl.org/content/article/1069740.html.

Gammer, Moshe. *Muslim Resistance to the Tsar: Shamil and the Conquest of Chechnya and Daghestan.* London: Frank Cass & Co. Ltd., 1994.

---

[130] Trenin and Malashenko, *Russia's Restless Frontier,* 87.
[131] Ibid., 72.
[132] Ibid., 88.

————. *The Lone Wolf and the Bear, Three Centuries of Chechen Defiance of Russian Rule.* Pittsburgh, PA: University of Pittsburgh Press, 2006.

Glinkina, Svetlana, and Dorothy Rosenberg. "The Socioeconomic Roots of Conflict in the Caucasus." *Journal of International Development* 15, no. 4 (2003): 519.

Higgins, Andrew. "Profile: Dzhokhar Dudayev: Lone Wolf of Grozny," The Independent. January 22, 1995. http://www.independent.co.uk/opinion/profile-dzhokhar-dudayev-lone-wolf-of-grozny-1569145.html.

Kramer, Mark. "Guerrilla Warfare, Counterinsurgency and Terrorism in the North Caucasus: The Military Dimension of the Russian-Chechen Conflict." *Europe-Asia Studies* 57, no. 2 (2005): 210.

LaFraniere, Sharon. "How Jihad Made Its Way to Chechnya, Secular Separatist Movement Transformed by Militant Vanguard." *The Washington Post*, April 26, 2003, A01.

Lanskoy, Miriam. "Daghestan and Chechnya: The Wahhabi Challenge to the State." *SAIS Review* XXII, no. 2 (Summer-Fall 2002): 183–194.

Lapidus, Gail W. "Contested Sovereignty: The Tragedy of Chechnya." *International Security* 23, no. 1 (Summer 1998): 10.

Menon, Rajan, and Graham Fuller. "Russia's Ruinous Chechen War." *Foreign Affairs* 79, no. 2 (2000): 39.

Moore, Cerwyn, and Paul Tumelty. "Assessing Unholy Alliances in Chechnya: From Communism and Nationalism to Islamism and Salafism." *Journal of Communist Studies and Transition Politics* 25, no. 1 (2009): 87.

————. "Foreign Fighters and the Case of Chechnya: A Critical Assessment." *Studies in Conflict and Terrorism* 31, no. 5 (2008): 416.

Obituary: Chechen Rebel Khattab," *BBC News*, April 26, 2002. http://news.bbc.co.uk/2/hi/europe/1952053.stm.

Office of the Special Representative of the Secretary-General for Children and Armed Conflict. "Special Representative for Children and Armed Conflict Concludes Russian Federation Trip; Welcomes Assurances on Voluntary Return of Displaced Chechen Populations." United Nations press release, June 24, 2002. http://www.un.org/children/conflict/english/pr/2002-06-2447.html.

Oliker, Olga. *Russia's Chechen Wars 1994-2000: Lessons from Urban Combat.* Santa Monica, CA: RAND Corporation, 2001.

"Remembering Stalin's Deportations." *BBC News*, February 23, 2004. http://news.bbc.co.uk/2/hi/3509933.stm.

Smick, Elisabeth. "Backgrounder: The Chechen Separatist Movement" Council on Foreign Relations, July 18, 2006. http://www.cfr.org/publication/11121/chechen_separatist_movement.html.

Sumner, Dianne L. "Success of Terrorism in War: The Case of Chechnya." Master's thesis, Naval Postgraduate School, 1999.

Takeda, Yuka. "Economic Growth and its Effect on Poverty Reduction in Russia." Global COE Hi-Stat Discussion Paper Series 075, Institute of Economic Research, Hitotsubashi University, July 2009. http://gcoe.ier.hit-u.ac.jp/research/discussion/2008/pdf/gd09-075.pdf.

Technical Cooperation Department, The Food and Agriculture Organization of the United Nations (FAO). "The FAO Component of the Inter-agency Consolidated Appeals 2005." Rome: Emergency Operations and Rehabilitation Division, 2004). http://www.fao.org/docrep/007/y5805e/y5805e06.htm#TopOfPage.

"Terrorist Organization Profile: Black Widows." National Consortium for The Study of Terrorism and the Response to Terrorism (START), University of Maryland. http://www.start.umd.edu/start/data_collections/tops/terrorist_organization_profile.asp?id=3971.

Tishkov, Valery. *Chechnya, Life in a War-Torn Society.* Berkeley, CA: University of California Press, 2004.

Trenin, Dmitri, and Aleksei Malashenko. *Russia's Restless Frontier, the Chechnya Factor in Post-Soviet Russia.* Washington, DC: The Brookings Institution Press, 2004.

Vidino, Lorenzo. "How Chechnya Became a Breeding Ground for Terror." *Middle East Quarterly* XII, no. 3 (Summer 2005): 57–66.

Wilhelmsen, Julie. "Between a Rock and a Hard Place: The Islamisation of the Chechen Separatist Movement." *Europe-Asia Studies* 57, no. 1 (January 2005): 36.

Wood, Tony. *Chechnya, The Case for Independence.* New York: New Left Books/Verso, 2007.

Zelkina, Anna. "Jihad in the Name of God: Shaykh Shamil as the Religious Leader of the Caucasus." *Central Asian Survey* 21, no. 3 (2002): 249–264.

HIZBOLLAH: 1982–2009

Shana Marshall

SYNOPSIS[1]

Hizbollah is a Lebanese political and militant organization that emerged in the early 1980s in response to the Israeli invasion of Lebanon. This Shi'a-based insurgent movement was driven by a historical narrative of repression and was the recipient of sustained support from Iran. The movement generated broad domestic support as a result of its military success against the technically superior Israeli forces and the scope of the social services it provided to the impoverished residents of South Lebanon. Hizbollah has employed a broad spectrum of military capabilities ranging from suicide bombers to medium-range rockets, with Israel Defense Forces (IDF) and Israeli citizens being the primary targets of its attacks. Through its engagement of the political process in Lebanon, Hizbollah has established itself as a legitimate political force within the country and, therefore, enjoys the dual benefits of military and political influence.

TIMELINE

March 1978	Israeli invasion of Lebanon. Withdrawal in June 1978.
June 1982	Israel conducts second large-scale invasion of Lebanon. Hizbollah conducts a campaign of small hit-and-run attacks, including suicide bombings.
October 23, 1983	Bombing of US Marine Barracks and French troops in Beirut.
1992	Hizbollah participates in Lebanese elections.
July 1993	Israel launches Seven Day War after the killing of seven of its soldiers by Hizbollah.

---

[1] There are several reliable synopses that detail different aspects of Hizbollah and its operations. These include Joseph Elie Algha, *The Shifts in Hizbullah's Ideology: Religious Ideology, Political Ideology and Political Program* (Amsterdam: Amsterdam University Press, 2006); Ahmad Nizar Hamzeh, *In the Path of Hizbullah.* (Syracuse, NY: Syracuse University Press, 2004); Augustus R. Norton, *Hezbollah: A Short History* (Princeton, NJ: Princeton University Press, 2007); and Amal Saad-Ghorayeb, *Hizbullah: Politics and Religion* (Sterling, VA: Pluto Press, 2002).

April 1996	In response to Hizbollah rocket attacks on southern Israel, the IDF conducts Operation Grapes of Wrath.
October 2000	Three Israeli soldiers are kidnapped and killed by Hizbollah inside of Israel.
July–August 2006	During the 2006 Lebanon War, Hizbollah makes heavy use of short-range rocket strikes into Israel.

THE ENVIRONMENT OF THE REVOLUTION

PHYSICAL ENVIRONMENT

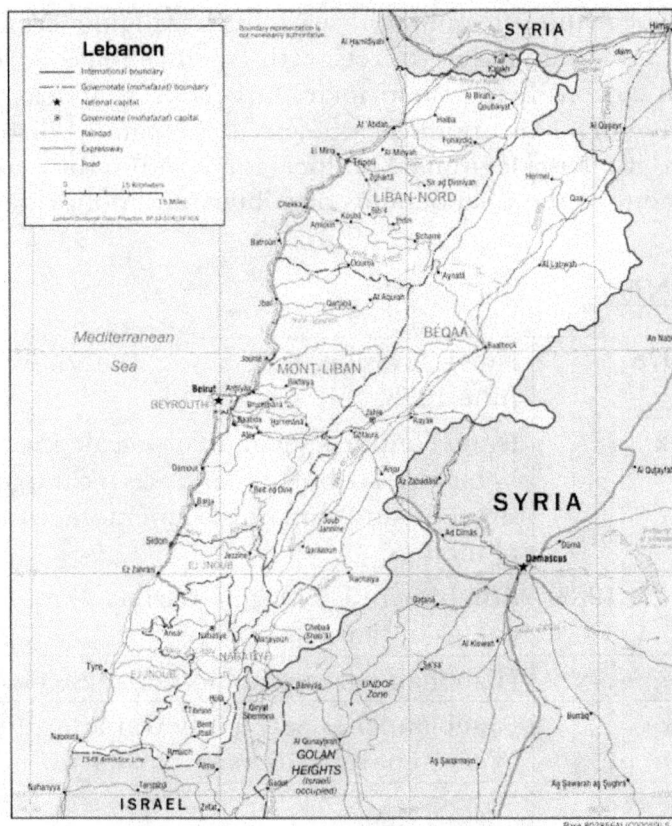

Figure 1. Map of Lebanon.[2]

---

[2]   Perry-Castañeda Library Map Collection, "Lebanon (Political) 2002 Map," The University of Texas at Austin, accessed March 14, 2011, http://www.lib.utexas.edu/maps/middle_east_and_asia/lebanon_pol_2002.jpg.

Lebanon is bordered by Israel to the south, Syria to the east and north, and the Mediterranean Sea on the west. The country is small; at roughly 4,000 square miles, it is only three-fourths the size of Connecticut. The landscape of Lebanon is dominated by two mountain ranges, the Lebanon and the Anti-Lebanon, which run lengthwise through the narrow country, with the Bekaa (Biqa) Valley lying in between. Approximately one-third of the land is arable, with one-fifth currently being cultivated.[3] The population is concentrated most densely along the coastal areas, especially in the capital city of Beirut and the major population centers of Sidon and Tyre to the south and Tripoli to the north. The Litani River, which flows south from the Bekaa Valley before turning sharply westward toward the Mediterranean, is a major source of water and also provides for irrigation and hydroelectricity. The phrase "Belt of Misery" describes the southern suburbs of the capital, which are inundated with thousands of Shi'ite and Palestinian refugees fleeing the south. The near-constant conflict means that public works, building standards, and regulations are rarely enforced, and the area is best described as a slum. "South Lebanon," from where Hizbollah draws most of its support, is generally considered the area bordered by the Litani River Gorge to the north, the Mediterranean to the west, the Bekaa Valley on the east, and Israel to the south. Green, hilly, and dotted with deep valleys, the terrain is inhospitable to large, armored vehicles.[4] Many of the region's villages are situated on hilltops, providing Hizbollah fighters with clear fields of fire and ample cover against ground attacks.[5]

CULTURAL AND DEMOGRAPHIC ENVIRONMENT

The most powerful foundation of cultural identity in modern Lebanon is sectarian affiliation, with the predominant groups falling under two broad categories of either Muslim (Sunni, Shi'ite, Druze)

---

[3] "Agriculture and Food—Lebanon: EarthTrends Country Profiles," EarthTrends: The Environmental Information Portal of the World Resources Institute, accessed September 5, 2010, http://earthtrends.wri.org/pdf_library/country_profiles/agr_cou_422.pdf. For information of commodities, see FAOSTAT, accessed September 5, 2010, http://faostat.fao.org/DesktopDefault.aspx?PageID=339&lang=en&country=121.

[4] Andrew Exum, "Hizballah at War: A Military Assessment," *Policy Focus* no. 63 (The Washington Institute, December 2006), www.washingtoninstitute.org/pubPDFs/PolicyFocus63.pdf, 3.

[5] Ibid. Exum points out that operations on this hilly terrain, which require dismounted infantry, are much different than the operations the IDF have traditionally carried out in other regional campaigns. Additionally, the terrain renders Israel's technological advantage mostly useless.

or Christian (Maronite, Greek, Orthodox, Catholic).[6] Although exact population figures do not exist, Muslims account for roughly 70% of the overall population of about four million, and Christians account for the remaining 30%. The presence of approximately 400,000 Palestinian refugees, most of whom remain in camps, is a further strain on this delicate demographic map.[7] About 60% of the Muslim population is Shi'a, with the remainder being Sunni and Druze.[8] Maronites account for roughly 75% of the Christian population. The country's sects are distributed largely along geographic lines, a phenomenon that was further reinforced by the civil war of 1975–1990. The various natural obstacles of Lebanon—mountain ranges, fast-flowing rivers, and climatic extremes—facilitate the isolation of factions based on clan, ethnic, and religious ties.[9] The south, which shares a border with Israel, is predominantly Shi'ite, whereas the mountains have traditionally been inhabited by Christians in the north and Druze in the south. The far north, around Tripoli, has traditionally been a Sunni population center.

---

[6] Ethnically, the Lebanese state is very homogenous; roughly 95% of the population is Arab. Kurds, Alawites, and Ismaelis are also present, as are Armenians (the only major non-Arab population) although in smaller numbers.

[7] Estimate from "Lebanon Camp Profiles," United Nations Relief and Works Agency for Palestine Refugees in the Near East (UNRWA), accessed September 5, 2010, http://www.unrwa.org/etemplate.php?id=73.

[8] Shi'a and Sunni Muslims consider the Druze a heretical sect and do not recognize members as fellow Muslims.

[9] Charles Winslow, *Lebanon: War and Politics in a Fragmented Society* (London: Routledge, 1996) states: "As a home for ancient coastal settlements; as Phoenician city states; entrepôt centers (under Persian, Greek Roman and Byzantine rule); as a haven for dissentients (during the Umayyad and Abbasid periods); and as semi-independent chieftaincies (under the Egyptian Mamluk and Ottoman Turkish Sultanates), the mountains and coast of Lebanon have often operated politically as separate entities. The mountains, many of them tree covered until the late nineteenth century, gave water and protection to their inhabitants. Because of them, historical Lebanon has served as a refuge for a great variety of groups, sects, and individuals who have had to flee the larger systems nearby. The *rawasab* (residue) of other peoples and cultures have discovered the independence of the mountain and have been stubborn to keep it. The result has been to pack a great deal of diversity into a small area."

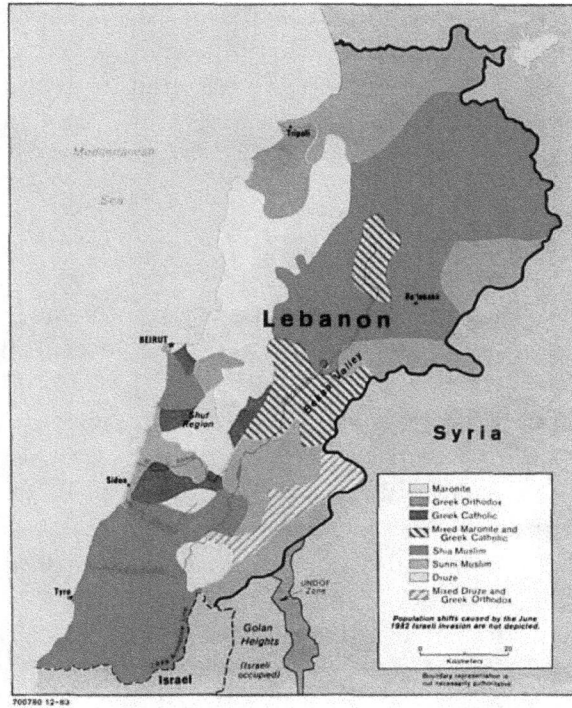

Figure 2. Distribution of religious groups in Lebanon in 1983.[10]

In addition to the unique nonconformist character of many of the groups that sought refuge in the area's mountains (Maronites were considered Christian heretics, and members of the Shi'ite and Druze sects were similarly labeled Muslim heretics by their Sunni rulers), this exclusionary identification has been reinforced by decades (if not centuries) of communal conflict. The configuration of sectarian identity is highly exclusionary, as it is based not only on religious affiliation but also on kinship ties.[11] Consequently, individuals cannot move between these categories in the same way that individuals in other societies might "convert" and therefore adopt a new religious identity, nor can they simultaneously have membership in multiple groups with crosscutting allegiances.[12] The extent of control this identity exerts over such things as social status and social mobility makes it highly similar to caste systems like those found in India.[13] This

[10] Perry-Castañeda Library Map Collection, "Lebanon - Distribution of Religious Groups 1983," The University of Texas at Austin, accessed March 14, 2011, http://www.lib. utexas.edu/maps/middle_east_and_asia/lebanon_religions_83.jpg.

[11] Simon Haddad, "The Political Transformation of the Maronites of Lebanon: From Dominance to Accommodation," *Nationalism and Ethnic Politics* 8, no. 2 (Summer 2002): 5.

[12] An analogous example might be an African American Baptist, who is both "black" and "Christian," therefore belonging simultaneously to two identity groups with equally powerful narratives.

[13] Haddad, "The Political Transformation," 5.

sectarian identity provides a strong sense of internal cohesion and acts as a social reference point, but it also limits an individual's contact with others and the kind of occupation open to that individual.[14] Sectarian identity is further reinforced by formal institutional arrangements, as well as by social customs.

SOCIOECONOMIC ENVIRONMENT

Agriculture accounts for approximately 10% of the gross domestic product and employs roughly 11% of the overall workforce in Lebanon, although the majority of rural households engage in agricultural activity at least part time. However, agriculture is the sole source of income for nearly half of the population of South Lebanon, and repeated conflicts on the southern border with Israel have been extremely disruptive to the local population.[15] The wide variation in topography and climate allows for the production of temperate and tropical crops. Fruits and vegetables, which require high inputs of labor, capital, and water resources, are largely grown along the coast, whereas agriculture in the Bekaa Valley is dominated by staples such as potatoes, tomatoes, and sugar beets, as well as hashish, which is a major cash crop. Cereals and olives are grown in the north, and wheat, tobacco, and figs are cultivated in the south. Unlike in many states of the region, minerals are scarce in Lebanon and are mined only for domestic consumption, not for export.

Throughout most of Lebanon's history, there was a severe bias in favor of the economic center—that is the Christian and Sunni elite. The only real exception to this was under President Shihab (1958–1967), a general who embarked on a systematic campaign to deliver the social justice he and his technocratic circle of advisers felt was really driving the state's debilitating sectarianism.[16] After the conflict of 1958, Shihab instituted significant changes meant to redress socioeconomic inequality, including imposition of an equal distribution of high-level administrative posts between Christians and Muslims (a large boost for the latter) and a dramatic increase in government spending. He also founded a central bank (to facilitate state regulation of the economy)

---

[14] Ibid., 6.

[15] "Country Information: Lebanon," Food and Agriculture Organization (FAO) of the United Nations, accessed August 15, 2010, http://www.fao.org/emergencies/country_ information/list/middleeast/lebanon/en/. Farmers not only lose their harvest during conflict but must also wait until unexploded munitions and mines are removed before returning to work; livestock populations have also been decimated by repeated conflict, and animal husbandry has most likely been in constant decline since the civil war began in 1975.

[16] This section is based on William W. Harris, *Faces of Lebanon: Sects, Wars, and Global Extensions*, (Princeton, NJ: Markus Wiener Publishers, 1996), 146–149.

and agencies for planning, statistics, development, and social security, and he dedicated substantial resources to public works projects such as roads, schools, and irrigation schemes. To accomplish these projects, Shihab used the domestic security services to weaken the political power of the ruling elite clans, who opposed these economic policies because their primary beneficiary was the poor Shi'ite periphery. This socioeconomic experiment came to an end roughly a decade later, when the commercial and landed elite finally put aside their sectarian differences to pursue their common interest in laissez-faire economics. The ensuing programs of economic liberalization eventually earned Lebanon the title of "Switzerland of the Middle East," especially for its banking secrecy laws and loose financial regulations, notably the absence of restrictions on movement of capital. It soon became a regional finance hub where Gulf monarchs, socialist dictators, and nervous bourgeoisie spirited away their fortunes. Although this inflow was interrupted by the civil war, the rush of liquid assets benefited well-connected elites, who used their access to capital and their ability to bypass the already weak bureaucracy to construct high-end shopping districts and luxury hotels adjacent to the slums and bombed-out buildings, thus further underscoring the division between Lebanon's economic elite and those on the periphery.

HISTORICAL FACTORS

The history of the Shi'a in Lebanon is suffused with a heritage of collective suffering that has a distinct communal character yet also reflects the grievances of the global Shi'a community.[17] Centuries of persecution by majority-Sunni empires, as well as the contemporary Shi'a community's minority status in most of the states where communities do exist, create a rich history with elements that provide a powerful basis for political mobilization. However, Shi'a traditions and Shi'a leaders historically promote political quiescence, even submission, to perceived tyranny and injustice.

The Shi'a of Lebanon had no powerful patron and instead comprised the class of "hewers of wood and drawers of water."[18] In the great struggle between the (Sunni) Ottoman Empire and the (Shi'a) Safavid Empire, the latter originating in what is modern-day Iran, Lebanon's Shi'a had the misfortune of being a religious minority in an empire at war with its sectarian brethren. They were under near-constant military assault from Ottoman officialdom, the ravages of

---

[17] Amal Saad-Ghorayeb, *Hizbullah: Politics and Religion* (Sterling, VA: Pluto Press, 2002).

[18] Fouad Ajami, "Lebanon and its Inheritors," *Foreign Affairs* 63, no. 4 (Spring 1985): 778–799.

which were partly to blame for the principles of political passivity the community cultivated in the intervening centuries.[19] As pan-Arab nationalism and leftist political ideologies swept the region, they also influenced Shi'a youth in great numbers. The Baath Party, numerous Communist parties, and the Syrian Socialist Nationalist Party (SSNP) appealed to the economic and social grievances that were dormant in Shi'a political awareness for decades. Yet, in a country so defined by sect, kinship, and clan, ideological parties based on more abstract notions of class were difficult to assimilate. As a result, many political platforms based on class grievances were rewritten using the language of sectarian identity, cloaking goals such as social justice and equality in a distinctly Shi'a language. Growing rates of urbanization, literacy, and exposure to printed material, as well as large inflows of worker remittances, increased the educational attainment of many Shi'a, but these changes were not mirrored by increasing employment opportunities.[20]

In addition to these factors, there were additional phenomena that contributed to both the sectarian conflict and the rise of Hizbollah. These included the post-World War I process of dismantling the Ottoman Empire, the establishment of the confessional system[21] of political representation, and the huge influxes of Palestinian refugees, which sparked the fifteen-year civil war that lasted from 1975 to 1990. During the post-World War I process of dismantling the Ottoman Empire, the victorious European powers divided the spoils between themselves with an eye more toward maintaining equilibrium between their colonial possessions than toward creating viable nation-states based on ethnic and religious divisions. Although the French sheltered and protected the Maronite population in the new state of Lebanon, often violently, the state's Muslim populations identified more with the anti-French and anti-British independence movements sweeping through the rest of the Arab world.[22] In this context, the declaration of the Lebanese state's independence (from the French) in 1943 was partly a strategic move to ensure that the state was not absorbed by Syria.[23] The National Pact, as it was known, included an agreement to abandon both Western and Syrian allegiances, as well as a statement that Christians constituted a majority of the population according to the census of 1932. Although this was probably not true—even in the

---

[19] Ibid.

[20] Ibid.

[21] A confessional state is a system of government in which there is a proportional allocation of political seats and governmental billets based on religious or ethnic groups.

[22] Hamzeh, *In the Path of Hizbullah*. French forces (with Maronite volunteers) bombed Shi'ite villages, destroyed militias, and forced capitulation of community leaders.

[23] Haddad, "The Political Transformation," 5.

1930s and 1940s when the census was taken and the pact agreed to—Muslim leaders saw this as a necessary concession to allay Christian fears and complete the withdrawal of the Maronite's French patron.[24]

The confessional system was enshrined in two agreements, the National Pact (unwritten) of 1943 and, later, the Taif Accords that brought an end to the civil war. The National Pact gave Christians the presidency and a guaranteed parliamentary majority while reserving the positions of prime minister and speaker of the Parliament for the Sunnis and Shi'a, respectively. The Taif Accords redressed the fundamental representative inequalities (although still without a formal census) by weakening the constitutional powers of the Maronite presidency and granting Muslims and Christians guaranteed equal representation in Parliament. Although the final agreement provided the technical basis for a new government, it did little to address the underlying issues that would contribute to future outbreaks of violence; for example, it did not include an accurate gauge of the current demographic distribution or the fate of civil war-era militant leaders. In most cases these leaders—many of whom were responsible for civilian massacres and other crimes—achieved formal amnesty, and they (or their family members and closest affiliates) continued to dominate the political scene. In the context of the negotiations leading up to the Taif agreement, the Maronite's ever-shrinking demographic status made them reactionary and uncompromising, whereas Muslims' rising majority-status emboldened their community leaders, causing them to overreach and be excessive in their demands. This excess was exacerbated by the ideology of Arab nationalism that was sweeping the region in the 1960s, as well as by the huge influxes of Palestinian refugees, who framed their own struggle in the pan-Arab context and further complicated sensitive demographic issues.[25]

In addition to sectarian tensions, which had simmered since independence until breaking out into full-scale conflict, Lebanon was also a staging ground for continued conflict between Israel and the Palestinians. In 1969, the Cairo Agreement, signed by the commander of the Lebanese Army and Yasser Arafat, established the legitimacy of Palestinian guerrilla activity (against Israel) in Southeastern Lebanon and ensured that the Lebanese government would not act to restrain the PLO's activities. The expulsion of the Palestine Liberation Organization (PLO) from Jordan in 1971 made the Lebanese arena even more crucial to the PLO, and afterward much of the group's activity was carried out from bases on Lebanese territory. However,

---

[24] Ibid., 7.
[25] Ibid., 8.

the weakness of the Lebanese state provoked both Palestinian and Israeli retaliation. In 1973, armed clashes between the Lebanese Army and Palestinian fighters broke out as the Lebanese government was unable (or unwilling) to prevent Israeli retaliation, and five years later, in 1978, Israel launched a full-scale invasion of Southern Lebanon to end Palestinian incursions into Israeli territory. The Lebanese military proved unable to reign in the PLO—even as the PLO's activities drew increasingly destructive responses from Israel. The demonstrated weakness of the state military led many sects and prominent political families to intensify their efforts to build up their own militias.[26] The scales finally tipped toward civil war in 1975 when a spate of (successful and failed) political assassinations and large-scale reprisals against unarmed civilians soon turned into generalized fighting. Initially, fighting was largely limited to Palestinians and Phalangists (right-wing Christian militias controlled by Bachir Gemayel), but it then spread to more general Christian versus Muslim violence.

For the ensuing fifteen years, Lebanon's confessional communities targeted one another even as Syria, Israel, the United States, and the PLO joined the fighting, marked by a dizzying array of temporary alliances and broken agreements. Much of the fighting took place in Beirut and the Shi'ite population centers in the south where the PLO was launching attacks against Israel. Nearly a quarter of a million were believed to have died, nearly one-fourth of the population was injured, and the economy collapsed almost completely. Each confessional group had at least one militia, although throughout much of the war several armed groups claiming to represent their sectarian communities were in direct competition with one another, carrying out reprisals against their own populations. Hizbollah coalesced several years into the fighting, around 1982, as the violence migrated from Beirut into the south and as fighting between the PLO and Israel escalated, culminating in a second Israeli invasion. Even the 1989 Taif Accords, which brought an end to the fighting by guaranteeing Christians and Muslims equal representation and making the (Sunni) prime minister and (Shi'ite) speaker responsible to the legislature rather than the (Maronite) president, did little to directly address Shi'a grievances, which were distinct from those of Sunni Muslims. The accords themselves were possible because there was no clear victor in the fifteen-year civil war and because the census necessary to provide an objective basis for constructing representative institutions was too dangerous and destabilizing to conduct. Thus,

---

[26] John Keegan, "Shedding Light on Lebanon," *The Atlantic*, April 1984.

although Shi'ites were given more political power and representation, it was still far short of their actual demographic weight.

It was during this time that an influential Imam (Musa Al-Sadr) from a notable religious family traveled to Lebanon from Iran, intent on mobilizing the Shi'a population outside the confines of either the leftist (nonreligious) parties or the few dominant feudal families who ruled the community in pursuit of their own narrow interests. Al-Sadr turned Shi'ite history and ceremony from a collection of passive ritual lamentations into calls to action, and by the mid-1970s, he had succeeded in establishing Amal, a Shi'ite militia and precursor to what Hizbollah would later become.[27] Al-Sadr worked hard to lure Shi'a recruits away from the PLO and its secular-leftist Lebanese allies, who he accused of using the Shi'a as disposable "canon fodder."[28] Although he lent rhetorical and ideological support to Palestinian aspirations, Al-Sadr insisted that he was unwilling to expose the already poor and marginalized Shi'a of the south to additional suffering, and in 1976, when it became clear the Shi'a would bear the full brunt of fighting between the PLO and Israel, Al-Sadr threw his support behind Syria, which intervened on behalf of the Maronite Christians in order to weaken the PLO and its leftist (and Lebanese nationalist) allies.[29] Al-Sadr was also less antagonistic toward the Maronite Christians because he believed they were driven to violence by an existential fear rooted in their minority status and their own historical experience with persecution.[30] Yet he also criticized the Christian political establishment for its gross neglect of the southern region of Lebanon and for its campaign of repression against poor Shi'a.

GOVERNING ENVIRONMENT

Although Lebanon was technically a democracy for the thirty years prior to the outbreak of civil war in 1975, political representation in the state was highly skewed in favor of the Maronite community

---

[27] The following statement is indicative of Al-Sadr's successful use of leftist themes to mobilize the formerly passive Shi'ite population: "Whenever the poor involve themselves in a social revolution it is a confirmation that injustice is not predestined." Norton, *Hezbollah: A Short History*, 18. Al-Sadr disappeared in 1978 on a trip to Libya, probably a victim of President Gaddafi.

[28] Indeed, more Shi'a died in the early stages of the civil war (before Hizbollah was founded) than any other sect. The coalition of parties fighting the Christians—the Lebanese National Movement—was led by the Druze Kamal Jumblatt, who Al-Sadr accused of a willingness to "combat the Christians [down] to the last Shi'a." Pakradouni (1983), 106. Cited in Norton, *Hezbollah: A Short History*, 19.

[29] Ibid., 20.

[30] Ibid., 19.

and, to a lesser extent, the Sunni elite.[31] Although independence was granted (under French protection) in 1926, the French continued to practice de facto rule through their Maronite Christian allies, principally in order to prevent possible unification into a single state of "greater Syria," which the Christians saw as an existential threat and the French saw as a potentially ungovernable state composed of competing demographic groups.[32] This combined French–Maronite rule proved oppressive and led to frequent clashes with the state's many confessional groups. Subsequent decades saw attempts to redress what many non-Christians saw as institutional obstacles to their representation (notably the National Pact "Mithaq Al-Watani" of 1943, which enshrined representation according to the flawed 1932 census), as well as conflicts that foreshadowed the bloody civil war that would begin in 1975 (mainly the 1958 civil conflict). The conflict in 1958, only halted by direct US intervention, later proved to be a sign that the legitimacy of the earlier pact was dissolving. The severe underrepresentation of the Shi'ite community in Parliament, as well as government and civil service jobs, combined with a massive population shift (Shi'ite numbers tripled between 1956 and 1975 from 250,000 to 750,000; by the 1980s it was the largest group, at 1.4 million compared to 800,000 each for Sunni and Maronite), required more than marginal tinkering within the existing institutional design.[33] The inflexibility of the confessional system, which did not allow for gradual changes in representation in response to demographic realities, created a political powder keg. Resources were also allotted to sect-based resource networks, which meant that the much larger Shi'a population received significantly less (per capita) state assistance.[34] Meanwhile, the economic growth that did take place was not in the agricultural or industrial sectors, which is where most of the Shi'ite population labored. The fact that the Shi'a members of Parliament came overwhelmingly from the landed elite, and were completely unbeholden to their poor constituents, exacerbated these problems.

Syria was also a major player on the Lebanese political scene. Syrian influence and control was nearly ubiquitous after 1976, when troops entered the country at the official request of the Maronite president, who reluctantly turned to Damascus after appeals for a second US intervention (after 1958) went unanswered.[35] The Christians, for their part, feared that the combined numerical superiority of the

---

[31]   Haddad, "The Political Transformation."

[32]   Keegan, "Shedding Light on Lebanon," 4.

[33]   Hamzeh, *In the Path of Hezbollah*, 13.

[34]   Laura Deeb, "Hizballah: A Primer," *Middle East Report*, July 31, 2006.

[35]   Haddad, "The Political Transformation," 32.

Palestinian fighters along with their Shi'a and Druze allies would finally succeed in ending Christian political dominance. The Syrians feared the same, although their ultimate concern was that the rise in radical (leftist, pro-Palestinian) Muslim opposition would pull Syria into a war with Israel, which they would quickly lose.[36] During the civil war, Syrian troops fought many parties, Christian and Muslim, and the most apparent pattern of alliances between the warring parties was the frequency with which they shifted. This reflected Syria's overall strategy, which was to maintain a rough balance between all the warring factions. Preventing the dominance of any single group allowed Syria to dictate Lebanese politics and extract important resources from its neighbor's economy.[37] Thus, when the Maronites again appeared to be on the cusp of reestablishing their dominance (achieved with Israeli assistance) in the late 1980s, Syria acted to sabotage their rise by aiding the anti-Maronite opposition. In 1991, at the conclusion of the civil war and after the signing of the Taif Accords, the Lebanese and Syrian regimes signed the "Treaty of Brotherhood, Cooperation, and Coordination." This treaty formalized the Syrian presence in Lebanon (considered illegal by many), asserted that conditions in Lebanon should not be allowed to destabilize neighboring Syria, and stated that Syria would refrain from interfering in Lebanon's foreign affairs. The Syrian relationship was defined by a massive military and intelligence presence, economic penetration, the relocation of a large Syrian civilian community—laborers, businessmen, etc.—to Lebanon, control over the Lebanese military, and the intense screening of candidates for domestic office.[38]

Because of the range and depth of interested parties (both domestic sectarian groups and foreign parties), as well as the absence of adequate "checks and balances," domestic politics in Lebanon resembled a patronage-based system. Consequently, the awarding of political office depended on personal loyalties rather than competence. These underlying conditions facilitated the communal politicization that provided the organizational scaffolding on which Hizbollah could be built.[39] In this regard, the Shi'a "awakening" had more in common with the politicization of other Lebanese sects than with the universal Islamic Revival.[40] Still, individual actors—both religious and secular—played important roles in this mobilization,

---

[36] As'ad Abu-Khalil, "Ideology and Practice of Hezbollah in Lebanon: Islamization of Leninist Organizational Principles," *Middle Eastern Studies* 27, no. 3 (July 1991): 7.

[37] Ibid.

[38] Haddad, "The Political Transformation," 15.

[39] Saad-Ghorayeb, *Hizbullah: Politics and Religion*, 9.

[40] Ibid.

notably Palestinian activists, leftist groups (including Druze), and the Iranian cleric Imam Musa Al-Sadr.[41] Al-Sadr, who came to Lebanon from Iran in 1957, founded Harakat Al-Mahrumin (Movement of the Dispossessed) in 1974, which eventually grew into the Afwaj Al-Muqawama Al-Lubnaniyya, "Battalions of the Lebanese Resistance," better known by its acronym, Amal. Unlike Hizbollah, Amal appealed more to middle-class Shi'a who were frustrated with the political elite and railed against the brutality of the PLO guerrillas who were engaging in their own occupation of the south.[42] The disappearance of Al-Sadr in 1978—and the decision by Amal's leadership to participate in the Salvation Committee[43]—contributed to the establishment of Hizbollah, which took advantage of the leadership vacuum left by Al-Sadr's absence and its apparent collusion with the Maronite Christians.[44] The emergent narrative was that the Shi'a, rather than the Sunni elite, should lead the struggle against the Maronite political establishment.[45]

WEAKNESSES OF THE PREREVOLUTIONARY ENVIRONMENT AND CATALYSTS

Although Hizbollah was certainly forged in the fires of Lebanon's fifteen-year civil war and became a central player in both attacks on foreign targets and the war's communal violence, it was not one of the original belligerent parties that sparked the conflict. This unfortunate designation belonged to Christian and Palestinian militias whose fighting quickly engulfed Lebanon's already divided society. As formal Palestinian organizations and their leftist allies were sidelined, their place in the sectarian fabric of Lebanese society was largely usurped by Shi'a organizations. Hizbollah was unique among those organizations that appealed to Shi'a communal sympathies in that it was both a religious party (as opposed to the secular Organization of Communist

---

[41] Abu-Khalil, "Ideology and Practice of Hezbollah in Lebanon."

[42] Ironically, although Amal was founded by an Iranian-born cleric and its members were originally trained by Fatah (PLO), Amal would later be distinguished from Hizbollah both by its political distance from Tehran and its opposition to the PLO's presence in South Lebanon. Norton, *Hezbollah: A Short History*, 17–22.

[43] The Salvation Committee had been formed by Lebanese President Elias Sarkis to bring the state's warring militias to the negotiating table amidst the Israeli siege of Beirut. Although Amal had tacitly welcomed the Israeli invasion—which presented the greatest possibility of expelling the PLO fighters from Lebanon—many younger, more radical members within Amal saw participation in the Salvation Committee as bringing the organization too close to Israel and the United States. Many defectors joined Hizbollah, and some others broke away to form Islamic Amal. Ibid., 23.

[44] Emmanuel Karagiannis, "Hezbollah as a Social Movement Organization: A Framing Approach," *Mediterranean Politics Journal* 14, no. 3 (2009): 365–383.

[45] Abu-Khalil, "Ideology and Practice of Hezbollah in Lebanon."

Action and Lebanese Communist Party, both with majority Shi'ite membership) and pro-Palestinian (as opposed to Imam Musa Al-Sadr's Amal movement, which quickly distanced itself from the Palestinian resistance).[46] For this reason, the Israeli invasions (in 1978 and again in 1982), the Iranian Revolution, and the decline in Amal's popularity served as important proximate causes leading to the formation of Hizbollah. Indeed, it was probable that no single factor alone would have been sufficient because the process of building up local defenses against an Israeli invasion provided the necessary organizational scaffolding while the demonstration effect of a successful Shi'ite Revolution in Iran provided the material for religious mobilization.[47] Moreover, many Shi'ite religious groups and associations formed in the years immediately following the Iranian Revolution in 1979, and some of these groups later merged with Hizbollah.[48] However, these groups would likely have remained independent and fairly marginal in the absence of a second Israeli invasion.[49]

The environment of increased militancy also contributed to the group's formation, especially vis-à-vis heightened Maronite political mobilization. This militancy was part of a larger cycle of social and economic dislocation in which the Shi'a community had been continuously caught. This cycle formed a major part of the community's political consciousness and included the dynastic struggles between regional empires that frequently ravaged their population; the creation of the Lebanese state, which had included significant anti-Shi'a violence by the French colonial forces and their Maronite allies; the civil conflicts of 1958 and 1975–1990; and, finally, the Israeli invasions.[50] The 1978 invasion killed 2,000 Shi'a and displaced another 250,000.[51] The leftist movements' continued support for the PLO despite the havoc visited on Shi'a villagers in the south convinced many that these movements were going in the wrong direction. Thus, when Amal fighters engaged the PLO in the

---

[46] Shortly after the outbreak of civil war, Al-Sadr withdrew his Amal movement from cooperation with the (secular) Lebanese National Movement (LNM), itself composed of leftists and Arab nationalists who supported Palestinian liberation. This move lost Al-Sadr a great deal of support, as many Shi'ites remained loyal to the LNM. Hamzeh, *In the Path of Hezbollah*, 14.

[47] See Abu-Khalil, "Ideology and Practice of Hezbollah." See also As'ad Abu-Khalil, "The Incoherence of Islamic Fundamentalism: Arab Islamic Thought at the End of the Twentieth Century," *Middle East Journal* 48, no. 4 (Autumn 1994).

[48] Husayn Fadlallah was an important religious figure at this time.

[49] Saad-Ghorayeb, *Hizbullah: Politics and Religion*, 11. The author cites Nasrallah [Hizbollah's leader] as saying, "Had the enemy not taken this step [invasion], I do not know whether something called Hezbollah would have been born. I doubt it."

[50] Hamzeh, *In the Path of Hezbollah*, 14.

[51] Deeb, "Hizballah: A Primer."

south, they gained significant popularity.[52] Waves of evictions (such as the eviction of 100,000 Shi'ites from Nab'a in 1976) also brought a large proportion of the Shi'a into close proximity with each other, frequently in Beirut's southern suburbs, which facilitated their capacity for organization.[53] However, the second invasion of 1982, when Israeli troops made it all the way to West Beirut, was the most powerful catalyst. Tens of thousands of Lebanese were killed and nearly half a million displaced.[54] Israel's "Iron Fist" policy, which included bombing Shi'ite villages in the south, destroying large tracts of farmland, imposing curfews, disconnecting utilities, and blockading villages, caused a mass exodus of poor Shi'a from the south into suburban slums around Beirut.[55] Still, perhaps the most powerfully symbolic events of the invasion were the massacres at the Sabra and Shatila refugee camps in 1982, perpetrated by right-wing Christian militias that had been granted access to the camps by the IDF, who had been left in charge of administering security. These attacks occurred despite promises from the United States that refugee populations would be protected in return for the PLO's withdrawal from Lebanon. Approximately one-fourth of the civilians killed were Lebanese Shi'a who had fled the fighting in the south. Ironically, Israel's success in pushing the Palestinians out of the southern battlefield simultaneously demonstrated the weakness of the Maronites (who had to rely on an Israeli invasion in order to weaken the Palestinians and the Arab nationalist opposition) and allowed for the emergence of distinctly Lebanese resistance groups, including Hizbollah.[56]

In addition to the demonstration effect provided by the Iranian Revolution, the cohort of young Lebanese Shi'ites who had studied

---

[52]   This gain was eventually reversed, however, when Amal leader Nabih Berri agreed to participate in the National Salvation Committee (NSC) in 1982, which caused a rift in the group. Disgruntled members (including future Hizbollah leader Hassan Nasrallah) later joined Hizbollah. The stated goal of the NSC, which included the pro-Israeli Maronite leader Bachir Gemayel and his militia the Lebanese Forces, was to replace the PLO presence in West Beirut with the Lebanese Army. Amal's declining popularity, fed also by its increasing involvement in patronage politics and detachment from the primary issue of poverty, directly benefited Hizbollah. Many Amal defectors joined with the Pasdaran and other Islamic resistance movements to establish the "Committee of Nine," a decision-making council composed of three ex-Amal members, three clerics, and three individuals from the Committee Supportive of the Islamic Revolution. Hizbollah became the umbrella movement that absorbed these disparate groups, which would ultimately include Amal defectors, Islamic Amal, individual clerics and their followers, members of Lebanese Da'wa, the Association of Muslim Students, the Association of Muslim Ulama in Lebanon, and the Committee Supportive of the Islamic Revolution. Saad-Ghorayeb, *Hizbullah: Politics and Religion*.

[53]   Hamzeh, *In the Path of Hezbollah*.

[54]   Deeb, "Hizballah: A Primer."

[55]   Saad-Ghorayeb, *Hizbullah: Politics and Religion*, 11.

[56]   Saad-Ghorayeb, *Hizbullah: Politics and Religion*, 10.

under radical Shi'ite ideologues—including Khomeini—in Najaf (Iraq) in the 1960s and 1970s also included many of the individuals who would later form the inner circle of Hizbollah. These youths had been expelled by the Baath Party in Iraq and, upon returning to Lebanon, established the Da'wa ("Call") Party and the Lebanese Muslim Students Union.[57] Iran's dispatch of 1,500 Revolutionary Guards (Pasdaran) to the Bekaa after the 1982 invasion also contributed directly to the formation of Hizbollah, as previous efforts to infiltrate existing groups (such as Amal) were abandoned in favor of bringing all resistance groups into a single organization. The entry of the Iranian Pasdaran was facilitated by Syria, which actively opposed any independent Lebanese peace agreement with Israel because an agreement would ultimately hamper Syrian efforts to recover the Golan Heights.[58] Indeed, by 1982, Syria's near-total control over the actions of the Lebanese central government allowed it to thwart attempts by Lebanese politicians to broker peace with Israel.[59]

Finally, there were also a number of symbolic events that lent a sense of urgency to existing organizational efforts in the Shi'a community and most likely helped the movements' leaders not only mobilize support but also overcome ideological obstacles that might have otherwise scuttled the formation of Hizbollah. These events include the disruption of an Ashura ceremony (the most significant annual religious event in Shi'a Islam) in 1983 by an Israeli military convoy, which killed two Lebanese civilians; the assassination of Shaykh Raghib Harb, a leading figure of Islamic resistance in South Lebanon, in 1984;[60] and the mass detention of many Southern Lebanese in Israeli-operated prisons.[61]

---

[57]   Ibid., 13.

[58]   The Golan Heights were captured by Israel in the 1967 war. Peace between Israel and Lebanon would remove the threat posed by Hizbollah and other militant groups, who could stage attacks on Israel's northern border, without provoking the sort of response that Syrian military action would engender. If the threat of these attacks were removed, it is unlikely the Israelis would consider yielding back this territory as part of a future peace agreement with Syria.

[59]   Haddad, "The Political Transformation," 16.

[60]   Daniel Byman, "Do Targeted Killings Work?" *Foreign Affairs* 85, no. 2 (2006).

[61]   The Ansar prison camp was reported to have detained half of the south's male population at some time between 1982 and its closing in 1985. The transfer by bus of 1,200 blindfolded prisoners from this prison into Israel in 1985 (in contravention of the Geneva Conventions) also caused unrest.

FORM AND CHARACTERISTICS OF THE REVOLUTION

OBJECTIVES AND GOALS

Hizbollah had both long- and short-term goals, as well as religious and political objectives.[62] Although it was not an expression of the group's ideology, Hizbollah's "Open Letter" of 1985 outlined specific actions the group sought to take in response to circumstances at the time.[63] In the letter, they made broad pledges of continued resistance to the United States and its European allies as well as to UNIFIL[64] and Israel. But they did so with explicit reference to events, actors, territories, and political forces in Lebanon, yielding quite a detailed document. Events referenced in the letter included the US intervention in the Lebanon civil war, the massacres at the Sabra and Shatila refugee camps, and the mass evictions carried out in South Lebanon. The explicit objectives stated therein included: (1) "to expel the Americans, the French, and their allies definitely from Lebanon, putting an end to any colonialist entity on our land"; (2) "to submit the Phalanges to a just power and bring them all to justice for the crimes they have perpetrated against Muslims and Christians"[65]; and (3) "to permit all the sons of our people to determine their future and to choose in all the liberty the form of government they desire." Although the organization periodically distanced itself from any commitment to a theocratic state, this original document did include an appeal to the Lebanese people to install an Islamic government: "We call upon all of them to pick the option of Islamic government which, alone, is capable of guaranteeing justice and liberty for all. Only an Islamic regime can stop any further tentative attempts of imperialistic infiltration into our country."

---

[62] For a comprehensive treatment see Joseph Elie Alagha, *The Shifts in Hizbullah's Ideology: Religious Ideology, Political Ideology and Political Program* (Amsterdam: Amsterdam University Press, 2006).

[63] Hamzeh, *In the Path of Hezbollah*, 27.

[64] UNIFIL is the United Nations (UN) Interim Force in Lebanon, which patrols the zone in Southern Lebanon previously occupied by Israel.

[65] The Phalangists were the most powerful Christian militia in operation during the Lebanese civil war and thus engaged in most of the fighting against the Shi'a leftist groups and the PLO. In addition to their role in the bus massacre of 1975, which directly contributed to the onset of hostilities, they are most well known for their role in the massacres at Sabra and Shatila. The group's violent role during the civil war led to its overall decline through much of the 1980s and 1990s, but it regained some popularity after it participated in anti-Syrian demonstrations (the Cedar Revolution) that led to its participation in the pro-Western March 14 Alliance.

Whether and how Hizbollah would seek to establish an Islamic state in Lebanon is an issue of considerable debate. Hizbollah's stated long-term goal was to establish an Islamic order through peaceful means and thus bring about the fulfillment of God's promise of a just order. In the short term, achievement of this goal required the institution of majority-rule democracy and abolition of the sectarian/confessional political system, which, given the Shi'ite demographic advantage, translated into a Shi'ite-controlled government.[66] Reform of Lebanon's current electoral system was, therefore, one of the group's most enduring goals. However, significant changes in the electoral system have historically been made only in the wake of major hostilities—for example, after the civil war of the 1980s and the sectarian violence in Beirut in 2007–2008.[67] Although these incremental changes conferred additional political power on the Shi'a and their political representatives (notably Hizbollah, but also Amal), sectarian voting patterns showed little sign of surmounting existing cleavages.

Opposition to Israel was also an enduring part of Hizbollah's program. The vast majority of Hizbollah's military activities targeted Israeli defense forces, especially after the end of the civil war.[68] It was unclear to what extent Hizbollah would be willing to coexist alongside an Israeli state that had made peace with the Palestinians. Although some of the group's statements stressed the necessity of the destruction of Israel, Hizbollah's actual behavior was far more pragmatic. For example, Hizbollah held indirect talks with Israel in 1996 and 2004 and participated in several prisoner exchanges.[69] The group also accepted a UN-brokered cease-fire that ended the 2006 war with Israel, despite an explicit statement in its 1985 program that it would accept "no treaty, no cease fire, and no peace agreements,

---

[66] "The main problem in the Lebanese political system, which prevents its reform, development, and constant updating, is political sectarianism. The fact that the Lebanese political system was established on a sectarian basis constitutes in itself a strong constraint to the achievement of true democracy where an elected majority can govern and an elected minority can oppose, opening the door for a proper circulation of power between the loyalty and the opposition or the various political coalitions. Thus, abolishing sectarianism is a basic condition for the implementation of the majority–minority rule." (Press conference given by Hassan Nasrallah, November 30, 2009).

[67] Smaller groups, such as the Armenians, have traditionally thrown their weight behind one of the major confessional groups. In this way, they secure political patronage for their community while avoiding outright electoral competition, which their small numbers would not support.

[68] Many of the attacks for which Hizbollah has gained notoriety were carried out against US and Western targets who intervened during the Lebanese civil war (including the bombing of the US Marine barracks in Beirut in 1983 and the subsequent kidnapping of Western hostages).

[69] Deeb, "Hizballah: A Primer."

whether separate or consolidated." Some of Hizbollah's operations against Israel (such as the kidnapping of two IDF soldiers, which sparked the 2006 war) have had specific goals: (1) the return of Lebanese prisoners being held in Israeli prisons and (2) the return of the Shebaa Farms, a fifteen-square-mile parcel of land near the Golan Heights.[70] These concrete and largely divisible set of goals (i.e., goals that can be partially satisfied) differentiated Hizbollah from groups such as Al Qaeda or Jamaa Islamiyya. Whereas Hizbollah could enter into negotiations in hopes of making incremental progress toward achieving its ultimate aims (and thus viewed political participation as a viable strategy), Al Qaeda could not make accommodations with its adversaries because the group's ultimate goal was the destruction of its adversaries.[71]

LEADERSHIP AND ORGANIZATIONAL STRUCTURE

Hizbollah was simultaneously a political party that participated openly in the Lebanese electoral system and a paramilitary organization. The group had roughly 100,000 active supporters, about half as many party members, and several thousand fighters.[72] The number of "elite" fighters that received advanced weapons training was probably much smaller—perhaps as few as 1,000.[73] The organization itself was separated into military and political wings, which included political and administrative councils, military and security organs, and specific service subunits including a social unit, a health unit, an education unit, an information unit, a syndicate unit, an external relations unit, a finance unit, and an engagement and coordination unit.[74] The group's leadership was composed of a seven-member Shura Council, which originally operated underground but then became more open. This collective leadership model was unusual for resistance groups, which frequently operated under a

---

[70] Israel claims that the Shebaa Farms are part of the Golan Heights (which it has occupied since the 1967 war) and, therefore, Syrian, not Lebanese, territory. Syria and Lebanon insist the Shebaa Farms are Lebanese territory, which would complicate Israel's claim to continue occupying the territory. See Human Rights Watch, "Why They Died," September 6, 2007, http://www.hrw.org/en/reports/2007/09/05/why-they-died.

[71] Daniel G. Arce and Todd Sandler, "Terrorist Spectaculars: Backlash Attacks and the Focus of Intelligence," *Journal of Conflict Resolution* 54, no. 2 (2010): 356.

[72] Estimates are from Adam Shatz, "In Search of Hizbullah," *New York Review of Books* 51 (April 29, 2004). However, as was made clear by the 2006 war with Israel, many villagers in South Lebanon took up arms and fought directly alongside Hizbollah fighters, although they were not members or active supporters of the organization. See Exum, "Hizballah at War: A Military Assessment."

[73] Exum, "Hizballah at War: A Military Assessment," 5.

[74] Hamzeh, *In the Path of Hezbollah*, 45–65.

single charismatic personality, although as secretary general of the party, Hassan Nasrallah demonstrated some of the characteristics of a charismatic leader and became a center of gravity for the party.[75]

The Shura Council was elected by the Central Council, a 200-member group of founders and cadres that served for three-year terms. The leadership was mostly clergy; laypersons had to demonstrate sufficient commitment to the principle of *wilayet al-faqih*, as well as possess skills in other fields, such as health, social affairs, finance, or information technology. The councils decided on issues of administration, planning, and policy making. Their decisions were final and binding and were taken either unanimously or by majority vote. In the case of deadlock, the Al-Wali Al-Faqih (i.e., Iranian Ayatollahs) cast the decisive vote. The Jihad Council made decisions at the level of strategy, including those on armed operations and political/social activities. Tactical and operational decisions were left to the party's military wing. The Jihad Council was probably headed by the party's secretary general and likely included a number of ground commanders and possibly members of Iran's Revolutionary Guard.[76] Hizbollah also had a large number of political organs, including a Politburo and a Parliamentary Council.[77] In addition to its aboveground political activities, the group had several security organs that provided security for the organization and its primary constituencies.[78]

---

[75] Ibid., 45–48. The Council on Foreign Relations provides a brief background on Nasrallah at "Backgrounder: Profile: Hassan Nasrallah," accessed August 15, 2010, http://www.cfr.org/publication/11132/profile.html.

[76] Ibid., 69.

[77] The Politburo, whose members serve in an advisory capacity, consists of eleven to fourteen members, both clerical and lay. The Politburo in turn has its own committees; among the most significant are a Cultural Committee, a Palestinian Affairs Committee, and a Security Zone Committee (so named for the Israeli security zone in the south for which it was responsible). The primary duty of the Cultural Committee is to resist the normalization of relations with Israel. It targets teachers, professors, journalists, and businessmen to encourage them to resist Israeli and American educational and cultural influences. The Palestinian Affairs Committee was tasked with strengthening ties with Palestinian resistance groups (primarily Hamas and Islamic Jihad), as well as coordinating activities and services in Lebanon's refugee camps. Finally, the Security Zone Committee (itself disbanded after the 2000 Israeli withdrawal) liaised between displaced southerners and the party leadership, also helping to relocate people to their villages and homes if fighting forces them to flee.

[78] The least secretive of these is the Engagement and Coordination Unit, which is under direct authority of the political wing and handles normal security matters, such as mediation in disputes between party members and other actors, including the government, and the investigation of crimes such as murder, robbery, theft, spying, etc. The Security Unit is similar to the Engagement and Coordination Unit but is under the control of the party's military wing and reports directly to the Shura Council. It is also the most discreet of all the party's units. It is divided into two parts: (1) Party security, tasked with preventing infiltration and dissension; and (2) External security, which conducts counterintelligence operations.

Hizbollah's primary fighting force was the Islamic Resistance, which consisted of both a combat section and an enforcement and recruitment section that provided ideological indoctrination to newly recruited fighters. To avoid infiltration, there was a high level of autonomy between the different sections. Although statements made by the Hizbollah leadership attributed strategic decision making to Al-Wali Al-Faqih, tactical and operational decisions were made by the party leadership.[79] For example, the struggle against Israel was declared a legitimate project, but how to prosecute this campaign was largely left to Hizbollah leaders.[80]

Hizbollah's military wing differed from most Arab armies in that it had a decentralized command structure, which allowed subordinates to exercise independent initiative and respond more quickly to changing circumstances. This decentralization of command and location was demonstrated in the 2006 war with Israel when many fighters were civilians assembled by local village leaders.[81] The number of fighters Hizbollah had at its disposal is unclear, especially given the number of villagers in the south who may have mobilized in defense of their homes but who might have been unwilling to take up arms in support of other Hizbollah activities. Some propose estimates of 5,000 active fighters and 3,000 security personnel, but recent military activities and rallies would suggest much larger numbers—perhaps 20,000 fighters and 5,000 security personnel.[82]

COMMUNICATIONS

Hizbollah possessed both a sophisticated narrative and an advanced physical infrastructure for disseminating its narrative. This narrative centered on a powerful and coherent exposition of the Islamic concept of "struggle" (jihad), which permeated the group's speeches and written materials. Believers must undertake a greater struggle, controlling themselves by purging the body and mind of evil; this struggle imparted conviction and bravery, which fed into success in the lesser struggle that was taken against the worldly enemies of Islam.[83] The group's media outlets constituted the largest information

---

[79] Hamzeh, *In the Path of Hezbollah.* See in particular excerpts from the author's interviews with Sheikh Naim Qasim, Hizbollah's deputy secretary-general, 34–35. It may be of interest to note that Qasim was passed over for the position of secretary-general in favor of current Secretary-General Hassan Nasrallah, his younger and less-experienced counterpart. Many analysts believe this was due to Nasrallah's closer ties to Tehran.

[80] Ibid., 34.

[81] Ibid., 72.

[82] Ibid., 74.

[83] Ibid., 37.

network of any regional political party and included a satellite channel *Al-Manar* (the beacon), with more than ten million viewers, as well as four licensed radio stations and five licensed newspapers, all of which garnered substantial receipts through advertising sales.[84] *Al-Manar*, available throughout the Arab world via satellite, had a corporate atmosphere with state-of-the-art editing and production equipment, as well as a team of foreign correspondents located throughout the Middle East, Europe, and North America.[85] Unlike much of the Western media, which abided by a taboo on depicting grisly images of the dead and wounded, *Al-Manar* openly showed such images (although it is worth noting that many other regional media outlets did as well). Such material routinely included maimed and dead children and disembodied limbs, and even live feeds (during the 2006 war) of civilians being shelled during Israeli air raids.

Al-Manar's pioneering use of footage from actual battles with Israeli soldiers has been widely credited with helping reverse regional feelings of impotence in the struggle with Israel.[86] The channel also targeted Israeli audiences with psychological operations aimed at demoralizing the public; messages were based on Hizbollah's understanding that Israeli society was incapable and unwilling to absorb large casualties because of its tightly knit social fabric.[87] The use of Google maps technology in 2006 allowed the station to pinpoint particular locales in Northern Israel that were being targeted, creating substantial anxiety for residents. In addition, the station also launched the "Who's Next" campaign, which displayed a continuously updated photo gallery of the latest IDF casualties followed by the image of a question mark superimposed over an empty silhouette.[88] Its post-2006 investment in longer-range antennae allowed the station to send its signal as far into Israel as Haifa, the Jewish state's third-largest city. In addition to targeting parties directly involved in the conflict, Hizbollah also crafted messages for parties that had influence over regional conflicts, including the United States and neighboring Arab countries.

Finally, cyberspace had also become an important arena for Hizbollah's communication strategy. The organization's technicians routinely hijacked servers and websites to transmit their own information (while also ensuring that ordinary traffic was not

---

[84] Ibid., 59.

[85] Ibid.

[86] Ibid.

[87] Amir Kulick, "The Next War with Hizbollah: Strategic Assessment" (Tel Aviv: The Institute for National Security Studies, 2007).

[88] Hamzeh, *In the Path of Hezbollah*, 59.

disrupted and that their hijacking activities remained unnoticed).[89] In addition to using the web to transmit information between group members, Hizbollah used the web to transmit information to the media. Because Hizbollah had an extensive presence on the web and the information available on its websites was credible, the information was used by Israeli journalists and other correspondents covering the region. When several Israeli teenagers launched a cyber attack on a number of Hizbollah sites (as well as those of Hamas and the Palestinian Authority) in 2000, the response was intense. Hackers struck the Knesset (Israeli Parliament), the Ministry of Foreign Affairs, and an IDF site and later also targeted the Israeli Prime Minister's Office, the Bank of Israel, and the Tel Aviv Stock Exchange.[90]

Hizbollah's battlefield communications network was crucial to the group's military successes in the 2006 war with Israel. Official reports indicate that the group was successful in rebuffing Israeli efforts to jam their communications systems south of the Litani (jamming signals are restricted in range to relatively small areas). In addition, reports suggest that the group may have possessed the capability to disrupt some Israeli communications.[91] Despite a relatively high level of technological sophistication, Hizbollah used landlines (primarily copper cable, which is highly susceptible to jamming and tapping). Most of these lines were merely laid next to existing utility lines (both public and private), enabling the group to use existing infrastructure and link far-flung outposts and offices. Yet, the vulnerability of these lines meant that the organization did not rely on them except as a secondary or emergency methods of communication. Increasingly, the group laid fiber optic landlines to serve their headquarters, television and radio stations, military compounds, and mobile rocket launching facilities; they oped for fiber optic lines because this material is immune to many of the deficiencies of copper wiring.[92]

Mobile communications technology—primarily cell phones, which were inexpensive, portable, and lightweight, but also satellite phones—was the group's most common method of communication, despite its vulnerability. During the conflict, Hizbollah utilized an elaborate system of radio call signs and a closed cellular phone system; they designed the latter to handle short message service (SMS) and

---

[89] "Lebanon: Hezbollah's Communication Network," *Stratfor* (May 9, 2008), accessed August 15, 2010, http://intellibriefs.blogspot.com/2008/05/lebanon-hezbollahs-communication.html.

[90] See Hasan M. Al-Rizzo, "The Undeclared Cyberspace War between Hezbollah and Israel," *Contemporary Arab Affairs* 1, no. 3 (2008).

[91] "Lebanon: Hezbollah's Communication Network."

[92] VoIP (Voice over Internet Protocol) transmitted by fiber optic cable would be a particularly secure form of communication available to the group. Ibid.

e-mail as the primary formats of information exchange.[93] Because jammers could disrupt the flow of signals only in a small area and were also limited by mountainous terrain (the Lebanese theater being both large and mountainous), Hizbollah operatives were largely able to evade Israel's efforts, which were concentrated in a few high-value areas, to deny the flow of communication. Hizbollah's Secretary General Hassan Nasrallah underscored the centrality of the group's communication network in 2008 when, as the Western-backed Lebanese government declared the organization's media assets both "illegal" and an "attack on Lebanese sovereignty," he declared these assets to be the group's single most important weapon and stated that any disruption perpetrated by government authorities to the network would be an act of war.

METHODS OF ACTION AND VIOLENCE

The majority of Hizbollah's military activities centered on their guerrilla campaign against the IDF in Southern Lebanon (1982–2000). The campaign focused on low-intensity attacks targeting a small number of IDF soldiers. Although it was not overwhelmingly successful from a tactical point of view, the campaign did achieve the group's strategic objectives, forcing Israel to increase the number of forces deployed to the area, to build additional military installations, and to spend large sums of money to supply the South Lebanese Army (SLA). Attacks on individual IDF soldiers were not designed to capture land but were an end in themselves; they constituted the core of the group's psychological operations aimed at demoralizing the Israeli occupation forces and their SLA collaborators.[94] These operations drove Israeli decision makers to abandon their original strategic vision, which focused on building up an indigenous SLA force that would police the security belt on Israel's behalf.[95] Overall, the extended guerrilla campaign was credited with ultimately forcing Israel's unilateral withdrawal from the south.

Hizbollah fighters were also very strategic in identifying suitable targets, often choosing them on the basis of their political impact rather than how they would affect battlefield operations. Early attacks often targeted reservists (as opposed to regular army members) in an

---

[93] Exum, "Hizballah at War: A Military Assessment," 5.

[94] Schleifer, 6.

[95] The SLA was originally a Christian militia formed to defend specific Christian villages against attack by PLO forces and their Lebanese allies. The vast majority of their weapons and training came from the IDF, which saw the SLA as a useful source of indigenous manpower. Ibid.

effort to undermine public support for the occupation. Because Israel conscripts all citizens into military service, the public response to the loss of reservists, who are less seasoned than their regular counterparts, was particularly acute.[96] This eventually forced the Israeli government to change their policy to declare that only members of the regular army would be stationed in Lebanon. The close-knit character of Israeli society and its history of ransoming prisoners of war made this an especially effective strategy. To this end, Hizbollah also targeted officers (as opposed to rank-and-file soldiers) and planted mines as close to the Israeli border as possible to increase the visibility such attacks would garner among the Israeli public.[97]

Hizbollah also took revenge on those Lebanese it accused of collaborating with Israel, often utilizing family connections to put pressure on soldiers serving in the SLA.[98] In addition to producing much of the high-quality intelligence that played a key role in the success of Hizbollah's military operations, it also exacerbated the SLA's already low morale.[99] Hizbollah's ability to collect intelligence was also facilitated by the SLA's high command, which treated rank-and-file recruits (especially Shi'a recruits) very poorly. In addition, the commitment of SLA fighters was also eroded by the popular perception that the SLA served as "Israel's sandbags" because SLA troops manned the frontline outposts while the Israeli soldiers operated from the better-protected positions to the rear.[100]

Although Hizbollah's successful exploitation of suicide attacks contributed to the popular association of militant Islam with suicide missions, the vast majority (81%) of suicide attacks during the Israeli occupation (1982–1986) were carried out by Christians or affiliates of secular or leftist parties,[101] and only twelve of Hizbollah's attacks involved the intentional death of a party operative.[102] Hizbollah was credited with inspiring the use of similar attacks by other ethno-

---

[96] Ibid.

[97] Ibid., 8.

[98] SLA members were mostly Maronite Christians but still included a significant minority of Muslim and Druze soldiers that Hizbollah could target.

[99] Schleifer, 4.

[100] Nicholas Blanford, "The Quandary of an SLA Amnesty," *The Daily Star* (Beirut), August 16, 2005. The SLA often imprisoned the family members of would-be recruits who initially refused to join, holding them as collateral against defection (Schleifer, 5). Like other parties to the conflict, the SLA committed a wide range of atrocities, including the systematic torture of civilians and captured militants and the detention of women and children, primarily at Khiam prison. The compensation provided to SLA soldiers was also extremely low, making it easy for Hizbollah to provide a superior financial incentive to those willing to defect (Schleifer, 8).

[101] Robert Pape, *Dying to Win: The Strategic Logic of Suicide Terrorism* (New York: Random House, 2005).

[102] Deeb, "Hizballah: A Primer."

nationalist groups because its 1983 attacks precipitated the withdrawal of both the French and American forces.[103] The number of suicide attacks dropped off precipitously after the Israeli withdrawal to the "security zone" in 1985, and attacks on Western targets largely ended with the civil war in 1991.[104] These early methods of kidnapping Westerners, initiating suicide attacks, and bombing high-profile targets were increasingly set aside as the dynamics of the conflict shifted. Subsequent attacks largely concentrated on Israeli military targets in the south of Lebanon and Katyusha rocket attacks on residential areas in Northern Israel.

Mounting losses of Israeli soldiers eroded public support for the occupation, and in 1999, Ehud Barak, who was a candidate for prime minister at the time, promised to bring the troops home if he were elected. Barak won the election, and the IDF withdrew from many of its forward military outposts. Hizbollah simultaneously ramped up the intensity of attacks (including the assassination of an SLA commander), and on the eve of Israel's May 2000 withdrawal, many of the SLA's brigades abandoned their posts and fled across the border into Israel.[105] Others turned themselves over to Hizbollah or the Lebanese police, and many were later tried in military courts on charges of treason.

From 2000 until the onset of the Israel–Lebanon war in 2006, military activities between IDF and Hizbollah forces were restricted in their intensity and were characterized mostly by tit-for-tat exchanges with few casualties. Hizbollah's military tactics were primarily rocket and mortar attacks targeting Northern Israel, as well as cross-border raids and kidnappings of Israeli soldiers. The International Committee of the Red Cross (ICRC) was generally denied access to kidnapped soldiers, and the Hizbollah party leaders routinely refused to confirm the fate of the soldiers.[106] Israeli tactics consisted primarily of artillery fire and air strikes in Southern Lebanon. Israeli forces occasionally targeted large electrical and industrial infrastructure or waged strikes against Syrian targets (such as radar stations) inside Lebanon. In

---

[103] These attacks killed 241 Americans and fifty-eight French nationals.

[104] A US State Department report on terrorism issued in 2001 stated that Hizbollah had not attacked any US interests in Lebanon since the conclusion of the civil war ten years before.

[105] Although some SLA members (mostly low-level recruits who had been press-ganged into the SLA ranks or joined out of economic necessity) returned to Lebanon after Hizbollah promised them amnesty, about 2,400 remained in Israel, where they received compensation packages (an $8,800 minimum, with bonuses for number of years served) and citizenship. Blanford, "The Quandary of an SLA Amnesty."

[106] Exum, "Hizballah at War: A Military Assessment," 2. Also see Human Rights Watch, "Why They Died," accessed August 15, 2010, http://www.hrw.org/en/reports/2007/09/05/why-they-died.

addition, the IDF allegedly assassinated several high-profile militants and Islamist spiritual leaders.[107] During this period (and after the 2006 war) the IDF remained in control of the disputed territory known as the Shebaa Farms, which lies along the border between Lebanon and Syria's occupied Golan Heights. The year 2000 also marked the beginning of the second Palestinian *intifada* (uprising), which saw Israeli Prime Minister Ariel Sharon reassert control over occupied West Bank territory that the previous Israeli government had forfeited as part of the Oslo Accords.[108] Because the timing of the second *intifada* largely coincided with Israel's withdrawal from Lebanon, it is difficult to determine what Hizbollah's behavior might have been in the absence of what many party members considered a direct provocation. Hizbollah intensified its attacks (launching daily rockets against IDF targets) after Israel launched Operation Defensive Shield—its largest military incursion in the West Bank since 1967. The operation, in retaliation for suicide bombings that killed roughly 400 Israelis over the previous eighteen months (Israeli countermeasures produced roughly 1,200 Palestinian casualties during the same period), trapped PLO leader Yasser Arafat in his Ramallah compound and laid siege to much of the West Bank. Despite the intensification of artillery attacks, Hizbollah stopped short of a large-scale mobilization, which party Secretary Hassan Nasrallah claimed must be preserved for retaliation in the event that the Israeli government attempted to expel the Palestinians from the West Bank and Gaza.[109]

The 2006 Lebanon war, although nominally fought between Hizbollah and Israel, extended well beyond the traditional battleground of South Lebanon. During the conflict, Hizbollah used primarily conventional tactics, in the context of a war of attrition, presuming that Israeli society would not tolerate massive casualties.[110] This strategy necessitated the maintenance of a constant barrage of rocket fire (averaging 150–200 rockets per day) to reinforce the perception that the Israeli campaign was ineffective at weakening the group's offensive

---

[107] Important examples include Ghalib Awali, a Hizbollah military commander who was killed in a car bombing in 2004; Sheikh Ahmad Yassin (the quadriplegic Hamas founder and an important spiritual figure for many Hizbollah members) who was killed, along with his bodyguard and nine bystanders, by a missile fired from an Israeli gunship in 2004; and Mahmoud Al-Majzoub (leader of Palestinian Islamic Jihad), who was killed by a car bomb in Sidon, Lebanon, in 2005.

[108] In addition to Hizbollah's attacks on Israeli targets, several other Arab governments also ramped up hostilities, including oil embargoes and downgrading of diplomatic relations, in response to Israel's policies in the occupied Palestinian territories.

[109] Paul Wachter, "Hezbollah: Lebanon's Paper Tiger," *Salon*, April 10, 2002, http://www.salon.com/news/feature/2002/04/10/Lebanon.

[110] Kulick, "The Next War with Hizbollah: Strategic Assessment."

capabilities.[111] Hizbollah's arsenal consisted primarily of Katyusha rockets, known for their distinctive screeching noise. Because these rockets lacked a guidance system and could only be outfitted with relatively small warheads, they were most effective when launched in highly concentrated numbers.[112] Hizbollah launched approximately 4,000 rockets into Israel during the fighting in 2006, with roughly 25% of them landing in populated areas, killing forty-three civilians.[113] To facilitate these operations, Hizbollah ensured that each unit was self-sufficient and prestocked with adequate supplies, making any Israeli attack on supply routes or large weapons depots irrelevant.[114] However, this approach also resulted in the isolation of each unit, which, because of the ubiquitous Israeli air presence, prevented fighters from communicating with or supporting nearby units.[115] To provide fortified cover and clear lines of sight, Hizbollah located these bunkers and launch sites, whose size and complexity surprised even IDF intelligence, in and around villages (traditionally situated on hilltops in South Lebanon).[116] These underground stations, which launched both short- and medium-range rockets, used pneumatic lifts to bring launchers up from underground and were often so well camouflaged that they were able to function from behind IDF lines as ground troops advanced through the south. In areas where villages or population centers were sparse, Hizbollah constructed extensive fighting positions with large and sophisticated bunker systems that included electrical wiring and ample provisions, often very close to IDF and UNIFIL positions.[117] Hizbollah's concentration on rocket launching sites and underground bunkers, as opposed to infantry or mobile anti-tank capabilities, suggested that the group anticipated an Israeli response composed primarily of air strikes rather than ground

---

[111] Exum, "Hizballah at War: A Military Assessment."

[112] "Hezbollah's Rocket Force," *BBC News*, July 18, 2006, http://news.bbc.co.uk/2/hi/middle_east/5187974.stm. Subsequent press reports and internal IDF investigations demonstrate that most Israeli military planners dismissed the import of these rockets, and although few of the rockets produced any civilian casualties, many urban centers and infrastructure sites in Northern Israel were paralyzed during the thirty-four-day conflict. Scott Wilson, "Israeli War Plan Had No Exit Strategy," *Washington Post*, October 21, 2006. Although Katyushas have a maximum range of only twenty-five kilometers, Hizbollah most likely also possessed several longer-range models, including the Iranian-built Fajr-3, Fajr-5, and ZelZal-2, with ranges of forty-five, seventy-five, and one hundred to four hundred kilometers, respectively. Even though many of the group's long-range launchers were hit in the first few hours of the war, they were able to extend their reach into Northern Israel, reaching as far as Haifa.

[113] Uzi Rubin, "Hizballah's Rocket Campaign Against Northern Israel," *Jerusalem Issue Brief* 6, no. 10 (August 31, 2006).

[114] Exum, "Hizballah at War: A Military Assessment," 10.

[115] Ibid.

[116] Ibid., 3.

[117] Ibid., 3–4.

attacks.[118] Although the conventional Israeli plan for an attack in Southern Lebanon focused on a ground invasion force of four army divisions, the military's chief of staff (the first air force general to be in that position) instead emphasized a combination of air power and special forces troops, hoping that strikes on major infrastructure targets would erode Hizbollah's support among the Christian and Sunni population.[119] Because the sites were underground, however, Israeli pilots were unable to spot them from the air, and pilots lacked the ground-intelligence necessary to pinpoint the locations of the sites.[120]

METHODS OF RECRUITMENT

Because Hizbollah's identity was inextricably linked with Lebanon's communal politics, the group had a built-in advantage in terms of recruiting. It was by far the most legitimate organization and was able to draw on the loyalties of Lebanon's most populous sect. Within the organization, different units appealed to different segments of Lebanon's Shi'ite community. In terms of formal recruitment activities, the Syndicate Unit (within the Executive Council) aimed to increase the party's presence in professional syndicates, recruiting from among Lebanese professionals (university faculty, students, lawyers, doctors, etc.).[121] Similarly, the External Relations Unit (also within the Executive Council) provided outreach to other political parties, government institutions, and nongovernmental organizations, often sending official representatives to attend the local meetings of these groups.[122] Of course, recruitment activities were not restricted to Shi'ites, and in 1997, Hizbollah created the multi-confessional brigade, which attracted Lebanese youth from across the sectarian spectrum.[123] The group's extensive network of centers for military training and its provision of social services and even forums for

---

[118] Kulick, "The Next War with Hizbollah: Strategic Assessment."

[119] Wilson, "Israeli War Plan Had No Exit Strategy." A report commissioned by the US Air Force concluded that, although air power remains the most flexible means for targeting irregular armies, Israel's indiscriminate and excessive bombing of civilian infrastructure sites was counterproductive. William M. Arkin, *Divining Victory: Airpower in the 2006 Israel-Hezbollah War* (United States: Air University Press, 2007).

[120] Exum, "Hizballah at War: A Military Assessment," 4.

[121] Hamzeh, *In the Path of Hezbollah*, 62.

[122] Ibid.

[123] Alagha, *The Shifts in Hizbullah's Ideology*, 169.

networking and socializing all helped to bring potential members into the party's orbit.[124]

Hizbollah provided multiple outlets for action, including activities aimed at defending the south against potential Israeli invasion, efforts to reform the state's political institutions in order to provide the Shi'a population with a greater voice in government, programs to provide services to the poor, and religious and cultural education programs designed to foster cohesion among the group's target population. This variety provided would-be recruits with multiple opportunities to contribute based on their personal strengths and experiences. This flexibility was reflected in its large number of party supporters, estimated at 200,000, the largest of any single political entity in Lebanon.[125]

METHODS OF SUSTAINMENT

In addition to its deep recruiting pool, Hizbollah was also able to draw on a wide range of financial sources, including funding from Iran, sympathetic donors in other Arab countries, and the group's own extensive business interests (licit and illicit). The extent to which the group benefited financially from illegal trade (in drugs, diamonds, and other contraband) was unclear, but at least some connections were uncovered between the group and various criminal enterprises. The sizeable Lebanese expatriate community in South America made the region a hub for generating finance, and in 2008, a drug probe found that the Lebanese operator of an enormous cocaine money-laundering operation in Colombia donated a portion of his proceeds to Hizbollah.[126] Estimates of Iranian resources flowing to Hizbollah varied from tens of millions of dollars a year to $1 billion a year and included money; hardware; training provided for military and "resistance" activities; the services of Iranian engineers, doctors and other professionals; and financial services designed to help the group evade international sanctions. However, most of these funds came from private foundations and charitable organizations in Iran or from

---

[124] The phenomenon of temporary marriage, or *mutaa*, and the venues necessary for bringing together potential participants, has created a thriving space for Hizbollah to recruit from among Lebanon's young adults. The connections (both social and political) these young adults are able to make through these centers bring them closer to the group's orbit and help spread the group's resistance narrative. Hanin Ghaddar, "The Militarization of Sex: The Story of Hezbollah's Halal Hookups," *Foreign Policy*, November 25, 2009, http://www.foreignpolicy.com/articles/2009/11/25/the_militarization_of_sex?page=0,1.

[125] Hamzeh, *In the Path of Hezbollah*, 74.

[126] Chris Kraul and Sebastian Rotella, "Drug Probe Finds Hezbollah Link," *Los Angeles Times*, October 22, 2008, http://articles.latimes.com/2008/oct/22/world/fg-cocainering22.

Iran's Revolutionary Guard. Thus, intelligence services were largely outside the purview of the Iranian Ministry of Finance or the office of Iran's president.[127] Financing from Arab donors to Hizbollah was primarily in the form of tithes (the portion of income that believers are required to donate to charitable causes under Shi'ite religious law), as well as donations from individuals, groups, small businesses, and banks in the Arab world and among the Shi'ite international community.[128] Hizbollah also received significant income from the group's domestic business chains, which included supermarkets, gas stations, department stores, restaurants, construction companies, and travel agencies, as well as offshore companies, banks, and currency exchanges. Most of this money was held in Tehran banks in order to prevent seizure of the group's assets; these Iranian institutions also operated under the legal strictures of Islamic finance, which provided added religious legitimacy.[129] In addition, several Lebanese financial institutions acted as intermediaries between Hizbollah and mainstream banks, facilitated by extremely lax transparency and oversight of the financial sector.[130] During the 2006 war, Israel bombed as many as twelve banks, including two very large ones, Al Baraka and Fransabank, as well as the home of a bank manager.[131]

METHODS OF OBTAINING LEGITIMACY

Hizbollah's ability to adapt was a key component of its success. The organization was both a revolutionary group, fervently defending its right to maintain an armed militia and its dedicated goal of liberating Palestine, and a parliamentary player, increasing its level of participation in the formal political system and orchestrating electoral alliances with other parties. This dual nature was a critical component of the group's identity and legitimacy, illustrating its need to maintain armed resistance even amidst a strategy of political

---

[127] Hamzeh, *In the Path of Hezbollah*, 63.

[128] Ibid., 64.

[129] Islamic law forbids Muslims from earning interest (the proceeds of idle capital), which is considered usury by religious scholars. However, in practice, these banks often function similarly to non-*shari'a*-compliant banks, merely using some intermediary asset to provide a degree of separation so that the interest payment does not pass directly from the borrower to the lender. Not surprisingly, Islamic jurists in the Gulf countries, where finance is most plentiful and there is the most potential for earning interest, have introduced or approved many of these creative instruments. Hamzeh, *In the Path of Hezbollah*, 64.

[130] These include *Bayt Al-Mal* (House of Money) and the Yousser Company for Finance and Investment of Lebanon.

[131] Adam Ciralsky and Lisa Myers, "Hezbollah Banks Under Attack in Lebanon," *NBC News*, July 25, 2006, http://www.msnbc.msn.com/id/14015377/.

accommodation.[132] In the most literal sense, Israel's presence in the disputed Shebaa Farms territory supported Hizbollah's claim that it must retain its arms for purposes of "national resistance." When Israel withdrew to the security zone in South Lebanon, the Taif Accords recognized Hizbollah's right to maintain a militia because Israel was still occupying Lebanese territory; when Israel withdrew in 2000, Hizbollah equated the IDF presence in the Shebaa Farms with its earlier presence in the "security zone," allowing Hizbollah to cling to its earlier justification. The group then framed its military arsenal as a crucial factor in deterring further Israeli aggression, characterizing the group's strength as a benefit in which all Lebanese could share equally.[133] This was an especially effective justification because it suggested that Hizbollah would not only restrict itself to defensive operations but that it would also aim to alleviate Christian, Sunni, and Druze fears that were exacerbated by the factional violence that broke out during the previous election cycle.[134]

A considerable amount of Hizbollah's historic legitimacy stemmed from its positions on issues that largely contradicted the words and/ or deeds of the region's other political leaders, especially on issues of normalizing relations with Israel, extending aid to the Palestinians, adhering to espoused religious principles, promoting regional unity, and relying on Western actors and institutions for political and economic assistance. Moreover, Hizbollah's popularity was propelled by its success against Israel in the 2006 war. The group's ability to repel Israeli ground forces while still launching a constant barrage of missiles into Northern Israel demonstrated that a non-state group of highly committed fighters could achieve more on the battlefield than the region's governments had achieved (with the full political and economic backing of the United States) in decades.[135]

---

[132] Mohammed Ayoob, *The Many Faces of Political Islam: Religion and Politics in the Muslim World* (Ann Arbor, MI: University of Michigan Press, 2009), 123.

[133] At an August 14 rally attended by members of most Lebanese political factions and marking the three-year anniversary of the 2006 war, Nasrallah stated (in a speech broadcast on large television screens), "You might ask, 'Do we have the power to prevent a war?' I will reply, 'Yes, there is a very real possibility that, if we cooperate with one another as Lebanese, we will be able to prevent Israel from launching a war against Lebanon.' I stress to you that there will be surprises in any new war with Israel, God willing. By saying this to the Israelis, we can deter and prevent them. Let them think a million times before waging a war on Lebanon. Let them look for other ways to confront us, but not war." Cited in Mohamad Bazzi, "Lebanon's Shadow Government: How Hezbollah Wins by Losing," *Foreign Affairs* (September 11, 2009).

[134] Bazzi, "Lebanon's Shadow Government: How Hizbullah Wins by Losing."

[135] Neil MacFarquhar, "Tide of Arab Opinion Turns to Support for Hezbollah," *New York Times* (July 28, 2006), http://www.nytimes.com/2006/07/28/world/middleeast/28arabs.html.

EXTERNAL SUPPORT

Like many territories in the developing world, Lebanon became a battlefield for both regional power struggles and rich-country rivalries. In addition to France, the United States, Israel, and Syria, all of which intervened directly to protect their own geostrategic interests, there was also a significant amount of indirect activity aimed at shifting the dynamics of the conflict. Hizbollah owed its very existence to Iran's efforts to export its revolutionary ideology and, by association, to Iraq's invasion of Iran and the ensuing war, which stifled Tehran's grander plans of expansion and forced it to rely more fully on Hizbollah as its primary agent.[136] Iran sent 1,500 Revolutionary Guard members to train Shi'ite fighters during the Israeli invasion of 1982, and although Hizbollah was very capable and well trained in the use of small arms, the Revolutionary Guard trained Hizbollah fighters on more sophisticated weapons systems, such as medium-range rockets and anti-tank missiles, and provided the party with tens of millions of dollars in money and equipment annually.[137] Reportedly, Iran heavily influenced Hizbollah's decision to maintain its armed wing after the 2000 withdrawal of Israel, rather than turning its full attention toward increasing power in the Lebanese domestic political system.[138] Hizbollah's leadership made no secret of these associations, instead justifying them based on numerous rationales, including the argument that by consulting with other regional players, the group's own decisions were more comprehensive and effective.[139] By the time of the 2006 war, however, it became clear that Hizbollah had gained significant financial and operational independence from Tehran.[140]

Whereas Iranian support was concrete and highly visible, Syrian support to Hizbollah was less so and consisted primarily of logistical

---

[136] Fouad Ajami, "Lebanon and its Inheritors."

[137] Exum, "Hizballah at War: A Military Assessment," 5. However, financing patterns often depend on the composition of Iran's leadership; support was cut by as much as 70% under previous reformist administrations in Iran (Presidents Rafsanjani and Khatami), although it was never fully removed. Hamzeh, *In the Path of Hezbollah*, 63.

[138] Shatz, "In Search of Hizbullah."

[139] Alagha, *The Shifts in Hizbullah's Ideology*, 172.

[140] Although Iranian finance certainly facilitates Hizbollah's social outreach activities, Shi'ite businesspeople in Lebanon, South America, West Africa, and the United States provide an increasing proportion of funding. Observers in Lebanon during the 2006 war also reported that IDF sources saw no indication that any of the group's activities stemmed from direction provided by either Iran or Syria. Graham Fuller, "The Hizballah-Iran Connection: Model for Sunni Resistance," *The Washington Quarterly* (Winter 2006–2007), 142.

and symbolic support.[141] Furthermore, while Iran remained an unwavering supporter of Hizbollah, Syrian assistance varied according to the status of its peace talks with Israel. As these talks continued to stagnate, and as Hizbollah's influence in Lebanese politics continued to expand, Syrian support became less important.[142]

COUNTERMEASURES TAKEN BY THE GOVERNMENT

Numerous state entities, including Israel, the United States, and successive Lebanese governments but also other regional players such as Saudi Arabia, engaged in activities aimed to weaken Hizbollah militarily, economically, and politically. Israel's activities were the most extensive and varied and included targeted assassinations of party members (Abbas Musawi and Imad Mughniyeh), the recruitment of domestic spy rings, conventional military attacks (especially the 1993 "Operation Accountability," the 1996 "Grapes of Wrath" campaign, and the 2006 War), coordinated efforts to delegitimize the group in international forums and among the broader Lebanese public (reflected in IDF 2006 aerial campaign targeting public infrastructure),[143] calibrated displays of Israeli military presence (air force flyovers and sonic booms), and the withholding of maps detailing unexploded ordnance (mines and cluster bombs) from previous conflicts.[144] Israel (and the United States) also made efforts to restrict the flow of weapons into Iran and Syria, hoping to cut off Hizbollah's major supply lines, in addition to stepping up efforts to cripple Hizbollah's access to funds by banning many of the charitable

---

[141] The central component of this support has been Syria's assistance with transferring Russian-made (and Iranian-purchased) weapons to Hizbollah. See Exum, "Hizballah at War: A Military Assessment," 5.

[142] Current support is largely restricted to logistical and organizational support facilitating the movement of weapons and supplies from Iran through Syrian-controlled territory and into Lebanon. In fact, it may be that the Syrian leadership now has more to gain from its association with Hizbollah than vice versa.

[143] This was most visible in the IDF's attacks on regions in Lebanon where the population was strongly anti-Hizbollah (such as Christian Achrafiyya, Amsheet and Jounieh); Israeli Prime Minister Ehud Olmert echoed this strategy of collective punishment, "This morning's events [kidnapping of Israeli soldiers] were not a terrorist attack, but the action of a sovereign state that attacked Israel for no reason and without provocation . . . Lebanon is responsible and Lebanon will bear the consequences of its actions." See Israel Ministry of Foreign Affairs, "PM Olmert: Lebanon Is Responsible and will Bear the Consequences," accessed August 15, 2010, http://www.mfa.gov.il/MFA/Government/Communiques/2006/PM+Olmert+-+Lebanon+is+responsible+and+will+bear+the+conseque nces+12-Jul-2006.htm.

[144] Barak Ravid, "Lebanon to UN: Israel Breached Truce Deal Hundreds of Times," *Haaretz*, November 1, 2007, http://www.haaretz.com/news/lebanon-to-un-israel-breached-truce-deal-hundreds-of-times-1.232334.

organizations that transferred funds to Hizbollah.[145] The United States also placed Hizbollah's television station *Al-Manar* on the Terrorism Exclusion List, meaning that individuals working with or providing financing for the station could be deported, and blocked its satellite signal. However, *Al-Manar* feeds were widely available online.[146]

The actions of successive Lebanese governments were largely inconsistent in terms of responding to Hizbollah, both because the parties that made up these governments had disparate interests and because countering Hizbollah's militia capacity required quite different tactics than countering its growing popularity. As the organization strengthened, Hizbollah's claim as the legitimate representative of the Shi'a population and other communal groups within the governing structure made political alliance and accommodation, rather than confrontation, the cornerstone of their electoral strategies.

SHORT- AND LONG-TERM EFFECTS

CHANGES IN THE ENVIRONMENT

The impact of the long civil war on Lebanese society has been extremely pernicious. Centuries-old struggles over socioeconomic issues and the division of concrete material goods and political offices transformed over time into a battle over sectarian identity—an existential fight in which compromise was complicated by the indivisibility of communal membership.[147] Both proximate (Syria, Israel, Iran, and Egypt) and distant (the United States and France) political entities amplified these divisions in pursuit of their own geopolitical aspirations, thus increasing the intractability of the conflict.[148] Repeat foreign interventions did little to resolve the underlying conditions that contributed to renewed conflict, instead opting for incremental alterations of the status quo ante. Rather, the ever-present specter of renewed conflict produced an environment where powerful actors relentlessly pursued short-term economic

---

[145] This was partially successful in at least one case. When Jihad Al-Bina ("The Struggle to Build") was blacklisted in the United States, making it illegal for any US companies or banks to do business with the organization, Hizbollah had to create a new organization Al-Wadaa ("The Promise") to carry out much of the post-2006 war reconstruction effort in the south.

[146] Jeremy M. Sharp, et al., "Lebanon: The Israel-Hamas-Hezbollah Conflict," Congressional Research Service, September 15, 2006, http://www.fas.org/sgp/crs/mideast/RL33566.pdf, 23.

[147] Samir Khalaf, *Civil and Uncivil Violence in Lebanon: A History of the Internationalization of Communal Conflict* (New York Columbia University Press, 2004).

[148] Ibid.

and political gains in order to consume immediately (what they may rightly perceive as) their fleeting benefits.[149] The sanctity of property that did not "belong" to a particular community (public goods, such as the environment or infrastructure) and rules that did not emanate from that community (national laws) were violated with impunity. These factors complicated the redressing of grievances (such as the pursuit of legal damages resulting from unlawful activity) and exacerbated resentment between groups. Decades of state policy promoting systematic discrimination on the basis of sectarian identity also meant that issues ordinarily defined by class identity or political ideology (which cut across ethno-religious lines in other states) were fundamentally communal battles.

Endemic political and economic uncertainty also increased the currency of kinship and religious ties, eroding opportunities for cross-communal interaction. This intragroup isolation facilitated both individual and group violence, as any particular individual was identified first and foremost by his or her sectarian association. Years of fighting also contributed to spatial segregation, yielding neighborhoods, public spaces (including universities), and entire regions where the opportunities for intercommunal interaction were minimal. These practical manifestations of the severe power imbalance in Lebanese society ensured Hizbollah, with its vast social service apparatus and highly visible rural presence, a continued place in Lebanese politics.

Hizbollah's "resistance" activities also provided the group with its other major source of resilience, and despite Israel's withdrawal from the security zone in South Lebanon in 2000, the Hizbollah narrative only grew more powerful. The outbreak of the second *intifada* in Palestine, the 2006 War in Lebanon, and the 2008–2009 Gaza War have nearly mooted the issue of disarming Hizbollah, even though the Security Council Resolution (1701) that ended the war called for the disarmament of all militias in Lebanon. However, even politicians in the anti-Hizbollah camp publicly recognized the group's right to maintain its militia, and Hizbollah has been integrated into the system of realpolitik that has defined Lebanon for decades.

CHANGES IN GOVERNMENT

Although Hizbollah has continued to gain power within the Lebanese state, both electorally and in non-state institutions, many factors combine to prevent the group from gaining full control over

---

[149] Ibid.

the government or a monopoly over the state's coercive apparatus (military and police). First, despite the fact that more Lebanese cast votes for Hizbollah's March 8 coalition in both of the elections prior to 2010, the group shares power with other powerful coalition members who do not necessarily share Hizbollah's political, economic, or religious agendas. Hizbollah's Christian ally Michel Aoun, leader of the Free Patriotic Movement, failed to fully split the vote of his co-religionists in the most recent election, and many Christians remained with the March 14 alliance.[150] This failure prevented Hizbollah from expanding its previous electoral gains, as the support of the Shi'a population alone could not propel Hizbollah to a parliamentary majority. A second factor preventing Hizbollah from gaining full control of the government are the electoral rules regarding the allotment of both parliamentary seats and cabinet positions. The result of these rules has been an intensely divided government often paralyzed by gridlock. This division came to a head in 2006, when Hizbollah and its allies left their cabinet posts, demanding a veto over government decisions. The Lebanese government was therefore temporarily deadlocked, as the number of members of parliament in attendance was insufficient to conduct most government business.[151] Observers likened the stalemate to the condition prevailing near the end of the fifteen-year civil war, which saw the formation of two rival governments, each claiming to be the legitimate representative of the Lebanese state.[152]

Despite some political gains in 2009 by the political coalition opposed to Hizbollah, many important factors, such as incumbent advantage, remained entrenched in the system. Indeed, many of the political elites continued to hail from the same dynasties that have dominated Lebanese politics for decades, especially within the minority Christian and Druze communities.[153] This reality not only prevented the measured, gradual deconfessionalization of Lebanese politics, but it also exacerbated bureaucratic inertia, as agencies loyal to particular families or incumbent politicians attempted to stifle the activities of one another. This favoritism also drove political leaders to weaken any institution, such as the Constitutional Court, or

---

[150] "March 14 Bloc Wins Lebanon Election," *Al Jazeera,* June 8, 2009, http://english.aljazeera.net/news/middleeast/2009/06/20096813424442589.html.

[151] Hugh Macleod, "Hezbollah Recruits Thousands in Lebanon Crisis," *Telegraph,* November 25, 2007, http://www.telegraph.co.uk/news/worldnews/1570478/Hezbollah-recruits-thousands-in-Lebanon-crisis.html.

[152] Ibid.

[153] "March 14 Bloc Wins Lebanon Election."

governing mechanism that sought to operate outside the control of any single faction.[154]

CHANGES IN POLICY

The most noticeable changes in policy in Lebanon have been increasing accommodations toward Hizbollah, both from Lebanon's political actors and from foreign governments. Hizbollah's primary adversary in Lebanon, the March 14 Alliance, has taken an increasingly pragmatic approach, including an August 2008 statement issued by Lebanon's cabinet declaring "the right of Lebanon's people, army, and resistance [i.e., Hizbollah] to liberate the Israeli-occupied Shebaa Farms, Kafar Shuba Hills, and the Lebanese section of Ghajar village, and defend the country using all legal and possible means,"[155] which was followed by the formal recognition of Hizbollah's right to maintain a militia (in December 2009) and formal rejections of Israeli and US claims that Syria transferred Scud missiles to Hizbollah (in April 2010). Michel Aoun's Free Patriotic Movement, which is allied with Hizbollah in the March 8 coalition, has also reframed its language regarding "foreign opposition," replacing perennial references to Damascus's stranglehold over Lebanese politics with references to the United States and Saudi Arabia. Although Israel and the United States have increased sanctions against Hizbollah in the aftermath of the 2006 conflict, and amid growing concerns over Iran's influence in the region, Hizbollah is no longer designated as a terrorist group by most Arab or European governments.[156] Even the Gulf monarchies, which are always suspicious of popular movements that could incite their own significant Shi'ite populations to violence, have been more practical in their relations with Hizbollah.

---

[154] "Lebanon's Elections: Avoiding a New Cycle of Confrontation," International Crisis Group, June 4, 2009, http://www.crisisgroup.org/en/regions/middle-east-north-africa/iraq-syria-lebanon/lebanon/087-lebanons-elections-avoiding-a-new-cycle-of-confrontation.aspx.

[155] Nafez Qawas, "Berri Summons Parliament to vote on Policy Statement," *Daily Star* (Beirut). Of course some political leaders in the March 8 alliance either insisted this reference did not pertain to Hizbollah, or claimed that it applied equally to all of the state's sectarian militias. However, the weakness of both these militias and Lebanon's official army left little doubt as to which actor the statement was referring.

[156] Australia and the United Kingdom distinguish between Hizbollah's political apparatus and its military wing, labeling only the latter as a "terrorist organization," while European Union policy requires that all members agree before a group can be placed on the terrorist list. This has proven to be a sore spot for transatlantic relations, and both the US government and private actors continue to pressure Europe's governments to add Hizbollah to the list.

CHANGES IN THE REVOLUTIONARY MOVEMENT

During the course of its insurgency, Hizbollah has periodically oscillated between religious commitment and political pragmatism. In 1989, the ratio of clerics to laity in the Shura Council was six clerics to one layperson; after the Taif Accords, efforts to be more representative of the population brought in more laity, with four clerics and three laypeople. However, by the time of the 2001 elections, there was a shift back to the pre-1989 focus on clerics in an effort to isolate some party elements who wanted to relax the group's religious focus and concentrate on political goals.[157] These changes represented a calculated balancing of Hizbollah's political aspirations with its religious agenda, as well as the selective use of resistance versus engagement in Lebanon's political system.[158]

On the military front, the 2006 war was viewed as a huge success for Hizbollah, but the presence of 10,000 Lebanese soldiers and 12,000 UN soldiers in the areas bordering Israel has complicated Hizbollah efforts to rearm (rebuild bunkers, rocket launchers, etc.).[159] Hizbollah appears to be focusing on rebuilding its capabilities closer to civilian centers in order to avoid having to pass through open space and risk being seen. It is also focusing on upgrading its long-range rockets to deter Israel from targeting Beirut in the future because previous IDF attacks on Beirut damaged the group's credibility with the non-Shi'a population in Lebanon and also damaged some of organization's logistical capabilities. In 2009, it was estimated that Hizbollah had between 40,000 and 80,000 rockets in its arsenal.[160] In addition, Hizbollah has built up its anti-tank capabilities by ordering more anti-tank missiles, such as the ones suspected of piercing a Merkava tank in 2006.[161] Despite these successes and advances, Hizbollah must adapt its strategy to deal with the reality that Israel is no longer in Lebanon, and as such, Hizbollah must decide which path to take: continue to fight Israel (not in the context of a Lebanese occupation, but in the context of its occupation of Palestine) or focus on domestic Lebanese politics.

---

[157] Hamzeh, *In the Path of Hezbollah*, 45.
[158] Shatz, "In Search of Hizbullah."
[159] Kulick, "The Next War with Hizbollah: Strategic Assessment."
[160] Bazzi, "Lebanon's Shadow Government: How Hizbullah Wins by Losing."
[161] Kulick, "The Next War with Hizbollah: Strategic Assessment."

BIBLIOGRAPHY

Abu-Khalil, As'ad. "Ideology and Practice of Hezbollah in Lebanon: Islamization of Leninist Organizational Principles." *Middle Eastern Studies* 27, no. 3 (July 1991).

———. "The Incoherence of Islamic Fundamentalism: Arab Islamic Thought at the End of the Twentieth Century." *Middle East Journal* 48, no. 4 (Autumn 1994).

Ajami, Fouad. "Lebanon and its Inheritors." *Foreign Affairs* 63, no. 4 (Spring 1985): 778–799.

Alagha, Joseph Elie. *The Shifts in Hizbullah's Ideology: Religious Ideology, Political Ideology and Political Program.* Amsterdam: Amsterdam University Press, 2006.

Al-Rizzo, Hasan M. "The Undeclared Cyberspace War between Hezbollah and Israel." *Contemporary Arab Affairs* 1, no. 3 (2008).

Arce, Daniel G., and Sandler, Todd. "Terrorist Spectaculars: Backlash Attacks and the Focus of Intelligence." *Journal of Conflict Resolution* 54, no. 2 (2010): 356.

Arkin, William M. *Divining Victory: Airpower in the 2006 Israel-Hezbollah War.* United States Air Force: Air University Press, 2007. http://www.nytimes.com/2007/10/14/world/middleeast/14airwar.html.

Ayoob, Mohammed. *The Many Faces of Political Islam: Religion and Politics in the Muslim World.* Ann Arbor, MI: University of Michigan Press, 2009.

Bazzi, Mohamad. "Lebanon's Shadow Government: How Hezbollah Wins by Losing." *Foreign Affairs* (2009).

Blanford, Nicholas. "The Quandary of an SLA Amnesty." *The Daily Star* (Beirut), August 16, 2005.

Byman, Daniel. "Do Targeted Killings Work?" *Foreign Affairs* 85, no. 2 (2006).

Ciralsky, Adam, and Myers, Lisa. "Hezbollah Banks Under Attack in Lebanon." *NBC News,* July 25, 2006. http://www.msnbc.msn.com/id/14015377/.

Deeb, Laura. "Hizballah: A Primer." *Middle East Report* (July 31, 2006), http://www.merip.org/mero/mero073106.html.

Exum, Andrew. "Hizballah at War: A Military Assessment." *Policy Focus* no. 63 (The Washington Institute, December 2006), www.washingtoninstitute.org/pubPDFs/PolicyFocus63.pdf

Fuller, Graham. "The Hizballah-Iran Connection: Model for Sunni Resistance." *The Washington Quarterly* (Winter 2006–2007): 142.

Ghaddar, Hanin. "The Militarization of Sex: The Story of Hizbullah's Halal Hookups." *Foreign Policy*, November 25, 2009. http://www.foreignpolicy.com/articles/2009/11/25/the_militarization_of_sex?page=0,1.

Haddad, Simon. *The Palestinian Impasse in Lebanon: The Politics of Refugee Integration.* Portland, OR: Sussex Academy Press, 2003.

———. "The Political Transformation of the Maronites of Lebanon: From Dominance to Accommodation." *Nationalism and Ethnic Politics* 8, no. 2 (Summer 2002): 27–50.

Harris, William W. *Faces of Lebanon: Sects, Wars, and Global Extensions.* Princeton, NJ: Markus Wiener Publishers, 1996.

Hamzeh, Ahmad Nizar. *In the Path of Hizbullah.* Syracuse, NY: Syracuse University Press, 2004.

"Hezbollah's Rocket Force," *BBC News*, July 18, 2006. http://news.bbc.co.uk/2/hi/middle_east/5187974.stm.

Karagiannis, Emmanuel. "Hezbollah as a Social Movement Organization: A Framing Approach." *Mediterranean Politics Journal* 14, no. 3 (2009): 365–383.

Keegan, John. "Shedding Light on Lebanon." *The Atlantic*, April 1984.

Khalaf, Samir. *Civil and Uncivil Violence in Lebanon: A History of the Internationalization of Communal Conflict.* New York: Columbia University Press, 2004.

Kraul, Chris, and Sebastian Rotella. "Drug Probe Finds Hezbollah Link." *Los Angeles Times*, October 22, 2008. http://articles.latimes.com/2008/oct/22/world/fg-cocainering22.

Kulick, Amir. "The Next War with Hizbollah: Strategic Assessment." Tel Aviv: The Institute for National Security Studies 10, December 2007.

"Lebanon: Hezbollah's Communication Network." *Stratfor* (May 9, 2008), http://intellibriefs.blogspot.com/2008/05/lebanon-hezbollahs-communication.html.

MacFarquhar, Neil. "Tide of Arab Opinion Turns to Support for Hizbullah." *New York Times*, July 28, 2006. http://www.nytimes.com/2006/07/28/world/middleeast/28arabs.html.

Macleod, Hugh. "Hezbollah Recruits Thousands in Lebanon Crisis," *Telegraph*, November 25, 2007. http://www.telegraph.co.uk/news/worldnews/1570478/Hezbollah-recruits-thousands-in-Lebanon-crisis.html.

Norton, Augustus R. *Hezbollah: A Short History.* Princeton, NJ: Princeton University Press, 2007.

Pape, Robert. *Dying to Win: The Strategic Logic of Suicide Terrorism.* New York: Random House, 2005.

Ravid, Barak. "Lebanon to UN: Israel Breached Truce Deal Hundreds of Times," *Haaretz*, November 1, 2007. http://www.haaretz.com/news/lebanon-to-un-israel-breached-truce-deal-hundreds-of-times-1.232334.

Rubin, Uzi. "Hizballah's Rocket Campaign Against Northern Israel." *Jerusalem Issue Brief* 6, no. 10 (August 31, 2006), http://www.jcpa.org/brief/brief006-10.htm.

Saad-Ghorayeb, Amal. *Hizbullah: Politics and Religion*. Sterling, VA.: Pluto Press, 2002.

Sharp, Jeremy M., et al., "Lebanon: The Israel-Hamas-Hezbollah Conflict," Congressional Research Service, September 15, 2006, http://www.fas.org/sgp/crs/mideast/RL33566.pdf.

Shatz, Adam. "In Search of Hizbullah." *New York Review of Books* 51 (April 29, 2004).

United Nations Relief and Works Agency for Palestine Refugees in the Near East (UNRWA). "Lebanon Camp Profiles." http://www.unrwa.org/etemplate.php?id=73.

Wachter, Paul. "Hizbullah: Lebanon's Paper Tiger." *Salon*, April 10, 2002. http://www.salon.com/news/feature/2002/04/10/Lebanon.

Wilson, Scott. "Israeli War Plan Had No Exit Strategy." *Washington Post*, October 21, 2006. http://www.washingtonpost.com/wp-dyn/content/article/2006/10/20/AR2006102001688_3.html.

Winslow, Charles. *Lebanon: War and Politics in a Fragmented Society*. London: Routledge, 1996.

HIZBUL MUJAHIDEEN

Maegen Nix and Summer Newton

SYNOPSIS

The Hizbul Mujahideen insurgency emerged in Kashmir in 1989 as the armed wing of the political party Jamaat-e-Islami (JEI) with the objective of attaining unification of Kashmir with Pakistan. Using tactics similar to those employed by the Afghan Mujahideen against the Soviets, the Hizbul Mujahideen tried to elevate India's economic, military, and political costs for retaining control of Kashmir until these costs became prohibitive to India. The Hizbul Mujahideen conducted kidnappings, assassinations, bombings, and hit-and-run tactics against government and military officials in the Kashmir Valley, and the group also expanded its area of operations beyond the Valley in order to overtax the Indian army. Aided by the sponsorship of the Pakistani Inter-Services Intelligence (ISI), the Hizbul Mujahideen replaced the Jammu Kashmir Liberation Front (JKLF), which sought Kashmir's complete independence from both India and Pakistan, as the dominant militant group in the region. By 1995, however, Wahhabi organizations also entered into the insurgency through the support of Pakistan, and these organizations began to achieve greater military success through more aggressive operations. These Wahhabi groups also alienated the population because of rising civilian casualties and the escalation in India's response to the groups' heavy-handed attacks. Friction between Hizbul Mujahideen commanders in the field and their political leadership in Pakistan, as well as the impression that the Hizbul Mujahideen was becoming a pawn in the Indo-Pakistani struggle over Kashmir, eventually weakened the resilience of the organization and decreased its external support. The movement was further fractured by disagreements over opportunities for peace negotiations with India, leading to the expulsion and assassination of key leaders within the Hizbul Mujahideen. Despite these numerous setbacks, the Hizbul Mujahideen movement remained an active insurgency and, in 2008, reiterated its call for jihad against India.

TIMELINE

August 1947	Indian and Pakistani independence granted from Britain.
October 1947	Kashmir formally accedes to the new Indian Union.
November 1947	Outbreak of the 1947 Kashmir War between India and Pakistan.
January 1949	Cease-fire signed between India and Pakistan. The United Nations Commission for India and Pakistan (UNCIP) issues a resolution stating that the fate of Jammu and Kashmir will be decided by a free and impartial plebiscite.
July 1952	Sheikh Abdullah signs the Delhi Agreement that gives Kashmir autonomy within India.
1962	India war with China over disputed territory in Ladakh region.
1965–1966	War between India and Pakistan.
1972	Simla Agreement formally establishes the Line of Control.
1987	In a purportedly rigged election, Farooq Abdullah wins as head of the National Conference-Congress Party alliance over Syed Salahuddin.
September 1989	Hizbul Mujahideen emerges as militant arm of JEI.
1990–1991	New Indian legislation grants judicial protection to security forces in Kashmir, as well as the right to detain suspected insurgents.
1990–2001	Successful Hizbul Mujahideen insurgency campaign throughout Kashmir, with thousands of Kashmiris receiving training and equipment across the border in Pakistan. Increased sponsorship from the ISI and an inflow of more extreme Islamic militant groups.
July 2000	Abdul Majid Dar enters into cease-fire negotiations with India and is removed from command.
March 23, 2003	Abdul Majid Dar killed by gunman at his home. Syed Salahuddin continues as the head of both the Hizbul Mujahideen and the United Jihad Council.

THE ENVIRONMENT OF THE REVOLUTION

PHYSICAL ENVIRONMENT

Kashmir generally refers to a northwestern state in India, officially referred to as Jammu and Kashmir, or Indian-controlled Jammu and Kashmir (IJK). Historically, the term Kashmir denoted a geographical region extending to the Pakistani-administered Azad Kashmir[1] and Gilgit-Baltistan (or the Northern Areas). Kashmir is the site of the world's largest and most militarized territorial dispute, with portions under the de facto administrative control of China (Aksai Chin), India (Jammu and Kashmir), and Pakistan (Azad Kashmir and Northern Areas). IJK consists of three regions, the Kashmir Valley (or the Valley), Jammu, and Ladakh, covering a total area of 100,948 square kilometers. The Line of Control, or LoC, is a military control line that separates Azad Kashmir from the IJK.[2] Lying between Azad Kashmir, south Jammu, and the Valley is the Pir Panjal mountain range, which, despite its ruggedness, is home to a migratory population. Traffic between Azad Kashmir and IJK is through the mountain passes of the Pir Panjal.[3]

The capital of Kashmir during the summer months is in Srinagar, located in the Valley, but moves to Jammu City, located in Jammu, during the winter months. Ironically, although Kashmir has heavily influenced Indian and Pakistani foreign policy for more than four decades, IJK comprises only 6.7% of the landmass of the overall country of Indian.[4] The Valley, sitting astride the upper Jhelum River, is the smallest region at 15,948 square kilometers. It is enclosed within two mountain ranges, the Himalayas and the Pir Panjal. Primarily arid and mountainous, most of the arable land capable of supporting intensive agriculture in IJK is within the Valley. The Valley's fabled beauty made it a tourist hotspot in Central Asia until the onset of the conflict.[5] Jammu, the second-largest region at 26,000 square kilometers, is primarily hilly and has a number of dense

---

[1] "Azad" here means "free."

[2] There is also a "Line of Actual Control," which separates the IJK from the Chinese-controlled territory of Aksai Chin.

[3] Kashmir Study Group website, accessed April 12, 2010, http://kashmirstudygroup.com/.

[4] Reeta Chowdhari Tremblay, "Nation, Identity, and the Intervening Role of the State: A Study of the Secessionist Movement in Kashmir," *Pacific Affairs* 69, no. 1 (1996–1997): 471–497.

[5] Terestia C. Schaffer, *Kashmir: The Economics of Peace Building* (Washington, DC: Center for Strategic & International Studies Press, 2005), 2.

forests, moderately arable valleys, and narrow strips of land capable of supporting intensive agriculture.[6] The topography of Ladakh, the largest of the three regions, covering nearly half the state at 59,000 square kilometers, includes some of the world's highest mountains, the Himalayas in the south and the Kunlun range in the north, and plateaus. The arid region makes life difficult for most of the inhabitants. The region supports only small-scale agriculture along with some sheep, goat, and yak herding.[7]

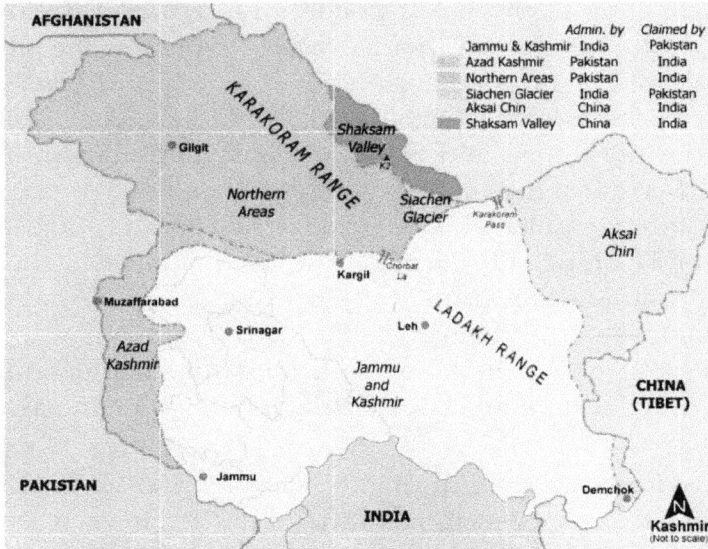

Figure 1. Map of Kashmir.[8]

CULTURAL AND DEMOGRAPHIC ENVIRONMENT

IJK is a diverse state, housing numerous ethnicities, languages, and religions. Potential cleavages across ethnicities, religions, and regions do not necessarily overlap. As of 1981, the population of Kashmir was 5.9 million, with only one-fifth residing in urban areas.[9] The Valley, undoubtedly because of its favorable geographical features, was the most densely populated region in IJK at 3.1 million. Jammu followed a close second at 2.7 million. Rugged Ladakh, the least populated of the three regions, had a sparse 134,000 inhabitants. Before the conflict,

---

[6] Ibid.

[7] Ibid.

[8] "File:Kashmir 2007.JPG." *Wikipedia*, http://en.wikipedia.org/wiki/File:Kashmir_2007.JPG.

[9] India was not able to conduct the 1991 census in Kashmir because of the conflict. Statistics on urban population are accessible at the official website of the Jammu and Kashmir Government, India, accessed on April 12, 2010, http://jammukashmir.nic.in/profile/facts.htm.

the Valley, a predominantly Muslim, Kashmiri-speaking region, was 95% Muslim (mostly Sunni) and 4% Hindu, or Kashmiri Pandit. After the insurgency began and many Hindus were killed by militants, the Pandit population left *en masse*, settling in refugee camps in Jammu and Ladakh.[10] Jammu, a predominantly Hindu, Dogri-speaking region, was 66% Hindu and 30% Muslim (mostly Sunni). Ladakh, bordering Tibet, was 50% Buddhist and 46% Muslim (mostly Shi'a). Before the accession of the princely state of Jammu and Kashmir to India in 1947, the minority Dogra Hindus ruled the state. After Jammu and Kashmir acceded, Sheikh Abdullah's National Conference, a mass-based party concentrated in the Valley, took control of the government for many of the succeeding decades. The shifting center of power, including the changing role of the Indian government in Kashmir affairs, sparked an amalgam of inter-regional, inter-ethnolinguistic, and later, inter-religious rivalries.[11]

The identity of ethnic Kashmiris is underscored by nearly 5,000 years of history. Consolidated in the culturally rich, religiously diverse, and isolated Valley, Kashmiriyat identity is based on ethnolinguistic similarities among the Kashmiri-speaking people of the Valley and its connected regions, to include both Hindus and Muslims. Indeed, the Islamic traditions of the Valley, influenced heavily by tolerant Sufism, are markedly autonomous from larger Islamic traditions and, in some cases, overlap with Hinduism, sharing shrines, saints,[12] and other religious practices. Kashmiriyat "suffocated," in part, when the Hindu minority fled the Valley during the early years of the insurgency after assassinations of Pandit officials.[13] Despite the Kashmiriyat demands (both Muslim and Hindu) for an independent Kashmir, the vision of sovereignty advocated by several Islamic groups in Kashmir made it difficult for many Hindus to support these separatist efforts, and they instead increasingly turned to India and the Hindu nationalist Bharatiya Janata Party (BJP) to realize their political goals. In turn, strong regional patriotism made accession to the monolithic, orthodox Muslim Pakistan undesirable among proindependence factions.[14]

---

[10] For an account of the flight of the Pandits, see Victoria Schofield, *Kashmir in Conflict: India, Pakistan and the Unending War* (New York: I. B. Tauris, 2003), 151–152.

[11] Reeta Chowdhari Tremblay, "Kashmir: The Valley's Political Dynamics," *Contemporary South Asia* 4, no. 1 (1995): 79.

[12] The term "saint" is often considered a Christian term but it is also used within Islam, Hinduism, and other religions to denote people of high religious stature.

[13] Balraj Puri, "Kashmiriyat: The Vitality of Kashmiri Identity," *Contemporary South Asia* 4, no. 1 (1995): 55.

[14] Sumantra Bose, *Kashmir: Roots of Conflict, Paths to Peace* (Cambridge, MA: Harvard University Press, 2003), 131.

SOCIOECONOMIC ENVIRONMENT

IJK was, and remains, a socioeconomically underdeveloped state, even in comparison with the rest of India. It is therefore likely that the insurgency in Kashmir was "motivated by more than political considerations." Growth rates in the gross state domestic product (GSDP) lagged behind growth in population, leading to increased poverty and unemployment. By 2002, more than a third of the population lived on less than a dollar a day. Poverty in Kashmir was also worse than in several neighboring states in India. Agriculture accounted for 37.8% of GSDP, the largest contributor to the economy. Second to agriculture was tourism, accounting for 10–20% of GSDP. The contribution of industry to the GSDP was minimal. After the onset of conflict in the early 1990s, employment dropped drastically.[15] The number of total workers dropped from 44.3% of the population in 1981 to 36.6% in 2001, affecting urban, rural, and youth populations alike.[16] In addition, in 1981 literacy rates for Kashmiris hovered around 26%, but substantially higher rates were registered in urban areas than in rural areas and rates for men were higher than those for women.[17]

After its accession to India in 1947, Kashmir was designated a "special category status" by New Delhi, meaning that 90% of the funds allocated to Kashmir by the central government were grants rather than loans. This policy is considered one of the reasons that Kashmir failed to develop its economic base, and it also gave the state no impetus to develop its tax base, resulting in Kashmir being one of the lowest-taxed states in India. In the 1970s, India changed its policy toward Kashmir and required that 50% of the funds Kashmir received from the central government be in the form of loans. As a result of the policy, like in many countries in Africa, the cost of servicing the loans absorbed the bulk of resources needed for productive investment. Servicing fees on the loan of a single rupee cost the Kashmiri state a staggering 5.35 rupees. For this reason, many militants cited the burdensome loans as evidence that Kashmir was a "colony" of India.[18]

---

[15] Shahid Javed Burki, *Kashmir: A Problem in Search of a Solution* (Washington, DC: United States Institute of Peace, 2007), 15.

[16] Debidatta Aurobinda Mahapatra, "Conflict and Development in Kashmir: Challenges and Opportunities," in *Proceedings of the International Conference, Center for International Studies and Cooperation* (Kathmandu, Nepal: SAHAKARYA Project, March 2007), 72; and Burki, *Kashmir: A Problem in Search of a Solution*, 17.

[17] Government of India, Department of Education website, accessed on April 12, 2010, http://www.education.nic.in/cd50years/12/8I/72/8I720J01.htm.

[18] Mahapatra, "Conflict and Development in Kashmir," 72.

Overall, since 1947, the Indian government has regularly intervened in Kashmir's economy because the central government was the source of most of the state's annual revenues. Government funds and initiatives to spur economic development and ease poverty, such as Sheikh Abdullah's redistribution of land after 1947 or a series of expenditures and programs in the 1980s to boost agricultural productivity and ease inequality, met with less than optimal outcomes, often due to poor management and rampant corruption. A rent-seeking government, combined with dismal economic prospects even for the urban educated population, led to the alienation of many Kashmiris.[19]

HISTORICAL FACTORS

From 1846 to 1947, Kashmir was a princely state[20] ruled by a Dogra (Hindu) family. Although Hindus formed a minority in the state, the majority-Muslim population was severely disadvantaged both politically and socioeconomically.[21] Attempts to educate selected members of the Muslim community in the early twentieth century created a more politically savvy Muslim elite that later established the National Conference party under the leadership of the charismatic schoolteacher Sheikh Abdullah.[22] Starting with a brief stint in the 1950s and then continuing again in the 1970s, Abdullah and the National Conference ruled Kashmir until his death in 1982. The National Conference advocated democratic governance with heavy socialist economic influences and a strong Kashmiri patriotism.[23] Although the National Conference was ostensibly secular and inclusive of Hindus and Muslims, it maintained a significant Islamic appearance.[24] Fighting for an independent Kashmir emancipated from Hindu Dogra rule, the National Conference was remarkably successful in gaining popular support in the 1940s and mobilizing the population in favor of its policies.[25]

Although the ruling Dogra king (the "Maharaja") executed the day-to-day governance of Jammu and Kashmir during the period

---

[19] Siddhartha Prakash, "The Political Economy of Kashmir since 1947," *Contemporary South Asia* 9, no. 3 (2000): 315–337.

[20] During the British colonial period in India, a "princely state" was ruled by a local ruler or king with a certain degree of autonomy from British interference.

[21] Bose, *Kashmir: Roots of Conflict*, 16–18.

[22] Ibid., 18–19. The first revolt against the Dogra rulers, a "historic day in the annals of Sringar," occurred on July 13, 1931.

[23] Ibid., 25.

[24] Ibid., 24.

[25] Ibid., 23.

of colonial rule, Britain maintained "paramountcy" over the local rulers.[26] When Britain disestablished the British Indian Empire in 1947 and created the dominions of India and Pakistan, the individual princely states of the empire were allowed to choose which of the new dominions they would join. This provided the Maharaja Singh in Kashmir with the choice to accede to India or Pakistan. With its substantial population of Hindus and Muslims, as well as trade relations, Kashmir had close ties with both of the new republics. Maharaja Singh, therefore, flirted with both governments but ultimately leaned heavily toward India.[27] Pakistan forced the Maharaja's hand, however, when Pashtun tribesmen from Pakistan crossed the border into Kashmir and attempted to decide the matter through force. The Maharaja requested aid from the Indian military to repulse the Pashtun invasion, but Indian officials refused until the matter of accession was settled. So, in October 1947, Kashmir officially acceded to India in exchange for military aid.[28]

At this time, Indian Prime Minister Jawaharlal Nehru and his personal friend Sheikh Abdullah promised that the residents of Kashmir could decide the matter of accession for themselves in a plebiscite supervised by the United Nations. For various reasons, the plebiscite was never held, in part because of conflict during the late 1940s. Moreover, the Indian military was only partially successful in repulsing Pakistanis from Kashmir. After a truce in 1949, India retained nearly two-thirds control of Kashmir (IJK), including the important Kashmir Valley, about 139,000 square kilometers of the total 223,000 square kilometers. Pakistan acquired the remainder, Azad Kashmir ("Free" Kashmir, or AJK). The two Kashmiris were eventually divided by the Line of Control (LoC), so named after the 1972 conflict between India and Pakistan, with the LoC changing only marginally through the succeeding several decades. The failure to hold the plebiscite and allow the Kashmiris to determine their own fate detracted from the legitimacy of the state for both pro-Pakistani and pro-independence factions in Kashmir,[29] and so, the Kashmir dispute was born.

Until 1989, when popular support for insurgency movements began, it was largely the Pakistani and Indian governments that played tug-of-war with the region. But the increased emphasis on the

---

[26] Ibid., 30. In a paramount system, multiple levels of legal jurisdiction may exist and operate independently, but when differences occur, the senior entity (in this case, the British) always has the final authority.

[27] Around the time of the Pakistan–Indian split, the population of Kashmir was 77% Muslim. Bose, *Kashmir: Roots of Conflict*, 31.

[28] Ibid., 33–36. Kashmir: Nuclear Flashpoint websitetime, http://www.kashmirlibrary.org/kashmir_timeline/kashmir_chapters/1947.shtml.

[29] Bose, *Kashmir: Roots of Conflict*, 38–42.

religious dimensions of the conflict escalated with Lashkar-e-Taiba and other foreign militants using Wahhabist doctrine to mobilize and frame the conflict, thus creating within Kashmir cleavages other than the obvious Hindu–Muslim divide. From roughly 1990 to 1995, the conflict took root in the ethnically distinct Valley, home to the raucous autonomist politics of the National Conference and the Plebiscite Front.[30] The conflict there became an extension of the old National Conference and Plebiscite Front politics but was led by a new generation of political activists. When the conflict expanded into Jammu, it reached areas dominated by an ethnolinguistically similar group, whereas other Muslim-dominant areas in Ladakh, composed of non-Kashmiri Gujjars and Rajputs, experienced minimal conflict. In this sense, the religious context of the Kashmiri conflict was initially a backdrop to the larger ethnolinguistic dynamics within the region, but the eventual rise of orthodox Islam and the introduction of transnational Islamic insurgency in the region altered the identity politics of the Kashmir conflict.[31]

GOVERNING ENVIRONMENT

The insurgency in Kashmir, and in an intimate fashion, the rise of the Hizbul Mujahideen, is a story of the failure of the political process in Kashmir. Indigenous political elites and the central government in India, with brief interludes, failed to address the grievances of the population through political channels. In 1950, with Article 370, Prime Minister Nehru's government institutionalized Kashmiri "uniqueness" in the federal union of India, offering a political compromise between proaccession and proindependence forces. Article 370 limited India's jurisdiction over Kashmir to matters of defense, foreign policy, and communications. Even on those subjects, the central government had to act in consultation with the Kashmiri government. On other matters, the central government had to work through the Kashmiri legislative assembly.[32] In the ensuing decades, Kashmiri autonomy, vis-à-vis the Indian government, steadily eroded. This steady erosion, and the Indian powers described in Article 370, served to fuel the flames of insurgency that exploded in 1989.

---

[30] See the *Governing Environment* section for an extended discussion of these parties.

[31] Bose, *Kashmir: Roots of Conflict*, 118 and 130.

[32] Ibid., 59.

After the Maharaja's[33] accession to India, Abdullah's National Conference took the reins of governmental power in Kashmir. Hidden during the period of mass mobilization, Abdullah's authoritarian tendencies, and his intransigence toward India, gradually emerged once he came to power. Although advocating a secular democracy held in check by a robust constitution, Abdullah paid lip service to democracy, and his policies belied his ostensible democratic aspirations. In 1951, he convened a Constituent Assembly, against the advice of India and the United Nations, rather than holding a plebiscite on the future of Kashmir. The National Conference's slogan was "One Leader, One Party, One Programme."[34] Despite his initially strong ties with India and Prime Minister Nehru, who supported Abdullah's socialist policies, Abdullah eventually made threats of Kashmiri separation. In 1953, Abdullah was arrested and imprisoned, and he remained in jail for most of the next two decades until he signed the Delhi Accord in 1975.

While further integrative measures with India were taken by the succeeding governments of Chief Ministers Bakshi Ghulam Mohammedand G. M. Sadiq in Kashmir, both governments were careful to emphasize "the special status Jammu and Kashmir enjoyed in the Indian Federation,"[35] thus providing a legal channel for the demands of the more extremist fringe that still advocated Kashmiri separatism. Furthermore, both chief ministers carefully avoided participation in Indian mainstream politics, a stance the central government respected. The Plebiscite Front, which was founded by a close colleague of Abdullah's, Mirza Afzal Beg, acted as a front for Abdullah and his separatist demands during his period of incarceration. Although electoral fraud was rampant, the Plebiscite Front gained several seats in elections and commanded significant popular sympathy. The National Conference became a branch of the Indian National Congress under Sadiq in the late 1960s and was renamed the Jammu and Kashmir Branch of the Indian National Congress. Overall, the integrative politics during the period from 1956

---

[33] The Maharaja was eased out of a governing role in Kashmir by 1949. Initially granted a ceremonial head-of-state role, in the 1960s the royal family's ceremonial role was replaced with a governor appointed by India.

[34] Bose, *Kashmir: Roots of Conflict*, 56.

[35] Tremblay, "Kashmir: The Valley's Political Dynamics," 79. Integrative measures included the application to Kashmir of Articles 356 and 357 of the Indian constitution. The former allowed the Indian Parliament to impose president's rule on Kashmir, whereas the latter granted the Indian Parliament the authority to grant to the president the power of the State Assembly. Moreover, during this period the Indian government also reserved for itself the right to interfere in Kashmiri elections in cases of "irregularities."

to 1974 were tempered with "the myth of Kashmiri distinctness."[36] The Bakshi and Sadiq regimes aided in the maintenance of interregional harmony, offering tokens to both the separatist and integrationist factions in society. Both chief ministers offered the Jammu region significant representation in their cabinets, helping to offset potential grievances against the Valley-dominated regime.[37]

In 1975, Sheikh Abdullah was released from prison. As a condition of his release, he signed an accord with Prime Minister Indira Gandhi, declaring the state of Kashmir as a constituent part of the Indian union ruled by Article 370, and also relinquishing his previous demands for secession that led to his imprisonment. The legislative assembly unanimously elected Abdullah to the chief ministership after his release. In response to the integrationist policies of his predecessors, Abdullah refused to join the Indian National Congress and, with the help of Beg, revived the National Conference. Abdullah's rule at the Kashmiri helm, from 1975 until his death in 1982, was characterized by the revival of the National Conference, which absorbed the Plebiscite Front and the assertion of Kashmiri autonomy. However, his populism was accompanied by the rise of a single-party state, alienating the extremist fringe of the Plebiscite Front, primarily in the Kashmir Valley, and the prointegrationist forces of the Jammu region. The latter would turn to Prime Minister Indira Gandhi's Congress party to represent their interests rather than indigenous ones. Likewise, after Beg and the Plebiscite Front joined forces with the National Conference, the secessionists "were suddenly deprived of both a political apparatus and a charismatic leader."[38] Although Abdullah sharply limited their scope of action, the pro-Pakistan and Muslim JEI[39] garnered more sympathy, eventually backing the loose coalition, the Muslim United Front (MUF), after Abdullah's death. Thus, the narrowing of political space in Abdullah's government contributed

---

[36] Another scholar, Bose, highlights only the integrative policies during this period and the lack of a viable, meaningful political opposition. He believes that this period's contribution to the onset of insurgency in the 1980s is found in its democratic deficit. Tremblay, on the other hand, points to the break of a careful balance between the separatist and prointegration demands of the population during Farooq Abdullah's government (this break is discussed in greater detail below). See Bose, *Kashmir: Roots of Conflict*, and Tremblay, "Kashmir: The Valley's Political Dynamics," 79.

[37] The JKLF, operating by at least 1971, if not earlier, received little to no popular support during this time. Moreover, during the 1965 India and Pakistan conflict over the disputed Kashmir region, Kashmiris themselves, especially Muslims, did not participate in the conflict at the level that Pakistan had anticipated. Bose, *Kashmir: Roots of Conflict*, 84.

[38] Tremblay, "Kashmir: The Valley's Political Dynamics," 79.

[39] Please see the *Leadership and Organizational Structure* section for more information on JEI.

to a sharpening of regional (Jammu versus Kashmir) and religious (Hindu versus Muslim) polarization.[40]

Abdullah designated his son, Farooq Abdullah, who had little political experience, as his successor. Mixing dynastic and democratic politics, after his father's death in 1982, Farooq took over as chief minister. Farooq revived the integrationist policies of chief ministers Bakshi and Sadiq. His regime began in an unfavorable political environment; "in the absence of an effective party apparatus, most of the oppositional forces had temporarily withdrawn themselves into a political solitude, a sharp polarization on a Hindu-Muslim basis had emerged between the Jammu region and the Valley, and the authoritative allocation of political and economic resources on the basis of a large-scale patronage system appears to have been ineffective in fulfilling the demands of a large section of the educated unemployed."[41] Additionally, Indira Gandhi and her Congress party on the national scene recaptured power shortly after Farooq's accession. To consolidate her party's power in the 1984 election, she adopted a strategy of "majoritarian mobilization," rousing the population against the secessionist Muslim and Sikh minorities. The Kashmir elections in late 1984 were to be the initial test case for her new strategy. The Congress party was successful in the Hindu-majority Jammu region, but it is not surprising that the party evoked little response from Kashmiris in the Muslim-majority Valley.[42] Gandhi's activities, and the renewed presence of the Congress party in Kashmiri politics, exacerbated regional and religious cleavages in the region, solidifying the secessionist movement in the region, encouraging opposition against the Kashmir state and India alike, and increasing religious tensions.[43]

Farooq's response to the interventionist policies of Indira Gandhi and the Congress party also contributed to the tense environment that saw emergence of an indigenous insurgency movement in the late 1980s, including the Hizbul Mujahideen. After an initial, failed attempt to create an alliance between the Congress party and the National Conference, each party contested the elections separately. To oppose the interventionist policies of the Congress party, Farooq aligned the National Conference with anti-Congress parties in other states of India. While his antagonistic relationship to the Congress would seemingly ingratiate Farooq with anti-Indian factions in society, his participation in mainstream Indian politics served to

---

[40] Tremblay, *Kashmir: The Valley's Political Dynamics.*

[41] Ibid.

[42] Bose, *Kashmir: Roots of Conflict*, 90–91.

[43] Tremblay, "Kashmir: The Valley's Political Dynamics," 79.

integrate Kashmir within the Indian union, eroding its autonomy and "distinctness" with the federal union. Even if Farooq's emergence on the Indian political scene contested the dominant party, it was still a welcome development as Farooq became "more than a regional figure" and signaled the "deeper political integration of Kashmir within the Union."[44]

Not content with a limited share of power in the Kashmir, the Congress government in India manufactured a takeover of Farooq's government in 1984 when a number of National Conference legislators quit the party and formed a new group with the support of Congress party members in the legislative assembly. At the center of the conspiracy was G. M. Shah, Abdullah's son-in-law, who became chief minister for several years. Farooq maintained that the plan originated with Prime Minister Gandhi. Protests erupted in the capital city, Srinagar, and throughout the Valley but were suppressed by the paramilitary police, the Central Reserve Police Force. To keep a lid on further eruptions of popular discontent, Shah instituted a curfew for his first seventy-two days in office, and he became known as the "curfew chief minister." The Congress government dismissed Shah and his government, using Article 356 (citing a breakdown of law and order), after communal riots broke out in the Valley and the state was unable to control the riots. The central government once again installed Farooq as chief minister pending the Assembly elections in 1987. Farooq allied with the Congress party, the latter of which would contest seats in Jammu while the National Conference contested seats in the Valley. Farooq's rapprochement with the Congress party was known as the Farooq Accord, or the Farooq–Rajiv Accord. Part of the Accord was New Delhi's promise to supply substantial funds to Kashmir to address the ailing economy and the army of the qualified but unemployed.[45] In 1975, when Abdullah Sr. made the proverbial deal with the devil (Indira Gandhi), he had the popular support and legitimacy as the "lion of Kashmir" to withstand the political fallout from his agreement. His son, however, did not have the political clout or legitimacy to lead the state in uncomfortable directions or to withstand the political fallout from his agreement with Rajiv Gandhi's[46] government.

---

[44] As quoted in Bose, *Kashmir: Roots of Conflict*, 91.

[45] Schofield, *Kashmir in Conflict*, 145.

[46] Prime Minister Indira Gandhi was assassinated by her Sikh bodyguards in 1984. Her son, Rajiv Gandhi, and the Congress party swept the polls in the subsequent elections.

WEAKNESSES OF THE PREREVOLUTIONARY ENVIRONMENT AND CATALYSTS

Farooq's decision to ally with the Congress party was seen as a "cowardly capitulation," evoking "contempt and hostility" among many Kashmiris, especially in the Valley.[47] As a result, many Kashmiris joined a loose umbrella coalition, the MUF, to contest the National Conference–Congress alliance in the upcoming elections. The MUF was an "anti-establishment" group with no real unifying ideology other than its animosity toward a moribund National Conference that represented a "narrow political elite." The group consisted of a wide spectrum of society, including educated youth, the uneducated working class, and farmers, among others. If nothing else, it signaled that the once-great National Conference, supported by a groundswell of popular support for regional autonomy and Kashmiri dignity at its inception, continued its quest for political power in a Valley sharply divided between a party machine that churned out the "traditional" vote and hundreds of thousands of citizens joining politics for the first time to give voice to their disapproval for the Abdullah dynasty, rampant government corruption, and dismal socioeconomic prospects. A massive wave of popular support for the MUF rocked the Valley. Its call for responsible government and "regional Kashmiri pride" was heard by many, and the MUF attracted an army of thousands of volunteer workers. It appeared that a successful, legal, constitutional opposition to the National Conference–Congress-dominated government was on the horizon.

It can be speculated that it was the hope raised by the MUF's political campaign in an ostensibly democratic country (indeed, India's sense of nationhood was intimately informed with secular democratic governance) that made the blatant electoral fraud of the Indian government in the 1987 elections such an explosive event.[48] The contest for one seat of the Assembly in particular would have important ramifications for the ensuing insurgency. Two men from the Amirakadal district of the capital city Srinagar competed for the same Assembly seat. One, Yusuf Shah, a political activist of the Pakistan-based JEI who ran under the MUF banner, tried, and failed, to win a seat in the previous three elections. His opponent, Ghulam Mohiuddin Shah, ran on the National Conference ticket. Voter turnout was heavy, and when the votes began to be counted, it was readily apparent that Yusuf Shah had won the seat by a landslide. The other Shah left the election center dejected, apparently the loser, while the

---

[47] Bose, *Kashmir: Roots of Conflict*, 93.
[48] Bose, *Kashmir: Roots of Conflict*, 101.

victorious Shah remained. In an interesting twist of events, presiding officials escorted Ghulam Shah back to the center and declared him the official winner. Now the loser, Yusuf Shah, along with his campaign manager, Yasin Malik, was arrested and imprisoned without charges or trial for the rest of 1987. Yusuf Shah adopted the *nom de guerre* "Syed Salahuddin" and became the bane of counterinsurgency forces as the commander-in-chief of Hizbul Mujahideen in the early 1990s. Yasin Malik, the campaign manager, became one of the quartet of "freedom fighters" that symbolized the JKLF after training and procuring weapons in Azad Kashmir.

The tale of the two Shahs was repeated during the 1987 elections throughout the Valley and in isolated spots in the Jammu region. The voter turnout, at nearly 75% throughout the state and almost 80% in the Valley, was the highest ever recorded in Kashmir.[49] Farooq vehemently denied charges of vote rigging, but evidence suggests that fraud was widespread. Reports trickled in that gangs forcibly took over poll stations, entire ballot boxes were prestamped for the National Conference, citizens were intimidated from voting, and government officials stopped ballot counts as soon as the opposition candidate appeared to gain a lead. Despite receiving 32% of the votes, even by official counts, the MUF captured just four of the seventy-six seats in the Assembly. The National Conference, led by Chief Minister Farooq Abdullah, formed the new government with the overwhelming majority.[50]

Any residual respect the bulk of Kashmiris, particularly those in the Valley, may have had for the political process fled their hearts and minds after the 1987 election. One top MUF leader, whose electoral "victory" in the Handwara district was likewise stolen, said, "If people are not allowed to vote, where will their venom go but into expression of anti-national sentiment?"[51] Having failed to address their grievances through legal, democratic channels, for many, it was time to take up the gun.[52] The educated, but unemployed, neither part of the swollen bureaucracy or the elite, alienated youth, turned to politico-religious organizations such as the People's Conference, Awami National Conference, and the JEI as the broader MUF coalition fell apart. Farooq's government was on increasingly shaky ground, failing to

---

[49] Schofield, *Kashmir in Conflict*, 137.

[50] Bose, *Kashmir: Roots of Conflict*, 49.

[51] From Abdul Ghani Lone, who was later assassinated by pro-Pakistani militants in 2002, as quoted in Bose, *Kashmir: Roots of Conflict*, 94.

[52] Schofield, *Kashmir in Conflict*, 138 and 146.

deliver on key electoral promises, especially economic development.[53] The expected New Delhi largesse promised by Prime Minister Rajiv as part of the accord with the National Conference was not forthcoming. Farooq claimed that without it, he could not create jobs, eliminate corruption, or invest in infrastructure.[54] The emphasis, which would later change as the insurgency progressed, was social justice through the establishment of an independent, sovereign Kashmiri state.[55]

In 1989 a number of insurgent groups began to gain ground in the Valley, some supporting independence and others unification with Pakistan. The Jammu Kashmir Liberation Front (JKLF), a proindependence group, enjoyed the most popular support. Others, such as Hizbul Mujahideen, a pro-Pakistan group, were militant wings of parties that had formed the now-defunct MUF.[56] During the year, many *bandhs/hartals*, or strikes, were held, eating up nearly a third of all working days. In addition, "processions, bomb blasts, acts of arson, selected killings of political workers and intelligence workers,[57] attacks on security forces, and the latter's violent response of cordoned-off searches that led to arrests and indiscriminate firing" rocked Kashmir in 1989, especially after massive demonstrations against the publication of Salman Rushdie's book *The Satanic Verses*.[58] Secessionist-nationalist groups seized on any pretext possible to attack the Kashmir and Indian governments. Groups constructed a new public events calendar, celebrating events associated with Islamic societies (such as the anniversary of Ayatollah Khomeini's death) and *shaheeds* (martyrs, or militants who lost their lives battling security forces). Any events associated with the Indian government or the National Conference, by contrast, were "commemorated" with Valley-wide blackouts, enforced by bombs and targeted assassinations. Various professional groups and civil servants supported militant activities, leading to a near-complete breakdown of civil administration in Kashmir.[59]

Militants, primarily the proindependence JKLF, contributed to the further breakdown of political order in Kashmir by boycotting the 1989 Assembly elections. The strategy was to withdraw public support

---

[53] Farooq had also promised to curtail the move from the Valley capital of Srinagar to Jammu (in the Jammu region) during the winter months. Tremblay, "Kashmir: The Valley's Political Dynamics," 79.

[54] Schofield, *Kashmir in Conflict*, 145.

[55] Ibid., 138–139.

[56] Ibid., 145.

[57] The JKLF, the dominant militant group at this time, killed approximately one hundred political and intelligence workers, about 75% of whom were Muslim and the remaining 25% Hindu. Bose, *Kashmir: Roots of Conflict*, 108.

[58] Tremblay, "Kashmir: The Valley's Political Dynamics," 79.

[59] Ibid.

for the government and break the hegemonic power of the National Conference. Initially, Farooq's speeches and rallies were boycotted. In early August, militants demanded that party members publicly terminate their party membership. A Kashmir daily newspaper dedicated a special section for the "declaration of dissociation," listing all those who had withdrawn from the National Conference. In late August, the JKLF killed a leading National Conference party member, resulting in a mass desertion of the National Conference party.

For years, high levels of Kashmiri participation in elections, up to a high of 75% of voter turnout in 1987, was construed by Indian officials and internationally as a measure of the legitimacy of the Kashmiri state and its place in the Indian federal union. Secessionist groups, therefore, urged and intimidated Kashmiris to boycott the 1989 elections (held in Srinagar) *en masse*. Voter turnout for the elections was astonishingly low—only 2% among the Srinagar constituency. A majority of the polling workers also refused to show up for work at the polling stations. Farooq's attempts to bolster support for and the legitimacy of the National Conference were unsuccessful.[60] The 1989 elections were the last to be held in Kashmir until 1996, and the breakdown of the political and civil order of Kashmir at the cusp of 1990 was near complete.

After JKLF militants kidnapped the daughter of a Kashmiri appointed to a ministership in the Indian government in December 1989, who was later released unharmed in return for the release of jailed militants, New Delhi appointed Jagmohan as governor and charged him with restoring law and order in the troubled state. JKLF supporters poured into the streets after the release of the militants, celebrating their victory against the government. During the celebration, the police killed several JKLF supporters. In response, in late January 1990 protestors took to the street once again aggrieved by the slaying of unarmed demonstrators. In one of the worst massacres in recent Kashmiri history, the Central Reserve Police Force (CRPF) and Kashmiri police opened fire on the crowd, killing 130 unarmed protestors. Jagmohan arrived in Kashmir shortly after the incident, so he was not involved in the slayings, but his response to the incident helped to further alienate the population, encouraging militancy.[61]

---

[60] Among other tactics, Farooq attempted grassroots mobilization after the National Conference's mass rallies were attended poorly, released twenty-five militants, promised a substantial relief package, censored the press, and apologized for the shootings of protestors by paramilitary forces. Tremblay, "Kashmir: The Valley Political Dynamics."

[61] Edward Desmond, "The Insurgency in Kashmir (1989–1991)," *Contemporary South Asia* 4, no. 1 (1995): 5.

More than the relatively simple denial of civil and political rights that characterized the Kashmiri government for more than four decades, the events of 1990, when Governor Jagmohan and the Indian government stepped up their counterinsurgency efforts, developed into a pronounced human rights crisis[62]—there were rampant abuses such as unarmed protestors shot indiscriminately, arrests without trial, and the rape and torture of prisoners. Jagmohan whitewashed the security forces' role in human rights violations, laying the blame for atrocities at the feet of "terrorist" forces. In February, he also dissolved the Assembly. Combined with the severe, indiscriminate harassment of the population, whereby all citizens were treated as potential suspects, the January massacre, and Jagmohan's draconian policies, support for the JKLF skyrocketed.[63] The unpopular governor was finally dismissed by New Delhi after the CRPF opened fire on a funeral procession for an important religious leader, killing fifty-five. Jagmohan was a boon to the militants, aiding in the transformation of a low-intensity conflict into a "full-blown revolt."[64]

However, it was JKLF, an ostensibly secular, proindependence movement, that dominated the field at the onset of the insurgency. For several reasons, including heavy counterinsurgency measures by the Indian government, Pakistani involvement, and infighting among various insurgent movements, Hizbul Mujahideen emerged as the most effective guerrilla organization on the ground in Kashmir in the early to mid-1990s until the "fidayeen phase" of the insurgency when foreign jihadists, including militants such as Lashkar-e-Taiba, took control of the field.

Despite theories at the time that it orchestrated the uprising, Pakistan, like India, was caught by surprise by the scope and intensity of the uprising.[65] Nevertheless, once the insurgency began, Pakistan's Inter-Services Intelligence (ISI) agency played a large role in shaping the nature of the insurgency. Support for the uprising was strong in Pakistan, particularly in military and intelligence circles where many believed Kashmir had been "stolen" from Pakistan because of the failure of New Delhi to initiate a plebiscite after Kashmir's accession to India. In addition, many were still smarting from the 1971 conflict against India in which they lost East Pakistan.[66] Along the LoC, Pakistan

---

[62] See the Human Rights Watch report on abuses by both Indian security forces and militants for a more detailed account. Meenakshi Ganguly, "Everyone Lives in Fear: Patterns of Impunity in Jammu and Kashmir," *Human Rights Watch* 18, no. 11 (2006).

[63] Desmond, "The Insurgency in Kashmir (1989–1991)," 5.

[64] Ibid.

[65] Ibid.

[66] East Pakistan is now the country of Bangladesh.

facilitated the infiltration of insurgents into Kashmir in early 1990, providing heavy artillery and mortar fire to ease their passage. Although ISI efforts initially aided the JKLF, Pakistani military and intelligence shunted aside the JKLF in favor of Hizbul Mujahideen, a pro-Pakistan Islamist insurgent group founded in 1989 by Ashan Dar as the militant wing of JEI. The ISI took an active role in the marginalization of the JKLF by "harassing its leadership, hijacking recruits, and inducing or coercing JKLF members to join Hizbul Mujahideen, or any of the many smaller, pro-accession, Islamic groups."[67] The ISI also cut off the JKLF's money, weapons, and training.[68]

Because of the early effectiveness of the JKLF, the Indian government's counterinsurgency measures fell heavily on it, further decreasing its capacity. Moreover, militants in Hizbul Mujahideen itself were active in targeting the JKLF. Clashes between the groups accounted for more than a dozen deaths in 1991. Together, these factors ensured that Hizbul Mujahideen, beginning as early as 1991, was the predominant insurgency movement on the ground in Kashmir.[69]

FORM AND CHARACTERISTICS OF THE REVOLUTION

OBJECTIVES AND GOALS

The Hizbul Mujahideen formed from indigenous Kashmiri members in 1989 to serve as the militant wing of the political party JEI. Once Jammu and Kashmir's Indian-led government was initially stunned by the popular Kashmiri rebellion in 1990, Pakistan's ISI, JEI's benefactor, directed Hizbul Mujahideen to counter the JKLF, Kashmir's alternative secular organization.[70] While the JKLF advocated Kashmiri independence from India and Pakistan, Hizbul Mujahideen promoted Kashmiri unification with Pakistan. "At the start of the 1990s, the (broader) movement quickly dissipated in a struggle for domination among different groups, and what had begun

---

[67] Ibid.

[68] Ibid.

[69] Ibid.

[70] Pakistan initially enabled the JKLF to gain power. By 1990, however, "where Pakistan was concerned, the 'Kashmiri card' had served its purpose and, if allowed to reach its logical conclusion, might backfire, especially since the JKLF's goal was 'independence and re-unification of the divided Kashmir and *not* accession to Pakistan.' Consequently, Pakistan decided to 'curb the *independence* sentiment in the Valley." Navnita Chadha Behera, *Demystifying Kashmir* (Washington, DC: Brookings Institution Press, 2006), 151.

as an ethnic conflict was given a religious identity by the ISI, which promoted religiously oriented groups."[71] Pakistan also encouraged intergroup fighting. Whereas initially the neighboring patron supported numerous groups within Jammu and Kashmir, hoping to mobilize the indigenous population, it soon shifted its philosophy to favor Hizbul Mujahideen and changed the discourse of the liberation. For a short time, the Hizbul Mujahideen was considered the most effective organization to attain Pakistani goals.

During the 1990s, the rhetoric of the Kashmiri liberation movement transitioned from that of an ethno-political conflict, which inspired frequent popular participation, to that of a struggle based on Islamization and a religious jihad. Whereas ethno-political rhetoric deemed social, religious, and political autonomy as the center of perceived grievances, Hizbul Mujahideen considered its plight neither a guerrilla war nor a national movement. Rather, efforts against India represented a holy war.[72] Pakistan's eventual injection of additional external Wahhabi terrorist organizations was meant to augment and supersede Hizbul Mujahideen. In many ways, the Kashmiri Sufi population became alienated from the Wahhabi perspective, caught between the Indian government and the terrorist organizations active within Kashmir. Pakistani terrorist organizations, such as Lashkar-e-Taiba, came to dominate Hizbul Mujahideen, inspiring a failed attempt by the Hizbul Mujahideen leadership living within Jammu and Kashmir to break away from their organizational counterparts living within Pakistan and dominated by the ISI.

LEADERSHIP AND ORGANIZATIONAL STRUCTURE

Hizbul Mujahideen exists within a broad resistance network that supports numerous regional and international actors. It was created as the militant wing of JEI at the request of Pakistan's external intelligence agency, the ISI. "The Jamaat-e-Islami is an Islamist party similar to the Arab Muslim Brotherhood, with which it has both ideological and organizational links. It has operated over the decades as a political party, a social welfare organization, a pan-Islamic network and the sponsor of militant groups fighting in Afghanistan and Kashmir."[73] JEI served as the ideological center for Hizbul Mujahideen and the militant group received direction, training, and funding through JEI assets.

---

[71] Wajahat Habibullah, *My Kashmir: Conflict and the Prospects of Enduring Peace* (Washington, DC: United States Institute of Peace Press, 2008), 66.

[72] Behera, *Demystifying Kashmir*, 154.

[73] Husain Haqqani, "The Ideologies of South Asian Jihadi Groups," *Current Trends in Islamist Ideology* 1 (April 2005): 15.

In addition to participation in JEI, the Hizbul Mujahideen was a member of two additional regional organizations. For example, "an alliance of 26 political, social and religious organizations, the All Party Hurriyat Conference (APHC) was formed on March 9, 1993 as a political front to further the cause of Kashmiri separatism. The amalgam has been consistently promoted by Pakistan in the latter's quest to establish legitimacy over its claim on the Indian State of Jammu and Kashmir."[74] JEI provided a representative who serves as one of seven on the executive council of the APHC, which may provide a political front for terrorist campaigns in the state. Counter arguments state that "the creation of the APHC was in fact an admission by the separatist political leadership that the course of violence was proving counterproductive."[75] Similar to the APHC, the United Jihad Council—or Muttahida Jihad Council (MJC)—was formed by the ISI in 1994. Whereas the APHC comprises Kashmiri-based groups, the MJC includes members external to Kashmir, with as many as fifteen organizations affiliated with the MJC in early 1999, including Lashkar-e-Taiba, Hizbul Mujahideen, Harkat-ul-Mujahideen, Al-Badar, and Tehrik-i-Jihad.[76]

The Hizbul Mujahideen as a singular organization was divided into two major components, of which the JEI had direct control.[77] The first component included the operational command responsible for maintaining five regional divisions within Kashmir itself. Each division oversaw supporting districts. The Central Division was based around Srinagar; the Northern Division covered the region of Kupwara-Bandipora-Baramulla; the Southern Division command included Anantnag and Pulwama; the Chenab Division encompassed Doda district and Gool in the Udhampur district; and the Pir Panjal Division commanded the districts of Rajouri and Poonch.[78] The second component of the Hizbul Mujahideen organization included the central command within Pakistan, responsible for political liaison and administrative control. "Each wing of the organization had a leader for military and ideological training, intelligence, supplies,

---

[74] "All Party Hurriyat Conference (APHC)," *Kashmiri Herald* 1, no. 12, May 2002, http://www.kashmirherald.com/profiles/hurriyat.html.

[75] Habibullah, *My Kashmir*, 82.

[76] John Pike, "United Jihad Council, Muttahida Jihad Council (MJC)," Federation of American Scientists (FAS), updated October 25, 2999, http://www.fas.org/irp/world/para/mjc.htm.

[77] Ali Chaudhry, "In the Spotlight: Hizb-ul-Mujahideen (HM)," Center for Defense Information, August 6, 2004, http://www.cdi.org/program/document.cfm?documentid=2363.

[78] "Hizb-ul-Mujahideen," South Asian Terrorism Portal, accessed October 8, 2010, http://www.satp.org/satporgtp/countries/india/states/jandk/terrorist_outfits/hizbul_mujahideen.htm.

logistics, and finances. All positions together formed the nucleus of the larger body, the Majlis-i-Shoora, which was the central command of the Hizbul-Mujahideen."[79]

Syed Salahuddin, a participant in the Afghan war against the Soviets, was the supreme commander of the 1,500-member Hizbul Mujahideen organization headquartered in Muzaffarabad, the capital of Azad Kashmir, or Pakistan-administered Kashmir.[80] Many of the Hizbul Mujahideen members joined the organization having combat experience in Afghanistan. "Its front organizations included the Jamiat-ul-Tulba, the student wing, and the Dukhtaran-e-Millat (Daughters of the Faith), the women's wing, which also ran the Islamic Relief Committee and the Islamic Blood Bank."[81]

COMMUNICATIONS

The Hizbul Mujahideen ran its own news agency, Kashmir Press International, and it also employed an official spokesperson for the organization.[82] Lead commanders and even Salahuddin himself gave interviews to the press in order to communicate the group's message regionally and abroad. Such open conversations allowed Hizbul Mujahideen to address grievances, discuss capabilities, weigh in on Indian and Pakistani political decisions, and speak to both supporters and adversaries.[83] Communications within the Hizbul Mujahideen were also facilitated by membership within larger organizations, such as the Muttahida Jihad Council (MJC). With respect to a more tactical communications perspective, the use of wireless communication grew steadily in Kashmir after 1992. "This equipment allowed the insurgents to build an extensive communications infrastructure across the region. By the late 1990s, there were reports of militants using satellite phones and high-end wireless sets capable of encrypted communication."[84]

---

[79] Behera, *Demystifying Kashmir*, 317. See Tahir Amin, *Mass Resistance in Kashmir: Origins, Evolution, Options* (Islamabad: Institute of Policy Studies, 1996), 92. Also Alifuddin Tarabi, *Hizbul Mujahideen: The Principles and Struggle*, translated from Urdu (n.d.); and Shamshul Haq, *Hizbul Mujahideen: Its Background and Struggle*, translated from Urdu (Rawalpindi: Markaz Matruit Kashmir, 1994).

[80] "Hizb-ul-Mujahideen," South Asian Terrorism Portal.

[81] Behera, *Demystifying Kashmir*, 154.

[82] "Hizb-ul-Mujahideen," South Asian Terrorism Portal.

[83] Owais Tohid, "Interview: Syed Salahuddin," *Newsline*, June 12, 2003, http://www.newslinemagazine.com/2003/06/interview-syed-salahuddin/.

[84] Gerald Meyerle, "Death by a Thousand Cuts: The Dynamics of Protracted Insurgency in Kashmir and Sri Lanka" (PhD dissertation, University of Virginia, 2008), 182.

METHODS OF ACTION AND VIOLENCE

The Hizbul Mujahideen began violent activity in the context of a larger ethnic uprising within Kashmir in 1989 and solidified its organizational constitution in 1990. These first two years of the insurgency were the most violent and the most effective against the Indian military and government. "Inspired by the Afghan model, the Hizbul Mujahideen's strategy was to make the economic, military, and political costs of retaining Kashmir too prohibitive for India. This had two components: to raise military costs by tying down large numbers of the Indian army in the Valley; and to extend the area of operations to other parts of the state."[85] Hizbul's indigenous Kashmiri insurgents primarily utilized military measures against government officials and military targets by using targeted assassinations, kidnappings, bombings, and hit-and-run tactics with weapons such as the AK-47 assault rifle, sniper rifles, hand grenades, rifle-propelled grenades, and additional small arms.[86] By late 1991–1992, Hizbul Mujahideen was successful at establishing liberated zones throughout Kashmir.

During the early to mid-1990s, Hizbul Mujahideen grew focused on minimizing the capabilities of the JKLF, as well as continuing strikes on Indian outposts and targets.[87] During this time, the local population still supported the movement. Police revolted against the Indian authority in support of Kashmiri militants, and businesses frequently participated in mass demonstrations against Indian security forces. "For years, Hizbul Mujahideen avoided the terrorist label by taking care that its target in Kashmir could be identified as military, as opposed to civilian, targets. Even the United States acknowledged this distinction and spared Hizbul Mujahideen from designation as a terrorist group."[88]

By 1995, however, Wahhabi organizations entered into the insurgency, aided by the local knowledge and capabilities of the Hizbul Mujahideen. In combination, "insurgents with high-level military training, weapons, and communications equipment began launching sophisticated commando-style assaults on heavily guarded government installations beginning in 1999."[89] Wahhabi-sponsored attacks became vicious and less discriminating against the local

---

[85] Behera, *Demystifying Kashmir*, 154.

[86] Ibid.

[87] Pakistan was equally as effective in neutralizing the JKLF by withholding weapons, cadres, and funds from the secular organization. Ibid., 166.

[88] Husain Haqqani, "The Ideologies of South Asian Jihadi Groups," 18.

[89] Meyerle, "Death by a Thousand Cuts," 181.

population, such as a Harkat-ul-Ansar bombing in July of 1995 that killed twenty civilians and injured forty-five in Jammu City.

Although the Kashmir and Pakistan-based Hizbul Mujahideen leadership started to strain internally during the late 1990s, Hizbul Mujahideen tactics did not significantly change. The insurgency exhibited periods of surging and then recouping but continued to push toward destabilizing the Indian government. In 1996, however, despite increased violence, indications hinted that Hizbul Mujahideen and other militant groups were giving ground to Indian forces in urban settings. Threats to interrupt Hindu holidays and national elections were left unfulfilled. In response, Wahhabi organizations began to turn to attacks against smaller Hindu villages in southern Kashmir, likely with some support from Hizbul Mujahideen.

In part, the militant resistance against India within Kashmir did atrophy in the latter half of the 1990s. Kashmiri residents began to resist Wahhabi tactics and the Sunni ideologies presented by ISI-backed organizations. During 1998, however, the Pakistani military—in conjunction with the United Jihad Council, or Muttahida Jihad Council—made preparations to draw a good deal of the Indian force to the border separating Indian and Pakistani territory. "In early 1999, troops of Pakistan's Northern Light Infantry, in the garb of Kashmiri militants, crossed the Line of Control and occupied strategic mountain peaks in Mushkoh Valley, Dras, Kargil, and Batalik sectors of Lakakh."[90] Hizbul Mujahideen participated in the fighting, although the primary militant organization was Lashkar-e-Taiba. "The goal was two-fold: to spark a limited war between India and Pakistan that would force India to negotiate, and to inject new life into the Kashmir insurgency by demonstrating that militants were capable of engaging the Indian army directly in an all-out war."[91]

Operations in the border region, however, did not go according to Pakistan's intentions and the Indian military proved extremely effective in countering the attack. Even so, a spokesman for the MJC and Hizbul Mujahideen told the media in Muzaffarabad that "neither Pakistan nor any other country could compel them to vacate the territory they 'liberated' from Indian occupation."[92] In direct

---

[90] Behera, *Demystifying Kashmir*, 84.

[91] Meyerle, "Death by a Thousand Cuts," 165.

[92] Muddassir Rizvi, "Sharif Faces New Front at Home" *Asia Times,* July 8, 1999, http://www.atimes.com/ind-pak/AG08Df01.html.

mediation with the United States, however, the Pakistani government agreed to pull back its troops and directed local militants to leave.[93]

Soon after the 1999 Kargil war ended, the rift within the Hizbul Mujahideen came out into the open. Abdul Majid Dar, the group's Kashmiri military commander, unilaterally called a cease-fire in order to initiate negotiations with the Indian government in July of 2000. The shift in Kashmir-based policy reflected the differences between the Hizbul Mujahideen leadership in Pakistan and the Hizbul Mujahideen commanders in the Kashmir Valley. The operational commanders felt that they were taking the risks even as the political leaders lived in comfort in Pakistan while taking direction from their ISI handlers. In addition, the field commanders believed that the mission and objectives of the Hizbul Mujahideen were being subordinated to the objectives and interests of Pakistan.[94]

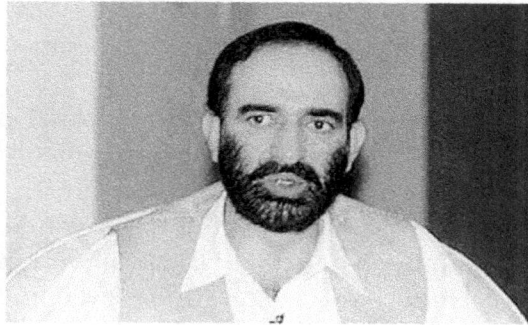

Figure 2. Abdul Majid Dar.[95]

Although India accepted and reciprocated Dar's motion, the cease-fire actually increased the level of violence within Kashmir as Pakistani-based militant groups retaliated in order to undermine the peace process and prove their operational dominance. Salahuddin, in coordination with the APHC, additionally sabotaged the process by marginalizing Dar's credibility and adding the stipulation that Pakistan be included within the negotiations.[96] Dar exposed weakness within the Hizbul Mujahideen, and Salahuddin eventually expelled him from the group. The Kashmiri leader maintained a strong following,

---

[93] This decision caused considerable criticism of the Pakistani government and contributed to Pakistan's military coup of October 1999. John Pike, "United Jihad Council, Muttahida Jihad Council (MJC)."

[94] Sudha Ramachandran, "Now It's India's Move After Kashmir Split," *Asia Times*, May 10, 2002, http://www.atimes.com/ind-pak/DE10Df03.html.

[95] Jammu and Kashmir website, http://www.jammu-kashmir.com/media/majid-dar.jpg.

[96] Owais Tohid, "Interview: Syed Salahuddin."

however, and was eventually assassinated at his home in March 2003.[97] In addition, "his associates were accused of assisting Indian security forces, and many of his loyalists were killed by cadres of the Salahuddin group.[98] Before his death, it was expected that Dar would announce a parallel organization claiming to be the "real Hizbul Mujahideen." Although the majority of Kashmiri leaders did not defect, after 9/11, additional commanders supportive of negotiations with India "felt that they did not want to be painted with the same brush as the jihadi organizations based in Afghanistan and Pakistan."[99] Some reporting goes as far as to say that Hizbul Mujahideen did in fact split.[100] Ultimately, Hizbul Mujahideen never fully recovered its full reputation and influence, and its subsequent operations after the Kargil war demonstrated that Hizbul was no longer the primary force behind the insurgency, even though it continued to play a supporting role.

METHODS OF RECRUITMENT

Hizbul Mujahideen recruitment from a general perspective was motivated by Indian repression of the Islamic population and the belief among many youth that all other options for seeking independence had failed and that the use of violence was their final option.[101] The fact that Hizbul Mujahideen presented the local population with what appeared to be a Kashmiri-led organization played an additional part in attracting support within the region. "Several members of the Hizb admitted that their allegiance to the organization has more to do with access to weapons and training than with a commitment to Hizb's aim of acceding to Pakistan."[102]

A second motivational factor that affected recruitment was unintentionally supported by the media. "The decline and fall of the Soviet Union, so graphically depicted by the media—particularly by the popular BBC TV, which was perceived as impartial by Kashmiris untrusting of propaganda—provided hope for a similar disintegration

---

[97] Mukhtar Ahmad, "Abdul Majid Dar Shot Dead," *Rediff*, March 23, 2003, http://www.rediff.com/news/2003/mar/23jk.htm.

[98] "Hizb-ul-Mujahideen," South Asian Terrorism Portal.

[99] Sudha Ramachandran, "Now It's India's Move After Kashmir Split."

[100] "We have launched our own faction of Hizbul Mujahideen," Tufail Ahmed, a former operational chief of the Hizb and Dar supporter, said in a March 27, 2003, report. Ahmed is the younger brother of Zafar Abdul Fateh, who was expelled along with Majid Dar by Salahuddin in May 2002. He claimed that commanders of the new faction on both sides of the border had "unanimously" appointed Ahmed Yasin as their "chief commander." "Around 40 per cent of the Hizb activists are with us," claimed Ahmed. See "Hizb-ul-Mujahideen," South Asian Terrorism Portal.

[101] Habibullah, *My Kashmir*, 66.

[102] Behera, *Demystifying Kashmir*, 151.

in India."[103] This perception was strengthened in 1989 when India's Congress party was replaced by a coalition of parties that was considered unstable and vulnerable to collapse.[104] These perceptions, in addition to the early momentum gained by the Hizbul Mujahideen movement, positively impacted Hizbul Mujahideen recruitment.

Outside of Kashmir, JEI opened camps in Pakistan to provide a recruiting and training base for the movement and to also collect funding to support the jihad.[105] JEI also recruited from an extensive network of mosques and madrasas within Kashmir. Overall since the 1970s, JEI funded approximately 2,500–3,000 facilities that were used to teach its interpretation of Sunni Islam, literature, and culture.[106] Finally, recruits for Hizbul Mujahideen were also drawn from numerous other Islamic organizations, such as Islami Jamiat Tulba, Jamiat Tulba Arabiya, Jamaat-i-Islami, National Labour Federation, and Shabab Milli.[107]

METHODS OF SUSTAINMENT

Hizbul Mujahideen received the most support between 1989 and 1999 from the Pakistani ISI, the Pakistani military, and JEI.[108] For example, JEI collected funds from 7,200 collection boxes placed throughout Pakistan to gather money; JEI also fostered several forums for this purpose in Pakistan and abroad, "including the Kashmir Fund, Kashmir Security Fund, Al-Khidmat Foundation, Martyrs of Islam Foundation, and Islamic Mission of the United Kingdom."[109] In addition to charitable funding and underground *hawala* networks, through which money was embezzled, the Pakistani ISI spent up to $60–80 million per year to support Kashmiri militants. "By the late 1990s, recruits from Pakistan were receiving an estimated $2,500 to $5,000 per year for fighting in Kashmir—far more than what regular

---

[103] Habibullah, *My Kashmir*, 66.

[104] Ibid., 66.

[105] Behera, *Demystifying Kashmir*, 154.

[106] Ibid.

[107] Ibid., 161.

[108] In support of current operations, narcotics trafficking played an increased role in organizational sustainment. Nihar Nayak. "Drug Trafficking in Punjab: An ISI Game Plan to Revive Militancy" Institute of Peace and Conflict Studies, June 24, 2003, http://www.ipcs.org/article/terrorism/drug-trafficking-in-punjab-an-isi-game-plan-to-revive-1066.html.

[109] Behera, *Demystifying Kashmir*, 161.

soldiers receive."[110] The military also offered weapons training to recruits in Pakistan and areas surrounding Kashmir.

METHODS OF OBTAINING LEGITIMACY

Begun in the midst of a popular movement by a repressed population, the Hizbul Mujahideen soon became the most viable option for Kashmiri retaliation and rebellion against the Indian government. This, in addition to limited early success, provided the group with an aura of legitimacy from which their popularity was strengthened. For some years the organization was able to successfully continue this perception as a justifiable liberation force in the eyes of Kashmiri citizens and the international community. Remaining focused on government and military targets, Kashmiri locals provided the movement with support and information. After 1995, however, more extremist Wahhabi organizations entered into Hizbul Mujahideen operations, and the local Sufi population began to be more skeptical of the militant movement in general. Even its own Kashmiri-based leadership started to question the authority emerging from Pakistan. Although JEI justified violent action on the basis of fundamental Islamic ideologies, the increased intensity of action and rhetoric imported by Pakistani-based groups represented the beginning of a loss of legitimacy for the Sunni group, even though its leaders, such as Syed Ali Shah Geelani, attempted to gloss over the discrepancies between the Sunni–Sufi divide.[111]

EXTERNAL SUPPORT

The insurgency in Kashmir grew as a product of Pakistani politics and a cognizant decision to utilize jihad against India. Formulated on an Afghan model that was used against the former Soviet Union, its "guerrilla strategy had two central features: it strived to maintain a low threshold and to invoke the principle of "plausible deniability," portraying the insurgency as an *intifada* or a "freedom struggle" to which Pakistan extended only moral and political support. A low

---

[110] Meyerle, "Death by a Thousand Cuts," 182. For this section, Meyerle cites a number of articles: "Recovery of Explosive Material in Jammu and Kashmir," Institute for Conflict Management, data table; Smith 1998, 3; Ganguly 1997, 40-41; Schofield, *Kashmir in Conflict*, 176; Rizwan Zeb, "Pakistan and the Jihadi Groups in the Kashmir Conflict," in, *Kashmir: New Voices, New Approaches*, ed. Waheguru Pal Singh Sidu (Boulder, CO: L. Rienner Publishers, 2006), 68; *India Today*, May 15, 1994; "Kashmir's Hawala Scandal," *Frontline*, September 5, 1997; "What the Intelligence Bureau Knew," *Frontline*, August 23, 1997; "Plans, Strategies," *Frontline*, November 1, 1996.

[111] Behera, *Demystifying Kashmir*, 152–153.

threshold meant supplying cadre, funds, and weapons without raising the ante on the conventional military front."[112] Whereas in some insurgent relationships, an external state sponsors funds or intercedes between two previously established warring factions, in the case of Kashmir, Pakistan's government enabled the creation of the Hizbul Mujahideen to serve its needs against India. Over time, however, Hizbul Mujahideen became ancillary to Wahhabi groups such as the Harkat-ul-Ansar and the Lashkar-e-Taiba, much as the JKLF became less important once Pakistan determined that Hizbul Mujahideen proved a better asset.

From a contemporary perspective, the Wahhabi terrorist organizations and many members of the ISI became an uncontrollable liability for the Pakistani government, especially since 9/11 and the advent of US interests in Pakistan. Lashkar-e-Taiba, HUA, and ISI have grown overly independent over that last decade, even striking out against targets within Pakistan and greater India. Lacking the need for external government support, these groups grew more independent, turning toward a pan-Islamic message that obviates the need for the state, hoping instead to establish a larger Islamic caliphate.

COUNTERMEASURES TAKEN BY THE GOVERNMENT

The strong feeling of marginalization of Kashmiri Muslims by the Hindu majority within India, combined with the "frequently ruthless Indian response to Kashmiri demands," spurred on popular rebellion and "cost about 100,000 lives, mainly of Kashmiri Muslims and of these the majority at the hands of the Indian Army and paramilitary forces."[113] In fact, the indiscriminate behavior of Indian military and paramilitary forces within the region served as an asset to Hizbul Mujahideen. For example, Indian security forces routinely fired openly into civilian Kashmiri crowds suspected of containing Hizbul Mujahideen militants. Three hundred unarmed demonstrators were killed by police forces over the course of a three-day period in January of 1990.[114] That July, the Indian central government passed new legislation that gave broad powers to the security forces in Kashmir, to include protecting them from prosecution for their actions and also allowing them to suspend the individual rights of suspected insurgents.

---

[112]   Ibid., 82.

[113]   Murtaza Shibli, "Kashmir: Islam, Identity and Insurgency (with Case Study: Hizbul Mujahideen)," Kashmir Affairs, January 2009.

[114]   Bose, *Kashmir: Roots of Conflict*, 109.

These new authorities led to accusations of illegal detentions, torture, and even detainee deaths in 1990 and 1991.[115]

India deployed hundreds of thousands of forces from numerous departments and agencies into Kashmir over the course of the insurgency within Kashmir. Often, communication between these groups hindered progress and effective counterinsurgent activity. Personnel deployed to major urban areas, as well as static checkpoints between cities. These checkpoints drew heavy fire because they were easy targets; their supply lines were also effectively targeted by Hizbul forces during resupply operations. In addition, India's military was trained for conventional war against Pakistan, not for large-scale insurgency operations. Routine cordon-and-search operations took into account civilian presence, and Indian action in general made little to no effort to respond to civilian needs or satisfy the requirements of basic stability operations. Indian officers were also short on troops and had to cover large areas of the thousand-mile LoC with insufficient numbers of soldiers.[116]

By the mid to late 1990s, Indian forces proved successful in infiltrating and accessing the chain of command structure for Hizbul Mujahideen, a fairly hierarchical organization, unlike its cellular counterparts such as Lashkar-e-Taiba. Surrendered Kashmiri militants were used to infiltrate and gain access to membership, communications, transportation routes, and safe houses. With this information, paramilitary units were able to pick off members of the insurgent leadership. "Several surrendered militants [had] also been absorbed into the police as well as units that were specifically fighting militancy in J&K."[117]

Another organization that initially proved useful to the Indian counterinsurgency was the Ikhwanis, a Kashmiri group attacked and marginalized by the Hizbul Mujahideen when it also attacked the JKLF. An ethnic Kashmir group, the Ikhwanis fostered a deep suspicion and hatred for the Islamic militant groups and were often the victims of violence by these groups.[118] At first the Indian government armed and provided cover for Ikhwani counterinsurgents. By 1998, however, they were stripped of their cover and labeled a liability for the Indian government because of their own terror campaigns against the Kashmiri people. "They were brutal in their methods to elicit information from the locals. They used their weapons to fight the militancy but

---

[115] Meyerle, "Death by a Thousand Cuts," 148.

[116] Meyerle, "Death by a Thousand Cuts," 155.

[117] Sudha Ramachandran, "The Downside to India's Kashmir 'Friendlies,'" *Asia Times*, September 26, 2003, http://www.atimes.com/atimes/South_Asia/EI26Df07.html.

[118] Ibid.

gradually they used it to settle personal scores, to extort and to further their individual interests. The role of the Ikhwanis in turning the tide against militancy was substantial. But their contribution has not been acknowledged enough by Delhi, prompting some to accuse India of not doing enough to protect its own in the Valley."[119]

SHORT- AND LONG-TERM EFFECTS

CHANGES IN THE ENVIRONMENT

The question of Kashmiri independence cannot be removed from the larger political confrontation between India and Pakistan. It also cannot be removed from the political environment internal to Pakistan's government, one subject to a larger international system. When the insurgency first began, Pakistan promised Kashmiri militants that their fight would be the first wave of a larger Pakistani offensive within Kashmir. This, however, never developed. Rather, Pakistan found utility in continuing the primary effort of proxy support and determined that autonomy from India was not as advantageous to Pakistani interests as Kashmir's accession to Pakistan. Thus, Pakistan manipulated the environment to marginalize Kashmiri militant groups that were unsupportive of this endeavor, putting weight and resources behind Hizbul Mujahideen at the expense of organizations such as the JKLF. Then, in the mid-1990s, the JEI lost prestige within Pakistan to the Wahhabi Deobandi faction. Pakistan came to believe that their control of the Kashmiri situation would increase with the development of Pakistani-based militant Wahhabi groups, such as the Lashkar-e-Taiba. "With the induction of Pakistan-based jihadi organizations, the Kashmiri component—its cadre, ideology, and political goals—became eclipsed."[120]

Until the late 1990s, Pakistan continued to use militant forces to engage India within Kashmir. In mid-May of 1998, however, India conducted two nuclear tests and Pakistan reciprocated with its own tests roughly two weeks later on May 28. State relations suffered, and tension produced by the tests caused the conflict to spill across the border. During the summer, there was a marked deterioration in relations between India and Pakistan and in May the number of border crossings in Kashmir was reported to be the highest ever.[121] Pakistan's

---

[119]   Ibid.

[120]   Behera, *Demystifying Kashmir*, 158.

[121]   Sten Widmalm, *Kashmir In Comparative Perspective, Democracy and Violent Separatism in India* (Oxford: Oxford University Press, 2006), 140–141.

initiation of the Kargil war in 1999 furthered the deteriorating relationship between the two states and also served to embolden and legitimize participating Wahhabi-based terrorist organizations. Its loss reconfirmed its belief that the best option for pressuring India was through the use of low-intensity conflict in Kashmir and that the sponsorship of the Kashmir insurgency—as well as insurgent activities in other areas of India—could provide Pakistan with an effective military option against India short of a full military confrontation.[122]

The viability of Hizbul Mujahideen in Kashmir was influenced by a number of external and internal variables, including the increased dominance of Wahhabi terrorist organizations and the leadership divisions within Hizbul Mujahideen. These conditions were compounded after the events of 9/11, including Pakistan's growing relationship with the United States, which led to US pressure on Pakistan to limit Hizbul Mujahideen's operations in both Kashmir and India.[123] Since 2000, Hizbul Mujahideen has maintained an operational capacity and is still capable of conducting attacks within Kashmir. At the same time, its leadership within the Kashmiri independence movement has become marginalized, and its direction has become increasingly influenced by its political leaders in Pakistan.[124]

CHANGES IN GOVERNMENT

Both Pakistan and India experienced major changes in government and national leadership between 1990 and 2000. Pakistan's loss at the battle of Kargil in 1999 prompted internal criticism and dissatisfaction, and its 1998 nuclear tests prompted a series of destabilizing economic sanctions; in combination, these conditions contributed to the successful coup led by Pakistan's chief of the army staff, General Pervez Musharraf.[125] Changes within the Indian government were less volatile as security forces battled numerous insurgent groups within Kashmir, as well as other regions within India.[126] In late 2002, the Indian Army finally called for a major change in the organization structure for the forces engaged in Kashmir by recommending in a

---

[122] Behera, *Demystifying Kashmir*, 87.

[123] The Hizbul Mujahideen is one of the thirty-two organizations proscribed under the Prevention of Terrorism Act, 2002. Haqqani, "The Ideologies of South Asian Jihadi Groups," 18.

[124] Widmalm, *Kashmir In Comparative Perspective*, 136.

[125] Ibid., 145.

[126] Sumit Ganguly, "Explaining the Kashmir Insurgency: Political Mobilization and Institutional Decay," *International Security* 21, no. 2 (Fall 1996).

classified document that all police and paramilitary forces be brought under the command of the Army for counterinsurgency operations.[127]

CHANGES IN POLICY

Indian military and paramilitary forces in Kashmir did not effectively adapt to counterinsurgency conditions. "Despite extensive experience fighting insurgencies in its northeastern states since the 1950s, Pujab in the 1980s and early 1990s, and in Sri Lanka, the Indian Army remained a conventional fighting force oriented primarily towards defense against Pakistan and China; it developed no new doctrine as a result of its experience in Kashmir."[128] Cordon-and-search practices continued as the primary method of confronting militants and general brutality has been curtailed only slightly. By the late 1990s, India was successful in developing the Rashtriya Rifles, a professional counterinsurgency force that could match the tactics and operational mobility of the insurgents;[129] however, these new counterinsurgency capabilities were never effectively integrated into the overall government strategy that continued to focus on large-unit operations and the physical control of terrain.[130] Through 1999, the Indian military still lacked new organizational and operational approaches to counterinsurgency in Kashmir and continued to have insufficient coordination between the military and security agencies involved in Kashmir. Moreover, the focus of the Indian Army was still on the potential for conventional warfare against Pakistan and not the real-time, day-to-day security risks in Kashmir.[131]

CHANGES IN THE REVOLUTIONARY MOVEMENT

Two broad thematic changes occurred between 1989 and 2000 within the Hizbul Mujahideen revolutionary movement. First, its popular support began to lose strength within Kashmir and decrease in status within Pakistan by the mid to late 1990s. "Military violence was most effective in 1989–90. At the peak of militancy in 1990, there was a naïve belief that the Indian state would soon 'withdraw,'

---

[127] Meyerle, "Death by a Thousand Cuts," 179. The document was called the "J&K Strategy for Resolution of Internal Conflict."

[128] Meyerle, "Death by a Thousand Cuts," 180. Here he cites the following article: Rajesh Rajagopalan, "Restoring Normalcy: The Evolution of the Indian Army's Counterinsurgency Doctrine," *Small Wars and Insurgencies* 11, no. 1 (Spring 2000): 62–64.

[129] Ibid., 178.

[130] Ibid., 178 and 180. Here he cites the following article: Rajesh Rajagopalan, "Restoring Normalcy: The Evolution of the Indian Army's Counterinsurgency Doctrine."

[131] Ibid., 179.

granting Kashmir independence, or that Pakistan would attack India to liberate Kashmir."[132] Armed conflict, however, did not yield the Hizbul Mujahideen's intended results, and support within Kashmir began to wane, making recruitment within Kashmir more challenging. Pakistan, whether attributing this trend to fatigue or finding it more convenient to support Pakistani-based groups, began to send foreign mercenaries, Afghan veterans, and Pakistanis into the Valley, which slowly eroded the support base of the Hizbul Mujahideen.[133]

The second major trend that developed within the movement during this time focused on the discrepancies between the Kashmiri-based leadership and Hizbul Mujahideen's Pakistani-based leadership. When Abdul Majid Dar made the move to disassociate from the Hizbul Mujahideen leadership in Pakistan, he represented the resentment and alienation of the resident Kashmiri membership. "The bulk of the midlevel command in the Hizbul Mujahideen's south and central Kashmir divisions threw their weight behind the expelled leader. Zafar Fateh remarked: 'Hizb is not anybody's handmaiden . . . Those who are sitting across [Pakistan-occupied Kashmir] cannot claim to be representatives of Kashmir and the organization as they have no understanding of the ground situation.' "[134] Since 2000, Hizbul Mujahideen Pakistani-led leadership has attempted to purge the organization of dissent, move closer to the ISI, and stay relevant as US pressure on Pakistan has cut off much of Hizbul Mujahideen's funding. Since 2008, however, Hizbul Mujahideen has again called for the conduct of jihad against India.[135]

BIBLIOGRAPHY

Amin, Tahir. *Mass Resistance in Kashmir: Origins, Evolution, Options.* Islamabad: Institute of Policy Studies, 1996.

Behera, Navnita Chadha. *Demystifying Kashmir.* Washington, DC: Brookings Institution Press, 2006.

Bose, Sumantra. *Kashmir: Roots of Conflict, Paths to Peace.* Cambridge, MA: Harvard University Press, 2003.

Burki, Shahid Javed. *Kashmir: A Problem in Search of a Solution.* Washington, DC: United States Institute of Peace, 2007.

Desmond, Edward. "The Insurgency in Kashmir (1989–1991)." *Contemporary South Asia* 4, no. 1 (1995).

---

[132] Behera, *Demystifying Kashmir,* 167.

[133] Ibid., 155–156.

[134] Behera, *Demystifying Kashmir,* 158.

[135] Shibli, "Kashmir," 261.

Ganguly, Meenakshi. ""Everyone Lives in Fear" Patterns of Impunity in Jammu and Kashmir." *Human Rights Watch* 18, no. 11(C) (2006).

Ganguly, Sumit. "Explaining the Kashmir Insurgency: Political Mobilization and Institutional Decay," *International Security* 21, no. 2 (Fall 1996).

Habibullah, Wajahat. *My Kashmir: Conflict and the Prospects of Enduring Peace.* Washington, DC: United States Institute of Peace Press, 2008.

Haq, Shamshul. *Hizbul Mujahideen: Its Background and Struggle,* translated from Urdu. Rawalpindi, Pakistan: Markaz Matruit Kashmir, May 1994.

Haqqani, Husain. "The Ideologies of South Asian Jihadi Groups." *Current Trends in Islamist Ideology* 1 (April 2005): 15.

Mahapatra, Debidatta Aurobinda. "Conflict and Development in Kashmir: Challenges and Opportunities." In Proceedings of the International Conference, Center for International Studies and Cooperation. Kathmandu, Nepal: SAHAKARYA Project March 2007.

Meyerle, Gerald. "Death by a Thousand Cuts: The Dynamics of Protracted Insurgency in Kashmir and Sri Lanka." PhD dissertation, University of Virginia, 2008.

Prakash, Siddhartha. "The Political Economy of Kashmir since 1947." *Contemporary South Asia* 9, no. 3 (2000): 315–337.

Puri, Balraj. "Kashmiriyat: The Vitality of Kashmiri Identity." *Contemporary South Asia* 4, no. 1 (1995).

Samad, Yunas. "Kashmir and the Imagining of Pakistan." *Contemporary South Asia* 4, no. 1 (1995).

Schaffer, Terestia C. *Kashmir: The Economics of Peace Building.* Washington, DC: Center for Strategic and International Studies Press, 2005.

Schofield, Victoria. *Kashmir in Conflict: India, Pakistan and the Unending War.* London and New York: I. B. Tauris, 2003.

Shah, Mehtab Ali. "The Kashmir Problem: A View from Four Provinces of Pakistan." *Contemporary South Asia* 4, no. 1 (1995).

Shibli, Murtaza. "Kashmir: Islam, Identity and Insurgency (with Case Study: Hizbul Mujahideen)." *Kashmir Affairs,* January 2009.

Tarabi, Alifuddin. *Hizbul Mujahideen: The Principles and Struggle,* translated from Urdu (n.d.).

Tremblay, Reeta Chowdhari. "Kashmir: The Valley's Political Dynamics." *Contemporary South Asia* 4, no. 1 (1995).

————. "Nation, Identity, and the Intervening Role of the State: A Study of the Secessionist Movement in Kashmir." *Pacific Affairs* 69, no. 1 (1996–1997): 471–497.

Widmalm, Sten. *Kashmir In Comparative Perspective, Democracy and Violent Separatism in India.* Oxford: Oxford University Press, 2006.

Zeb, Rizwan. "Pakistan and the Jihadi Groups in the Kashmir Conflict." In *Kashmir: New Voices, New Approaches,* edited by Waheguru Pal Singh Sidu. Boulder, CO: L. Rienner Publishers, 2006.

SECTION IV

·················

REVOLUTION BASED ON
RELIGIOUS FUNDAMENTALISM

GENERAL DISCUSSION

At the time of this writing, the idea of a religious fundamentalist mind-set as a distinct category of "revolutionary type" is still under debate. Due to the topical interest, we have separated those cases that explicitly concern themselves with the creation of a dedicated religiously driven government system and fight against secular government power. There are certainly other issues that enter into the objectives and motivations of the groups described here (the Palestinian/Israel issue, US foreign military presence, etc.), but we will leave it to the reader to decide whether we have made a case for a separable category.

The three cases in this section are explicitly motivated by the establishment of an Islamic-driven legal/government system, and the reestablishment of the global caliphate. They reject Western-style government as an appropriate system and reject Western political, social, and cultural influences. The category of a religious fundamentalist-driven revolution is not meant to be exclusive to this exclusively Islamic vision. There are other smaller examples of groups looking to establish either Christian or Jewish hard-line religiously driven government systems. However, to date, they have not been significant enough to use as an educational tool about insurgencies and revolutions.

Three factors are important to stress with the religiously motivated revolution. The first two specifically involve religious issues. First is the ideologically driven nature of the revolution, its necessity, its legitimacy, its purity, and its eventual outcome as foretold in religious texts. This ideology is a very specific, particular interpretation that is not widely held within the greater religious community but defines the group's social identity and evokes a strong affinity within the group. In the past decade, a great deal has been written about the fundamental basis of Al Qaeda's motivations and its use of Quranic scholarship and pronouncements to build its case overall, as well as to justify (religiously) its specific operations. *Fatwas*—religious opinions of Islamic law that provide the justification for types of operations (e.g., suicide bombings, incidental deaths of Muslims in the prosecution of an attack on infidels)—are important for legitimacy of the group both internally and within the greater Islamic community. The narrative structure setup since Qutb on the comprehensiveness and importance of jihad provides both motivation for individual participation, but also glory and eventual success as a destined outcome.

The second factor is highly related to the first, being dependent on the interpretation of religious texts and doctrine, and involves the definition of the "infidel" or outsiders. As discussed in previous sections, the definition of the "other" or "out-group" is important in the set up of a revolutionary environment and allows for advanced radicalism to occur. The three cases presented here all show how the definition of the threat—whether it be secular government power, Western influence within the nation, or a "far-enemy" like the United States—drives the strategy of the groups, their selection of operations, and their definition of success.

The last distinction to make with this category is their view of the opposing government and their willingness to accept political solutions. Like in the ethnic category, political solutions do not easily solve these types of revolutions, but here it is their ideological purity that disallows them to negotiate or share power. Religious fervor is a strong and constant motivator, especially when driven by a narrative of predestined victory.

EGYPTIAN ISLAMIC JIHAD (EIJ)

Jason Spitaletta

SYNOPSIS

Egyptian Islamic Jihad (EIJ) was a terrorist group that emerged from the larger Islamic movement in Egypt in the latter half of the twentieth century and was considered the first element of the modern global Salafist jihad. This Islamic political organization emerged in 1928 as the Muslim Brotherhood, serving as a counterweight first to the colonial government and then to the secular Egyptian government under Presidents Nasser, Sadat, and Mubarak. The Muslim Brotherhood was officially opposed to violence as a means of conveying its objective, which was to affirm the Quran and Sunnah as the sole reference point for ordering the lives of the Muslim individual, family, community, and state. This renunciation of violence precipitated the development of groups who espoused a more aggressive means of obtaining similar goals. Notable among those are EIJ and the Egyptian Islamic Group (EIG). The latter emerged as an umbrella organization for the proliferation of Islamic student groups that emerged in the 1970s, whereas the former was a more secretive and elitist organization that considered itself the ideological vanguard of the global Salafist jihad movement of the late twentieth and early twenty-first centuries. EIJ, although not as prolific in terms of the number and boldness of its offensive actions as EIG, under the leadership of Dr. Ayman Zawahiri in the late 1980s and 1990s, transformed from a domestic threat to the Egyptian regime to a global terrorist concern. EIJ members were responsible for terrorist operations, including assassinations, suicide bombings, and psychological warfare in Egypt, Afghanistan, Pakistan, Sudan, Tanzania, and Kenya. By 1997, EIG had formerly renounced violence while EIJ, operating outside Egypt, continued to further its objectives through a strategic partnership with Osama bin Laden's Al Qaeda, ultimately merging in 2001.

TIMELINE

1922	The Muslim Brotherhood is founded in Egypt by Hassan al-Banna.
1949	Al-Banna is assassinated by the Egyptian government.
1951	Qutb returns from the United States and joins the Muslim Brotherhood.
1952	The Egyptian Revolution of 1952, led by the Free Officers movement, executed a coup d'etat deposing King Farouk.
1953	All Muslim Brotherhood activities are banned by the Egyptian government after the Brotherhood insists that Egypt be governed under *shari'a* law.
1954	Qutb is arrested twice and eventually sentenced to 25 years of hard labor; while in prison, he writes *Milestones*.
1966	Qutb is hanged for his alleged involvement in a Muslim Brotherhood conspiracy.
1967	Israel defeats the combined forces of Egypt, Jordan, and Syria, gaining control of the Sinai Peninsula, the Gaza Strip, the West Bank, East Jerusalem, and the Golan Heights.
1970	Nasser dies; Sadat assumes the presidency.
1973	A joint surprise attack by Egypt and Syria against Israel enjoyed early tactical success but did not result in a decisive victory. The war itself laid the groundwork for the ensuing peace process and, ultimately, the Camp David Accords.
	The Gama'a al-Islamiyya, or EIG (initially encouraged by the Sadat regime), is formed.
1977	Islamists dominate political and social life in Egyptian universities and begin printing literature (including that of Qutb).
1978	The Sadat regime strips EIG of its victories in the student elections, further stoking tension among the regime, student populace, and broader Islamist sentiment.
1979	Signature of the Camp David Accords.
	Abdel Salam Faraj, author of *The Neglected Duty*, unites several small jihadi groups under his leadership, signifying the beginning of the group known as Tanzim al-Jihad from which later sprang EIJ.
	The Soviet Union invades Afghanistan to support the ruling communist government against a loose confederate of tribal militias.

1981	During a Shura, leaders of the two branches of the Tanzim al-Jihad (Faraj and Zuhdi) agree to merge their groups (now EIG) under the leadership of Sheikh Umar Abd-al Rahman, who agrees to be the mufti.
	The regime bans EIG and other Islamist groups; numerous Brotherhood members are imprisoned
	Islamists (led by the brother of imprisoned Mohammed al-Islambuli, a member of Tanzim al-Jihad) assassinate Egyptian President Anwar Sadat.
	Hosni Mubarak is elected president.
1984	Sheikh Abdullah Azzam and Osama bin Laden create the Mekhtab al-Khidemat (MAK) or Services Bureau to facilitate administrative problems for foreign Muslim fighters.
1986	EIG and EIJ members travel to Pakistan and Afghanistan to participate in jihad along with the other Afghan Arabs.
1992	EIG and EIJ commence terrorist attacks in Egypt against the Mubarak regime.
1993	The first bombing of the World Trade Center in New York occurs.
1997	The Luxor massacre occurs: Six operatives from EIG and an element of EIJ kill sixty-three tourists in Egypt; this is the last terrorist event attributed to EIG.
	EIG, disappointed with Al Qaeda's inability to help Rahman (now imprisoned in the United States) and the decision to target the United States rather than Egypt, declares a unilateral cease-fire.
2000	Sheikh Omar withdraws his support for the EIG peace initiative.
2001	In June, Al Qaeda and EIJ merge, forming "Qaeda al-Jihad."

THE ENVIRONMENT OF THE REVOLUTION

PHYSICAL ENVIRONMENT

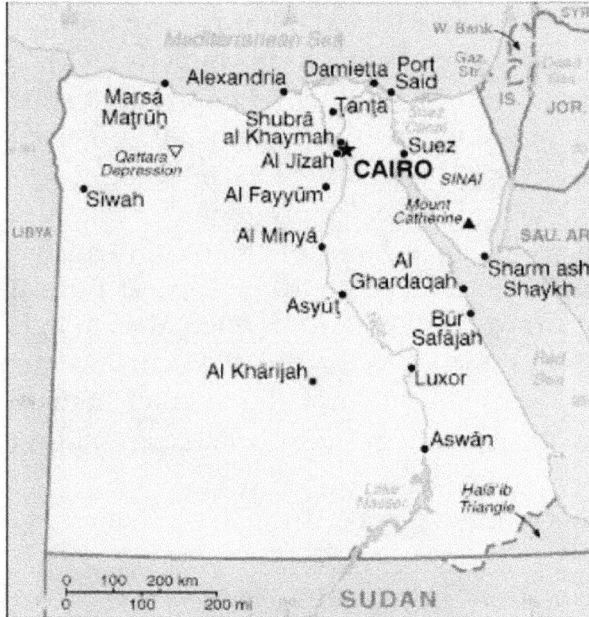

Figure 1. Map of Egypt.[1]

Egypt's geography is defined by the regularity and richness of the annual Nile River flood, coupled with the semi-isolation provided by deserts to the east and west. Roughly the size of Texas and New Mexico combined; the country covers 1,001,449 square kilometers. Egypt is located in northeastern Africa and includes the Sinai Peninsula. Its natural boundaries include more than 2,900 kilometers of coastline along the Mediterranean Sea, the Gulf of Suez, the Gulf of Aqaba, and the Red Sea.[2] Neighboring countries include Israel, Libya, Sudan, and the Gaza Strip.[3] Outside of the major urban centers of Cairo and Alexandria, small communities clustered around oases and historic trade and transportation routes are distributed throughout the desert regions.[4]

---

[1] Central Intelligence Agency, "Egypt," *The World Factbook*, accessed March 14, 2011, https://www.cia.gov/library/publications/the-world-factbook/maps/maptemplate_eg.html.

[2] Central Intelligence Agency, "Egypt," *The World Factbook*, accessed August 23, 2010, https://www.cia.gov/library/publications/the-world-factbook/geos/eg.html.

[3] H. C. Metz, *Egypt: A Country Study* (Washington, DC: Federal Research Division. Library of Congress, 1990), accessed August 23, 2010, http://lcweb2.loc.gov/frd/cs/egtoc.html.

[4] Ibid.

CULTURAL AND DEMOGRAPHIC ENVIRONMENT

Egypt is the most populous country in the Arab world and the second most populous on the African continent. The majority of its 79 million people are concentrated in Cairo and Alexandria as well as along the banks of the Nile, the Nile Delta, and the Suez Canal. These regions are among the most densely populated in the world, containing an average of more than 3,820 persons per square mile compared to 181 persons per square mile for the country as a whole.

Egyptians are a relatively homogeneous people of Hamitic origin with Mediterranean and Arab influences in the north and Nubian influences in the south. Ethnic minorities include a small number of Bedouin Arab nomads in the eastern and western deserts and in the Sinai, as well as some 50,000–100,000 Nubians clustered along the Nile in upper or southern Egypt.[5]

During the 1980s and 1990s, the literacy rate of the adult population was approximately 58%. Education, which is compulsory from ages six to fifteen, is free for both secondary and tertiary institutions. By 1994, 87% of Egyptian children had entered primary school, and enrollment rates continue to rise. Major universities include Cairo University, Alexandria University, and Al-Azhar University, one of the world's major centers of Islamic learning whose history spans millennia.[6] Beginning in the 1970s and continuing today, Al-Azhar University has served as a friction point between the increasingly secular government and various Islamic groups who opposed those policies, with religious scholars, or *ulema*, often caught in the middle.[7]

SOCIOECONOMIC ENVIRONMENT

Egypt's economy was highly centralized during the rule of former President Gamal Abdel Nasser but became more accessible to the Egyptian financial elite and open to foreign investment under former Presidents Anwar Sadat and Mohamed Hosni Mubarak. During the 1960s, Sadat encouraged Western financial investment but understood that Western investment would not be forthcoming until there was peace between Egypt and Israel, the Soviet influence was eliminated, and the climate became more favorable to Western capitalism.[8]

---

[5] Ibid.

[6] Ibid.

[7] G. Abdo, *No God but God: Egypt and the Triumph of Islam* (Oxford, UK: Oxford University Press, 2000).

[8] Metz, *Egypt: A Country Study.*

However, the years between 1967 and 1974—the final years of Gamal Abdel Nasser's presidency and the early part of Anwar Sadat's— were lean times, with growth rates of only about 3.3%. The slowdown was caused by many factors, including agricultural and industrial stagnation and the costs of the June 1967 War. Investments, which were a crucial factor for growth, also plummeted and only recovered in 1975 after dramatic increases in oil prices. Beginning in 1974, the Sadat government introduced a series of laws intended to restore and promote private ownership of previously socialized sectors of the economy. Like many oil-producing counties, Egypt benefitted from the oil boom but suffered during the subsequent slump.[9] In this period, agriculture as a percentage of gross domestic product (GDP) dropped from approximately 33% in the early 1950s to approximately 15% by the middle of the 1970s. Likewise, manufacturing declined from approximately 15% to 12%. By contrast, industry as a percentage of GDP increased from approximately 13% to 35%. The rise was almost solely attributed to energy-related activity, especially oil drilling. The inability to adapt and transform the manufacturing industry was a primary contributor to the Egyptian economy's inability to become self-sustaining and to its reliance on oil and external financing.[10]

Beginning in the 1980s, the Mubarak government encouraged internal migration to newly irrigated land reclaimed from the desert. However, the proportion of the population living in rural areas has continued to decrease as people move to the cities in search of employment and a higher standard of living.[11]

HISTORICAL FACTORS

The Islamic movement in Egypt can be traced to The Society of Muslim Brothers, more popularly known as the Muslim Brotherhood, which was founded by Hassan al-Banna in 1928 as an attempt to turn

---

[9] Ibid. "The 1973 oil crisis started in October, when the members of Organization of Arab Petroleum Exporting Countries or the OAPEC (consisting of the Arab members of OPEC, plus Egypt, Syria and Tunisia) proclaimed an oil embargo in response to the U.S. decision to support Israeli military during the Yom Kippur war. The crises lasted until March 1974. The embargo disrupted supply, causing an economic recession, and exacerbated a rift within NATO causing some nations to disassociate themselves from the U.S. Middle East policy. Independently, OPEC members decided to use their leverage over the world price setting mechanism for oil to stabilize their incomes by raising world oil prices. Industrialized economies relied on crude oil, and OPEC was their predominant supplier. Because of the dramatic inflation experienced during this period, a some economists have theorized the price increases were to blame, however, the causality is not necessarily widely accepted."

[10] Ibid.

[11] Ibid.

Egypt away from secularism and toward an Islamic government based on *shari'a*[12] and traditional Muslim principles. Al-Banna believed that applying the Salafist[13] interpretation of Islam to modern exigencies was the antidote to Western domination and a solution to the political, social, and economic problems that plagued Egypt and the Muslim world.[14] Initially serving in opposition to colonial occupation, the organization has continued to advocate Islamic principles, often in opposition to the secular Egyptian governments under Presidents Nasser, Sadat, and Mubarak. Successive Egyptian administrations have targeted the group for its often illegal oppositional activities.[15]

Since 1954, the Muslim Brotherhood has served as a religious charitable and educational institution, having been banned as a political party by Nasser. In the 1940s and early 1950s, the Muslim Brotherhood appealed primarily to urban civil servants and white- and blue-collar workers. Al-Banna was executed by the Egyptian government in 1949 for an alleged assassination attempt on the Egyptian prime minister,[16] and his pioneering organization was significantly dismantled as a result.[17] However, his enduring legacy was a reminder to all Muslims that jihad against unbelievers is an obligation of all Muslims and that obligation extended beyond the defense of Muslim lands but also served as a means "to safeguard the mission of spreading Islam."[18]

In 1950, a year after its leader was hanged; the Muslim Brotherhood instigated an uprising against the British, whose lingering occupation of the Suez Canal zone enraged Islamists and secular nationalists alike.

---

[12] *Shari'a*, or the Islamic law, "is the collection of prohibitions and regulations derived from the Qur'an and Traditions [Sunna]. The term covers a more comprehensive area than is commonly understood by the term 'law' in English translation. [It] encompasses but it's not limited to matters of law as it is understood in the West. . . . shari'a is perceived as infallible legislation for almost all aspects of human existence and so governs the seemingly disparate realms of religious belief, practice, and observance of law. It extends to matters of administration, justice, morality, ritual washing, dispensation of property and political treaties." This is the sense in which Sayyid Qutb, discussed below, understood and applied this concept. Roxanne Leslie Euben, *Enemy in the Mirror: Islamic Fundamentalism and the Limits of Modern Rationalism: A Work of Comparative Political Theory* (Princeton, NJ: Princeton University Press, 1999), 239.

[13] The term Wahhabism is used interchangeably with Salafism, although those who follow this particular variant of Islam view the term Wahhabbism as pejorative. Salaf means "to follow," as in following the ways of the Prophet Muhammad and his companions. Wahhab refers to the eighteenth-century Muslim scholar Muhammad bin Abd al-Wahhab, who advocated a return to the religious and social principles that prevailed during the Prophet's lifetime.

[14] D. C. Eikmeier, "Qutbism: An Ideology of Islamic-Fascism," *Parameters* (2007).

[15] Richard P. Mitchell, *The Society of the Muslim Brothers* (New York: Oxford University Press, 1993; 1969).

[16] Ibid.

[17] Mitchell, *The Society of the Muslim Brothers*.

[18] Eikmeier, "Qutbism: An Ideology of Islamic-Fascism."

That same year, Sayyid Qutb, a well-educated literary critic in Cairo, returned to Egypt after spending two years at Colorado State College of Education under the sponsorship of the Egyptian Ministry of Public education. Qutb left Egypt a secular nationalist with an opposition to British occupation and returned to his homeland a radical Islamist. His writings stoked the revolutionary fire within disillusioned young Muslims who sought a more active role in returning Egypt to the center of the Islamic world. It was Qutb's unique Salafist interpretation, and not al-Banna's, that would provide the ideological foundation of the late twentieth-century global Salafist insurgency.

In January 1952, the Muslim Brotherhood organized a protest in response to the British massacre of 50 Egyptian policemen. Mobs set fire to any establishment representing the British presence, resulting in thirty deaths, more than 700 buildings destroyed, and 12,000 Egyptians left homeless.[19] On July 23, 1952, the Free Officer's Movement, led by Army Colonel Gamal Abdel Nasser, deposed King Farouk in a bloodless coup.[20] The 1952 revolution was a seminal event in Egyptian political history. Its implications were felt internationally as Egypt positioned itself more aggressively as the leader of the modernizing Muslim world and domestically as the relationship between the Nasser government and supporters of the Islamic movement would soon sour as a result of Nasser reneging on an alleged agreement with the Muslim Brotherhood. A struggle for the influence over and support of the populace developed immediately between the Free Officer's Movement, which enjoyed the support of the Army, and the Muslim Brotherhood, which had a significant presence in the mosques and growing popular support but lacked the weapons and training necessary for a violent confrontation.[21]

Throughout the 1950s and 1960s, political Islam served as an intellectual and ideological counterweight to the uniquely Egyptian blend of Arab nationalism and socialism espoused by the Nasser government. Nasser created the Arab Socialist Union to be the sole political party and as a means of gathering the support of the masses, a policy that reclassified the Muslim Brotherhood from a political party to a socioreligious organization. The ideological war over Egypt's future reached a climax on the night of October 26, 1954, when a Muslim Brother, Abdul Munim Abdul Rauf, attempted to assassinate

---

[19] Lawrence Wright, *The Looming Tower: Al Qaeda and the Road to 9/11* (New York: Knopf, 2006).

[20] Paul A. Jureidini et al., *A Casebook on Insurgency and Revolutionary Warfare: 23 Summary Accounts* (Washington, DC: American University, Special Operations Research Office, 1962).

[21] Mitchell, *The Society of the Muslim Brothers.*

Nasser as he spoke before a crowd in Alexandria; six conspirators were quickly executed, and more than 1,000 Muslim Brothers, including Qutb, were imprisoned. Qutb's imprisonment, and the accompanying physical and psychological torture that he endured, appeared to aid in the articulation of his revolutionary Salafist doctrine that would inspire readers to wage jihad for generations. His most famous work, translated as *Milestones* or *Signposts Along the Road*, was published in Cairo in 1964. It was quickly banned, and anyone caught with a copy was charged with sedition.[22] Qutb was released from prison the same year *Milestones* was published.

In 1965, Qutb was arrested again and charged with conspiracy to overthrow the government; he was hanged in August 1966. Authorities hoped that the execution would extinguish the Islamist threat to the Nasser regime. However, the tactic did not have the desired effect.[23] The execution drove many Islamists underground, and clandestine cells began to form among the restless, disillusioned, and disaffected youth, especially students.[24] The Qutbist legacy was not only the binary worldview (or one that perceived the world starkly divided between good and evil, with the former being true Islam and the latter being the state of ignorance, *jahiliyya*, in which the rest of the world existed) but also the resurgence of *takfir*[25] (or declaration of apostasy) authority pronounced by the *umma* (Muslim community) and no longer solely the prerogative of the *ulema* (religious scholars), *imam* (prayer leader), or *shari'a* court.[26] Qutb's teachings were embodied by a number of young Muslims who opposed both the government and the Muslim Brotherhood's perceived appeasement of the regime.

In the eleven years leading up to the June 1967 War, also known as the Arab–Israeli War and the Six-Day War, the military had been intensively trained and outfitted with new Soviet weapons and equipment, which led to a false sense of capability. Despite indications of an imminent attack, the Israelis still took Egypt by surprise on June 5, and, within three hours, the Israelis had destroyed 300 Egyptian

---

[22] Ibid.

[23] Ibid.

[24] Marc Sageman, *Understanding Terrorist Networks* (Philadelphia: University of Pennsylvania Press, 2004).

[25] The traditional sentence for apostasy under *shari'a* was execution, amputation, or expulsion, thus requiring stringent evidence for such accusations and often requiring an Islamic court or a religious leader to pronounce a fatwa (religious edict) of *takfir*. This subordination or decentralization of authority to declare *takfir* now empowered devout Muslims to rationalize not only judgment of their fellow Muslims but also violent punishment. Jarret M. Brachman, *Global Jihadism: Theory and Practice* (New York: Taylor & Francis, 2008).

[26] Devin R. Springer, L. Regens, and David N. Edger, *Islamic Radicalism and Global Jihad* (Washington, DC: Georgetown University Press, 2009).

combat aircraft, including all of Egypt's thirty long-range bombers. Although it lasted less than a week, the conflict was disastrous for the Egyptians and the greater Middle East. Egypt suffered the loss of 10,000 soldiers and 1,500 officers, the capture of 5,000 soldiers and 500 officers, and the destruction of 80% of its military equipment.[27] The rapidity and efficiency with which Israel was not only able to subdue the Egyptian forces but also recapture the Jewish quarter of Old Jerusalem left the Middle East shocked and many of its Arab citizens humiliated.

When Nasser died of a heart attack in 1970, he was succeeded by Anwar Sadat, who had been appointed to the vice presidency in 1969. Sadat, who had a better appreciation than his predecessor for the increasingly dominant role of the Islamic movement on the Egyptian political landscape, further reduced the military's influence in government. However, Sadat was careful to protect the career interests of professional soldiers and to provide for the material requirements of the military. He was also careful not to alienate the Islamic movement, initially supporting its growing presence on university campuses. The Islamic groups that began on campus evolved into EIG, where their political prominence and perceived threat to the Egyptian government grew.[28] In the latter half of 1972, there were large-scale student riots, and some journalists came out publicly in support of the students.

In October 1973, during the Muslim holy month of Ramadan and the Jewish holy day of Yom Kippur, Egyptian forces launched a successful surprise attack across the Suez Canal, while Syria executed a simultaneous attack on Israel. On October 22, the United Nations Security Council passed Resolution 338, calling for a cease-fire by all parties within twelve hours in the positions they occupied.[29] Militarily, the result was a stalemate; however, Sadat's supporters viewed the conflict favorably because it helped recover some of the collective pride lost in the 1967 war.[30]

To establish his political legitimacy, Sadat attempted to make peace with the broader Islamic movement and the Muslim Brotherhood in particular. Sadat offered the Brothers a deal in return for their political support: They were allowed to preach and to advocate

---

[27] Michael Scheuer, *Through Our Enemies' Eyes: Osama Bin Laden, Radical Islam, and the Future of America* (Washington, DC: Potomac Books, Inc., 2007).

[28] Gilles Kepel, *The Roots of Radical Islam* (London: Saqi, 2005).

[29] Scheuer, *Through Our Enemies' Eyes: Osama Bin Laden, Radical Islam, and the Future of America.*

[30] The 1973 victory would also increase Egypt's advantage in future negotiations with Israel, a notion that Sadat did not express publicly.

peacefully, although their status as a political party would not be restored. His misperception or underestimation of the Qutbist ideology among the loosely, if at all, connected clandestine groups, which would evolve into the terrorist organizations EIG and EIJ, later served to undermine the influence of the Brotherhood and funnel the recently released Islamist prisoners to these smaller groups.[31] The Muslim Brotherhood's reemergence as a political force after Sadat's attempt at reconciliation coincided with the proliferation of Islamic groups that believed that the Brotherhood was too accommodating to the secular regime. The individual groups, too plentiful to enumerate and sometimes clandestine, were often composed of Muslim Brothers but were not ideologically or operationally homogeneous with their parent organization. Some espoused the violent overthrow of the government, while others simply advocated living a devout life of rigorous observance of religious practices. Offshoots of the Brotherhood, two of which—Egyptian Islamic Group (EIG) and Egyptian Islamic Jihad (EIJ)—emerged in the late 1970s, were more militant and espoused violence to overthrow the government and establish an Islamic state in Egypt.

Arguably, the most geopolitically significant issue in the Middle East during this period was the issue of a negotiated peace with Israel. In December 1977, Egypt and Israel began peace negotiations in Cairo that continued on and off over the next several months and were initially beset with difficulties. However, in September 1978, US President Jimmy Carter announced that the Camp David Accords[32] had been signed. The Camp David Accords made Sadat a hero in Europe and the United States, and although the reaction in Egypt was generally favorable, there was opposition from leftists and the Muslim Brotherhood, especially from its more militant members. In the larger Arab world, Sadat was almost universally condemned. The Arab states suspended all official aid and severed diplomatic relations, and Egypt was expelled from the Arab League, which it was instrumental in founding.[33]

GOVERNING ENVIRONMENT

The Egyptian Constitution provides for a powerful executive with authority vested in an elected president whose term runs for six years and who is empowered to appoint one or more vice presidents, a prime

---

[31] Wright, *The Looming Tower: Al Qaeda and the Road to 9/11.*

[32] The Camp David Accords included the "Framework for Peace in the Middle East" and the "Framework for the Conclusion of a Peace Treaty between Israel and Egypt."

[33] Ibid.

minister, and a cabinet. Egypt's legislative body, the People's Assembly, has 454 members, with 444 popularly elected and ten appointed by the president. The assembly sits for a five-year term but can be dissolved earlier by the president. There is also a 264-member Shura Council, with eighty-eight appointed members and 176 elected for a six-year term. At the local level, authority is exercised by governors and mayors appointed by the central government and by popularly elected local councils.[34] The country is divided into twenty-six governorates.

Egypt's judicial system is based on European, primarily French, legal concepts and methods. Under the Mubarak government, the courts demonstrated increasing independence, and the principles of due process and judicial review gained greater respect. The legal code is derived largely from the Napoleonic Code. Marriage and personal status are primarily based on the religious law of the individual concerned, which for most Egyptians is *shari'a*.

The military became one of the most important factors in Egyptian politics after the overthrow of the monarchy in 1952. Nasser appointed members of the officer corps to senior positions in the bureaucracy and public sector to help implement his social revolution. In the later years of the Nasser regime, fewer military figures occupied high government posts, and even fewer held posts during the Sadat and Mubarak regimes.[35]

WEAKNESSES OF THE PREREVOLUTIONARY ENVIRONMENT AND CATALYSTS

The conditions that gave rise to EIG and EIJ had less to do with the increasing economic deprivation and alienation felt by the Egyptian populace than the Islamic elites' dissatisfaction with an increasingly secular regime.[36]

The combination of stunted economic development and the defeat in the 1967 war led Egyptian students, at the time the strongest adherents to and proponents of the Nasser ideology, to become increasingly disillusioned. The campus protests in the 1960s advocating Arab socialism evolved into Islamic rhetoric in the 1970s. Upon assuming the presidency, Sadat initially supported the rise of Islamic associations, the most prominent being EIG, as a counterbalance to leftist influence among students. Despite the surplus of academically

---

[34] Metz, *Egypt: A Country Study.*

[35] Wright, *The Looming Tower: Al Qaeda and the Road to 9/11.*

[36] Gilles Kepel, *Muslim Extremism in Egypt: The Prophet and Pharaoh* (Berkeley, CA: University of California Press, 1985).

trained engineers and the scarcity of technicians, Sadat increased the provincial enrollment into Egyptian universities, unintentionally creating a cadre of dissatisfied and disaffected engineers who could not find adequate professional employment.[37] EIG spread rapidly and started to dominate campus life and have a strong influence over university faculties and administrations. Although the total number of activists was believed to be several hundred thousand, the membership in clandestine organizations was small, with estimates ranging from 3,000 to 20,000; violent activists numbered even fewer at 1,000.[38] By the late 1970s, the numerous underground groups became cognizant of other groups and the similarity of their goals. They slowly began to establish networks among the community of interest.[39] In April 1974, Saleh Seriya, a Palestinian academic, lead the Islamic Liberation Organization in a failed coup attempt at the Technical Military Academy, killing eleven and wounding twenty-seven.[40] Many young Egyptians, who had long since lost faith in their leaders, found meaning and identity in the clandestine Islamic groups and, as a result, were galvanized by the 1967 war. Their primary target remained the near enemy, the secularizing Egyptian government.[41]

By the late 1970s and early 1980s, with the older generation occupying most of the government bureaucratic positions and the guarantee of a government job weakening, many Egyptian technocrats found employment in the oil-producing Gulf states. Those who elected to remain in Egypt often selected unemployment over menial, although reasonably well-paying, job opportunities. In addition, because of the difficulties in finding employment, marriage was unlikely for many young men, resulting in an unfulfilled sense of purpose for a growing number who subsequently found solace and dignity in the anti-materialist Islamic morality prevalent in mosques and the growing number of Islamic social organizations.[42] This collective frustration, although palpable, only set the conditions for radicalization; the more decisive factor was the authoritarian government's repression of this increasingly militant network of Islamic groups.[43] Nowhere was this repression more evident than in the series of prisons and concentration

---

[37] Ibid.

[38] Metz, *Egypt: A Country Study.*

[39] Sageman, *Understanding Terrorist Networks.*

[40] Ibid.

[41] Fawaz A. Gerges, *The Far Enemy: Why Jihad Went Global* (New York: Cambridge University Press, 2005).

[42] Diego Gambetta and Steffen Hertog, "Engineers of Jihad," Sociology Working Papers, University of Oxford, 2007, accessed August 23, 2010, http://www.nuff.ox.ac.uk/ users/gambetta/engineers%20of%20jihad.pdf.

[43] Ibid.

camps used by Nasser and Sadat to silence and weaken the networks of opposition groups from the Muslim Brotherhood, to the left-leaning socialist organizations, and, ultimately, to the militant Islamic groups such as EIG and EIJ.

In September 1981, Sadat ordered the largest arrest of his opponents when at least 1,500 people, many of whom were Muslim Brothers, including supreme guide Umar Tilmasani, were imprisoned. Sadat also withdrew his recognition of the Coptic pope and banished him to a desert monastery, arresting several bishops and priests. Sadat ordered the closing of the *Ash Shaab* (*The People*) newspaper and officially banned EIG and other Islamic groups.[44] The imprisonment of many prominent Islamic leaders of the Muslim Brotherhood, EIG, and EIJ would further stoke the fires of militant Islamic sentiment.

FORM AND CHARACTERISTICS OF THE REVOLUTION

OBJECTIVES AND GOALS

The goals of the Muslim Brotherhood have evolved somewhat in response to changing local and global policies and geopolitical landscapes. Remaining consistent, however, has been the reinstatement of the Islamic caliphate, which was dissolved along with the Ottoman Empire in 1923; the strengthening and administrative discipline of its internal organization; and the mobilization of the greater community of Muslim believers, or *umma*, through social, religious, and political outreach and civic action.

The goal of both EIG and EIJ was to overthrow what they classified as an apostate Egyptian regime with swift, violent action and apply Islamic rule, which constituted the imposition of *shari'a* and a return to a governing body of *ulema*.[45] Both groups promised a purer society for Egyptian Muslims, free of external influences and the internal corruption and poverty that plagued them. These objectives reflected both the fundamentalism and the pragmatism of the groups, blending the Qutbist Salafism with an appreciation for the environmental restrictions resulting from the combination of the densely populated urbanized landscape and wide-open deserts

---

[44] Kepel, *Muslim Extremism in Egypt: The Prophet and Pharaoh.*

[45] Montasser al-Zayyat, *The Road to Al-Qaeda: The Story of Bin Laden's Right-Hand Man* (London; Sterling, VA: Pluto Press, 2004).

with an aggressive government security service.[46] Both groups were cognizant of the government's superiority of presence in the urban areas and the infeasibility of protracted operations in the harsh desert. Therefore, to achieve the desired objective of instituting an Islamic government, required action was planned covertly and executed quickly and violently. The primary difference between EIG and EIJ collectively and the Muslim Brotherhood was the former's preference to Islamize the Egyptian government and society from the top down by deposing the apostate ruler using violence, while the latter focused on accomplishing the same goal from the bottom up through a broader sociopolitical movement.[47] EIG and EIJ can be further differentiated from the Muslim Brotherhood in the latter's renunciation of violence and the former's reliance upon it. Both groups favored sudden precise action over a protracted guerrilla revolution that the arid topography of Egypt's vast deserts would not accommodate.[48]

While EIG was initially conceived as an Islamic opposition to communism, both EIG and EIJ adopted goals that expanded to include the local regime.[49] What has been retrospectively classified as the founding manifesto and the de facto operational manual of EIJ was written by an electrical engineer, Muhammad abd-al-Salam Faraj. In *The Neglected Duty* or *The Absent Obligation*, Faraj began by stating, "Jihad for God's cause . . . has been neglected by the Ulema of this age."[50] He went on to expand the interpretation of jihad as a violent struggle that is a duty incumbent on all Muslims. Faraj's argument was an extension of Qutb's with a more explicit advocacy for the consideration of jihad as the sixth pillar of Islam, approximating a heretofore-considered Shi'a belief. Faraj's advocacy entailed the establishment of an Islamic vanguard, an elite cadre of pious Muslims with either academic or military credentials that made them pillars of society. The responsibility of this vanguard was to serve as a model for elites in other Muslim nations to emulate. He made the initial classification of the "near enemy" and the "far enemy," subordinating all Islamic goals to the fight against local apostates.[51] There was no

---

[46] In *Milestones,* Qutb wrote "The defense of the homeland is not the ultimate objective of the Islamic movement of jihad, but it is a means of establishing the Divine authority within it so that it becomes the headquarters for the movement of Islam, which is then to be carried throughout the earth to the whole of mankind, as the object of this religion is all humanity and its sphere of action is the whole earth." Brachman, *Global Jihadism: Theory and Practice.*

[47] Gerges, *The Far Enemy: Why Jihad Went Global.*

[48] Ibid.

[49] Kepel, *The Roots of Radical Islam.*

[50] Johannes J. G. Jansen, *The Neglected Duty: The Creed of Sadat's Assassins and Islamic Resurgence in the Middle East* (New York: Macmillan, 1986).

[51] Gerges, *The Far Enemy: Why Jihad Went Global.*

written EIG counterpart to Faraj's work. Instead, the goals were articulated through the Friday sermons of the group's spiritual leader, Sheikh Umar Abd-al Rahman.[52] This reflected not only the more overt activities of EIG but also their potential recruitment pool, which was often younger, less educated, and less financially stable than the smaller membership of EIJ. In 1987, once Zawahiri had assumed the duties of EIJ emir, a pamphlet entitled "The Inevitability of Confrontation" ranked EIJ's goals in the following order: toppling the impious ruler who has abandoned religion; fighting any Muslim community that deserts Islam; reestablishing the caliphate and installing a caliph; and liberating the homeland, freeing the captives, and spreading religion.[53]

In small groups in general and jihadist groups in particular, individual personalities and leadership styles had a greater impact on instilling discipline, cohesion, and esprit de corps than did ideology, and this is evident in the differences between EIG and EIJ.[54] The Cairo group (which later emerged as EIJ) was seconded by future leader and current Al Qaeda deputy Dr. Ayman Al-Zawahiri. The Said group (which later emerged as EIG) was led by a Shura but deferred to their ideological guide, Sheikh Umar Abd-al Rahman (who came to be known as the Blind Sheikh). The dispute was ostensibly over the propriety of a blind man leading a militant organization, although it was more likely a result of the perceived threat of the religious credentials Rahman had that the abrasively negativistic Zawahiri did not.[55] From the late 1970s through the late 1990s, the groups had similar, if not congruent objectives and goals, with the primary difference being the operational tempo of violent action, because EIJ preferred a more systematic approach with deliberate planning and detailed surveillance before execution. EIG and EIJ can be further differentiated by EIG's formal renunciation of violence in 1997, while EIJ expanded their objectives beyond Egypt.[56] Although Zawahiri believed that the immediate focus should be on the near enemy, he thought the success of the movement was to serve as a vanguard[57]

---

[52] Rahman (blind since childhood from diabetes) would come to be known as "The Blind Sheikh" in the West and is currently incarcerated in the United States for his role in the 1993 World Trade Center bombing.

[53] Ibid.

[54] Ibid.

[55] Zayyat, *The Road to Al-Qaeda: The Story of Bin Laden's Right-Hand Man.*

[56] By the late 1980s, when he had assumed the duties of emir, Zawahiri (who initially believed that the immediate focus should be on the near enemy—the Egyptian government) thought the success of the movement was to serve as a vanguard and model for Salafists around to world to emulate, thereby restoring the prestige the Islamic world had lost.

[57] The term "vanguard" is a direct reference to Qutb's *Milestones.*

and model for Islamists to emulate, thereby restoring the prestige it had lost.[58]

Until the mid-1990s, the Egyptian jihadist movements limited the targets of their operations to Egypt and its leaders and consciously avoided provoking the West (namely the United States), considering such actions as politically, more than morally, counterproductive. At some point in the mid to late 1990s, as EIJ and their senior leaders became increasingly integrated into bin Laden's Al Qaeda, and as EIJ became increasingly dependent on Al Qaeda's financial resources, they shifted their objectives and target selection from the near enemy to the far enemy.[59] Faraj's concept of the "near enemy" was the apostate Egyptian regime, while the "far enemy" were those forces (individual and governments) who stood in the way of Islam's expansion. While the "near enemy" remained the same to EIJ and EIG, the "far enemy" was more specifically defined as the United States, which was believed to be responsible for the exploitation of oil-producing Muslim countries at the expense of their Islamic ideals. This shift in targeting coincided with a shift in credit claimed for operations, specifically the 1998 attacks on the US embassies in Kenya and Tanzania, transferring from EIJ to Al Qaeda.

LEADERSHIP AND ORGANIZATIONAL STRUCTURE

The Muslim Brotherhood is led by a Supreme Guide within an Executive Office who reports to a Shura Council that overseas and plans the general policies and programs of the group and the General Organization Conference. During certain periods in its history, the Brotherhood is said to have employed a secret apparatus,[60] or a small cell, that operated independently of the Conference and Shura Council, and possibly without the knowledge of the Supreme Guide, to conduct terrorist operations in support of broader Brotherhood political objectives.[61]

EIJ and EIG briefly merged in 1981, although differences in support and structure remained. Muhammad Uthman Isma'il,[62] a Sadat associate and attorney, was considered to have had a prominent

---

[58] Ibid.

[59] Gerges, *The Far Enemy: Why Jihad Went Global.*

[60] It is often the secret apparatus of the Brotherhood that was accused of violent acts, including the assassinations of Egyptian Prime Minister Mahmoud Fahmi an-Nukrashi Pasha in 1948 and President Sadat in 1981.

[61] Kepel, *Muslim Extremism in Egypt: The Prophet and Pharaoh.*

[62] Muhammad Uthman Isma'il was appointed governor of Asyut in 1973 and served until 1982, when he was removed by Mubarak.

role in EIG's emergence and an influential Islamic voice in Cairo from late 1971.[63] The Qutb-inspired jihadist groups that rose to prominence in the early 1980s could be divided into two predominant factions; the group that was to become EIG had its basis in the poorer districts of Cairo and Port Said,[64] while the group that was to become EIJ had its roots in the wealthier districts of Cairo. The latter group was formed in 1979 when Faraj united four to six clandestine jihadi cells under his leadership, including one that was still underground and not yet operational led by Zawahiri and two army officers.[65] During a Shura in late spring 1981, leaders of the two branches (Karam Zuhdi of EIG and Faraj of EIJ) agreed to merge their groups under the leadership of Sheikh Umar Abd-al Rahman, who agreed to be the *mufti*.[66] Rahman, an alumnus and former professor at Al-Azhar University in Cairo, became one of the most outspoken clerics to denounce Egypt's secularism. Although respected by both groups, he did not have a unanimous mandate to serve as the group's emir.[67] The Zuhdi faction was dominated by students (64%), while Faraj's group was composed of only 43% students, with the rest primarily professionals with academic backgrounds in engineering, mathematics, medicine, or the military. Zuhdi's group enjoyed the support of the population and were afforded freedom of movement through their communities, which were historically resistant to the Egyptian government and safe havens along the Maoist model. The Cairo groups perceived their operational environment to be controlled by the enemy and operated in an underground cellular network.[68] Zawahiri's cell[69] and, later, EIJ as a whole, employed a blind-cell structure like that of the Leninist Communist Party where members in one cell did not know the identities or activities of those in another, so that if one member was captured or compromised, he would not be able to endanger more than a few people.[70]

---

[63] Kepel, *The Roots of Radical Islam.*

[64] Jansen, *The Neglected Duty: The Creed of Sadat's Assassins and Islamic Resurgence in the Middle East.*

[65] Sageman, *Understanding Terrorist Networks.*

[66] A mufti is a Sunni Islamic scholar who is an interpreter of Islamic law.

[67] Zayyat, *The Road to Al-Qaeda: The Story of Bin Laden's Right-Hand Man.*

[68] Sageman, *Understanding Terrorist Networks.*

[69] In this environment (Cairo), the combination of Zawahiri's obsessive compulsive (detail-oriented, rigid enforcement of challenges and passwords; conscious avoidance of pattern setting, etc.) and paranoid (assumed the group was under surveillance, exhibited caution in recruitment, attempted to avoid large congregations of jihadists, etc.) leadership traits served him and his colleagues well in avoiding the Egyptian law-enforcement and counterintelligence apparatus.

[70] Ibid.

In the wake of the October 1981 Sadat assassination, the Egyptian police arrested much of the EIG and EIJ leadership, including Rahman, Zuhdi, Faraj, Zumur, Zawahiri, and roughly 300 others[71]. The post-assassination imprisonments under Mubarak saw an irreparable rupture between the Cairo element and the Said element over the former's opposition to Rahman's leadership due to his disability.[72] The dispute resulted in the splitting of the groups into EIJ, which was run by Faraj and seconded by Zumur, with Zawahiri serving as the public face of the organization, and EIG, which was ruled by a Shura. Debate exists as to whether the EIJ cells were formally dissolved while many of its members were in prison.[73] However, the operational capacity of both EIJ and EIG was significantly degraded in the period between the Sadat assassination and the mid-1980s.

During the mid-1980s, many EIJ and EIG members relocated to Peshawar, Pakistan, to escape the Egyptian government and to participate in the Afghan resistance to the Soviet Union's occupation. Although some EIJ members recognized the still-imprisoned Zumur as emir, others began to consider Sayyed Imam al-Sharif, also known as "Dr. Fadl," EIJ's leader. Sharif, a physician who was administering to the refugees from Afghanistan under the Red Crescent, emphasized the importance of the Qutbist ideology and the condemnation of those who deviated from it. Sharif had established a series of guesthouses and training camps in Peshawar to receive, stage, and prepare to move and integrate young Muslims from the Middle East into the Afghan resistance. His was a parallel effort to that of Osama bin Laden and his spiritual adviser (a role that Zawahiri would eventually assume), Abdullah Azzam, who had established Mekhtab al-Khidemat (MAK) or Services Bureau to facilitate administrative problems for foreign Muslim fighters.[74] At some point during the mid-1980s, Sharif's academic, professional, and ideologically like-minded colleague, Zawahiri,[75] became the recognized leader of EIJ.[76] The opportunity to select, train, and indoctrinate operatives from the pool of Egyptians seeking to martyr themselves in Afghanistan enabled Zawahiri to slowly unite the still disparate cells of EIJ and build capability.[77]

---

[71] Zayyat, *The Road to Al-Qaeda: The Story of Bin Laden's Right-Hand Man.*

[72] Springer, Regens, and Edger, *Islamic Radicalism and Global Jihad.*

[73] Gerges, *The Far Enemy: Why Jihad Went Global.*

[74] Wright, *The Looming Tower: Al Qaeda and the Road to 9/11.*

[75] Both Zawahiri and Imam al-Sharif were pious, high-minded, prideful, and rigid in their views. They evaluated matters of religion as a series of immutable rules established by God, a mind-set representative of the well-educated engineers and technocrats who occupied the higher ranks of Salafists.

[76] Ibid.

[77] Gerges, *The Far Enemy: Why Jihad Went Global.*

In 1987, conflict erupted between the two groups; EIG was now Saudi funded and headed by Mohammed Shawqi al-Islambuli. He criticized Zawahiri's financial mismanagement, exacerbating the rift and alienating the Gulf countries, which henceforth channeled the majority of their financial support to EIG. In 1992, the conflict between EIG and EIJ reached the stage of mutual accusations of apostasy and individual assassination attempts. Zawahiri emerged as the winner, largely because of bin Laden's support[78] and because of the murder of Abdullah Azzam, the spiritual leader of bin Laden, which some attribute to EIJ.[79] At its pinnacle in the early 1990s, EIG is estimated to have had 10,000 full-time members, of which approximately 800 were Egyptian veterans of the war in Afghanistan.[80]

During summer 1999, members of EIJ were becoming uncomfortable with Zawahiri's growing ties to Osama bin Laden and Al Qaeda. As a result, Zawahiri was ousted as their leader. EIJ briefly turned the leadership over to Thartwat Shehada, who took control with the intent of refocusing the group's efforts on operations in Egypt.[81] However, Shehada did not have the financial backing to achieve this goal and by spring 2001, Zawahiri was able to reassert control over the group.[82]

In June 2001, although difficult to distinguish for years, Al Qaeda and EIJ merged, forming "Qaeda al-Jihad";[83] hereafter, all activities and operations of former EIJ members were done under bin Laden's banner. At this point, Zawahiri was presumed to be the deputy to bin Laden and the leader of EIJ. Other notable Egyptians in Qaeda al-Jihad include Mustafa Abu al-Yazid (who had close ties to both bin Laden and Zawahiri), Saif al-Adel (believed to be under house arrest in Iran), Abdullah Ahmed Abdullah (chief financial officer of Al Qaeda), and Mustafa Abu al-Yazid (an original member of Al Qaeda's leadership council and adviser to bin Laden for more than a decade). Both al-Yazid and deputy leader Zawahiri served time in prison in the early 1980s for their role as conspirators in the 1981 assassination of Egyptian

---

[78] Zayyat, *The Road to Al-Qaeda: The Story of Bin Laden's Right-Hand Man.*

[79] Sageman, *Understanding Terrorist Networks.*

[80] Carla Liberatore, "Islamic Fundamentalism in Egypt: U.S. Policy Recommendations" National War College, 1994, accessed August 23, 2010, http://www.dtic.mil/cgi-bin/GetTR Doc?AD=ADA440564&Location=U2&doc=GetTRDoc.pdf.

[81] Wright, *The Looming Tower: Al Qaeda and the Road to 9/11.*

[82] After 2001, the Al Qaeda–EIJ partnership was organized into four committees (each headed by an emir or leader): military, religion, finance, and the media. Although it is unclear whether this was patterned after existing EIJ structure, the organization was presumed to be Zawahiri's suggestion.

[83] Zayyat, *The Road to Al-Qaeda: The Story of Bin Laden's Right-Hand Man.*

President Anwar Sadat. Both Adel and Abdullah are still wanted for their role in the 1998 US embassy bombings in Kenya and Tanzania.[84]

METHODS OF ACTION AND VIOLENCE

The Sadat regime's imprisonment of many prominent Islamic leaders of the Muslim Brotherhood, EIG, and EIJ in early October 1981 served as the final catalyst to the growing resentment toward the government, the results of which were evidenced on October 6, when assailants, led by Army Lieutenant Khalid al-Islambuli, brother of imprisoned EIG member Mohammed al-Islambuli, assassinated Sadat while he was attending a military parade commemorating the Egyptian action in the 1973 Yom Kippur War.[85] Although numerous cells were aware of the impending threat, there was no organized movement to coordinate a coup to exploit the assassination. However, an EIG faction did organize an insurrection in Asyut, Upper Egypt and took control of the city for three days until they were defeated by paratroopers who arrived from Cairo. Ultimately, sixty-eight policemen and soldiers were killed in the fighting, and the surviving perpetrators were imprisoned, although their sentences of three years were light in comparison to their counterparts implicated in the Sadat assassination. The Sadat assassination led to the election of Vice President Hosni Mubarak by plebiscite followed by a swift and severe governmental crackdown on all Islamic groups.[86]

Figure 2. Assassination of Egyptian President Anwar Sadat on October 6, 1981.[87]

---

[84] Jayshree Bajoria and Greg Bruno, "al-Qaeda (a.k.a. al-Qaida, al-Qa'ida)" (New York: Council on Foreign Relations), accessed August 23, 2010, http://www.cfr.org/publication/9126/alqaeda_aka_alqaida_alqaida.html#p1.

[85] Kepel, *Muslim Extremism in Egypt: The Prophet and Pharaoh.*

[86] Sageman, *Understanding Terrorist Networks.*

[87] History Commons, accessed March 14, 2011, http://www.historycommons.org/context.jsp?item=a84mohamedcia#a84mohamedcia.

The subsequent incarceration by Mubarak's regime served not only to take many Islamic opponents to the government off of Egyptian streets but also to harden the resolve of the most radically minded individuals. Many traveled via Pakistan to Afghanistan to participate in the resistance movement, which was beginning to solicit volunteers from throughout the Muslim world to wage jihad in support of their Afghan brethren. Both Rahman, who was acquitted in the Sadat assassination but later expelled from Egypt, and Zawahiri, who had served three years on a weapons charge and traveled to Saudi Arabia and then Pakistan as a physician working for the Red Crescent, encountered Osama bin Laden[88] and Abdullah Azzam in Peshawar, Pakistan. Azzam had established MAK while Sharif established jihad camps in Peshawar.[89] Rahman eventually traveled to the United States in July 1990 to assume the leadership of MAK's New York City office, which operated out of the Al-Farooq Mosque in Brooklyn.[90] Zawahiri's experience with the mujahidin resistance helped transform EIJ by convincing him of the need for a secure base to train operatives and from which to launch attacks, as well as the primacy of violence, particularly martyrdom operations, as a means of achieving their political objectives.

During the 1980s, both EIG and EIJ initially saw their capability diminish in the wake of the Sadat assassination but then slowly build as many gained valuable training and experience in the Soviet-Afghan war. By the early 1990s, the decade-long Egyptian respite from domestic terrorism was about to end. From 1990 until 1997, the Egyptian government waged a low-intensity war of attrition with EIG (which accounted for 90% of the attacks)[91] and EIJ resulting in about 1,300 casualties, billions of dollars in damage to the tourist industry, and significant costs to relations between state and society.[92]

In June 1992, Faraj Fawda, a prominent Egyptian writer and human rights activist who had open criticized Islamic fundamentalist ideology, was declared an apostate and foe of Islam and subsequently shot to death by two EIG operatives. Testifying at their trial, an Al-Azhar scholar declared their actions to be sanctioned under *shari'a*

---

[88] Both Zawahiri and bin Laden were scions of prominent families, well educated, and wealthy (although Zawahiri was better educated academically and theologically, bin Laden was certain the wealthier of the two); both were members of the educated classes, intensely pious, quiet-spoken, and politically stifled by the regimes in their own countries (Egypt and Saudi Arabia respectively).

[89] Ibid.

[90] Wright, *The Looming Tower: Al Qaeda and the Road to 9/11.*

[91] Gerges, *The Far Enemy: Why Jihad Went Global.*

[92] Fawaz A. Gerges, "The End of the Islamist Insurgency?: Costs and Prospects," *The Middle East Journal* 54, no. 4 (2000): 592–612.

because Fawda was declared an apostate, and the assailants were implementing a justified punishment.[93] This use of a fatwa or religious ruling to sanction violence against civilians, specifically Muslim targets, before the execution of the operation was to serve as a Qutbist-group model that eventually extended to Al Qaeda. The fatwas generated much debate within Islamic circles and have continued to be points of contention and legitimacy among many Islamic terrorist groups.

In 1993, the insurgent groups turned to attacks on US soil. On February 26, in what was to be the first attack, EIG, EIJ, and/or Al Qaeda operatives parked a truck loaded with a 1,500-pound urea nitrate–hydrogen gas-enhanced bomb in the parking garage of the World Trade Center's Tower One with the intent of knocking it into Tower Two and killing 250,000 people. The attack was planned by a group of conspirators with ties to Al Qaeda, EIG, and EIJ, including Ramzi Yousef (nephew of senior Al Qaeda leader Khalid Sheikh Mohammed), Mahmud Abouhalima, Mohammad Salameh, Nidal Ayyad, Abdul Rahman Yasin, Eyad Ismoil, and Ahmad Ajaj. The blast shook but failed to destroy either tower, killing six, injuring 1,042 and causing $300 million in property damage.[94]

Domestic attacks in Egypt also mounted, including the first use of suicide tactics the same year. In April, EIG claimed responsibility for an assassination attempt against a motorcade of the Egyptian information minister, injuring the minister and his bodyguard.[95] Later, in August, EIJ operatives attempted to assassinate the Egyptian interior minister, who was leading a crackdown on Islamic militants and their terror campaign. A bomb-laden motorcycle exploded next to the minister's car, killing the bomber and his accomplice, but not the minister. The failed attack marked the first time that Sunni Islamists had made use of suicide in terrorism, a technique made famous by Hizbollah in Lebanon.[96]

Shortly thereafter, the membership director for EIJ was arrested in Egypt; as a result, approximately 800 members of EIJ were arrested, effectively decimating the group's presence in Egypt. Most of the remaining members were in scattered cells in other countries. In November, EIJ made another bombing attempt, this time with the goal of killing Egypt's prime minister, Atef Sidqi. A car bomb exploded close to a girls' school in Cairo as the minister was chauffeured in

---

[93] A. B. Soage, "Faraj Fawda, Or the Cost of Freedom of Expression," *The Middle East Review of Foreign Affairs* 11, no. 2 (2007).

[94] Wright, *The Looming Tower: Al Qaeda and the Road to 9/11.*

[95] American University, "Terrorism in the Middle East," accessed August 23, 2010, http://www1.american.edu/TED/hpages/terror/islamic.htm.

[96] Ibid.

his armored car. Sidqi was unhurt, but the explosion injured twenty-one people and killed a young schoolgirl, Shayma Abdel-Halim, significantly degrading popular support for EIJ even among militant groups. The assassination attempt on Sidqi was preceded by two years of aggressive operations by EIG operatives, at times with the support of EIJ, that had killed 240. The resulting police crackdown was facilitated by anonymous help from Egyptian citizens who disapproved of the tactics employed, which resulted in the arrest of 280 EIJ members, of whom six were later executed.[97] By the end of 1994, EIJ had become increasingly dependent on Al Qaeda (whose inner circle was well represented by Egyptians) for both training and financial support because most of its members were reportedly on bin Laden's payroll.

In June 1995, another failed assassination attempt caused a greater setback for EIJ. Together with EIG, and with help from Sudanese intelligence, EIJ plotted the assassination of President Hosni Mubarak during a visit to Ethiopia. The attack was planned a year in advance by Mustafa Hamza, a senior Egyptian member of Al Qaeda and commander of the military branch of EIG. Despite thoughtful planning and detailed surveillance and reconnaissance, the operatives were tactically impatient and were undone by minor errors, including unfamiliarity with a rocket-propelled grenade launcher, and Mubarak's armored limousine. This resulted in much criticism of Zawahiri's operational leadership. Several months later, in November, EIJ and EIG used a suicide truck bomb to destroy the Egyptian embassy in Islamabad, Pakistan. Two dismounted operatives approached the embassy and executed the security detail with a combination of small-arms fire and hand grenades. Then a taxi loaded with 250 pounds of explosives breached the gate and the driver detonated the bomb, destroying both the vehicle and the gate. Minutes later, a second vehicle carrying a larger bomb was detonated alongside the embassy, which destroyed the building and damaged the adjacent Japanese and Indonesian embassies and a nearby bank. The attack killed sixteen, injured sixty, and served as a prototype for the 1998 Al Qaeda attacks on the two American embassies in Africa.[98]

In 1996, bin Laden issued a fatwa declaring war against the "Americans Occupying the Land of the Two Holy Places (Expel the Infidels from the Arab Peninsula)," which signified the point at which the targeting focus of EIJ changed from the near enemy, the Egyptian government, to the far enemy, the United States. Zawahiri's public support of the fatwa created fallout among EIJ members, including

[97] Ibid.
[98] Ibid.

his brother, who were not consulted.[99] Some EIJ members became affiliated with EIG, while others simply disengaged from the cause. A small cadre remained with Zawahiri and bin Laden, who now operated in Sudan.

In 1997, EIG renounced violence, reportedly because of a deal struck with the Egyptian government in return for a release of its jailed members.[100] From a US prison, Sheikh Rahman (incarcerated because of the 1993 bombing of the World Trade Center in Manhattan) initially gave his blessing to the deal, while Zawahiri publicly condemned the decision. That November, operatives from EIG who had not agreed with the renunciation of violence and an element of EIJ called Jihad Talaat al-Fath or "Holy War of the Vanguard of the Conquest" killed sixty-three tourists at the Temple of Hatshepsut, Egypt. During the operation, dubbed the "Luxor massacre," six assailants dressed in police uniforms systematically shot and stabbed to death fifty-eight foreign tourists and four Egyptians. The audacity and depravity of the attack stunned Egyptian society, ruined tourism for years, and destroyed much of the remaining popular support for EIG and EIJ in Egypt. The day after the attack, EIG's Taha claimed that the attackers intended only to take the tourists hostage, despite the evidence of the systematic execution of the attack. Others denied Islamist involvement completely. Rahman blamed the Israeli Mossad, while Zawahiri maintained that the Egyptian police were responsible. This operation was to be the last kinetic action attributed to either EIG or its members.[101]

The next year, in 1998, Zawahiri issued a joint fatwa with bin Laden and Taha under the title "World Islamic Front Against Jews and Crusaders," sanctioning the ruling that "to kill the Americans and their allies—civilians and military—is an individual duty for every Muslim who can do it in any country in which it is possible to do it."[102] EIJ membership had dwindled significantly, with estimates of only forty members, all of whom operated outside Egypt.[103] The "World Islamic Front" was essentially an amalgamation of Al Qaeda, EIJ, and other transnational jihadi organizations because the groups were operationally intertwined to a degree that made distinguishing among them nearly impossible. That August, the ramifications of the fatwa were soon evidenced by near simultaneous explosions at the US embassies in Dar es Salaam, Tanzania, and Nairobi, Kenya, killing

[99]   Gerges, *The Far Enemy: Why Jihad Went Global.*
[100] Wright, *The Looming Tower: Al Qaeda and the Road to 9/11.*
[101] Springer, Regens, and Edger, *Islamic Radicalism and Global Jihad.*
[102] Ibid.
[103] Wright, *The Looming Tower: Al Qaeda and the Road to 9/11.*

eleven and injuring eighty-six in Dar es Salaam and killing 212 and injuring 4,000 in Nairobi.[104] That attack placed Zawahiri, along with bin Laden, on the US Federal Bureau of Investigation's (FBI's) Ten Most Wanted Fugitives list; shortly thereafter, Zawahiri was sentenced in absentia to death by the Egyptian government.[105]

Soon afterward, Zawahiri performed his first public act as part of the Al Qaeda strategic communications effort when he conducted a telephone interview with a Pakistani reporter where he conveyed a message from bin Laden denying involvement in the embassy bombings yet encouraging Muslims to continue to wage jihad against the occupiers of the two holy places.[106] With the official merging of Al Qaeda and EIJ in June 2001, forming "Qaeda al-Jihad," terrorist recruiting, planning, and operations involving EIJ members were instigated under the guise and leadership of the new organization.

METHODS OF RECRUITMENT

The recruitment techniques of EIJ, EIG, and individual leaders evolved over the course of their operations. Early in its campus recruiting and social outreach programs during the 1970s, EIG employed a tactic whereby the potential recruiting pool was subjected to a single, undifferentiated message to which some, but not all, were expected to respond positively.[107] After the Sadat assassination, EIJ employed a different recruiting tactic where recruiters used an incremental approach. With this tactic, recruiters were to select individuals who were potentially vulnerable and therefore ripe for recruitment and could easily yield to significant transformation in identity and motivation. Once identified, recruiters started potential operatives at one end of the process and then vetted, trained, and transformed them into dedicated group members over time. Recruits demonstrated their commitment through knowledge of radical Islam, such as the doctrines of Qutb, Faraj, and Zawahiri, and the use of violence to achieve the stated goals.[108] EIJ, under Zawahiri's leadership, relied on yet another method of recruitment, dispatching Afghan veterans to locales in Egypt to recruit quality fighters with military or engineering backgrounds in order to contribute operationally to

---

[104] Sageman, *Understanding Terrorist Networks.*

[105] Ibid.

[106] Ibid.

[107] Gerwehr and Daly refer to this model as "the Net." Scott Gerwehr and Sara Daly, "Al-Qaida: Terrorist Selection and Recruitment," in *The McGraw Hill Security Handbook*, ed. David G. Kamien (New York: McGraw Hill, 2006).

[108] Ibid.

the mujahidin resistance.[109] This type of approach was typically used when a target population was insular or a subcomponent of a larger group that would actively deny terrorist recruitment. Trusted agents, like the Afghan veterans, were then inserted into the target population to rally potential recruits through direct, personal appeals, leveraging individual charisma and the persuasive strength of the source's credibility.[110] When the EIG's leadership operated outside of Egypt in the 1990s and was denied access to domestic populations by the government, recruitment became more problematic. Here, recruiters would provide the context and often the tools for individuals to seek out the groups.[111]

In addition to the methods of recruiting described above, EIJ, EIG, and Muslim Brotherhood recruitment can also be analyzed according to the environment in which the organizations operated. Within the aforementioned models, Gerwehr and Daly also proposed four recruiting approaches that were used interchangeably, depending on the environment, including public and proximate, public and mediated, private and proximate, and private and mediated.[112] All three groups used a tactic that observers call the "public and proximate" whereby recruiters commingle with the target population and make appeals individually or in small groups irrespective of any opposition or observation by authorities.[113] The groups used this approach for generations in Egyptian prisons and concentration camps to great success.

Another approach favored by each group, but in varying degrees, was described as "public and private mediated," which was often exploited mass media and marketing tactics, techniques, and procedures and modern information technology to convey their message.[114] This approach combined overt messaging through a variety of media with personal reinforcement by a group representative. The Muslim Brotherhood historically used print media in various forms to further their message, while EIG used the tactic to a lesser degree, often preferring Friday sermons at the mosques or public gatherings as

---

[109] Youssef H. Aboul-Encin, "Ayman Al-Zawahiri: The Ideologue of Modern Islamic Militancy" *The Counterproliferation Papers, Future Warfare Series,* no. 21 (Montgomery, AL: USAF Counterproliferation Center, Air University, Maxwell Air Force Base, 2004), accessed August 23, 2010, http://www.dtic.mil/cgi-bin/GetTRDoc?AD=ADA446154&Location=U2&doc=GetTRDoc.pdf.

[110] This model is known as "Infection." Gerwehr and Daly, "Al-Qaida: Terrorist Selection and Recruitment."

[111] This model is known as the "Seed Crystal." Ibid.

[112] Ibid.

[113] Ibid.

[114] Ibid.

better venues. EIJ's use of print media is best described as propaganda or psychological warfare where recruiting was not necessarily the primary intent, although it was a desired higher-order effect. Several EIJ members distributed Zawahiri's writings during the late 1980s and early 1990s in Egypt.[115] The writings, distinguished from other jihadist literature by their fine printing and distinctive yellow covers, served to elevate Zawahiri's status in the eyes of current EIJ members and legitimize the message to potential recruits.[116] Use of the Internet as a recruiting medium in which EIJ conveyed its message and targeted recruits has been evident. However, the tactic did not reach maturity or operational effectiveness until after the 2001 merger with Al Qaeda. Through its beginnings as an unfederated network of clandestine cells, EIJ's preferred recruiting approach was private and proximate. In this context, pitches were made in intimate settings with the explicit intent of avoiding observation. This technique leveraged the influential power of conformity and relied heavily on personal appeals tailored specifically to a targeted individual, often using peers and/or relatives when making the pitch.[117]

Individuals with academic backgrounds, such as science, engineering, and medicine, were strongly overrepresented among Islamic movements in the Muslim world.[118] In comparison to EIG, EIJ recruits were often older, with stable family lives, careers, or academic backgrounds in science, engineering, or the military.[119] Many of the Islamic groups that existed in Egypt in the twentieth century were led by individuals who possessed a technical education and were composed of individuals of high motivation and achievement.[120] Technical professionals were included among the ranks of the Muslim Brotherhood from its inception, but individuals from elite faculties seem to be a rarity. As groups like EIG and EIJ began to emerge from the Muslim Brotherhood in the 1960s and 1970s, they were often led by individuals with backgrounds in science and engineering. In the Egyptian school system, acceptance and subsequent enrollment in an engineering program indicated above-average intellect and ambition, while earning an engineering degree carried both technical and social status.[121] The mechanistic rigidity often required to master the engineering discipline in an academic setting seemed to carry over

---

[115] Aboul-Enein, "Ayman Al-Zawahiri: The Ideologue of Modern Islamic Militancy."
[116] Ibid.
[117] Gerwehr and Daly, "Al-Qaida: Terrorist Selection and Recruitment."
[118] Gambetta and Hertog, "Engineers of Jihad."
[119] Sageman, *Understanding Terrorist Networks*.
[120] Gambetta and Hertog, "Engineers of Jihad."
[121] Ibid.

well to the selective Salafist interpretation of the Quran and hadiths, making the technocratic social circles fertile recruiting ground for the professional elite of both EIJ and EIG.[122]

Because EIG began as a provider of social services (housing, transportation, tutoring, mentoring, etc.) on college campuses, they served as a conduit for young Muslim men seeking both intellectual and spiritual guidance to local mosques and Muslim organizations, including both the Muslim Brotherhood and EIG. EIG favored a public and proximate approach whereby prominent members, such as civic leaders or clerics, would serve as recruiters, openly espousing the benefits of affiliation.[123] This occurred in full view and often with the tacit support, if not overt consent, of the Sadat regime. EIG activists were often university students or recent graduates, including rural-to-urban migrants and urban middle-class youth whose fathers were middle-level government employees or professionals. Their fields of study—medicine, engineering, military science, and pharmacy—were among the most highly competitive and prestigious disciplines in the university system. The rank-and-file members of EIG came from the middle class, the lower-middle class, and the urban working class.

By contrast, EIJ, under Zawahiri's guidance, favored a predominantly private and proximate recruiting technique. Zawahiri's requirement for secrecy, along with the combination of a highly capable and, at times, brutal counterintelligence apparatus of the Egyptian government, resulted in rather slow organizational growth because of the requirement to thoroughly vet and slowly incorporate potential members.[124] The clandestine groups united by Faraj, namely Zawahiri's cell, favored a military coup, and thus attempted to infiltrate the military by identifying like-minded military officers either in the mosques or through their affiliation with the Muslim Brotherhood. During the late 1970s, EIJ was careful to approach only certain types of individuals as potential recruits. Members would appeal to potential recruits based on their prior personal knowledge of, or relationship with, the individual and often did so out of the public eye so as to avoid local authorities.[125] Recruits were typically well educated, most with academic backgrounds in science, math, and/or engineering, and pious Muslims often coming from the ranks of the broad membership of the Muslim Brotherhood. EIJ's membership in the 1980s benefited from bin Laden and Assam's MAK and their efforts across the Middle East to recruit, train, and equip "Arab Afghans" to

---

[122] Ibid.
[123] Gerwehr and Daly, "Al-Qaida: Terrorist Selection and Recruitment."
[124] Sageman, *Understanding Terrorist Networks*.
[125] Gerwehr and Daly, "Al-Qaida: Terrorist Selection and Recruitment."

support the mujahidin resistance to the Soviet occupation. Mosques, schools, and boardinghouses throughout Egypt served as recruiting stations with the support, although not the same level of financial assistance as in Saudi Arabia, of the government.[126] Some recruits, particularly those with military experience, were enmeshed into the leadership hierarchy of what was to become Al Qaeda and remained in central Asia and later Sudan with Zawahiri, while others returned to Egypt after serving with implicit instructions to sustain the level of violence and action that Zawahiri believed was necessary to achieve the group's objectives.

EIG's domestic recruiting efforts fared better than EIJ during the 1980s and early 1990s, and their membership, although always significantly larger than EIJ—estimated in the tens of thousands compared to hundreds—remained strong until the Egyptian law-enforcement and counterintelligence apparatus began to systematically dismantle the networks in the mid-1990s.[127]

METHODS OF SUSTAINMENT

Little is known about the funding sources for the initial set of clandestine groups that arose in late-1970s Egypt. However, because many members were from prominent families and were educated professionals, it is hypothesized that the groups were financially self-sustaining.[128] In the 1970s, EIG and EIJ adopted a conventional guerrilla approach to weapons procurement and caching by attempting to co-opt Egyptian military and/or police officers to gain access to stores and retaining hardware and ammunition after engagements with security forces.

As EIJ evolved, operational funding seemed a perpetual constraint that necessitated strategic alliance, most notably with MAK and later Al Qaeda. EIJ's relative lack of financial stability in comparison with the Muslim Brotherhood in Egypt was also a recurring theme in EIJ, particularly Zawahiri's, literature.[129] In the mid-1980s, while Zawahiri continued to run EIJ from Afghanistan, he found his job rather difficult because of both the financial mismanagement of his operatives and the robust Egyptian counterintelligence and law-

---

[126] National Commission on Terrorist Attacks Upon the United States, *The 9/11 Commission Report: Final Report on the National Commission on Terrorist Attacks upon the United States* (New York: W. W. Norton, 2004).

[127] Gerges, *The Far Enemy: Why Jihad Went Global.*

[128] Sageman, *Understanding Terrorist Networks.*

[129] N. Raphaeli, "Ayman Muhammad Rabi' Al-Zawahiri: The Making of an Arch-Terrorist," *Terrorism and Political Violence* 14, no. 4 (2002): 1–22.

enforcement apparatus.[130] In 1995, at an EIJ meeting in Yemen, Zawahiri announced that the organization was out of funds and urged his subordinates to become self-sufficient, causing many to serve in the employ of Islamic relief organizations and use their salaries to fund their activities.[131] From this point until its formal merger with Al Qaeda, EIJ received most of its funding from bin Laden.[132]

METHODS OF OBTAINING LEGITIMACY

EIJ drew its legitimacy from a Salafist interpretation of Islam viewed through a Qutbist takfiri lens. EIJ saw its organization not only as the vanguard of Qutb's vision for an Islamic revolution but also as an entity with the requisite political and religious authority to declare all those who did not meet their requirements for piety essentially non-Muslim, regardless of what the individuals professed to believe. The Qutbist worldview was binary, whereby Islam represented all that was good and pure in human society, and all else was *jahiliyya*.[133] To Qutb, the entire modern world, including Muslim societies like Egypt, was *jahiliyya*, and the only way for a society to move from that state of ignorance to a higher state of being was for the pious among them to wage jihad.[134] Because social and political systems in both the Islamic and non-Islamic world were considered pagan, it was the individual Muslim's obligation to change this state of affairs by means of preaching and jihad through a vanguard effort of the truly committed, with the ultimate goal being the elimination of suffering and oppression of the Muslim people through the return of God's rule on Earth.[135] Qutb did not believe that piety alone would secure an individual's place in paradise, but that it was the obligation of those able to dedicate their lives to jihad. EIJ operationalized Qutbism by exemplifying piety in its members, as well as the organization's oppositional stance against the Egyptian government. Nowhere was this more apparent than in EIJ's commitment to conducting attacks against Egyptian and Western

---

[130] Zayyat, *The Road to Al-Qaeda: The Story of Bin Laden's Right-Hand Man.*

[131] Ibid.

[132] Gerges, *The Far Enemy: Why Jihad Went Global.*

[133] *Jahiliyya* is an Islamic concept of ignorance of divine guidance or the state of ignorance of the guidance from God. The concept is applied to the pre-Islamic Arab world before the revelation of the Quran to Muhammad. Qutb extended the concept to include all those (including the Egyptian government) who did not practice his rigidly adherent Salafist variant of Islam.

[134] Springer, Regens, and Edger, *Islamic Radicalism and Global Jihad.*

[135] Brachman, *Global Jihadism: Theory and Practice.*

interests, despite the increasing difficulty and relative lack of success in the early to mid-1990s.[136]

Unlike the more elitist and clandestine approach taken by EIJ, EIG attempted to appeal to a broader demographic of Muslims through social outreach and engagement along with their anti-regime actions. Their civic action paralleled that of the Muslim Brotherhood, although their criticisms of the Egyptian regime were more vociferous and at times critical of the Muslim Brotherhood's appeasement of and cooperation with the regime. EIJ's failed assassination attempt on Prime Minister Sidqi, which resulted in the death of a young girl, further alienated the secretive group, whose principal leadership had not resided in Egypt for years, from the population. EIJ made little effort to garner the favor or support of the Egyptian populace, and their insular policies did not allow for a transparent articulation of their objectives as a counter-narrative to that promulgated by the Egyptian regime. By the mid-1990s, Egyptian authorities had identified many key nodes in the EIG and EIJ support networks, including their family members, business associates, and friends who were not directly involved in operations against the government. As these individuals were arrested, interviewed, interrogated, and tracked, the popular view of the groups transformed from an Islamic alternative to the oppressive secular regime to criminal organizations. Although not a populist movement, EIG understood the requirement for broader popular support to the Islamic cause, which likely precipitated their 1997 renunciation of violence.[137] However, the renunciation may have been the result of a deal struck with the Egyptian government in return for a release of its jailed members.

EXTERNAL SUPPORT

The origins of many Islamic economic and financial regulatory organizations date back to the Muslim Brotherhood's development of political, economic, and financial infrastructures that enabled fulfillment of their religious obligations. Al-Banna viewed finance as a critical weapon in undermining the infidels and reestablishing the Islamic caliphate. To do so, he believed that Muslims must create an independent Islamic financial system that would parallel and later supersede the Western economy.[138] Al-Banna's successors set

---

[136] Gerges, *The Far Enemy: Why Jihad Went Global.*

[137] Ibid.

[138] Terror Finance Blog, "The Muslim Brotherhood New International Order," accessed August 23, 2010, http://www.terrorfinance.org/the_terror_finance_blog/2007/10/the-muslim-brot.html.

his theories and practices into motion, developing uniquely Islamic terminology and mechanisms to advance the Brotherhood's system of faith, as well as their unique financial apparatuses. Nasser negated the Brotherhood's attempt to establish an Islamic banking system during the mass arrests in 1964. Saudi Arabia welcomed this Egyptian dissident idea, and in 1961, King Saud bin Abdul Aziz funded the Brotherhood's establishment of the Islamic University in Medina to proselytize their fundamentalist Islamic ideology.

In 1962, the Brotherhood convinced the king to launch a global financial joint venture that established numerous charitable foundations around the world, which would become the cornerstone and spread Islam (and later fund terrorist operations) worldwide. The first to be established were the Muslim World League and Rabitta al-Alam al-Islami, which united Islamic radicals from more than twenty nations. In 1978, the kingdom backed another Brotherhood initiative, the International Islamic Relief Organization (IIRO), an entity that has been implicated in funding organizations such as Al Qaeda and Hamas. Most Muslim nations collect mandatory Islamic charity (zakat is the Third Pillar of Islam, an annual wealth tax for charitable purposes) of about 2.5% from Muslim institutions and companies.[139] Zakat is proscribed to go to those less fortunate. However, the Brotherhood has determined that those engaged in jihad against the enemies of Islam are entitled to benefit from the charitable offering. The interpretation that modern jihad is a serious, purposefully organized work intended to rebuild Islamic society and state and to implement the Islamic way of life in the political, cultural, and economic domains is widely accepted among Muslims and is thus viewed as a legitimate recipient of zakat.[140]

Both EIG and EIJ likely used these mechanisms to their advantage. EIG often sought donations during Friday prayers at mosques in order to facilitate their social outreach programs. It is unknown how much of this funding went toward illicit activity; however, it is presumed that the preponderance of the collected funds were reinvested into public and legal organizational activities. The extent of the group's aid from outside of Egypt is not known, although the Egyptian government has claimed that both Iran and Saudi Arabia have provided financial and material support to EIJ.[141] Given the socialization of the tactic, it may be assumed that EIJ obtained some funding through various Islamic nongovernmental organizations, cover businesses, and zakat funding

---

[139] Ibid.

[140] Ibid.

[141] Scheuer, *Through Our Enemies' Eyes: Osama Bin Laden, Radical Islam, and the Future of America.*

operations, and possibly, although not likely, criminal acts.[142] The most evident external support to EIJ was the symbiosis with Al Qaeda and their increasing dependence on that organization. Few exact figures exist; however, from 1996 to 1997, EIJ received more than $5,000 per month[143] from Al Qaeda.[144]

COUNTERMEASURES TAKEN BY THE GOVERNMENT

The Egyptian law-enforcement and corrections apparatus had been historically brutal when dealing with threats to domestic security and the political authority of the regime.[145] A deviation from this pattern was evident briefly in the early 1970s after Sadat succeeded Nasser and attempted to mollify the Muslim Brotherhood. To limit their criticism of his regime, Sadat pardoned many imprisoned Islamists. Both Qutb and Zawahiri's accounts of their respective imprisonments two decades apart indicate that the torture and horrific conditions suffered by Islamists while imprisoned only further radicalized many, while often reinforcing the rather loosely federated relationships in the process. Internal security was the responsibility of three intelligence organizations: General Intelligence, which was attached to the presidency; Military Intelligence, which was attached to the Ministry of Defense; and the General Directorate for State Security Investigations (GDSSI), which was under direct control of the interior minister. The GDSSI was accused of torturing Islamic extremists to extract confessions. In 1986, forty GDSSI officers went on trial for 422 charges of torture brought by EIJ defendants,[146] although each was exonerated in 1988.

During the 1980s, the Egyptian government supported the efforts of the Muslim Brotherhood and the more militant Islamic groups (namely EIG and EIJ) to recruit, encourage, and, to a lesser degree, facilitate the movement of young Muslim males from Egypt to Pakistan, where they were received, housed, and trained by Azzam and bin Laden's MAK and sent on to Afghanistan to support the Afghan

---

[142] Sageman, *Understanding Terrorist Networks*.

[143] In late 1996, Zawahiri traveled clandestinely to a number of former Soviet Caucasian republics (including Chechnya), and in December, he was arrested by a Russian patrol in Dagestan. He was released in May, having stuck to his cover story without the Russians determining his identity, but he was chastised by Al Qaeda members for his carelessness and saw the subsidy (paid to EIJ by Al Qaeda) lowered to $5,000 during his imprisonment, resulting in an even closer operational relationship between the two groups as EIJ members were asked to fund their own operations.

[144] Ibid.

[145] Kepel, *Muslim Extremism in Egypt: The Prophet and Pharaoh*.

[146] Zayyat, *The Road to Al-Qaeda: The Story of Bin Laden's Right-Hand Man*.

mujahidin resistance of the Soviet occupation.[147] Under Mubarak, the tendency of military intelligence to encroach on civilian security functions had been curbed, although there was a mutually supporting relationship.[148] The Egyptian intelligence service had a highly capable human intelligence apparatus and has used a variety of tactics, techniques, and procedures, ranging from clandestine penetration to document and media exploitation, as counterterrorism operations.[149]

The Sadat and Mubarak regimes attempted to alleviate Egypt's constantly problematic employment crisis and urban overcrowding by modernizing their economic system and manufacturing infrastructure. However, because this required external support (the United States Agency for International Development [USAID] spent $858 million on local development programs in Egypt, most of it on water, sewer, and transportation projects in rural areas),[150] the policies further weakened the government in the eyes of the Islamic movement.

In the early 1990s, the Mubarak regime tried to seize the moral initiative from EIG and EIJ by retreating from secular politics and culture through the Islamization of the sociopolitical space.[151] The state-run education and media systems were more publicly "Islamic" and openly sought to cultivate the image of Mubarak as a devout Muslim who was not only strong enough to protect Egypt from its attackers but also compassionate enough to provide for the average citizen. By the mid-1990s, Egyptian authorities had identified many key nodes in the EIG and EIJ support networks, including their family members, business associates, and friends who were not directly involved in operations against the government. As these individuals were arrested, interrogated, and tracked, the populace became, if no longer tacitly supportive of the groups, outright allies of the government.[152]

The US State Department credits Egyptian counterterrorism efforts and frequent extremist crackdowns as instrumental in reducing domestic EIJ operations.[153] The Egyptian security and police services were effective in reducing the operational capability of EIJ in Egypt, and attacks that were reliably attributed to the group declined by the 1990s. After the 1997 Luxor massacre, the Egyptian army was

---

[147] Gerges, *The Far Enemy: Why Jihad Went Global*.

[148] Metz, *Egypt: A Country Study*.

[149] Scheuer, *Through Our Enemies' Eyes: Osama Bin Laden, Radical Islam, and the Future of America*.

[150] Liberatore, "Islamic Fundamentalism in Egypt: U.S. Policy Recommendations."

[151] Gerges, "The End of the Islamist Insurgency?: Costs and Prospects."

[152] Gerges, *The Far Enemy: Why Jihad Went Global*.

[153] Scheuer, *Through Our Enemies' Eyes: Osama Bin Laden, Radical Islam, and the Future of America*.

deployed to augment the security apparatus in many tourist areas, serving as both a deterrent to potential attackers and a reassurance to Egyptians and foreigners alike.[154] After 1998, US intervention on Egypt's behalf in extradition requests of EIG and EIJ members from other states in Europe, Africa, Latin America, and the Middle East and their subsequent rendition and interrogation proved helpful in identifying the remainder of the parent networks and organizational structure in Egypt.[155] The Mubarak government also began rewarding EIG for its renunciation of violence through prisoner releases and releasing pressure on the non-member support network, further incentivizing the transformation of the organization.

SHORT- AND LONG-TERM EFFECTS

CHANGES IN THE ENVIRONMENT

Two potential explanations for the failure of the Islamist revolutionary groups in Egypt (EIG and EIJ) to achieve their operational goals in Egypt are their loss of popular support and their inability to effectively integrate their operations.[156] Victims of campaigns against the Egyptian state from 1990 to 1997 totaled more than 1,300 and included the head of the counterterrorism police, a speaker of parliament, dozens of European tourists and Egyptian bystanders, and more than 100 Egyptian police.[157] Popular support for Islamist causes among the Egyptian populace waned with the collateral damage inflicted on innocents by both EIJ and EIG terrorist operations. The death of the young girl during the failed assassination attempt on former Prime Minister Sidqi so outraged Egyptian public opinion that Zawahiri, in a rare display of a conciliatory tone, issued a statement expressing regret of her loss as an unintended consequence of the Islamic resistance.[158] The Muslim Brotherhood's condemnation of the terrorist attacks, particularly the 1997 Luxor massacre, shifted the perceived legitimacy of Islamic organizations from EIG and EIJ back to the Muslim Brotherhood. Both EIG and EIJ placed more emphasis on capturing political power than cultivating popular support and misinterpreted the geopolitical environment and the external influences of the Egyptian government and society.[159]

---

[154] Gerges, "The End of the Islamist Insurgency?: Costs and Prospects."
[155] Ibid.
[156] Gerges, *The Far Enemy: Why Jihad Went Global.*
[157] Gerges, "The End of the Islamist Insurgency?: Costs and Prospects."
[158] Aboul-Enein, "Ayman Al-Zawahiri: The Ideologue of Modern Islamic Militancy."
[159] Gerges, "The End of the Islamist Insurgency?: Costs and Prospects."

Although their respective efforts to Islamize society through violent overthrow of the government were never achieved, their efforts resulted in the unintended consequence of reestablishing the political viability of the Muslim Brotherhood in Egypt as a mainstream Islamic activist organization that continues to influence domestic policy through its advocacy of welfare programs, public morality, and government accountability.[160] The Muslim Brotherhood, although still active, has done relatively little to improve the economic and social conditions of Egyptians from lower socioeconomic status groups. This same demographic, which once comprised the ranks of EIG and EIJ, is still marginalized as the distribution of wealth in Egypt has made the distinction between haves and have-nots more prominent in recent years.[161]

CHANGES IN GOVERNMENT

Although the Islamic terrorist organizations in Egypt, including EIJ and EIG, were often a focus of the Egyptian security apparatus, particularly in the wake of the high-profile attacks, no dramatic changes in organizational structure can be attributed to either group. The goal of overthrowing the secular Egyptian government and replacing it with an Islamic emirate had not come to fruition by the 2001 EIJ–Al Qaeda merger.

EIG and EIJ's relevance to the Egyptian state was the exposure of its "weakness, rigidity, and closeness of the political system and highlighting the pervasiveness of corruption, the government's dependency on foreign handouts, and its inept microeconomic policies."[162] Although the Muslim Brotherhood made similar attempts to make the failures and priorities of the government more transparent, their tactics lacked the sensationalism of terrorism. Although this enabled the government's tolerance of the Brotherhood's existence, it allowed the secular government to placate the group without improving the quality of life for the average Egyptian. US aid to Egypt not only reinforced the Islamic groups' antipathy for the regime but also helped to defeat the more militant of those groups.[163]

Egypt participated in the coalition to expel Iraqi forces from Kuwait in 1990–1991, a decision that brought moderate condemnation from the Islamic groups within the country but ultimately resulted in greater

---

[160] Ibid.
[161] Abdo, *No God but God: Egypt and the Triumph of Islam.*
[162] Gerges, "The End of the Islamist Insurgency?: Costs and Prospects."
[163] Ibid.

material and military-to-military training support from the United States. With the increased international attention of transnational jihadist groups, including the remnants of EIJ, after the 9/11 Al Qaeda attacks on the United States, Egypt's strategic, operational, and tactical alliance with the United States has grown stronger.

CHANGES IN POLICY

The successive post-1952 Egyptian administrations were weary of Islamist threats and thus employed a robust and, at times, brutal law-enforcement and counterintelligence capability to hinder the political and operational influence of these groups. While waxing and waning with the threat, the vigilance by the government and its distrust of the populace did not change appreciably due to EIJ operations. With increased international, notably US, cooperation after the post-9/11 declaration of the Global War on Terrorism, the freedom of movement of Islamist terrorist organizations has been further degraded within Egypt. However, the sentiment that initially led to the original support of groups like EIJ is still prevalent. With the dissemination capabilities brought about by advances in information technology, there is still a threat of individual and small-group radicalization leading to a resurgent capability of Islamic organizations in opposition to the Egyptian regime.

CHANGES IN THE REVOLUTIONARY MOVEMENT

The Muslim Brotherhood, formally banned in Egypt since 1954, is still permitted by the government to operate somewhat openly, although it is subject to mass arrests, the most recent in February 2010. The organization has affiliates across the world and still advocates an Islamic state under Islamic law, in opposition to secular, authoritarian regimes. The regional and social differences among the memberships of EIG and EIJ were never overcome by their leadership, many of whom were either imprisoned in Egypt or abroad or led in absentia.

One scholar identifies the first stage of the global Salafist jihad as EIJ, exemplified by Faraj's message of the forgotten duty of jihad against the near enemy, followed by Azzam's global expansion of the defensive jihad, most notably in Afghanistan in the 1980s, and finally, by bin Laden and Zawahiri's transition to a global jihad offensive against the far enemy under the guise of Al Qaeda and its affiliated movements.[164]

---

[164] Sageman, *Understanding Terrorist Networks*.

The most dramatic change in the EIJ movement was the shift from targeting the Egyptian government to a series of harassing, although nonetheless effective, terrorist operations directed at the United States. The debate over the near versus far enemy seems to have changed in 1990 after bin Laden was rebuked by Saudi Arabia when he offered to employ his mujahidin against Saddam Hussein's army after it had annexed Kuwait.[165] Whether out of design or necessity, the decision to move from the near enemy to the far enemy disillusioned some EIJ members and motivated others to pursue their objectives on targets in other nations. The June 2001 merger between EIJ and Al Qaeda, forming "Qaeda al-Jihad," effectively ended the operational role of EIJ. However, many Egyptian members of Al Qaeda are still considered to belong to both organizations and are often identified as such.[166]

After years of violent confrontation, EIG and the remnants of EIJ that did not affiliate with Al Qaeda accepted the failure to achieve their objectives and recognized the deleterious effects of their violence on the society and peoples they sought to liberate.[167] From this acknowledgment, there has arisen an Egyptian jihadist revisionism that renounces violence. The revisionism recalibrates antigovernment sentiments while employing the same Islamist narrative once used to incite the populace to now advocate nonviolent social and political activism.[168] Although numerous factors have prevented the remnants of EIG, EIJ, and the Muslim Brotherhood from seeing their social policies come to fruition, the current state of political discourse among these three groups is much closer to the activist zone along the political mobilization continuum than the radical terrorist.[169]

---

[165] Gerges, *The Far Enemy: Why Jihad Went Global.*

[166] Zayyat, *The Road to Al-Qaeda: The Story of Bin Laden's Right-Hand Man.*

[167] Amr Hamzawy and Sarah Grebowski, "From Violence to Moderation: Al-Jama'a al-Islamiya al-Jihad" *Carnegie Papers*, no. 20 (Washington, DC: Carnegie Endowment for International Peace, 2010), accessed August 23, 2010, http://carnegieendowment.org/files/violence_moderation.pdf.

[168] Ibid.

[169] Ibid.

BIBLIOGRAPHY

Abdo, G. *No God but God: Egypt and the Triumph of Islam.* Oxford, UK: Oxford University Press, 2000.

Aboul-Enein, Youssef H. "Ayman Al-Zawahiri: The Ideologue of Modern Islamic Militancy." *The Counterproliferation Papers, Future Warfare Series,* no. 21. Montgomery, AL: Air University, Maxwell Air Force Base.

Bajoria, Jayshree, and Greg Bruno. "al-Qaeda (a.k.a. al-Qaida, al-Qa'ida)." New York: Council on Foreign Relations. Accessed August 23, 2010. http://www.cfr.org/publication/9126/alqaeda_aka_alqaida_alqaida.html#p1.

Brachman, Jarret M. *Global Jihadism: Theory and Practice. Cass Series on Political Violence.* New York: Taylor & Francis, 2008.

Eikmeier, D. C. "Qutbism: An Ideology of Islamic-Fascism." *Parameters* (2007): 85.

Euben, Roxanne Leslie. *Enemy in the Mirror: Islamic Fundamentalism and the Limits of Modern Rationalism: A Work of Comparative Political Theory.* Princeton, NJ: Princeton University Press, 1999.

Gambetta, Diego, and Steffen Hertog. "Engineers of Jihad." Oxford, UK: University of Oxford.

Gerges, Fawaz A. "The End of the Islamist Insurgency?: Costs and Prospects." *The Middle East Journal* 54, no. 4 (2000): 592–612.

———. *The Far Enemy: Why Jihad Went Global.* New York: Cambridge University Press, 2005.

Gerwehr, Scott, and Sara Daly. "Al-Qaida: Terrorist Selection and Recruitment." In *The McGraw Hill Security Handbook.* Edited by David G. Kamien. New York: McGraw Hill, 2006.

Hamzawy, Amr, and Sarah Grebowski. "From Violence to Moderation: Al-Jama'a al-Islamiya al-Jihad." *Carnegie Papers,* no. 20. Washington, DC: Carnegie Endowment for International Peace.

Jansen, Johannes J. G. *The Neglected Duty: The Creed of Sadat's Assassins and Islamic Resurgence in the Middle East.* New York: Macmillan, 1986.

Jureidini, Paul A., et al. *A Casebook on Insurgency and Revolutionary Warfare: 23 Summary Accounts.* Washington, DC: American University, Special Operations Research Office, 1962.

Kepel, Gilles. *Muslim Extremism in Egypt: The Prophet and Pharaoh.* Berkeley, CA: University of California Press, 1985.

———. *The Roots of Radical Islam.* London: Saqi, 2005.

Liberatore, Carla. "Islamic Fundamentalism in Egypt: U.S. Policy Recommendations." Washington, DC: National War College.

Metz, H. C. *Egypt: A Country Study.* Washington, DC: Federal Research Division, Library of Congress, 1990. Accessed August 23, 2010. http://lcweb2.loc.gov/frd/cs/egtoc.html.

Mitchell, Richard P. *The Society of the Muslim Brothers.* New York: Oxford University Press, 1993; 1969.

National Commission on Terrorist Attacks Upon the United States. *The 9/11 Commission Report: Final Report on the National Commission on Terrorist Attacks upon the United States.* New York: W. W. Norton, 2004.

Raphaeli, N. "Ayman Muhammad Rabi' Al-Zawahiri: The Making of an Arch-Terrorist." *Terrorism and Political Violence* 14, no. 4 (2002): 1–22.

Sageman, Marc. *Understanding Terrorist Networks.* Philadelphia: University of Pennsylvania Press, 2004.

Scheuer, Michael. *Through Our Enemies' Eyes: Osama Bin Laden, Radical Islam, and the Future of America.* Washington, DC: Potomac Books, Inc., 2007.

Soage, A. B. "Faraj Fawda, Or the Cost of Freedom of Expression." *The Middle East Review of Foreign Affairs* 11, no. 2 (2007).

Springer, Devin R., L. Regens, and David N. Edger. *Islamic Radicalism and Global Jihad.* Washington, DC: Georgetown University Press, 2009.

Wright, Lawrence. *The Looming Tower: Al Qaeda and the Road to 9.11.* New York: Knopf, 2006.

al-Zayyat, Montasser. *The Road to Al-Qaeda: The Story of Bin Laden's Right-Hand Man.* London: Pluto Press, 2004.

TALIBAN: 1994–2009

Sanaz Mirazei

SYNOPSIS

The Taliban insurgency took place in two periods: the first from 1994 to 1996 against the ruling Mujahidin and again from 2001 to 2009.[1] In the first period, the Taliban took advantage of the weak centralization after the Mujahidin came to power to quickly gain control of the country, starting with attacks in the south and moving north toward Kabul. The Taliban ruled Afghanistan from 1996 until 2001, when in the aftermath of 9/11, the United States sought to find Osama bin Laden, who was suspected to have been given refuge by the Taliban government. After many failed attempts by the United Nations and the United States to convince the Taliban government to hand over bin Laden, the United States and Britain issued attacks on the Taliban. Employing an Islamic narrative with strong influence from Wahhabism, the Taliban capitalized on the use of religion and a promise to return to a pure, uncorrupt time in order to gain recruits for their war against both the international coalition forces and the new Afghan government installed in 2002. Using a variety of guerrilla warfare tactics and capitalizing on considerable opium revenue, the Taliban insurgency, led primarily by Mullah Mohammed Omar, remained strong even until 2010.

TIMELINE

1996	The Taliban seize Kabul from the Mujahidin.
1997	The Taliban government is recognized by Pakistan and Saudi Arabia.
1999	The United States imposes sanctions and an air embargo on the Taliban government to pressure them to hand over Osama bin Laden.
March 2001	The Taliban blow up the Buddha statues in Bamiyan in defiance of the international community.

---

[1] As of 2010, the insurgency is still ongoing, but for the purposes of this chapter, we have stopped the observation at 2009.

October 2001	The US and Britain launch air strikes against Afghanistan after the Taliban's refusal to hand over bin Laden.
November 2001	US and British-led opposition forces capture Mazar-e Sharif, Kabul, and other important cities, leading to the Taliban's fall.
December 2001	Agreement among Afghan groups is reached to establish an interim government.
	Hamid Karzai is sworn in as the interim president.
September 2002	The Loya Jirga elects Karzai, extending his term until 2004.
January 2003	The Taliban begin their military campaign in Helmand and Zabul Provinces.
June 2003	Mullah Omar reorganizes the Taliban leadership by adding a ten-man leadership council.
June to August 2003	The Taliban initiates one or two attacks every other day, killing more than 220 Afghan soldiers and civilians.
	The North Atlantic Treaty Organization (NATO) takes control of security in Kabul.
2003	The Taliban is able to secure 80% of Zabul.
2005	The Taliban begins cooperation with Al Qaeda.
May to June 2006	Many are killed in battles between the Taliban and Afghan and coalition forces in the south.
July 2006	International Security Assistance Force (ISAF)-led NATO troops take over military leadership in the south.
October 2006	NATO relieves the US-led coalition force of the responsibility for security across the entire country as they take command.
2007	Several top Taliban leaders are killed, forcing the group to once again restrategize, including Mullah Dadullah, its most senior military commander.

THE ENVIRONMENT OF THE REVOLUTION

PHYSICAL ENVIRONMENT

Figure 1. Map of Afghanistan.[2]

Afghanistan lies east of Iran, north and west of Pakistan, south of the central Asian states of Uzbekistan, Turkmenistan, and Tajikistan, and southwest of China, encompassing approximately 652,290 square kilometers, roughly the size of Texas. Afghanistan features a rugged, mountainous terrain with a major mountain range, the Hindu Kush, that runs northeast to southwest and divides the northern provinces from the rest of the country. Flanking the Hindu Kush mountains are fertile but isolated valleys, as well as deserts and river valleys. A landlocked country, Afghanistan primarily relies on water from surrounding rivers. Having limited natural freshwater resources and inadequate supplies of potable water are sources of environmental concern for the country. The climate is arid to semi-arid.

Afghanistan is divided into eleven geographic zones. The first six—the Wakhan Corridor-Pamir Knot, Badakhshan, Central Mountains, Eastern Mountains, Northern Mountains and Foothills, and Southern Mountains and Foothills—are located in the mountainous region of the Hindu Kush. The remaining five are mostly plains and desert: the

---

[2] Central Intelligence Agency, "Afghanistan," *The World Factbook*, https://www.cia.gov/library/publications/the-world-factbook/geos/af.html.

Turkistan Plains, Herat-Farah Lowlands, Sistan Basin-Hilmand Valley, Western Stony Desert, and Southwestern Sandy Desert.

CULTURAL AND DEMOGRAPHIC ENVIRONMENT

Afghanistan has a population of about 17,000,000, with almost 1,000,000 of the population having died in the war up until 2001 and a refugee population of close to 5,000,000.[3] Afghanistan represents a culturally diverse country with several, often clashing, ethnic and linguistic groups. Organized by tribal and kin lineages that mobilize the people both politically and economically, Afghanistan's deep ethnic divisions have been the source of several domestic conflicts.[4] Afghanistan's patrilineal society consists of about 40–50% Pashtun, 25% Tajik, 9% Uzbek, 12–15% Hazara, and minor ethnic groups, including Chahar Aimaks, Turkmen, Baloch, and others.[5] Corresponding to the different ethnic groups, there are also several spoken languages in Afghanistan. About 50% speak Pashtu, 35% speak Dari, 11% speak Turkic languages (primarily the Uzbek and Turkmen), and about 4% speak thirty minor languages (primarily Balochi and Pashai).[6]

The majority of Afghans are Muslim, representing about 99% of the population.[7] There are a few thousand Hindus and Sikhs, plus small numbers of Armenian Christians and Jews in the major cities.[8] Roughly 85% of the Muslims are Sunnis of the Hanafi school. The rest are Shi'a, most Twelver Shi'a, like those of Iran.[9]

With a long history of patrilineal organization, this male-dominated society is rooted in the Islamic tradition, which dominates Afghanistan. The male gender controls both the private and public spheres and few rights are given to women, who are subjected to forced covering of their bodies by head-to-toe burqas, arranged marriages, and other Islamic practices. Women are precluded from political life and are deprived of the opportunity to get an education.[10] Instead, they are usually married at very young ages and expected to stay at home, raise families, and maintain the honor of the family, tribe, and

---

[3] Nasreen Ghufran, "The Taliban and the Civil War Entanglement in Afghanistan," *Asian Survey* 41, no. 3 (May, 2001): 462–487.

[4] Barnett Rubin, *The Fragmentation of Afghanistan: State Formation and Collapse in the International System* (New Haven, CT: Yale University Press, 2002).

[5] Ibid.

[6] Ibid.

[7] Ibid.

[8] Ibid.

[9] Ibid.

[10] Ahmed Rashid, *Taliban: Militant Islam, Oil, and Fundamentalism in Central Asia* (New Haven, CT: Yale University Press, 2000).

kin group. These practices were especially apparent during Taliban rule. In addition to combating the corruption that was rampant in Afghanistan, another major goal of the Taliban was to reinforce Islam in the everyday lives of Afghans. As such, they "issued decrees in which they required men to wear turbans, beards, short hair, and shalwar kameez."[11] The Taliban also banned "music, games, any representation of the human or animal form, and entertainment including television, chess, kites, cards, etc."[12]

SOCIOECONOMIC ENVIRONMENT

Poppy cultivation in Afghanistan increased in the mid-1970s, with more than one-half of Afghanistan's provinces involved in the poppy growth and cultivation.[13] Furthermore, by the late 1980s, millions of dollars of foreign aid had ended and the political fragmentation[14] left in Afghanistan led local Mujahidin commanders to turn to criminal activities, including heroin production, trafficking, extortion, and kidnapping for ransom. Devastation from the Soviet war, combined with the increasing spread of political Islam, hindered the Afghan government's ability to establish the legitimacy it needed to construct and maintain effective control of its people. This lack of political authority led Mujahidin commanders to evolve into warlords with their own local gendarmes, effectively splitting Afghanistan into several different fiefdoms.[15] Opium production was close to 3,400 tons by 2002 and 3,600 tons in 2003, making up 75% of the world's heroin.[16]

As poppy cultivation continued, smuggling also produced a viable source of income for the Afghan people. By the late 1990s, a "World Bank study estimates that the smuggling trade between Pakistan and Afghanistan alone amounted to $2.5 billion in 1997, equivalent to more than half of Afghanistan's estimated GDP." This smuggling had adverse effects on the local economy, as "factories [could not]

---

[11] Ibid.

[12] Peter Marsden, *The Taliban: War and Religion in Afghanistan* (London: Zed Books, 2002).

[13] Ibid.

[14] At this point, the Mujahidin were not a unified movement; instead, they were regionally fragmented, with the commanders of each regional Mujahidin group taking on a new role as warlords. The Mujahidin were not able to reach a sustainable power-sharing agreement.

[15] Rasul Bakhsh Rais, *Recovering the Frontier State: War, Ethnicity, and the State in Afghanistan* (Lanham, MD: Lexington Books, 2008).

[16] Marsden, *The Taliban: War and Religion in Afghanistan*; Larry P. Goodson, "Afghanistan in 2003: The Taliban Resurface and a New Constitution Is Born," *Asian Survey* 44, no. 1 (January 2004): 14–22.

compete with smuggled, foreign-made, duty-free consumer goods," and led to losses in revenue from customs and sales taxes. Although officially the Afghan government and local industry were harmed by the smuggling economy, the Taliban were able to use the lack of security in the region to effectively tax the smuggling trade, making it the Taliban's "second largest source of income after drugs."[17]

However the financial benefits from the drug trade were not shared evenly among the public. According to the United Nations Drug Control Program (UNDCP), "farmers received less than 1% per cent of the total profits generated by the opium trade; another 2.5 per cent remained in Afghanistan and Pakistan in the hands of dealers."[18] In addition, the Afghan people continued to suffer from poverty, with average salaries around one to three US dollars a month, creating further dependency on the United Nations and other aid agencies for food. In 1998, Afghanistan witnessed two massive earthquakes that exacerbated the already poor state of the Afghan people. Furthermore, the Taliban engaged in increasingly confrontational rhetoric and actions against the international community, instituting repressive laws against nongovernmental organizations (NGOs) that sought to provide relief for the suffering Afghan people. In 2001, Afghanistan was not only a security threat but also the "world's worst humanitarian disaster zone."[19] The country suffered from four years of nationwide drought, "destroying seventy percent of the livestock and making fifty percent of the land uncultivable."[20] As a result, the country faced mass starvation, forcing one of the largest refugee populations in the world, with close to 3.6 million refugees outside of Afghanistan and 800,000 internally displaced.[21]

Attempting to transition from the war economy built on heroin production, smuggling, and other criminal activities, there were a few positive developments in the Afghan economy by 2003. First, Afghanistan began transitioning to a new currency. Second, Afghanistan is strategically located with access to natural gas, giving the government an opportunity to reap the benefits of this scarce resource. However, understanding that Afghanistan had been in a constant state of war for the last twenty years, it became faced with donor fatigue and fear of continued violence as several NGOs and

---

[17] Ahmed Rashid, "The Taliban: Exporting Extremism," *Foreign Affairs* 78, no. 6 (November 1999): 22–35.

[18] Ibid.

[19] Ahmed Rashid, *Descent into Chaos: The US and the Failure of Nation Building in Pakistan, Afghanistan, and Central Asia* (New York: Viking, 2008).

[20] Ibid.

[21] Ibid.

international donors pulled out, leaving "Afghanistan with little capacity for redevelopment."[22]

HISTORICAL FACTORS

Political Islam offered a concrete framework with which the Afghans were familiar, understood, and practiced for as long as their history. As political Islam gathered a more domestic following, communist counterparts with support from the Daoud regime also drew significant attention. In 1977, the two major communist groups, Khalq and Parcham, reunited once again after ten years of separation.[23] However, this coalition was rather short-lived. When the Saur Revolution took place in 1978, the Khalq-Parcham party faced very little resistance as the Daoud regime expunged all of the opposition and made political parties and organization illegal. However, power struggles between the two factions in the summer of 1978 as a result of the inability of the regime to effectively deal with the rising domestic unrest, Islamic jihad, and tense relations with the Soviets caused the party to split once again, with the Khalq faction winning.[24] Beginning in 1979, with the Soviet reinforcement of troops, the Afghan Mujahidin, the first group to embody political Islam in Afghanistan, was able to effectively organize an armed resistance, largely from Pakistan where the party's headquarters and leaders were based. The Mujahidin sought to remove the Soviet-backed government from power, and, additionally, to establish an Islamic government for the country. Waging a multidimensional war and strategically using the geography, the Mujahidin, who seemed at first to be at a disadvantage, were able to expel the Soviets from their homeland by acquiring significant foreign backing from the United States, Saudi Arabia, and other countries. As a result of their insurgency, the Mujahidin established and led a highly fragmented decentralized Islamic government in Afghanistan from 1992 until 1996.

After the Soviet retreat, local Mujahidin commanders took advantage of the new transitioning government and were able to divide Afghanistan into separate fiefdoms controlling their respective areas. Lacking a strong centralized government and distraught after years of destructive Soviet bombing campaigns, the Mujahidin commanders set up base as warlords engaging in the illicit drug trade,

---

[22] Goodson, "Afghanistan in 2003."

[23] Nazif M. Shahrani and Robert L. Canfield, *Revolutions and Rebellions in Afghanistan: Anthropological Perspectives* (Berkeley, CA: Institute of International Studies, University of California, 1984); Rashid, "The Taliban: Exporting Extremism."

[24] Ibid.

smuggling, and ruling their respective provinces with little regard for the new Afghanistan. Simultaneously, ethnic Pashtuns became more and more dissatisfied with the new primarily Tajik government exacerbating divides in the Mujahidin. The traditionalist Mujahidin and the fundamentalist Mujahidin became wearier of their ethnic and ideological differences as they were more and more incapable of coming to a power-sharing agreement for the new government. All of these factors led the Pashtuns to lend support to the Taliban, a largely Pashtun Islamic fundamentalist organization championed by Mullah Omar and other *ulama* with heavy influence from the Egyptian Islamic Jihad and other jihadi movements. The Taliban, which began as a "small, spontaneous group" in 1994,[25] is an almost exclusively Sunni organization with devotion to the principles of Islam and the *shari'a*[26] rule of law.[27] This idea was appealing to Pashtuns tired of the chaos and warlordism that was rampant in the country at this time. An offshoot of the fundamentalist Mujahidin, the Taliban infused political Islam learned from the Egyptian Islamic Jihad with Saudi Arabian Wahhabism.

The Taliban began their rise to power in October 1994 when close to 200 fighters gathered at the Durand line border on the Pakistan side in an attempt to secure the commercial road from Kandahar to Pakistan. On November 4, a thirty-truck Pakistani convoy was captured by warlords near Kandahar. Soon thereafter, the Taliban crossed the border and captured Spin Buldak, a town in Kandahar Province in southern Afghanistan, from Hikmatyar, one of the main Mujahidin leaders. Capturing a large quantity of arms from Spin Buldak, they went forward to Kandahar, where they freed the Pakistani commercial caravan from local armed groups who controlled the road and were able to convince Mullah Najibullah, a prominent commander in Kandahar, to surrender. Once again, the Taliban secured a depot of arms, armored vehicles, artillery pieces, six MiG-21 fighters, and several helicopters, giving them a military advantage in the area. As a result of securing these two strategic positions and their weapons acquisitions, they successfully disarmed local armed groups, secured the military garrison and administration in Kandahar, and gained control of the commercial road between Kandahar and Pakistan, maintaining security on that road for goods and travelers. This was their first demonstration to the Afghan people that they were committed to combating corruption and establishing order in Afghanistan—a

---

[25] Marsden, *The Taliban: War and Religion in Afghanistan.*

[26] *Shari'a* is the main document on which Islamic law is based and is sometimes more generally referred to as the Islamic legal system.

[27] Ibid.

goal that was admired by the war-torn Afghan people. Afterward, the Taliban established their administration in Kandahar and began to organize their military campaign for the surrounding towns and provinces.

Shortly thereafter, they secured and controlled Helmand Province, a center for opium production.[28] Moving against the forces of Ghaffar Akhunzadeh, a clan leader of the opium-rich Helmand Province, the Taliban were faced with fierce resistance. However, the Taliban successfully captured the province by January 1995 by playing smaller drug warlords and bribing others against Akhunzadeh. Afterward, they led their efforts to the west, reaching Dilaram on the Kandahar-Herat highway. While maintaining their successes in the southwest, they began to move north toward Kabul. The Taliban were able to progress easily through the Pashtun belt, facing more surrender than resistance. Their strategy was unusual in that they were regularly successful in avoiding direct fighting. Instead, they would send a delegation of *ulama*, a group of educated Islamic legal scholars, to talk to local militia commanders, asking them to implement *shari'a* law and promote peace by surrendering their weapons and ammunition. The *ulama* tried to convince the militia commanders that these arms were actually part of the national treasury and because Afghanistan has a single government, they should give them up. If the commander agreed, then there was no fighting. If they rejected, then the Taliban would send a second delegation that included religious tribal elders and a Taliban representative. If this second delegation also failed, then they would engage in violent, armed attacks.[29]

As they moved toward the north, they were faced with major warlords and deep political and ethnic divisions. In January 1995, the Taliban opposition came together to attack the government led by President Rabbani in Kabul. In the north, Hikmatyar allied with the Uzbek warlord General Rashid Dostam, as well as Hazaras living in Kabul. By February 1995, the Taliban had continued their strategy to the northeast, capturing Hikmatyar's headquarters at Charasyab, a town close to Kabul, and opened the roads to Kabul. After several months of a Hikmatyar-imposed blockade of Kabul, this Taliban move was quite popular with the public because it enabled food convoys to finally reach the capital. The Taliban continued northward, reinforced by new recruits won after freeing the roads to Kabul and showing the people that they were capable of keeping their promises for order and ending corruption. The Taliban were able to capture Urozgan

---

[28] Neamatollah Nojumi, *The Rise of the Taliban in Afghanistan: Mass Mobilization, Civil War, and the Future of the Region* (New York: Palgrave, 2002).

[29] Ghufran, "The Taliban and the Civil War," 462–487.

and Zabul Provinces in the north without a shot being fired. Given their lucrative backing from the opium trade and transport business, in addition to foreign funding from Saudi Arabia and Pakistan, the Taliban were able in many cases to bribe commanders and armed groups into surrender.[30]

Although the Taliban were successful in the Pashtun belt, the minority ethnic Mujahidin groups did not buy into the Taliban's claim to represent Islamic unity above ethnicity and were very much aware of the Pashtun composition of the group. The Mujahidin leaders in power—especially the ruling coalition of Tajiks, including Massoud[31]—were not going to simply allow the Taliban to take over. Unlike southern Afghanistan, which is represented mostly by Pashtuns, outside of the Pashtun belt, the Taliban were not seen as peacemakers but rather as a force to revive Pashtunism.[32] These minority groups, especially the Tajiks, Uzbeks, and Hazaras, were ready to defend their territory and claims for regional autonomy by any means necessary, starting yet another civil war in the country as the Taliban ascended to power in 1996. To balance the rising Taliban threat, the Mujahidin, or jihadi, warlords formed a coalition with the Hazaras and Uzbeks named the Northern Alliance. Massoud, remembered as the Lion of Panjshir and a successful commander during the Soviet war, was able to push the Taliban out of Kabul after a bloody fight that killed hundreds of Taliban. "This was the first major battle that the Taliban had fought and lost. Their weak military structure and poor tactics ensured their defeat at the hands of Massoud's more experienced fighters."[33]

This event did not suppress the Taliban's determination. They continued their quest to the west in Herat. In March, after heavy fighting, the Taliban gained control of Nimroz and Farah, two provinces guarded by Ismael Khan, and made their way to Shindad, a former Soviet airbase in the south of Herat. In response, the Kabul government sent Massoud to help Ismael Khan in the fight against the Taliban. As part of their strategy to quell Taliban power, Massoud launched a punishing attack, airlifting 2,000 of his experienced Tajik fighters against the Taliban from Kabul in March 1995. By the end of March, the Taliban had been pushed out of Shindad, suffering another major loss. After the Taliban retreat, Massoud also committed

---

[30] Ahmed Rashid, *Taliban Islam, Oil and the New Great Game in Central Asia* (London: Tauris, 2002).

[31] Massoud was one of the strongest and most brutal warlords. Named the Lion of Panjshir, he was a Tajik commander who was very difficult to break down. He wielded a lot of power and respect, especially by the Tajiks in Afghanistan and during the Mujahidin insurgency.

[32] Rashid, "Taliban: Militant Islam, Oil, and Fundamentalism."

[33] Ibid.

atrocities against the citizens of Kabul. As a result, President Rabbani was able to consolidate his power in Herat and Kabul and regained control of six provinces around Kabul from the Taliban.[34] However, Ismael Khan largely miscalculated the effects of the Taliban's defeat. Because he believed that the Taliban were no longer a viable threat, he led an ill-prepared offensive against them in Helmand Province, which was under Taliban control. Given the poor timing and overexertion of his forces in the hostile area, what Khan failed to consider was that the Taliban had spent that summer restrategizing, rebuilding their forces, and reformulating their command structure, making them a much stronger adversary. As a result of these initiatives and new recruits from Pakistan, the Taliban sent 25,000 men to fight against Khan's forces, defeating them in two days. The Taliban forced Khan out of Herat as they occupied that city and made plans for Kabul. This marked the beginning of the end for the Rabbani government.[35]

In the next year, an important decision was to be made by the leaders of the two adversaries: How would the Taliban led by Mullah Omar win Kabul, and how would the government with Massoud at its military head win Kandahar, the Taliban base? Already winning the trust of Pashtuns, the Taliban were concerned with the response of non-Pashtuns, who make up a large percentage of Kabul, especially after the confrontation with Massoud, and the northern provinces loyal to the Mujahidin leaders. In an effort to help legitimize the Taliban's plans, Mullah Omar was named "Amir-ul Momineen" or "Commander of the Faithful"—a title that gave the movement legitimacy unlike any Mujahidin leader could gain.[36] However, this move was not enough to secure the capital. As a result, the Taliban engaged in rocket attacks backed by ground assaults in Kabul and the surrounding cities throughout the year. After several months of deliberation by the Taliban, the Mujahidin and interested foreign parties—chiefly the United States and Pakistan—met to determine the fate of Kabul. On June 26, 1996, Hikmatyar went to Kabul to take his post as prime minister. In direct response to this act, the Taliban launched an intense rocket attack on Kabul, killing sixty-one people and injuring more than one hundred. The Taliban then continued to move on to Jalalabad, a main stronghold for the Mujahidin leaders, on August 25, 1996, leading to the disintegration of the Shura council. The speed of the offensive led the Taliban to take control of Jalalabad on September 11, 1996, and Kabul on September 26, 1996, when Massoud decided to withdraw from the capital, marking the beginning

---

[34] Ibid.
[35] Ibid.
[36] Ibid.

of Taliban rule of Afghanistan. To demonstrate their new authority over the country, on September 27, 1996, the Taliban brutally killed Najibullah, who had ruled from 1986 to 1992. From 1996, when the Taliban essentially ruled Afghanistan, until 2001, which marked the US-led invasion of Afghanistan, the Taliban and Massoud's forces continued fighting, losing and regaining control of several different provinces.[37]

GOVERNING ENVIRONMENT

Afghanistan has been ruled by several different governments over the last one hundred years. Traditionally, the country has been organized by two power structures. The first is the local government administration, which is directed from the central government. The second is the tribal or village structures within each region.[38] Mohammad Daoud attempted to maintain these two separate power structures, allowing local systems to keep their forms of self-government, which tended to be tribal institutions.[39] After the Daoud regime, these two power structures came into conflict as the old regime's royal family elites, which held most of the official government power, clashed with the local village leaders or khans.[40] This tension resulted in an attempt to consolidate the two power structures into one communist government.

From 1979 to 1989, Afghanistan was led by a Soviet-style Communist government until the Soviets withdrew in 1989, when the Mujahidin led the new government. On April 27, 1978, in what is known as the Saur Revolution,[41] the People's Democratic Party of Afghanistan (PDPA)[42] overthrew the Daoud government in a bloody coup, installing a new communist government named the Democratic Republic of Afghanistan. A few months later, a revolt in eastern Afghanistan gave way to a larger countrywide insurgency. Facing increasing domestic pressure from the insurgency, the new government in Afghanistan signed a bilateral treaty of friendship and cooperation with the Soviet Union in December. As a result of this treaty, Soviet military assistance greatly increased, causing the fledgling regime to become more dependent on this foreign assistance to combat the spreading insurgency as the Afghan army began to collapse.

---

[37] Ibid.

[38] Shahrani and Canfield, *Revolutions and Rebellions in Afghanistan.*

[39] Rubin, *The Fragmentation of Afghanistan.*

[40] Ibid.

[41] Saur is April in the Afghan language.

[42] The People's Democratic Party of Afghanistan is also referred to as Khalq-Parcham.

In May 1986, Babrak Karmal, a communist president of Afghanistan who was supported by the Soviet regime, was replaced by Najibullah, who launched the "national reconciliation" policy to draw more support from traditionalist Mujahidin. However, the Najibullah government began to weaken between 1989 and 1992 when it lost support from the northern militia. By 1990, the Mujahidin had established an Islamic government in Afghanistan. The new constitution proclaimed that Islam was the national religion and that *shari'a* law would rule the country. The 1987 and 1990 versions of the new constitution allowed for "institutionalized representation at the local level and for some devolution of control over administration to these local bodies."[43] Local Mujahidin groups were incorporated into the new government, with much of the eastern part of Afghanistan still under Mujahidin rule. The Mujahidin factions maintained relative autonomy in their respective areas.

As Najibullah was planning to leave Afghanistan, which was set to be ruled by the Mujahidin in April 1992, the initially peaceful transfer of power turned violent when Hikmatyar resorted to arms and a violent overthrow. Although the Mujahidin factions attempted to form an alliance, largely to attract more funds from foreign powers, political differences and grievances with respect to the power-sharing led to a new arrangement. A new power-sharing deal was formed keeping Burhanuddin Rabbani, the Tajik leader of Jamiati Islami Afghanistan (JIA-Islamic Society of Afghanistan), as the new interim president, and appointing Hikmatyar as the prime minister. However, shortly thereafter Rabbani was shot, causing concern for Hikmatyar's safety. As the fighting in Kabul continued in the wake of the dissolution of Najibullah's regime, Massoud, the lead commander of JIA, came to have a significant role in the new government.

In 1992, leaders of the exiled Mujahidin in Pakistan created the Islamic Jihad Council to assume power in Kabul, appointing Sibghatullah Mojaddedi as an interim chair for two months until the Mujahidin decided on who should be represented in the ten-member leadership council. From 1992 to 1996, Afghanistan outside of Kabul was essentially ruled as separate fiefdoms, where the local Mujahidin commanders served as warlords controlling those areas. The Mujahidin government of the Islamic State of Afghanistan, as it was now called, was merely an extension of the Afghan Interim Government. As the Mujahidin led the new Afghan government, the ethnic composition of the different groups became more salient. This new council was to be led by Rabbani for the next four months. However, the new

---

[43] Rubin, *The Fragmentation of Afghanistan*, 174.

government, led by Rabbani and his leading commander, Massoud, both ethnic Tajiks, did not sit well with the Pashtuns, who made up most of Afghanistan's ethnic composition, because they felt they had lost political sovereignty over their people.[44] This development also resulted in increased support for the Taliban, who took over Kabul in 1996, implementing an Islamic government and strict compliance with *shari'a* law.

In 1997, the Taliban were officially recognized as legitimate rulers by Pakistan and Saudi Arabia; however, most other countries continued to regard Rabbani as head of state. During the Taliban's rule, the United States launched several offensives in Afghanistan in response to several terrorist attacks led by Osama bin Laden and Al Qaeda. Although these attacks were not aimed at the Taliban, from 1998 to 2001, the United States and United Nations put pressure on the Taliban in the form of air embargoes and financial sanctions in order for them to hand over Osama bin Laden. In response to the 9/11 attacks on the United States, the United States and Britain launched air strikes against Afghanistan in October 2001, after continued Taliban refusal to hand over bin Laden. By November of that year, the coalition forces led by the United States and Britain captured Mazar-e Sharif, Kabul, and other important cities, leading to the fall of the Taliban.[45] By December 5, 2001, the Afghan people agreed on an interim government in Bonn. Two days later, the Taliban relinquished their last stronghold in Kandahar, and on December 22, 2001, Hamid Karzai, a Pashtun royalist, was sworn into office as the interim president. Karzai extended his presidency in June 2002 when the Loya Jirga (the grand council of Afghan tribal leaders) elected him as the head of state until 2004. The Loya Jirga adopted a new constitution in 2004 allowing for a strong presidency, and Karzai was once again elected president in 2004.

WEAKNESSES OF THE PREREVOLUTIONARY ENVIRONMENT AND CATALYSTS

Perhaps the greatest catalyst for the Taliban insurgency was the inability of the Mujahidin government to effectively consolidate its power. During the Mujahidin insurgency, the high level of factionalization and differences among the leaders of the various groups led to the inability to reach a long-lasting power-sharing agreement. As a result, there was almost constant fighting in some

---

[44] Rais, *Recovering the Frontier State*, 44.
[45] BBC timeline.

areas, including brutality against their own citizens. This political vacuum, in addition to the public's concern for stability and safety, made the power transition to the Taliban much easier.[46] Furthermore, the Afghan people were tired of the constant fighting, chaos, and disorder in their country. The already poor economic situation was exacerbated by the harsh Soviet tactics that left many towns and cities completely demolished, and as foreign governments and NGOs that had provided a significant amount of funds to the Mujahidin leaders witnessed donor fatigue and heavily decreased their support, the new Afghan leadership was left with few solutions to the mounting economic problems.

With the loss of millions of dollars of foreign aid and the continuing political and ethnic fragmentation in Afghanistan, local Mujahidin commanders turned to criminal activities, including heroin production, trafficking, extortion, and kidnapping for ransom with the arms that were in their possession from the insurgency. As a result, the Afghan people were left economically worse off than during the communist regime, were disillusioned from the Mujahidin promises of order and an Islamic system that would represent the majority religion, and, in the absence of a foreign invader against which they could rally, once again became cognizant of the ethnic differences that had divided the country for so many years. Given that the new Mujahidin government consisted in large part of Tajiks, the Pashtun majority felt unrepresented in this new government. With a government that did not ethnically represent the majority of the population, the devastation from the war, and the increasing spread of political Islam, the Afghan government was not able to establish the legitimacy it needed to construct and maintain effective control of its people. This lack of political authority led these commanders to evolve into warlords with their own local gendarmes, which effectively split Afghanistan into several different fiefdoms. All of these factors led the Pashtuns and non-Pashtuns to support the Taliban to demand order and stability in their war-torn country.

FORM AND CHARACTERISTICS OF THE REVOLUTION

OBJECTIVES AND GOALS

In the post-2001 Taliban insurgency, or the Taliban after the US invasion, the group's main objective was to regain order and stability in

---

[46] Marsden, *The Taliban: War and Religion in Afghanistan.*

Afghanistan through a strong government reflected in the principles of Islam while respecting and upholding *shari'a.*[47] In an attempt to recreate the time of Prophet Muhammad 1,400 years ago, an initial meeting of the *ulama* was convened to determine how the Taliban should rule. In this meeting, they reinforced the idea that the "sharia does not allow politics of political parties," which is why they do not pay their soldiers; rather, they provide them only with food, clothes, and the weapons they need to survive.[48] Accordingly, several directives were ordered to support their main objectives, including a prohibition against "mistreating the population, forcibly taking personal weapons, taking children to conduct jihad, punishment by maiming, forcing people to pay donations, searching homes, kidnapping people for money."[49] After losing power in 1996, the Taliban's political program also included removing the foreign powers and reestablishing the Islamic Emirate.[50] Throughout its conception, the Taliban remained committed to *shari'a*, the basic principles of Islam, and devotion to jihad as every man's right. However, for some of the Taliban leaders, their devotion to Islamic rule of law provides a cover that enables them to reach their own personal economic goals.

LEADERSHIP AND ORGANIZATIONAL STRUCTURE[51]

The Taliban rules with a shadow government approach, observing strong adherence to *shari'a* law. The Taliban's shadow government approach aims to be an important counterweight to the Afghan government, seeking to provide the justice, security, and dispute resolution that the Afghan public has craved after decades of war and unrest. Through this shadow governance approach, the Taliban has attempted to fill the vacuum of corruption and neglect with hope,

---

[47] Jeffrey Dressler and Carl Forsberg, *The Quetta Shura Taliban in Southern Afghanistan: Organization, Operation, and Shadow Governance* (Washington, DC: Institute for the Study of War, 2010), accessed September 8, 2010, http://www.understandingwar.org/report/quetta-shura-taliban-southern-afghanistan; Marsden, *The Taliban: War and Religion in Afghanistan.*

[48] Rashid, *Descent into Chaos.*

[49] Michael Flynn, "State of the Insurgency: Trends, Intentions, and Objectives," accessed September 8, 2010, http://coincentral.wordpress.com/2010/02/03/state-of-the-insurgency-2010/.

[50] Thomas Ruttig, *The Other Side, Dimensions of the Afghan Insurgency: Causes, Actors – and Approaches to Talks* (Afghanistan Analysts Network, 2009), accessed September 8, 2010, http://aan-afghanistan.com/index.asp?id=114.

[51] For a comprehensive chart on the Taliban organization, see National Security Archive, *The Taliban Biography: The Structure and Leadership of the Taliban 1996–2002* (2009). Also see *The Long War Journal*, In Pictures: The Taliban Leadership, http://www.longwarjournal.org/multimedia/Taliban-Leaders-Jan2008/index.html.

order, and stability, which has bolstered not only the Taliban's following but also their legitimacy as an alternate form of government.[52]

The Taliban's organizational structure consisted of a *rahbari shura*, or leadership council, and the *majlis al-shura*, made up of four regional military councils and ten committees.[53] The *rahbari shura* controlled all other subdivisions of the group and was led by Mullah Mohammad Omar, who was a member of the Mujahidin group Islamic Party Khalis[54] and named "Amir Al-Mu'minin" or Leader of the Faithful.[55] The *rahbari shura* was also referred to as the Quetta Shura Taliban (QST), named after its base in Quetta, Pakistan. The QST controlled and led all of the Taliban subdivisions, including four regional military *shuras* and ten committees. Although the official leader of the QST is Mullah Omar, after the 9/11 attacks when he and Mullah Abdul Ghani Baradar retreated to the Quetta base in Pakistan, he entrusted Mullah Baradar with the day-to-day operations of the Taliban. Until February 2010, when he was captured by Pakistani officials, Mullah Baradar acted as the second man in charge of the QST.[56] From the beginning of the Taliban movement in late 2003, Baradar provided the logistics, supplies, and ammunition to Taliban fighters inside Afghanistan.[57] Mullah Omar and Taliban members continue to refer to themselves "as the Islamic Emirate of Afghanistan, despite being removed from power in 2001."[58]

---

[52] Dressler and Forsberg, *The Quetta Shura Taliban.*

[53] Ibid.; Bill Ruggio, "The Afghan Taliban's Top Leaders," *The Long War Journal,* accessed September 8, 2010, http://www.longwarjournal.org/archives/2010/02/the_talibans_top_lea.php#ixzz0gvJMKftH.

[54] Marsden, *The Taliban: War and Religion in Afghanistan.*

[55] Dressler and Forsberg, *The Quetta Shura Taliban in Southern Afghanistan.*

[56] Ibid.

[57] Ahmed Rashid, "Ahmed Rashid Offers an Update on the Taliban," *NPR,* February 17, 2010, accessed September 8, 2010, http://www.npr.org/templates/transcript/transcript.php?storyId=123777455.

[58] Dressler and Forsberg, *The Quetta Shura Taliban in Southern Afghanistan.*

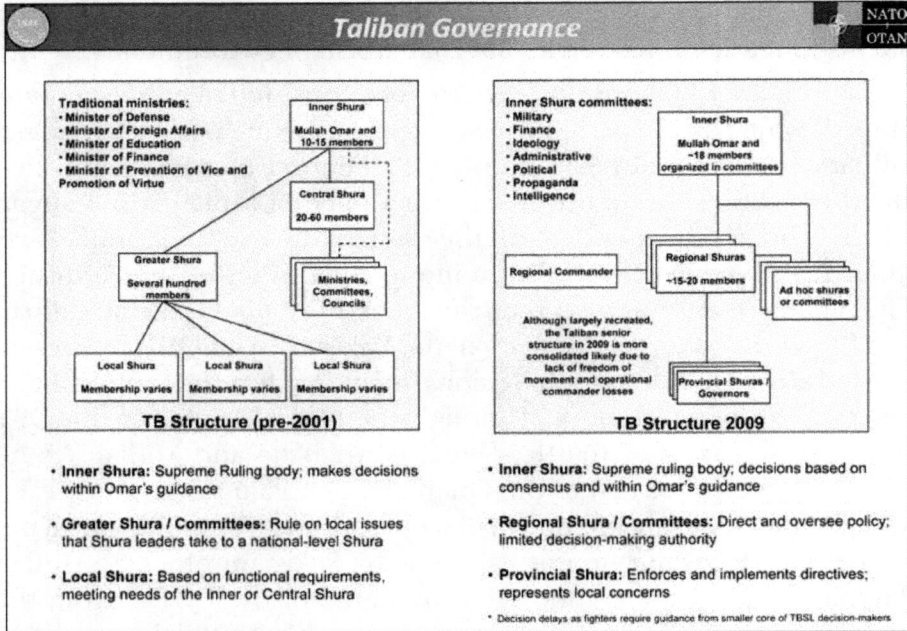

Figure 2. Organizational structure of the Taliban.[59]

The four military *shuras* correspond to the four main geographical areas of operations and are the Quetta, Peshawar, Miramshah, and Gerdi Jangal—named after their bases. The Quetta Regional Military *Shura*, led by Hafez Majid, controls activities in the southwestern areas of Afghanistan. The Peshawar Regional Military *Shura*, led originally by Maulavi Abdul Kabir before his arrest in early 2010, is now under the direction of Abdul Latif Mansour and operates in the eastern and northeastern areas of Afghanistan. In these areas, they are met by fierce resistance by the Mujahidin leaders and Northern Alliance members. The third *shura* is the Miramshah Regional Military *Shura*, which is based in North Waziristan and led by Siraj Haqqani, the son of Jalaluddin Haqqani. This *Shura* directs Taliban activities in southeastern Afghanistan. Finally, the Gerdi Jangal Regional Military *Shura* in Baluchistan is led by Mullah Adbul Zakir and runs operations in Helmand Province and, sometimes, Nimroz Province.[60] An interesting organizational characteristic of the Taliban's leadership is that even the top leaders often go back and forth from their duties as administrators and military commanders, enabling a somewhat flexible structure.[61] In addition to the leadership council and four military *shuras*, there are ten committees: military, *ulema*, finance,

[59] Flynn, "State of the Insurgency," 7.
[60] Ruggio, "The Afghan Taliban's Top Leaders."
[61] Ghufran, "The Taliban and the Civil War Entanglement," 462–487.

political affairs, culture and information, interior affairs, prisoners and refugees, education, recruitment, and repatriation.[62]

COMMUNICATIONS

Aware of US communication interception strategies, the Taliban's main source of communication depended on messengers, which reflect traditional Afghan practice to make sure a planned operation remains secure.[63] In addition to traditional messengers, the Taliban also used "night letters" to communicate, including written communications, pamphlets and underground newspapers, and communal messages. These "night letters" were used not only between Taliban members but also between the Taliban and non-Taliban Afghans. The latter use of "night letters" often contained death threats to anyone who dealt with foreigners and as a way to maintain control the public.

The Taliban also relied on radios and mobile phones even though they were more susceptible to interception. However, because the Taliban were weary of civilian informants, they banned cell phones since they were a means for private citizens to call in tips on the Taliban's plans. As a result, in March 2008, the Taliban ordered cell phone companies in Kandahar "to suspend service from five in the evening to seven in the morning so that the Taliban could operate safely during those hours."[64]

METHODS OF ACTION AND VIOLENCE

Unlike the relatively peaceful strategy used in their initial ascension to power from 1994 to 1996, after the US invasion in 2001 and losing power, the Taliban began to employ more violent force in their strategy. They led several offensives that were orchestrated from their base in Quetta. The QST focused mostly on activity in the southern provinces of Helmand and Kandahar where the narcotics trade dominated, providing serious cash flow to the movement and posing a challenge to coalition efforts aimed at bringing stability in the region.[65]

In southern Afghanistan, the QST was organized functionally with both indigenous fighting units and foreign fighters. While QST

---

[62] Ruggio, "The Afghan Taliban's Top Leaders."
[63] Rowan Scarborough, "Taliban Outwits US Eavesdroppers," *Human Events* 65, no. 7 (2009).
[64] Dressler and Forsberg, *The Quetta Shura Taliban in Southern Afghanistan.*
[65] Ibid.

commanders planned and led both offensive and defensive operations against the coalition and Afghan forces, facilitators managed logistical elements. Senior-level commanders were responsible for both. As highlighted in the *Leadership and Organizational Structure* section, the QST operated within a hierarchical chain of command, receiving orders from its base in Quetta.[66] These operational orders were planned offensives designed and led by senior leaders in the group, who made adjustments to the plans as the campaign unfolded, including requesting reinforcements in critical enemy terrain.[67]

Even though QST operations at the district level were fought by local indigenous people, they were subject to orders from senior Taliban commanders—mostly based in Quetta. Given the distance from the planning to the theater of action, the central command had weaker control over operations at the village level. Given the nature of their operations, lower-level commanders and small-unit leaders may have had more autonomy in their operations. These small units normally consisted of between eight and twelve men who typically planted improvised explosive devices (IEDs), engaged in small-scale attacks on coalition and Afghan patrols, and gathered intelligence on the interactions of locals with coalition and Afghan forces.[68] Because these units had a good understanding of coalition and Afghan lines of operation, they had the ability to launch attacks without specific orders to do so.[69] Taliban weaponry consisted primarily of small arms weapons and ammunition, in addition to IED and homemade explosive (HME) materials and technology.

Especially in the heavily populated districts in Kandahar and Helmand, the Taliban had active strategic intimidation campaigns to reduce government and coalition activity and to exert a fear campaign against locals who considered cooperating with these forces. However, these Taliban intimidation campaigns were not only limited to attacks on the government; after 2006, violence was also used against ordinary citizens to increase Taliban influence in contested areas. For instance, "in December of 2006 they executed twenty-six men in a Taliban-dominated village west of Kandahar City for cooperating with ISAF troops."[70] The Taliban then displayed the headless bodies publicly to warn locals that they would face a similar fate if they collaborated with the coalition or government forces. Although the public display of headless corpses continues, in 2008, Mullah Omar banned the

---

[66] Ibid.

[67] Ibid.

[68] Ibid.

[69] Ibid.

[70] Ibid.

beheading of informants and suggested firing squads instead.[71] In fact, since mid-2008, Omar consistently banned violence against civilians, even though the Taliban did engage in such activity. This illustrates how little control Omar had on the local Taliban.

In 2001, the insurgency illustrated its more confrontational style against the Western coalition forces when, at the encouragement of Osama bin Laden, Mullah Omar ordered the destruction of two ancient Buddha statues in the Bamiyan Valley in response to increasing pressure from the United States to hand over bin Laden.[72] By 2003, the Taliban began their military campaign, launching guerrilla attacks in Helmand and Zabul Provinces, even with a scarce US presence.[73] At the end of January 2003, the first major battle took place near Spin Buldak when close to eighty Taliban were surprised by US forces and dozens of Taliban were killed. In response, the Taliban[74] attacked a US compound in Bagram with rockets and mortars in February.[75]

After the coalition forces led by the United States refused to negotiate with the Taliban on the principle that the United States does not negotiate with terrorists, the Taliban increased pressure on the international presence in Afghanistan. They issued a new policy making it almost impossible for the United Nations and other aid agencies to provide relief to the Afghan people. In a stark rejection of Western influence and assistance, the Taliban attacked Western-run hospitals, refused to cooperate with a UN-led polio immunization campaign for children, and harassed female aid workers. Toward the beginning of the conflict, the Taliban detained fourteen people working for a German aid agency—eight of whom were Westerners—and accused them of promoting Christianity, which in Afghanistan is punishable by death by the Afghan government.[76] On March 27, 2003, the armed Taliban also held up a jeep convoy of International Committee of the Red Cross members, killing a foreign worker and setting fire to his corpse.[77] On April 8, an Italian tourist was killed in Zabul Province. This pattern of behavior continued in 2007 when the Taliban kidnapped a group of South Korean Christian charity workers, killing two of those workers and freeing the rest over the next six weeks.

---

[71] Ibid.

[72] Rashid, *Descent into Chaos.*

[73] Ibid.

[74] Other warlords and anti-coalition forces also launched rocket attacks.

[75] Ibid.

[76] Ibid.

[77] Ibid.

In June 2003, in response to mounting pressure by the Afghan government on Pakistan to turn in Taliban leaders based in Quetta, Mullah Omar reorganized the Taliban leadership by adding a ten-man leadership council. This reorganization enabled the Taliban to carry out more coordinated attacks on soft targets. That summer, the Taliban initiated one or two attacks every other day, killing more than 220 Afghan soldiers and civilians. On August 13, 2003, fifty people were killed in multiple attacks led by the Taliban. This increase in violence led the United Nations to suspend travel in the south, and aid agencies left the dangerous Kandahar and Helmand Provinces.[78]

Zabul Province in southern Afghanistan, which provided an entry point for the Taliban from Baluchistan, became another important area of conflict. The Taliban wanted to secure this strategic location for a base area. However, Zabul was also a key location for coalition forces, who wanted to provide security on the Kabul-Kandahar highway. As a result, the US forces launched Operation Mountain Viper in September to push out close to 500 Taliban led by Dadullah. Although previously this sort of offensive would have deterred the Taliban, this time they mounted a defense. After nine days of heavy air and artillery bombardment that killed more than 100 Taliban, the Taliban were able to secure 80% of Zabul by the winter of 2003. They were also successful in pushing back the Loya Jirga from October to December because of the increased fighting and worsening security situation in the south.[79]

In June 2004, the Taliban took advantage of the US focus on Iraq and Al Qaeda and assassinated twenty-four Afghan officials and killed fourteen foreigners. In the first six months of 2004, the number of US troops killed doubled compared to 2003.[80] The Taliban slept in mosques during the day, and at night, they would try to "persuade, bribe, or terrorize farmers into helping them kill US troops."[81] At this point in the conflict, two of the four southern provinces were under Taliban control. In an effort to stall the elections slated for 2004, the Taliban continued their fear campaigns, attacking several voter registration teams and issuing death threats to others. However, these efforts were ultimately unsuccessful because a large percentage of the Afghan public registered and participated in the elections. Much of the fighting between 2004 and 2008 was aimed at the Taliban, who still sought to maintain control of Helmand, Zabul, Kandahar, and Uruzgan Provinces. The Taliban continued their tactics of "stand-up battles,

---

[78] Ibid.
[79] Ibid.
[80] Ibid.
[81] Ibid.

suicide attacks, ambushes, roadside explosions, and the assassination of aid workers."[82] These initiatives were bolstered through cooperation with Al Qaeda in 2005, which, in effect, internationalized the conflict. In 2007, several top Taliban leaders were killed, forcing the group to once again restrategize.

In 2008, the Taliban assassinated the deputy to the National Directorate of Security (NDS) and launched two failed attacks on NDS headquarters in August and September 2009. The Taliban continued to target the NDS by attacking family members of those within the institution.[83] In addition, the Taliban targeted Afghan police commanders by engaging in several suicide bombings in Kandahar and by killing a police chief in Helmand in 2008. Furthermore, from 2008 to 2009, the Taliban conducted armed attacks against police officers and governors, using guns, suicide bombings, grenades, and roadside bombs. As these tactics enabled more control of these southern provinces, the Taliban also began to target tribal leaders, militia commanders, government officials, pro-government clerics, and security forces that were protecting the targeted individuals. By January 2009, the Taliban had assassinated 24 members of the 150 member *ulema* council who actively opposed Taliban propaganda.[84] The QST also assassinated several leading clerics, as well as leading government officials.

METHODS OF RECRUITMENT

The Taliban focused their recruitment on a wide range of the local population for many reasons, including "tribal identity, resentment of local government officials, or resentment of International Security Assistance Force (ISAF) forces."[85] Given the high level of frustration with the corrupt Mujahidin government, a skewed tribal representation in the government, and rampant poverty in the post-Soviet war era, the Taliban offered an attractive alternative to many people. The Taliban used several different strategies to attract new followers, including propaganda on Islamic jihad as a religious obligation, financial incentives, security offerings, revenge, and foreign fighters.

To draw recruits, the Taliban engaged in active propaganda, with Taliban mullahs arguing that it was the religious duty of locals to support the insurgency. In the beginning stage of the insurgency, the

---

[82] Ibid.
[83] Dressler and Forsberg, *The Quetta Shura Taliban in Southern Afghanistan.*
[84] Ibid.
[85] Ibid.

Taliban were the "orphans of the war;"[86] they had no memory of tribe or of a peaceful Afghanistan. All they had seen was war and conflict. They lived only in the present—not the past or the future. "Their simple belief in a messianic, puritan Islam, which had been drummed into them by simple village mullahs, was the only prop they could hold on to and which gave their lives some meaning."[87] Even though Afghanistan is fragmented along ethnolinguistic lines, religion does provide a sense of unity. However, although the purpose of the propaganda was to recruit new followers on the basis of ideology and *jihad,* not all Taliban fighters were drawn to the organization for these reasons. Instead, and especially in the post-2001 insurgency, financial motivations were significant methods of recruitment for the Taliban. Some estimates suggest that "the average foot soldier is paid between $100-150 a month, while cell commanders make considerably more, approximately $350 a month," and these soldiers are normally "deployed for only short temporary service."[88]

The Taliban focused their efforts on establishing recruitment bases in the areas that they controlled. In their 2006 rulebook, it is noted that "A Taliban commander is permitted to extend an invitation to all Afghans who support infidels so that they may convert to the true Islam."[89] This was particularly appealing to new constituencies when the Taliban established control in a new area. Security offerings were made to new recruits in the sense that they would be protected from execution.

Another effective recruitment strategy was providing people with an opportunity to seek revenge against the weak, corrupt Mujahidin government and against the United States and its allied North Atlantic Treaty Organization (NATO) forces. The Taliban represented something that the government did not—it promised and delivered order and stability. This was one of the most attractive features of the movement. In addition to the financial and security guarantees, the Taliban enabled their followers to fight against the corrupt government, which only handed them false promises, as well as foreign forces that bombed several towns and continued the foreign occupation of the war-torn country. Moreover, many Afghans felt alienated because, in the initial insurgency, the Americans not only targeted the Taliban but essentially all Pashtuns.[90]

---

[86] Rashid, *Descent into Chaos.*

[87] Ibid.

[88] Dressler and Forsberg, *The Quetta Shura Taliban in Southern Afghanistan.*

[89] Ibid.

[90] Rashid, "Ahmed Rashid Offers an Update on the Taliban."

After 2008, the Taliban also resorted to foreign fighters as a source of recruitment. Most of these fighters were drawn from refugee camps, orphanages, and madrasas in Pakistan and Baluchistan.[91,92] Many new recruits were orphans, and others were madrasa students or came from refugee camps. "The male brotherhood offered these youngsters not just a religious cause to fight for, but a whole way of life to fully embrace and make their existence meaningful."[93] The Taliban were able to secure a following from these individuals regardless of their tribal identities.[94] In fact, for several of the Taliban offensives, they were able to draw thousands of Punjab—not Pashtun—followers from Pakistan.[95]

METHODS OF SUSTAINMENT

Unlike the Mujahidin, which drew most of its support from foreign funding, the early Taliban insurgency was largely sustained by domestic sources capitalizing on poppy cultivation; tax collection; transport; and weapons, munitions, tanks, and helicopters accumulated from new towns under their control.[96] The Taliban's strategy of securing the commercial trade routes between Afghanistan, then controlling Helmand Province, the locus for the bulk of the opiate development, greatly enabled them to carry out their mission. After the Soviet withdrawal, the Afghan people were able to once again concentrate on reviving their economy by focusing on agriculture and trade. However, instead of growing food crops and positive forms of economic development, the lack of governance and the lucrative benefits from opiate production caused much of the renewed production to be turned into poppy cultivation, heroin refining, and smuggling.[97] These illicit activities were initially controlled by Mujahidin warlords, but as the Taliban advanced and were able to either negotiate or win battles in these provinces, they began to reap the benefits of this illegal trade. Although Islam forbids the production of intoxicants, because the opiate production was meant for trade and use by *kafirs*, or unbelievers, it was initially allowed to continue. However, Mullah Omar contemplated placing a ban on the production of these substances, provided the United States and United Nations gave the

---

[91] Dressler and Forsberg, *The Quetta Shura Taliban in Southern Afghanistan*.

[92] Marsden, *The Taliban*.

[93] Olivier Roy, "Rivalries and Power Plays in Afghanistan: The Taliban, the Shari'a, and the Pipeline," *Middle East Report* 202, Winter 1996.

[94] Ibid.

[95] Rashid, "The Taliban: Exporting Extremism," 22–35.

[96] Ibid.

[97] Rubin, *The Fragmentation of Afghanistan*.

Taliban international recognition—a move that the international community was not ready to make.

Because the international community did not accept their offer and because the profits from opiate production were so appealing, replacing this production with cash crops did not make sense for the Taliban. Instead, they formalized the drug economy in order to collect taxes.[98] Heroin production became a large part of Taliban sustenance. According to the UNDCP, in 1999, Afghanistan produced 4,600 metric tons of opium, doubling its production from the previous year. By 2000, Afghanistan produced "three times more opium than the rest of the world put together."[99] Of this production, 96% is grown in Taliban-controlled areas, which makes the Taliban the largest heroin producer in the world. From this production, "the Taliban collect a twenty percent tax from opium dealers and transporters; money that goes straight to the Taliban war chest."[100] In addition to controlling most of the opiate development and trade, the Taliban also received support from *zakat*, a local tax that varies from place to place, and *ushr*, an Islamic tithe of ten percent.[101] Although *zakat* is typically a tax of about 2.5%, the Taliban had no qualms charging 20% on the opiates trade. In addition to the tax on opiates, individual governors also imposed their own taxes.[102] It is estimated that by 2000, the Taliban were bringing in at least $20 million in taxes and more from the opiates trade.[103]

METHODS OF OBTAINING LEGITIMACY

The Taliban established their legitimacy from three main sources: Islam, Pashtun identity, and effective shadow governance.[104] Similar to the Mujahidin, the Taliban were able to garner support and establish legitimacy by illustrating their devotion to Islam. After more than 10 years of war and foreign occupation, Islam was a unifying force that was representative of the Afghan people and culture, regardless of tribe or ethnolinguistic allegiance; as many as 99% of the Afghan public is Muslim. The Taliban movement was further legitimized when Mullah Omar, the leader of the QST, was named Amir-ul Momineen or the

---

[98] Rashid, "The Taliban: Exporting Extremism," 22–35.

[99] Ibid.

[100] Ibid.

[101] Dressler and Forsberg, *The Quetta Shura Taliban in Southern Afghanistan.*

[102] Rashid, *Descent into Chaos.*

[103] Ibid.

[104] The shadow governance was especially important in the later Taliban insurgency from 2001 to 2010.

"The Leader of the Faithful" on April 4, 1996.[105] It has been suggested that the Taliban's "distinctive white turbans and obvious religious fervour and purity, lend them an almost supernatural aura."[106]

The Taliban's Pashtun identity also bolstered the group's legitimacy—especially in the southern Pashtun belt in Afghanistan. Although the Mujahidin also used Islam as a foundation for their movement, many Afghans felt ethnically isolated and misrepresented by the largely Tajik ethnic representation of the Mujahidin in office. The ethnolinguistic representation of Afghans in office did not accurately reflect the majority of Pashtuns in Afghanistan. Thus, the Pashtun identity of the Taliban provided not only a recruiting motivation but also legitimacy for the movement.[107] The Taliban have been able to penetrate the fabric of the country in areas with large Pashtun representation and in traditionally non-Pashtun areas.[108]

Against the backdrop of several failed governments and broken promises from the Mujahidin, the effectiveness of the Taliban's shadow governance also provided legitimacy for the movement. By establishing military-backed *shari'a* courts in their areas of control, the Taliban effectively brought order to Kandahar, which also earned them considerable popularity.[109] Not only were the Taliban better able to deliver on their promises of providing and enforcing justice, but they were also able to influence the population more effectively than the local or national government.[110] However, part of this influence was derived from the Taliban's "sophisticated, multi-pronged campaign of intimidation designed to dissuade the population from cooperating with the coalition and Afghan government, and not necessarily from the esteem that many had for them."[111]

However, as the war between the Taliban and the coalition forces continued, the Taliban lost much of the legitimacy they had gained during their ascent to power. As they engaged in heavy intimidation campaigns aimed not only at the coalition and Afghan forces but also at civilians, the appeal of the Taliban wore off for many of their followers. The Taliban lost touch with the Pashtun nationalism that had attracted many at the onset of their insurgency. Moreover, as attacks became more violent, several Taliban leaders were influenced by Al Qaeda-style terror tactics, such as burning down schools and

---

[105] Rashid, "The Taliban: Exporting Extremism," 22–35.

[106] Marsden, *The Taliban*.

[107] Roy, "Rivalries and Power Plays in Afghanistan."

[108] Rashid, "Ahmed Rashid Offers an Update on the Taliban."

[109] Dressler and Forsberg, *The Quetta Shura Taliban in Southern Afghanistan*.

[110] Ibid.

[111] Ibid.

clinics, killing, and mutilating aid workers, and these were far from the group's initial goals to restore stability and order to Afghanistan. Instead of building and maintaining support and legitimacy through their mission and actions, the Taliban instead increasingly turned to campaigns of fear to force loyalty to the movement.[112]

EXTERNAL SUPPORT

Although the Taliban were able to secure much of their sustainment from domestic sources, their movement was also supported externally, mainly by Pakistan and Saudi Arabia, especially after they controlled the Afghan government from 1996 to 2001. Pakistan has been a major force in providing support to the Taliban. Not only has it lobbied other countries for cash and arms supplies, but it also has enabled the Taliban to run their operations from Pakistan. Because of Pakistan's direct intervention, the Taliban were afforded tanks, aircraft, and effective telecommunications.[113] Moreover, Pakistan also provided "a new telephone and wireless network for the Taliban, refurbished Kandahar airport and helped with spare parts and armaments for the Taliban's air force, while continuing to provide food, fuel, and ammunition, including rockets."[114] Support that the Taliban garnered while in power was available to them in the post-2001 insurgency.

Weary of the external support that Pakistan—a traditional US ally—was providing to the Taliban and to Osama bin Laden, Al Qaeda, and other terrorist groups, the United States mobilized international support for UN Resolution 1333, which "imposed a complete arms ban on the Taliban and closing of training camps, as well as a seizure of Taliban assets outside Afghanistan."[115] In response, Pakistan's Inter-Service Intelligence established the Afghan Defense Council to resist UN pressure and register support for the Taliban.[116] After the development of the Council, Pakistan continued to provide the Taliban with arms. As a result, the UN Security Council passed Resolution 1363 authorizing enforcement monitors on the Pakistan–Afghanistan borders to uphold the UN arms embargo. However, both the Taliban and their Pakistani supporters threatened to kill any UN monitors.[117] Pakistan also lobbied Saudi Arabia and the United Arab Emirates, who were sympathetic to the Taliban's Wahhabi rhetoric,

---

[112] Rashid, *Descent into Chaos.*
[113] Roy, "Rivalries and Power Plays in Afghanistan."
[114] Rashid, *Descent into Chaos.*
[115] Ibid.
[116] Ibid.
[117] Ibid.

on the Taliban's behalf, resulting in support from these countries in the form of fuel, money, pick-up trucks, and recognition for their movement.[118] This external support was delivered to the Taliban via couriers and *hawalas*.[119] After 2004, cooperation between the Taliban and Al Qaeda became more evident as the conflict became more internationalized and confrontational.

COUNTERMEASURES TAKEN BY THE GOVERNMENT

In the aftermath of the 9/11 attacks, the United States was determined to take action against those responsible. Although at first, and throughout the initial insurgency, the United States' main target was not the Taliban but Osama bin Laden and Al Qaeda, who they suspected were responsible for the attacks, as the conflict progressed, the Taliban became an increasingly important target. The United States continued to maintain that it would not negotiate with terrorists and refused to negotiate with Mullah Omar, even though he was open to such talks. The United States initially pursued a "warlord" strategy, with Central Intelligence Agency (CIA) budgets of $1 million, to pay off warlords in order to gain control of Taliban-concentrated areas.[120] After the Taliban failed to hand over bin Laden, the United States launched several major air strikes against Taliban strongholds in Kabul, Kandahar, and Jalalabad on October 7, 2001. As fighting between the Taliban and the US and Afghan forces continued, ISAF was created in December 2001, after NATO forces joined the United States to manage the counterinsurgency and establish the new Afghan government. The new Afghan government also took countermeasures against the Taliban, arresting those who contravened informal or local amnesties and detaining them in Afghan or American prisons, such as Bagram or Guantanamo.[121] At the end of January 2003, eighty Taliban were surprised by US forces near Spin Buldak. During the twelve-hour battle, the United States dropped twenty-one bombs on Taliban

---

[118]   Ibid.

[119]   Flynn, "State of the Insurgency." Hawala is a remittance system in which money is transferred without actually being moved. This parallel remittance system often relies on the extensive use of connections, sometimes within a family or regional network. Starting in India, it is commonly practiced in Asia and the Middle East. Because hawala does not make use of traditional banking systems to transfer money, they are sometimes flagged for money laundering and terrorist financing, although they are not always used for these purposes. For more information on hawala, see "Money Laundering – the Hawala Alternative Remittance System," accessed September 14, 2010, http://www.interpol.int/public/financialcrime/moneylaundering/hawala/default.asp.

[120]   Rashid, *Descent into Chaos*.

[121]   Matt Waldman, *Golden Surrender? The Risks, Challenges, and Implications of Reintegration in Afghanistan* (Afghanistan Analysts Network, 2010).

forces, killing dozens.[122] In an effort to show good faith to the coalition partners, President Karzai seemed concerned with the Taliban advancements against the counterinsurgency, and, in April 2003, he visited Islamabad and urged President Musharraf to arrest Taliban leaders living in Quetta, giving him a list of the leaders' names.[123]

By August 11, 2003, NATO took the lead in managing the ISAF operation.[124] In support of the Afghan government, ISAF's main role was to conduct operations "to reduce the capability and will of the insurgency, support the growth in capacity and capability of the Afghan National Security Forces (ANSF), and facilitate improvements in governance and socioeconomic development, in order to provide a secure environment for sustainable stability that is observable to the population."[125] ISAF continued to expand its operations in the following years, providing security for about half of Afghanistan's territory by 2005. By summer 2004, Lieutenant General David Barno, the new head of US forces in Afghanistan, "introduced new counterinsurgency tactics involving small groups of US soldiers living in villages to win hearts and minds and collect better intelligence."[126] Meanwhile, Zabul Province, a significant entry point for the Taliban from Baluchistan in southern Afghanistan, became another important area of conflict. The Taliban wanted to secure this strategic location for a base area; however, this area was also essential for coalition forces who wanted to provide peace and stability on the Kabul-Kandahar highway. As a result, the US forces launched Operation Mountain Viper to push out close to 500 Taliban led by Dadullah.[127]

As fighting continued after July 2006, ISAF-led NATO troops took over military leadership in the south. As the coalition forces sought to increase government control in areas of strong Taliban influence, fierce fighting ensued. In October 2006, NATO relieved the US-led coalition force of the responsibility for security across the entire country as they took command after three years of hard fighting. Meanwhile, Pakistan illustrated its commitment to the counterinsurgency by arresting Mullah Obaidullah Akhund, the third most senior member of the Taliban's leadership council. Soon thereafter, NATO and Afghan forces launched Operation Achilles, which was their largest offensive to date against the Taliban in the south. Heavy fighting

---

[122] Rashid, *Descent into Chaos.*

[123] Ibid.

[124] International Security Assistance Force Afghanistan, "History," accessed September 8, 2010, http://www.isaf.nato.int/en/our-history/.

[125] Ibid.

[126] Rashid, *Descent into Chaos.*

[127] Ibid.

continued in Helmand Province while a controversy over an Italian negotiation with the Taliban—releasing five rebels in exchange for kidnapped reporter Daniele Mastrogiacomo—took place. Although the Italians were successful in the trade, the reporter's Afghan driver and translator were beheaded. In the last six months of 2006, coalition forces launched 2,100 air strikes.[128] By 2006, "NATO forces in Afghanistan had grown from thirty-two thousand to forty-five thousand troops, but only one third were available for fighting."[129] NATO forces' initiatives showed progress when, in 2007, the Taliban's most senior military commander, Mullah Dadullah, was killed during fighting with coalition forces.

SHORT- AND LONG-TERM EFFECTS

CHANGES IN THE ENVIRONMENT

Significant attempts from the United States and other foreign development agencies continue funding projects aimed at instituting a peaceful, democratic Afghanistan. Given that more than $4 billion has been spent on development programs in Afghanistan since 2002, "USAID provides the largest bilateral civilian assistance program to Afghanistan," by promoting private-sector-based economic growth, establishing good governance programs, and providing basic needs to the people.[130]

CHANGES IN GOVERNMENT

As the insurgency progresses, the Taliban continue to challenge the Karzai government. Until the coalition forces and Afghan government are able to quell the Taliban insurgency, and the Karzai government acts as a responsive, transparent, and representative alternative, the Taliban will still have a large influence in the governance of Afghanistan. Measures have been taken to bolster the transparency and legitimacy of the Karzai government in order to dissuade potential followers from lending more support to the Taliban. However, given the infant state of the new government combined with continued foreign presence, Afghanistan is still far from reaching this ideal type.

---

[128] Ibid.

[129] Ibid.

[130] USAID Afghanistan, "About USAID Afghanistan," accessed January 10, 2010, http://afghanistan.usaid.gov/en/Page.About.aspx.

CHANGES IN POLICY

Faced with the ongoing struggle, the Afghan government continues to push for policies that make Taliban operations more difficult. However, given that the Taliban are integral to the fabric of the country, these initiatives are challenged by the creativity and adaptability of the Taliban. The Taliban seek to carry on promoting justice and offering an Islamic alternative to secular state policies.

CHANGES IN THE REVOLUTIONARY MOVEMENT

Since February 2010, in order to prove allegiance to the coalition forces, Pakistan has taken a firmer stance against the Taliban, capturing and detaining several of the group's top leaders. The Taliban's top military commander, Mullah Abdul Ghani Baradar, was arrested in a joint CIA–Pakistani operation in Pakistan. Baradar, second only to Mullah Omar, heads the Taliban's military council and coordinates the movement's military operations throughout the south and southwest of Afghanistan.[131] This arrest, in addition to the arrests of Mullah Abdul Kabir, a regional commander based in Peshawar, and Agha Jan Mutassim, the Taliban's former finance minister, has posed a threat to the insurgency movement as they are forced to restructure their efforts and take on new leadership.[132] It has been suggested that perhaps these more moderate Taliban leaders will now be replaced with hardliners, which could pose a serious change to the revolutionary movement and produce an even more dangerous Taliban.[133] In addition, there is evidence to suggest that the Taliban, especially in the later part of their insurgency, have been cooperating more with other terrorist networks, including Al Qaeda and the Pakistani Taliban.

BIBLIOGRAPHY

Ahmad, Munir. "Taliban's Top Military Commander Captured." *Associated Press*, February 16, 2010.

De Montesquiou, Alfred. "US Marines Airdropped into Taliban-Held Territory." *Netscape News*, February 19, 2010.

---

[131] Munir Ahmad, "Taliban's Top Military Commander Captured," *Associated Press*, February 16, 2010, accessed September 8, 2010, http://www.afghannews.net/index.php?action=show&type=news&id=3619.

[132] Jackie Northam, "Pakistan Arrests Throw Afghan Taliban into Disarray," *NPR*, March 9, 2010, accessed September 8, 2010, http://www.npr.org/templates/story/story.php?storyId=124501310.

[133] Ibid.

Dressler, Jeffrey, and Carl Forsberg. *The Quetta Shura Taliban in Southern Afghanistan: Organization, Operation, and Shadow Governance.* Washington, DC: Institute for the Study of War, 2010.

Flynn, Michael. "State of the Insurgency: Trends, Intentions, and Objectives." Accessed September 8, 2010. http://coincentral. wordpress.com/2010/02/03/state-of-the-insurgency-2010/.

Ghufran, Nasreen. "The Taliban and the Civil War Entanglement in Afghanistan." *Asian Survey* 41, no. 3 (May 2001): 462–487.

Goodson, Larry P. "Afghanistan in 2003: The Taliban Resurface and a New Constitution Is Born." *Asian Survey* 44, no. 1 (January 2004): 14–22.

International Security Assistance Force Afghanistan. "History." Accessed September 8, 2010. http://www.isaf.nato.int/en/our-history/.

Marsden, Peter. *The Taliban: War and Religion in Afghanistan.* London: Zed Books, 2002.

"Money Laundering – the Hawala Alternative Remittance System." Accessed September 14, 2010. http://www.interpol.int/public/ financialcrime/moneylaundering/hawala/default.asp.

National Security Archive. *The Taliban Biography: The Structure and Leadership of the Taliban 1996–2002.* National Security Archive Electronic Briefing Book no. 295, 2009.

Nojumi, Neamatollah. *The Rise of the Taliban in Afghanistan: Mass Mobilization, Civil War, and the Future of the Region.* New York: Palgrave, 2002.

Northam, Jackie. "Pakistan Arrests Throw Afghan Taliban into Disarray." *NPR,* March 9, 2010.

Rais, Rasul Bakhsh. *Recovering the Frontier State: War, Ethnicity, and the State in Afghanistan.* Lanham, MD: Lexington Books, 2008.

Rashid, Ahmed. "Ahmed Rashid Offers an Update on the Taliban." *NPR,* February 17, 2010.

———. *Descent into Chaos: The US and the Failure of Nation Building in Pakistan, Afghanistan, and Central Asia.* New York: Viking, 2008.

———. "The Taliban: Exporting Extremism." *Foreign Affairs* 78, no. 6 (November 1999): 22–35.

———. *Taliban: Militant Islam, Oil, and Fundamentalism in Central Asia.* New Haven, CT: Yale University Press, 2000.

Rashid, Ahmed. *Taliban Islam, Oil and the New Great Game in Central Asia.* London: Tauris, 2002.

Roy, Olivier. "Rivalries and Power Plays in Afghanistan: The Taliban, the Shari'a, and the Pipeline." *Middle East Report* 202 (Winter 1996).

Rubin, Barnett. *The Fragmentation of Afghanistan: State Formation and Collapse in the International System.* New Haven, CT: Yale University Press, 2002.

Ruggio, Bill. "The Afghan Taliban's Top Leaders." *The Long War Journal.* Accessed September 8, 2010. http://www.longwarjournal.org/archives/2010/02/the_talibans_top_lea.php#ixzz0gvJMKftH.

Ruttig, Thomas. *The Other Side, Dimensions of the Afghan Insurgency: Causes, Actors – and Approaches to Talks*: Afghanistan Analysts Network, 2009. Accessed September 8, 2010. http://aan-afghanistan.com/index.asp?id=114.

Scarborough, Rowan. "Taliban Outwits U.S. Eavesdroppers." *Human Events* 65, no. 7 (2009).

Shahrani, Nazif M., and Robert L. Canfield. *Revolutions and Rebellions in Afghanistan: Anthropological Perspectives.* Berkeley, CA: Institute of International Studies, University of California, 1984.

Waldman, Matt. *Golden Surrender? The Risks, Challenges, and Implications of Reintegration in Afghanistan*: Afghanistan Analysts Network, 2010.

AL QAEDA: 1988–2001

Jason Spitaletta and Shana Marshall

SYNOPSIS

Al Qaeda (the base) is headed by Osama bin Laden, son of a Yemeni laborer turned Saudi construction magnate, and his deputy, Dr. Ayman Al-Zawahiri, an Egyptian physician and former leader of Egyptian Islamic Jihad (EIJ). Al Qaeda was originally established to facilitate the integration of young Arabs into the anti-Soviet resistance in Afghanistan, but after the Soviet withdrawal in 1989, the organization evolved into the vanguard of a global fundamentalist insurgency seeking to unite the international Muslim community under a united political and religious authority. At the most basic level of analysis, Al Qaeda represents the internationalization of Islamic militancy, facilitated by technological change, but ultimately driven by the failure of earlier Islamist movements to bring about serious political change in their own domestic environments.

Notable incidents attributed to Al Qaeda include the 1992 bombings of two hotels in Yemen; the 1993 bombing of the World Trade Center in New York City; the 1998 bombings of two United States (US) embassies in Tanzania and Kenya; the 2000 bombing of the USS *Cole* in Yemen; and the 2001 attacks that destroyed the World Trade Center in New York and damaged the headquarters of the US Department of Defense in Washington, DC. After the 2001 attacks, Al Qaeda became less of a concrete organization and more of an operational philosophy providing inspiration, direction, and resources for like-minded individuals. The proliferation of violent fundamentalist groups that have claimed association with Al Qaeda or adopted the Al Qaeda name yet bear only loose tangible connections to the group's original leadership, financial networks, or strategic planning apparatus demonstrates the longevity of Al Qaeda's revolutionary principles. For the purposes of the present analysis, the Al Qaeda case ends in 2001 when US and Northern Alliance forces drove the organization from Afghanistan.

TIMELINE

1922	Ottoman Empire comes to a close upon the conclusion of World War I.
1928	Muslim Brotherhood is founded in Egypt by Hassan Al-Banna.
1967	Israel defeats the combined forces of Egypt, Jordan, and Syria, gaining control of the Sinai Peninsula, the Gaza Strip, the West Bank, East Jerusalem, and the Golan Heights.
1973	A joint surprise attack by Egypt and Syria against Israel enjoyed early tactical success but did not result in a decisive victory.
1979	Signing of the Camp David Accords.
	Soviet Union invades Afghanistan to support the ruling communist government against a loose confederate of tribal militias.
1985	Sheik Abdullah Azzam and Osama bin Laden create the *Maktab Al Khidamat* (MAK) or Services Bureau to address administrative problems for foreign Muslim fighters.
1987	Soviet Spetsnaz attack a *jihadi* training camp at Ali Kheyl in Jaji, Khost Province, Afghanistan, named *Masada* (The Lion's Den); with bin Laden (who served with great distinction) present, the small group repels the Soviets and bin Laden's legend grows.
1988	On February 15, the last Soviet soldiers leave Afghanistan.
1990	On August 2, Iraq invades and annexes Kuwait.
1992	Bin Laden travels from Jeddah, Saudi Arabia, to Kabul, Afghanistan, to help stop the intertribal militia conflict; the trip ultimately terminates in Khartoum, Sudan, where the Al Qaeda inner circle had established a base of operations.
	First bombing of World Trade Center in New York.
1994	Al Qaeda elements are ordered to leave the Sudan
	bin Laden and approximately 150 Al Qaeda members and their families depart Khartoum for Afghanistan.
1995	Bin Laden establishes a close relationship with the Taliban leader, Mullah Mohammed Omar.
1996	Al Qaeda issues a *fatwa* declaring war against the "Americans Occupying the Land of the Two Holy Places (Expel the Infidels from the Arab Peninsula)."
1998	Near-simultaneous explosions at the US embassies in Dar es Salaam, Tanzania, and Nairobi, Kenya.

2000	Bombing of the USS *Cole*.
2001	In June, Al Qaeda and EIJ merge, forming "Qaeda al-Jihad."
	9/11 attacks.
	United States declares a "Global War on Terrorism."
	US and coalition forces launch Operation Enduring Freedom.
	Taliban vacates Kabul under coalition and Northern Alliance pressure.

THE ENVIRONMENT OF THE REVOLUTION

PHYSICAL ENVIRONMENT

The physical environments in which Al Qaeda operated were diverse and shifted over time as the organization responded to pressures and opportunities in the international system and as the group's own objectives changed.[1] These environments included weakly governed states characterized by poor infrastructure, civil conflict, and rugged terrain (Somalia, Sudan, Afghanistan); states with strong, repressive central governments, powerful traditions of religious resistance, and high levels of inequality (Saudi Arabia, Egypt); as well as the wealthy and technologically advanced capitals of Western countries (London, Madrid, New York). The diversity of locations reflected Al Qaeda's conceptualization of a two-front war, the first against the nominally Islamic regimes of the Middle East and Asia, considered apostates by the Al Qaeda leadership, and the second against the Western governments that provided those regimes with military, economic, and political support.[2] The group's decentralized structure, exploitation of modern technology, and use of traditional institutions, such as unofficial moneylenders and charitable organizations, aided in its geographic dispersal. In addition, the economic and social impacts of globalization also helped the group to draw resources from disgruntled communities across the globe.[3]

---

[1] One example of the group's response to changing circumstances on the ground was its decision to reroute militants to the Philippines for training after security along the Afghanistan–Pakistan border increased in the aftermath of the embassy bombings in East Africa in 1998. Rohan Gunaratna, *Inside Al Qaeda: Global Network of Terror* (New York: Berkley Books, 2003).

[2] Ibid.

[3] In addition to citing the repression and violence visited on Muslim communities by Western regimes and indigenous governments, the Al Qaeda leadership has also cited the devastating impacts of climate change (which have disproportionately impacted developing countries more susceptible to dramatic weather patterns and changes in temperature), as well as the exploitative behavior of multinational corporations headquartered in the

Additionally, the disintegration of the Soviet Union sparked conflicts that impacted Muslim populations in many diverse locales, including the former Yugoslav Republic and Chechnya. The diverse physical environments in which it operated allowed Al Qaeda to draw upon the comparative advantage offered by each location: poorly governed and sparsely populated locations, such as Afghanistan, afforded space for training militants and concealing high-level leadership figures; the traditional, religious institutions that permeated society in many Middle East and Asian countries provided fund-raising opportunities and recruitment networks; and the broad political and civil liberties present in Western capitals offered the group's leaders the freedom to advocate on behalf of their cause, especially that of discrediting their native "apostate" governments while also accessing the resources provided by large Muslim diaspora communities.[4]

CULTURAL AND DEMOGRAPHIC ENVIRONMENT

Like the physical environment in which Al Qaeda operated, the cultural and demographic environments were also diverse. Although the Middle East and Asia continue to be broadly conceived as traditional societies with conservative social norms, adherence to these strictures can vary enormously according to a number of factors, including generation, social class, religion, and rural or urban settings. Linkages based on identity, like kinship or sect, strongly influenced relations between individuals on questions of marriage, business, and many other transactions. Family name or ties to a certain geographic region within a specific state were also likely to be used as shortcuts for identifying another's social status, such as occupation or education.

Yet, despite the strong current of traditionalism, public opinion polls demonstrated overwhelming support for universal principles, such as democratic governance, human and women's rights, media freedom,

---

Western countries, as cause for launching attacks against the West. Indeed, many militants' analyses of the inequities of the global system are very sophisticated. Mohammad Atta, one of the 9/11 hijackers, reportedly pointed out the adverse impact on nutrition and self-sufficiency brought about by the Egyptian government's emphasis on growing cash crops for export, which diverted peasant resources away from subsistence farming and toward growing fruits and vegetables for export to Europe. Agricultural subsidies for domestic farmers in Europe meant that Egyptian farmers did not get a fair price for their exports, the proceeds from which went to buy staples imported from the West that were less nutritious than the native cereals the farmers had been growing under earlier subsistence schemes. Jason Burke, *Al-Qaeda: Casting a Shadow of Terror* (New York: IB Tauris & Co LTD, 2003).

[4] See especially Allison Pargeter, "North African Immigrants in Europe and Political Violence," *Studies in Conflict and Terrorism* 29, no. 8 (2006), 731–737, and Peter R. Neumann, *Joining Al-Qaeda: Jihadist Recruitment in Europe*, Adelphi Series (New York: Routledge for the International Institute for Strategic Studies, 2008), 71.

and religious tolerance. However, the experience of these principles in the region has been imperfect. Democracy has become synonymous with elections,[5] the results of which are predetermined either by legal maneuverings on behalf of incumbents or by outright fraud, while the issue of women's rights has likewise been used strategically by regimes to control public displays of piety perceived as supportive of opposition religious parties. Al Qaeda's leadership was skilled at demonstrating the practical shortcomings of these "foreign principles," which they claimed were only effective in weakening Arab and Muslim unity rather than contributing to goals of social justice or development.[6]

The rise of mass consumption and the revolution in communications brought Western popular culture into the region's most isolated corners. Human trafficking and drug abuse, facilitated by technological advances in transportation and finance, were broadly perceived as consequences of this cultural Westernization. Moderate and radical religious opposition groups long advocated addressing these concerns through a return to the region's religious and cultural roots. Many regimes responded by adopting some measure of religious window-dressing, like inserting Quranic language into the constitution or financing the construction of religious institutions, but the superficial character of these efforts meant they frequently backfired. Because Islam was viewed by its adherents as a unitary system that provided guidance in both public and everyday life, many believed that politics should also be governed by religious principles and that the pervasive corruption of their governments was attributable to a decline in religiosity. As a result, many reformists advocated a return to the premodern principles governing life during the time of the Prophet Muhammad.[7] To a large degree, Middle Eastern and Asian peoples are more religiously observant today than during much of the twentieth century when the alternative ideologies of nationalism and socialism held more sway. This social transformation greatly emboldened religious activists and encouraged them to challenge the political system.

---

[5] Without institutional checks and balances, separation of powers, or a free press, elections tend to produce a system based purely on patronage, where individuals provide electoral support in exchange for private remuneration. This is characteristic of most elections held in the Middle East, where a small number of elites "deliver" the votes of their dependents. As a result, the equitable distribution of many public goods, such as health care, education, and infrastructure, is noticeably lacking, leaving the concept of democracy an empty one for many in the region.

[6] Foreign Broadcast Information Services, "Compilation of Usama Bin Ladin Statements 1994 - January 2004," accessed August 23, 2010, www.fas.org/irp/world/para/ubl-fbis.pdf.

[7] The Prophet Muhammad was born in 570 CE in Mecca, modern-day Saudi Arabia.

As religion became more central in many societies, it also became more radical, owing in part to Saudi government efforts to spread the particular form of Islam, Salafism, practiced by the ruling family.[8] Partially to preempt the rising popularity of Shi'a Islam in the wake of Iran's 1979 revolution, which established a populist theocratic government that contrasted sharply with the ostentatious wealth of the Saud dynasty, the Saudi government embarked on a massive scheme to finance Salafi religious schools and organizations around the world.[9] Given the unavailability and/or low quality of existing secular schools throughout much of the developing world, these Saudi-funded schools filled an important void. They also displaced local institutions that promoted less radical forms of Islam such as Sufism, which was traditionally practiced throughout much of central Asia. Because these Saudi-funded schools frequently served the populations of remote and poor areas, central governments could exert little authority over them. Ultimately, they helped create large pools of young, relatively marginalized youth conversant in the religious principles used by radical opposition groups. Islam, as an oppositional platform, also underwent a significant change during the mid-twentieth century when activists in Egypt, Iran, and Pakistan succeeded in synthesizing the principles of religion and political resistance.[10] Religious authorities, who owed their positions to the ruling elite, had historically promoted a vision of Islam that rejected political activism in favor of personal piety.

SOCIOECONOMIC ENVIRONMENT

The Muslim world is home to some of the world's wealthiest populations, such as the Gulf countries of Qatar, Kuwait, and the

---

[8] The term Wahhabism is used interchangeably with Salafism, although those who follow this particular variant of Islam view the term Wahhabism as pejorative. Salaf means "to follow," as in following the ways of the Prophet Muhammad and his companions. Wahab refers to the eighteenth-century Muslim scholar Muhammad bin Abd al-Wahhab, who advocated a return to the religious and social principles that prevailed during the Prophet's lifetime. Al-Wahhab is considered a vital player in the Al-Saud dynasty's successful consolidation of power on the Arabian Peninsula.

[9] For an in-depth examination of how Saudi and Iranian competition played out in much of central Asia, including Afghanistan, see Mohammed E. Ahrari and James Beal, *The New Great Game in Muslim Central Asia* (Washington DC: National Defense University, 1996). Saudi Arabia offered not only sectarian affinity (since the vast majority of Muslims worldwide are Sunni) but also significant economic assistance, including access to the Islamic Development Bank. Iran, on the other hand, appealed to central Asian populations because of ethnic and linguistic ties as well as its more populist version of Islam.

[10] See especially Gilles Kepel, *Jihad: The Trail of Political Islam* (Cambridge: Harvard University Press, 2002). The most significant thinkers in this regard would be Qutb (Egypt), Khomeini (Iran), and Mawdudi (Pakistan).

United Arab Emirates, as well as some of its poorest, like Egypt and Yemen. These economic distinctions have special significance for many religious activists, including Al Qaeda's leadership, who believe their struggle to unite the Muslim community is confounded by the concentration of wealth in the hands of the few.[11] The frugality displayed by leading Al Qaeda figures contrasted sharply with the corruption and decadence associated with the traditional state-sponsored clergy. Income inequality in most Muslim countries is greater than that in Western Europe but falls below measures for the United States and most of the developing countries as well.[12] The elite classes consist of both fully Westernized elites as well as those that maintain the vestiges of tradition, whether in dress, language, or social customs, while also participating in modern institutions, such as financial markets. The perception, whether true or false, that these elites are tools of Western imperialism was a particularly potent aspect of Al Qaeda's ideology.[13]

Like that in most of developing countries, the Muslim world's experience with modernity and globalization has been disruptive and unbalanced. In many cases, certain aspects of globalization were introduced, like the lowering of trade barriers and integration of new technologies, but without policies to cushion their impact on the less fortunate, such as social safety nets and accountability for capital flows. The promise of social mobility and economic freedom promised by liberal capitalism has largely proven a failure in the region. Access to external markets, credit, and employment is available only to the wealthy and politically connected, while austerity programs have stripped public sector workers and peasants of the subsidies and social programs that once mitigated the economic insecurity they faced because of low incomes.

Religion provided a "goal-replacement" mechanism for many of those individuals who have been left out of this modernization

---

[11] Bin Laden stated that Western support for corrupt leaders in the Muslim world allows the latter to "steal our community's wealth and fortunes and sell them to you [Western countries] for a cheap price, so that a few of the elite may indulge themselves whilst the general population starves to death." Foreign Broadcast Information Services, "Compilation of Usama Bin Ladin Statements." For example, the oil-exporting countries of the Arab Gulf are home to only 11% of the region's population but control 46% of gross domestic product. See Ali Abdel Gadir Ali, *Globalization and Inequality in the Arab Region* (Kuwait: Arab Planning Institute, 2003).

[12] See the UN Human Development Reports, which provide measures of income inequality for most countries: http://hdr.undp.org/en/ (accessed August 23, 2010).

[13] See Osama bin Laden, "Open Letter to King Fahd," accessed August 23, 2010, http://en.wikisource.org/wiki/An_Open_Letter_to_King_Fahd_on_the_Occasion_of_the_Recent_Cabinet_Reshuffle.

process.[14] Many of the dispossessed were young adults—members of the region's "youth bulge." In 2007, nearly 60% of the population in the Middle East was under 25 years old. The combination of young, largely unemployed (or underemployed) youth,[15] social dislocation, and the disruption of previous economic arrangements has channeled much of the population's discontent into religious extremism. By tapping into these specific sources of frustration, Al Qaeda and other groups effectively recruited from across the socioeconomic spectrum.

HISTORICAL FACTORS

The rise of Al Qaeda was facilitated by both broad historical trends and more specific events. Negative fallout from US and European policy in the region and the violent repression of religiously oriented political opposition by authoritarian regimes in the Middle East and Asia were two important general trends. These are interrelated, in actual terms and in the judgment of Al Qaeda's leadership, because much of the repression visited upon the religious opposition would have been impossible in the absence of significant military aid and political support from Western governments. US and European history in the region was characterized largely by a procession of colonial governments (especially the British in Egypt, Palestine, and the Indian subcontinent and the French in North Africa and Indochina); monopolization of the region's natural resources (notably the oil industry in Iran and the Suez Canal in Egypt); the perceived injudicious use of its human population (including the conscription of colonial populations during World War I and World War II); abrogated agreements (including the unfulfilled British pledge to support a united Arab nation in exchange for assistance in fighting the Ottoman Turks);[16] and actions that either supported or destabilized

---

[14] The intense antimodern slant of many current militant groups reflects the deeply disorienting method by which modern concepts and technologies have been introduced in many countries. Earlier leading figures of the Islamic opposition (Afghani, Abduh, Al-Banna, Mawdudi) stressed the necessity of integrating religious principles into modern society, but the accelerated pace of change that ensued triggered a much more reactionary form of religious opposition that ultimately manifested itself in groups like Al Qaeda and the Taliban.

[15] Although unemployment is a major issue in the region, underemployment is equally detrimental. The rate of university graduation in the region is fairly high—because public institutions in many places, such as Egypt, are relatively accessible—but graduates end up working in low-skilled jobs because the training provided by these institutions is out of sync with the demands of the domestic economy. See Jeff Defferios, "Youth Unemployment: Mideast 'Ticking Time Bomb,'" *CNN*, March 10, 2010, http://www.cnn.com/2010/WORLD/meast/03/12/bahrain.youth.unemployment/index.html.

[16] Bin Laden referenced the Sykes–Picot treaty explicitly in his justification of the internationalization of Al Qaeda, that is, the shift from targeting "apostate" Muslim

regimes according to their pro-West orientation regardless of the regimes' illiberal character (including the 1953 Central Intelligence Agency [CIA]-led coup against the democratically elected Prime Minister Mossadeq of Iran in favor of the Shah).

The most visible reminder of the impact that US policy had on the region was the conflict between the Israelis and the Palestinians, which is still central to the grievance list of Al Qaeda and most other religious movements in the Muslim world.[17] The group's leadership saw the Jewish presence not only as a religious affront and a physical threat to the Palestinians, but also as a symbol of the weakness of Arab leaders and the ability of Western governments to dictate the region's future. The war of 1967 resulted in the extension of Israeli control over significant additional territory, including the Gaza Strip, the West Bank, and the Golan Heights, which are still in dispute. Later, in 1973, emergency US military assistance would prove crucial to Israel's success in fending off a surprise attack by Egypt and Syria and would solidify the US–Israel connection so unpopular among the Muslim public. This hastened the declining legitimacy of Arab nationalism and socialism while paving the way for the increasing salience of religious rhetoric, especially vis-à-vis the Israeli–Palestinian conflict.

The methods regional leaders used to deal with political opposition contributed to further radicalization and directly swelled the ranks of Al Qaeda. In the early to mid-twentieth century, when communism and socialism were powerful mobilizing frameworks throughout the developing world, secular regimes in the Middle East frequently cultivated religious groups as a counterweight to the more powerful secular opposition groups.[18] Later, as religious movements grew in popularity, the region's leaders cracked down hard on dissenters, torturing and imprisoning thousands. The prisons became potential powder kegs, providing incubation sites for radicalization and

---

governments to targeting the West: "it is essential to hit the main enemy who divided the ummah into small and little countries and pushed it for the last few decades into a state of confusion." Burke, *Al-Qaeda: Casting a Shadow of Terror*.

[17] Al Qaeda coalesced largely in the shadow of increasing tensions between Israel and the Palestinians: the planning sessions between bin Laden, Zawahiri, and Azzam in 1988 on what direction the "Arab Afghans" should take following the Soviet pullout were roughly coincident with the first Intifada (Palestinian uprising) of 1987–1993. The subsequent failure of the Oslo Accords and the increasing influence of extremist elements in both the Israeli and Palestinian governments mark a low point in the peace process.

[18] Anthony Cordesman, of the Center for Strategic and International Studies, has confirmed that Israeli officials funneled money to Hamas militants in the 1970s to weaken the Palestine Liberation Organization and provide an avenue for agents to infiltrate the secular opposition, and that the Egyptian authorities have engaged in similar tactics as well. Pakistani leaders also encouraged religious extremism in an effort to consolidate power after their separation from Hindu-India.

networking among extremists.[19] Many regimes chose to release large numbers of prisoners on the condition that they travel to Afghanistan to aid in the resistance against the Soviets.[20]

It is difficult to underestimate the impact of the Soviet invasion of Afghanistan on the formation of Al Qaeda. A well-organized system, financed primarily by Saudi authorities, was established to assist prisoners and other religious zealots in making the journey.[21] Thousands of "Afghan Arabs," as they came to be known, were drawn to the fight, and many of these were transported, organized, and trained by the leaders of the *Maktab Al Khidamat* (Services Bureau), the organization that would later develop into Al Qaeda. After the Soviets withdrew in 1989, some Afghan Arabs remained, but many traveled to other conflict sites, such as Chechnya, not only to support the global religious struggle but also because they were unwelcome in their home countries.[22] Many who did return joined in opposition activities against their own governments and ultimately either fled to Europe, knowing that they were under state surveillance and likely to be arrested, or found themselves in prison.[23] Those who escaped to Western capitals became an important asset in the global religious struggle because they had access to advanced communications infrastructure and the freedom to use it to promote their message.[24]

The Soviet withdrawal from Afghanistan emboldened the Al Qaeda leadership who believed their success in Afghanistan could be replicated in other Muslim territories.[25] The Soviet withdrawal also

---

[19] Marc Sageman, author of *Understanding Terror Networks* (Philadelphia: University of Pennsylvania Press, 2004), is the most well-known proponent of the idea that the social bonds created through these networks, rather than historical grievances or behavioral disorders, are the primary motivating force behind individual acts of terrorism.

[20] This expedition gave religious militants an outlet for their fervor while also removing a serious threat to the incumbent regime. Walter Laqueur, *No End to War: Terrorism in the Twenty-First Century* (New York: Continuum, 2003).

[21] Pargeter, "North African Immigrants in Europe and Political Violence," 731–737.

[22] The existence of so many conflicts, spread across a wide geographic space, facilitated the entry of many individuals into the militant apparatus. When enhanced security in one region made movement difficult (as was the case on the Afghanistan–Pakistan border after the 1998 bombings in East Africa), Al Qaeda's leadership reached out to movements located elsewhere that could house and train militants (in this case the Philippines offered an alternate location). Gunaratna, *Inside Al Qaeda*.

[23] Pargeter, *North African Immigrants in Europe and Political Violence*, 731–737. Many regimes sent members of their intelligence branches abroad to track down and kill those who had fought in Afghanistan, and in 2000, Pakistani President Musharraf began rounding up Arab fighters in and around Peshawar, where the Services Bureau was headquartered, and handing them over to their home governments.

[24] Based in London, the Advice and Reformation Committee, which bin Laden helped form in 1994, was adept at using the country's media freedom to propagate its message.

[25] Iran's successful Islamic Revolution in 1979 and Hizbollah's use of a suicide truck bomb to drive the American and Israeli militaries out of Beirut were other successful efforts initiated by Islamic militants.

left a power vacuum, which the Taliban eventually filled, creating an ideologically and strategically favorable base of operations for Al Qaeda. As Soviet power waned, the influence of secular and leftist opposition groups throughout the Middle East and Asia lost increasing ground to religious movements. In addition, the weakening of many central governments, now cut off from Soviet support, contributed to rising tensions between ethnic and religious groups. The ensuing conflicts not only played into Al Qaeda's narrative about the Muslim struggle as a continuation of the medieval Crusades, but also provided ample opportunities to train fighters and recruit from among those who were hardest hit by the violence and upheaval.[26]

GOVERNING ENVIRONMENT

The governing environment in most Muslim-majority countries has been characterized by authoritarianism, limiting the political participation of ordinary citizens while giving political elites wide scope to pursue their own interests. Because most civil organizations, such as labor unions, nongovernmental organizations, and literary societies, were either infiltrated or controlled by the government, religious organizations were the only institutions that remained outside direct state control. Thus, it was through these institutions that opposition activists channeled their antiregime activities. In addition, the state's coercive capacity was unrivaled. Many of the military and police forces in the region were often leftovers of European colonial governments, as in North Africa and much of the Levant, or the product of massive Western programs to train and equip them, as in much of the Arab Gulf. Attacks launched by religious militants increased in number and severity throughout the 1980s and 1990s. Rather than simply reflecting the growing strength of the groups, the scope and frequency of the attacks were indicative of a strategic reaction to the continuing isolation of the opposition from normal politics.[27] States responded

---

[26] Al Qaeda's leadership frequently invoked the image of Christian Crusaders in their public statements. See World Islamic Front, "Jihad Against Jews and Crusaders," accessed August 23, 2010, http://www.fas.org/irp/world/para/docs/980223-fatwa. htm. This statement is an early example of Al Qaeda's outreach efforts, which appealed to besieged Muslim populations around the world (in Somalia, Palestine, Pakistan, Indonesia, the Philippines, and Africa) to aid in the global movement to establish a single Islamic government to rule over all the world's Muslims. Although Al Qaeda has succeeded in recruiting individuals from among these populations to carry out specific attacks in its name, the leadership of these disparate opposition movements continues to focus on national aspirations, such as political autonomy, rather than the pursuit of a global caliphate.

[27] This interpretation is supported by numerous scholars, including Olivier Roy and Gilles Kepel. Notable attacks during this period include attacks against tourist sites in Egypt

with indiscriminate repression, which further radicalized many religious movements and drove even nonviolent religious opposition figures underground. Many dissidents from Jordan, Egypt, Syria, and other Muslim countries fled to Saudi Arabia, where religious fundamentalists did not need to fear prosecution from the state for their religious and political views.

This permissive environment in Saudi Arabia contributed to the rise of Al Qaeda just as the repressive environment did in other Muslim-majority states. The absence of any meaningful regulation of religious charities or the Kingdom's financial infrastructure gave Al Qaeda a relatively free hand in terms of raising and transferring funds. But, because the Saudi dynasty's legitimacy rested on its family's stewardship of the Islamic faith, it could not easily restrict citizens' practice of that religion, regardless of how radical such practices were, without undermining its claim to the throne. Other states also found themselves beholden to religious extremists for various reasons. The Sudanese government offered Al Qaeda's leaders sanctuary in return for badly needed investment and infrastructure projects,[28] while the Afghan Taliban, and the anti-Soviet fighters before them, benefited immensely from financial and military support provided by Arab militants. Many of the resources of Al Qaeda and other militant groups went to development projects, which employed thousands of the otherwise unemployed and provided basic necessities for the families of militants and those inadvertently killed in regional conflicts. The ability of these groups to fill this void, especially when compared with the immense wealth of many Muslim political elites and the inability of international financial institutions to alleviate poverty, exposed the corruption and failure of the region's governments.[29]

(1997 attack in Luxor); assassinations (Egyptian President Anwar Sadat in 1981; Lebanese President Gemayel in 1982); attacks on US military installations in Saudi Arabia and Yemen (Khobar Towers in 1996; USS *Cole* bombing in 2000); attacks on state property by North African groups (hijacking of Air France plane in 1994); and attacks on US embassies (in Lebanon in 1982; in Kenya and Tanzania in 1998).

[28] After coming to power in 1989, the Islamic government in Sudan was desperate to demonstrate its ability to bring development to the region. It invited veterans of the Afghan war to establish bases in the country, and bin Laden spent tens of billions of dollars building a road across the desert, financing a new airport, and providing hard currency to the Sudanese government during a series of fiscal crises. Burke, *Al-Qaeda: Casting a Shadow of Terror.*

[29] The religious opposition frequently focuses on the importance of the "moral administration" of the economy—especially the cronyism, nepotism, and waste that exist alongside extreme poverty. See the Saudi CDLR's (Committee for the Defense of Legitimate Rights) 1991 "Letter of Demands" and its 1992 "Memorandum of Advice" to the Saudi King.

WEAKNESSES OF THE PREREVOLUTIONARY ENVIRONMENT AND CATALYSTS

Against the backdrop of historical circumstances that contributed to the formation of Al Qaeda, including the Soviet invasion of Afghanistan, the demonstrated success of Islamic activism in Iran, Lebanon, and Afghanistan, the negative fallout from US and European foreign policy, and the continuing conflict between Israel and Palestine, there were also more immediate contributing factors. These included the US policy during the Soviet war in Afghanistan of allowing Pakistani authorities to choose the Afghan recipients of US and Saudi funds, which benefitted the most extreme elements of the resistance; the stationing of US troops in Saudi Arabia to repel Saddam Hussein in Iraq; and finally, the broad shift in focus away from the "near enemy" of Israel, Egypt, and the other "apostate" governments toward the "far enemy," or Western governments.

The US government funneled some $3 billion into Afghanistan to aid the resistance to the Soviet invasion.[30] Although bin Laden and the Arab Afghans had their own sources of funding and did not benefit directly from US dollars, the fact that the extremist Afghan factions that shared more in common with bin Laden's doctrinal fundamentalism were strengthened by this influx of resources certainly made bin Laden's transition to Afghanistan, and the consolidation of Al Qaeda, much easier. The preference for extremists was essentially a policy of the Pakistani regime, which viewed these Afghans as crucial allies in any potential conflict with India, although these were not the most competent fighters available among the Afghan resistance.[31]

Like the conflict in Afghanistan, Iraq's tumultuous history figured centrally into much of bin Laden's rhetoric. He frequently cited the humanitarian impact of the United Nations (UN) sanctions against Iraq as evidence of Western disregard for human life,[32] but more importantly, bin Laden saw the stationing of US troops in Saudi

---

[30] John Rollins, *Al Qaeda and Affiliates: Historical Perspective, Global Presence, and Implications for US Policy* (Congressional Research Service, 2010).

[31] Burke, *Al-Qaeda: Casting a Shadow of Terror,* 74. See also Steve Coll, *Ghost Wars: The Secret History of the CIA, Afghanistan, and Bin Laden from the Soviet Invasion Until September 10, 2001* (New York: Penguin, 2004), 225.

[32] Although the death toll bin Laden cited, 1.5 million, was extremely inflated, a Columbia University study, widely viewed as the most authoritative study on sanction-related child mortality in Iraq, estimated that about 350,000 excess childhood deaths were caused by sanctions on the regime between 1990 and 2000. However, the Iraqi regime bears significant blame for this as well because it rejected many attempts by the Security Council to redress the humanitarian impact of the sanctions.

Arabia, home to Islam's holiest sites, as an unforgiveable offense.[33] The Royal Family's dismissal of bin Laden's offer to provide security using his hardened Arab Afghan fighting force, rather than receiving the American military delegation, compounded his indignation. The Americans' increasing role in the region, signified by repeated bombing campaigns in Iraq, the United States' central role in sanctions on the regime, the increasing visibility of US military forces, and the United States' bankrolling of many despotic Arab regimes, had now largely outstripped the residual good will it had earned by abstaining from colonization and supporting the Afghan resistance to the Soviets. This shift certainly facilitated bin Laden's efforts to convince his partners to move Al Qaeda away from a focus on domestic "apostate" regimes and in the direction of confronting the "far enemy"—the United States.

FORM AND CHARACTERISTICS OF THE REVOLUTION

OBJECTIVES AND GOALS

Al Qaeda represents the extremist fringe of the global political Islamist movement[34] and the culmination of a religious struggle viewed in evolutionary terms. Early phases of this struggle were characterized by groups of dissident Muslims actively seeking to overthrow their own regimes, which they viewed as "apostate" governments.[35] This "defensive *jihad*" was then expanded beyond the domestic arena to the entire globe. Proponents of the radical doctrine argued that it was necessary for Muslims to fight alongside their co-religionists wherever apostate governments existed. This was exemplified by the participation of many Arab fighters in the Afghan struggle against the Soviets. Lastly came the transition to an offensive *jihad* against the far enemy—Europe and the United States.[36]

The phases correspond roughly to the schools of thought prevalent among Islamic militants during different times. The first phase was associated with the charter of EIJ, titled *The Neglected Duty*, written by Mohammad Abdel Salam Farraj. The second phase, epitomized by the writings of Sheik Abdullah Azzam, was strongly influenced by the

---

[33] It was widely believed at the time that the Iraqi military—then the world's fourth largest—would make quick work of Kuwait and subsequently invade neighboring Saudi Arabia.

[34] Ibid.

[35] The Egyptian, Jordanian, and Saudi governments were frequent targets—as were many North African regimes—because of their close ties to the United States.

[36] Sageman, *Understanding Terror Networks*.

Soviet invasion of Afghanistan and the continued occupation of the Palestinian territories. The third phase, offensive *jihad*, has long been present in the writings of political Islamists but became Al Qaeda's guiding principle when Azzam was killed and bin Laden secured singular control over the group's future direction.[37] Realizing that Western military and political support for regional governments was preventing Islamists from achieving their goals of radical change, bin Laden adopted a strategy of targeting Western governments, hoping to cut off these sources of support.

Although Al Qaeda's grievances were essentially political, they were articulated in religious terminology.[38] The organization declared both broad and more limited goals. The organization's most expansive goal was to unite the contemporary Muslim community into a single political and religious unit—a Caliphate—governed by a descendant of the Prophet Muhammad.[39] Many Islamist groups diverge from Al Qaeda on this point. Although most militant groups wanted substantial reform of their governments according to religious principles, far fewer believed the best way to achieve this is through a unification scheme. The group's more short-term goals included the withdrawal of American troops from Saudi Arabia; practical reforms in tax law, monetary policy, and sanitation in the Kingdom; the lifting of sanctions on Iraq; and an end to the oppression of Muslims in Palestine, Kashmir, and Chechnya.[40]

LEADERSHIP AND ORGANIZATIONAL STRUCTURE

Al Qaeda was less an organization than a decentralized network. During the 1990s, the group had approximately twenty operational cells throughout the world,[41] with estimates ranging up to 70,000

---

[37] There is some speculation that bin Laden was responsible for Azzam's —the latter was an obstacle to bin Laden's desire to shift Al Qaeda's focus to Western targets.

[38] "[T]he grievances they are seeking to resolve are not in any way metaphysical . . . In their manifestos they refer to real events and real people and what are perceived to be real problems . . . While bin Laden's discourse may be based on an interpretation of Islamic history, his power is derived from playing on the current social, economic and political problems of the Muslim World. Just because a lack of graduate employment, decent housing, social mobility, food, etc. is explained by an individual through reference to a religion does not make it a religious grievance. It remains a political grievance articulated with reference to a particular religious worldview." Burke, *Al-Qaeda: Casting a Shadow of Terror.*

[39] Yassin Musharbash, "The Future of Terrorism: What Al Qaeda Really Wants," *Speigel Online,* December 8, 2005, http://www.spiegel.de/international/0,1518,369448,00.html.

[40] Burke, *Al-Qaeda: Casting a Shadow of Terror.*

[41] Peter L. Bergen, *Holy War, Inc.: Inside the Secret World of Osama bin Laden* (New York: Free Press, 2001).

members in more than 60 countries by 2001.[42] The leadership did not necessarily exert command and control over tactical planning and operations, instead emphasizing local initiative, flexibility, and robustness.[43] While the core leadership retained responsibility for the ideological direction and strategic messaging of the movement, when necessary, they drew upon the leaders of regional nodes to serve in positions of high command.

Though the exact structure of Al Qaeda is still unknown, information acquired from former members provided US authorities with a rough picture of how the group was organized.[44] Al Qaeda was administered by a *Shura* (consultative) council that discussed and approved major actions, including terrorist operations.[45] It is a structure similar to that of a holding company with a core management group (Al Qaeda) controlling partial or complete interests in other companies (Al Qaeda affiliates).[46] This innovative structure made Al Qaeda the first truly multinational terrorist group with a global strike capability.[47] The core of the organization was composed of a dozen or so militants, primarily Egyptians who had previously been held as political prisoners and traveled to Afghanistan to fight against the Soviets.[48] This inner cadre was surrounded by an outer layer of one hundred or so highly motivated and well-trained loyalists from throughout the Muslim world.[49]

The Al Qaeda core was augmented by a set of specialized committees with a tailored set of goals, missions, and budgets. The Military Committee was responsible for training operatives, acquiring weapons, and planning attacks.[50] The Finance and Business

[42] Gunaratna, *Inside Al Qaeda.*

[43] Sageman, *Understanding Al Qaeda Networks,* 27.

[44] Jarret M. Brachman, *Global Jihadism: Theory and Practice,* Case Series on Political Violence (New York: Taylor & Francis, 2008).

[45] Jayshree Bajoria and Greg Bruno, "Al-Qaeda (a.k.a. Al-Qaida, Al-Qa'ida," Council on Foreign Relations, accessed August 23, 2010, http://www.cfr.org/publication/9126/alqaeda_aka_alqaida_alqaida.html#p1.

[46] Bergen, *Holy War, Inc.*

[47] Gunaratna, *Inside Al Qaeda.*

[48] Sageman, *Understanding Al Qaeda Networks,* 20. http://www.usini.org/prog_01072004.html (accessed August 23, 2010).

[49] Angel Rabasa et al., *Beyond Al-Qaeda: Part I: The Global Jihadist Movement* (Santa Monica, CA: RAND Corporation, 2006); Lawrence Wright, *The Looming Tower: Al-Qaeda and the Road to 9/11* (New York: Knopf, 2006).

[50] The most operationally viable component of the military committee (and Al Qaeda writ large) was the 055 Brigade. At its peak in 2001, the 055 Brigade had an estimated 2,000 soldiers and officers comprising Arabs, central Asians, and south Asians, including Chechens, Bosnians, and Uighurs from western China. These included veterans of the Afghan resistance, many of whom had remained in Afghanistan, as well as a second generation of younger, better-educated recruits. The 055 Brigade fought with the Taliban against US and Northern Alliance forces in 2002 and suffered significant losses. Although

Committee, composed of professional bankers, accountants, and financiers, funded the recruitment and training of operatives, often through the *hawala*[51] banking system, providing airline tickets and false passports, issuing paychecks, and overseeing the group's vast network of businesses as well as dealing with large organizational issues, such as developing financial resources to meet Al Qaeda's payroll and fund its operations and those of its various affiliates.[52] The Foreign Purchases Committee was responsible for acquiring weapons and other technical equipment.[53] The *Shari'a* Committee determined whether particular courses of action conformed to established Islamic law.[54] The *Fatwa* (formal legal opinion) Committee issued religious edicts, while the Media Committee supplied video and audio materials.[55]

The roots of Al Qaeda lay in a number of different organizations, including the *Maktab al-Khidamat* (MAK) or Services Bureau, a clearinghouse established to facilitate the recruitment, transportation, organization, training, and equipping of Arabs to support the Afghan resistance.[56] Established by Abdullah Yusuf Azzam, a Palestinian scholar of Islamic law, and Osama bin Laden in Peshawar, Pakistan, in

---

the brigade did not possess a capability to operate outside of the region, individual members were permitted to join terrorist cells if selected to do so. Wright, *The Looming Tower*. See also Gunaratna, *Inside Al Qaeda*.

[51] *Hawala* is an alternative or parallel remittance system outside of, or parallel to, traditional banking or financial channels. It was developed in India, before the introduction of Western banking practices, and is currently a major remittance system used around the world. The components of *hawala* that distinguish it from other remittance systems are trust and the extensive use of connections, such as family relationships or regional affiliations. Unlike traditional banking, *hawala* makes minimal use of any sort of negotiable instrument. Transfers of money take place on the basis of communications between members of a network of *hawaladars*, or *hawala* dealers. INTERPOL website, accessed August 23, 2010, http://www.interpol.int/public/FinancialCrime/MoneyLaundering/hawala/default.asp.

[52] Steve Kiser, "Financing Terror: An Analysis and Simulation to Affect Al Qaeda's Financial Infrastructures" (PhD diss., Pardee RAND Graduate School, 2005).

[53] Ibid.

[54] Wright, *The Looming Tower*.

[55] Ibid.

[56] There is a long history of Arab migration into Afghanistan, much of it following the Russian Revolution of 1917 when Arabs living throughout central Asia fled encroaching Soviet control for the religious freedom afforded them in Afghanistan. Later, under Stalin, many Muslims of Arab and non-Arab descent living in the Caucasus were deported en masse to Siberia and central Asia. The current Islamist resistance in Chechnya owes much to these earlier Soviet policies. Although this resistance was originally led by indigenous fighters, it has received substantial financial and human support from Muslims throughout the Middle East, where many deportees settled and integrated into the local communities. Therefore, to suggest that the Arabs who supported the anti-Soviet resistance are religious mercenaries focused only on global *jihad* and without personal ties to Muslims living outside the Middle East is misleading. More likely, the intense migration and intermarriage between Muslims of Arab and non-Arab descent throughout central Asia, the Caucasus, and the Middle East created strong personal ties that supported the narrative of religious *jihad* and resistance to foreign occupation.

1984, the MAK also attracted other militant leaders.[57] It consisted of a network of international recruiting offices, bank accounts, and safe houses and was also responsible for the construction of paramilitary camps for the training of militants and the fortifications used by Arab fighters.[58] Between 1982 and 1992, estimates report approximately 35,000 foreign fighters contributed to the Afghan effort, though there were probably never more than 2,000 in Afghanistan at any one time. The MAK was responsible for training approximately 12,000–15,000 of those fighters, with approximately 4,000 remaining connected through either chain of command or ideological affinity after the conflict.[59]

Osama bin Laden was the group's undisputed leader, to whom the newly admitted swore an oath of allegiance.[60] Although few of the Afghan Arabs or later Al Qaeda recruits had direct contact with bin Laden,[61] he was able to build a popular following throughout the Islamic world and continues to be regarded as the supreme symbol of resistance to US imperialism.[62] Though bin Laden was principally a financier, logistician, and facilitator, he also retains some distinction as a military commander. In 1987, Soviet Special Forces attacked a training camp in Afghanistan. Bin Laden and a small group of fighters repelled the attack, earning bin Laden a reputation for tactical prowess.[63] Bin Laden's principal ideological adviser, and the organization's Deputy Chief of Operations, was Ayman Al-Zawahiri.[64] Zawahiri provided the scriptural and juridical substance for bin

---

[57] Azzam, who earned a PhD from Cairo's Al-Azhar University, was a very charismatic character and played a central role in crafting the narrative of resistance that drew thousands of Arabs to the Afghan cause. His previous combat experience (he fought the Israelis in 1967), combined with his religious credentials (his education and his connections with the family of Sayyid Qutb, an important ideological leader of the early Muslim Brotherhood), made him a particularly appealing figure to bin Laden. Omar Abd Al-Rahman, an Egyptian militant also educated at Cairo's Al-Azhar University and a key ideological figure for both Al-Jama'at Al-Islamiyya and the EIJ, also used MAK's resources to contribute to the Afghan resistance.

[58] Wright, *The Looming Tower.* Azzam originally established the MAK and later persuaded bin Laden to join. Bin Laden used his family's relationship with the Saudi Royal Family to support the effort overtly—through a strategic communications plan—and covertly, eventually matching US financial contributions to the resistance. Michael Scheuer, *Through Our Enemies' Eyes: Osama bin Laden, Radical Islam, and the Future of America* (Washington, DC: Potomac Books, Inc., 2008).

[59] Bergen, *Holy War, Inc.*

[60] Gunaratna, *Inside Al Qaeda.*

[61] Bergen, *Holy War, Inc.*

[62] Gunaratna, *Inside Al Qaeda.*

[63] Sageman, *Understanding Terror Networks.*

[64] Zawahiri, a trained surgeon, was born into a pious, middle-class Egyptian family. A key figure in EIJ, he spent time in prison in Egypt on suspicion of being involved in the assassination of Egyptian President Anwar Sadat in 1981.

Laden's more general political ideals.[65] Al Qaeda's senior leadership (top aides, media representatives, military advisers, etc.) were more likely to have technical and professional backgrounds—in business, public administration, law, engineering, or medicine—than religious ones.[66]

COMMUNICATIONS

Throughout the 1990s, Al Qaeda used satellite phones and computers to organize and maintain plans and faxed copies of religious rulings issued by bin Laden throughout the Muslim world and Europe where they were picked up by Arabic-language media outlets.[67] The group also exploited informal, traditional forms of communications, such as pre-existing social bonds, to transfer information.[68] Ultimately, its communications infrastructure and operations made extensive use of electronic media for mobilization, communication, fund-raising, and planning attacks. Al Qaeda successfully combined multimedia propaganda with advanced communications technologies to integrate, professionalize, and disseminate a highly sophisticated message.[69] Through its Media Committee, Al Qaeda was able to centralize the group's strategic messaging content yet decentralize its distribution, simultaneously increasing the group's target audience while retaining thematic integrity.

A number of themes repeatedly continued to appear in Al Qaeda's messaging: that the oppressive regimes persecuting Muslims (Egypt, Saudi Arabia, Pakistan, etc.) are in league with the United States; since the abolition of the caliphate in 1924, the "Crusaders" have worked to prevent true believers from establishing an Islamic state; Christendom, together with world Jewry, is seeking to destroy Islam; the United States has created "an ocean of oppression, injustice, slaughter and plunder" and has thus merited responses such as the 9/11 attacks; the economy is the US center of gravity and is acutely vulnerable; and contributing in some way to violent, defensive *jihad* is the solemn obligation of every Muslim. These themes were circulated via the Internet, in books and pamphlets, and through videotapes and

---

[65] Wright, *The Looming Tower.*

[66] This is as much a product of changes in the region's educational system as it is indicative of psychological links between certain occupations and support for terrorist tactics. Bergen, *Holy War, Inc.*

[67] Ibid.

[68] Sageman. *Understanding Al Qaeda Networks*, 27, http://www.usini.org/prog_01072004.html.

[69] Rabasa et al., *Beyond Al-Qaeda.*

audiotapes in which bin Laden and Zawahiri expounded on various ideological and political issues, as well as current events.[70]

METHODS OF ACTION AND VIOLENCE

Al Qaeda operations were distinguished by their audacity; thoughtful and deliberate planning; thorough and detailed reconnaissance/surveillance; a stringent emphasis on training; and efforts by the leadership to secure religious sanction through the issuing of *fatwas*.[71] Several manuals, often multivolume ones including content from US military doctrinal publications, were disseminated by hand at training camps and, later, electronically. By designing specialized courses and constructing secret camps to train volunteers for martyrdom operations, Al Qaeda institutionalized and formalized the tactics, techniques, and procedures of suicide terrorism.[72]

The first attack attributed to Al Qaeda took place in 1992, when bombs were detonated at the Mövenpick and Goldmohur hotels in Yemen.[73] Although American soldiers transiting Yemen on a UN mission to Somalia were the presumed target, the attacks killed only two civilians.[74] The bombings, largely unnoticed in the United States, were symbolic of Al Qaeda's shifting operational philosophy, mainly the acceptance of civilian deaths in operations targeting military assets.[75] Two *fatwas* referencing Ibn Taymiyyah (a thirteenth-century scholar and ideological forefather of Salafism, who sanctioned resistance by any means during the period of the Mongol invasion) were issued to justify the killings according to Islamic law.[76] Bin Laden later claimed responsibility for arming the Somali factions that battled US forces there in October 1993, killing eighteen US service members in Mogadishu, though there is little tangible evidence to suggest a direct support role.[77]

A scant year later, in 1993, Al Qaeda operatives parked a rental truck loaded with a 1,500-pound explosive in the parking garage of the World Trade Center hoping to damage the first tower sufficiently to send it crashing into the second tower, with a casualty estimate of

---

[70] Ibid.

[71] Gunaratna, *Inside Al Qaeda*.

[72] Ibid.

[73] For a complete timeline of Al Qaeda operations, see http://news.bbc.co.uk/2/hi/3618762.stm (accessed August 23, 2010).

[74] Wright, *The Looming Tower*.

[75] Scheuer, *Through Our Enemies' Eyes*.

[76] Wright, *The Looming Tower*.

[77] Rollins, *Al Qaeda and Affiliates*, 6.

about 250,000.[78] Although the blast did not destroy either tower, it shook both, killing six and injuring 1,042. The following year, four individuals were convicted of conspiracy, explosive destruction of property, and interstate transportation of explosives, and in 1997 both the mastermind, Ramzi Yousef, and the driver of the truck were also convicted.[79]

Several years later, in 1996, Al Qaeda announced its intention to expel foreign troops and interests from what they considered Muslim territory. Bin Laden issued a *fatwa* entitled "Declaration of War Against Americans Occupying the Land of the Two Holy Places," a public declaration of war against the US and any of its allies, and began to refocus the organization's resources toward large-scale psychological operations.[80] The *fatwa* represented an overall shift in focus on behalf of the EIJ and the other cells comprising Al Qaeda from the near enemy, or Muslim "apostate" governments, to the far enemy, or the United States.[81] The *fatwa* also cemented the operational and strategic relationship that had been evolving between bin Laden's Al Qaeda and Zawahiri's EIJ. As a result, some EIJ members left to join the Egyptian Islamic Group, which engaged in an ambitious and highly active campaign against the Egyptian government in the 1990s, while others simply disengaged from the cause. A small cadre remained with Zawahiri and bin Laden, who were now operating out of Sudan.[82]

In 1998, Zawahiri issued a joint *fatwa* with bin Laden and Rifi Taha (Egyptian Islamic Group) under the title "World Islamic Front Against Jews and Crusaders," which read:

> The ruling to kill the Americans and their allies—civilians and military—is an individual duty for every Muslim who can do it in any country in which it is possible to do it, in order to liberate the al-Aqsa Mosque[83] and the holy mosque from their grip, and in order for their armies to move out of all the lands of Islam, defeated and unable to threaten any Muslim. This is in accordance with the words of Almighty Allah, 'and fight the pagans all

---

[78] Wright, *The Looming Tower.*

[79] Ibid.

[80] Scheuer, *Through Our Enemies' Eyes.*

[81] Fawaz A. Gerges, *The Far Enemy: Why Jihad Went Global* (New York: Cambridge University Press, 2005).

[82] Ibid. In 1991, after public statements against the Saudi government for harboring American troops, bin Laden traveled from Jeddah, Saudi Arabia, to Kabul, Afghanistan, to help stop the intertribal militia conflict; his trip ultimately terminated in Khartoum, Sudan, where the Al Qaeda inner circle established a base of operations.

[83] The Al Aqsa Mosque, considered Islam's third holiest site, is located in East Jerusalem, the proposed capital of an eventual Palestinian state.

together as they fight you all together,' and 'fight them until there is no more tumult or oppression, and there prevail justice and faith in Allah.'[84]

The near-simultaneous explosions at the US embassies in Tanzania and Kenya followed six months later, killing 223 and injuring more than 4,000.[85] The attack placed Zawahiri on the list of the US FBI's Ten Most Wanted, alongside bin Laden, and Zawahiri was shortly thereafter sentenced to death in absentia by the Egyptian government.[86]

In 2000, Al Qaeda operatives attacked the missile destroyer USS *Cole* while it was refueling at a port in Yemen. The attack, a suicide mission using a skiff packed with approximately 1,000 pounds of explosives, killed seventeen US servicemen and damaged the vessel. Then, in 2001, nineteen Al Qaeda operatives simultaneously hijacked four commercial airliners, flying two into the World Trade Center towers in New York City while another crashed into the Department of Defense in Washington, D.C. The fourth—destined for the US Capitol building—crashed in Pennsylvania.[87] The attacks killed nearly 3,000 altogether.[88] The hijackers received flight training in the United States and conducted extensive surveillance and drills. The operatives commandeered the aircraft by using improvised weapons (box cutters and/or small knives) to subdue flight attendants, gain access to the cockpits, and execute some of the aircrew.

METHODS OF RECRUITMENT

Al Qaeda used many pre-existing social institutions, such as mosques, schools, and boardinghouses, as recruiting stations.[89] The organization was also adept at using media to publicize successful missions, which facilitated recruitment, fund-raising, and status. The long history of recruiting, transporting, training, and equipping fighters to support the Afghan resistance to the Soviet occupation meant that significant support for these operations already existed within many communities. Many of the necessary channels were

---

[84] As quoted in Scheuer, *Through Our Enemies' Eyes*.

[85] Sageman, *Understanding Terror Networks*.

[86] Ibid.

[87] At 8:46 am, American Airlines Flight 11 struck the World Trade Center's North Tower; at 9:03 am, United Airlines Flight 175 struck the South Tower; at 9:37 am, American Airlines Flight 77 struck the Pentagon; and at 10:03 am, United Airlines Flight 93 (whose intended target was the US Capitol building) crashed near Shanksville, Pennsylvania.

[88] Wright, *The Looming Tower*.

[89] National Commission on Terrorist Attacks upon the United States, *The 9/11 Commission Report: Final Report on the National Commission on Terrorist Attacks upon the United States* (New York, W. W. Norton, 2004).

already established.[90] Although money does not appear to play a dominant role in recruitment—many militants came from middle and upper class backgrounds and thus were not engaging in militancy to earn a livelihood—many members did draw monthly paychecks, and the organization frequently made severance payments to the families of those who had died fighting or while carrying out operations.[91]

Al Qaeda did not necessarily use a centralized recruiting approach, often allowing subordinate leaders to determine the most appropriate method of recruitment for a given operational need.[92] In general terms, recruiters made subtle contact initially. As contact increased in intensity, the varied components of an individual's previous identity, such as occupation, education, or membership in different political or social groups, weakened until they were displaced by an identity as a militant. Accordingly, the recruiter sought out individuals who already possessed a weak identity, for example, those who were unemployed or underemployed[93] or did not possess strong roles in a family or community network.[94]

Al Qaeda's recruiting methods varied both across time and across the varied groups that eventually formed the organization. The loosely federated groups that would form Al Qaeda in 1988 used recruiting techniques designed for targeting individuals in larger populations that would actively subvert group recruitment. This approach was often used when targeted recruits were in the existing security apparatus, such as the police or military forces, from which Al Qaeda consistently drew recruits. Al Qaeda's preferred recruiting approaches were often attempts made in intimate settings with the explicit intent of avoiding observation. This technique leveraged the influential power of conformity and relied heavily on personal appeals tailored specifically for a targeted individual, often using peers and/or relatives to make the pitch.[95] Use of the Internet as a recruiting medium through which Al Qaeda conveyed its message and targeted recruits did not reach maturity or operational effectiveness until after the 2001 merger with EIJ. Al Qaeda typically measured the progress of individual recruitment by evaluating the individual's commitment to Salafist

---

[90] Gerges, *The Far Enemy.*

[91] Bergen, *Holy War, Inc.*

[92] Sageman, *Understanding Al Qaeda Networks* (2004), 22. http://www.usini.org/prog_01072004.html.

[93] Such as an individual with a law degree who is unable to find suitable work and is reduced to being a shopkeeper. Not only is this individual's self-worth reduced by virtue of not securing a job, but they also lack consistent contact with a familiar peer group.

[94] Scott Gerwehr and Sara Daly, "Al-Qaida: Terrorist Selection and Recruitment," in *The McGraw Hill Security Handbook*, ed. David G. Kamien (New York: McGraw Hill, 2006).

[95] Ibid.

principles through their demonstrated knowledge of that particular interpretation of Islam, as well as their willingness to use violence to further the universal observance of these principles.[96] However, this approach could have backfired if individuals misunderstood or were offended by the violent means, and the justifications offered for those means, used to achieve the group's goals and objectives.

METHODS OF SUSTAINMENT

The 9/11 Commission estimated that Al Qaeda operations cost $30 million per year prior to 9/11.[97] Swiss intelligence estimates ranged from $250 to $500 million, while the Australian government placed the total above $250 million and the British placed it around $280–$300 million. The most recent analyses suggest the figure is closer to $30–$35 million.

Al Qaeda's sophisticated methods of sustainment date back to the MAK in the late 1980s when Azzam and bin Laden started creating camps inside Afghanistan to better prepare Arab recruits for more active roles in combating Soviet forces. The leadership of many militant groups (including those of EIJ and Gamaat Al-Islamiyya), as well as state sponsors such as Saudi Arabia, contributed significant financial and physical resources to supplement MAK's activities. In Saudi Arabia, Prince Turki bin Faisal bin Abdul-Aziz coordinated the efforts of as many as twenty charities set up for the express purpose of funding the Arab resistance fighters, channeling as much as $2 billion to the effort.[98]

Al Qaeda used an extremely sophisticated, complex, and resilient money-generating and -transferring network.[99] The organization's highly resilient financial infrastructure spanned the globe with fund-raising operations, various types of accounts, and financiers on every continent in approximately one hundred countries.[100] The versatility of Al Qaeda's financial infrastructure was primarily due to its compartmentalized structure. Sources of funding were kept separate from the cells to which Al Qaeda distributed money, and high priority was assigned to financial training and management, as well as to the sustained generation and investment of funds. The Al Qaeda Finance and Business Committee managed the group's resources across four

---

[96] Ibid.

[97] National Commission on Terrorist Attacks upon the United States, *The 9/11 Commission Report*.

[98] Kiser, "Financing Terror."

[99] Gunaratna, *Inside Al Qaeda*.

[100] Kiser, "Financing Terror."

continents. To move funds clandestinely from source to recipient, Al Qaeda's financial network disguised the identities of both parties and established several legitimate institutions, including state and privately owned charities, banks, and companies through which to funnel funds.[101] Revenues went directly to central headquarters. With the exception of some operational expenses, cells were typically expected to be self-sufficient. Operational cells could deploy without much information regarding the larger organization's underlying financial network.[102]

Al Qaeda favored charities as a primary source of income. Charitable contributions are a sizeable source of financing for humanitarian, educational, and foreign aid activities in Muslim countries. Donations were largely cyclical, peaking during the months of Ramadan, suggesting funds came from *zakat* and other obligatory charitable activities, often from unsuspecting religious institutions.[103] The prominence of charity exploitation by Islamist groups can be partially explained by religious principles. One of the five pillars of Islam (*zakat* or tithing) requires Muslims to provide a small percentage of their incomes to help the poor.[104] Al Qaeda effectively exploited this principle for their own goals by merging it with militant operations to develop an effective line of persuasion that resonated with the more militant elements in the Muslim world. Bin Laden reportedly said that "Muslims and Muslim merchants in particular should give their *zakat* and their money in support of this state [Afghanistan] . . . where followers of Islam can embrace the Prophet of God."[105]

Al Qaeda used two different approaches in redirecting funds from charities to its own organization. The first involved creating, subsuming, or collaborating with a cooperative charitable organization. Donors to these charities may or may not have known the true nature of the organization to which they were contributing. Once donations were made to the charity, money was then sent to the network's headquarters. The charities would print false documentation for the benefit of donors, typically showing the money had been spent on humanitarian causes. The second, less common, and riskier approach involved Al Qaeda operatives infiltrating unwitting charities with the intent of having those charities send funds to support the group's

---

[101] Gunaratna, *Inside Al Qaeda.*

[102] Kiser, "Financing Terror."

[103] National Commission on Terrorist Attacks upon the United States, *The 9/11 Commission Report.*

[104] Kiser, "Financing Terror."

[105] Ibid.

efforts in various parts of the globe.[106] Most donors to these types of benevolent organizations probably did not know the organization had been infiltrated. However, some donors may intentionally have provided funds to such charities, providing an additional layer of plausible deniability to hide the true intentions of their funding activities.

Al Qaeda was adept at exploiting the environment of limited financial regulation prevalent in the Middle East and many Asian states but was also able to make use of formal financial institutions (including wire transfers and bank accounts). Like other components of their logistical and operational networks, redundancies were incorporated to avoid interruption, and nodes were replaced frequently to avoid detection and surveillance.[107] A presence was established in most countries using indigenous or migrant Muslim communities. Although Al Qaeda had no single, central financial repository, it could never operate in isolation because mounting a terrorist operation required financial and technical logistical support that often had to be in place years in advance. In the Middle East, especially in the Gulf states, Al Qaeda had a great deal of covert support among the public and received practical help from Islamic philanthropists and foundations, particularly from the United Arab Emirates and Saudi Arabia.[108]

The exploitation of existing commercial, financial, and transportation systems and institutions is common practice in organized crime and was implemented adeptly by Al Qaeda. The *hawala* system, an informal network of money brokers originally established to facilitate long-distance trade beginning around the eighth century, was one such system.[109] MAK relied heavily on the *hawala* system to move money from its satellite entities (recruiters, arms dealers, logisticians) to the training camps during the Afghan resistance, driven partly by the weakness and undependability of the existing financial and banking system.[110] *Hawaladars* used formal financial institutions and couriers to transfer funds, but the process was not subject to substantial government oversight or record keeping, which was kept in short-hand and only for brief periods. This enabled operatives to access funds without opening an account. Couriers came from inside the network and were chosen for certain characteristics

---

[106] Ibid.

[107] National Commission on Terrorist Attacks upon the United States, *The 9/11 Commission Report*.

[108] Gunaratna, *Inside Al Qaeda*.

[109] National Commission on Terrorist Attacks upon the United States, *The 9/11 Commission Report*.

[110] Ibid.

that facilitated their movement (ethnicity, documentation, language skills), but they were rarely privy to operational details.[111]

METHODS OF OBTAINING LEGITIMACY

Al Qaeda's strength and appeal did not lay solely in its sophisticated theological discourse, but also in its ability to comprehend, co-opt, and exploit modern grievances. This narrative combination resonated with extremists and moderates alike, regardless of whether an individual approved of the means by which Al Qaeda sought to accomplish its goals. The psychological simplification (or splitting) of complex, multi-faceted issues into binary arguments of good versus evil is a common political practice and has been used effectively by Islamist theorists throughout the twentieth century. Al Qaeda's leadership was not composed of highly trained religious scholars, and their religious rhetoric was far from complex or nuanced, making it broadly accessible across the socioeconomic spectrum. The specific messages within the larger narrative rarely focused on citing authoritative texts (beyond selective interpretations of previous theorists reinforced by Quranic quotes without context) but rather relied on the application of general religious or ethical principles to modern political and social problems.[112]

Osama bin Laden's personal credibility was among the highest of any modern Muslim leader. His ascetic lifestyle contrasted sharply with those of the state-sponsored clergy, as well as those of many previous movement leaders, who amassed significant personal fortunes.[113] Although the textual and visual narrative of bin Laden as the warrior–scholar has been carefully crafted, it does rest on a body of proven exploits. Although he lacked many of the serious religious, academic, and military credentials of other movement leaders, his image as a billionaire's son who forsook wealth and comfort for the austerity and deprivation of the life of a militant was admirable to many. The congruity between Al Qaeda's message and actions, particularly

---

[111]   Ibid.

[112]   Dale C. Eikmeier, "Qutbism: An Ideology of Islamic-Fascism," *Parameters* 37 (Spring 2007), 85.

[113]   The fortune of the Saudi Royal Family, which claims legitimacy as the custodians of Islam's two holiest sites, Mecca and Medina, comes wholly from the family's control over that nation's oil resources. Pan-Arab leaders such as Gaddafi, Hussein, and Nasser also gained significant personal wealth through their control over state resources. More recently, Salah Ezzedine, a Lebanese billionaire and supporter of Hizbollah, was charged with fraud after investigators revealed that he was operating a Ponzi scheme. The massive profits he made while custodian of Hizbollah's finances contrasts sharply with the extreme poverty of most Lebanese Shi'ites and the social welfare programs Hizbollah provides to them.

when juxtaposed with the incongruity between the words and deeds of most other actors, including foreign governments, international organizations, and their own domestic governments, was refreshing to many Muslims who were locked out of the small circle of elites able to live richly off the largesse of corrupt regimes. The idea of a courageous and pious vanguard standing up to political, military, and economic superpowers, and achieving a degree of success, was a profoundly empowering narrative for many.

EXTERNAL SUPPORT

Al Qaeda did not appear to have an overt state sponsor along the Cold War model. However, at various points in its history, bin Laden exploited his personal relationships with members of the Saudi Royal Family and other government elites for the financial and operational survival of the organization. The 9/11 Commission Report indicates no significant state support for Al Qaeda.[114] But, during Operation Cyclone, the CIA provided funds to the Pakistani Inter-Services Intelligence directorate, which then distributed the funds, some of which may have reached the MAK.[115] At two points in the 1990s, Al Qaeda did rely upon the tacit permission, if not protection, of a state or ruling power. While in Sudan (1991–1996), bin Laden worked closely with Dr. Hassan al-Turabi, the head of the National Islamic Front and key adviser to President Umar Hassan Ahmad al-Bashir. Unlike Al Qaeda's remote camps in Afghanistan, bin Laden's base in Sudan provided Al Qaeda with greater access to the international community, facilitating the operations of their legitimate and illicit business enterprises. In return, bin Laden's construction enterprise built a new highway from Khartoum to Port Sudan at a discounted rate.[116] Bin Laden also purchased wide tracts of depressed real estate, where he established small-scale manufacturing and agricultural operations. These business operations supplied additional income but also served as fronts to acquire weapons, explosives, and other equipment. After his departure from Sudan, bin Laden paid Mullah Mohammed Omar and his Taliban organization (then the de facto government of Afghanistan) approximately $10–$20 million per year for safe passage. The Taliban also received weapons, vehicles,

---

[114] National Commission on Terrorist Attacks upon the United States, *The 9/11 Commission Report*.

[115] Kiser, "Financing Terror."

[116] National Commission on Terrorist Attacks upon the United States, *The 9/11 Commission Report*.

commodities, and money for social projects from Al Qaeda.[117] These much-needed resources made the Taliban government reluctant to turn over bin Laden to international authorities; however, the government was under pressure to expel him in return for recognition by the UN.

The aforementioned mechanisms of sustainment, primarily the exploitation of charitable organizations and legitimate businesses, provided more support to Al Qaeda than any single external organization, whether state or non-state. Al Qaeda also enhanced its capabilities by acquiring and integrating existing militant organizations. Al Qaeda had been a supporter of EIJ for years when in June 2001, the two organizations merged.[118] Thereafter, all activities and operations of former EIJ members were carried out under bin Laden's banner. Al Qaeda may also have formed a strategic partnership with other militant groups, including possibly Hizbollah, from whom it may have received technical assistance, training, and intelligence. Israeli intelligence also believes that Al Qaeda infiltrated the Palestinian Occupied Territories with the support of Hamas.[119]

COUNTERMEASURES TAKEN BY THE GOVERNMENT

In 1996, the CIA established a bin Laden Issue Station—code named "Alec Station"—to gather, analyze, and disseminate intelligence on, and plan operations against, bin Laden.[120] In cooperation with military and law enforcement entities, the station considered numerous contingency plans to capture bin Laden. However, none were implemented. Their efforts were largely informed by the defection of Jamal Ahmed al-Fadl, who surrendered to the US embassy in Eritrea after embezzling more than $100,000 from Al Qaeda's business ventures.[121] Some of the targeting information used in the 1998 cruise middle attacks on Al Qaeda training camps in Sudan and Afghanistan in the wake of the US embassy bombings in Kenya and Tanzania came from Alec Station. By the spring of 2000, Alec Station supported the growing number of unmanned aerial vehicle (UAV)

---

[117] Ibid.

[118] Montasser al-Zayyat, *The Road to Al-Qaeda: The Story of Bin Laden's Right-Hand Man*, Critical Studies on Islam (London: Pluto Press, 2004); Coll, *Ghost Wars*.

[119] Gunaratna, *Inside Al Qaeda*.

[120] National Commission on Terrorist Attacks upon the United States, *The 9/11 Commission Report: Final Report on the National Commission on Terrorist Attacks upon the United States*.

[121] Scheuer, *Through Our Enemies' Eyes*.

surveillance and reconnaissance flights searching for bin Laden in southeastern Afghanistan.[122]

The 9/11 attacks resulted in a considerable shift in the countermeasures taken by the United States and its allies. On October 7, 2001, the United States declared a "Global War on Terrorism" in order to protect the citizens of the United States and allies, to protect the business interests of the United States and allies at home and abroad, to break up terrorist cells in the United States, and to disrupt the activities of the international network of terrorist organizations made up of a number of groups under the umbrella of Al Qaeda. The US and coalition partners commenced with Operation Enduring Freedom to "find Osama bin Laden and other high-ranking Al Qaeda members and put them on trial, to destroy the whole organization of Al Qaeda, and to remove the Taliban regime which supported and gave safe harbor to Al Qaeda."[123] The military invasion of Afghanistan and subsequent advance was rapid and effective, and by November 12, 2001, the Taliban vacated Kabul under coalition and Northern Alliance pressure. By December, the Al Qaeda leadership was isolated in a cave complex near the Tora Bora mountains but ultimately escaped, likely to Pakistan's Northwest Frontier Province.[124] Since its inception, Operation Enduring Freedom has expanded to include ongoing operations in the Middle East, central and southeast Asia, Africa, and South America.

Through increased efforts and cooperation, law enforcement elements have detected and disrupted Al Qaeda cells in the United Kingdom, the United States, Italy, France, Spain, Germany, Albania, Uganda, Saudi Arabia, Yemen, Indonesia, and elsewhere since 2001.[125] Efforts include aggressive financial intelligence gathering to track the flow of financing; freezing assets; interrogation of suspects with multinational military, intelligence, and law enforcement cooperation; increased paramilitary training for law enforcement officers; and exponential increases in local, state, and federal counterterrorism budgets throughout the world. However, Al Qaeda's financial infrastructure has proven more difficult for agents to penetrate than the traditional institutions and channels used to investigate and prosecute white-collar financial crimes. Al Qaeda financing was often

---

[122] Coll, *Ghost Wars.*

[123] John F. Kerry, "Tora Bora Revisited: How We Failed to Get Bin Laden and Why It Matters Today," A Report to Members of the Committee on Foreign Relations United States Senate, 111th Congress, 1st Session (Washington, DC: US Government Printing Office, 2009).

[124] Ibid.

[125] Bajoria and Bruno, "Al-Qaeda."

tied to legitimate charities that provided genuine humanitarian relief, largely because regulations to enhance transparency or accountability in financial transactions were almost completely absent from most of the countries that Al Qaeda used to raise and transfer funds.[126]

Efforts to disrupt militant operations have also been hampered by the tension between the missions of different government agencies, especially with respect to the contrast between the evidence required to incarcerate individuals versus what is required to disrupt their operations. The acquisition of documentary evidence from foreign governments and institutions is likely to tip off operatives, making legal pursuit an obstacle to continued intelligence gathering. Many analysts have concluded that tracking and disrupting terrorist finances has little impact because operations are relatively inexpensive to carry out. The use of underground financial networks makes freezing assets far more complicated, if not impossible, and staunch resistance from private financial institutions to reforming laws regarding secrecy, confidentiality, and reporting requirements has been a significant obstacle.[127]

In addition to military and intelligence operations, the governments of some predominantly Muslim countries have initiated rehabilitation or "reeducation" programs directed at young militants.[128] Inmates at a series of modern facilities in Saudi Arabia (some of which have, ironically, been built by the bin Laden group) are treated less like criminals and more like impressionable youth who, having been led astray, receive correction of their theological misunderstandings from Islamic scholars. The programs also address possible psychological needs and emotional weaknesses that might have rendered the youth susceptible to the militant narrative through postincarceration social welfare programs. These programs are designed to facilitate the integration of the reformed militant back into Saudi society by helping him find both a job and a wife. Enrollment in the *Munasaha* (Advising) program is not voluntary, and Human Rights Watch has complained that some participants have been detained for lengthy periods without trial or access to legal counsel. However, treatment is not considered harsh by Middle Eastern standards, and the Saudi Ministry of the Interior claims a 0% recidivism rate. The Saudi government has also intensified efforts to turn public sympathies

---

[126] National Commission on Terrorist Attacks upon the United States, *The 9/11 Commission Report.*

[127] Ibid.

[128] Under Dr. Abdulrahman al-Hadlag, the general director of the Ideological Security Directorate at the Saudi Arabian Ministry of the Interior, Saudi Arabia has instituted its own rehabilitation programs to steer young (18–36 years old) Saudi militants away from terrorist operations.

away from terrorist groups through public information campaigns, including employing prominent clerics to take public stands against Al Qaeda and its affiliated movements. In late 2007, Saudi Mufti Sheik Abdulaziz Bin Abdullah Bin Mohammed al-Sheikh issued a *fatwa* prohibiting Saudi youth from traveling overseas to wage militant campaigns. The Saudi Ministry of Islamic Affairs also initiated an online program called Serenity to fight militancy by drawing recruiters into ideological debates with moderate-minded clerics.[129]

SHORT- AND LONG-TERM EFFECTS

CHANGES IN THE ENVIRONMENT

Although Al Qaeda's dramatic operations have cast the current era as one dominated by terrorism, politically motivated violence targeting civilians and governments dates back to the first century CE.[130] In fact, the most notable changes in the environment during the period 1988–2001 include an overall *reduction* in the frequency of attacks, the majority of which were perpetrated in Latin America, as well as a shift away from state-sponsored terrorism, which had originated primarily in Eastern Europe and the Middle East.[131] The post-2001 period looks less rosy, with reports from the US State Department showing an increase in "significant" acts of terrorism (those that caused death, serious injury, or major property damage)[132] and reports from the US National Counterterrorism Center demonstrating increases in the number of fatalities from terrorist incidents every year from 2003 to 2008.[133]

---

[129] See NewAgeIslam.com, November 12, 2008.

[130] Most historians agree that terrorism dates at least to the first century CE with the Jewish *Sicarii*, who targeted both their Roman occupiers and those coreligionists they believed to be insufficiently anti-Roman (*sicarii* comes from the Latin word for dagger—members assassinated their targets using daggers hidden in their robes and afterward blended away into the crowd). The Muslim Hassassin—or "assassins"—a Shi'ite sect that carried out assassinations against the elite of neighboring empires, emerged a few centuries later.

[131] For a pre-9/11 examination of the history of terrorism and state responses, as well as theoretical and definitional issues, see Grant Wardlaw, *Political Terrorism: Theories, Tactics, and Countermeasures* (New York: University of Cambridge, 1989). For a review of the literature and pre-/post-9/11 distinctions, see Isabelle Duyvesteyn, "How New Is the New Terrorism?" *Studies in Conflict and Terrorism* 27, no. 5 (2004), 439–454.

[132] Alan B. Krueger and David Laitin, "Faulty Terror Report Card," *Washington Post*, May 17, 2004.

[133] Although much of this increase is due to incidents in Iraq, other states are also experiencing large increases in terrorist incidents, including Afghanistan, Pakistan, India, Thailand, and several African states (Somalia, Kenya, Niger). Even when deaths in Iraq are excluded, the number of fatalities trends upward, suggesting that the frequency of terrorist

The extraordinary character of the 9/11 attacks provoked an unprecedented response from the US government. Less than a month after the attacks, the US invaded Afghanistan, where the Taliban regime had been harboring Al Qaeda's leadership. Although Al Qaeda, in its original incarnation, was seriously weakened by this invasion and subsequent US operations targeting operatives in Pakistan and elsewhere, Al Qaeda's *essence* has proved quite resilient. Today, many militant groups carry out attacks under the Al Qaeda banner without necessarily possessing significant financial or operational links to the original group. It may be this contagion effect that is Al Qaeda's most lasting legacy. A 2010 Congressional Research Service report concluded that Al Qaeda, "has transformed into a diffuse global network and philosophical movement composed of dispersed nodes with varying degrees of independence."[134]

This proliferation of loosely affiliated groups—and the string of high-profile attacks that occurred in the aftermath of 9/11[135]—have also provided many authoritarian regimes with a free hand to repress legitimate and nonviolent dissent, citing dubious linkages between Al Qaeda and domestic opposition forces. Because the US government acts as gatekeeper to a significant portion of the global economic, political, and military infrastructure cooperation with the United States, counterterrorism efforts have become an important commodity for developing countries.[136] Their cooperation with the United States on counterterrorism measures not only brings them privileges, such as military aid, favorable International Monetary Fund lending, or World Bank development projects, but also allows them to deflect criticism of their own policies by placing them under the auspices of cooperation with the United States. One result, whether real or perceived, is US complicity with the repression, torture, and deaths

---

attacks is not solely explained by the Iraqi case. Myra Williamson, *Terrorism, War, and International Law: The Legality and the Use of Force Against Afghanistan in 2001* (Burlington, VT: Ashgate Publishing, 2009).

[134] John Rollins, Liana Sun Wyler, and Seth Rosen, *International Terrorism and Transnational Crime: Security Threats, US Policy, and Considerations for Congress* (Washington, DC: Congressional Research Service, 2010).

[135] Large attacks include the London subway bombings (2005), the Madrid train bombings (2004), the nightclub bombing in Bali, Indonesia (2002), and the hostage crisis in a Moscow theater (2002). Al Qaeda-affiliated groups are also blamed for numerous post-9/11 attacks in Morocco, Turkey, Tunisia, Pakistan, Kenya, Saudi Arabia, Madrid, London, Indonesia, and Algeria. For a list of attacks, see the Council on Foreign Relation's Backgrounder on Al Qaeda: http://www.cfr.org/publication/9126/#p8.

[136] The US position on the UN Security Council, its unrivaled supremacy in military technology, and its hefty voting weight within the International Monetary Fund and World Bank make the United States a virtual gatekeeper for any country wanting access to any of these global resources.

of antiregime activists and insurgents in many countries.[137] Another implication is increased domestic support for authoritarian regimes that have resisted pressure to cooperate with the United States, like Syria, and until the disputed elections in 2009, Iran as well.[138] This relationship allows foreign regimes to weaken opposition to their rule but has damaged US standing in the international community.[139]

CHANGES IN GOVERNMENT

There are numerous changes in government that resulted either directly or indirectly from the 9/11 attacks. These include the removal of the Taliban regime in Afghanistan; the electoral defeats of Pakistani President Pervez Musharraf and Spanish Prime Minister José María Aznar, whose cooperation with the United States in the Global War on Terror had made both leaders extremely unpopular;[140] fundamental changes in intelligence and law enforcement cooperation between many states, notably the United States and the Gulf States, where much of Al Qaeda's financial support originated; and lastly the concentration of power in the executive branches of many governments, including the imposition of martial and emergency law in many places.

CHANGES IN POLICY

The US declaration of the Global War on Terror was probably the single largest policy shift to come out of Al Qaeda's attacks on 9/11. Efforts by the international community to address the phenomenon of "failed states," whose governance structures have collapsed and left them unable to administer rules of law or basic public services,

---

[137] See Salon.com, October 1, 2008. "Almost everyone I spoke with assumed — whether true or not — that the United States backed the arbitrary arrest and unlawful rendition of men like Ishmael and the still-detained Kenyans . . . [W]hen US officials interrogate rendition victims who are being held incommunicado, the United States becomes complicit in the abuse. The US is funding the Ethiopian military, supporting its activities in Somalia, and training Kenyan security forces in counterterrorism — so as US-backed military and police forces in the region brutalize their domestic opponents in the name of fighting terrorism, the United States is often blamed."

[138] See Flynt Leverett, "Illusion and Reality," *The American Prospect*, accessed August 23, 2010, http://www.prospect.org/cs/articles?articleId=11859.

[139] See the American Political Science Association's task force report: "U.S. Standing in the World: Causes, Consequences, and the Future." http://www.apsanet.org/content_59477.cfm.

[140] Aznar's electoral defeat came just weeks after the Madrid train bombings—and his opponent pledged to withdraw Spanish troops from Iraq. Richard Wike, "Musharaff's Support Shrinks, Even as More Pakistanis Reject Terrorism . . . and the U.S.," *Pew Global Attitudes Project*, accessed August 23, 2010, http://pewresearch.org/pubs/561/pakistan-terrorism.

have also intensified because many observers believe these conditions create safe havens for the formation and operation of militant organizations.[141]

Physical security centered on transportation facilities and critical infrastructure has also been increased, especially as analysts continue to explore the possibility that Al Qaeda or a related group will seek to attack power plants, electrical grids, or nuclear sites.[142] Government surveillance of individuals, including electronic communications, has intensified since 9/11, and several programs in the United States have been scrapped because of pressure from civil activists.[143] The United States has also dramatically increased military aid and training assistance to foreign governments cooperating with the Global War on Terror initiatives, including many recipients that are accused of human rights violations, such as Pakistan, the Philippines, Indonesia, and Yemen. This policy trajectory, combined with highly publicized incidents of detainee abuse and death at numerous US facilities, including Guantanamo Bay detention camp in Cuba, Abu Ghraib prison in Iraq, and Bagram Theater Internment Facility in Afghanistan,[144] as well as the CIA's use of extraordinary rendition to transfer detainees to alternate sites where they were allegedly tortured,[145] has intensified

---

[141] Max Boot, "Pirates, Terrorism, and Failed States," *Wall Street Journal*, December 9, 2008, http://online.wsj.com/article/SB122869822798786931.html. For an opposing view, see Anna Simmons and David Tucker, "The Misleading Problem of Failed States: A 'Socio-Geography' of Terrorism in the Post-9/11 Era," *Third World Quarterly* 28, no. 2 (2007), 387–401. Simmons and Tucker argue that most international terrorists, and the organizations that support their activities, do not in fact come from failed states.

[142] John D. Moteff, *Critical Infrastructures: Background, Policy, and Implementation* (Congressional Research Service, 2011), accessed August 23, 2010, http://www.fas.org/sgp/crs/homesec/RL30153.pdf.

[143] Terminated programs include CAPPS II (Computer Assisted Passenger Prescreening System), which would have assigned individual travelers a color-coded risk rating based on personal information provided by the airlines, as well as Operation TIPS (Terrorism Information and Prevention System), which would have provided a direct link to relevant law enforcement agencies for those service workers, such as telephone repairmen, letter carriers, and truck drivers, who frequently enter individual homes and might observe suspicious activity. For an overview of the post-9/11 struggle between security and civil liberties, see Daniel B. Prieto, *War About Terror: Civil Liberties and National Security: After 9/11: A CFR Working Paper* (Council on Foreign Relations Press, 2009), accessed August 23, 2010, http://www.cfr.org/publication/18373/war_about_terror.html.

[144] See the 2004 CIA Office of Inspector General's Counterterrorism Detention and Interrogation Activities Report. Also see an American Civil Liberties Union (ACLU) compilation of autopsy reports prepared by US military doctors, acquired through a Freedom of Information Act request, which shows fourteen cases of homicide: http://action.aclu.org/torturefoia/released/102405/. Alternate sources, including reports by Human Rights First, claim that more than one hundred detainees died during interrogation, nearly half of which were homicides: http://www.humanrightsfirst.org/us_law/etn/dic/index.aspx.

[145] A 2004 CIA memo to the Department of Justice, obtained by the ACLU through a Freedom of Information Act request, details the rendition process: http://www.aclu.org/

scrutiny of the legality of US actions. The realization that a significant number of these detainees were apprehended on the basis of false information provided by local informants, either because of personal vendettas or in exchange for bounties, has not only caused political blowback but also probably complicated the collection of genuine intelligence.[146]

Most recently, the utilization of UAVs to assassinate individual Al Qaeda operatives demonstrates another significant policy shift. Although this tactic too has been criticized on both legal and ethical grounds,[147] it has also been highly successful, killing hundreds of militants without risking US casualties. Despite this high head count, some officials believe that the intelligence functions of the CIA, which operates the drone program, are being neglected in favor of what is fundamentally a military operation.[148]

CHANGES IN THE REVOLUTIONARY MOVEMENT

Although Al Qaeda's core leadership is probably still training operatives, recruiting, and engaging in strategic communication from a base in northwest Pakistan, the group's real power centers are instead affiliates active in places like Yemen and Somalia.[149] By some estimates, Al Qaeda has autonomous underground cells in some one hundred countries.[150] This apparent decentralization not only makes tracking affiliated militants and uncovering their operations more

---

torturefoia/released/082409/olcremand/2004olc97.pdf. Also see the 2007 report by the International Committee of the Red Cross: http://www.nybooks.com/icrc-report.pdf.

[146] Tom Lasseter, "Day 1: America's Prison for Terrorists Often Held the Wrong Men," *McClatchy Newspapers*, June 15, 2008, accessed August 23, 2010, http://www.mcclatchydc.com/2008/06/15/v-print/38773/day-1-americas-prison-for-terrorists.html. Stern and Weiner examine the impact of some precautionary measures taken by Western governments to protect against terrorism and suggest that more extensive risk analysis must be undertaken in planning counterterrorism operations so the negative consequences of incidents, such as false imprisonment, do not outweigh whatever positive progress is achieved. Jessica Stern and Jonathan B. Weiner, "Precaution Against Terrorism," *Journal of Risk Research* 9, no. 4 (2006), 393–447.

[147] The legal issue is whether UAV assassinations violate a 1976 Executive Order signed by President Ford banning US intelligence forces from engaging in assassinations; ethical issues revolve mostly around the rate of civilian casualties. See Jane Mayer, "The Predator War," *The New Yorker*, October 6, 2009, http://www.newyorker.com/reporting/2009/10/26/091026fa_fact_mayer.

[148] Siobhan Gorman and Jonathan Weisman, "Drones Kill Suspects in CIA Suicide Bombing," *Wall Street Journal*, March 18, 2010.

[149] Rollins, Wyler, and Rosen, *International Terrorism and Transnational Crime*. Notable affiliate groups include Al Qaeda in the Arabian Peninsula, Al Qaeda in Iraq, Al Qaeda in the Islamic Maghreb, Al Shabaab, and Abu Sayyaf.

[150] Bajoria and Bruno, "Al-Qaeda"; Rollins, Wyler, and Rosen (*International Terrorism and Transnational Crime*) provide a figure of seventy.

difficult, because there are no longer central "nodes" of individuals or institutions where significant activity is taking place, but also increases the probability that cells will recruit on their own. Yet, these cells probably do not possess the capability to carry out a "large, catastrophic operation" on the scale of 9/11.[151] Despite this dramatic change in the group's overall structure, its strategic objectives remain the same. Al Qaeda's recruitment strategies have also undergone significant changes. Intelligence reports that have appeared in media outlets suggest that Al Qaeda is recruiting in Western prisons, inner cities, among immigrant populations in the United States and Muslim converts in France, and, increasingly, among women.[152]

BIBLIOGRAPHY

Ahrari, Mohammed, and James Beal. *The New Great Game in Muslim Central Asia.* Washington, DC: National Defense University, 1996.

Ali, Ali Abdel Gadir. *Globalization and Inequality in the Arab Region.* Kuwait: Arab Planning Institute, 2003.

Bajoria, Jayshree, and Greg Bruno. "Al-Qaeda (a.k.a. Al-Qaida, Al-Qa'ida)." Council on Foreign Relations. Accessed August 23, 2010. http://www.cfr.org/publication/9126/alqaeda_aka_alqaida_alqaida.html#p1.

Bergen, Peter L. *Holy War, Inc.: Inside the Secret World of Osama bin Laden.* New York: Free Press, 2001.

bin Laden, Osama. "Open Letter to King Fahd." Accessed August 23, 2010. http://en.wikisource.org/wiki/An_Open_Letter_to_King_Fahd_on_the_Occasion_of_the_Recent_Cabinet_Reshuffle.

Boot, Max. "Pirates, Terrorism, and Failed States." *Wall Street Journal,* December 9, 2008. http://online.wsj.com/article/SB122869822798786931.html.

Brachman, Jarret M. *Global Jihadism: Theory and Practice.* Case Series on Political Violence. New York: Taylor & Francis, 2008.

Burke, Jason. *Al-Qaeda: Casting a Shadow of Terror.* New York: IB Tauris & Co LTD, 2003.

Coll, Steve. *Ghost Wars: The Secret History of the CIA, Afghanistan, and Bin Laden from the Soviet Invasion Until September 10, 2001.* New York: Penguin, 2004.

Defferios, Jeff. "Youth Unemployment: Mideast 'Ticking Time Bomb.'" *CNN,* March 10, 2010. http://www.cnn.com/2010/WORLD/meast/03/12/bahrain.youth.unemployment/index.html.

---

[151] Ibid.

[152] Jessica Stern, "When Bombers Are Women," *Washington Post.* December 18, 2003.

Duyvesteyn, Isabelle. "How New Is the New Terrorism?" *Studies in Conflict and Terrorism* 27, no. 5 (2004): 439–454.

Eikmeier, Dale C. "Qutbism: An Ideology of Islamic-Fascism," *Parameters* 37 (Spring 2007): 85.

Foreign Broadcast Information Services. "Compilation of Usama Bin Ladin Statements 1994 - January 2004." Accessed August 23, 2010. www.fas.org/irp/world/para/ubl-fbis.pdf.

Gerges, Fawaz A. *The Far Enemy: Why Jihad Went Global.* New York: Cambridge University Press, 2005.

Gerwehr, Scott, and Sara Daly. "Al-Qaida: Terrorist Selection and Recruitment." In *The McGraw Hill Security Handbook,* edited by David G. Kamien. New York: McGraw Hill, 2006.

Gorman, Siobhan, and Jonathan Weisman. "Drones Kill Suspects in CIA Suicide Bombing." *Wall Street Journal,* March 18, 2010.

Gunaratna, Rohan. *Inside Al Qaeda: Global Network of Terror.* New York: Berkley Books, 2003.

Kepel, Gilles. *Jihad: The Trail of Political Islam.* Cambridge: Harvard University Press, 2002.

Kerry, John F. "Tora Bora Revisited: How We Failed to Get Bin Laden and Why It Matters Today." A Report to Members of the Committee on Foreign Relations United States Senate, 111th Congress, 1st Session. Washington, DC: US Government Printing Office, 2009.

Kiser, Steve. "Financing Terror: An Analysis and Simulation to Affect Al Qaeda's Financial Infrastructures." PhD diss., Pardee RAND Graduate School, 2005.

Krueger, Alan B., and David Laitin. "Faulty Terror Report Card." *Washington Post,* May 17, 2004.

Laqueur, Walter. *No End to War: Terrorism in the Twenty-First Century.* New York: Continuum, 2003.

Lasseter, Tom. "Day 1: America's Prison for Terrorists Often Held the Wrong Men." *McClatchy Newspapers,* June 15, 2008.

Leverett, Flynt. "Illusion and Reality." *The American Prospect,* August 23, 2006. http://www.prospect.org/cs/articles?articleId=11859.

Mayer, Jane. "The Predator War." *The New Yorker,* October 6, 2009. http://www.newyorker.com/reporting/2009/10/26/091026fa_fact_mayer.

Moteff, John D. *Critical Infrastructures: Background, Policy, and Implementation.* Congressional Research Service, 2011. Accessed August 23, 2010. http://www.fas.org/sgp/crs/homesec/RL30153.pdf.

Musharbash, Yassin. "The Future of Terrorism: What Al Qaeda really Wants." *Speigel Online,* December 8, 2005.

National Commission on Terrorist Attacks upon the United States. *The 9/11 Commission Report: Final Report on the National Commission on Terrorist Attacks upon the United States.* New York: W. W. Norton, 2004.

Neumann, Peter R. *Joining Al-Qaeda: Jihadist Recruitment in Europe.* Adelphi Series. New York: Routledge for the International Institute for Strategic Studies, 2008.

Pargeter, Allison. "North African Immigrants in Europe and Political Violence." *Studies in Conflict and Terrorism* 29, no. 8 (2006): 731–737.

Prieto, Daniel B. *War About Terror: Civil Liberties and National Security: After 9/11: A CFR Working Paper.* Council on Foreign Relations Press, 2009. Accessed August 23, 2010. http://www.cfr.org/publication/18373/war_about_terror.htm.

Rabasa, Angel, Peter Chalk, Kim Cragin, Sara A. Daly, Heather S. Gregg, Theodore W. Karasik, Kevin A. O'Brien, and William Rosenau. *Beyond Al-Qaeda: Part I: The Global Jihadist Movement.* Santa Monica, CA: RAND Corporation, 2006.

Rollins, John. *Al Qaeda and Affiliates: Historical Perspective, Global Presence, and Implications for US Policy.* Congressional Research Service, 2010.

Rollins, John, Liana Sun Wyler, and Seth Rosen. *International Terrorism and Transnational Crime: Security Threats, US Policy, and Considerations for Congress.* Washington, DC: Congressional Research Service, 2010.

Sageman, Marc. *Understanding Terror Networks.* Philadelphia: University of Pennsylvania Press, 2004.

Scheuer, Michael. *Through Our Enemies' Eyes: Osama bin Laden, Radical Islam, and the Future of America.* Washington, DC: Potomac Books, Inc., 2007.

Shahin, Sultan. "Deprogamming Jihadists: A Fascinating Saudi Endeavor." *NewAgeIslam.com,* November 12, 2008.

Simmons, Anna, and David Tucker. "The Misleading Problem of Failed States: A 'Socio-Geography' of Terrorism in the Post-9/11 Era." *Third World Quarterly* 28, no. 2 (2007): 387–401.

Stern, Jessica. "When Bombers Are Women." *Washington Post,* December 18, 2003.

Stern, Jessica, and Jonathan B. Weiner. "Precaution Against Terrorism." *Journal of Risk Research* 9, no. 4 (2006): 393–447.

Wardlaw, Grant. *Political Terrorism: Theories, Tactics, and Countermeasures.* New York: University of Cambridge, 1989.

Wike, Richard. "Musharaff's Support Shrinks, Even as More Pakistanis Reject Terrorism . . . and the U.S." *Pew Global Attitudes Project*, November 7, 2007. http://pewresearch.org/pubs/561/pakistan-terrorism.

Williamson, Myra. *Terrorism, War, and International Law: The Legality and the Use of Force Against Afghanistan in 2001.* Burlington, VT: Ashgate Publishing, 2009.

World Islamic Front. "Jihad Against Jews and Crusaders." Accessed August 23, 2010. http://www.fas.org/irp/world/para/docs/980223-fatwa.htm.

Wright, Lawrence. *The Looming Tower: Al-Qaeda and the Road to 9/11.* New York: Knopf, 2006.

al-Zayyat, Montasser. *The Road to Al-Qaeda: The Story of Bin Laden's Right-Hand Man.* Critical Studies on Islam. London: Pluto Press, 2004.

SECTION V

....

REVOLUTION FOR
MODERNIZATION OR REFORM

GENERAL DISCUSSION

The last category is the most loosely bound in terms of being a "type" of insurgency or revolution. However, we do believe that there are elements within the included case studies that point to important trends and aspects that are not covered elsewhere within this volume, and we believe that we can make some general observations about what may be triggering these distinctions. The broad umbrella terms of "modernization" and "reform" are used to denote both the motivations of these cases as well as a variance from the strict notions of "classical insurgencies," where arms are lifted and violence ensues for the pure purpose of bringing about a change in the governmental system. Two of the four cases in this section explore how the reform of the government may be more of a side objective rather than the main one, and the remaining cases are ones in which nonviolent means were chosen to render a revolution of government. Strictly speaking, these cases would not have been included in the original Special Operations Research Office (SORO) Casebook.

In most of the included cases, the weakening power of the state heavily contributes to the fomenting of the movements, primarily through the disintegration of services and elements of power. In these cases, however, instead of a mature ideological alternative being available for revolution, looser or nonpolitical factors contribute to the growth of a movement. In two cases, these factors are economic concerns and a growth in criminal elements shaping the new landscape. In the case of Solidarity, a labor union movement combined with the weakening of both the internal and external communist structure led to an eventual overthrow of the government, even though this was not the original motivation of the union.

This looser style of revolutionary adherence affects the strength of the popular support as well, giving rise to a more fluid support structure and greater flexibility in terms of enticing occasional support. As has been seen with other mass-motivated revolutions, critical mass of popular support combined with a weakened government can push the revolution quickly from low-level operations to victory, often without clear indicators when it is happening. The velvet "Orange" revolution in the Ukraine is used to show how quickly massive support can lead to dramatic results.

MOVEMENT FOR THE EMANCIPATION OF THE NIGER DELTA (MEND)

Jerry Conley

SYNOPSIS

The root causes of the Niger Delta insurgency are well known. Violence, underdevelopment, environmental damage, and failure to establish credible state and local government institutions have contributed to mounting public frustration at the slow pace of change under the country's nascent democracy, which is dogged by endemic corruption and misadministration inherited from its military predecessors.[1] In response to the public's growing resentment of the government, in 2006 the Movement for the Emancipation of the Niger Delta (MEND) formed as an umbrella organization representing several independent militias to fight for "total control" of the Niger Delta's oil wealth and to reverse the exploitation of the region's natural resources by foreign companies and people from other parts of Nigeria.[2] Feeding on the impoverished state of the Nigerian public, MEND recruited mostly less-educated young males, offering them an ability to protect their community and ethnic groups, an opportunity to fight oil companies and the government over political and economic marginalization, and financial gain, among other reasons. Through a combination of kidnappings, pipeline and other bombs, and other serious criminal activities, MEND's struggle continued into 2010 in order to establish a redistribution of Nigeria's oil wealth to the public.

TIMELINE

1939	Britain restructures Nigeria along three main regions that align with the three largest ethnic groups, resulting in the political isolation of the Ijaw, the fourth largest ethnic block in Nigeria.
October 1, 1960	Nigeria gains its independence from the United Kingdom.

---

[1] International Crisis Group, "The Swamps of Insurgency: Nigeria's Delta Unrest," *Africa Report No. 115*, August 3, 2006, i.

[2] "Nigeria's Shadowy Oil Rebels," *BBC News*, April 20, 2006. This complaint about exploitation of the Niger Delta by people from other parts of the country would continue with the March 2010 bombing of the Warri post-amnesty dialogue. See Abdullahi Yahaya Bello, Muhammad Bello, and Monday Osayande, "Why Bombs Returned to Niger Delta," *Daily Trust Online*, March 19, 2010, http://www.news.dailytrust.com.

1960–1973	Rapid expansion of the oil industry in Nigeria increases oil production from 5 million barrels per year to 600 million barrels per year.
1963–1999	Nigeria has eleven changes of government (seven by coup).
February 1966	Isaac Adaka Boro leads the first significant Ijaw armed insurrection in the Niger Delta against the federal government.
February 1999	End of military rule in Nigeria enables the rise of security forces that are controlled by local politicians and business interests.
1999 and 2003	Nationwide elections provide a venue for new partnerships among political, business, criminal, and militant groups.
September 2004	The Niger Delta Peoples Volunteer Force begins "Operation Locust Feast" against the oil industry in the Niger Delta.
Late 2005	A series of militant meetings in Delta State leads to the creation of MEND and the coordination of oil bunkering, the purchase of advanced weaponry, and the targeting of oil infrastructure.
December 2005	MEND takes credit for its first attack (on the Royal Dutch Shell's Opodo pipeline in Delta State). This attack is followed a month later by attacks on an oil flow station and the kidnapping of four foreign oil workers from Shell's offshore E.A. oil field.
April 2006	Oil production in the Niger Delta is reported down by 25%.
2006–2007	MEND conducts dozens of attacks and kidnappings against the oil industry and foreign oil workers in the Niger Delta.
June 20, 2008	MEND attacks the Shell-operated Bonga oil platform that is 120 kilometers offshore, shutting down 10% of Nigeria's oil production.
September 2008	MEND announces that it is beginning an all-out "oil war" in the Niger Delta and targets key oil infrastructure. A week later, MEND announces a cease-fire due to pressure from Ijaw leaders.
May 15, 2009	Joint Task Force commences major military operations in the Niger Delta against MEND camps and villages.

July 11, 2009	MEND attacks the Atlas Cove Jetty in Lagos, a major oil hub.
October 25, 2009	MEND announces a unilateral truce and accepts the government proposal for reintegration.
March 15, 2010	Two bombs explode outside of a Nigerian government complex in the city of Warri at the beginning of a high-level dialogue on post-cease-fire planning for militants from the Niger Delta.
May 5, 2010	President Yar'Adua dies after a period of failing health, putting the Niger Delta peace negotiations in jeopardy.

THE ENVIRONMENT OF THE REVOLUTION

PHYSICAL ENVIRONMENT

The name "Nigeria" was first used in 1897 by the editor of *The Times* to describe to readers the combined region of the Niger River and its surrounding "area" in West Africa. This name was adopted seventeen years later by Frederick Lugard, the husband of the editor, when he became the first governor general of the British colony that covered this region.[3] Located on the Gulf of Guinea, Nigeria borders Benin to the west, Niger to the north, Chad to the northeast, and Cameroon to the east. With a total size of 923,768 square kilometers, Nigeria is approximately twice the size of California and the thirty-second largest country in the world.[4] It has five major geographical zones: the Niger-Benue River Valley; a stepped plateau along the northern border; a mountainous region in the east; a low coastal zone to the southwest; and hills and low plateaus north of this coastal zone. Nigeria experiences an arid climate in the northern regions, a tropical climate in the central area of the country, and an equatorial climate to the south. The Benue tributary flows approximately 1,400 kilometers from Cameroon in the east and empties into the 4,000-kilometer-long Niger River, which traverses Guinea, Mali, Niger, and Benin and is the longest river system in West Africa. The second major river system in the country is the Yobe River, which passes along the northeastern border with Niger and empties into Lake Chad. The highest point in Nigeria is Chappal Waddi in the eastern mountain range.[5]

---

[3] International Crisis Group, "Nigeria: Want in the Midst of Plenty," *Africa Report No. 113*, July 19, 2006.

[4] Library of Congress, *Country Profile: Nigeria*, 2008, accessed January 14, 2010, http://memory.loc.gov/frd/cs/profiles/Nigeria.pdf; Central Intelligence Agency, "Nigeria," *The World Factbook*, accessed February 4, 2010, https://www.cia.gov/library/publications/the-world-factbook/geos/ni.html.

[5] Ibid.

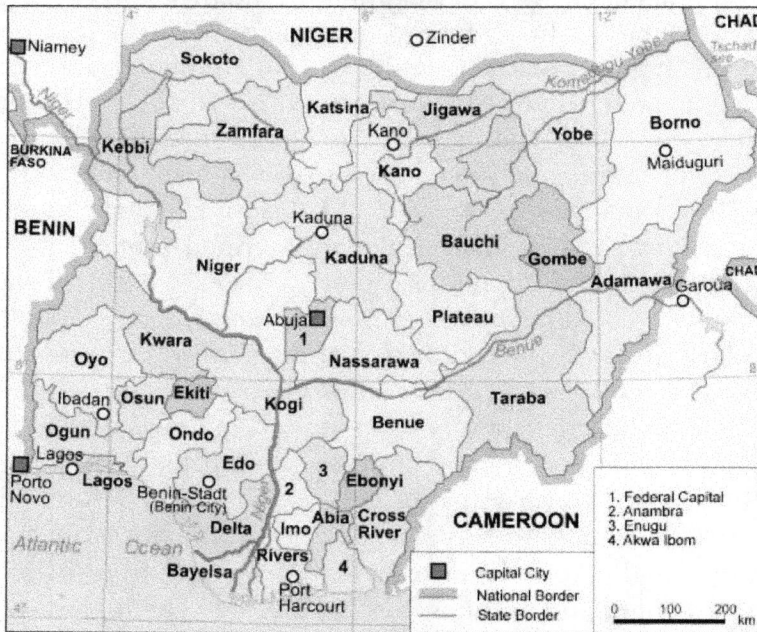

Figure 1. Maps of Nigeria and its states.[6]

---

[6] (Top) Reproduced with permission from Jennifer M. Hazen and Jonas Horner, "Small Arms, Armed Violence, and Insecurity in Nigeria: The Niger Delta in Perspective," *Small Arms Survey*, Occasional Paper 20 (Geneva, 2007), xvi (© Small Arms Survey/ MAP*grafix*); (bottom) "File:Nigeria political.png," *Wikipedia*, accessed March 14, 2011, http://en.wikipedia.org/wiki/File:Nigeria_political.png.

Nigeria has approximately 193,200 kilometers of roads, of which 32,000 kilometers fall within the federal roads network. A vast majority of these roads are in poor condition, and in 2004, the Federal Roads Maintenance Agency (FERMA) began a major repair effort for the federal roads. As part of a twenty-year strategic plan to improve its 3,505-kilometer narrow-gauge rail system, Nigeria signed an $8.3 billion contract in November 2006 with the China Civil Engineering and Construction Corporation. This plan included the construction of 8,000 kilometers of standard-gauge rail lines to include a link between Lagos and Kano in the North and a possible link between Port Harcourt and Jos.[7] The rivers and waterways in Nigeria serve as a major transportation link for commerce within the country, and there are approximately 8,600 kilometers of inland waterways.

CULTURAL AND DEMOGRAPHIC ENVIRONMENT

The overall population of Nigeria is estimated at 149 million people, making it the eighth most populous country in the world and the most populated country on the African continent, and resulting in a ratio of one of every six Africans being Nigerian. Approximately 42% of the population is fourteen years old or younger, with another 55% between the ages of fifteen and sixty-four, leaving only 3% of the population above the age of sixty-four. The annual population growth rate for Nigeria is 1.99%, with approximately 48% of the population living in urban centers[8] and an unusually high urbanization rate of 5.3% per year.[9] Nigeria's poor health care system and living conditions contribute to an average life expectancy of only 47.8 years and an infant mortality rate of 93.93 deaths per 1,000 live births, or almost one in ten deaths per births.[10]

Nigeria has more than 250 ethno-linguistic groups, with four of these groups comprising major ethno-regional clusters in the country and representing 70% of the population—the Hausa-Fulani in the north (29%), the Yoruba in the southwest (21%), the Igbo or Ibo in the southeast (10%), and the Ijaw in the Niger Delta (10%).[11] Of these four, rivalries among the Hausa-Fulani, the Yoruba, and the Igbo

---

[7] *Country Profile: Nigeria*

[8] Central Intelligence Agency, "Nigeria," *The World Factbook*, accessed March 4, 2010.

[9] *Country Profile: Nigeria.*

[10] Ibid. Inadequate infrastructure is a key reason for poor health conditions in Nigeria, with access to safe drinking water available to only 72% of urban residents and 49% of rural residents and access to adequate sanitation available to only 48% of urban residents and 30% of rural residents (statistics from 2002).

[11] Central Intelligence Agency, "Nigeria," *The World Factbook*, accessed March 4, 2010; "Nigeria: Want in the Midst of Plenty," *Africa Report No. 113*; *Country Profile: Nigeria.*

have dominated postcolonial politics within Nigeria and have served as a major source of instability and ethnic violence. At the center of this ethnic tension is the concept of "indigeneity" in which natives of a subregion of the country discriminate against "outsiders" and ethnic groups from other parts of the country.[12] Nigeria's population is approximately 50% Islamic and 40% Christian, with the Muslim population mostly in the north and Christians living in the south. The official language of Nigeria is English.

SOCIOECONOMIC ENVIRONMENT

Nigeria's socioeconomic challenges were best described in July 1975 by its military dictator, General Yakubu Gowon: "Want in the midst of plenty."[13] Blessed with an abundance of natural resources, Nigeria has struggled to establish an equitable and sustainable economy and political system that ensures that these resources are properly accessed, maintained, and distributed to meet the needs of its rapidly growing population. The central dynamic in these socioeconomic factors is Nigeria's robust natural gas and petroleum reserves, estimated at 182 trillion cubic feet (seventh in the world) and 36.2 billion barrels (eighth in the world), respectively. In the 1970s, national policies led to Nigeria's economy becoming completely dependent on its oil and natural gas sector to the detriment of other sectors, such as agriculture, which used to be the foundation of the economy.[14] It has been suggested that "Oil-related rents (royalties, taxes, oil export earnings, interests on joint venture investments, etc.) are the lifeblood of Nigeria's economy."[15] Despite the vast financial potential offered by the export of petroleum, corruption permeates this entire sector in Nigeria and results in 80% of oil revenues benefiting only 1% of the population, with approximately 55% of the population living on less

---

[12] This type of ethnic violence is at the heart of the ongoing massacres in northern Nigeria around the city of Jos and elsewhere. Portrayed in the media as religious conflicts between Christians and Muslims, religion is just one aspect of the ethnic identity of the competing factions.

[13] "Nigeria: Want in the Midst of Plenty," *Africa Report No. 113*. Economists have also used the phrase "paradox of plenty" for this phenomena, while activists in Africa often use the phrase the "curse of oil." *Country Profile: Nigeria*.

[14] An interesting divergence from this trend occurred in 2008 when the agricultural sector was 42% of Nigeria's gross domestic product (GDP) because of the insurgency-related decrease in oil production in the Niger Delta. World Bank, *Nigeria: Country Brief*, 2009, accessed February 4, 2010, http://go.worldbank.org/FIIOT240K0. Irrelevant of its share of GDP, agriculture employs approximately 70% of Nigeria's workforce.

[15] Kenneth Omeje, "Oil Conflict and Accumulation Politics in Nigeria," *ECSP Report*, Issue 12, 44–49, http://www.wilsoncenter.org/topics/pubs/Omeje12.pdf.

than one dollar per day. The 2009 Human Development Index (HDI) by the United Nations ranked Nigeria 158 out of 182 countries.[16]

Described by its own National Planning Commission in 2004 as "dysfunctional," Nigeria's education system is government-funded but not compulsory at any level. Attendance is highest at the primary school level of education (grades 1–6), with 59% of girls and 68% of boys enrolled, but drops to 23% for girls and 28% for boys after the completion of junior secondary school (grades 7–9) when the students are enrolled in secondary school (grades 10–12). For those Nigerians who do pursue a university education, a 2007 study found that 40% of undergraduate students "were more interested in leaving Nigeria as a way of gaining social status than in seeking gainful employment at home."[17] Overall, the adult literacy rate in Nigeria is 69.1%, with 78.2% of men and 60.1 % of women considered literate.[18]

Another significant socioeconomic challenge for Nigeria involves the tremendous amount of environmental degradation that has occurred in the country over the last three decades. The rapid growth of Nigeria's urban population and its overall industrial expansion have led to a significant waste-management crisis linked to the dumping of waste, open pit burning, improper construction of landfills, automobile emissions, oil spills, natural gas flaring, etc.[19] As discussed below, environmental degradation in the Niger Delta due to the petroleum sector is a key catalyst for the emergence of numerous insurgent groups in the region.[20] In addition, deforestation has emerged as a major issue across Nigeria, with the land area covered by forests reduced by almost 50% over the last twenty years.[21]

A microcosm of Nigeria's national socioeconomic ills can be found within the Niger Delta region of the country. With approximately 20 million people and twenty ethno-linguistic groups, the Niger Delta is the center of Nigeria's petroleum sector, accounting for 90% of the country's oil and gas export earnings and approximately 70% of federal revenues generated by taxes and fees.[22] The region is

---

[16] Factors included in the HDI ranking are life expectancy, access to health services, access to safe water, access to sanitation, adult literacy, gross national product (GNP) per capita, Real GNP per capita, and calorie supply as a percentage of daily requirements.

[17] *Human Development Report 2009* (New York: United Nations Development Programme, 2009).

[18] *Nigeria: Country Brief; Country Profile: Nigeria.*

[19] Ibid.

[20] Because of their large onshore and nearshore operations, Shell and Chevron Texaco are considered to be some of the primary causes of environmental degradation in the Niger Delta and thus are often the targets of insurgent attacks and kidnapping. Omeje, *Oil Conflict and Accumulation Politics in Nigeria.*

[21] *Country Profile: Nigeria*

[22] Omeje, "Oil Conflict and Accumulation Politics in Nigeria."

composed of nine of Nigeria's thirty-six states, covers approximately 5,600 square miles, and is bounded by the Benin River in the west, the Imo River in the east, and Atlantic Ocean to the south.[23] A large majority of this region is swamp and inundated terrain with a belt of mangrove trees up to 10 kilometers deep along the Atlantic Ocean. The region experiences approximately 234 days a year of significant cloud coverage and 100 days a year with thunderstorms. The two main urban centers are Warri in the west and Port Harcourt in the east, with 1,600 additional autonomous communities spread across the region.[24]

HISTORICAL FACTORS

Before the arrival of Portuguese explorers in 1471, and subsequently Dutch and British traders, the region that now comprises Nigeria was a series of diverse kingdoms and empires that were based on ethnic and tribal lines, with some dating back to the first millennium.[25] Islam had also spread throughout the northern region of Nigeria, with much of this area and neighboring parts of Niger and Cameroon being consolidated under a single Islamic government—the Sokoto Caliphate—after the holy wars of 1804–1808.[26] The establishment of permanent European trading posts and ports coincided with the commencement of a slave trade[27] in the region and the introduction of formal government, education, and economic practices, as well as Christianity, to the ethnic groups with whom they traded—primarily in the south. These European influences exacerbated a north–south divide within the region that was codified under the British Empire.

In 1885, Great Britain declared West Africa as a British "sphere of interest"; Northern and Southern Nigeria were established as two British protectorates in 1901 and then as a unified colony and protectorate in 1914. Despite this amalgamation of Northern and Southern Nigeria, the British authorities continued to govern these two regions in distinctly different manners, with "indirect rule" being used in the north to empower local political and religious institutions that

---

[23] Brian Lionberger, "Emerging Requirements for U.S. Counterinsurgency: An Examination of the Insurgency in the Niger Delta Region" (US Army Command and General Staff College, 2007).

[24] Ibid.

[25] The precolonial kingdoms included the Kano, Katsina, Fulani, and Hausa in northern Nigeria; the Nok in central Nigeria; and Ife, Oyo, Nri, and Benin in southern and southwest Nigeria.

[26] *Country Profile: Nigeria.*

[27] In 1916, the British passed the Slavery Ordinance, which outlawed slavery and slave trading in Nigeria.

favored conservative institutions to the detriment of social progress.[28] In an attempt to address the imbalanced distribution of power among Nigeria's ethnic groups and also streamline trading practices—not necessarily complimentary tasks—the British restructured Nigeria as three regions in 1939, with the Northern Region dominated by the Hausa-Fulani and the cash crops of cotton and groundnuts, the Western Region dominated by the Yoruba and the trade of cocoa, and the Eastern Region dominated by the Igbo and the production of palm oil. With these three ethnic blocks representing approximately two-thirds of Nigeria's population and agriculture accounting for more than 64% of the nation's gross domestic product (GDP) during the 1950s, a power-sharing environment emerged for the allocation of political power and control of Nigeria's limited wealth.[29] This delicate balance would be undermined during the 1960s, however, with the granting of national independence and the emergence of Nigeria's oil wealth.

On October 1, 1960, Nigeria became an independent country within the British Commonwealth. Despite the creation and updating of its constitution in order to promote democratic institutions and practices, the next four decades would be dominated by military rule and the changing of governments initiated through coups rather than elections. The artificial attempt to pull together 250 ethno-linguistic groups within one Nigerian national identity failed to materialize during this period and was further hampered by the tremendous influx of oil wealth into the country, which raised the stakes for those in positions of power. Between 1960 and 1973, oil output increased from 5 million barrels to more than 600 million barrels a year. This led to consequential government oil-revenue increases from $250 million in 1970 to $2.1 billion in 1972 and $11.2 billion by 1974. However, the massive influx of revenue into the government accounts during this period did not trickle down into improved quality of life for average

---

[28] "Nigeria: Want in the Midst of Plenty," *Africa Report No. 113.*
[29] Ibid. The emergence of these three strong ethnic and political blocks became a major topic of concern for minority ethnic groups in Nigeria and led the British government in 1957 to commission a panel led by Harry Willink, the former Vice Chancellor of Cambridge University, to explore this topic. See http://nigerianwiki.com/wiki/Willink_Commission.

citizens, as per capita income fell from $250 in 1965 to $212 in 2004 and the percentage of Nigerians living on less than one dollar per day increased from 36% in 1970 to 70% in 2000. Finally, the magnitude of petro-corruption within Nigeria was best underscored by a World Bank estimate that at least $100 billion of Nigeria's $400 billion in oil revenues from 1970 until 2010 have "gone missing."[30]

As the center of Nigeria's oil industry,[31] the Niger Delta has experienced a significant degree of postindependence conflict over ethnic tensions, land-ownership rights, the struggle for political representation, the control of natural resources, and the associated protection of the environment. Although the Nigerian constitution guarantees the civil rights of all citizens, it also provides for "indigenes" rights that allow for preferential treatment of members of ethnic groups that are considered to be the original settlers of an area.[32] For minority ethnic groups in a region, this "indigenes" clause places the burden of proof on them to convince the majority ruling ethnic elite to grant certain local status and rights to the minority. In the Niger Delta, where ownership and control of certain areas and townships can translate directly into petroleum rents,[33] there is no incentive for recognizing special status for minority groups. In particular, the Ijaw, which is the fourth-largest ethnic group in Nigeria with approximately 10% of the population, became a vocal critic of the three-region structure established by the British because it placed the Ijaw as a minority in both the Western Region and Eastern Region. Moreover, this minority status continued into the postcolonial period, with the Ijaw finding that it had minority status in both state and local areas, to include the city of Warri, because of indigeneity claims by two other ethnic groups—the Itsekiri and the Urhobo.[34] This minority status for the Ijaw resulted in denied access to "political appointments and

---

[30] Michael Watts, "Petro-Insurgency or Criminal Syndicate? Conflict & Violence in the Niger Delta," *Review of African Political Economy* 114 (2007): 641; "Nigeria: Want in the Midst of Plenty," *Africa Report No. 113*, 7.

[31] Ibid., 639. With a population of approximately 28 million people, the Niger Delta has ten export terminals, 275 flow stations, ten gas plants, four refineries, 606 oil fields, 5,284 wells, and 7,000 kilometers of pipeline.

[32] Michael Watts, ed., *Curse of the Black Gold: 50 years of Oil in the Niger Delta* (Brooklyn, NY, PowerHouse Books, 2009): 40–42.

[33] In 1978, the Land Use Decree was issued by the Obasanjo administration and stated that all subsoil minerals—to include oil—belonged to all of the people of Nigeria and not just those people who lived on the land. This allowed greater federal control of the oil resources in the Niger Delta but led to the current crisis over revenue sharing between federal and state and between state and local.

[34] Ukoha Ukiwo, "From 'Pirates' to 'Militants': A Historical Perspective on Anti-State and Anti-Oil Company Mobilization Among the Ijaw of Warri, Western Niger Delta," *African Affairs*, 106, no. 425 (2007): 591. Warri was in fact the Itsekiri homeland, but Ijaw requests for the establishment of autonomous divisions within the Warri area were denied.

public service employment, employment and contract opportunities in the oil industry, and status of traditional rulers."[35]

This exploitation of the resources of Niger Delta as well as the perceived repression of the Ijaw people led to the first significant armed insurrection in the Niger Delta against the federal government. Led by Isaac Adaka Boro, who formed an Ijaw group called the Niger Delta Volunteer Force (NDVF), an independent republic was declared on February 23, 1966, but federal forces quickly put down this uprising in twelve days.[36] Although unsuccessful in achieving autonomy or independence for the Ijaw or the Niger Delta, Boro's actions are still cited today by insurgent leaders in the region as inspiration for their ongoing struggle.[37]

With the end of military rule in Nigeria in 1999, an unexpected trend emerged in which the strong security institutions that were employed by the military to enforce stability across the country quickly gave way to regional, state, and local security forces that were controlled by politicians and business interests.[38] In many cases, the private militias were used to control or remove political opponents and often were structured along ethnic lines, with a patronage relationship between the political and business leaders and the leadership of the militia force.[39] The militias played a prominent role in the 1999, 2003, and 2007 Nigerian elections. The local militia groups also began to collect security and/or consulting fees from oil firms as a means of ensuring the safety and protection of the firms' facilities and employees in the area, including the prevention of oil

---

[35] Ibid., 595. As a specific example, since the establishment of the Warri Local Government Area (LGA) in 1976, only Itsekiri have been elected chairman of the LGA.

[36] A short autobiography by Boro can be found online. See Major Isaac Adaka Boro, *The Twelve Day Revolution.* http://www.adakaboro.org. Boro would be granted amnesty by the Nigerian government and join the Nigerian army as a major, where he would ironically die while fighting in support of the federal government against the secession Biafran movement, although some have questioned the circumstances of his 1968 death.

[37] Judith Burdin Asuni, *Understanding the Armed Groups of the Niger Delta* (New York, Council on Foreign Relations, 2009), 5.

[38] An example of the military leadership using force to quell political and ethnic violence can be seen in Operation Weite in January 1966, which involved a series of political assassinations as well as the Biafra War from 1966 to 1970 that involved an attempted secession from Nigeria and resulted in more than 2 million deaths due to conflict, disease, and starvation. It should also be underscored, however, that most of the military coups from 1963 to 1993 had some element of ethnic struggle in them as a catalyst. "Nigeria: Want in the Midst of Plenty," *Africa Report No. 113,* 6. The Nigerian government was also highly criticized for the 1995 trial and execution of Ken Saro-Wiwa, who was the leader of the Movement for the Survival of the Ogoni People (MOSOP), which employed peaceful means to protest environmental damage and abject poverty related to the oil industry. Saro-Wiwa considered his activism to be a peaceful extension of the 1966 revolt by Boro.

[39] Hazen and Horner, "Small Arms, Armed Violence," 11, accessed February 4, 2010, http://www.smallarmssurvey.org/files/sas/publications/o_papers.html.

bunkering activities.[40] Within a few years, however, the composition and objectives of these militia groups began to vary, with some becoming more independent of their political patrons and more focused on the tremendous financial rewards and power that could be gained through criminal activities. This has produced four general categories of armed groups now operating in the Niger Delta region: ethnic militias, criminal gangs, vigilante groups, and confraternity/cult groups (see the following table).[41]

	Ethnic Militias	Criminal Gangs	Vigilante Groups	Confraternities/Cults
Purpose	Aims are to redress grievances and injustices and protect and defend the rights of the ethnic group	Economic gain	Provide security to communities; provide law-and-order services in areas in which police presence is minimal; provide economic opportunities to members	Self-enrichment and defending territory
Membership	Ethnic group; other sympathetic ethnic groups	Unemployed youth	Community organization	Confraternities; students; cults; unemployed youth
Support Base	Typically grassroots organizations receiving widespread support; able to mobilize more widely	Members; politicians	Community support; community funding through dues; many receive government support	Members; alliances with other armed groups; politicians

[40] Bunkering involves the theft of oil by tapping into pipelines and pumping the oil into barges for delivery and sale to ships waiting out at sea. A complex and international trafficking effort, bunkering costs the oil companies and Nigeria billions of dollars a year in lost revenues. In many cases, the oil companies were paying a "security fee" to militias as a bribe to keep the same militia group from stealing from the oil company's pipelines.

[41] For a more detailed discussion on ethnic militias, see Austine Ikelegbe, "State, Ethnic Militias, and Conflict in Nigeria," *Canadian Journal of African Studies* 39, no. 3 (2005): 490–516. Nigerian confraternities were similar to those same organizations that existed in Europe in the eighth century and conducted charitable work and worship then slowly transitioned toward political influence, the disciplining of members, and the establishment of educational institutions. Rather than being affiliated with churches, however, Nigerian confraternities were like US fraternities and were linked to universities. As such, they became embroiled in the violence that engulfed many Nigerian universities in the 1970s, and some eventually became surrogate armed groups for the Nigerian military and even controlled drug trading in their territory. Three well-known confraternities were the Supreme Vikings, Black Axe, and the Klansmen Konfraternity. Stephen Davis, *The Potential for Peace and Reconciliation in the Niger Delta* (Coventry, UK: Coventry Cathedral, 2009), accessed March 19, 2010, http://www.adakaboro.org/ndmiscreports/doc_download/31-the-potential-for-peace-and-reconciliation-in-the-niger-delta; Asuni, "Understanding the Armed Groups of the Niger Delta," 8–10; Hazen and Horner, "Small Arms, Armed Violence," 74–75.

	Ethnic Militias	Criminal Gangs	Vigilante Groups	Confraternities/ Cults
Area of Operations	Communities of ethnic groups; also across states where ethnic group is dominant	Dominate particular neighbor-hoods; local-ized area of operations	Localized area of operations, usually at community level	Confraternities tend to be on campus, similar to US fraternities; cults operate off campus, tend to be more violent of the two; localized area of operations
Main Activities	Defense of eth-nic group rights, might include: political protests, attacks on politi-cians, attacks on oil pipelines, kidnapping, oil bunkering	Engage in armed rob-bery and other crimi-nal activity	Activities aimed at community security; some-times administer physical punish-ments to suspects or take law into own hands; some work with police to enforce law and order	Control and de-fend territory; drug trafficking; oil bunkering; reputation for being brutal and secretive with elaborate rituals for initiation
Arms	Paramilitary groups; of all armed groups, best trained, armed, organized; usually armed with sophisticated weapons	Not all are armed	Not all are armed	Not all are armed, but most are armed; prospec-tive members must demonstrate bravery and ability to use weapons
Examples	MEND Niger Delta Peo-ple's Volunteer Force (NDPVF) Federated Niger Delta Ijaw Com-munities (FNDIC)	Area Boys Yandaba groups	Bakassi Boys Anambra State Vigilante Service O'odua People's Congress (OPC)	Niger Delta Vigi-lante (NDV)/Ice-lander Deebam Deewell Greenlander Outlaws

Based on Hazen and Horner, "Small Arms, Armed Violence," 74–75.

GOVERNING ENVIRONMENT

Upon gaining its independence from Great Britain in 1960, Nigeria experienced almost four decades of military rule intermingled with coups and short periods of civilian rule before the transition to a sustained civilian rule in 1999. During this period, the assignment of jurisdictional boundaries and authorities varied and often cut across ethnic and tribal areas.[42] Throughout the country's history, Nigeria

---

[42] At the time of independence, Nigeria was divided into only three states and 131 LGAs before the cascading of federalism led to this same landmass eventually becoming thirty-six states and 774 LGAs. "Nigeria: Want in the Midst of Plenty," *Africa Report No. 113*, 3.

has struggled with "how to structure the state so that every ethnic or religious group and every Nigerian as an individual becomes a stakeholder," which is also referred to as the "National Question."[43]

Nigeria is now a federal republic composed of thirty-six states and three main branches of government: the executive, legislative, and judicial. The executive branch includes the president, the vice president, and the Federal Executive Council (FEC). The National Assembly is a bicameral legislature with 109 Senators and 360 Representatives.[44] Supreme Court judges are appointed by the president and confirmed by the Senate. The legal system in Nigeria is somewhat convoluted, with a combination of English common law and statutory law applied by the federal and state courts and customary law and Islamic law (*shari'a*) recognized by local courts in the south and north, respectively.[45] Below the federal level, each of the thirty-six states has a governor who is limited to two four-year terms. The third tier of government is composed of 774 LGAs that are governed by a council, and this council receives a monthly stipend from federal accounts to support local services and governance.

Put into effect on May 29, 1999, the fourth constitution of Nigeria is modeled after the US Constitution with its three branches of government and its separation of powers, but the federal control of finances is considered to usurp the authorities and powers of the states.[46] More importantly, the application of the US model in Nigeria after decades of postcolonial military rule has yet to bring about the

---

[43] Ibid., 1

[44] Senate membership is composed of three senators from each state and one senator representing the capital territory, while seats within in the House of Representatives are allocated based on population. Elections are held every four years, and the president and members of the National Assembly are limited to two four-year terms. *Country Profile: Nigeria.* Despite these constitutional rules, however, Nigeria's political parties have long observed a practice of rotating the presidency among the six geopolitical regions and in 1999 entered into a pact in which the presidency would be rotated between the north and the south (e.g., the Muslim and Christian halves of the country). This was observed with the election of Yar'Adua (a northerner) after eight years of the Obasanjo presidential administration (a southerner) but may lead to political violence in 2011 with Goodluck Jonathan (a southerner) currently serving as acting president due to the failing health of Yar'Adua. Hazen and Horner, "Small Arms, Armed Violence," 9.

[45] Ibid.

[46] In the lead up to independence, unusually extensive consultations were conducted across Nigeria with stakeholders at the local, regional, and national levels in order to determine the most appropriate constitutional structure for the new nation. The draft 1951 "MacPherson Constitution" and draft 1954 "Lyttleton Constitution" reflected the strong belief at the time that federalism would provide the desired degree of regional autonomy while still giving the central authorities control over interregional and international affairs. "Nigeria: Want in the Midst of Plenty," *Africa Report No. 113*, 5. The 1999 Constitution also provides a legal quandary by presenting Nigeria as a secular state with personal freedoms yet also permitting Muslims in the north to follow Islamic law (*shari'a*).

democratic freedoms and equal opportunities that were originally envisioned. Specifically, a culture of corruption within government bureaucracies remains and—some have argued—has increased with democracy:

> 'Democracy' has—forgive the redundancy—democratised corruption. Under the military, corruption was a quasi-monopoly; it was tightly controlled by a small cohort. Under our 'democracy,' the need to cultivate political support and immunity means that the loot has to circulate.[47]

In addition, there was an underlying assumption that the removal of military control and the introduction of democratic institutions would help ease the plight of Nigeria's poor and impoverished because it was perceived that "pro-poor reform" would bolster and make more equitable the control and distribution of resources. The past decade has demonstrated, however, that the personal benefits that the ruling elite received from their "highly personalised, discretionary use of resources" far outweighed the minimal political benefit that might be gained through the promotion of more equitable resource allocation.[48]

WEAKNESSES OF THE PREREVOLUTIONARY ENVIRONMENT AND CATALYSTS

The challenges of poverty, corruption, control of oil rents, unemployed youth, environmental degradation, and the rise of local militia organizations created a perfect storm for the rise of general armed insurgency in the Niger Delta and specifically for the emergence of MEND as an umbrella organization for several of these militia organizations. A 2009 report by the Coventry Cathedral on the prospects for peace and reconciliation in the Niger Delta finds that:

> This endemic corruption and abject poverty in the Niger Delta provides an environment that favours criminal activity including small arms dealings, money laundering, and large-scale oil theft that, if left unchecked, will continue the social disintegration of the Niger Delta, hinder the economic growth of Nigeria and threaten the political stability of key states and the Federation. These factors provide the funding,

---

[47] Moses Ochonu, "The Failures of Nigerian Democracy," *Pampazuka News*, March 18, 2010, 474, accessed February 4, 2010, http://pambazuka.org/en/category/features/63116.

[48] Hazen and Horner, "Small Arms, Armed Violence," 10.

weapons, and militia needed to precipitate and sustain conflict, as well as undermine society's ability to prevent or recover from conflict.[49]

After the transition from military rule to democracy in 1999, many unemployed young males in the Niger Delta were drawn into ethnic and militia organizations to serve as strongmen for competing political factions who were scrambling for power after the end of military rule.[50] One specific group to emerge was the Ijaw Youth Council (IYC), which was created by the Ijaw National Congress (INC) in December 1998 to support Ijaw candidates within the People's Democratic Party (PDP) who were running for office in the Niger Delta.[51] The IYC and other similar groups were armed by their political patrons and engaged in harassment activities and sometimes assassinations against non-PDP candidates.[52] The true power and autonomy of these militia groups would emerge after the 2003 elections, however, when they retained their weapons and were able to branch out into new criminal enterprises such as oil bunkering. The significant revenues that were generated by these criminal activities enabled the militia groups to purchase additional and more sophisticated weapons and also gave them the opportunity for complete independence from their patrons, although in some cases the politicians were directly involved in the criminal enterprises themselves.[53]

Within this context of the 2003 elections, two key militia leaders emerged who would play a prominent role in the expansion of the influence of armed groups in the Niger Delta region as well as the eventual emergence of MEND. Mujahid Dokubo-Asari became the leader of the IYC in 2001 with the assistance of Governor Peter Odili

---

[49] Davis, *The Potential for Peace and Reconciliation in the Niger Delta*, 36.

[50] The PDP emerged in 1999 as the dominant political organization in Nigeria and saw its ranks swell with new members who were often previous adversaries and now all competing for the same, limited number of electoral seats. *Nigeria's MEND: Connecting the Dots* (Austin, TX: STRATFOR Global Intelligence, 2009), 3, accessed February 14, 2010, www.stratfor.com.

[51] Ibid., 2–3.

[52] Asuni, "Understanding the Armed Groups of the Niger Delta," 13–15. As one Human Rights Watch report stated, "One cult member described a meeting in Government House in Port Harcourt just prior to the April 14 polls during which he saw government officials hand out between N5 million and N10 million ($38,000 to $77,000) to several different cult groups in return for their assisting or simply accepting the PDP's plans to rig the polls."

[53] Hazen and Horner, "Small Arms, Armed Violence," 13–15.

from Rivers State, who was Asari's patron.[54] For two years during the lead-up to the 2003 elections, Asari served as a strongman for his political patron and also enjoyed political protection for his illegal oil bunkering activities. But Asari had a falling out with Governor Odili, and in response, Odili shifted his support to Ateke Tom, who was Asari's deputy.[55] The resulting split between two influential militant leaders (Asari and Ateke Tom) led to direct fighting between their organizations and also turned Asari against the government of Rivers State.[56] In response to increased armed resistance from Ateke Tom, state military forces under Governor Odili, and federal military forces, Asari announced in September 2004 that his Niger Delta People's Volunteer Force (NDPVF) was commencing "Operation Locust Feast" against the oil industry in the Niger Delta, which contributed to global oil prices rising to about $50 a barrel for the first time.[57]

Figure 2. MEND logo.[58]

---

[54] Asari replaced Felix Tuodolo as IYC president. Tuodolo was sponsored by Ijaw Chief Edwin Clark, who was the head of the INC, which created the IYC. As the leader of the Ijaw tribe in the Niger Delta, Chief Clark operates behind the scenes but commands strong influence and respect. As governor of the region's wealthiest oil-producing state, Odili sought independence from Clark's influence, which led to his patronage of Asari and short-term influence within the IYC. See *Nigeria's MEND: Connecting the Dots*; Ochereome Nnanna, "Nigeria: Clark, Niger Delta Overlord?," *Vanguard*, October 8, 2007, accessed February 4, 2010, http://intellibriefs.blogspot.com/2007/10/nigeria-clark-niger-delta-overlord.htm.

[55] Ateke Tom was the head of the Icelanders, which conducted many of the bunkering activities from which Asari initially benefited before they became foes. Upon falling under the patronage of Governor Odili, Ateke Tom changed the name of the Icelanders to the Niger Delta Vigilantes (NDV). *Nigeria's MEND: Odili, Asari and the NDPVF* (Austin, TX: STRATFOR Global Intelligence, 2009); Hazen and Horner, "Small Arms, Armed Violence."

[56] At this time, Rivers State was experiencing the bulk of violence among the armed groups in the region, although the other two major oil-producing states in the Niger Delta (Bayelsa and Delta State) would also see a surge in violence as turf battles over oil bunkering rights emerged.

[57] Michael Peel, "Into the Heart of the Niger Delta Oil War," *Financial Times*, September 12, 2009; Asuni, *Understanding the Armed Groups of the Niger Delta*.

[58] National Mirror, accessed March 14, 2011, http://nationalmirroronline.net/news/4914.html.

One month later, Nigerian President Olusegun Obasanjo called Asari and Ateke Tom to the capital in order to negotiate a cease-fire. As part of the cease-fire agreement, militia leaders would receive $2,800 for each weapon turned in by their group.[59] This arms-for-money agreement and cease-fire significantly reduced the amount of direct fighting between the two groups, but it did little to control the rising interest in oil bunkering and other criminal activities because a resupply of arms was readily available. Both Asari and Ateke Tom were soon arrested (Asari in September 2005 for treason after a speech he made at a conference and Ateke Tom in November 2004 for the murder of a rival gang member). Although Ateke Tom would be freed after he shifted blame for the murder to his own deputy, Asari would be sentenced to jail and serve eighteen months. By the time he was released, MEND was a prominent insurgent force within the Niger Delta and composed of many of his former militia associates.[60]

In December 2005, less than a month after Asari began his prison sentence, MEND took credit for an attack on Royal Dutch Shell's Opodo pipeline in Delta State. This attack was followed by two more on January 10, 2006, that involved blowing up a flow station for the Trans Ramos oil pipeline in Bayelsa State as well as the kidnapping of four foreign oil workers from Shell's offshore E.A. oil field.[61] Although it was not uncommon for new armed groups to emerge in the Niger Delta, what surprised analysts and security experts about MEND was the tactical proficiency of its attacks, the diverse type of these attacks, and the broad geographic area in which MEND appeared to operate. By April 2006, MEND had selectively targeted key choke points in several oil pipelines, continued kidnapping of foreign oil workers, and conducted a car bomb attack on a military barracks in Port Harcourt. Within this first four months of MEND operations, oil output from the Niger Delta region was reported to be down by 25%.[62]

The escalation of violence between the forces loyal to Asari and those loyal to Ateke Tom was by no means the single source of conflict within the Niger Delta during this period. What is noteworthy about this conflict, however, is that it represented all of the key environmental

---

[59] $2,800 breaks down to $1,000 from the federal government and $1,800 from Rivers State.

[60] *Nigeria's MEND: Odili, Asari and the NDPVF.*

[61] The hostages were captured by Farah and Boyloaf's groups and taken to Tompolo's camp, where it took six days to determine what to do with them and the list of demands. Upon payment of a ransom by Bayelsa State, the hostages were released on January 30, which set an ongoing pattern for the safe return of MEND hostages. This January 2006 attack on the Shell E.A. field would actually remove the field from operation for more than three years until July 2009.

[62] "Nigeria's Shadowy Oil Rebels."

elements and catalysts that merged into the overall MEND insurgency, including the struggle for control of the "rentier space" for oil production,[63] the use of armed militia groups by politicians for private security and political persuasion, the frequent splitting of allegiances within these groups based on personal opportunities, the targeting of international oil companies as a ready source of income and as the cause of environmental degradation, and the rapid expansion of local and state power struggles into national and international affairs. In addition, from a counterinsurgency perspective, jurisdictions, authorities, and capabilities were diverse and inconsistent among the local, state, federal, and international actors who bore the brunt of these attacks and thus significantly complicated the development of any coordinated strategy or policy.

FORM AND CHARACTERISTICS OF THE REVOLUTION

OBJECTIVES AND GOALS

Several key events in 2005 triggered the forming of alliances and the merging of operational capabilities among several armed groups in the Niger Delta under the MEND banner. The first of these events was a decision by Asari to accept an invitation from a militant leader named Tompolo to move to Delta State in order to escape the attacks from Ateke and the government troops in Rivers State. This coincided with the arrest on corruption charges of Chief Alamieyeseigha, the governor of Bayelsa State, which was viewed as a politically motivated action by Nigerian President Obasanjo against an Ijaw chief who was also the leader of the only Ijaw-majority state in Nigeria. The subsequent arrest of Asari in September 2005 and the arrest in Rivers State of another group leader ("Olo") in November 2005 prompted the leaders of two key militant organizations (Farah Dagogo and Boyloaf) to leave Rivers State for the safety of Delta State.[64] The presence of these significant militant groups and leaders in Delta State prompted Tompolo to convene a series of meetings in late 2005 that led to the creation of MEND and operational planning concerning bunkering

---

[63] The "rentier space" is "a term encompassing the acquisition, control, and disposition of oil and oil-related resources, including the financial benefits derived from them." Omeje, "Oil Conflict and Accumulation Politics in Nigeria."

[64] Asari was not arrested in Delta State; rather, he was captured when he went to Abuja for a supposed meeting with the president.

syndicates, the purchase of advanced weaponry, and the targeting of oil infrastructure.[65]

An additional aspect of MEND that surprised observers was that the true identity of its leadership was not known at the time, most likely because of security concerns after the recent arrest of Asari. It was suspected that former members of Asari's NDPVF served as a key component of this new organization, and this belief was bolstered by that fact that one of MEND's initial demands was for the release of Asari.[66] But it also became quickly apparent to observers that MEND was not a clearly defined group but rather an umbrella organization for multiple militant, criminal, and cult groups across the Niger Delta, which explained some of the flexibility and broad geographic reach of this new organization.

In April 2006, during one of its first media interviews, MEND stated that they were fighting for "total control" of the Niger Delta's oil wealth and to reverse the exploitation of the region's natural resources by foreign companies and people from other parts of Nigeria.[67] The MEND representative also stated that all oil companies and Nigerians who were not from the Niger Delta should immediately leave the area. This was a repeat of a threat initially made by e-mail in January 2006:

> It must be clear that the Nigerian government cannot
> protect your workers or assets. Leave our land while you
> can or die in it. Our aim is to totally destroy the capacity
> of the Nigerian government to export oil.[68]

Although these demands were not new for a militia group from the Niger Delta to make, their impact was more significant since the MEND appeared to have more proficiency and potentially better insider-knowledge for its attacks. However, within a month of these initial demands, MEND would conduct another set of kidnappings and issue revised demands that were more moderate and better articulated: a demand for 50% (rather than 100%) of oil revenues from the Niger Delta to be allocated back to the region, an increase in political participation for the people of the region, increased

---

[65] Attending these meetings were militants from the Federated Niger Delta Ijaw Council, NDPVF, and cult members from the Greenlanders and KKK. Asuni, *Understanding the Armed Groups of the Niger Delta.*

[66] Ukiwo, "From 'Pirates' to 'Militants'," 587–610.

[67] *Nigeria's Shadowy Oil Rebels.* This complaint about exploitation of the Niger Delta by people from other parts of the country would continue with the March 2010 bombing of the Warri post-amnesty dialogue. See Bello, Bello, and Osayande, "Why Bombs Returned to Niger Delta."

[68] Quoted in Daniel Howden, "Nigeria: Shell May Pull Out of Niger Delta After 17 Die in Boat Raid," *The Independent (UK)*, January 17, 2006, accessed February 4, 2010, http://www.corpwatch.org/article.php?id=13121.

involvement in the oil and gas industry, development of the region's economy and infrastructure; and a reduction in federal military presence in the region.[69]

LEADERSHIP AND ORGANIZATIONAL STRUCTURE

In the four years since MEND first announced its presence in the Niger Delta, numerous individuals have communicated with the media and presented themselves as part of MEND leadership, but the exact hierarchy of the organization – if there is one – is unknown. Among the better known names within MEND are Government Ekpemupolo (aka Tompolo) who led the Federated Niger Delta Ijaw Communities (FNDIC) and held security contracts with multiple foreign oil companies; Prince Farah Ipalibo (aka Farah Dagogo), and Victor Ben Ebikabowei (aka Boyloaf). As of 2007, Henry Okah was the most recent and most visible leader of MEND though he lived in South Africa and his career as an arms merchant has called into question his true interest in the Niger Delta despite his origins being in Bayelsa State.[70]

MEND is composed of mostly young Ijaw men; the number of people under arms is uncertain, although estimates range from a few hundred to a few thousand.[71] One West Africa scholar noted that "MEND seems to be led by more enlightened and sophisticated men than most of the groups in the past."[72] MEND has also been described as a "franchise operation," which allows it to adjust its membership and tactics to its operating environment but also results in people switching affiliation between MEND and other groups depending on their personal opportunities.[73]

---

[69] Asuni, *Understanding the Armed Groups of the Niger Delta*

[70] Okah was arrested in Angola in November 2007 on arms smuggling charges and deported to Nigeria, where he remained incarcerated until July 2009. MEND critics of Okah point out as proof of his limited leadership role that the organization continued to function while he was in prison. Proponents say Okah spends very little time in South Africa but moved his family there for safety and that he is very well regarded within MEND as the "Overall Master" or "Oga" (boss). See Daniel Alabrah, "Mystery World of MEND Leader, Henry Okah; Untold Story of His Life and Struggle," *Daily Sun Online*, November 22, 2009, accessed February 4, 2010, http://www.sunnewsonline.com.

[71] A 2007 study conducted for the Delta State government found that, including MEND, there were forty-eight armed groups operating in Delta State with more than 25,000 members and 10,000 weapons. Other experts estimate the number of armed group members at closer to 60,000. Asuni, *Understanding the Armed Groups of the Niger Delta*.

[72] Nnamdi K. Obasi, West Africa senior analyst at the International Crisis Group, quoted in Stephanie Hanson, *MEND: The Niger Delta's Umbrella Militant Group* (New York: Council on Foreign Relations, 2007), http://www.cfr.org/publication/12920.

[73] Ibid.

Within a few months of forming, MEND already began to break into three main factions because of rivalries and the greed of individual leaders. The "Western MEND" operated out of Delta State under Tompolo and was based on his FNDIC. The "Eastern MEND" operated primarily in Rivers State, was led by Farah Dagogo, and was composed of fighters from the Niger Delta Strike Force (NDSF) as well as other groups, especially the Outlaws under George Soboma, who controlled Port Harcourt. Its focus was primarily on kidnapping for the revenue. The "Central MEND" was led by Boyloaf and operated in Bayelsa State, and this faction appeared to have the closest link to Henry Okah.[74]

Finally, tensions surrounding recent MEND negotiations with the federal government led to an open airing in the media of an internal power struggle between supporters of Henry Okah (including Farah Dagogo and Boyloaf) and those of Tompolo and another "creek general," John Togo.[75] After the expiration of an amnesty offer from President Yar'Adua in October 2009, the Tompolo and Togo faction, which accepted the amnesty offer, dismissed Henry Okah and his negotiating team as a "paper general without a battalion" and demanded that the federal government stop negotiating with the Okah-appointed "Aaron Team."[76] As of 2010, the negotiations are still under way, and the Aaron Team remains in place.

COMMUNICATIONS

MEND has several audiences with which it communicates, including residents of local Niger Delta communities; the leadership of local, state, and federal government organizations; the international community; and international oil companies. Because the government controls most means of broadcast communication (e.g., television

---

[74] Asuni, *Understanding the Armed Groups of the Niger Delta*

[75] Creek general is a term of respect for a militant leader who has operational forces in the creeks of the Niger Delta.

[76] The Aaron Team is the group of negotiators that Henry Okah proposed to the federal government and is composed of a former Nigerian Chief of General Staff, Vice Admiral Okhai Mike Akhigbe (retired), Professor and Nobel Laureate Wole Soyinka, Major General Luke Kakadu Aprezi (retired), Dr. Sabella Ogbobode Abidde, Ph.D., and Mr. Amagbe Denzel Kentebe. During a direct meeting with President Yar'Adua, the team was also accompanied by Henry Okah and Farah Dagogo. Emma Amanze, "MEND's Aaron Team Seeks Inputs from Nigerians," *Vanguard Online*, December 5, 2009, accessed February 4, 2010, http://www.vanguardngr.com/2009/12/05/mend's-aaron-team-seeks-inputs-from-nigerians/. The Tompolo faction was reportedly also angry with Henry Okah for not showing enough appreciation for their help in securing his release from prison in July 2009 after his arrest in Angola in September 2007 on arms smuggling charges. "Crisis Rocks MEND," *The Neighborhood*, November 19, 2009.

and radio), MEND relies heavily on the use of the Internet and e-mail for its outreach efforts. Communication and interviews are conducted primarily by e-mail, and one spokesperson (Jomo Gbomo) has emerged as a constant voice for the organization, although he only communicates via e-mail.[77] This virtual connection also provides a layer of security for MEND leadership. However, MEND has also been extremely effective in engaging the media and nurturing relationships with certain reporters in order to effectively disseminate its message both domestically and abroad. This point was underscored after MEND bombed a post-amnesty conference in March 2010 in Warri that was sponsored by the Vanguard newspaper. Governor Timipre Sylva stated that everyone was shocked by this attack because the event was coordinated by a media group, and "All of us took few things for granted because we felt that MEND and journalists have been friends for some time."[78] MEND has also used websites for archiving media stories about its activities (http://mendnigerdelta.com) and to solicit input from citizens about local grievances and demands (www. theaaronteam.org).

METHODS OF ACTION AND VIOLENCE

In the four years of its existence, MEND has conducted a broad range of attacks across the Niger Delta that supported revenue-generating activities, such as oil bunkering and kidnapping, while also pursuing its stated goal of sharply reducing oil production in the Niger Delta. Unlike other militia or criminal groups, MEND has also demonstrated a willingness to directly engage the Nigerian military both in combat operations, in order to protect base camps, and in retaliatory strikes. Moreover, MEND has conducted attacks more than 100 kilometers out at sea against oil rigs[79] and is also linked to a February 2009 attack on the Equatorial Guinea presidential palace in Malabo that was purportedly planned as a robbery.[80] Between 2006 and 2008, MEND engaged in kidnapping, pipeline and other bombings, as well as other serious criminal activities. Kidnappings peaked in 2007 with close to forty-six incidents, about 50% more than in the previous and following years. While in 2006–2007, there were

---

[77] Some observers have speculated that Henry Okah and Jomo Gbomo are the same person. Asuni, *Understanding the Armed Groups of the Niger Delta*.

[78] "Gov Sylva Condemns Warri Bomb Attack," *Yenagoa Glory*, March 17, 2010.

[79] "Militants in Nigeria Attack Offshore Oil Rig, Cut Production," *Voice of America*, June 19, 2008. This shocking attack in 2008 against Shell's Bonga oil field cut oil production for the company by 200,000 barrels a day.

[80] "Nigerians Jailed After Attack on Equatorial Guinea Presidential Palace," *Agence France Press*, April 5, 2010.

relatively few pipeline bombs (about four incidents per year), in 2008, MEND placed twenty pipeline bombs. Other bombs, although present, seem to be a less dominant strategy, with only four incidents over the three-year period. However, serious criminal activity rose steadily over the three-year period, with a large jump from 2006 (three incidents) to 2007 (twenty incidents) to 2008 (thirty-two incidents).[81]

Although MEND is credited with being perhaps the most tactically proficient armed group operating in the Niger Delta, it is by no means the only group in operation. An array of criminal groups, vigilante organizations, politically motivated groups, and even corrupt governmental groups have caused a swath of death and destruction in the region. These other groups and organizations account for the vast majority of attacks and criminal events across the Niger Delta, with MEND being linked to only about one-in-seven to one-third of the attacks conducted in its first three years of operations.

Overall, the selection of targets by MEND shows a preference to avoid civilian casualties accompanied by a willingness to destroy critical, debilitating components of the Nigerian oil infrastructure. After the March 2010 car bombing at the post-amnesty dialogue in Warri, MEND released a statement saying it had only detonated two of the three car bombs because it wanted to avoid casualties when people fleeing from the first two bombs gathered in the vicinity of the third undetonated bomb.[82] However, as MEND becomes more daring, the scope of the attacks makes it more likely that civilians will be killed. This was the case with the July 11, 2009, attack on the Atlas Cove Jetty that killed five workers while also destroying a major Nigerian oil hub during MEND's first attack in Lagos.

METHODS OF RECRUITMENT

There are few detailed studies that explain the reasons why and/or the method by which young males join up with MEND and other militia organizations in the Niger Delta, but the data that do exist show that socioeconomic factors and anger toward corrupt government are key factors. A vast majority of Niger Delta fighters are males between the ages of twenty and thirty-nine; they are unemployed, and they are

---

[81] Bergen Risk Solutions, *Niger Delta Security Briefing*, 2008, accessed February 4, 2010, www.bergenrisksolutions.com.

[82] Ahamefula Ogbu, "MEND: We Stopped Third Explosion to Save Lives," *This Day Online*, March 17, 2010, accessed February 4, 2010, http://www.thisdaylive.com. The car bombs were also placed across the road from the meeting venue, although that is likely due to security blocking access to the meeting site. Despite the claim to prevent casualties, media reports indicate that three civilians walking through the area were possibly killed by the bombs.

not married but often have dependent children. The fighters often come from broken homes and lack role models in their lives. Many are drug users—some are also drug dealers—and few are literate, although leaders often are educated to a secondary or tertiary level. With no economic resources, they are completely dependent on their leaders for financial support, food, and shelter.[83]

When questioned in 2006–2007 about their reasons for joining a Niger Delta militant group, these members cited numerous reasons, including their desire to protect their community and ethnic group; concerns over personal safety after threats from other groups or government agencies; a desire to fight oil companies and the government over political and economic marginalization; job offers from politicians related to vote-rigging, voter intimidation, and attacks on opponents; financial gain through criminal activity; peer pressure; and seeking power and influence to offset low self-esteem. Follow-up questioning in 2008 revealed that a desire to avenge the death of a family member or friend as well as direct coercion from existing members of armed groups also played a role in recruitment, as did the ever-present desire for financial gain.[84] It was reported that a trained militant could earn N50,000 a month (approximately $330), which is well above the income level for an educated youth working in the formal economy.[85]

A separate survey in 2007 found that only 5% of the surveyed Niger Delta population was satisfied with the status quo and that 36.23% had a "willingness or propensity to take up arms against the state." Anti-state sentiment has also been fueled by heavy-handed military operations against civilian villages by the federal government's Joint Task Force during the course of its anti-militia operations. A federal attack on the towns of Odi and Odiama in Bayelsa State in November 1999 to search for militant groups destroyed the towns and caused international rebuke.[86] In addition, MEND often conducts its boat raids while flying long white flags that symbolize Egbesu (an Ijaw water spirit representing peace) or red flags (representing the Ijaw fighting spirit) as a means of drawing upon local Ijaw support for the group.[87] The bombing of an Ijaw community in February 2006 by the Joint Task Force is often cited as a key catalyst to the emergence of MEND, with fighters from NDPVF and other groups joining with MEND in order

---

[83] Asuni, *Understanding the Armed Groups of the Niger Delta*.
[84] Ibid.
[85] Watts, "Petro-Insurgency or Criminal Syndicate," 637–660.
[86] Asuni, *Understanding the Armed Groups of the Niger Delta*.
[87] Watts, *Curse of the Black Gold: 50 Years of Oil in the Niger Delta*.

to fight the Joint Task Force.[88] Finally, unlike most armed groups that had ethnic origins, MEND expanded its potential recruiting pool by allowing all ethnic groups to join its membership.[89]

METHODS OF SUSTAINMENT

MEND sustains its operations through the support of local communities as well as the financial gains from bunkering and kidnapping. Some militant leaders, such as Tompolo, have been extremely effective at generating local support by fostering an image of freedom fighters, and this has given the militants not only safe harbor in these communities but also access to the armories that many of these villages have in response to ethnic conflicts in the region.[90] Moreover, in response to huge financial losses experienced by the oil companies, the companies have increased their payments to militant leaders for protection and surveillance services, with these payments often presented as "community development" funding.[91] This substantial influx of cash has allowed the militant groups to import large caches of weapons; one intercepted shipment from Ukraine in August 2007 contained "950 AK assault rifles, 150 under-barrel grenade launchers to suit AK-47, one million rounds of 7.62 mm ammunition, 475,000 rounds of 4.45 mm ammunition, 500 pistols, 300,000 rounds of ammunition for the pistols, 8,000 hand grenades, 200 RPG launchers, 1,000 RPG projectiles, 500 kg of TNT, plus various mortars and mines."[92] Usually, the arms shipments originate in Turkey, Liberia, Cameroon, South Africa, Ukraine, and the Ivory Coast and come in on tankers that exchange the weapons and ammunition for the bunkered oil. Although there are an estimated 1–3 million small arms and light weapons in circulation in Nigeria—mostly in the hands of civilian villagers rather than militants—the density of these weapons is so high in the Niger Delta region that militia leaders from northern Nigeria who are involved in the ethnic and religious fighting near Jos have been seen shopping for weapons in the Niger Delta.[93]

---

[88] Hazen and Horner, "Small Arms, Armed Violence."

[89] Angela Kariuki, *The Niger Delta: A History of Insecurity* (Pretoria: Consultancy Africa Intelligence, 2009).

[90] Asuni, *Understanding the Armed Groups of the Niger Delta.*

[91] "The Swamps of Insurgency," *Africa Report No. 115.*

[92] Stephen Davis, "Arms, Arms and More Arms," *NEXT Community Online,* 2009.

[93] Omafume Amuran, "Jos Crisis: Muslim Leaders Shop for Arms in the Niger Delta," *Niger Delta Standard,* March 17, 2010.

METHODS OF OBTAINING LEGITIMACY

Legitimacy for MEND has been provided by the continuous rounds of negotiations, cease-fires, and amnesty offers between the group and the federal and state governments. This ability to force government officials to the negotiation table comes from the willingness of MEND to target critical components of Nigeria's oil infrastructure as opposed to less critical, symbolic nodes.[94] The result has been a drop in oil production in the Niger Delta of approximately 25–33%, thus depriving the Nigerian government, as well as state coffers, of billions of dollars in revenue since MEND declared its oil war in 2007. The traditional, armed response by the federal and state authorities has also proven ineffective because MEND has demonstrated its ability to fight these military forces, resulting in the double-digit loss of government forces in many engagements. MEND seems less inclined to completely stop its operations based on payoffs and bribes from government and private sector officials, although members of MEND are certainly not immune from collaborating with government officials and soldiers during bunkering operations, kidnappings, etc. Taken together, the military capability and effective, aggressive targeting by MEND gives it unique leverage with the federal and state governments.

EXTERNAL SUPPORT

In addition to the availability of external arms supply routes into the Niger Delta, it is believed that mercenary elements from South Africa and other locations are providing training support in the Niger Delta to MEND as well as other armed groups. The exact extent of this support is not well documented, but the linkages are becoming more apparent with the discovery of mercenaries from Cameroon and Equatorial Guinea found in Niger Delta militant camps as well as the extension of MEND operations into Equatorial Guinea.[95] Another source of external support continues to be developmental aid for Niger Delta villages and communities that comes from foreign donors and international oil companies but is siphoned off to support local militia elements.

---

[94] Many militant groups are hesitant to completely disable sections of the oil infrastructure because this will interfere with the flow of oil necessary to sustain bunkering operations. In 2007, however, Ijaw leaders are reported to have put pressure on MEND leaders to reduce the destructiveness of their attacks because the financial consequences were also impacting their villages.

[95] "Mystery Over E Guinea Gun Battle," *BBC News*, February 18, 2009, accessed February 4, 2010, http://news.bbc.co.uk/2/hi/africa/7894651.stm.

COUNTERMEASURES TAKEN BY THE GOVERNMENT

By 2010, the countermeasures taken by state and federal governments to control the actions of MEND have been largely ineffective. This is due to the fact that MEND is not a single, homogenous entity; thus, the acceptance of certain government offers—be they financial, amnesty, resource reallocation, developmental, etc.—by one faction leader or group of leaders within MEND will not be comprehensive enough to satisfy all leaders and factions within MEND. This is the ongoing case with the Aaron Team, which is composed of well-regarded, neutral Nigerians who were selected to mediate between MEND and the federal government; however, the fact that one faction leader (Henry Okah) was instrumental in putting this team together undermines the overall credibility of this expert group. State and federal governments have also not been able to form a sufficient military dragnet around militant camps and support bases because of the difficult environment of the Delta basin, the proficiency of the fighters, and the support of local communities. Even if government forces were able to capture and remove a militant leader, experience with the detention of Asari, Okah, and others shows that the militant organizations are self-generating and that a new leader would quickly emerge to replace one who is killed or incarcerated. Finally, several previous attempts at a government-brokered amnesty deal with MEND and other militants have failed because of corruption within the process involving the skimming of funds by federal and military officials and also the failure of the militant leaders who received these funds to distribute them among the fighters or the affected communities.[96] In addition, some domestic and international observers believe that for many government officials and militants alike, the financial benefits of the status quo are significant enough that it is in their personal interest to prevent sustained peace in the region. As such, absent the ability or willingness to remove the core causes of the militant and criminal activities, the government is left with two options—negotiation and armed force—neither of which has proven effective for sustained peace with MEND.[97]

---

[96] The use of bribes to secure a short-term peace has been successful before, however, especially during the run-up to an election. This was the case in Rivers State in October 2006 when the state government "is alleged to have released over N15 million each to the Outlaws (led by Saboma George) and Icelanders (led by Ateke Tom) as an incentive to the two groups to stop fighting during Governor Odili's attempts to gain the PDP presidential nomination." Asuni, *Understanding the Armed Groups of the Niger Delta.*

[97] Davis, *The Potential for Peace and Reconciliation in the Niger Delta.*

SHORT- AND LONG-TERM EFFECTS

CHANGES IN THE ENVIRONMENT

The overarching factors that gave rise to MEND in late 2005 continue to be present in the region. Unless addressed, the core issues that impact quality of life and human development opportunities in the Niger Delta will likely remain central catalysts for anti-government sentiment and the fielding of armed groups (militia, criminal, cult, and vigilante) in order to meet individual needs and/or the needs of the community and family.

CHANGES IN GOVERNMENT

In 2010, the Nigerian federal government was balanced upon a constitutional crisis concerning the failing health of President Yar'Adua and the role of acting president being filled by Vice President Goodluck Jonathan.[98] Initial efforts to block the rise of Jonathan to the president's seat came from the FEC as well as the National Assembly.[99] With the eventual judicial decision to allow Jonathan to replace a living but ill president, the Nigerian democratic system demonstrated unusual resilience, although significant concern surrounds the 2011 national elections. At the center of this concern is the fact that Yar'Adua is from the north but had secured political allegiances in the south, too. Per electoral custom, the next Nigerian president should be from the south, and Goodluck Jonathan, who is the former governor of Bayelsa State in the Niger Delta, could be positioned to take this next presidential slot although he does not have the necessary allegiances in the north. These political uncertainties have severely undermined the ongoing federal amnesty negotiations with MEND due to the fact that President Yar'Adua had gained the confidence of MEND faction leaders before his health failed. With the future of the federal government in question, mistrust of federal military leadership, and ongoing disputes with the governors at the state levels, it is unlikely that MEND can achieve a unified voice for itself or find a unified representative voice for the federal government with which it can negotiate peace requirements.

---

[98] Although still alive, Yar'Adua was under medical care in Saudi Arabia for almost two months in late 2009 and completely removed from the day-to-day decision-making within Nigeria.

[99] Davo Benson and Gbenga Oke, "Can Jonathan be His Own Man?" *Vanguard Online,* February 18, 2010, accessed February 4, 2010, http://www.vanguardngr.com/2010/02/18/can-jonathan-be-his-own-man.

CHANGES IN POLICY

With the existing legacy of corruption and the financial pull of petroleum dollars heavily influencing the implementation of policy within Nigeria, any significant policy changes are unlikely to be attempted before the 2011 elections. In the interim, however, some symbolic efforts at peace and stability will likely be attempted in order to please the electorate, but based on the experiences during the 1999, 2003, and 2007 elections, these symbolic policy changes will probably not be sustainable.

CHANGES IN THE REVOLUTIONARY MOVEMENT

The tensions that were created within the MEND hierarchy during the waning days of the 2009 amnesty highlight differences of opinion and power struggles between the leadership of the three main MEND factions. At the center of this dispute is the desire of "creek generals" to remain at the center of negotiations and have a seat at the negotiating table.

In addition, the Joint Revolutionary Council (JRC), which was formed in 2006 as an umbrella organization for the Reformed NDPVF, the Martyrs Brigade, the Outlaws, and elements of MEND,[100] emerged in 2010 as a vocal critic of MEND, thus raising the potential of a power struggle between these two "umbrella" organizations for Niger Delta militants. Specifically, the JRC was highly critical of MEND after the March 2010 bombing at the Warri post-amnesty dialogue and called the bombing the "handiwork of a dementia inflicted cabal who has cunningly infiltrated the just and noble struggle for the liberation and emancipation of the Ijaw and Niger Delta struggle."[101] For its part, MEND was critical of the JRC only a week before the Warri bombing, after the JRC claimed credit for a pipeline attack. In response, MEND told the media, "The Nigerian government has not seen anything yet. This (the Agip attack) is just child's play. When we begin, they will know we are not joking."[102] Finally, the JRC has been very vocal during the political power struggle surrounding the appointment of Goodluck Jonathan as acting president. This prompted MEND to tell the media that:

---

[100] "MEND Dissociates Self from Anarchy Threat," *The Neighborhood,* December 1, 2009; Ogbu, "MEND: We Stopped Third Explosion to Save Lives."

[101] Ibid.

[102] Daniel Alabrah, "Militants Resume Attacks; 'Expect More Hostilities'," *NBF News,* March 7, 2010, http://m.thenigerianvoice.com/mobile/12939/1/militants-resume-attacks-expect-more-hostilities.html; Dokubo Asari, "I'm a Militant Forever," *Daily Sun Online,* March 7, 2010, http://www.sunnewsonline.com.

As far as we know, this group exists only in cyberspace and there is no collaboration of any sort with this imaginary group as claimed in the misleading statement. The Movement for the Emancipation of the Niger Delta is apolitical and will continue to remain focused in the fight for land ownership and resource control for the impoverished people of the Niger Delta. We will neither act in haste nor be drawn into political speculation.[103]

This apolitical claim by MEND may be called into question with the approach of the 2011 elections and the potential challenge from the JRC.

BIBLIOGRAPHY

Alabrah, Daniel. "Mystery World of MEND Leader, Henry Okah; Untold Story of His Life and Struggle." *Daily Sun Online,* November 22, 2009.

———. "Militants Resume Attacks; 'Expect More Hostilities.'" *NBF News,* March 7, 2010. Accessed February 4, 2010. http://m.thenigerianvoice.com/mobile/12939/1/militants-resume-attacks-expect-more-hostilities.html.

Amanze, Emma. "MEND's Aaron Team Seeks Inputs from Nigerians." *Vanguard Online,* December 5, 2009.

Amuran, Omafume. "Jos Crisis: Muslim Leaders Shop for Arms in the Niger Delta." *Niger Delta Standard,* March 17, 2010.

Asari, Dokubo. "I'm a Militant Forever." *Daily Sun Online,* March 7, 2010. Accessed February 4, 2010. http://www.sunnewsonline.com.

Asuni, Judith Burdin. *Understanding the Armed Groups of the Niger Delta.* New York: Council on Foreign Relations, 2009.

Bello, Abdullahi Yahaya, Muhammad Bello, and Monday Osayande. "Why Bombs Returned to Niger Delta." *Daily Trust Online,* March 19, 2010.

Benson, Davo, and Gbenga Oke. "Can Jonathan be His Own Man?" *Vanguard Online,* February 18, 2010. Accessed February 4, 2010. http://www.vanguardngr.com/2010/02/18/can-jonathan-be-his-own-man.

Bergen Risk Solutions. *Niger Delta Security Briefing,* 2008. Accessed February 4, 2010. www.bergenrisksolutions.com.

---

[103] Amanze, "MEND's Aaron Team Seeks Inputs from Nigerians."

Central Intelligence Agency. "Nigeria." *The World Factbook.* Accessed February 4, 2010. https://www.cia.gov/library/publications/the-world-factbook/geos/ni.html.

"Crisis Rocks MEND." *The Neighborhood,* November 19, 2009.

Davis, Stephen. "Arms, Arms and More Arms." *NEXT Community Online,* 2009.

———. *The Potential for Peace and Reconciliation in the Niger Delta.* Coventry, UK: Coventry Cathedral, 2009.

"Gov Sylva Condemns Warri Bomb Attack," *Yenagoa Glory,* March 17, 2010.

Ikelegbe, Austine. "State, Ethnic Militias, and Conflict in Nigeria." *Canadian Journal of African Studies* 39, no. 3 (2005): 490–516.

International Crisis Group. "Nigeria: Want in the Midst of Plenty." *Africa Report No. 113,* 2006.

International Crisis Group. "The Swamps of Insurgency: Nigeria's Delta Unrest." *Africa Report No. 115,* 2006.

Hanson, Stephanie. *MEND: The Niger Delta's Umbrella Militant Group.* New York: Council on Foreign Relations, 2007.

Hazen, Jennifer M., and Jonas Horner. *Small Arms, Armed Violence, and Insecurity in Nigeria: The Niger Delta in Perspective.* Geneva: Small Arms Survey, Occasional Paper 20, 2007.

Howden, Daniel. "Nigeria: Shell May Pull Out of Niger Delta After 17 Die in Boat Raid." *The Independent (UK),* January 17, 2006.

Human Development Report 2009. New York: United Nations Development Programme, 2009.

Kariuki, Angela. *The Niger Delta: A History of Insecurity.* Pretoria, South Africa: Consultancy Africa Intelligence, 2009.

Library of Congress/Federal Research Division. *Country Profile: Nigeria.* July 2008, 17. Accessed January 14, 2010. http://memory.loc.gov/frd/cs/profiles/Nigeria.pdf.

Lionberger, Brian. "Emerging Requirements for U.S. Counterinsurgency: An Examination of the Insurgency in the Niger Delta Region." U.S. Army Command and General Staff College, 2007.

"MEND Dissociates Self from Anarchy Threat." *The Neighborhood,* December 1, 2009.

"Militants in Nigeria Attack Offshore Oil Rig, Cut Production." *Voice of America,* June 19, 2008.

"Mystery Over E Guinea Gun Battle." *BBC News,* February 18, 2009. Accessed February 4, 2010. http://news.bbc.co.uk/2/hi/africa/7894651.stm.

Nigeria: Country Brief. Washington, DC: World Bank, 2009.

"Nigerians Jailed After Attack on Equatorial Guinea Presidential Palace." *Agence France Press*, April 5, 2010.

Nigeria's MEND: Connecting the Dots. Austin, TX: STRATFOR Global Intelligence, 2009.

Nigeria's MEND: Odili, Asari and the NDPVF. Austin, TX: STRATFOR Global Intelligence, 2009.

"Nigeria's Shadowy Oil Rebels." *BBC News,* April 20, 2006.

Nnanna, Ochereome. "Nigeria: Clark, Niger Delta Overlord?" *Vanguard,* October 8, 2007.

Ochonu, Moses. "The Failures of Nigerian Democracy," *Pampazuka News,* March 18, 2010, 474. Accessed February 4, 2010. http://pambazuka.org/en/category/features/63116.

Ogbu, Ahamefula. "MEND: We Stopped Third Explosion to Save Lives." *This Day Online,* March 17, 2010.

Omeje, Kenneth. "Oil Conflict and Accumulation Politics in Nigeria." *Environmental Change and Security Program Report, Issue 12.* Woodrow Wilson International Center for Scholars, 2006–2007.

Peel, Michael. "Into the Heart of the Niger Delta Oil War." *Financial Times,* September 12, 2009.

Ukiwo, Ukoha. "From 'Pirates' to 'Militants': A Historical Perspective on Anti-State and Anti-Oil Company Mobilization Among the Ijaw of Warri, Western Niger Delta." *African Affairs* 106, no. 425 (2007): 587–610.

Watts, Michael, ed. *Curse of the Black Gold: 50 Years of Oil in the Niger Delta.* 2009 ed. Brooklyn, NY: PowerHouse Books.

———. "Petro-Insurgency or Criminal Syndicate? Conflict & Violence in the Niger Delta." *Review of African Political Economy* 114 (2007): 637–660.

REVOLUTIONARY UNITED FRONT (RUF)— SIERRA LEONE

Jerry Conley

SYNOPSIS

From March 1991 through January 2002, the country of Sierra Leone was embroiled in a civil war against an insurgent movement that was driven by financial interests and sustained by external actors. Although casualty estimates range from 20,000 to 120,000 civilian deaths, the Revolutionary United Front (RUF) insurgency is known more for the thousands of amputations conducted against the population and the central role of "blood diamonds." Government efforts to counter the insurgency were limited by a poorly trained and equipped army, as well as two decades of corruption that undermined the security and public safety institutions of government. These weaknesses led the government of Sierra Leone to rely on regional and international peacekeeping forces, a British military intervention, as well as the hiring of mercenary forces. In addition, a series of coups and elections during the decade-long civil war resulted in four different heads of government trying to coordinate countermeasures against the RUF. As the civil war concluded in 2002, the main quality of life factors that initially sowed the seeds for the rise of the RUF remained in place, especially the presence of abundant natural resources and a large number of impoverished, unemployed, and uneducated young males.

TIMELINE

April 27, 1961	United Kingdom grants independence to Sierra Leone.
1968–1985	President Siaka Stevens leads a one-party socialist state that is described as a "17-year plague of locusts."
1971	Attempted coup against President Stevens. Corporal Foday Sankoh is sentenced to seven years for not reporting the coup plot.
1987	A group of Sierra Leoneans that includes Foday Sankoh, expelled student leaders, and recruits receives training in Libya. Sankoh meets Charles Taylor, who is also undergoing training.

December 25, 1989	Liberian civil war commences with 150 rebel fighters entering Liberia from Côte d'Ivoire.
1990–1991	Training of RUF rebels takes place in Liberia.
Mar 23, 1991	RUF-National Patriotic Front of Liberia (NPFL) forces enter eastern Sierra Leone from Liberia, signaling the commencement of the RUF insurgency.
March 1991 to November 1993	Phase I of the RUF insurgency in eastern Sierra Leone.
April 29, 1992	Coup removes General Joseph Momoh as president and establishes the National Provisional Ruling Council (NRPC) under Captain Valentine Strasser as the national leadership of Sierra Leone.
November 1993 to March 1997	Phase II of the RUF insurgency with use of hit-and-run tactics in eastern and central Sierra Leone.
March 1996	Ahmed Kabbah elected president of Sierra Leone.
November 30, 1996	A peace accord is signed between President Kabbah and Foday Sankoh.
May 25, 1997	A coup by the Armed Forces Revolutionary Council (AFRC) and Major Johnny Paul Koroma established a ruling AFRC-RUF junta.
May 1997 to January 2002	Phase III of the RUF insurgency included direct assaults on Freetown and western Sierra Leone.
March 1998	President Kabbah is returned to power by the Nigerian-led Economic Community of West African States Monitoring Group (ECOMOG).
January 1999	Rebels lay siege to Freetown during Operation No Living Thing.
July 7, 1999	President Kabbah and Foday Sankoh sign Lome accord.
May 2000	During disarmament process, numerous occurrences of rebels refusing to surrender their arms and capturing United Nations (UN) peacekeepers.
September 2000	British expeditionary forces rescued British peacekeepers.
May 16, 2001	Rebels and militia forces agreed to enter disarmament process.
May 19, 2002	President Kabbah is declared the winner of the first post-conflict national elections.

THE ENVIRONMENT OF THE REVOLUTION

PHYSICAL ENVIRONMENT

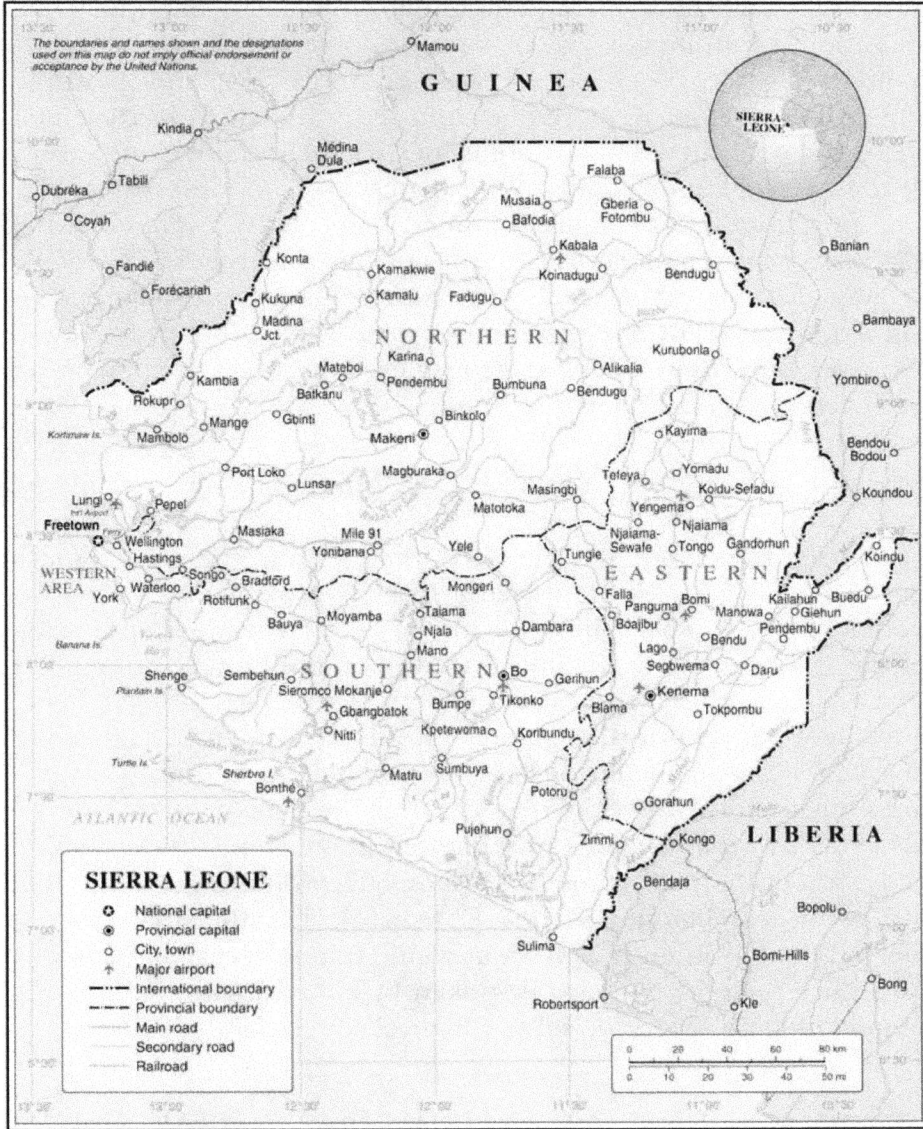

Figure 1. Map of Sierra Leone.[1]

---

[1] Sierra Leone, UN Map No. 3902 Rev. 5 (January 2004), www.un.org/Depts/
Cartographic/map/profile/sierrale.pdf. Reprinted with permission from the United
Nations.

Sierra Leone is a West African country that borders the North Atlantic Ocean and is located between Guinea and Liberia. The country's terrain includes mangrove swamps along the coastline, woodlands and rainforests along the higher regions to the north and east, and a plateau region and mountains in the east. Slightly smaller than South Carolina at 71,740 square kilometers, Sierra Leone has a hot, humid tropical climate with a typical rainy season during the months of May to December and a dry season from December to April. With an average rainfall of 195 inches along the coastal region, Sierra Leone is one of the wettest areas in West Africa.[2] The transportation infrastructure is limited, with only 372 miles of paved highways, approximately 7,026 miles of unpaved roads, 52 miles of railway, 373 miles of navigable waterways, and only one major airport (Lungi, outside of Freetown) with a 10,600-foot paved runway. However, the port of Freetown is considered the largest natural harbor in Africa and one of the largest in the world.

CULTURAL AND DEMOGRAPHIC ENVIRONMENT

During the eleven-year period of the civil war, the population in Sierra Leone grew by almost 25% from approximately 4.3 million people in 1991 to 5.4 million people in 2002.[3] With fifteen different ethnic groups, Sierra Leone's ethnic composition is nonetheless dominated by two groups: the Mende in southeastern Sierra Leone and the Temne in northern and western Sierra Leone. Each group comprises approximately 30% of the population and also serves as a strong political force in regional and national politics. Diversity also exists in the religious mix for the Sierra Leonean population, with 30% of the population as Muslim, 30% with indigenous beliefs, 30% with no religion, and 10% Christian.[4] The official language of Sierra Leone is English with Mende and Temne vernaculars in their respective regions, but Krio is spoken by approximately 95% of the population and is considered the "native" tongue of the country.[5]

---

[2] Central Intelligence Agency, "Nigeria," *The World Factbook*, accessed September 16, 2010, https://www.cia.gov/library/publications/the-world-factbook/geos/sl.html.

[3] At the height of the civil war in 1997–1998, however, approximately 2 million people (more than one-third of the nation's population) were either internally displaced or in a refugee status in neighboring countries because of the hostilities. Ibid.

[4] According to 1990 CIA *World Factbook* estimates. The 2002 *World Factbook* would change these numbers to reflect 60% Muslim, 30% indigenous, and 10% Christian.

[5] Ibid.

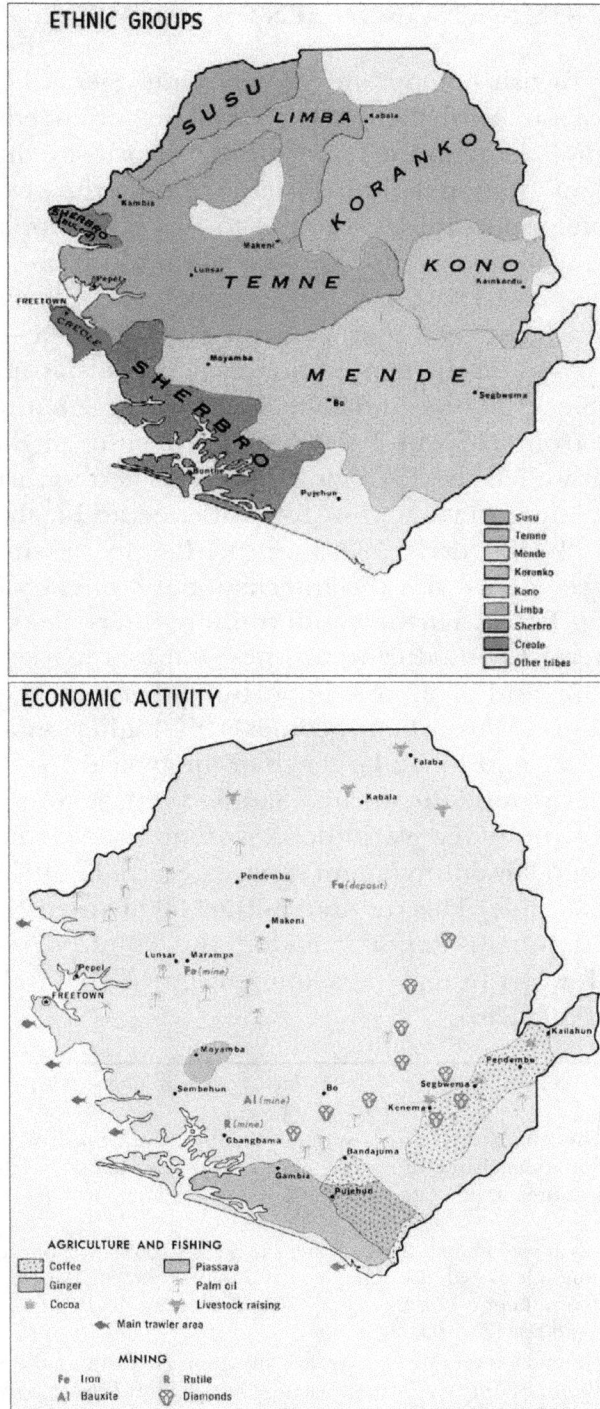

ETHNIC GROUPS

ECONOMIC ACTIVITY

Figure 2. Ethnic groups and economic activity.[6]

---

[6] These maps are from 1969 but provide an accurate depiction of the ethnic distribution and economic activities of Sierra Leone in the period leading up to the 1991 hostilities. Available online at The Perry-Castañeda Library Map Collection, The University

SOCIOECONOMIC ENVIRONMENT

A former British colony and protectorate, Sierra Leone gained independence on April 27, 1961. It then experienced decades of political turmoil, corruption, and ethnic favoritism that undercut the growth of national infrastructure and the expansion of socioeconomic opportunities. Central to this postcolonial period of economic stagnation was the corrupt administration of President Siaka Stevens (1968–1985) and the continuation of Stevens' policies by his chosen successor, General Joseph Momoh (1985–1992). With an economy based on the exportation of diamonds, iron ore, coffee, and cocoa,[7] Sierra Leone initially had a modest 4% annual growth rate in GNP from 1965 to 1973, but the decline in diamond and iron prices, as well as the 1973 global oil crisis, led to further foreign borrowing by the government of Sierra Leone and a sharp increase in inflation.[8] By the early 1980s, Sierra Leone was in default on numerous agreements with the International Monetary Fund (IMF) and the World Bank.[9] Further undermining internal socioeconomic stability in Sierra Leone during this period was a marked difference in quality of life and economic opportunity between the urban and rural populations, with urban income 410% higher on average than rural income, a trend much larger than the typical 50–100% seen in other countries around the world.[10] Saddled with a culture of political patronage and noneffective judicial, administrative, public health, educational, and law-enforcement services, Sierra Leone experienced a steady societal slide. This resulted in the 1990 United Nations (UN) Human Development Report ranking the country as having the world's fourth worst human development index (behind only Niger, Mali, and Burkina Faso).[11]

---

of Texas at Austin, accessed October 18, 2009, http://www.lib.utexas.edu/maps/sierra_leone.html.

[7] Although these natural resources were (and are) the largest sources of revenue in the country, approximately two-thirds of the working-age population is engaged in subsistence agriculture that generated one-third of gross national product (GNP) in 1990. Ibid.

[8] Inflation was approximately 2.1% during the period 1965–1973 but rose to 50% in the 1980s, with economic growth dropping from 4% to 0.7% during these same periods. Michael Chege, "Sierra Leone: The State that Came Back from the Dead," *The Washington Quarterly* 25, no. 3 (2002): 147–160.

[9] A detailed discussion on IMF and World Bank efforts to prop up the Sierra Leonean economy can be found in *National Integrity Systems: Country Study Report – Sierra Leone, 2004* (Berlin: Transparency International, 2004).

[10] *Human Development Indicators Report 1990* (New York: United Nations Development Programme, 1990), http://hdr.undp.org/en/reports/global/hdr1990/.

[11] Ibid. In the 2007/2008 report, Sierra Leone ranked last (#177) in the world according to the HDI. Factors included in the HDI ranking are life expectancy, access to health services, access to safe water, access to sanitation, adult literacy, GNP per

HISTORICAL FACTORS

In addition to the dominant culture of corruption and political patronage within postcolonial Sierra Leone, key pillars of government, such as an independent judiciary and law-enforcement organization, a fiscally responsible treasury, and a trained, disciplined military all eroded under the All People's Congress (APC) governments of Siaka Stevens and Joseph Momoh. Student protests in 1977 and 1984–1985, teacher and labor strikes, as well as external pressures from the IMF and other foreign donors finally forced the Momoh government in 1991 to appoint a National Constitutional Review Commission (NCRC). The recommendations of this commission and the results of a national referendum eventually led to the adoption of a new multiparty constitution in Sierra Leone in 1991.[12]

By this point, however, a small opposition movement had emerged that was composed of expelled student radicals[13] living in Ghana with connections to the Revolutionary Council of Libya and unemployed Sierra Leonean youth living in Liberia and in Sierra Leone. Drawn together by a mixture of individual aspiration and common cause against the corrupt rule of President Momoh, this small cadre benefited from the quiet sponsorship of neighboring governments. Among the student group was an activist named Alie Kabba, the student union president from Fourah Bay College (FBC), who espoused the Green Book teachings of Gaddafi and who went to Ghana and then Libya in 1987 with several other students who were expelled from FBC.[14] This

---

capita, real GNP per capita, and calorie supply as a percentage of daily requirements. In addition, a 2010 United Nations Development Programme (UNDP) report states that the unemployment rate for youth in Sierra Leone is currently at 60% and is combined with a low overall adult literacy rate of 31%. See "Youth in Sierra Leone Find Hope in Entrepreneurship" (New York: United Nations Development Programme), accessed September 16, 2010, http://content.undp.org/go/newsroom/2010/january/youth-in-sierra-leone-find-hope-in-entrepreneurship.en.

[12] Jimmy D. Kandeh, "Transition without Rupture: Sierra Leone's Transfer Election of 1996," *African Studies Review* 41, no. 2 (1998), 91–111.

[13] The enforcement of a one-party political system under the APC resulted in university-based radical groups and "study clubs" emerging as the only visible opposition to the government. After a series of protests and acts of civil disobedience by students at Fourah Bay College (FBC) in 1985, leaders of these groups were expelled from the college, with some transferring to Ghana in order to complete their studies. It was in this environment that the student radicals became acquainted with members of the Revolutionary Council of Libya, Charles Taylor from Liberia, and Corporal Foday Sankoh from Sierra Leone, who would emerge as the leader of the RUF. *Truth and Reconciliation Commission for Sierra Leone – Final Report*, Truth and Reconciliation Commission for Sierra Leone (TRC), 2004, www.sierra-leone.org/TRCDocuments.html.

[14] Ibrahim Abdullah, "Bush Path to Destruction: The Origin and Character of the Revolutionary United Front/Sierra Leone," *The Journal of Modern African Studies* 36, no. 2 (1998): 203–235; Lansana Gberie, *A Dirty War in West Africa: The RUF and the Destruction of Sierra Leone* (Bloomington, IL: Indiana University Press, 2005).

group was joined by other Sierra Leoneans in Libya, including Foday Sankoh, a former corporal in the Sierra Leone army who had served seven years in prison (1971–1978) for not warning officials about a 1971 coup plot against then-President Stevens. The more militant members of the opposition group, including Sankoh, eventually broke off from the more idealistic and less action-oriented student radicals and formed the core leadership of the RUF.

Another significant historical factor that paved the way for the rise of the RUF was the integration of lower-class, unemployed urban youth with their counterpart middle-class high school and university youth around the drug culture and social environment of the cities' *potes*.[15] The lack of employment opportunities for both the uneducated and the educated youth of the country fostered a breeding ground for social discontent and a gray market economy, which forged personal relationships that would become significant when these youth became the foundation for both the military and insurgent forces during the civil war.

A final critical factor that contributed to the rise of the RUF was the outbreak of a civil war in 1989 in neighboring Liberia between the ruling government of President Samuel Doe and the insurgent National Patriotic Front of Liberia (NPFL) forces under the operational leadership of Prince Yormeh Johnson and the political leadership of Charles Taylor.[16] Some members of the Sierra Leonean opposition movement living in Ghana and Libya joined the NPFL forces during the early campaigns against Doe, and this paved the way for personal relationships, as well as some basic training and operational experience for the fighters from Sierra Leone. As detailed below, the NPFL would directly participate in—and in some cases lead—the early RUF incursions into Sierra Leone. Moreover, an additional incentive for Charles Taylor to support the RUF insurgency centered on his anger toward President Momoh in Sierra Leone after Momoh allowed multinational military operations by the Economic Community of West African States (ECOWAS) to be based out of Freetown in support of Samuel Doe and against Taylor's NPFL forces.

---

[15] The *pote* is the historical name in Sierra Leone for "a popular peri-urban area of relaxation for unemployed youths" and became the center for the emerging drug and reggae culture of youth in Sierra Leone. Abdullah, "Bush Path to Destruction."

[16] Paul Richards, *Fighting for the Rain Forest: War, Youth and Resources in Sierra Leone* (Portsmouth, NH: Heinemann, 1996).

GOVERNING ENVIRONMENT

In the immediate years after independence, politics in Sierra Leone were based on a true Westminster-style constitution and a multiparty political system that was "open, competitive, representative, and accountable."[17] At the local and regional levels, however, the historical chieftaincy system remained in place and continued to be manipulated for personal gain across all levels of government, including the national level. The chieftaincy system encompassed— and continues to encompass—a "chief" as a local source of authority and often "the only visible element of government."[18] The underlying principle for the chieftaincy was based on the concept of a social contract in which the citizens of a village provide gifts, services, and other forms of compensation in return for the chief's supervision of the two primary tasks of the chiefdom: "collecting local taxes and maintaining security."[19] The administrative structure and authorities for the 149 chiefdoms in Sierra Leone were outlined in colonial-era legislation: the Tribal Authorities Ordinance of 1938, the Chiefdom Treasuries Act of 1938, and the Tribal Authorities (Amendment) Act of 1964. In addition, the continued existence of the chieftaincy system and the offices of the paramount chiefs were specifically guaranteed within Sierra Leone's 1991 constitution.[20] Under the British colonial system, however, a "district officer" was appointed above the chief and provided local oversight for the central government, thus making chiefs more accountable to national leaders. It was the creation of this artificial bureaucratic layer that further corrupted the chieftaincy system by injecting national politics and patronage into local affairs, thereby leading the Truth and Reconciliation Commission (TRC) for Sierra Leone to conclude:

---

[17] Abdullah Abraham, "Dancing with the Chameleon: Sierra Leone and the Elusive Quest for Peace," *Journal of Contemporary African Studies* 19, no. 2 (2001): 205–228.

[18] Paul Jackson, "Reshuffling an Old Deck of Cards? The Politics of Local Government Reform in Sierra Leone." *African Affairs* 106, no. 422 (2007): 95–111. There are several different levels of chief within Sierra Leone, with the greatest degree of authority and power residing with the paramount chief. A counter-argument to these negative views of the chieftaincy system in Sierra Leone is the perspective that this system is still highly regarded and accepted in rural areas and villages where section chiefs and headmen play a critical role in settling day-to-day, small disputes. See Edward Sawyer, "Remove Or Reform? A Case for (Restructuring) Chiefdom Governance in Post-Conflict Sierra Leone," *African Affairs* 107, no. 428 (2009): 387–403.

[19] Jackson, "Reshuffling an Old Deck of Cards," 95–111.

[20] "The institution of chieftaincy as established by customary law and usage and its non-abolition by legislation is hereby guaranteed and preserved." *The Constitution of Sierra Leone, 1991,* accessed September 16, 2010, http://www.sierra-leone.org/Laws/constitution1991.pdf.

> The Commission finds that the Colonial government manipulated the Chieftaincy system and, in so doing, undermined its legitimacy. The Chiefs became mere surrogates of the colonial government. They owed their loyalty to their colonial masters rather than to the people they were meant to serve.[21]

In addition to this inversion of the chieftaincy scope of responsibility (accountable to national leaders rather than local citizens), the centralization of political power at the national level by Siaka Stevens in 1978 with the declaration and enforcement of a one-party system (in favor of the APC) eliminated almost all remaining aspects of organized political opposition and true democracy in the country. The use of state-sponsored violence and youth gangs to influence local and national elections became a hallmark of the APC's time in office and would set a standard and ominous tone for youth-based "thuggery"[22] within cities and in rural locations across Sierra Leone.

WEAKNESSES OF THE PREREVOLUTIONARY ENVIRONMENT AND CATALYSTS

The previous discussion highlights numerous factors that emerged in postcolonial Sierra Leone to undermine the pillars of government and instigate public discontent and despair. As detailed below, these factors may have served as catalysts for the emergence of the RUF; however, this emergence did not constitute a "popular uprising" against the government. A popular uprising would imply broad public support for a revolutionary movement that had the interests of the common citizen in mind. Instead, individual opportunistic ventures by internal and external actors combined with a readily available pool of unemployed young males provided the key and necessary ingredients for the emergence of the RUF. The unemployed young males of society (often referred to as "lumpen youth") hailed from the urban centers, diamond-mining districts, and rural corners of Sierra Leone.

---

[21] *Truth and Reconciliation Commission for Sierra Leone – Final Report.*

[22] As Arthur Abraham describes, "the dangerous introduction of thuggery into the political landscape—the recruitment and mobilisation of unemployed or under-employed youths from urban and diamond-mining areas to browbeat, under the influence of drugs deliberately fed them, opponents in all parts of the country" would reemerge as a basic tool of the RUF in the 1990s. Moreover, "these 'lumpen' youths maintained clientelist relations with the party, and increasingly came to be recruited into the army." Abraham, "Dancing with the Chameleon," 205–228. The use of "thuggery" for political activities was by no means an exclusive tactic of the APC but is noted here because of its contribution to the APC's stranglehold on national power. On this point, see Maya M. Christensen and Mats Utas, "Mercenaries of Democracy: The 'Politricks' of Remobilized Combatants in the 2007 General Elections, Sierra Leone," *African Affairs* 107, no. 429 (2008): 515–539.

As Sierra Leone's economy continued to fail, its population continued to grow. These factors, along with continued government corruption, caused the ranks of the lumpen youth to steadily grow.[23] By the time of the 1991 RUF invasion into Sierra Leone from Liberia, unemployed, disillusioned youth were a critical component of not only the RUF fighting force but also the armed forces of Sierra Leone and the fighting forces of other insurgent groups and militia elements in the country. "Lumpen youth," therefore, became a common ingredient for all of the internal combatant forces during the decade-long civil war and often resulted in strange alliances and the switching of allegiances based on preexisting personal relationships and opportunism rather than loyalty to a higher cause or ideology.[24] In this regard, the chaos and violence that defined the Sierra Leone civil war directly reflected the day-to-day uncertainty, chaos, and attitude of self-preservation of the country's youth culture, while the failures of state governance provided a catalyst and opportunity for this youth culture to take action.[25]

FORM AND CHARACTERISTICS OF THE REVOLUTION

OBJECTIVES AND GOALS

Perhaps the greatest topic of uncertainty surrounding the RUF insurgency is the basic question of motive and why the armed struggle occurred when it did.[26] As detailed above, there were clear reasons for grievance against the ruling APC government as the level of corruption and the dwindling quality of life factors provided incentive for social discontent and revolt. However, it is not as easy to identify a specific triggering event or factor that led to the formation of the

---

[23] Ibrahim Abdullah, "Youth Culture and Rebellion: Understanding Sierra Leone's Wasted Decade," *Critical Arts* 16, no. 2 (2002): 28.

[24] The Truth and Reconciliation Commission highlighted this "factional fluidity" and "chameleonic tendencies" of the lumpen youth to change allegiances as an astonishing and unique characteristic of the Sierra Leone conflict.

[25] For additional background on the lumpen youth culture in Sierra Leone and its influence on the civil war, see Ismail Rashid, "Student Radicals, Lumpen Youth, and the Origins of the Revolutionary Groups in Sierra Leone, 1977–1996," in *Between Democracy and Terror: The Sierra Leone Civil War*, ed Ibrahim Abdullah (Dakar, Senegal: Codesria, 2000); and Lansana Gberie, "The 25 May Coup d'Etat in Sierra Leone: A Lumpen Revolt?" in Ibid.

[26] Gberie is perhaps more direct to the point when he writes "It is a mark of the mercenary character of the RUF's war that nearly ten years after it began observers were still struggling to find not just a coherent explanation for its remarkably brutal nature— demonstrated by the amputations—but also the motivation behind waging the war itself." Gberie, *A Dirty War in West Africa*.

RUF and the initiation of armed insurrection. Previous efforts to unseat an unpopular government were focused primarily in Freetown on key seats of governmental power and often took the shape of a military coup.[27] But in the late 1980s, the training, networking, and financial support provided first by Gaddafi and then by Charles Taylor enabled the gathering of a select group of Sierra Leoneans who were motivated and willing to take action. Perhaps most critically, however, the personal motives of Gaddafi and Taylor—as well as President Blaise Compaoré in Burkina Faso—would shape the design and implementation of the RUF offensive into Sierra Leone, with particular focus placed on controlling the diamond-rich regions in eastern Sierra Leone rather than the seat of governmental power in western Sierra Leone. In this regard, the supposed "revolutionary" character of the RUF took a subservient role to the commercialist objectives of the external sponsors and underscored the fact that the RUF was not a popular uprising or a "people's movement" despite the insurgency contributing to the eventual removal from power of the APC.[28]

LEADERSHIP AND ORGANIZATIONAL STRUCTURE

The identities of the key political and operational leaders of the RUF during the decade-long civil war in Sierra Leone seem to be well understood, although some uncertainty continues to surround the true lines of authority and command during this chaotic period.[29] Despite eyewitness accounts at the scenes of attacks that provide a vivid record of RUF tactical decision-making, the links between these tactical operations and the strategic leadership (and strategic motives) are more difficult to trace, especially due to a limited written record

---

[27] An exception was the student demonstrations at FBC in 1984–1985 that led to the expulsion of forty-one students. Among this group was a student activist named Alie Kabba who would eventually establish the seeds of a Sierra Leonean revolutionary movement in Libya under the sponsorship of Gaddafi. Ibid.

[28] In a 1996 interview with Lansana Gberie, Foday Sankoh—the leader of the RUF insurgency—responded to a question about the objective of the RUF campaign by stating, "I did not start a war. It is a people's struggle. The people rose up against the rotten APC system. Before it all started, everyone was crying for war. Everybody wanted the rotten APC to be overthrown. Now that there is a people's struggle against the system, why should I be blamed for it? You ask me about war aims. Everyone knows what Sierra Leoneans want: free education, free health care, proper use of our natural resources, provision of basic necessities which the politicians have denied them. That's what the people's struggle is all about." Ibid. As Gberie and others point out, however, the RUF initiated its insurgency without ever publicly issuing a manifesto or political agenda, and it would not be until five or six years into the civil war that this rhetoric of a people's war became more standardized among RUF leaders in attempts to appease a growing international audience. A copy of the RUF Manifest called "Footpaths to Democracy: Towards a New Sierra Leone" was released in 1995 and can be found online: http://www.sierra-leone.org/AFRC-RUF/footpaths.html.

[29] Key leaders within the RUF included Foday Sankoh, Rashid Mansaray, Sam Bockarie, Gibril Massaquoi, Denis Mindo, Isaac Mongor, and Issa Sesay.

of RUF battlefield orders. Furthermore, the weak communication infrastructure and mobile/foraging nature of the RUF rebel forces resulted in long periods of time in which specific instructions were not received from the senior RUF leadership.[30] However, the systematic and consistent manner in which the tactical operations were conducted—as well as the presence of senior RUF leaders at numerous villages and towns that were attacked by the group—demonstrated clear awareness, guidance, and endorsement by these senior leaders of the tactical operations that were occurring.[31]

Witness testimony to the Special Court for Sierra Leone identified three main groups of rebels that made up the RUF organization. The first group was called the "Vanguards," which encompassed the core group of RUF leaders and fighters who trained with Foday Sankoh in Liberia.[32] The second group within the RUF was the "Special Forces"—the rebels who received military training in Libya and, in most cases, also had combat experience fighting with the NPFL. This small group was often distinguished by the US camouflage uniforms they wore during the early stages of the civil war. The final group of rebels within the RUF were the "Junior Commandos" that included all of the fighters recruited (or abducted) and trained in Sierra Leone. Although these distinctions existed and seemed to be recognized by many members of the RUF, the three groups were intermingled and did not operate as unique, stand-alone combat units within the RUF. In this sense, the "Vanguard" and "Special Forces" labels primarily represented a status within the RUF hierarchy, as well as a means of identifying original members of the RUF insurgency.

Figure 3 is a hierarchical depiction of the chain of command within the RUF and is based on testimony provided by several witnesses during the proceedings of the Special Court for Sierra Leone.[33] Most likely influenced by the limited Western-style training that the Vanguards and Special Forces members received in Libya and Liberia, this type of organizational structure nonetheless demonstrates an attempt at defining authorities and responsibilities for members of the RUF.[34]

---

[30] The RUF relied heavily on the use of commercial radio networks to broadcast general guidance to their rebel forces throughout the country.

[31] An exception is Charles Taylor of Liberia, who was never seen entering Sierra Leone, although his logistical and financial support of the RUF is well documented.

[32] According to Gibril Massaquoi (former spokesperson for the RUF), the Vanguards also provided ideological training to members of the RUF in Liberia during the months leading up to the invasion of Sierra Leone. Testimony of Gibril Massaquoi, October 7, 2005, 12.

[33] Witness TF1-045, Case No. SCSL-2004-16-T (July 20, 2005; 9:17 a.m.).

[34] One witness to the Special Court for Sierra Leone described an incident in 1993 when a battalion commander was relieved of command for failure to follow the operational orders given to him by the battle group commander.

In terms of practicality and implementation, however, much of the operational decision-making rested with the battalion commanders and—if communications existed—included direction from the area commander or battle group commander. The result of these limited lines of communication, as well as the limited overall number of combatants within the RUF, was the execution of small-scale operations by "Target" units that ranged in size from a platoon or company-size group and were often led by a "Junior Commando" who had proven himself in combat. Because the "Target" units usually had much fewer fighters than the one hundred depicted in Figure 3, the RUF also tended to avoid force-on-force confrontations with the Sierra Leone Army and instead attacked "soft" civilian objectives, such as villages and towns.

In addition to its chain of command, the RUF had personnel designated to General staff positions (G-1, G-2, G-3, G-4, and G-5) as well as Special staff (S-1, etc.) positions. The G-1 was responsible for recruiting and training, the G-2 provided intelligence and counterintelligence, the G-3 was in charge of general administration, the G-4 coordinated materials and supplies, and the G-5 was in charge of civilian matters (such as ideology and coordinating relief supplies). These positions were considered to be executive positions and reported directly to the commander-in-chief of the RUF.[35] Finally, starting in 1993, the RUF leadership was advised by a "War Council" that included senior rebel commanders and some prominent civilians.[36]

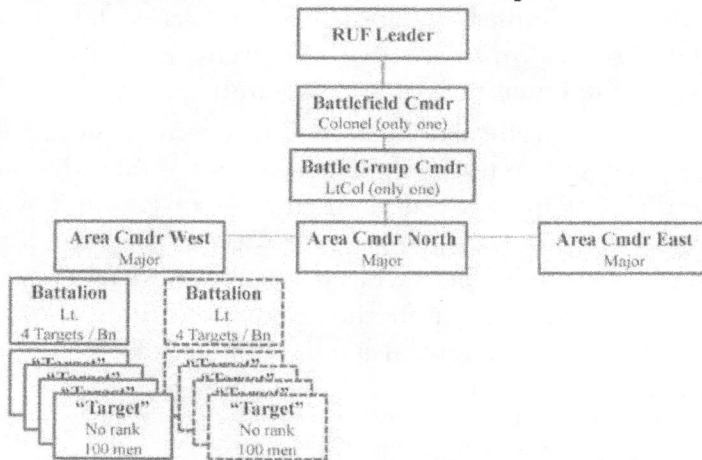

Figure 3. RUF organization.[37]

---

[35] Testimony of Witness TF1-168, Case No. SCSL-2004-15-T (April 3, 2006, 12:40 p.m., and March 31, 2006), 90–96. Note that although the names are the same, the specific functions of the General staff and Special staff do not mirror those of the US military. Court records show that this unnamed witness served as a G-5 for the RUF.

[36] Testimony of Witness TF1-168, Case No. SCSL-2004-15-T (April 3, 2006), 62–63.

[37] This diagram is generated from the courtroom testimony of numerous witnesses to the Special Court for Sierra Leone, www.sc-sl.org/.

Despite this potential appearance of military structure and discipline, the fact remains that the RUF was composed of mercenaries, student radicals, unemployed youth, discharged junior soldiers, and other people whose identity was based on opposition to "the system" as well as self-preservation. In this regard, it is no surprise that internal power struggles emerged almost immediately within the RUF leadership structure, to include two significant events in 1993. The first event involved the execution of Rashid Mansaray (second-in-command of the RUF) and forty other RUF members in Kailahun district by Foday Sankoh and Sam Bockarie, a senior RUF field commander, because of personal rivalries. The second event was the torture and execution of twenty-five members of the RUF's First Battalion (mostly Vanguards from northern Sierra Leone) in Pujehun district by Gibril Massaquoi, a RUF commander and soon-to-be RUF spokesperson, and other Mende from southern Sierra Leone in an attempt to shift the leadership structure of the RUF to those of southern ancestry.[38] These internal purges, as well as actual combat losses, eliminated some of the most popular and competent leaders within the movement and decimated the limited military expertise of the RUF during the early years of the conflict.

COMMUNICATIONS

Because of a limited technical infrastructure within Sierra Leone, the RUF relied on somewhat simplistic means of communication throughout the eleven years of the civil war. This included the use of "runners" to hand deliver messages between RUF leaders and also the use of some radio relay capabilities, such as field radios in Burkina Faso and Côte d'Ivoire, which allowed one-way messaging from Sankoh down to forces in the field. Given the primitive state of the RUF's communications, the group sometimes struggled to maintain communications with each other when operating in different towns. For example, from 1992 to 1994, the RUF forces operating in Kailahun and in Pujehun were effectively split from each other and had very limited communication with each other.[39]

Because the population was mostly illiterate, mass communication was conducted by the RUF through radio broadcasts. Specifically,

---

[38] *Truth and Reconciliation Commission for Sierra Leone – Final Report*. On this specific point, the TRC states: "The Commission finds that the majority of killings of key RUF commanders between 1991 and 1993 were attributable not to battlefield casualties, but to lethal manifestations of acrimony, rivalry and personal vendettas." paragraph 122.

[39] Special Court for Sierra Leone Testimony of Mr. Palmer, Witness TFI-168, Case No. SCSL-2004-15-T (April 3, 2006), 10. Rumors also persisted during this period that Sankoh had been wounded or killed.

the RUF made heavy use of the BBC Africa Service and employed interviews with its leaders in order to communicate to external and internal audiences. It was one such BBC interview with Sam Bockarie that enabled him to direct RUF forces to descend on Freetown for the commencement of "Operation No Living Thing" in January 2001. In the latter stages of the civil war, the RUF also received training and operational support from some South African mercenaries, and this assistance led to the use of the old British Slidex encryption code for important internal communications.[40] Finally, telephones were used intermittently by Sankoh to coordinate arms and diamond shipments, and he is reported to have had access to a satellite phone provided by Charles Taylor for the early years of the insurgency.

At a higher level, the notion of "communication" to the population that traditionally corresponds with an insurgency was not witnessed in Sierra Leone. The formal manifesto of the RUF was released four years after the initial incursions in the eastern province and provided little insight into its political agenda. More importantly, the repressed and impoverished population that the RUF was supposedly rescuing from the corruption of the APC was often the victim of RUF atrocities. Therefore, the RUF provided no concerted political message to the people of Sierra Leone and actually victimized this supposed audience during the course of the civil war.

METHODS OF ACTION AND VIOLENCE

The RUF insurgency covered a period of approximately eleven years (1991–2002) and involved three distinct operational periods that were defined by cease-fires and peace accords. Per an assessment by the TRC database of the total number of recorded violations by all combatant forces that occurred during the course of the civil war, the highest levels of violence were reported in 1991, 1995, and 1998–1999, and the lowest levels were in 1993, 1996, and 2000.[41]

---

[40] Al J. Venter, *War Dog: Fighting Other People's Wars.* (Drexel Hill, PA: Casemate, 2008).

[41] In addition to the RUF, high numbers of atrocities were committed by government forces, other rebel forces, and civil defense militias. Richard Conibere et al., "Statistical Appendix to the Report of the Truth and Reconciliation Commission of Sierra Leone," *A Report by the Benetech Human Rights Data Analysis Group and the American Bar Association Central European and Eurasian Law Initiative to the Truth and Reconciliation Commission* (2004). The three peaks on the graph coincide with the initiation of hostilities in 1991, a major RUF offensive in 1995, and the 1999 invasion of Freetown. The interviews by the TRC recorded seventeen types of violations that were conducted during the civil war: abduction, amputation, arbitrary detention, assault/beating, destruction of property, drugging, extortion, forced cannibalism, forced displacement, forced labor, forced recruitment, killing, looting, physical torture, rape, sexual abuse, and sexual slavery.

Violence was used as a tool of intimidation and was orchestrated at very high levels within the RUF organization in order to achieve near-term military objectives. During the harvest season, the RUF attacked farming villages and amputated the hands or arms of field workers as a warning to others that the crops should not be harvested, thus denying the government of this food supply. During the 1996 national elections, the RUF conducted "Operation Stop Elections" and amputated hands as a symbolic gesture of preventing people from voting. The RUF also routinely killed village elders and left their bodies or heads on public display as a warning to the rest of the village to not support government forces. This was complimented by "false flag" attacks in which the RUF wore Sierra Leone Army uniforms during some of their raids on villages, thus undermining public trust in the army. Finally, the RUF conducted large numbers of rapes and abductions against female civilians, although the specific numbers are difficult to verify because of under-reporting. In many cases, these atrocities were conducted by child soldiers within the RUF as a means of indoctrination and forced loyalty to the RUF.

If specific focus is placed on those violations believed to have been conducted by the RUF, as well as the location and year of those violations, the RUF insurgency seemed to transition from the east–southeast in the first phase of the conflict (March 1991 to November 1993), toward the southern region during the second phase (November 1993 to March 1997), and then finally into the north and west in the third and final phase of the civil war (March 1997 to January 2002).

Phase Zero—Pre-Invasion

Some of the most detailed information concerning the military preparations of the RUF in late 1990 and early 1991 comes from General John Tarnue, who at the time was the commander of training for the Liberian army and eventually served as the Commanding General of the Armed Forces of Liberia under Charles Taylor from 1999 to 2003.[42] Tarnue testified that in November 1990, he became aware of complaints by NPFL soldiers concerning the presence of "foreign soldiers" at Liberian training facilities, and these complaints prompted an inventory of all trainees that determined that ninety-six

---

[12] General Tarnue provided extensive testimony to the Special Court for Sierra Leone in 2005, and his perspective of events is extremely insightful because of the access he reportedly had to the inner circle of RUF planning as well as his use of Western military terminology during his testimony (General Tarnue received training at both Fort Benning and Fort Gordon early in his career). Pro-Taylor critics of his testimony argue that Tarnue provided the testimony in exchange for the safe relocation of himself and his family.

Sierra Leoneans were undergoing training at Konola Academy.[43] At approximately the same time, General Tarnue claimed that Charles Taylor introduced him to Foday Sankoh:

> This is Foday Sankoh. He's my friend, my old-time friend. We did our training in Libya. . . . Because of the inadequacies of the commandos squad that were trained along with him in Sierra Leone from Libya, I had decided to tell him—encourage him to have a military alliance that he, Foday Sankoh, together with his men would help me fight the war in Liberia to capture, and when Liberia is capture (sic), in return I would also provide manpower, ammunitions and other thing to be able to help him.
>
> Tarnue, you know that's what we call "susu" in Liberia.[44]

In February 1991, General Tarnue reported to Charles Taylor that the training of the ninety-six fighters from Sierra Leone was complete. Taylor sent trucks to have the Sierra Leoneans taken to Camp Naama, a former Liberian artillery base. There, they were joined with 150 NPFL fighters who were to serve as the "strikers"[45] for the offensive into Sierra Leone, as well as fifteen "special forces" personnel from Gambia.[46] Also joining this mix were two separately trained Sierra Leonean commando squads[47] as well as Foday Sankoh,

---

[43] When asked to describe the type of training provided to the Sierra Leoneans, Tarnue stated that "revolution is not like training people to become a professional soldier. Of course, the discipline was taught. We taught them discipline; we taught them drills and ceremonies and we taught tactics; we taught them weapons and we taught them courtesy and discipline, the physical fitness and then the CQ: Cover, concealment and camouflage." General John Tarnue testimony, October 4, 2004, 94.

[44] When asked by the prosecutor, Tarnue explained that "the word 'susu' in Liberia typically is just a colloquialism. It is something like: 'You help me first and when I succeed, then I will help you second.'" General John Tarnue testimony, October 4, 2004, 79–80. During his own trial for war crimes related to the Sierra Leone civil war, Charles Taylor denied supporting the RUF in its invasion of Sierra Leone. Rather, he claims to have provided limited support between August 1991 and May 1992 as the RUF and the NPFL were "jointly fighting a common enemy, happening to be ULIMO." July 14, 2009, Open Session testimony of Charles Taylor, The Hague. Transcript page 24329.

[45] The NPFL referred to "strikers" as their battle-hardened, experienced forces who would serve as the lead element of their offensive operations. Upon seizing an objective, the strikers would turn the objective over to the regular NPFL forces—often child soldiers—who would establish a defensive stance. These specific strike forces were brought to Camp Naama from Lofa County, Liberia, were they were engaged in combat operations.

[46] The Gambian special forces were reportedly out of Burkina Faso and fought within the NPFL organizational structure. It should be noted that the term "special forces" is often used to describe military personnel who were trained in another country and does not equate to these personnel having specialized military training or capabilities.

[47] These two squads are reported by General Tarnueto to have trained in Libya and had combat experience fighting alongside NPFL forces.

bringing the total number of initial fighters for the invasion into Sierra Leone to 292.[48]

On February 27, 1991, a meeting was held at Taylor's Executive Mansion at Gboveh Hill, Gbarnga (the NPFL headquarters). According to General Tarnue, Charles Taylor reported to the group that:

> I have already instructed the G4 to send six-man diesel trucks that were looted from Bong's mining company. I have already instructed to make available 100 AK-47 Kalashnikov weapons. I have already instructed the G4 to make 50 Berettas available. I have already instructed the G4 to make 20 RPGs available. I have already instructed the G4 to make ten LAR.[49]

Phase I (March 1991 to November 1993)

The RUF insurgency officially began on March 23, 1991, in the eastern region of Sierra Leone with RUF forces (augmented by the NPFL and Gambian mercenaries) coming out of Liberia. The initial attack was against the border towns of Bomaru and Sienga in Kailahun District followed by a withdrawal back into Liberia. Reports of this RUF-NPFL attack were initially dismissed by the Sierra Leone Army back in Freetown as a continuation of the ongoing cross-border looting raids that were being conducted by Taylor's forces.[50] Four days later, 300 heavily armed fighters occupied the town of Buedu and then moved on to Koindu, the main commercial center for the district. After a series of counter-attacks and modest success by the Sierra Leone Army (which numbered 3,000 soldiers total at the time), the RUF-NPFL forces opened up a second front by attacking towns and villages in the southern district of Pujehun a few weeks later.[51]

---

[48] General John Tarnue testimony, October 4, 2004, 99–100.

[49] General John Tarnue testimony, October 4, 2004, 105.

[50] On December 18, 1990, approximately 100 NPFL fighters crossed the border from Liberia and looted the village of Kissy-Tongay in eastern Sierra Leone; this was preceded by a separate looting attack by the NPFL in Kailahun district (Gberie, *A Dirty War in West Africa*). However, others—such as Brigadier Jusu Gottor, the Sierra Leone Army Chief of Staff—believed that the RUF incursions into Pujehun district were the continuation of the 1982 Ndorgborwusui crisis that involved a local uprising against an APC politician accused of rigging an election (Ibid.). Unlike previous NPFL looting excursions, however, this first RUF attack resulted in the death of an army major and lieutenant from the Sierra Leone army.

[51] Concerning the initial wave of atrocities conducted during the RUF insurgency, the TRC concluded that "The majority of violations attributed to the RUF in the period between March 1991 and September 1992 were in fact the acts of commandos fighting on behalf of the NPFL. In the Commission's view the NPFL faction, under the indisputable overall command of Charles Taylor, was chiefly responsible for the bulk of the abuse inflicted on the civilian populations of Pujehun and Kailahun Districts, in particular, during this period. The Commission finds further that the NPFL component of the

The expansion of the RUF insurgency to a two-pronged war in the eastern region significantly stretched the resources of the ill-equipped and poorly trained Sierra Leone Army.[52] Moreover, the RUF avoided direct engagements with the Sierra Leone Army and instead selected civilian targets. However, it was during this period that the United Liberation Movement of Liberia (ULIMO), which was made up of refugees from Liberia as well as the remnants of Doe's army, began offensive operations along the Liberian border against the NPFL and thus eventually provided a source of pressure on Taylor to recall his forces operating in Sierra Leone.[53] With limited interest in expanding its operations beyond the resource-rich eastern region and facing a poorly equipped and trained Sierra Leone Army, the RUF enjoyed relative freedom of movement and action during the first year of the conflict despite its own limited capabilities.[54]

---

initial incursion force that subsequently entered Sierra Leone outnumbered the RUF "vanguards" by at least four to one. The Commission finds that the NPFL forces were primarily responsible for the initial peak in brutality against civilians and, especially, against traditional and state authorities that were the hallmark of the first year of the conflict." Ibid. paragraph 382.

[52] It is believed that Stevens intentionally kept the army small in size and limited in terms of capabilities in order to reduce the likelihood of a coup.

[53] TRC survey data indicate that many of the Liberian rebels had departed Sierra Leone by the end of 1994. For those documented incidents of RUF violations, Liberians were involved in 78% in 1991, 69% in 1992, 21% in 1993, and 13% in 1994.

[54] Freedom of movement is a military term that means that a military force is able to operate freely in a certain area because they are unopposed or vastly superior to the opposing force. In this case, the RUF was facing a weak opponent even though the RUF's own military capabilities were fairly limited.

The boundaries and names shown and the designations used on this map do not imply official endorsement or acceptance by the United Nations.

Mamou

G U I N E A

SIERRA LEONE

Kindia

Médina Dula

Falaba

Dubréka Tabili

Musaia Gberia Fotombu

Banian

Coyah

Bafodia

Kabala

Konta

Fandié

Kamakwie Koinadugu Bendugu

Forécariah

Kukuna Kamalu Fadugu

Bambaya

Madina Jct.

N O R T H E R N

Karina Kuruborla

Mateboi Alikalia

Yombiro

Kambia Pendembu Bumbuna Bendugu

Batkanu

Rokupr Mange Gbinti Binkolo

Kayima

Kortimaw Is. Mambolo Makeni

Bendou Bodou

Port Loko Lunsar Magburaka Tefeya Yomadu

Koidu-Sefadu Koundou

Masingbi

Lungi Pepel Matotoka Yengema Njaiama

Freetown Masiaka Mile 91 Njaiama-Sewafe Tongo Gandorhun

Wellington Yonibana Yele Tungie

Hastings Koindu

WESTERN AREA Songo Bradford Mongeri E A S T E R N

York Waterloo Rotifunk Falla Bomi Kailahur Suedu

Panguma Manowa Gi n

Bauya Moyamba Taiama Boajibu Pendembu

Banana Is. Njala Dambara Bendu

Mano Lago

Shenge Sembehun Segbwema Daru

Plantain Is. Sieromco Mckanje Bo Gerihun Kenema

Gbangbatok Bumpe Tikonko Blama Tokpombu

Nitti Kpetewoma Koribundu

Turtle Is. Matru Sumbuya

Sherbro I. Bonthe Potoru Gorahun

Mar–Apr 1991

ATLANTIC OCEAN Pujehun Zimmi Kongo L I B E R I A

Bendaja

Bopolu

SIERRA LEONE

◎ National capital
◉ Provincial capital
○ City, town
✈ Major airport
–··– International boundary
–·– Provincial boundary
······ Main road
—— Secondary road
–···– Railroad

Sulima Bomi-Hills

Bong

Robertsport Kle

| 0 | 20 | 40 | 60 | 80 km |
| 0 | 10 | 20 | 30 | 40 | 50 mi |

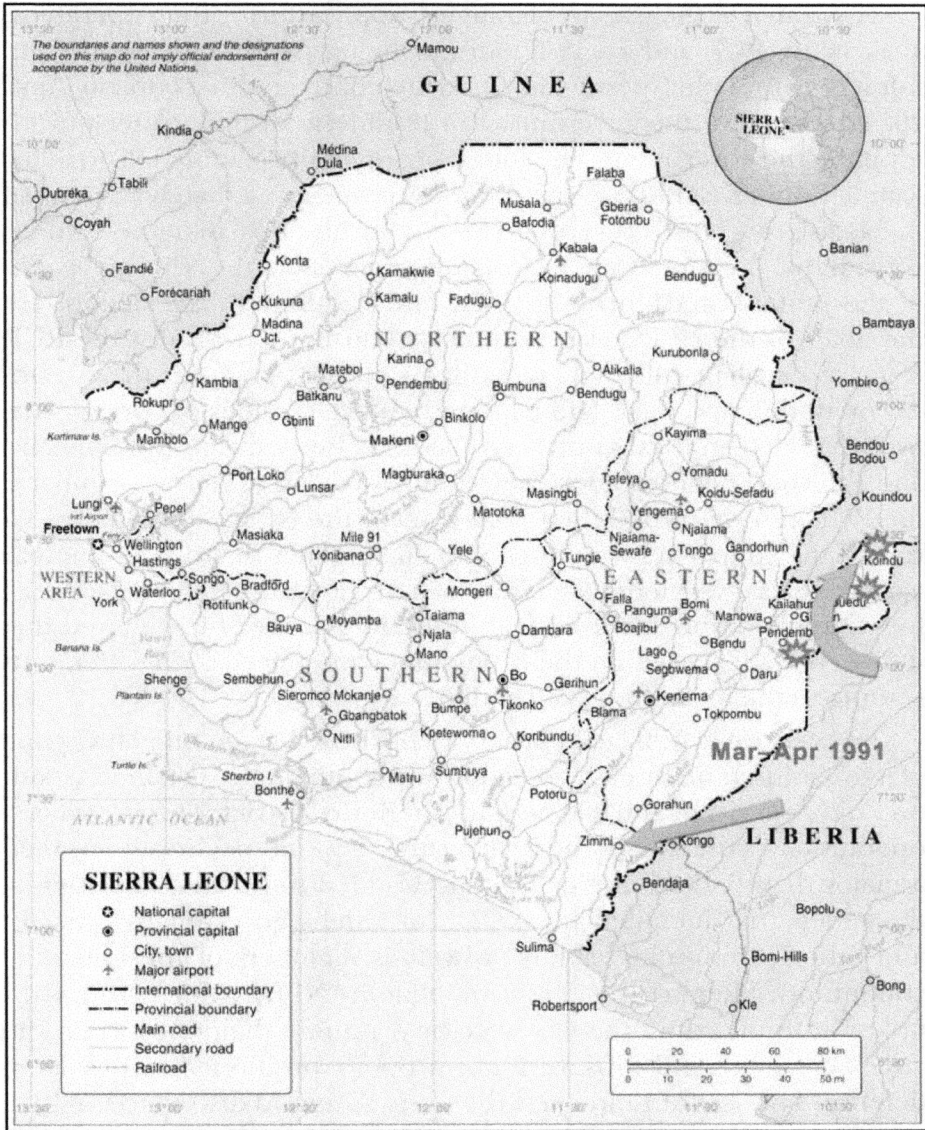

Figure 4. The initial RUF invasion of Sierra Leone from Liberia in March 1991.[55]

[55] Based on a UN map from www.un.org/Depts/Cartographic/map/profile/sierrale.pdf (accessed March 15, 2011).

Frustrated by the scarcity of military supplies during their operations against the RUF and angry about not being paid for three months, soldiers within the Sierra Leone Army conducted a coup on April 29, 1992, against the government of President Momoh. After quickly toppling the government, the soldiers formed the National Provisional Ruling Council (NPRC) to run the country and immediately suspended the legislature and other government institutions that the soldiers considered corrupt. Captain Valentine Strasser, the chosen leader of the junta, stated during his first radio address to the nation that the goals of the NPRC were to quickly end the war against the RUF, return to civilian rule, and rehabilitate the nation from the stresses of corruption and war.[56] After more than two decades of corrupt rule by the APC, the NPRC coup was welcomed by many Sierra Leoneans, and the young coup leaders were initially hailed as heroes. Similarly, the RUF also welcomed the initial news of the NPRC coup because it viewed the coup as a by-product of the insurgency that began one year earlier.[57] Strasser extended an offer of amnesty to the RUF in exchange for its unconditional surrender, but the RUF leadership believed that it deserved a place within the ruling junta and rejected this offer, thus leading to the resumption of hostilities.

From October 1992 to February 1993, the RUF expanded the scope of its operations in the eastern region and conducted a sustained assault on Koidu—a city of 200,000 people—in Kono District. This prolonged operation was the first by the RUF insurgency that specifically targeted a major diamond-mining center and also led to further atrocities by the rebels that included the killing of chiefs, government officials, and other community leaders, as well as members of the Lebanese community.[58] Reports also indicated that the NPRC soldiers who were tasked with guarding the city were busy mining diamonds when the attack occurred and thus were not properly positioned to provide a defense. This loss of Koidu and the Kono diamond district to the RUF caused great alarm across the country because it posed a potential loss of 60% of the total export revenue for Sierra Leone.

---

[56] Ibid.; *Truth and Reconciliation Commission for Sierra Leone – Final Report.*

[57] Gberie, *A Dirty War in West Africa.*

[58] Richards, *Fighting for the Rain Forest; Truth and Reconciliation Commission for Sierra Leone – Final Report.* Lebanese traders were targeted throughout the course of the civil war because of perceptions that their control of diamond trading in Sierra Leone symbolized foreign exploitation of the country's natural resources. There is also significant evidence that many community leaders were tortured and/or killed by the NPRC soldiers when they retook Koidu under the auspices that these prominent citizens had collaborated with the RUF. Because many of these citizens were known supporters of the overthrown APC government, however, it is more likely that their torture and/or death were related to their APC links rather than any RUF support.

In response to this sudden escalation in RUF activities, the NPRC launched "Operation Genesis" and sent many of its prominent military officers who were now living in Freetown back to the battlefront to fight the RUF. To augment its force, the NPRC recruited more than 1,000 boys under the age of fifteen who had been orphaned by the RUF attacks into the Sierra Leone Army, thus mirroring the RUF and NPFL practice of using child soldiers.[59] The NPRC also purchased advanced weaponry and communications equipment for the Sierra Leone Army from Russia, Romania, and Ukraine.[60] As a result of this operational surge by the Sierra Leone Army, the RUF forces operating in Kailahun (under the direct command of Foday Sankoh) and in Pujehun were effectively split from each other for almost two years, from 1992 to 1994.[61] The Sierra Leone Army succeeded in driving the RUF out of Kono District in late 1993 and into the rainforest region along the Liberia–Sierra Leone border. In December 1993, with the rebel forces apparently driven out of the diamond-mining centers and back across the border into Liberia, Captain Strasser and the NPRC declared a cease-fire in the civil war.[62]

Phase II (November 1993 to May 1997)

The second phase of the RUF insurgency began almost immediately upon the consolidation of the RUF forces in the rainforests of eastern Sierra Leone and involved a shift in tactics from the more conventional selection of clear targets, such as villages and towns, to the use of hit-and-run tactics, ambushes, and the frequent disguising of RUF forces in Sierra Leone Army uniforms that were taken from their victims.[63] This phase of the insurgency also witnessed the introduction of

---

[59] "Sierra Leone: Prisoners of War?" AFR 51/06/93 (London: Amnesty International, August 12, 1993), 1.

[60] Among the items purchased by the NPRC were Romanian SPG-9 rocket-propelled grenade launchers, 75–100 automatic grenade launchers, various light and heavy machine guns, various mortars from 60 mm to 120 mm, and approximately 1,000 rifles. Weapons purchased from Russia included BMP-2 armored personnel carriers. From Ukraine, the NPRC purchased two helicopters, an Mi-17 transport, and an Mi-24 gunship (piloted by a mercenary from South Africa). Berman, *Re-Armament in Sierra Leone.*

[61] Special Court for Sierra Leone Testimony of Mr. Palmer, Witness TFI-168, Case No. SCSL-2004-15-T (April 3, 2006), 10. Rumors also persisted during this period that Sankoh had been wounded or killed.

[62] By the time of the cease-fire, however, the NPRC was the subject of international and domestic condemnation for its use of child soldiers, as well as for a series of atrocities conducted by its soldiers, to include the execution of twenty-nine people (including a pregnant woman) in response to an alleged coup plot in December 1992.

[63] The TRC referred to this use of Sierra Leone Army uniforms by the RUF as "false flag" attacks; these attacks contributed significantly to the erosion of trust between the civilian population and the Sierra Leone Army, especially in light of the increasing occurrence of atrocities being conducted by the army. Mohamed Tarawallie (known as "Zino" or "CO Mohamed") was the RUF battlefield commander from 1994 to 1996 and is

private military companies into the conflict, as well as another change of national government in Freetown.

The cease-fire announced by Captain Strasser and the NPRC junta in late 1993 unfortunately coincided with the beginning of the annual "Zone 2" soccer tournament of West African countries, with Freetown being a host city for the major sporting event. Because of the cease-fire, many troops abandoned their positions on the battlefront in order to watch the popular soccer matches. In addition, the Sierra Leone Army soldiers were beginning to feel betrayed by Captain Strasser and the other NPRC members because many soldiers were not receiving their salaries, and there were growing rumors about the NPRC leaders leading lavish lifestyles and replicating the excesses previously demonstrated by the APC. Moreover, Strasser had forced the retirement of the twelve most senior officers in the Sierra Leone Army and followed this with the appointment (and rapid promotion) of his peers to the now vacant high posts.[64] These factors not only contributed to a sharp reduction in morale within the army but also degraded the effectiveness of the army and facilitated the rapid escalation of RUF operations from an eastern-centric insurgency to one that spread to all regions of the country. The government's ability to effectively employ its forces against the rebels was also hampered during this period by the rise of the "sobel" phenomenon in which Sierra Leone Army service members operated as soldiers during the day and as rebels at night. Moreover, there are dozens of documented cases in which soldiers and rebels coordinated the distribution of stolen items from villages and towns and even coordinated their illegal mine activities.[65]

In April 1994, the RUF conducted its first major operations in the northern region of Sierra Leone, and by the beginning of 1995, it had control of the three most important mining sites in the country and had its forces in position for an attack on Freetown.[66] This expansion of the RUF's area of operations and the loss of the three key mining

considered to be the main architect of the "false flag" operations. *Truth and Reconciliation Commission for Sierra Leone – Final Report*

[64] Gberie, *A Dirty War in West Africa.*

[65] The "sobel" phenomenon appears to be an extension of the lumpen youth dynamics and relationships that permeated both the military and the rebel forces. Detailed descriptions of "sobel" activities can be found in *Truth and Reconciliation Commission for Sierra Leone – Final Report*; "Special Court for Sierra Leone – Court Transcripts, 2009," www.sc-sl.org/.

[66] As David Francis explains, "The mining sites included a number of leased concessions in the Kono and Kenema diamond districts; the Swiss-owned Sierra Ore and Metal Company (SIEROMCO) bauxite mine at Mokanji; and the US-Australian-owned Sierra Rutile operations at Gbangbatok, both in the southern province. The RUF attack on these strategic resources was a major blow to the government's principal foreign exchange

sites put tremendous political and economic pressure on the NPRC leaders and led to their decision to look outside of Sierra Leone and seek the assistance of a private military company.

In January 1995, the government of Sierra Leone hired Gurkha Security Guards (GSG) to train the Sierra Leone Army in counterinsurgency (COIN) operations. GSG sent fifty-eight Gurkhas[67] and three managers—Robert MacKenzie (an American with combat experience with the US Army in Vietnam, the Rhodesian SAS, and the South African Defence Force), James Maynard (a former British Gurkha officer), and Andrew Myres (a former sergeant in the British Coldstream Guard).[68] The GSG personnel established a training base in Sierra Leone and began training the army in basic tactics and techniques, such as ambushing, hot pursuit, and evacuation. The GSG would later deny having any role in offensive combat operations but was reported to have supported the Sierra Leone Army when it fought the RUF at Mile 91 (a key strategic highway location from Freetown to Bo and Kenema) and at Camp Charlie, a major Sierra Leone Army logistics base. On February 24, 1995, however, MacKenzie, Myres, and nineteen other GSG and Sierra Leone Army personnel were killed in the Malal Hills during an engagement against the RUF, although the exact events leading up to this engagement remain unclear.[69] This loss of the GSG team leader and the refusal of the GSG to fulfill requests from the NPRC to directly engage in offensive combat operations against the RUF led to the closing of the GSG contract in April 1995 and a search for a different private military company that was willing to provide offensive combat forces.

Based on advice from the directors of Heritage Oil and Gas in the United Kingdom, Captain Strasser signed a contract with Executive Outcomes, a private military company based out of Pretoria, South Africa, that was composed mostly of Angolans and Namibians from the South African 32nd Battalion and led by former white officers from the 32nd Battalion. Because Strasser did not have the finances

---

earner." David J. Francis, "Mercenary Intervention in Sierra Leone: Providing National Security or International Exploitation," *Third World Quarterly* 20, no. 2 (1999): 319–338.

[67] Gurkhas are primarily from Nepal but also some areas in northern India. They got their reputation as exceptional soldiers while fighting for the British Empire but are still used as hired soldiers in India, Singapore, some places in Africa, and even by the United States at some overseas bases.

[68] Alex Vines, "Gurkhas and the Private Security Business in Africa," in *Peace, Profit Or Plunder? The Privatisation of Security in Worn-Torn African Societies,* eds Jackie Cilliers and Peggy Mason (Pretoria, South Africa: Institute for Security Studies, 1999), http://www.privatemilitary.org/Peace_Profit_or_Plunder.html.

[69] Reports vary on whether the GSG team was deliberately ambushed or whether they accidently came across a RUF training camp while they were scouting the Malal Hills to find a site for a live firing range.

to pay the $15 million cost of the contract, it was agreed that Heritage Oil and Gas would cover the costs of the operation in exchange for future mining concessions in Sierra Leone.[70] The tasking for Executive Outcomes within the contract included the restoration of internal security, the elimination of "terrorist enemies of the state," and assistance with revitalizing the economic health and investment opportunities of Sierra Leone. This latter point was a core operational priority for Executive Outcomes and led to offensive operations to regain control of key mining centers in Sierra Leone. In addition, Executive Outcomes provided convoy services and site security to Sierra Rutile and other foreign mining firms.

The operations by Executive Outcomes were immediately successful in pushing the RUF out of key resource regions in Sierra Leone. Executive Outcomes not only used advanced weaponry, to include two Mi-17 transport helicopters and a Hind Mi-24 gunship, but also armed and trained the Kamajor hunters from the regional Civil Defence Forces who became an extremely effective but controversial fighting force within Sierra Leone.

In the midst of these major military gains being led by Executive Outcomes, the NPRC prepared for national elections in March 1996 that returned civilian rule to the country. In January, Captain Strasser was ousted in a "Palace coup" as the leader of the NPRC and replaced by his deputy when Strasser tried to position himself as a "civilian" candidate in the elections.[71] On March 15, 1996, multiparty elections were held across the country, with the Sierra Leone People's Party emerging as the winner; Tejan Kabbah was elected as the nation's president.

Leveraging the military successes provided by Executive Outcomes, President Kabbah pushed for a cease-fire and negotiations with the RUF, and on November 30, 1996, the Abidjan Peace Agreement was signed in Côte d'Ivoire between President Kabbah and Foday Sankoh. As part of the peace negotiations, Kabbah agreed to remove all "foreign fighters" from the employment of the government of Sierra Leone, which effectively ended the role of Executive Outcomes in the conflict and simultaneously removed the main element that drove Sankoh and the RUF to the negotiation table. Further complicating President Kabbah's tenuous hold on national

---

[70] The intertwined corporate relationship of Executive Outcomes with Heritage Oil and Gas and various mining firms is discussed in detail in Khareen Pech, "Executive Outcomes – A Corporate Conquest," in Ibid.; Gberie, *A Dirty War in West Africa.* Heritage Oil and Gas used revenues from an Executive Outcomes operation in Angola to cover the costs of the Sierra Leone operation.

[71] Ibid.

power, he allowed the Kamajor Civil Defence Forces that were trained and armed by Executive Outcomes to be incorporated as a formal government security institution—even going so far as to name the Chief of the Kamajors (Hinja Norman) as the Deputy Minister of Defence—thus creating a rift between President Kabbah and the Sierra Leone Army.[72]

On May 25, 1997, fourteen months after being elected and less than six months after the Abidjan Peace Agreement, President Kabbah was overthrown in a coup by Sierra Leone Army soldiers who established the Armed Forces Revolutionary Council (AFRC) to rule the country. Under the leadership of Major Johnny Paul Koroma, the AFRC immediately established communications with Foday Sankoh and the RUF and invited the RUF to join the ruling junta.[73] Thus, the second phase of the RUF insurgency ended with a third change of government leadership in five years and the sudden emergence of the RUF within the ruling coalition.

---

[72] The Civil Defence Forces was dominated by Mende, while the Sierra Leone Army was composed primarily of northerners.

[73] The AFRC coup began with an attack on the Pademba Road prison to free Major John Paul Koroma and other plotters from an August 1996 coup attempt. This was followed by an attack on the State House and nearly a week of violence and theft across Freetown. Major Koroma contacted Foday Sankoh by telephone in May 1997 while Sankoh was under house arrest in Nigeria on weapons-smuggling charges.

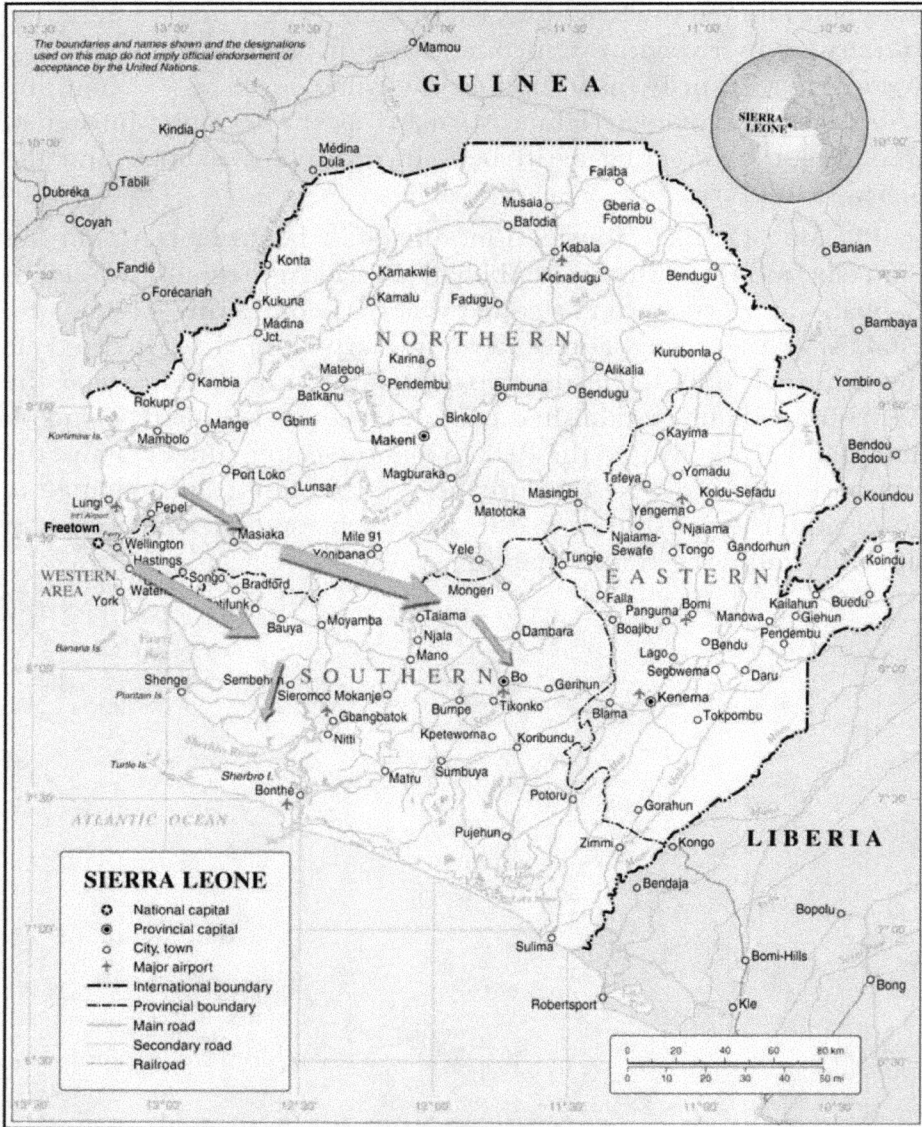

Figure 5. Offensives by executive outcomes, 1995–1996.[74]

---

[74] Based on a UN map from www.un.org/Depts/Cartographic/map/profile/ sierrale.pdf (accessed March 15, 2011); Larry J. Woods and Timothy R. Reese, "Military Interventions in Sierra Leone: Lessons from a Failed State." *The Long War Series Occasional Paper 28* (Fort Leavenworth, KS: Combat Studies Institute Press, 2008), 31.

Phase III (May 1997 to January 2002)

The third and final phase of the RUF insurgency is defined by internal power struggles within the RUF, as well as the complex and violent dynamics that emerged when the RUF and the AFRC formed what they called a "People's Army"[75] to fight the remnants of the Sierra Leone Army as well as international peacekeepers. Unlike the previous military junta (the NPRC), which initially had popular support after it ousted a corrupt and unpopular APC government, the AFRC was immediately viewed as a self-serving and violent organization more interested in personal gain than the judicious running of the country. After receiving the offer for the RUF to join the ruling junta, Foday Sankoh accepted the offer and gave instructions for the RUF forces to converge on Freetown. Under the operational leadership of Sam Bockarie, the RUF fighters began to enter Freetown and joined with the junta forces of the AFRC to establish defensive positions. This willful joining of forces between the RUF and the AFRC puzzled outside observers but became more transparent once the "sobel" and lumpen youth characteristics of the Sierra Leone Army were better understood. Specifically, many of the foot soldiers for both the RUF and the Sierra Leone Army came from the same geographic pool of young, unemployed, and under-educated males with their personal relationships continuing after they joined their respective organizations.

In late 1997, there was increased international pressure to have the joint AFRC-RUF junta removed from power, and in early 1998, a Nigerian-led Economic Community of West African States Monitoring Group (ECOMOG) force succeeded in retaking Freetown and returning President Kabbah from exile to his seat of power. The ECOMOG soldiers were also assisted by the Kamajor Civil Defence Forces, with both groups receiving training and weapons from Sandline International, a UK-based corporation affiliated with Executive Outcomes.[76] With little structure or faith remaining in the capabilities of the Sierra Leone Army, the reinstated President Kabbah had to rely on the ECOMOG forces as a surrogate national army. Although this

---

[75] This concept of a "People's Army" under the AFRC-RUF junta should not be confused with the traditional use of this term by some communist countries. In the case of the AFRC-RUF, this term was meant to imply a military force that represented the interests of the people, something that was quickly dismissed due to AFRC-RUF atrocities against the civilian population in Freetown.

[76] Several inquiries were conducted in the United Kingdom concerning the role of Sandline in supplying weapons to the Kabbah government and Civil Defence Forces because this was in violation of a blanket arms embargo on Sierra Leone. These inquiries showed a perception that the Kabbah government in exile was viewed by some parties as exempt from the embargo because it was the legitimate government of Sierra Leone.

provided some short-term security for the capital, the inability—or unwillingness—of ECOMOG to provide comprehensive security out into the countryside enabled the AFRC and the RUF to regroup and prepare an offensive to retake Freetown.[77]

In January 1999, AFRC-RUF rebels launched "Operation No Living Thing" against Freetown, which resulted in mass casualties and atrocities against the civilian population. Despite the presence of more than 15,000 Nigerian soldiers in Sierra Leone, the slow and steady advance on Freetown by the rebel forces[78] was never stopped, with the rebels capturing weapons and gaining confidence as they advanced.[79] Sweeping in from the east end of the city, the rebel forces went on a killing and looting spree for almost three weeks before they were eventually driven out of Freetown again by Nigerian and Civil Defence Forces. But during this period, more than 6,000 civilians were killed, thousands more were injured and mutilated, and more than 100,000 were driven from their homes. The indiscriminate savagery—yet calculated[80] nature—of the attacks was captured in numerous media accounts, as well as a Human Rights Watch report that detailed rebel units with specific names related to their method of attack, such as the Burn House Unit, Cut Hands Commando, and the Blood Shed Squad.[81]

With global attention now turned to Sierra Leone, Western governments sent envoys to Sierra Leone and began to pressure the Kabbah government to reach a peace settlement with the RUF.[82] On

---

[77] Ibid. The Nigerian ECOMOG forces were also accused of conducting extrajudicial killings of suspected AFRC members and pursuing their own financial interests in the diamond districts.

[78] The term "rebel force" is used here rather than RUF because this was a transition period as the AFRC soldiers became aligned under the leadership of the RUF field commanders.

[79] Among the key victories for the rebels before their attack on Freetown was the capture of Koidu, which reportedly held 50% of the ECOMOG arms stockpile. It has also been reported that the rebel forces were trained in infantry tactics by "white mercenaries" hired through Charles Taylor in exchange for diamonds, with the rebel tactics leading up to the attack on Freetown considered to be "textbook South African army tactics." Ibid.

[80] This was not just random violence. The selection of which civilians to kill may have been somewhat random, but the method of killing was systematic. Different rebel units had specific types of atrocities that they carried out against the civilian population. Ironically, during one of the trials after the war ended, a defense lawyer for one of the RUF defendants argued that the RUF could not have been responsible for the mass amputations in one village because the RUF rebels cut off hands at the forearms, and the amputations in the village were done above the elbows.

[81] *Getting Away with Murder, Mutilation, and Rape* (New York: Human Rights Watch, 1999), http://www.hrw.org/en/reports/1999/06/24/getting-away-murder-mutilation-and-rape.

[82] The US envoy to the peace talks (Reverend Jesse Jackson) raised considerable anger in Sierra Leone when he called Foday Sankoh a freedom fighter and compared him to Nelson Mandela, thus underscoring the lack of Western understanding of the conflict.

July 7, 1999, the Lomé Peace Accord was signed and included the release of Foday Sankoh from prison and his appointment in the government as the head of the Commission for the Management of Strategic Resources, National Reconstruction, and Development, thus putting him in charge of the allocation of the nation's diamond and mineral resources.[83] Having been separated from his RUF forces for almost two years while imprisoned, however, Sankoh no longer enjoyed control over all of the operational units because the RUF was now divided into two distinct groups—a political wing that sought a power-sharing arrangement with the government and a "combatant cadre" under Sam Bockarie that pushed to continue the insurgency.[84]

In addition to the political component of the Lomé Peace Accord (the power-sharing arrangement), the agreement also had a major military component that involved the disarmament of the rebel combatants. This second component became the most difficult to enact because of Sankoh's limited control over the operational RUF elements in the field. RUF commanders refused to comply with the disarmament provisions of the Lomé Peace Accord and instead resorted to numerous instances of attacks against UN peacekeepers who were sent in to monitor the disarmament process.[85] In May 2000, more than 500 UNAMSIL (United Nations Mission in Sierra Leone) peacekeepers were taken hostage and had their weapons and ammunition taken from them. This, and several similar events, underscored an aggressive posture by the RUF but also the operational inefficiencies of the UN forces. After a change in mandate, UNAMSIL began direct combat operations against the RUF, increased the number of peacekeepers in country, and had a change in its command structure.[86] In response to the hostage taking of British military members, the United Kingdom also deployed a contingent of Royal Marines, paratroops, and Special Forces who operated independently of the UNAMSIL operations but who provided decisive firepower during several key engagements with rebel forces. The enhancements to the UNAMSIL mission and forces, as well as the introduction of the highly capable British forces into Sierra Leone, coincided with the withdrawal of approximately 2,000

---

[83] The irony and hypocrisy of appointing Foday Sankoh as the highest government official responsible for the trade of diamonds in Sierra Leone cannot be overstated and obviously created tremendous criticism of the Western officials who forced the Lomé Peace process upon President Kabbah.

[84] *Truth and Reconciliation Commission for Sierra Leone – Final Report.*

[85] The UN peacekeepers were initially called UNOMSIL (United Nations Observer Mission in Sierra Leone) but became UNAMSIL (United Nations Mission in Sierra Leone) when their mandate became more operational.

[86] UNAMSIL is viewed as a major turning point in the credibility and reputation of UN peacekeeping. See *Report of the Panel on United Nations Peace Operations* (the "Brahimi Report"), New York, August 21, 2000.

RUF rebels back across the border into Liberia to support Charles Taylor, who was losing ground to anti-Taylor forces. These factors pushed the remaining RUF fighters into the disarmament process, with approximately 72,490 combatants (both RUF and Civil Defence Forces) disarmed, 42,000 weapons confiscated, and 1.2 million rounds of ammunition collected by January 2002.[87] That same month, the civil war in Sierra Leone was officially declared over, and national elections were conducted two months later.

METHODS OF RECRUITMENT

The RUF and other combatant forces during the Sierra Leone civil war employed forced recruitment at various stages during the insurgency in order to meet the manning requirements for the operations they were undertaking.[88] Surges in forced recruitment mirrored the three main phases of the RUF insurgency. When the TRC investigated the topic of forced recruitment and abduction, it found that:

> The RUF pioneered the policy of forced recruitment in the conflict. The RUF bore a marked proclivity towards abduction, abuse, and training of civilians for the purpose of creating commandos. It was the first armed group to practice forced recruitment and was responsible for the vast majority of the forced recruitment violations recorded by the Commission.[89]

The TRC also found, however, that some recruits willingly joined the RUF:

> In addition, the Commission finds that many young men joined the RUF voluntarily because they were disaffected. This trend demonstrates the centrality of bad governance, corruption, all forms of discrimination and the marginalisation of certain sectors of society among the causes of conflict in Sierra Leone. Historical ills and injustices had prepared the ground for someone of Foday Sankoh's manipulative ability to canvass among

---

[87] Gberie, *A Dirty War in West Africa*. As part of the Lomé Peace Accord, the Civil Defence Forces militia forces were also required to disarm.

[88] In August 1993, Amnesty International released a report that condemned the National Patriotic Revolutionary Council under Captain Strasser for having at least 1,000 boys under the age of fifteen—some as young as seven—serving in the Sierra Leone Army.

[89] *Truth and Reconciliation Commission for Sierra Leone – Final Report*, volume 2, chapter 2, paragraph 140.

the people and find scores of would-be RUF commandos who could be brought on board with relatively little persuasion.[90]

As alluded to earlier, the RUF and government forces obtained their recruits—those willing and unwilling—from the same pool of impoverished, unemployed youth, and this commonality undermined the combat effectiveness of the Sierra Leone Army, paved the way for the coup by the AFRC, and gave rise to the "sobel" phenomenon that was so unique to the Sierra Leone conflict. The TRC again underscored this significant issue with its finding that:

> There existed an astonishing *factional fluidity* among the different militias and armed groups. Overtly and covertly, gradually and suddenly, fighters switched sides or established new units on a scale unprecedented in any other conflict.[91] (emphasis added)

METHODS OF SUSTAINMENT

Throughout the insurgency, the RUF forces were sustained primarily through foraging and the execution of raids against villages in their areas of operation. Villagers were often captured and forced to serve as "human caravans" by carrying any food or possessions taken from the village by the RUF. Although arms and ammunition were often captured from government forces and later from peacekeeping forces, a dedicated and significant supply line of weapons and drugs through Burkina Faso, as well as Liberia, sustained the RUF and was paid for in diamonds.[92] Narcotics, such as cocaine and marijuana, were a critical and daily component of the insurgency because they enabled the rebels—especially the youth—to continue fighting when they were hungry and, more importantly, because they were used to increase or decrease the aggression of the fighters depending on the operational temperature. Reports exist of young rebel fighters entering Freetown

---

[90] Ibid., paragraph 141.

[91] Ibid., volume 2, chapter 1, paragraph 36.

[92] Dena Montague, "The Business of War and the Prospects for Peace in Sierra Leone," *Brown Journal of World Affairs* IX, no. 1 (2002): 229–238. The trading of diamonds for weapons and drugs was a natural extension of the illegal diamond-mining industry that already existed in Sierra Leone. Before the RUF insurgency, there was a well-established illicit trade in diamonds coming out of Sierra Leone. In 1970, approximately 2 million carats of diamonds were legally exported out of the country, and this number dropped to 48,000 carats by 1988. It is estimated that 97% of diamonds coming out of Sierra Leone in the 1980s were taken out of the country illegally. See Ian Smillie, Lansana Gberie, and Ralph Hazleton, *The Heart of the Matter: Sierra Leone, Diamonds, and Human Security* (Ottawa: Partnership Africa Canada, 2000)

during "Operation No Living Thing" with bandages on their heads covering the incisions where crack cocaine had been inserted under their skin.[93] For the RUF insurgency to continue for the duration and scope that it did, this external supply of weapons and narcotics was a necessity, as was the continuous mining of diamonds in order to purchase these supplies.

METHODS OF OBTAINING LEGITIMACY

The RUF demonstrated very little concern about obtaining legitimacy during the civil war, as demonstrated by the continuous targeting of the population that the RUF claimed it was liberating from a repressive government. With the exception of the sporadic release of poorly constructed political messages, the singular occurrence of legitimacy came with the 1999 Lomé Peace Accord, in which the RUF was given senior positions within the newly formed government as compensation for laying down its arms. This power-sharing arrangement lasted only a few months as it became apparent that the RUF leadership was divided with only a small fraction seeking peace and some level of legitimacy.[94]

EXTERNAL SUPPORT

The external support provided by Charles Taylor in Liberia, Gaddafi in Libya, Blaise Compaoré in Burkina Faso, and diamond business interests in Lebanon, Belgium, France, and Israel allowed for critical supplies and operational coordination in support of the RUF.[95] The motivation for this support varied to some degree based on personal relationships and the desire to foster Green Book ideology, but underlying most of these transactions were direct payments in diamonds or future concession rights to diamond fields. For this reason, the commercial and financial aspects of the RUF insurgency serve as key indicators of the strategic motives for the civil war and the rationale for sustained external support for more than a decade.

COUNTERMEASURES TAKEN BY THE GOVERNMENT

Decades of corruption, cronyism, and institutional decay, as well as the intentional degradation of military capabilities, left the

---

[93] Gberie, *A Dirty War in West Africa.*

[94] *Truth and Reconciliation Commission for Sierra Leone – Final Report*, paragraph 159.

[95] Smillie, Gberie, and Hazleton, *The Heart of the Matter.*

government and army of Sierra Leone ill-equipped to counter the unexpected rise of the RUF. When the insurgency began in 1991, the army was dangerously under-resourced because of years of neglect at the hands of the APC government. As President Momoh began to realize the threat posed by the RUF, he doubled the size of the army to 6,000 by recruiting lumpen youth as foot soldiers while keeping the membership of the officer corps based on a patronage system loyal to the APC.[96] Unable to field operationally proficient forces and equipment, Momoh's successor, Captain Strasser, turned to private military companies for support. Executive Outcomes proved to be extremely effective in countering the RUF insurgency and drove the rebels out of key strategic locations in the country. However, the subsequent removal of Executive Outcomes from Sierra Leone as a result of the cease-fire terms dictated by the RUF led the government of Sierra Leone to rely on the capabilities (and limitations) of ECOMOG and UN peacekeeping forces. In addition, four changes in government during the course of the civil war—as well as continued atrocities by government forces—all served to undermine internal efforts at counterinsurgency. In this regard, the eventual government countermeasures against the RUF were not indigenous to Sierra Leone or truly controlled by the government, but they were eventually effective. The delay in bringing these external forces to bear, however, resulted in tens of thousands of deaths within the country.

SHORT- AND LONG-TERM EFFECTS

CHANGES IN THE ENVIRONMENT

The scars from the RUF's insurgency were still fresh in Sierra Leone less than 10 years from the end of the conflict, but they exist in a nation that was already damaged from two decades of governmental corruption and abuse of the population. As of 2010, Sierra Leone is still ranked at or near the very bottom of global HDI scores, and the internal contributing factors that gave rise to the RUF remain very prevalent. Corruption remains rampant in the country, and the emergence of Sierra Leone as a new route in the drug trade between Latin America and Europe is troubling. Of greatest concern are the continued high unemployment rate among young men throughout the country and the reappearance of political "thuggery" in recent elections. Although measures to enhance the tracking and accountability of the global

---

[96] Gberie, *A Dirty War in West Africa*.

diamond trade have improved, Sierra Leone continues to have a significant level of illegal mining activity.[97]

CHANGES IN GOVERNMENT

The government of Sierra Leone surprised many outside observers with the successful and relatively peaceful democratic transition of power in 2007. The RUF, however, never established any sustained political support, although it did provide a political candidate for the 2002 national elections, Alimamy Pallo Bangura, who received less than 2% of the popular vote.

CHANGES IN THE REVOLUTIONARY MOVEMENT

The end of the civil war in Sierra Leone led to the end of the RUF because of the absence of any clear political agenda or plan that would sustain the organization during peacetime. As part of the peace process, the RUF was mandated to develop a political wing for the organization, but the well-documented atrocities that were committed by its forces made this an untenable goal. Absent the ability to fight and a peacetime agenda to implement, the RUF movement ceased to exist as a significant force in Sierra Leone.

BIBLIOGRAPHY

Abdullah, Ibrahim. *Between Democracy and Terror: The Sierra Leone Civil War.* Dakar, Senegal: Codesria, 2000.

———. "Bush Path to Destruction: The Origin and Character of the Revolutionary United Front/Sierra Leone." *The Journal of Modern African Studies* 36, no. 2 (1998): 203–235.

———. "Youth Culture and Rebellion: Understanding Sierra Leone's Wasted Decade." *Critical Arts* 16, no. 2 (2002): 28.

---

[97] In response to use of diamond smuggling to fuel several African conflicts, the Kimberley Process was established to provide an international mechanism for certifying and tracking diamonds as being "non-conflict" diamonds. According to the 2009 annual report from Partnership Africa Canada which tracks the progress of the Kimberley Process, "the Kimberley Process (KP), designed to halt and prevent the return of 'conflict diamonds,' is failing. On the topic of Sierra Leone, this 2009 report states that "There are two things everyone in Sierra Leone seems to agree on, namely that its diamonds are yet to benefit the country, and that it desperately needs a profound overhaul of its mining sector. That is where agreement ends." See Smillie, Gberie, and Hazleton, *The Heart of the Matter.*

Abraham, Abdullah. "Dancing with the Chameleon: Sierra Leone and the Elusive Quest for Peace." *Journal of Contemporary African Studies* 19, no. 2 (2001): 205–228.

Berman, Eric G. *Re-Armament in Sierra Leone: One Year after the Lome Peace Agreement.* Geneva: Small Arms Survey, 2000.

Chege, Michael. "Sierra Leone: The State that Came Back from the Dead." *The Washington Quarterly* 25, no. 3 (2002): 147–160.

Christensen, Maya M., and Mats Utas. "Mercenaries of Democracy: The 'Politricks' of Remobilized Combatants in the 2007 General Elections, Sierra Leone," *African Affairs* 107, no. 429 (2008): 515–539.

Central Intelligence Agency. "Sierra Leone." *The World Factbook.* Accessed September 16, 2010. https://www.cia.gov/library/publications/the-world-factbook/geos/sl.html.

Cilliers, Jackie, and Peggy Mason, eds. *Peace, Profit or Plunder? The Privatisation of Security in Worn-Torn African Societies.* Pretoria, South Africa: Institute for Security Studies, 1999.

Conibere, Richard, et al. "Statistical Appendix to the Report of the Truth and Reconciliation Commission of Sierra Leone," *A Report by the Benetech Human Rights Data Analysis Group and the American Bar Association Central European and Eurasian Law Initiative to the Truth and Reconciliation Commission,* 2004.

Francis, David J. "Mercenary Intervention in Sierra Leone: Providing National Security or International Exploitation." *Third World Quarterly* 20, no. 2 (1999): 319–338.

Gberie, Lansana. *A Dirty War in West Africa: The RUF and the Destruction of Sierra Leone.* Bloomington, IN: University Press, 2005.

Getting Away with Murder, Mutilation, and Rape. New York: Human Rights Watch, 1999.

Human Development Indicators Report 1990. New York: United Nations Development Programme, 1990. http://hdr.undp.org/en/reports/global/hdr1990/.

Jackson, Paul. "Reshuffling an Old Deck of Cards? The Politics of Local Government Reform in Sierra Leone." *African Affairs* 106, no. 422 (2007): 95–111.

Kandeh, Jimmy D. "Transition without Rupture: Sierra Leone's Transfer Election of 1996." *African Studies Review* 41, no. 2 (1998): 91–111.

Montague, Dena. "The Business of War and the Prospects for Peace in Sierra Leone." *Brown Journal of World Affairs* IX, no. 1 (2002): 229–238.

National Integrity Systems: Country Study Report – Sierra Leone, 2004. Berlin: Transparency International, 2004.

Report of the Panel on United Nations Peace Operations (the "Brahimi Report"). New York, August 21, 2000.

Richards, Paul. *Fighting for the Rain Forest: War, Youth and Resources in Sierra Leone.* Portsmouth, NH: Heinemann, 1996.

Sawyer, Edward. "Remove or Reform? A Case for (Restructuring) Chiefdom Governance in Post-Conflict Sierra Leone." *African Affairs* 107, no. 428 (2009): 387–403.

"Sierra Leone: Prisoners of War?" AFR 51/06/93. London: Amnesty International, August 12, 1993, 1.

Smillie, Ian, Lansana Gberie, and Ralph Hazleton. *The Heart of the Matter: Sierra Leone, Diamonds, and Human Security.* Ottawa: Partnership Africa Canada, 2000.

"Special Court for Sierra Leone – Court Transcripts, 2009." Accessed November 21, 2009. http://www.sc-sl.org/scsl/listcases.asp.

The Constitution of Sierra Leone, 1991. Accessed September 16, 2010. http://www.sierra-leone.org/Laws/constitution1991.pdf.

"Timeline Sierra Leone." Accessed September 16, 2010. http://timelines.ws/countries/SIERLEON.HTML.

Truth and Reconciliation Commission for Sierra Leone – Final Report. Truth and Reconciliation Commission for Sierra Leone, 2004.

Venter, Al J. *War Dog: Fighting Other People's Wars.* Drexel Hill, PA: Casemate, 2008.

Woods, Larry J., and Timothy R. Reese. "Military Interventions in Sierra Leone: Lessons from a Failed State." The Long War Series Occasional Paper 28. Fort Leavenworth, KS: Combat Studies Institute Press, 2008, 31.

"Youth in Sierra Leone Find Hope in Entrepreneurship." New York: United Nations Development Programme. Accessed September 16, 2010. http://content.undp.org/go/newsroom/2010/january/youth-in-sierra-leone-find-hope-in-entrepreneurship.en.

ORANGE REVOLUTION OF UKRAINE: 2004–2005

Jerry Conley

SYNOPSIS

The Orange Revolution took place during the 2004 presidential election in Ukraine and involved the mass mobilization of the population and the unification of key leaders and organization in order to prevent a fraudulent election. The promotion of nonviolent civic disobedience, as well as embracing the existing constitutional and institutional judicial and legislative structures within Ukraine, ensured the successful completion of a democratic electoral process.

TIMELINE

October 31, 2004	Voting in the presidential election gives Yushchenko a small 0.6% lead against Yanukovich, triggering a second-round ballot.
November 21, 2004	The second round of voting takes place after an interim period of rising tensions.
November 22, 2004	The Central Electoral Commission declares Yanukovich the winner. Yushchenko's supporters reject the result and gather in Kiev amid claims of vote-rigging. In the following days, the protests build, despite subzero temperatures.
November 24, 2004	The official results are published, giving Yanukovich 49.46% and Yushchenko 46.61%.
November 25, 2004	The Supreme Court suspends publication of the results while it examines the case, after the opposition appeals.
November 26, 2004	Yanukovich and Yushchenko hold talks and agree to seek a peaceful solution. Yushchenko demands a re-run of the vote. Meanwhile, Yushchenko's supporters lay siege to government buildings.
November 27, 2004	Parliamentary deputies declare the poll invalid and pass a symbolic, nonbinding vote of no-confidence in the electoral commission. Rival protests backing Yanukovich are held in his stronghold of Donetsk.

November 28, 2004	Eastern regions threaten to secede if Yushchenko is declared president.
November 29, 2004	The Supreme Court begins considering allegations of electoral abuses. Mr. Yanukovich says he might accept a re-vote in certain disputed areas.
November 30, 2004	Outgoing President Leonid Kuchma—who backed Yanukovich during the election campaign—says only fresh elections can resolve the stand-off.
December 1, 2004	Parliament narrowly passes a motion of no-confidence in the government on the second attempt, prompting opposition fireworks in Kiev, but Yanukovich dismisses the vote as illegal. Yushchenko agrees to lift a blockade on government buildings but asks supporters to remain on the streets.
December 2, 2004	Crisis talks to try to find a solution to the deadlock continue as parties await the decision of the Supreme Court.
December 3, 2004	The Supreme Court annuls the results of the second round of the elections, paving the way for fresh elections.
December 8, 2004	Parliament passes a wide-ranging reform bill, paving the way for a December 26 re-run of the disputed presidential election.
December 9, 2004	Government employees return to work after opposition demonstrators scale down their protest in Kiev.
December 11, 2004	Yushchenko's Vienna doctors confirm after exhaustive tests that he was poisoned with a form of deadly dioxin.
December 24, 2004	Campaigning ends at midnight, with both candidates saying they are confident of victory.
December 25, 2004	Constitutional Court strikes down reform restricting home voting; election officials say vote will proceed regardless.
December 27, 2004	With nearly all votes counted, Yushchenko's lead becomes unassailable, but Yanukovich says he will never concede defeat, claiming election abuses.
December 30, 2004	Supreme Court rejects all four complaints against the conduct of the presidential election lodged by Yanukovich. The Central Election Commission also rejects his appeal over the vote.
December 31, 2004	Yanukovich resigns as prime minister, saying he cannot work with people loyal to Yushchenko.

January 6, 2005	Supreme Court rejects an appeal by Yanukovich against the electoral commission's handling of the poll. The ex-prime minister had wanted the court to make the commission reexamine complaints about the election.
January 11, 2005	Electoral commission declares Yushchenko the official winner of the re-run presidential election with 51.99% of the vote. Yanukovich gets 44.2% but continues the legal battle.
January 16, 2005	Thousands of demonstrators rally in Yanukovich's home town, Donetsk, and elsewhere, to condemn Yushchenko's "anti-constitutional" election.
January 17, 2005	Supreme Court starts hearing Yanukovich's final appeal after he submitted 600 volumes of evidence indicating irregularities in the re-run election. All of his previous appeals have been rejected.
January 18, 2005	A ban on publication of the presidential election results is lifted by the Supreme Court—allowing them to be published in newspapers, making them legal.
January 20, 2005	Supreme Court rejects Yanukovich's final appeal against the result of the re-run election and declares Yushchenko the winner. Parliament votes to hold Yushchenko's inauguration. Yanukovich concedes that he has lost the election re-run to Yushchenko, telling supporters in his Donetsk stronghold: "The right of force has won against the force of the law."
January 23, 2005	Viktor Yushchenko is sworn in as Ukraine's new president ending the bruising election marathon. In taking the oath of office before parliament, Yushchenko said he would defend the unity of Ukraine.

THE ENVIRONMENT OF THE REVOLUTION

PHYSICAL ENVIRONMENT

Figure 1. Map of Ukraine.[1]

Sitting on the strategic crossroads between Europe and Asia, Ukraine is located in Eastern Europe with the Black Sea to the south; Romania, Moldova, Slovakia, and Poland on its western borders; Belarus to the north; and the Russian Federation to the north and east. Slightly smaller than the size of Texas, it is the second largest country in Europe at 603,550 square kilometers.[2] Ukraine has a temperate climate with higher summer temperatures along the Crimean coast and Black Sea and colder winters to the north.

---

[1] worldofmaps.net, "Map of Ukraine," accessed March 14, 2011, http://www.worldofmaps.net/en/europe/map-ukraine/map-administrative-divisions-ukraine.htm.

[2] Central Intelligence Agency, "Ukraine," *The World Factbook*, accessed September 30, 2010, https://www.cia.gov/library/publications/the-world-factbook/geos/up.html.

The republic is composed of twenty-four provinces called oblasts, one autonomous republic (Crimea), and two cities with oblast[3] status (Kiev and Sevastopol). Kiev is the capital of Ukraine and also its largest city, with more than 2.5 million residents. Sevastopol is the home of the Russian Black Sea Fleet under a special lease agreement, although Ukraine has indicated that it will let the lease expire in 2017.[4]

CULTURAL AND DEMOGRAPHIC ENVIRONMENT

Considered the breadbasket of the Soviet Union before the dissolution of the USSR on August 24, 1991, Ukraine has the twenty-seventh largest population in the world, at just under forty-six million people, yet also has one of the lowest population growth rates in the world, with its population shrinking at a negative rate of 0.632%.[5] The ethnic distribution of the country is a majority Ukrainian at 77.8%, with Russians as the second largest ethnic group at 17.3%. A large portion of this Russian minority lives in the eastern regions of Ukraine. The population is highly literate, with a 99.4% literacy rate and an average of fourteen years of primary, secondary, and tertiary education. The Ukraine has a diverse religious base, and according to 2006 estimates, about one-half of the population is Ukrainian Orthodox of the Kyiv Patriarchate, and 26.1% are Ukrainian Orthodox of the Moscow Patriarchate. Other religions practiced include Ukrainian Greek Catholic (8%), Ukrainian Autocephalous Orthodox (7.2%), Roman Catholic (2.2%), Protestant (2.2%), Jewish (0.6%), and other (3.2%).[6]

SOCIOECONOMIC ENVIRONMENT

Upon gaining its independence in 1991, Ukraine transitioned to a market economy but experienced a significant recession with

---

[3] An oblast is the term for an administrative area or zone within some Slavic countries as well as some former Soviet countries. For a city to have oblast status, it usually means that the city has autonomy from its surrounding region and has a direct administrative relationship with the national administration.

[4] Mark Franchetti, "Russia Fleet 'May Leave Ukraine'," *BBC News*, October 18, 2008, accessed April 12, 2010, http://news.bbc.co.uk/2/hi/europe/7677152.stm.

[5] In 2004, the population of Ukraine was estimated at 47.7 million. Ukraine's birth rate is ranked 202nd in the world at 9.6 births per 1,000 population, and its death rate is the thirteenth highest in the world at 15.81 deaths per 1,000 population. "Ukraine," *The World Factbook*.

[6] Ibid.

a 60% loss of its gross domestic product (GDP) from 1991 to 1999.[7] The country experienced a sharp economic recovery from 1999 to 2004, however, with GDP nearly doubling. This gave rise to a middle class in Kiev, as well as further riches to a small elite group of wealthy oligarchs.[8] In 2004, President Kuchma was in his second term and pledged to implement a variety of economic stimulus measures in order to promote investment and growth in Ukraine, including a reduction in the number of government agencies, streamlining of the regulatory process, creation and enforcement of a legal environment in Ukraine that would encourage entrepreneurs, and the enactment of a comprehensive tax reform in order to capture significant revenue losses to the gray economy. In 2004, approximately 23.4% of the workforce was employed in the agricultural sector, 41.4% in industry, and 35.1% in services.[9]

HISTORICAL FACTORS

After the breakup of the Soviet Union, Ukraine elected its first president, Leonid Kravchuk, on December 5, 1991, who "operated through elite bargaining" and whose political approach "retained a neo-Soviet style."[10] Although the Ukrainian constitution set presidential terms at five years, Kravchuk was forced into an early election cycle in 1994 because of the economic crisis that was overwhelming Ukraine, and he lost to his former prime minister, Leonid Kuchma. Kuchma ran on a campaign for closer relations with Russia and fighting the corruption and economic failures that defined the Kravchuk presidency. His tilt toward the important Russian voting block that dominated eastern Ukraine led to a very definitive East–West split in the electoral returns. Immediately upon assuming the presidency, however, it became apparent that Kuchma would not undo the corrupt practices set in place under Kravchuk but instead would accelerate them by leveraging his relationships with Ukraine's oligarchs in order to achieve his personal and political agendas.[11] At the center of this group of corrupt cronies was Pavlo Lazarenko who, between 1995 and

---

[7] World Bank Group, "Can Ukraine Avert a Financial Meltdown," *Beyond Transition Newsletter*, accessed September 30, 2010, http://web.archive.org/web/20110628222439/http://www.worldbank.org/html/prddr/trans/june1998/ukraine.htm.

[8] Adrian Karatnycky, "Ukraine's Orange Revolution," *Foreign Affairs* (2005).

[9] Central Intelligence Agency, "Ukraine," *The World Factbook*, accessed September 30, 2010, https://www.cia.gov/library/publications/the-world-factbook/geos/up.html.

[10] Andrew Wilson, *Ukraine's Orange Revolution* (New Haven, CT: Yale University Press, 2005).

[11] Under the Soviet Union, Kuchma had been the factory director for an ICBM plant and then the boss of Ukraine's "red director's union" after he was ousted as prime minister in 1993. Ibid., 38–40.

1997, served as first deputy prime minister and then prime minister under Kuchma and was accused of stealing hundreds of millions of dollars from the state, as well as from businesses through kickbacks and extortion.[12]

A variety of factors enabled this large-scale, unchecked corruption in Ukraine during the 1990s. Similar to what occurred in Russia during this same time frame, the rapid privatization of the former Soviet industries and enterprises provided tremendous opportunities for government insiders who could arrange extremely low-price sales of oil facilities, energy plants, steel mills, etc. Because of the high inflation rate in Ukraine, there was also heavy use of a barter system for cross-border financial transactions that enabled businesses to underreport the value of their business transactions. In addition, the tax system was highly corrupt, with oligarchs buying off tax inspectors to harass and fine their business competitors while guaranteeing protection of their own business interests.[13]

When Kuchma ran for reelection in 1999, he was increasingly unpopular because of the blatantly corrupt practices that surrounded his administration. However, the election was a rigged affair, with the three main opposing candidates in the race secretly sponsored by oligarchs and by Kuchma's own administration, thereby making them virtual, "cut-out" opponents who were predetermined to lose the election.[14] Despite winning this 1999 election, Kuchma lost his control and leverage over the wealthy oligarchs who no longer needed his political patronage and protection because of their own accumulated power and wealth.[15]

GOVERNING ENVIRONMENT

Ukraine is a republic with a legal system based on civil law and judicial review of legislative acts.[16] The chief of state is the president, who is elected by popular vote and can serve for two five-year terms.

---

[12]   In June 2004, Lazarenko was convicted in US district court of "fraud, conspiracy to launder money, money laundering, and transportation of stolen property." Karatnycky, *Ukraine's Orange Revolution.*

[13]   Ibid.

[14]   The only true potential competitors to Kuchma in 1999 were former Prime Minister and colleague Pavlo Lazarenko, Lazarenko's business partner Yuliya Tymoshenko, and Viktor Yushchenko. After his protégé, Vadym Hetman, was gunned down in an elevator, Yushchenko lost his desire to run for president and was convinced by Ukraine's leading oligarch—and Kuchma's campaign manager—not to run for president in exchange for being selected as prime minister when Kuchma won. Wilson, *Ukraine's Orange Revolution.*

[15]   Karatnycky, *Ukraine's Orange Revolution.*

[16]   Central Intelligence Agency, "Ukraine," *The World Factbook.*

The head of government is the prime minister. Cabinet ministers are appointed by the president and approved by the legislature. The Supreme Council is Ukraine's 450-member unicameral legislature with 225 seats allocated on a proportional basis to parties that garner at least 4% of the national vote and the remaining 225 seats elected by popular vote. All members serve for a four-year term.

WEAKNESSES OF THE PREREVOLUTIONARY ENVIRONMENT AND CATALYSTS

The beginning of the prerevolutionary period for the Orange Revolution is perhaps best marked by several key events after the 1999 presidential election, when public support for President Kuchma was openly waning and—more critically—when formal political and activist opposition movements to the president began to form. The first of these events was Kuchma's apparent role in authorizing the September 2000 murder of investigative journalist Georgy Gongadze, who founded the Internet site *Ukraine Pravda* and was a vocal critic of the Kuchma regime. Gongadze's headless body was found two months after he disappeared, and despite evidence that indicated that Kuchma was aware of the plot to kill Gongadze, the necessary legal procedures for impeaching Kuchma were stalled, and he was never formally charged.[17] Among the pieces of evidence against him at the time was the release in November 2000 of secret recordings made by Kuchma's bodyguard that exposed the extent to which Kuchma was involved in corruption across the state and that also included a conversation in which Kuchma said Gongadze must be silenced. These revelations led to the emergence of an opposition movement called "Ukraine without Kuchma" that started protests in Kiev and called for his impeachment, although eventually it failed. This movement began by staging a protest in Kiev on December 15, 2000, with approximately 20,000–30,000 people in attendance. A second wave of protests began on February 6, 2001, and continued until around March 9, with the authorities successfully instigating confrontations and violence from the protestors by "infiltrating fake nationalists from government-funded parties" into the tent city in Kiev, which justified a crackdown by the police.

At approximately the same time, Kuchma fired Yuliya Tymoshenko as the deputy prime minister for the fuel and energy sector on January 19, 2001, because of business squabbles she was having with other

---

[17] Karatnycky, *Ukraine's Orange Revolution*; Michael McFaul, "Ukraine Imports Democracy: External Influences on the Orange Revolution," *International Security* 32, no. 2 (Fall 2007).

oligarchs, and she was arrested one month later on charges of not paying taxes on gas that was smuggled into Ukraine when she was the president of a major Ukrainian energy company. This removal of Tymoshenko from the tight circle of the Kuchma administration enabled her to position for an anti-Kuchma political agenda. It appears to be no coincidence, therefore, that when Tymoshenko formed an organization called the National Salvation Forum on February 9, 2001, with the purpose of removing Kuchma from office, she was arrested four days later on tax evasion charges that dated back to 1995–1997.[18] She was released several weeks later.

A third key event during the prelude to the Orange Revolution that galvanized the opposition movement to Kuchma was the removal of his popular prime minister, Leonid Yushchenko. Yushchenko was the former chairman of the National Bank of Ukraine who developed a reputation for fighting corruption and for tightening economic regulations within Ukraine, especially the collection of taxes.[19] There was some concern that his reform efforts threatened the oligarchs and posed a risk to Kuchma himself. In addition, he formed an alliance with Tymoshenko, who was then the deputy prime minister for the fuel and energy sector and considered an insider among the oligarchs. Together, they recovered more than $1 billion in one year in tax revenues that had been siphoned off by the oil and energy oligarchs. Yushchenko was forced out of office in May 2001 by Kuchma.

Although the initial efforts of the "Ukraine without Kuchma" forum failed, the lessons learned and experiences gained from this initial anti-Kuchma movement proved critical in the prelude to the 2004 Orange Revolution. A central lesson was that the movement did not have a polarizing leader who could mobilize the masses. By removing both Tymoshenko and Yushchenko from his inner circle, Kuchma unwittingly provided two leaders who would form the core leadership of the Orange Revolution.

In preparing for the 2004 presidential elections, Yushchenko formed a political party called "Our Ukraine" that performed surprisingly well in the March 2002 parliamentary elections by capturing 31% of the parliamentary seats that were allocated by party-based portioning of the popular vote. Kuchma was nearing the completion of his constitutionally limited two terms in office and, therefore, searched for a replacement who would remain true to him and his associates and who was also capable of winning the 2004

---

[18] Wilson, *Ukraine's Orange Revolution*.
[19] Karatnycky, *Ukraine's Orange Revolution*, 35–52.

election. He selected Viktor Yanukovich, who was the governor of Donetsk oblast, and Kuchma made him prime minister in November 2002. Within a year of taking office, however, voters began to question the selection of Yanukovich as the potential successor to Kuchma when reports emerged about Yanukovich's criminal past, including a three-and-a-half-year prison sentence for assault and robbery.[20] With public support for Yanukovich in question and the presidential elections scheduled for late October 2004, Kuchma and his administration began a concerted effort in the spring of 2004 to discredit the two main political opponents to Yanukovich in the election: Yushchenko and Tymoshenko.

Figure 2. From left to right: Yushchenko, Tymoshenko, and Yanukovich.[21]

Among the tactics used by Kuchma were a significant pro-Yanukovich media campaign on government-controlled television networks, as well as a smear campaign against Yushchenko, who was now running as the primary presidential candidate for the opposition under his "Our Ukraine" party in conjunction with Tymoshenko, who agreed to accept the prime minister position if Yushchenko won the election. To interfere with the Yushchenko-Tymoshenko campaign effort, roadblocks were often placed in the way of their vehicles, and airplane landing rights were often denied as the campaign was en route to major rallies. Some of the activists from their coalition were also arrested on false charges, and students who campaigned for "Our Ukraine" were told they would be evicted from the dorms. The pro-Yanukovich group even went so far as to put disappearing ink in the pens that were being used for the ballots in districts that were known to be pro-Yushchenko.[22]

---

[20] Ibid.

[21] Edward Lozansky, "Ukraine: Five Years On," *Rusia Blog*, accessed March 14, 2011, www.russiablog.org/2010/01/ukraine-elections-2010-timoshenko-yushchenko-yanukovych.php.

[22] Ibid.

Finally, Yushchenko became seriously ill on September 6, 2004, less than two months before the election, and had to be flown out of the country for treatment. It was eventually determined that he was suffering from dioxin poisoning; although he would recover in time to continue the campaign, his face and body were permanently scarred by the effects of the poison.

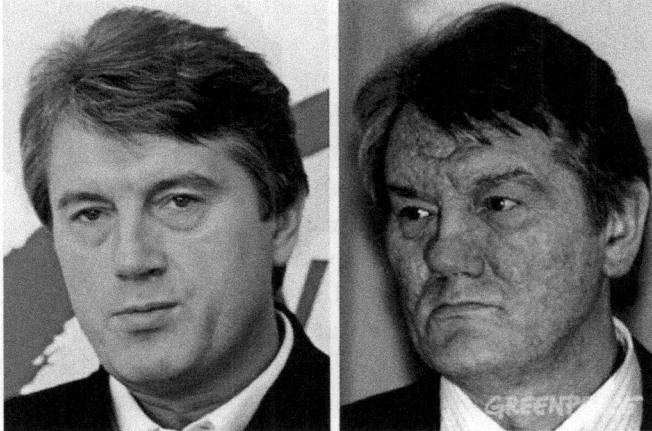

Figure 3. Yushchenko before and after the dioxin poisoning.[23]

When the first round of voting took place on October 31, 2004, Kuchma and his chosen successor assumed that the vote would be somewhat close but that Yanukovich would defeat the challenge from Yushchenko's "Our Ukraine" party.[24] To the dismay of Yanukovich and Kuchma, however, the level of support for Yushchenko was higher than anticipated, and the measures that were in place to control the vote were insufficient to swing it in favor of Yanukovich. When the final tally was released, Yushchenko had a slight lead over Yanukovich (39.87% to 39.26%, respectively), but because neither candidate received a necessary majority vote of the populace, a second-round "runoff" election between the two top candidates was scheduled for November 21. Although voting irregularities and potential vote rigging were reported during this first round, this was not considered significant enough to provide either candidate with a majority, so no formal challenges were filed.

On November 21, 2004, the presidential runoff election was held in Ukraine, with nonpartisan exit polling showing Yushchenko

---

[23]   Greenpeace, accessed April 12, 2010, http://www.greenpeace.org/new-zealand/en/campaigns/toxics/dioxin/impacts-on-health-environmen/.

[24]   This level of comfort by Kuchma appears to have been based primarily on the ability to control the vote because his approval rating just before the election was at 8%, with 62% of the population disapproving of his performance. Although he was not running for president, Yanukovich was viewed as his proxy. McFaul, *Ukraine Imports Democracy*.

with a 52% lead to Yanukovich's 43%. When the official results were released, however, Yanukovich was declared the winner with 49.5%, compared to 46.6% for Yushchenko.[25] Within hours, the details of the voting returns were being assessed, and a significant inconsistency was identified in the returns that were reported from the eastern regions of the country. Specifically, at the time that polling stations closed, the Central Election Commission (CEC) reported consistent voter turnout across the country, with approximately 78–80% turnout in all regions, including the eastern, Russian-speaking districts. However, by the next morning, these percentages were revised, most likely by Kuchma or his supporters, to show sharp increases in voter turnout in some regions, with large percentages of all voters in these regions casting their votes for Yanukovich. One glaring example was Yanukovich's eastern Donetsk region, where voter turnout increased from 78% to 96.2% overnight and 97% of all votes were recorded for Yanukovich. In some districts, voter turnout increased by as much as 40% from the turnout seen the previous month for the first round of the election. Overall, this sudden increase in voter turnout accounted for approximately 1.2 million additional votes, and with more than 90% of these votes going to Yanukovich, this surge was sufficient for his 800,000-vote margin of victory.[26]

---

[25] Wilson, *Ukraine's Orange Revolution.*
[26] Karatnycky, *Ukraine's Orange Revolution.*

Figure 4. October 31, 2004, Ukrainian presidential election results.[27]

27   (Top) "File:Ukraine Presidential Oct 2004 Vote (Yushchenko).png," *Wikimedia Commons*, accessed March 14, 2011, http://commons.wikimedia.org/wiki/File:Ukraine_Presidential_Oct_2004_Vote_(Yushchenko).png; (bottom) "File:Ukraine Presidential Oct 2004 Vote (Yanukovych).png," *Wikimedia Commons*, accessed March 14, 2011, http://commons.wikimedia.org/wiki/File:Ukraine_Presidential_Oct_2004_Vote_(Yanukovych).png.

This obvious and blatant manipulation of the electoral returns for the runoff vote served as the critical catalyst for the Orange Revolution, with hundreds of thousands of people immediately taking to the street in protest. How the government and the opposition responded over the next two months determined the outcome of the revolution.[28]

FORM AND CHARACTERISTICS OF THE REVOLUTION

OBJECTIVES AND GOALS

With the population already mobilizing in response to the well-documented manipulation of the vote, Yushchenko and his advisers from "Our Ukraine" chose two strategies to pursue. The first strategy was considered to be "revolutionary," and the second strategy was viewed as "constitutional and institutional."[29] For his revolutionary approach, Yushchenko went to parliament on November 22, declared himself president, and took the oath of office. In his capacity as "president," he instructed the military and security forces not to oppose the protesters, and he called for a nationwide general strike. Yushchenko also requested the allegiance of local governments and councils. For his constitutional and institutional appeal, Yushchenko knew he needed the support of the legislature and the judiciary if he was to have any checks and balances on whatever Kuchma and Yanukovich planned to do with executive powers. In the end, these two approaches worked, and the concern about a forced repression of the protestors was not realized. The extent of the vote manipulation was so extreme that parliament voted on November 27 to declare the election invalid.[30]

On December 26, 2004, a third election was conducted that was technically a repeat of the invalid November runoff election. With a watchful domestic and global audience now observing all aspects of the electoral return, Yushchenko received 52% of the votes and Yanukovich received 44%; the specific voter margin of victory was 2.2 million votes out of the 28 million votes cast.[31] Despite this clear margin of victory and the presence of thousands of independent international election monitors, Yanukovich challenged the election results, but his protests were eventually rejected by the electoral commission as well as the Supreme Court. Finally, on January 23, 2005, Leonid Yushchenko was formally sworn in as the third president of Ukraine.

---

[28] McFaul, *Ukraine Imports Democracy.*

[29] Karatnycky, *Ukraine's Orange Revolution.*

[30] Ibid.

[31] Ibid.

Ukrainian Presidential
Election December 26, 2004

Viktor Yushchenko

Ukrainian Presidential
Election December 26, 2004

Viktor Yanukovych

Figure 5. December 26, 2004, election results.[32]

[32] (Top) "File:Ukraine Presidential Dec 2004 Vote (Yushchenko).png," *Wikimedia Commons,* accessed March 14, 2011, http://commons.wikimedia.org/wiki/File:Ukraine_ Presidential_Dec_2004_Vote_(Yushchenko).png; (bottom) "File:Ukraine Presidential Dec 2004 Vote (Yanukovych).png," *Wikimedia Commons,* accessed March 14, 2011, http:// commons.wikimedia.org/wiki/File:Ukraine_Presidential_Dec_2004_Vote_(Yanukovych).png.

LEADERSHIP AND ORGANIZATIONAL STRUCTURE

Perhaps the most significant contributing factor to the success of the Orange Revolution was the role of experienced leadership within the movement. The direct lessons from the 2001 "Ukraine without Kuchma" campaign provided activists and political strategists with keen insight about mass mobilization, information management, and sustainment of a protest movement. In addition, lessons were gleaned from other "colored revolutions," such as Slovakia in 1998, Serbia in 2000, and Georgia in 2003, and the collaboration from the leaders in those movements with the leaders of the Orange Revolution is well documented. As one report stated, "The operation—engineering democracy through the ballot box and civil disobedience—is now so slick that the methods have matured into a template for winning other people's election."[33] The leadership of the movement also benefited significantly from the joining of forces between Yushchenko and Tymoshenko, a move facilitated by Kuchma's dismissal of both of them from his administration in 2001. Finally, the mobilization of civic youth organizations, such as "PORA!" ("It's Time!"), under the well-trained youth movement leaders from Serbia and Georgia, provided a level of maturity and nonviolence that was critical for the effective presentation of a united front against the sitting administration.

COMMUNICATIONS

At the time of the 2004 election, most of the national media outlets in Ukraine were controlled by or loyal to the Kuchma regime. Although there were several independent television networks, the pro-Kuchma oligarchs owned the major Ukrainian television channels in the country, and the popular Russian stations that played in the country provided positive coverage to Yanukovich.[34] Independent media outlets did exist, however—such as Channel 5 television, Radio Era, BBC, Voice of America, print newspapers, and internet news outlets, but all of these outlets had limited distribution or the population had limited access to them in 2004.[35] Although the Internet is often cited as a major enabler for the Orange Revolution, it is estimated that only 2–4% of the population had access to the Internet in late

---

[33] Ian Traynor, "U.S. Campaign Behind the Turmoil in Kiev," *The Guardian*, November 26, 2004.

[34] McFaul, *Ukraine Imports Democracy.*

[35] See Ibid.

2004.[36] Of this limited percentage, however, a large number of the users were located in Kiev (the center of the protest movement), and, more importantly, the users of the Internet provided an amplifying capability by reading or distributing by text message or cell phones key pieces of information that they found on the Internet. The Internet site *Ukrainska Pravda*[37] was especially popular because it provided exit polling data as well as stories about fraud allegations. More importantly, it also provided logistical data concerning protests, and it had 350,000 readers and 1 million hits a day. The Internet and cell phones were also used to distribute satire as an effective way of poking fun at Kuchma and Yanukovich while breaking up the monotony of the long days of protesting in the squares. Some observers have noted that the popular culture of satire in Ukraine made the emergence of "viral satire" during the Orange Revolution a critical component of organizational unity and morale.[38] Finally, although their pre-vote influence may have been limited, independent media became critical in the mobilization of the population after the vote. This was especially true of Channel 5, a television station that was purchased by an ally of Yushchenko in 2003. Despite the fact that it had only eight million viewers and a signal that reached only 30% of the country, its live, twenty-four-hour coverage of the election results and subsequent protests raised its rating from thirteenth to third in the nation.[39]

An unintended consequence of the attempts by the government to block Yushchenko's access to the national media was that it forced him to develop a grassroots outreach effort involving meetings across the country. In July, August, and September, he and his representatives traveled across the country and held five to six meetings a day in towns and cities in central and eastern Ukraine. It was in these settings that the organization's experience and lessons learned (as well as external guidance) proved critical as Yushchenko stuck to a singular messaging campaign of criticizing Kuchma and his administration rather than proposing a broad range of policy changes.[40] Yushchenko also kept his message positive with the slogan "*Tak!*" ("Yes") as well as the use of the bright color orange. These gatherings not only brought together crowds of tens of thousands but also led to the creation of networks of

---

[36] Josh Goldstein, *The Role of Digital Networked Technologies in the Ukrainian Orange Revolution*, Publication No. 2007-14 (Cambridge, MA: Berkman Center for Internet and Society at Harvard University, 2007), accessed April 12, 2010, http://cyber.law.harvard.edu/publications/2007/The_Role_of_Digital_Networked_Technologies_in_the_Ukranian_Orange_Revolution.

[37] http://www.pravda.com.ua/.

[38] Ibid.

[39] McFaul, *Ukraine Imports Democracy*; Karatnycky, *Ukraine's Orange Revolution*.

[40] McFaul, *Ukraine Imports Democracy*.

activists and civic organizations that would be crucial for orchestrating mass protests a few months later.[41]

METHODS OF ACTION AND VIOLENCE

The primary and most effective method of action during the Orange Revolution was the ability to mobilize hundreds of thousands of people—and in some cases a million people—as a show of solidarity against the sitting government while simultaneously avoiding the use or provocation of violence. On November 22, 2004, the day after the fraudulent vote count, approximately 500,000 people gathered in Maidan Nezalezhnosti (Independence Square) in Kiev and marched to the headquarters of the Ukrainian parliament while carrying orange flags; many dressed in orange as well. This image was shown across the country and around the world and, just as critically, sent a very clear message to the members of parliament who would vote a few days later to void the election results. For the next two months, this ability to draw together huge crowds—often in freezing temperatures, concentrating their presence at key points and times, while avoiding the use of violence[42]—would be a key success of the Orange Revolution.

Figure 6. The "Orange Revolution" in Maidan Nezalezhnosti (Independence Square), Kiev.[43]

---

[41] Karatnycky, *Ukraine's Orange Revolution.*

[42] It has also been observed that the large size of the crowds may have deterred the use of force by governmental forces because they were so significantly outnumbered.

[43] "File:Orange revolution kyiv.jpg," *Wikimedia Commons,* accessed March 14, 2011, http://commons.wikimedia.org/wiki/File:Orange_revolution_kyiv.jpg.

METHODS OF RECRUITMENT

In the run-up to the 2004 election, the "Our Ukraine" campaign conducted an extensive "get out the vote" campaign that involved nongovernmental organizations, youth civic organizations, and party gatherings to raise awareness about the upcoming election and the need to vote. In many cases, there was an emphasis not on the need to vote for Yushchenko but rather on the need to participate in the democratic process. This was due to an underlying belief that new voters were more likely to be pro-Yushchenko anyway as opposed to being in favor of the status quo.[44] This "get out the vote" effort was considered to be a success, as demonstrated by the large number of people who did vote in the first and second rounds of the election (80% and 77%, respectively).

METHODS OF SUSTAINMENT

The protestors living in the tent cities relied both on contributions from local civilians and on support from local government organizations to sustain them. This included Mayor Omelchenko in Kiev, who provided logistical support, food, and sanitation, and also opened government buildings to shelter out-of-town protestors when the weather was cold.[45] In addition, many protestors were from the local area and, thus, were able to take breaks to refresh and nourish themselves and provide relief for fellow protestors.

METHODS OF OBTAINING LEGITIMACY

In addition to the use of massive, nonviolent protests, the emphasis on obtaining domestic and international legitimacy was critical to the success of the Orange Revolution. Specifically, the strategy of embracing the constitution and institutions of Ukraine provided firm footing during the legal wrangling that followed each round of voting. Final decisions were made by the Ukrainian parliament and Supreme Court, not by either political campaign. In addition, the ability to rapidly gather and distribute independent exit polling data allowed the Yushchenko camp to maintain momentum and undermine potential fraudulent efforts by Yanukovich.[46] This was also aided by the presence of 12,000 international election monitors from

---

[44] McFaul, *Ukraine Imports Democracy*.

[45] Ibid.

[46] Ibid.

throughout Europe, Asia, the United States, and Russia during the third round of elections.

EXTERNAL SUPPORT

Perhaps the most controversial or questioned aspect of the Orange Revolution is the role of external actors in supporting—or even bringing about—the change in government. The presence of "activists for hire" from Belgrade who led the "*Otpor*" (Resistance) youth movements against Serbian President Slobodan Milosevic is well known and clearly aided the planning and preparations for the 2004 election.[47] Less well understood is the degree to which pro-democracy organizations (e.g., the National Democratic Institute, the International Republican Institute, the United States Agency for International Development [USAID], Freedom House, and the Open Society Institute) that supported the process of open, free elections also directly contributed to campaign efforts against Yanukovich. The US government is reported to have spent at least $18 million in "election-related assistance efforts in Ukraine in the two years leading up to the 2004 presidential vote."[48]

A secondary form of support that was external to the Yushchenko campaign, but critical throughout the electoral process, was the involvement of Ukraine's Security Service, the Sluzhba Bezpeky Ukrayiny (SBU). Yushchenko's chief of staff received regular reports from a senior SBU official on threats and tricks that were emerging from the Yanukovich campaign, thus enabling both preparation and appropriate responses.[49] Perhaps more critically, the SBU reportedly played the leading role in calling the Ukrainian Internal Ministry to prevent it from having 10,000 troops put down the protest movement in Independence Square on November 28, 2004, after these forces were mobilized by their commander.

COUNTERMEASURES TAKEN BY THE GOVERNMENT

In the months leading up to the 2004 presidential election, the administration of Leonid Kuchma made numerous attempts to block and undermine the electoral campaign of Yushchenko. This included the airing of pro-Yanukovich advertisements on government-controlled television networks; the use of road-blocks and denied airplane

---

[47] Traynor, *US Campaign Behind the Turmoil in Kiev.*
[48] McFaul, *Ukraine Imports Democracy.*
[49] Karatnycky, *Ukraine's Orange Revolution.*

landing rights to prevent the Yushchenko-Tymoshenko campaign from reaching major rallies; the arresting of coalition activists on false charges; threatening to evict students involved in the campaign from their dorms; the use of disappearing ink in the pens being used to cast votes in pro-Yushchenko districts; and the mysterious dioxin poisoning of Yushchenko in September 2004.[50] Although these types of tactics were used with success against previous campaigns, they were unsuccessful against the overwhelming size and coordination of the Orange Revolution. Despite being an autocratic leader, Kuchma did not truly control all institutions of state power because of the rising power and influence of the oligarchs as well as independence within the judiciary and legislative branches of government.[51] This lack of complete governmental control—as well as Kuchma's restraint in the handling of the protestors—enabled the Yushchenko-Tymoshenko campaign to enact a change in government through the use of the existing political and constitutional processes.

SHORT- AND LONG-TERM EFFECTS

CHANGES IN THE ENVIRONMENT

The election of Leonid Yushchenko in 2004 was in response to rapid corruption within Ukraine and a perception that the country was leaning too far eastward toward Russia. Yushchenko attempted to correct this by being too "pro-West," which led to an ambitious attempt to join the North Atlantic Treaty Organization (NATO), an attempt to limit Russian access to the Black Sea, and support to Georgia during its war with Russia. In response, Russia took advantage of Ukraine's heavy reliance on Russian energy imports and adjusted oil and natural gas prices to reflect fair market values that were much higher than the bargain prices Ukraine had been paying. These events occurred while Ukraine was struggling with the global economic downturn. Because of its geographic position as a crossroads, Ukraine, therefore, had to try to strike a balance between the West and Russia.[52] The Western media quickly lost interest in Ukraine in the aftermath of the Orange Revolution, and a power struggle soon emerged between Yushchenko, Tymoshenko, and Yanukovich, especially with the constitutional shifting of power toward the prime minister.

---

[50] Ibid.

[51] McFaul, *Ukraine Imports Democracy*.

[52] Jonathan Debilde, "Presidential Elections in the Ukraine: The End of the Orange Revolution," *The Geopolitical and Conflict Report*, March 19, 2010.

CHANGES IN GOVERNMENT

In response to the Orange Revolution, the Ukrainian constitution was changed from a semi-presidential republic to a parliamentary democracy. This placed greater power in the hands of the prime minister but also complicated the development of national policy because of the need to rule by forming political coalitions. Yushchenko remained president for six years, but there were four prime ministers during five of these years (Tymoshenko twice, Yanukovich, and Yekhanuriv). On February 9, 2010, Yanukovich was elected the president of Ukraine during the second round of national elections over his main opponent, Tymoshenko.

CHANGES IN POLICY

In the prelude to the 2010 national elections, a circuit court in Kiev outlawed mass gatherings at Independence Square from January 9, 2010, to February 5, 2010, per a request from the mayor's office.[53]

CHANGES IN THE REVOLUTIONARY MOVEMENT

In the immediate aftermath of the Orange Revolution, the coalition that had formed to defeat Yanukovich and the Kuchma regime began to splinter. Eventually, it fell apart because of the personal motivations of individual politicians, parties, and youth organizations, and because of the absence of a unifying foe. Although it has been argued that the 2010 election of Yanukovich signified the end of the Orange Revolution, it has also been argued that the enforcement of a democratic process during the 2004 national elections was in the long-term interest of the country and that this "breakthrough" of democracy was a significant step for the country.[54]

OTHER EFFECTS

The Orange Revolution has been viewed by some analysts as a continuation of the color revolutions that began with Slovakia in 1998, Serbia in 2000, and Georgia in 2003. It is posited that after the lessons and experiences with Ukraine in 2004, this string continued with Kyrgyzstan in 2005.

---

[53] "Court Forbade Maydan After First Tour of Election: Ukraine News by UNIAN," accessed September 30, 2010, http://www.unian.net/eng/news/news-356613.html.

[54] McFaul, *Ukraine Imports Democracy*.

BIBLIOGRAPHY

Central Intelligence Agency. "Ukraine." *The World Factbook*. Accessed September 30, 2010. https://www.cia.gov/library/publications/the-world-factbook/geos/up.html.

"Court Forbade Maydan After First Tour of Election: Ukraine News by UNIAN." Accessed September 30, 2010. http://www.unian.net/eng/news/news-356613.html.

Debilde, Jonathan. "Presidential Elections in the Ukraine: The End of the Orange Revolution." *The Geopolitical and Conflict Report*. March 19, 2010. http://gcreport.com/index.php/analysis/162-presidential-elections-in-the-ukraine-the-end-of-the-orange-revolution.

Franchetti, Mark. "Russia Fleet 'May Leave Ukraine'." *BBC News*, October 18, 2008.

Goldstein, Josh. *The Role of Digital Networked Technologies in the Ukrainian Orange Revolution*. Publication No. 2007-14. Cambridge, MA: The Berkman Center for Internet and Society at Harvard University, 2007.

Karatnycky, Adrian. "Ukraine's Orange Revolution." *Foreign Affairs* (2005): 35–52.

McFaul, Michael. "Ukraine Imports Democracy: External Influences on the Orange Revolution." *International Security* 32, no. 2 (Fall 2007): 45–83.

"Timeline: Battle for Ukraine." *BBC News*, January 23, 2005. http://news.bbc.co.uk/2/hi/europe/4061253.stm.

Traynor, Ian. "US Campaign Behind the Turmoil in Kiev." *The Guardian*, November 26, 2004.

Wilson, Andrew. *Ukraine's Orange Revolution*. New Haven, CT: Yale University Press, 2005.

World Bank Group. "Can Ukraine Avert a Financial Meltdown." *Beyond Transition Newsletter*. Accessed September 30, 2010. http://web.archive.org/web/20110628222439/http://www.worldbank.org/html/prddr/trans/june1998/ukraine.htm

SOLIDARITY

Chuck Crossett and Summer Newton

SYNOPSIS

In late 1980, the Communist leadership of the People's Republic of Poland announced another large price hike for food in response to a weakening economy. Protests and strikes sprang up throughout the country, but the Lenin (Gdansk) shipyards captured the fervor and heart of the rising movement. In response to the strikes, the government allowed a workers' union, Solidarity, to form separate from the official government workers' union. Industrial laborers and various other trades soon constituted a large but fairly loose organization, with a farmer's version of Solidarity soon following. Coordinated strikes and demands then tested the government, pushing for further reforms and concessions. Sensing a dangerous slip of their power, the government declared martial law at the end of 1981 and outlawed the independent unions, arresting multiple tiers of leaders and activists. The heavy repression stopped the strikes and violence, but an underground movement continued. Martial law was lifted in 1983 and another union, under more restrictions, was allowed, but events and world attention (including a Nobel Peace Prize for Solidarity's leader, Lech Walesa) kept the momentum for the government's collapse alive. By 1988, negotiations resulted in Solidarity being allowed to field candidates for the upcoming elections, which they swept. A coalition government between Communist and union parties was forged, and reforms began to be instituted in 1989. Lech Walesa was elected as president of the Republic of Poland at the end of 1990.

TIMELINE

June 1976	Violent protests and strikes occur across Poland in response to announced food price hikes.
October 1978	Karol Wojtyla is elected as Pope John Paul II.
August 1980	Strikes begin at Lenin (Gdansk) shipyards and spread across Poland.

August to September 1980	Agreements are signed that meet many striker demands, including the right to unionize independent of the Communist Party's control.
September 1980	Solidarity is formed for nationwide industrial workers.
March 27 1981	A nationwide "warning strike" occurs in response to beatings of Solidarity activists.
May 1981	Rural Solidarity is formed, organizing farm workers.
December 1981	Martial law is declared and Solidarity leaders are arrested.
September 1982	Solidarity is outlawed and underground organizations form.
November 1982	Walesa is released.
July 1983	Martial law is lifted, but some restrictions remain.
October 1983	Lech Walesa is awarded the Nobel Peace Prize.
June 1984	Solidarity calls for a boycott of local elections.
March 1985	Mikhail Gorbachev becomes the leader of USSR Communist Party.
September 1986	Most political prisoners are granted general amnesty, and Walesa forms a new legal Solidarity union.
April 1988	New strikes erupt after a large food price hike.
August 1988	Larger strikes occur, forcing the government to negotiate.
November 1988	A televised debate is held between Walesa and Miodowicz, the head of official state sponsored trade union.
December 1988	A hundred-member Citizens' Committee is formed to prepare for negotiations.
February to April 1989	Round-table talks among the government, Solidarity, and other opposition groups result in a new government structure.
June 1989	Solidarity wins 35% of seats in *Sejm* (the maximum allowed by the Round-Table Agreement) and 99 of 100 Senate seats.
August 1989	Solidarity member appointed prime minister and a non-Communist coalition government formed
January 1990	The first wave of economic reforms occurs.
December 1990	Lech Walesa is elected president of the Republic of Poland.

THE ENVIRONMENT OF THE REVOLUTION

PHYSICAL ENVIRONMENT

Figure 1. Map of Poland.[1]

Poland is slightly smaller than the state of New Mexico, with its capital in Warsaw. The total area of Poland is 312,685 square kilometers. Lying in the eastern region of Europe, Poland is bordered by Belarus, the Czech Republic, Germany, Lithuania, Russia, Slovakia, and Ukraine.[2] Before its dissolution in 1990, Poland shared a large border with the USSR and was a key member of the Warsaw Pact, located between the Soviet Union and a divided Germany. Poland's geography is composed primarily of flat land and some mountainous areas along its southern border. The climate is temperate, with moderate to severe winter weather and frequent precipitation in the winter months.[3]

CULTURAL AND DEMOGRAPHIC ENVIRONMENT

As of 1991, Poland's population was an estimated 36.1 million. As a result of intensive urbanization after the onset of communist rule,

---

[1] Central Intelligence Agency, "Poland," *The World Factbook*, https://www.cia.gov/library/publications/the-world-factbook/geos/pl.html.

[2] Ibid.

[3] Ibid.

more than 60% of Polish residents resided in urban centers covering 6% of the total territory. Poland has 24 cities with more than 150,000 residents, with Warsaw and Lodz being the first and second largest cities. Literacy rates are high, around 98%.[4]

Poland experienced extreme demographic changes after World War II. Before World War II, Poland included substantial numbers of Belarusians, Germans, Jews, and Ukrainians. German and Jewish populations endured the sharpest reductions As a result of the Holocaust, Poland's Jews, representing 10% of the population, were reduced to less than 0.1% of the population—a loss of more than 3 million.[5] Postwar resettlements and border adjustments resulted in the additional loss of 2 million ethnic Germans and 500,000 Ukrainians, Belarusians, and Lithuanians. As a result, by 1989, approximately 98% of Polish citizens claimed Polish ethnicity. The ethnic homogenization of Polish society also reduced religious diversity. Previously, Jews, Protestants, and Greek Orthodox members resided in Poland, but reductions in ethnic populations left Poland a predominantly Roman Catholic society. In a 1991 government survey, 96% of the population claimed an affiliation with the Roman Catholic Church.[6]

At the post-World War II Yalta summit, the USSR was granted the Polish territories, a "liberation" that was particularly galling to Poles because Russia had a historical penchant for interfering in Polish politics. Moreover, mistrust of central authority was evident in Polish culture long before the communists took control in the 1940s. In the sixteenth century, Poland developed a unique parliamentary system requiring unanimity among its nobles. Russia and Prussia had previously divided Polish territories among themselves in a series of partitions that literally erased Poland from the map in 1795,[7] provoking Polish insurrections that were brutally suppressed by Russia. Polish identity, developing as it did in the shadow of her large and powerful neighbor, is "historically defined in opposition to Russia."[8] In addition to more distant antagonisms, Soviet forces invaded eastern Poland in 1939, deporting more than 1 million Poles to Siberia and killing 22,000 military officers in the Katyn massacre. Less than one-half of those deported returned to Poland. No country in Eastern Europe

---

[4] Glenn E. Curtis, ed., *Poland: A Country Study* (Washington, DC: Government Printing Office, 1992).

[5] A. Kemp-Welch, *Poland Under Communism: A Cold War History* (Cambridge, UK: Cambridge University Press, 2008), 158.

[6] Curtis, *Poland: A Country Study*.

[7] Poland did not reappear as a sovereign country until the end of World War I. The Second Polish Republic lasted from 1918 until the Nazi and Soviet invasion in 1939.

[8] Timothy Garton Ash, *The Polish Revolution: Solidarity* (New Haven, CT: Yale University Press, 2002).

was less prepared to endure Soviet rule. Stalin himself acknowledged that implementing Soviet-style communism in Poland was akin to "saddling a cow."[9]

Figure 2. Jerzy Popieluszko (1947–1984), a Catholic priest from Poland associated with the Solidarity union, was murdered by the communist internal intelligence agency, the Sluzba Bezpieczenstwa.[10]

After the implementation of communist rule in the Soviet bloc, religious institutions were marginalized. It was only in Poland that the church remained an independent actor and continued to exert tremendous influence over the people of Poland. The symbols and rituals of the Roman Catholic Church pervaded the scene of strikes. Primarily because of the astute leadership of Cardinal Wyszynski and Catholic intellectuals, the church in Poland began to champion basic human rights and political and civil liberties, providing the non-Catholic intelligentsia and the workers in Poland with a common vocabulary to oppose the state.[11] This was especially apparent after the student revolt of March 1968, when the church and the left developed

---

[9] Ibid., 4.

[10] "File:Jerzy Popieluszko.jpg," *Wikipedia*, accessed March 11, 2011, http://en.wikipedia.org/wiki/File:Jerzy_Popieluszko.jpg.

[11] Ibid., 20.

closer ties.[12] During periods of social unrest, the state often turned to the church to repair its relations with society, such as the release of Cardinal Wyszynski from prison after the 1956 de-Stalinization crisis. Church officials acted as mediators during the 1988–1989 negotiations, providing a trusted third-party assurance to both the state and society that the negotiations were conducted in good faith. Furthermore, the election of a Polish Pope in October 1978 and his subsequent triumphant visit to Poland the next year despite the reservations of the Communist regime were landmark events that left an indelible mark on Poland. During his roving visit, some twelve million Poles gathered to hear his sermons emphasizing the dignity of labor, human rights, and the need for reconciliation in a polarized world.[13] His visit became a "symbolic confrontation" with the regime, uniting and inspiring Poles for a "new round of political struggle."[14]

SOCIOECONOMIC ENVIRONMENT

This history of the Solidarity movement is encapsulated in a series of dates: 1956, 1968, 1970, 1976, 1980, and 1989. Each crisis, with exception of the student revolt in 1968, was precipitated in part by an acute economic crisis. From the demands of "bread and freedom" by workers in the city of Poznan in 1956 to wage increase demands in 1980, economic demands made by the opposition contained implicit, and sometimes explicit, political criticisms. A series of acute economic crises resulted in social and regime instability, and sometimes changed leadership, during the period of Communist rule.

After the 1956 strikes, the new First Secretary issued economic concessions, such as wage increases and subsidies, to bolster the legitimacy of the regime. In 1970, however, another round of price increases was announced right before Christmas, and workers took to the streets again. The next First Secretary, Gierek, introduced the "great leap" forward, which was the last comprehensive economic vision offered by the Communist Party. Because the enervated communist ideology could no longer garner popular support, the

---

[12] Leftist intellectuals, many of whom were former members of the Polish United Workers Party (PZPR), openly expressed anti-Catholic sentiments in the early years of the communist regime. After the events of March 1968, Catholic leaders laid down their grievances against the intellectuals and provided support. For an excellent account of the unlikely alliance between the church and the left, see Adam Michnik and David Ost, *The Church and the Left* (Chicago: University of Chicago Press, 1993).

[13] Kemp-Welch, *Poland Under Communism*, 228–229.

[14] Grzegorz Ekiert and Jan Kubik, *Rebellious Civil Society: Popular Protest and Democratic Consolidation in Poland, 1989–1993* (Ann Arbor, MI: University of Michigan Press, 1999), 38.

program was an attempt to secure that support by providing Polish citizens with material well-being.[15]

The promised Polish economic miracle rested on proposed intensive industrial development, importing technology from the West, paid for by Western credit.[16] The net result, the government argued, would be increased exports to Western markets, further developing the economy. Real wages were increased 22%, and price increases on staple foods were scaled back to pre-1970 prices.[17] The reforms were a spectacular failure. External debt skyrocketed, and collectivization decreased food production while artificially deflated prices increased demand. Runaway spending by managers in heavy industry led to more investments in fixed assets than originally projected, and the expected increase in productivity did not materialize to offset the expenditures.[18] The most apparent result of the reforms was raising the expectations of a generation that became accustomed to food and Western consumer goods on the shelves that the Communist regime, by the end of the 1970s, could no longer provide.

HISTORICAL FACTORS

Stalin's death in 1953 signaled an about face in communist policy throughout the Soviet bloc and prompted widespread agitation for change in Eastern Europe after Stalinism was denounced by the Communist Party of the Soviet Union (CPSU). Nikita Khrushchev, First Secretary of the CPSU and later Premier of the USSR, denounced Stalin, his cronies, and policies in the "Secret Speech" to the Twentieth Party Congress of the CPSU in 1956. Crises erupted throughout Eastern Europe during the de-Stalinization process.[19] Uncertainty regarding the future in the Polish ruling Communist Party—the Polish United Workers Party (PZPR or simply the "Party")—as well as within society

---

[15] Garton Ash, *The Polish Revolution*, 13.

[16] For an overview of Gierek's economic program, and its failure, see David Kemme, "The Polish Crisis: An Economic Overview," in *Polish Politics: The Edge of the Abyss*, eds. Jack Bielasiak and Maurice D. Simon (New York: Praeger, 1984).

[17] Ekiert and Kubik, *Rebellious Civil Society*, 37.

[18] Garton Ash, *The Polish Revolution*, 13–17.

[19] De-Stalinization was a process that began with Nikita Khrushchev's (the head of the Soviet Union) "Secret Speech" in 1956 after the death of Stalin in 1953. In the speech, Khrushchev denounced many of Stalin's policies and his crimes. Afterward, members of the ruling Communist parties of the Eastern Bloc, rank-and-file and elite alike, began openly debating the efficacy of Stalinism, including the forced collectivization of agriculture, heavily industrialized economies, rigid adherence to the Soviet model, and terror tactics. Such debate before Khrushchev's speech was violently quashed by the Soviets. The extent of de-Stalinization was limited and varied from country to country. For a more detailed description, see Ibid.

at large, combined with a dismal economy, sparked unrest in Poland. Reformist factions of the Party challenged the Stalinist establishment, questioning close economic ties with the USSR, reassessment of Stalinist domestic policies, and the re-evaluation of Stalinist forced industrialization policies.[20] In June, workers in the city of Poznan protested unrealistic production quotas, taxes, and food and housing shortages. In contrast to the Solidarity strikes, the Poznan workers channeled their grievances through the Party apparatus. The regime's harsh response left 100 dead.[21]

Challenged by both the internal dissension and mobilization of oppositional societal forces after a series of meetings, the Politburo elected Wladyslaw Gomulka as First Secretary. A victim of Stalinist terror, Gomulka had long advocated a socialist path more attenuated to local Polish circumstances, garnering him significant popular support for the purported reform he would bring to Poland.[22] To defuse political and social tensions, Gomulka granted economic and political concessions, including increase of real wages and the legalization of civil society organizations. The failure of Gomulka's regime to follow through with many of the above promises was apparent already in 1957. The "Polish October," as the period is called, was stillborn.[23] Although liberalization of political and economical life did not materialize, the Polish October did bequeath important legacies for the Solidarity movement. As a result of the mobilization of oppositional forces during the 1956 crisis, Poland emerged as the most "open" regime in the Soviet bloc. The "uncontrolled spaces" opened during the 1956 crisis became laboratories of political dissent, and society—particularly intellectuals—nurtured a culture of resistance and gained assurance that they could force concessions from the regime, all developments that aided popular resistance in the 1970s and 1980s.[24]

In 1956, it was primarily the workers who set the streets ablaze in Poznan, but other oppositional actors emerged in the 1960s: students and intellectuals. In 1968, a year that saw large-scale student protests around the world, Poland underwent another political crisis. The crisis

---

[20] Ekiert and Kubik, *Rebellious Civil Society*, 27.

[21] Ibid., 28.

[22] Gomulka was the first communist leader in Poland. He was replaced by the Stalinist stooge Bierut after his attempts to adopt a socialist path for Poland that responded to local circumstances. For instance, Gomulka never supported the collectivization of agriculture. Garton Ash, *The Polish Revolution*, 5.

[23] "Stillborn" is a term frequently used to describe this period, meaning that as soon as the Polish October reforms emerged, they were crushed by the Party. The reforms never came to any fruition; they were dead before they were really born.

[24] Ekiert and Kubik, *Rebellious Civil Society*, 31.

was precipitated by "the regime's retreat from the course of the Polish October, the authoritarian political style of the Gomulka leadership, and the intensification of social and economic problems in the 1960s."[25] The conflict was primarily a confrontation between the increasingly conservative regime and revisionist activists seeking reform of the communist system. Revisionists within the Party represented several strands of thought, from appreciation of the Yugoslavian industrial self-management model to a rethinking of holdover assumptions from the Stalinist era.[26] Students agitated for more democratic freedoms at Warsaw University in March 1968 but were assaulted and arrested by security forces.[27] More student revolts followed. Students also attempted to unite with workers, visiting, among other places, the Gdansk Shipyard. However, workers were largely unsympathetic and waited on the sidelines while students protested and endured state oppression. In turn, less than two years later, when workers hit the streets in protest, students returned the favor and were largely absent from protests. The unification, or "solidarity," of various social actors would have to wait until after 1976.[28]

The ramifications of the students' protests in 1968 across Eastern Europe were important for later developments in Poland. The opposition was not successful in securing any major concessions from the regime or the removal of Gomulka from office. The events signaled that the sectors of society that had secured the important concessions in 1956—the reformists within the Party and intellectuals—were a spent force. After reformers failed to gain control of the Party or persuade those in authority to listen, revisionists were either driven from the Party or marginalized. Those who were affected by the events of 1968 returned to politics to great effect in the late 1970s and early 1980s.

After 1956, Gomulka had toed the Soviet economic line and maintained a highly centralized economy focusing on heavy industry while stifling private enterprise. He ignored suggestions by Polish economists to introduce market mechanisms and reduce economic

---

[25] Jack Bielasiak, "Social Confrontation and Contrived Crisis: March 1968 in Poland," *East European Quarterly* 22, no. 1 (1988): 81–103.

[26] Kemp-Welch, *Poland Under Communism*, 133–135.

[27] Gomulka's regime issued propaganda claiming that the student protests were inspired by the Writer's Plenum (an intellectual organization), academics in the humanities, and Jews, despite the fact that the latter population was decimated after the Holocaust and emigration (from 10% of the population before World War II to less than 0.1% in 1967 – Poland lacked even a single qualified rabbi). As a result of the events of March 1968, many of the faculty were purged from Warsaw University. Protests by students and intellectuals continued after the sackings. Ibid., 152–160.

[28] Bielasiak, "Social Confrontation and Contrived Crisis."

centralization, a result of bureaucratic inertia, vested interests, and the limited imagination of the leadership.[29] Shortages of food and consumer goods, stagnant wages, and unsafe working conditions plagued workers. The primary economic benefit common Poles received, half of whose monthly budget went to food, was heavily subsidized food prices.[30] In December 1970, the regime opted to address, as they would several times until 1989, economic difficulties by slashing costly food subsidies rather than engaging in any major structural reforms.[31] The protests that followed resulted in Gomulka's fall from power.

The price increases were announced on Saturday, and when workers returned to their workplaces the following Monday, many went on strike in protest. Events escalated quickly at the Gdansk Shipyard and throughout the Gdansk region.[32] The demands of the striking workers were economic, although some nascent political demands emerged that would later be echoed and amplified by the massive 1980–1981 Solidarity strikes. Demands for the withdrawal of price increases were widespread, recognition of strike committees, and more importantly, the demand for autonomous organizations to represent workers' demands—trade unions.[33] Gomulka called for the use of force, dispatching soldiers under the false pretenses of a German invasion. Soldiers fired on the crowds, although some surrendered when they encountered Polish workers rather than German invaders.[34] The martyrdom of the workers shot at the gates of the Gdansk Shipyard began immediately. The workers, murdered by the purported "workers' state," "became the symbol of all their accumulated grievances."[35] The strikes spread to several other cities, including Szczecin and Gdynia. In the latter city, workers were shot as they attempted to return to work after the strikes ended.[36] The articulation and development of proto-political demands, along with the unification of workers, especially in the Gdansk Shipyard where the strikes of 1980–1981 took place, were crucial results of the 1970–1971 strikes.[37]

---

[29] Garton Ash, *The Polish Revolution*, 10–11.

[30] Kemp-Welch, *Poland Under Communism*, 180.

[31] Ibid., 180.

[32] Garton Ash, *The Polish Revolution*.

[33] Ekiert and Kubik, *Rebellious Civil Society*, 35.

[34] Kemp-Welch, *Poland Under Communism*, 180–185.

[35] Garton Ash, *The Polish Revolution*, 13. Lech Walesa, cofounder of Solidarity, helped plan the strike and was present at the shooting of the Gdansk workers.

[36] Kemp-Welch, *Poland Under Communism*, 186.

[37] Garton Ash, *The Polish Revolution*, 13.

Workers were not the only casualties of the strikes. Gomulka's regime, unable to muster a response to the political and economic crises, was ousted in December 1970. Gomulka's replacement as First Secretary was Edward Gierek. In a move that was part politics, part theater, Gierek visited the Warski and Gdansk Shipyards in January 1971 and, in an unprecedented nine-hour conversation with strikers, allowed workers to voice their grievances (including complaints about official trade unions, lies in the official media, and the inefficiency of the ruling class) and responded to their concerns.[38] He appealed to the workers for assistance in the monumental task of reinvigorating the economic and political life in Poland. The workers infamously shouted back, "Pomolemy! Pomolemy! We will help you, we will help you!" Workers gave the Party another shot to deliver Poland from ruin.[39]

GOVERNING ENVIRONMENT

As a communist state in the Soviet bloc, Poland was "subject to the monopolistic rule of the communist party."[40] Rank-and-file Party membership reached a high of 3.08 million in the early 1980s but dropped to a little over 2 million by 1987.[41] However, the size of the mass membership, especially in the 1980s when more than one-third of Poles left the Party—many to join Solidarity—is not a reliable indicator of the support it enjoyed in society. The executive organ of the Party, the Politburo, was composed of ten to twenty members and led by the First Secretary.[42] It was the primary political power holder in Poland—not state institutions—directing the political, economic, and social developments of the country. Rank-and-file members belonged to local Party organizations, most often located in their place of work, and were connected to the Party in a complex hierarchy.

The Party maintained its domination of Polish society through an extensive bureaucratic apparatus. It controlled the appointment of officials to institutions affecting everyday life, including managers in industry and commerce, publishers, newspaper editors, judges, trade union officials, university rectors, leaders of youth and women's

---

[38] At the shipyard, Gierek said, "When it was proposed that I take over the leadership of the Party, at first I thought I would refuse . . . I am only a worker like you . . . ." Ibid., 13.

[39] Ibid., 13.

[40] George Kolankiewicz and Paul G. Lewis, *Poland: Politics, Economics and Society* (London: Printer Publishers, 1988): 66.

[41] For a demographic breakdown of Party membership, see Ibid., 69.

[42] Communist regimes in the Soviet bloc were infamous for their stifling, convoluted bureaucratic structures. See Ibid., 72–75, for a flowchart and description of the organizational structure of the Party.

organizations, and a host of others. To fill the positions, the Party's local and central offices maintained extensive lists of positions and people fit to fill them. The Soviet term for the lists, *nomenklatura*, became the term to describe the class filling these important positions. By 1980, some 200,000–300,000 belonged to the *nomenklatura*, not including their families and dependents. The *nomenklatura* acted as an elite class receiving benefits and perks in return for their loyalty to the system.[43] Anger at this "ruling class" in an ostensibly egalitarian society informed the demands of oppositional actors.

The system of nomenklatura and the wide-ranging domination it provided the Party allowed it to dominate the organs of the state as well. Although periodic elections were held to fill the seats in the parliament (the *Sejm*), it had no capacity as a legislative body. Two additional parties were represented in the *Sejm*, the United Peasant Party and the Democratic Party, but both were subsidiaries of the PZPR and offered little to no opposition to its rule.[44] During several periods of crisis, however, including the 1980s when Solidarity exerted pressure on the government, the *Sejm* did act as a forum for debate.

Communists quashed early attempts of worker resistance, declaring strikes "logically impossible" because the ostensibly worker-owned economy meant that workers would only be striking "against themselves."[45] All trade unions were federated into one mass trade union (OPZZ) connected to the Party-state; all others were illegal.[46] Trade unions, furthermore, did not exist to represent the interests of the workers, but only as, per Lenin's famous phrase, "transmission belts" feeding Party directives to workers.[47] Lack of free trade unions amid the series of economic crises in the following decades was a sticking point for Solidarity activists.[48]

WEAKNESSES OF THE PREREVOLUTIONARY ENVIRONMENT AND CATALYSTS

The legitimacy of Gierek's regime hinged not on its espousal of communist ideology but on its ability to successfully deliver an increased standard of living for Polish citizens. Gierek promised more food and consumer goods in shops, secure employment, social security, and low prices for staple goods. His "great leap" economic reforms failed, and

---

[43] Garton Ash, *The Polish Revolution*, 6–7.
[44] Kolankiewicz and Lewis, *Poland: Politics, Economics and Society*, 82.
[45] Kemp-Welch, *Poland Under Communism*.
[46] Ibid., 35.
[47] Garton Ash, *The Polish Revolution*.
[48] Kemp-Welch, *Poland Under Communism*.

by the mid-1970s, the economy was unraveling. Gierek implemented price increases in 1976 that were greeted with the same response as 1970. Strikes began in more than 130 enterprises, with two massive strikes in the industrial cities of Radom and Ursus. The latter strikes turned violent, and, as in 1970, the regime responded brutally and decisively, aggressively dispersing crowds and beating, arresting, and dismissing workers.[49] Thousands of workers were fired, found guilty, and sentenced for varied crimes; several were killed; and many more were wounded. These ominous results were belied by several other more positive ones: Gierek's regime capitulated, rolling back the price increases, and the Committee for Workers' Defense (KOR), was founded.

A diverse group of intellectuals formed the KOR after learning that many of the workers subjected to state repression after 1976 did not have adequate resources to defend themselves against state charges.[50] The KOR demanded amnesty for workers targeted after the 1976 strikes and argued for an end to repression. In addition, it published information on regime persecution, acting as a watchdog for state repression. The founding of the KOR was the first important step in bridging the gap between workers and intellectuals. Collaboration between the two important groups had heretofore not materialized during previous political crises. In both 1968 and 1970, each group had largely waited on the sidelines while the other risked life and limb opposing the communist regime. As one observer noted, "Without this bridge, Solidarity would have developed, if at all, very differently."[51] The KOR included former communists and Party members, former prisoners of Stalin, writers, economists, and students from 1968. The KOR encouraged numerous intellectuals to join the opposition, helping Poland to develop the most sophisticated opposition counterculture in the Soviet bloc. It published journals, the most important of which was *Robotnik*, "The Worker," in an extensive underground publication network.[52]

The KOR was an incubator for a new opposition strategy, developed by the intellectuals Adam Michnik, Leszek Kolakowski, and Jacek Kuron. After 1968, attempts to reform the Party into a more democratic, responsive institution died. The geopolitical reality facing the oppositionists, however, precluded any attempts to overthrow the

[49] Ekiert and Kubik, *Rebellious Civil Society*.
[50] For a detailed description of the founding of the KOR, see Michael H. Bernhard, *The Origins of Democratization in Poland: Workers, Intellectuals, and Oppositional Politics, 1976–1980* (New York: Columbia University Press, 1993).
[51] Garton Ash, *The Polish Revolution*, 18.
[52] Ibid., 18.

existing regime because the USSR had limited tolerance for shifts to pluralism or capitalism. Kolakowski argued for hope for democratic political change by exerting "effective, slow, and gradual" social pressure on the regime.[53] Michnik and Kuron developed the strategy further, transforming Kolakowski's ideas into practical politics with the founding of the KOR.[54] As a result, the KOR expanded its mission in 1977 to reflect its aspirations, changing its name to the Committee for Social Self-Defense. The KOR encouraged civic activities in all areas of life.

Other independent organizations developed, some under the KOR umbrella. Approximately one-third of the population joined independent, professional, social, or political organizations from 1976 to 1981. After the declaration of martial law in 1981, one in five Poles had participated in at least one collective protest.[55] One of the most important organizations formed during this time was the Founding Committee of Free Trade Unions on the Coast in Gdansk in 1978.[56] Its earliest members, Andrzej Gwiazda, Bogdan Lis, and Lech Walesa, would lead the August 1980 strike in the Gdansk Shipyard. Also, in 1979, it published an issue of its newsletter that included a "Charter of Workers' Rights," prefiguring many of the demands made by Solidarity in 1980 and 1989, such as independent trade unions. The KOR, disseminating information through *Robotnik*, was able to translate the strategy into specific tactics, influencing workers while developing a nationwide opposition network, which together played a tremendous role in generalizing societal grievances, formulating

---

[53] Michael Bernhard, "Civil Society and Democratic Transition in East Central Europe," *Political Science Quarterly* 108, no. 2 (1993): 307–326.

[54] Ibid. For a description of the strategy, see Adam Michnik "New Evolutionism," *Letters from Prison and Other Essays* (Berkeley, CA: University of California Press, 1985): 135–148.

[55] Ekiert and Kubik, *Rebellious Civil Society*, 41.

[56] Although the KOR supported worker organization and independent trade unions, the organization was apprehensive with regard to what it saw as the premature declaration of the Founding Committee of Free Trade Unions, fearing police reprisal. Michael Bernhard, "Reinterpreting Solidarity," *Studies in Comparative Communism* 24, no. 3 (1991): 313–330.

remedies, and coordinating workers' actions.[57] Past collective protests were carried by one class, whether it was the students and intelligentsia in 1968 or the workers in 1970 and 1976. By 1979, the stage was set for the "tacit alliance of workers, intelligentsia, and Church," which was to grow into the Solidarity movement.[58]

By 1980, with the failure of Gierek's economic policies, the regime was set for yet another round of price increases. Poland's external debt had increased from $1.2 billion in 1971 to $20.5 billion in 1979. In servicing the debt rather than repaying it, Gierek's legacy to Poland sucked up more than 81% of annual export earnings. The cost of maintaining food subsidies had also skyrocketed because the prices did not reflect demand or current production costs. Major structural reforms were untenable according to communist ideology. A prisoner of communist inefficiency, the regime's stock answer was price increases. However, it was not only the increases themselves that so disturbed the population. Despite Gierek's promises of "consultation" with society, the increases were unilaterally imposed on a society with no input into their own economic future. Additionally, the increases were instituted through commercial shops, which sold the scarce goods at much higher prices, and Pewex shops selling Polish goods for hard currency were also introduced, both out of reach for most workers. In a society for whom egalitarianism was bread and butter, such notably non-egalitarian policies emphasized the illusory nature of the so-called "worker's" state.[59] Demands in 1980–1981, and later in 1988, included the abolishment of privileges for elite groups such as the police, militia, and *nomenklatura*. Stanislaw Kania replaced Gierek as a result of the subsequent strikes in 1980–1981.

---

[57] Garton Ash, *The Polish Revolution*, 23–24. Garton Ash argues that the KOR functioned much in the way that Lenin described a vanguard movement in his essay "What Is to be Done?", raising the political consciousness of proletariats in industrial centers. In doing so, Garton Ash supports an elite-driven thesis, that Solidarity, and Poland's revolutionary change, was largely the product of intelligentsia efforts. However, several Solidarity scholars, notably Roman Laba, *The Roots of Solidarity: A Political Sociology of Poland's Working-Class Democratization* (Princeton, NJ: Princeton University Press, 1991), and Lawrence Goodwyn, *Breaking the Barrier: The Rise of Solidarity in Poland* (New York: Oxford University Press, 1991), challenge Garton Ash's thesis, arguing that Solidarity—its tactics, strategies, and goals—originated with the workers themselves. The different theses are reviewed in Jan Kubik, "Who Done it: Workers, Intellectuals, or Someone Else? Controversy Over Solidarity's Origins and Social Compositions," *Theory and Society* 23, no. 3 (1994): 441–466.

[58] Garton Ash, *The Polish Revolution*, 24.

[59] Kemp-Welch, *Poland Under Communism*, 230–231.

FORM AND CHARACTERISTICS OF THE REVOLUTION

OBJECTIVES AND GOALS

The objectives and goals of the Solidarity movement were first and foremost circumscribed within the allowable limits set by the Soviet Union. The very real threat of a Soviet armed intervention, institutionalized in the Brezhnev Doctrine, loomed over all Solidarity activities. According to the Brezhnev Doctrine, any and all shifts away from socialism and toward capitalism within the Soviet Bloc were of concern to the CPSU. After its articulation in late 1968, it was used to retroactively justify the Soviet invasion of Czechoslovakia earlier that year and the invasion of Hungary in 1956. The Doctrine left little to no room for liberalization or independence for the ruling communist parties of the Warsaw Pact. In this regard, the denunciation of the Brezhnev Doctrine and Gorbachev's *perestroika* and *glasnost*[60] policies were important to the later success of the Solidarity movement. As a result, Solidarity activists and[61] intellectuals walked a very narrow line over shark-infested waters in their agitations for a legal trade union. Intellectual advisers to Solidarity "coined the term 'self-limiting revolution'" to describe its agenda—society would organize itself, but it would not make an explicit grab for power.[62] In the words of Lech Walesa, "The communists had to believe that our actions and aspirations did not threaten the foundation of the system. They could not know that after taking one finger, we would reach out in a moment for the entire hand under the right circumstances."[63] Beginning with the strikes of 1980, Solidarity activists developed a comprehensive set of political demands.

At the start of the August strike at the Lenin shipyard, the strike committee, headed by Lech Walesa, put forward five demands: reinstatement of Walesa and another fired worker who were activists for a free trade union, permission to build a monument for the

---

[60] *Perestroika* and *glasnost* are the words that Mikhail Gorbachev used to describe the policies he implemented to reform the Soviet system. Although both words have multiple connotations, *perestroika* was used to describe the extensive political, social, and economic restructuring undertaken by the CPSU. *Glasnost* indicated the Soviet's willingness to promote transparency and openness in the Soviet government.

[61] Lech Walesa, "Foreword," in *From Solidarity to Martial Law: The Polish Crisis of 1980–1981, a Documentary History*, eds. Andrzej Paczkowski and Malcolm Byrne (Budapest, Hungary: Central European University Press, 2008).

[62] Jan Repa, "Analysis: Solidarity's Legacy," *BBC News*, August 12, 2005, accessed July 28, 2010, http://news.bbc.co.uk/2/hi/europe/4142268.stm.

[63] Walesa, "Foreword," in *From Solidarity to Martial Law*, xv.

victims of the 1970 crackdown, a guarantee of no reprisals against the strikers, a wage increase, and an increase in family benefits. The factory agreed to the reinstatement of the two men but said that the rest of the demands were beyond their scope of authority.[64]

As the strike continued, members from the shipyard and other striking locations agreed on an expanded list of twenty-one demands.[65] At the top of the list were demands such as free trade unions independent of the Communist Party, the right to strike and security for the strikers, guarantees for free speech and non-censorship of independent publications, amnesty for those dismissed or arrested in the crackdowns of 1970 and 1976, access to the media for worker views, and publication of full and accurate economic data for debate on reforms. The broad range of issues showed how important the economic crises and its affect on the entire population of Poland were viewed. The usual concessions that the government gave to workers on a factory-by-factory basis did not quell the nationwide movement as they had before.[66]

Agreeing to an independent union and the right to strike was unique in the Soviet sphere. However, the Party delayed in implementing the terms of the agreement after the August strikes, refusing to register Solidarity as an official trade union for many months. Even as Solidarity was officially registered in November 1980, tensions mounted as meetings in Moscow and within the Warsaw Pact raised the specter of a Soviet invasion.[67]

Throughout 1981, the union continued its push for further implementation of the agreement, successfully registered another Solidarity union for farm workers, and grew its membership in local and regional organizations. As more protests and strikes continued around the country, Solidarity warned again that a full strike would result from any governmental crackdown against protesters or strikers. In December, the government declared martial law, calling it a "state of war," and the army surrounded factories with strikers, breaking them by starvation or force. The demand for an end to the martial state then became the major objective of the underground

---

[64] Maryjane Osa, Pastoral Mobilization and Contention: The Religious Foundations of the Solidarity Movement in Poland (New York: Routledge, 1989), 145.

[65] The statement was released by the MKS, the Interfactory Strike Committee, a progenitor of Solidarity.

[66] Douglas J. MacEachin, US intelligence and the Polish crisis: 1980–1981 (Washington, DC: Center for the Study of Intelligence, 2001), 5–8.

[67] Kemp-Welch, *Poland Under Communism,* 44.

movement, with the reestablishment of a legal union and release of all detainees being subparts of that demand.[68]

LEADERSHIP AND ORGANIZATIONAL STRUCTURE

The local organization of the strikes and loose national coordination meant that there were few who gained recognition as leaders of the Solidarity movement as a whole, and those who did became as much celebrities as leaders of a movement. The only true national figure was Lech Walesa, who had been fired by the Lenin Shipyard but returned to lead the strike along with Andrzej Gwiazda and Bogdan Lis.

The first critical decision faced by the proponents of an independent national labor union was how to organize themselves. One scholar notes that movements based on democratic values, such as Solidarity, often have difficulty reconciling those values with the "instrumental necessity of establishing organizational control."[69] Tension between that necessity and Solidarity's values was evident throughout the movement's activities. Did they require large centralized and hierarchical structures to coordinate a nationwide campaign, or could loosely coupled regional or local efforts provide the necessary momentum and pressure? Centralized government was enough of an anathema to the Polish population to avoid recreating it within the opposition, and the fear that such a centralized movement could be crushed by an overwhelming military and intelligence effort (either Polish or Soviet) pointed to the more plausible decentralized option.

Once the movement was forced underground during the period of martial law, it had to reestablish its structures and leadership in accordance with the new conditions. Only 20% of the Solidarity leadership avoided internment under the initial sweeps, but the underground was still heavily influenced by those in prison, with those who avoided capture acting as intermediaries between the rest and the underground movement. The underground movement became even more nonhierarchical, local, and autonomous. Strategy and communications were the most centralized functions at this stage, with the agreement that a passive resistance campaign should be followed, and many newspapers/leaflet publishers were established to spread messages across the nation. A Provisional Coordinating Committee (TKK) was established at the national level to issue demands, as

---

[68] George Sandford, *Military Rule in Poland: The Rebuilding of Communist Power, 1981–1983* (London: Croom Helm, 1986), 254.

[69] Elisabeth Crighton, "Resource Mobilization and Solidarity: Comparing Social Movements Across Regimes," in *Poland After Solidarity: Social Movements Versus the State*, ed. Bronislaw Misztal (New Brunswick, NJ: Transaction, Inc., 1985), 129.

well as a strategy document, "Statement on the Methods and Forms of Action." Regional structures (including Regional Coordinating Committees, or RKKs) emerged in Gdansk, Wroclaw, Warsaw, and Krakow—places where the crackdown had not completely decimated the organizational structure or caused the leadership to flee. Some of these regional structures also set up organizations—Secret Factory Committees (TKZs)—within the largest factories and workplaces.[70] The underground resisted efforts mostly at the local level through loosely connected and often re-forming organizations and networks.[71]

COMMUNICATIONS

Unlike most revolutionary organizations, Solidarity held its meetings openly and distributed its messages widely and overtly through networks in the beginnings of the movement, and then was forced underground as martial law was imposed. The National Committee debated and passed resolutions as a modern trade union would, operating as though its "enemy" would not take action by extreme military force or violence.[72] Negotiations between the government and the union's leaders were secret, both as the union was legalized and as demands grew through 1980–1981. A majority of Solidarity members felt that they were not sufficiently informed about the National Committee's activities, however.[73]

Expatriate journals such as *Kultura* (Paris) and *Aneks* (Sweden) allowed for the discussion of events and dissemination of communiqués and messages to the worldwide community during the period of martial law. Within Poland, the underground claimed that more than 1,500 regular publications appeared within the last half of 1982 that bypassed the government censor. Underground publishers also produced vast reprints of George Orwell's *1984* and other counter-themed works. Radio Free Europe was also crucially important to the spread of the opposition's version of events and messages to the Polish population.[74]

---

[70] Sanford, *Military Rule in Poland*, 256.

[71] Osa's research on the networks that made up the pre-legal and then legal Solidarity only extends up through the imposition of martial law. It does not cover the period of underground activity.

[72] Alain Touraine, *Solidarity: The Analysis of a Social Movement: Poland, 1980–1981* (Cambridge, UK: Cambridge University Press, 1983), 181–182.

[73] Jadwiga Staniszkis, *Poland's Self-Limiting Revolution* (Princeton, NJ: Princeton University Press, 1984), 77.

[74] Sanford, *Military Rule in Poland*, 256–258.

METHODS OF ACTION AND VIOLENCE

Another crucial decision faced the workers and union proponents in Poland during the 1970s and 1980s: the choice either to pursue an armed campaign against the Polish government or to continue the tradition of nonviolent protest as a means to reform. The choice seemed to hinge on the preference of the Solidarity leadership to avoid having to muster the necessary support to supply and organize such an opposition force, as well as the fear that the weight of the Soviet army could be called in to support the government (particularly during the period of martial law when it seemed likely that the Soviets would intervene to support the regime) and vastly overwhelm a small insurgent force. Therefore, a passive resistance campaign was initiated during the strike campaigns, as well as during the underground period of martial law. Oppositional activists, especially after 1976, focused on exerting social pressure on the regime to achieve democratic reforms. The protests, despite the massive numbers involved, the highly charged atmosphere, and the threat of a Soviet invasion, are noted for their nonviolence and the dignity with which they were carried out.

While the democratic opposition was gathering steam after the 1976 strikes and the formation of the KOR, the Polish economy entered another downward spiral that had the potential to capsize the regime. The strikes that occurred during 1980–1981 were different in several important respects from those that had occurred previously. While the previous strikes included proto-political demands, the 1980–1981 strikes evidenced a more notable concern for political rather than economic well-being. Second, the sheer scope of these strikes dwarfed those in the past. With nearly 10 million members at the onset of martial law in 1981, Solidarity's capacity for mobilizing and coordinating the Polish citizenry was nothing short of extraordinary. An estimated 8–9 million workers participated in the strikes, effectively halting economic and political life in Poland. Last, the concessions granted to Solidarity by the Party far exceeded those given in the past. Continuous public pressure resulted in the regime granting legal existence to Solidarity and trade unions in the Gdansk Agreements. Although the regime repudiated the concessions with the onset of martial law in December, the strikes set the stage for the regime's greater concessions in 1989.

Like the previous strikes in 1970 and 1976, the strikes of 1980–1981 were precipitated by an economic crisis. As Western creditors became wary, Gierek announced plans to cut the trade deficit by $1.3–1.5 billion, necessitating a reduction in food subsidies. Further exacerbating the problem were accompanying increased work quotas in some enterprises.[75] Workers began striking in opposition to the price increases in July 1980. In a strategy that had been successful previously, the management of the facilities in several cities offered the strikers a 10% wage increase, effectively ending some strikes. The divide-and-rule tactic would ostensibly prevent widespread, coordinated strikes as strikers' immediate economic needs were addressed.[76] However, in no small part thanks to the organization capabilities of the KOR, especially KOR member Jacek Kuron, workers throughout Poland were kept abreast of strike activity and government concessions to the strikes.[77] As news spread of government concessions, the strikes escalated rather than diminished.[78] Large strikes, including more than 18,000 workers and 177 workplaces in Lublin, began in July but ended after government economic concessions and promises of OPZZ reforms.[79]

Figure 3. Solidarity leader Lech Walesa in 1980.[80]

[75] Kemp-Welch, *Poland Under Communism*, 231.

[76] Garton Ash, *The Polish Revolution*, 33.

[77] Garton Ash notes that 1980 could have turned out very differently if the authorities had arrested thousands of KOR activists and cut some telephone lines. Instead, they initially pursued a policy of appeasement. Ibid., 34.

[78] Kemp-Welch, *Poland Under Communism*, 231–232.

[79] Ibid., 233–236.

[80] "Plik:Lech Walesa 1980.jpg," *Wikipedia*, accessed March 14, 2011, http://pl.wikipedia.org/w/index.php?title=Plik:Lech_Walesa_1980.jpg.

Strikes, hatched among the Free Trade Unionists, began in the Gdansk Shipyard in mid-August. Using the pretext of the dismissal of a shipyard worker, Anna Walentynowicz,[81] where the threat of price increases had failed, the strike began on August 14, 1980. Demands included the reinstatement of Walentynowicz, wage increases, family allowances equal to those of the police, and security for the strikers from persecution. Negotiations with the recently established strike committee dragged on, with government representatives concerned about the "escalation" of demands from pecuniary interests to political ones, including demands for free trade unions and release of political prisoners.[82] After the team made an attractive wage increase offer, the strike committee, including Lech Walesa, electrician and cofounder of Solidarity, agreed to halt the strike. As workers began streaming out the gates, however, they were met by representatives from other workplaces across Poland calling for a "solidarity" strike. Walesa, moved by the would-be strikers, called for a halt to the evacuation of the shipyard. After establishing an Interfactory Strike Committee (MKS), the Gdansk Shipyard strike was back on. This time the strike would not cease until the MKS under Walesa's charismatic leadership had wrung unprecedented concessions from the communist regime.[83]

Over the course of the strikes, the MKS came to represent the interests of workers of more than 400 enterprises across Poland. Government negotiators attempted to meet with striking groups individually, refusing to acknowledge the MKS, but it was successful in disciplining its members. The MKS developed a list of twenty-one demands; the most contentious, and the condition on which the MKS consistently refused to budge, was the right to free trade unions.[84] Despite Walesa's assurances to the contrary, the Politburo viewed the demand as an assault on the Party's political authority.[85] With the economy worsening daily and the determination of the strikers to extract political concessions mounting, Mieczyslaw Jagielski

---

[81] In an ironic twist, Walentynowicz was fired for collecting candle stubs from a nearby graveyard for a memorial for workers killed at the shipyard in December 1970. Garton Ash, *The Polish Revolution*, 38.

[82] The use of force against the strikers was discussed in the Politburo. The Ministry of Interior declared the social unrest a threat to domestic security and had plans to drop military commandos by helicopter into the shipyard if necessary. While the military believed force could destroy the counter-revolutionary forces in Gdansk, members of the Politburo were less certain. Stanislaw Kania, member of the Politburo and future First Secretary of the Party, held that a large-scale operation was unfeasible in the limited time available. He maintained that negotiations were the most viable option at this juncture. Kemp-Welch, *Poland Under Communism*, 240.

[83] Ibid., 237–242.

[84] For a description of additional demands, see Ibid., 248–253.

[85] Ibid., 261.

met with the Strike Presidium, representatives of the MKS aided by their intelligentsia advisers, to begin negotiations. After several days, Jagielski signed an agreement with the strikers on August 31, 1980, known as the Gdansk Agreement, which granted the MKS their primary demand: free trade unions.[86]

Despite the hopeful outcome of the Gdansk strikes, the regime recapitulated and enacted martial law in 1981 to pacify continuing social unrest and striking activity in the country. During earlier political crises, Gomulka and Gierek had been able to regain popular support for the Party. Facing a more cynical, politically savvy crowd waiting for the materialization of the promises made by the regime in the Gdansk Agreement, First Secretary Kania, replacing Gierek, needed to develop a policy to "get Poland back to work."[87] After delaying the initial registration process and arresting some Solidarity members, Solidarity reinitiated strikes that nearly led to Soviet intervention. Tensions increased within the Party itself as Kania's leadership was viewed as ineffective and incapable of extracting Poland from its political and economic crises. In February 1981, General Jaruzelski replaced Kania as First Secretary. Jaruzelski, stating that Poland had descended into "criminality and chaos," declared martial law. Civilian riot police systematically broke strikes throughout country, killing several and arresting hundreds.[88]

A period of stasis followed the 1981 martial law. Forced underground, Solidarity lost the vast majority of its membership, counting only 4,000 members in its underground movement.[89] A sense of impending crisis, however, did not desert Polish political and economic life. A widening chasm existed between state and society, but the state required the support of society for political and economic reforms as the crises in preceding decades indicated. Under Jaruzelski's leadership, after using force to reconsolidate Party authority in 1981, the Party failed to recapture popular support or develop reforms necessary to extract Poland from crisis. Paralyzed, over the course of the 1980s, the Party accepted the necessity of entering into a dialogue with society's chosen representative—Solidarity. Although both sides had their hardliners, it was the moderates who carried the day in 1989. Finally, the move toward a more pluralist Poland would have been dangerous, and arguably impossible, without Gorbachev's renunciation of the Brezhnev Doctrine and the reforms he initiated within the USSR.[90]

---

[86] Ibid., 243–268.

[87] Ibid., 276.

[88] Ibid.

[89] Ibid., 338.

[90] Ibid., 410–411.

Jaruzelski initiated moves toward consultation with prominent sectors of society, workers, and the church as early as 1987, but the Party insisted on maintaining its monocentric grip on power, refusing legalization of Solidarity until early 1989. The Party, which had already hemorrhaged members in the early 1980s,[91] especially workers and young people, many of whom had left to join Solidarity, was nearly paralyzed in a "technocratic–bureaucratic managerialism" that left little room for the innovation necessary to weather the crisis.[92] His initial offer to individuals of Solidarity, but not the group itself, to join the government was rebuffed by Walesa. The leaders of Solidarity stood firm, insisting that any partnership with the Party to extract the state from its economic crisis required the return of independent trade unions and freedom of association.[93]

Strikes during 1988 broke the impasse. After a summer truce following strikes in the spring of 1988 in response to yet another economic crisis and a round of price increases, in August, mine workers initiated strikes demanding the legalization of Solidarity. After the spring strikes, Jaruzelski had suggested Round-table talks to bring state and society representatives together to discuss economic and political reforms. After the August 1988 strikes, Jaruzelski made overtures to Walesa to bring him to the round-table talks. Months of negotiations followed, at one point nearly collapsing on the issue of Solidarity's legality. The church, which was present at all discussions, provided a third-party assurance that neither the Party nor Solidarity was selling their constituents down the river in a back-room deal, lending credence to the discussions among hardliners in both groups.[94] In January 1989, the Party voted to recognize and negotiate with Solidarity,[95] helped in part by a televised debate between Walesa and the chairman of the OPZZ. Walesa came off as moderate, dispelling years of state propaganda that presented him as a dangerous radical.[96]

---

[91] The Party lost more than one-third of its membership from 1980 to 1985. Kolankiewicz and Lewis, *Poland: Politics, Economics and Society*, 68.

[92] Kemp-Welch, *Poland Under Communism*, 343.

[93] Ibid., 346–347.

[94] Ibid., 370.

[95] Ibid., 387.

[96] Ibid., 379–380.

Figure 4. Strike at the Gdansk Shipyard in 1980.[97]

In February 1989, the round-table talks began. With the recognition of Solidarity off the agenda, a decision on parliamentary elections was the largest component of the talks. Minister of the Interior Kiszczak headed the communist delegation, and Walesa headed the Solidarity delegation. By April, the delegations signed an agreement allowing Solidarity to contest a limited number of seats in parliamentary elections, while still allowing the Party a majority of seats.[98] After elections were held in June, Solidarity swept the communist contenders from nearly all of the available contested seats in the *Sejm.*[99] After the elections, Solidarity was able to form an opposition coalition with smaller parties formerly allied with the PZPR. For the first time in communist history, a non-Communist Prime Minister, Tadeusz Mazowiecki, a Catholic intellectual, was elected. Demoralized

---

[97] http://libcom.org/history/1980-poland-mass-strikes; photo by T. Michalak.

[98] The Party was under the impression that it would retain considerable power after the elections, gaining legitimacy in "another guise." Ibid., 402.

[99] Solidarity candidates won 160 of the 161 seats in the *Sejm* that they were allowed to contest, as well as 92 of 100 seats in the Senate. Voter turnout for the first round of elections was 62%. The Party briefly debated the possibility of invalidating the elections. Ibid., 404. The monumental elections of 1989 received little attention in Western media. On the day of the first round of elections, another Communist country attracted the world's attention when the regime in China crushed protestors in Tiananmen Square.

and bereft of support from society or the dissolving USSR,[100] the PZPR held its last Party Congress in January. The Communist government in Poland was peacefully dismantled.

METHODS OF RECRUITMENT

The Solidarity movement did not form overnight. At the peak of its strength, Solidarity had organized 70% of the 14 million workers in the country. Not only do the numbers themselves astonish, but Solidarity membership also represented a broad-based coalition across varying social classes, occupational groups, and the rural–urban divide. Solidarity combined with various ancillary civil society organizations already operating in Poland, such as the Students' Solidarity Committee and Rural Solidarity, meaning there were few corners of Polish society that Solidarity did not touch. Solidarity also reached into the Party itself. The "radical, antibureaucratic rump" of many local Party offices contributed more than 700,000 members. Moreover, about one-third of the MKS belonged to the Party.[101]

Mobilization of Polish society emerged several years before the large strikes during the summer of 1980. The foundation and the practice of the KOR encouraged and inspired a wide variety of organizations to form and contest state policy.[102] Some of the groups that formed subsequent to KOR activities belonged to the same general flavor as the KOR, while others, by contrast, had ideological or political agendas that were notably different from those of the KOR. Not only were the groups ideologically diverse, but they were also geographically dispersed throughout the country.[103] Some 33.5 million men and women—one-third of the population—joined independent professional, social, and political organizations.[104] At the close of the 1970s, Poland had a remarkable organizationally competent and dispersed opposition that would form the societal infrastructure of support upon which Solidarity relied and drew many of its members. In this regard, the church and its lay organizations and a loose network of secular groups and clubs were particularly important.[105] The recruitment and mobilization of Solidarity members relied on existing networks, much like the civil rights movement in the United

---

[100]   Ibid., 410.

[101]   Crighton, *Resource Mobilization and Solidarity*, 121–125.

[102]   The "civic fever" reached even state-controlled organizations, affecting professional associations and youth organizations; even the PZPR itself began to democratize internally and change leadership.

[103]   Bernhard, *The Origins of Democratization in Poland*.

[104]   Ekiert and Kubik, *Rebellious Civil Society*, 40.

[105]   Crighton, *Resource Mobilization and Solidarity*, 121.

States did with the networks of black churches and historically black colleges in the South.

METHODS OF SUSTAINMENT

Although industrial unrest was a feature of Polish politics throughout Communist rule there, previous striking activity suffered from severe limitations and was incapable of exerting enough pressure on the regime to secure significant political and economic concessions. Before the Solidarity strikes, industrial action generated by price increases began with walkouts culminating in riots and burning down the local Party headquarters. After limited economic concessions from the regime, strikers went back to work. The cycle of striking activity typically lasted no more than a few days. As the objectives of oppositional actors in Poland transformed to those prescribed by Kuron, Michnik, and other intellectuals, utilizing social mobilization from below to exert pressure on the regime, the new breed of opposition following the food riots of 1976 required organizational discipline lacking in previous strikes.

In 1980, Solidarity activists addressed these limitations with the formation of factory strike committees. The good grassroots organization before the strikes, especially after the formation of the KOR in 1976, resulted in a nonviolent, tightly organized, well-coordinated series of strikes "that could be escalated as necessary to support Solidarity's negotiating position." However, after the legalization of Solidarity, intra-movement power conflicts divided the National Commission, regional presidiums, and local unions, in some instances leading to unauthorized strikes in various factories. After Jaruzelski implemented martial law, some elements within the movement blamed Solidarity leaders for insufficiently bridling the radical elements in the movement, resulting in a period of military rule.[106]

METHODS OF OBTAINING LEGITIMACY

One scholar notes that Solidarity was more than an organizer and mobilizer of Polish society; it was also a torch-bearer of collective identity. That identity was built on the combination of the concepts and democratic ideals of the intelligentsia's opposition movements in the 1970s with the social and ethical concerns addressed by the Roman Catholic Church, and above all, Pope John Paul II. Solidarity offered

---

[106] Ibid., 125.

society a very seductive alternative "vision" for government and society than that espoused by the communist regime, including notions of inalienable human rights, political rights, and nationalist values and traditions, offering a "powerful vision of reform and political change based on the self-organization of a democratic society against the post-totalitarian state."[107] Solidarity's powerful symbolic discourse enabled a moral consensus that resonated across a wide spectrum of society otherwise riven by social, economic, and occupational cleavages. It handily defeated the PZPR in the battle for "national symbols and legitimation claims."[108]

EXTERNAL SUPPORT

Solidarity's external support derived primarily from Western Europe, the United States, and the Roman Catholic Church. Despite the similar straits of workers in other countries in the Eastern Bloc, Solidarity did not form alliances with other nascent oppositional actors in the communist world. At a Solidarity Congress after its legalization, one worker did suggest such an alliance with Soviet workers that met with harsh repercussions for the Solidarity movement.

The moral authority and support of Polish Pope John Paul II is difficult to overestimate. Several prominent Solidarity activists, including Walesa himself, credit the Pope's election, visit to Poland, and commitment to human rights and peaceful protest to Communist powers as a critical factor in uniting and inspiring Poles to political action. However, the Pope's actual material support of Solidarity is contested. One scholar claimed that the Pope provided financial assistance to the movement and formed an alliance with the Central Intelligence Agency (CIA) to hasten the fall of communism in the Eastern bloc, but such notions are hotly contested.[109]

---

[107] Ekiert and Kubik, *Rebellious Civil Society*, 278, 41.

[108] Ibid., 41.

[109] Idesbald Goddeeris, "Solidarnosc, the Western World, and the End of the Cold War," *European Review* 16, no. 1 (2008): 55–64.

W SAMO POŁUDNIE
4 CZERWCA 1989

Figure 5. Solidarity poster for the 1989 election featuring American actor Gary Cooper from the iconic movie *High Noon*.[110]

There is no question, however, that Solidarity received ideological and material support from sectors of Western society, especially in the United States and France. Lane Kirkland, then president of the American labor union AFL-CIO (American Federation of Labor and Congress of Industrial Organizations), was a staunch supporter of Solidarity. He lobbied for Solidarity but met with resistance from both the Carter and Reagan administrations. Believing official US responses to be too weak and ineffectual, Kirkland created the Polish Workers Aid Fund in 1980 to raise money and supplies for Solidarity, sending more than $3 million to Solidarity over the course of several years after its inception.[111] In 1987, Congress took over, granting Solidarity several million dollars.[112] In France, fascination and fraternity with the Solidarity movement resulted in the establishment of the *Solidarité*

---

[110] The actor Gary Cooper, playing US marshal Will Kane in the iconic western movie *High Noon*, holds a Solidarity ballot in his hand and wears a Solidarity badge on his chest. The text on the bottom reads "It's High Noon, June 4, 1989." Created by artist Thomas Sarnecki, this poster was one of the most powerful images that emerged during the Solidarity movement. "The attraction of this figure lies in its ability to combine an American image, given all that America meant to Poland, historically and culturally, with the powerful Solidarity material. This poster hammered home the message that the June 4 elections offered a stark choice between two opponents and would have momentous consequences for Poland." Making the History of 1989: The Fall of Communism in Eastern Europe, accessed on July 12, 2010, http://chnm.gmu.edu/1989/.

[111] After Jaruzelski declared martial law, most of the funds that the AFL-CIO sent to Solidarity were government funds channeled through the National Endowment for Democracy. Ibid.

[112] Ibid.

France-Pologne, a French trade union still extant. The union aimed at collaboration between the two unions. During Solidarity's legal phase, Solidarity collaborators and their French counterparts had more than 100 joint meetings. In addition, *Solidarité* channeled humanitarian aid to Poland. Some French trade unionists even reportedly participated in the August strikes. The enactment of martial law in Poland prompted a wave of demonstrations in France in which hundreds of thousands participated. Martial law in Poland sparked the organization of numerous French trade unions into one umbrella organization that provided material support to the Solidarity-in-exile organization, which was composed of Solidarity expatriates who were abroad when martial law was declared, and to Solidarity's activities in Poland proper.

COUNTERMEASURES TAKEN BY THE GOVERNMENT

The Party had opportunities to implement countermeasures against Solidarity activities at numerous points, including during the initial phase of oppositional activity after the 1976 food riots when the networks that eventually supported Solidarity were forming, during the strikes themselves, and in the interim period after martial law and before the round-table talks. However, after the food riots of 1976 and the formation of the KOR, oppositional and other civil society activity flourished. It is noted that "[within] three years, Poland developed a whole opposition counter-culture without parallel in the Soviet bloc."[113] Repression of such activities was technically feasible given the resources available to Gierek's regime. A high-ranking member of the security forces on one occasion questioned why they were not targeting the system of extensive underground publications and was told that higher authorities would not give the go ahead. One observer offers several reasons for Gierek's reluctance to take decisive action against the opposition. First, the regime simply did not view the opposition as a serious threat, and they hoped tolerance would soften the stance of the intelligentsia to the regime. Second, the détente between the East and West moderated the regime's response. Gierek's regime was in desperate financial straits at this juncture, and with the Helsinki process in full swing, the Carter administration expressly established a linkage between economic assistance and human rights records. In 1977, President Carter praised Poland's respect for human rights and granted $200 million in credits to the country in the same breath. Thus, while KOR activists were harassed, arrested, and held

---

[113] Timothy Garton Ash, "Ten Years After," *New York Review of Books* 46, no. 18 (1999): 18.

for questioning, the activists would not be held longer than forty-eight hours until August 1980.[114]

As strikes began in July 1980 in response to increased food prices, Gierek initiated the tried-and-true Party tactic to break strikes: divide and conquer. Authorities quickly capitulated to strikers' economic and local demands, shipping in containers of meat when necessary. When workers in some areas received wage increases, they picked up their tools and went back to work. Rather than quieting industrial unrest by preventing large-scale coordination of strikers across the country, the regime's tactics only further fanned the flames. What confounded the regime was the speed with which information about the regime's capitulation crisscrossed the country despite the dead silence of government-controlled mass media. As a result, as soon as one strike was put out, another took its place. KOR networks transmitted information all over Poland, to the West, and to Western radio stations heard in Poland, especially Radio Free Europe, keeping millions of Poles abreast of strike developments. Typical regime appeasement countermeasures to industrial unrest—which divided and conquered strikers, putting them back to work in a matter of days—failed.[115] Police action against strike committee members, including threats to family members, residence raids, constant surveillance, and other attempts at intimidation, failed to break the strikes. Adopting a more menacing tone, one senior official aptly encapsulated the regime's position by declaring, "We will not give up power nor will we share it with anyone."[116] On the advice of local Party officials that the strikers were "tired but determined," the regime authorized negotiations with Solidarity.[117]

The negotiations primarily stuck to the issue of free trade movements. Met with resolute strikers supported by a critical mass of society on the union issue, the Party faced a crisis-point, with limited options to move forward to get Poles back to work without degrading the Party's political power. A "Party-state Crisis Staff," assembled to address potential measures, saw three possible options: (1) "administrative measures," or keeping any Gdansk Agreement vague and then recapitulating when the Party was no longer "under the gun," eventually resorting to martial law if necessary; (2) the use of military force to break strikers and restore order, including a Soviet invasion; and (3) a political struggle to win the propaganda war. The last of the three options won out for a number of reasons, including

---

[114]   Garton Ash, *The Polish Revolution*, 18–19.

[115]   Garton Ash, *Ten Years After*, 33–34.

[116]   Kemp-Welch, *Poland Under Communism*, 246.

[117]   Ibid., 247–248.

the influence of the church and Soviet reluctance to engage in an uncertain armed struggle after its exhausting efforts in Afghanistan. As a result, Jagielski signed the draft known as the Gdansk Agreement, which included the right to free trade unions, legalizing Solidarity.[118]

The signing of the Gdansk Agreement did not usher in a more democratic era in Poland. The regime dragged its feet in implementing the terms of the agreement, and Solidarity, likewise, continued to use strikes as a political tool to pressure the regime into fulfilling its end of the bargain. After failing to mend relations between state and society, Jaruzelski replaced Kania as First Secretary and declared martial law to break the continuing strikes. In the interim period when Solidarity was an underground movement, the Party faced its own crisis as its equally rigid bureaucratic structure and ideology offered little hope of addressing the acute economic crisis facing the country. Party members jumped ship in droves. Moreover, as history had proven to the regime, the Party could do little to mend Poland's economic woes without the support of society.

When social unrest broke out once again in 1988, the regime was left with few options to quell agitation in Poland. Using security forces to break the strikes, it was thought, would only inflame the situation further.[119] Additionally, the threshold of fear of the Polish public had risen considerably, and few were still afraid of the authorities. The threat of another period of martial law had little to no restraining impact on rebellious Poles. At the end of the decade, when Gorbachev's reforms in the USSR largely removed the Soviet "bogey,"[120] the moribund Party capitulated and sought out Walesa and Solidarity as negotiating partners in rebuilding Poland after granting Solidarity legality. By entering into negotiations with Solidarity, Jaruzelski hoped to make the popular movement "co-responsible" for political and economic reform. He noted, "The game is about swallowing up the opposition of our system, and their participation in (re)shaping it. This is a great historical experiment which—if it works—can have an importance

---

[118] Ibid., 262–268. In defense of his enactment of martial law, Jaruzelski has long claimed that he did so in order to prevent a Soviet invasion. However, many contest that such a threat existed in late 1981. Former high-ranking Soviet officials claim that although a Polish invasion was discussed, the idea was ultimately rejected. Repa, "Analysis: Solidarity's Legacy."

[119] Kemp-Welch, *Poland Under Communism.*, 357.

[120] Ibid., 360. In the latter half of the 1980s, Gorbachev made several momentous announcements. The first was a limited moratorium on nuclear testing followed by the signing of the Treaty on Intermediate-Range Nuclear Forces. Then, in December 1988, he announced in a speech at the UN that there would be a unilateral decrease of Soviet conventional forces and that the countries of the Warsaw Pact could decide their own political fate. Goddeeris, "Solidarnosc, the Western World, and the End of the Cold War."

extending beyond Poland's borders."[121] In allowing Solidarity to participate in the elections of 1989, the Party made a gamble that they still had enough traction with Polish society to maintain political power—one they ultimately lost.

SHORT- AND LONG-TERM EFFECTS

CHANGES IN THE ENVIRONMENT

In 2005, at the celebration of the twenty-fifth anniversary of Solidarity, the central motif was the slogan "Everything began in Gdansk" and an image of tumbling dominoes. Indeed, the cracks in the seemingly unassailable Soviet edifice appeared first in Gdansk. However, whether those cracks were the cause of the subsequent dissolution of the Warsaw Pact and the USSR is not clear.[122] What is clear is that after Poles shed their Communist regime, replacing it with a Western-style representative government, several surrounding countries followed suit, including Czechoslovakia, East Germany, and Hungary. The reunification of Germany and the removal of Communist Party control in nearly the whole of Eastern Europe significantly reordered the political and security environment in short order. The Warsaw Pact dissolved, and the North Atlantic Treaty Organization (NATO) quickly approached former Warsaw members for inclusion in its mission, redefining itself in the process. NATO adopted a new mission with the addition of the former countries that were to provide a buffer to a Western invasion of the USSR, a broader approach to security that featured more crisis management and conflict prevention than resisting Soviet military might.[123]

CHANGES IN GOVERNMENT

The parliamentary elections of mid-1989 were to fill the *Sejm* (lower house) and the Senate. The failure of the communists to produce a majority led to the coalition with Solidarity, and eventually the election of Lech Walesa as president in 1990. Political parties were numerous, with more than 100 having candidates and 20 gaining seats in one of the two bodies during the 1991 elections. No one party obtained more than 13% of the vote, and coalition governments have been common

---

[121] As quoted in Kemp-Welch, *Poland Under Communism*, 390.

[122] Goddeeris, "Solidarnosc, the Western World, and the End of the Cold War."

[123] William Drozdiak, "NATO Finds New Role, with Soviet Threat Gone," *The Washington Post*, November 2, 1991.

since. The government was run under temporary amendments to the former constitution of the Peoples' Republic, while a new constitution was finally adopted in 1997.

CHANGES IN POLICY

The move of Poland from a keystone of the Warsaw Pact to a member of NATO in the span of 20 years shows the dramatic shift in its relationship to the world order and demonstrates its prime role in the world's exit from the Cold War and the breakup of the USSR. Poland undertook a crash course in Western industrialization and private enterprise implemented by Finance Minister Leszek Balcerowicz.[124] In 2004, Poland joined the European Union, the same year as other former Soviet sphere countries, Slovenia, Estonia, Latvia, Lithuania, the Czech Republic, Slovakia, and Hungary.

CHANGES IN THE REVOLUTIONARY MOVEMENT

The largest change in the movement was its normalization into a legitimate and powerful political party, in part due to its being awarded status by the government, as well as its vast mass popularity and membership. The ability of Solidarity to win almost all of the seats in which it fielded a candidate during the first election shows how dissatisfied the people were with the government and how willing they were to accept the Solidarity union as a legitimate member of the political scene.

As Solidarity transitioned into an organization with a political agenda, differences in opinions became more apparent between the leadership, including a split between Mazowiecki and Walesa when Walesa decided to run for president. Walesa did not approve of the coalition between Solidarity and the Communists, so he ran against his former advisers with the slogan, "I don't want to, but I must."[125] Solidarity fell into a slump in the mid-1990s, garnering only single-digit percentages of the vote. Walesa was narrowly defeated in 1995 by the former Communist Aleksander Kwaniewski, proving that Solidarity had now moved from the successful and powerful union that toppled a government to a marginal party that could be thrown out of office when the people did not see the results they wanted. A broader

---

[124] Leszek Balcerowicz, "Transition to the Market Economy: Poland, 1989–93 in Comparative Perspective," *Economic Policy* 9, no. 19 (December 1994): 72–97.

[125] "Lech Walesa," Timothy Garton Ash, *Time*, April 13, 1998, accessed March 26, 2010, http://www.time.com/time/magazine/article/0,9171,988170,00.html.

coalition—Solidarity Electoral Action—emerged in the late 1990s, winning one-third of the vote in the 1997 parliamentary elections.[126]

BIBLIOGRAPHY

Balcerowicz, Leszek. "Transition to the Market Economy: Poland, 1989–93 in Comparative Perspective." *Economic Policy*, 9, no. 19 (December 1994): 72–97.

Bernhard, Michael. "Civil Society and Democratic Transition in East Central Europe." *Political Science Quarterly* 108, no. 2 (1993): 307–326.

———. "Reinterpreting Solidarity." *Studies in Comparative Communism* 24, no. 3 (1991): 313–330.

Bernhard, Michael H. *The Origins of Democratization in Poland: Workers, Intellectuals, and Oppositional Politics, 1976–1980.* New York: Columbia University Press, 1993.

Bielasiak, Jack. "Social Confrontation and Contrived Crisis: March 1968 in Poland." *East European Quarterly* 22, no. 1 (1988): 81–103.

Central Intelligence Agency. "Poland." *The World Factbook.* https://www.cia.gov/library/publications/the-world-factbook/geos/pl.html.

Crighton, Elisabeth. "Resource Mobilization and Solidarity: Comparing Social Movements Across Regimes." In *Poland After Solidarity: Social Movements Versus the State.* Edited by Bronislaw Misztal, 113–132. New Brunswick, NJ: Transaction, Inc., 1985.

Curtis, Glenn E., ed. *Poland: A Country Study.* Washington, DC: Government Printing Office, 1992.

Drozdiak, William, "NATO Finds New Role, with Soviet Threat Gone." *The Washington Post*, November 2, 1991.

Ekiert, Grzegorz, and Jan Kubik. *Rebellious Civil Society: Popular Protest and Democratic Consolidation in Poland, 1989–1993.* Ann Arbor, MI: University of Michigan Press, 1999.

Garton Ash, Timothy. "Ten Years After." *New York Review of Books* 46, no. 18 (1999).

———. *The Polish Revolution: Solidarity.* 3rd ed. New Haven, CT: Yale University Press, 2002.

———. "Lech Walesa," *Time*, April 13, 1998. http://www.time.com/time/magazine/article/0,9171,988170,00.html.

Goddeeris, Idesbald. "Solidarnosc, the Western World, and the End of the Cold War." *European Review* 16, no. 1 (2008): 55–64.

---

[126] Garton Ash, *The Polish Revolution*, 377.

Goodwyn, Lawrence. *Breaking the Barrier: The Rise of Solidarity in Poland.* New York: Oxford University Press, 1991.

Kemme, David. "The Polish Crisis: An Economic Overview." In *Polish Politics: The Edge of the Abyss.* Edited by Jack Bielasiak and Maurice D. Simon. New York: Praeger, 1984.

Kemp-Welch, A. *Poland Under Communism: A Cold War History.* Cambridge, UK: Cambridge University Press, 2008.

Kolankiewicz, George, and Paul G. Lewis. *Poland: Politics, Economics and Society.* London: Printer Publishers, 1988.

Kubik, Jan. "Who Done it: Workers, Intellectuals, or Someone Else? Controversy Over Solidarity's Origins and Social Compositions." *Theory and Society* 23, no. 3 (1994): 441–466.

Laba, Roman. *The Roots of Solidarity: A Political Sociology of Poland's Working-Class Democratization.* Princeton, NJ: Princeton University Press, 1991.

MacEachin, Douglas J. *US Intelligence and the Polish Crisis: 1980–1981.* Washington, DC: Center for the Study of Intelligence, 2001.

Michnik, Adam. *Letters from Prison and Other Essays.* Berkeley, CA: University of California Press, 1985.

Michnik, Adam, and David Ost. *The Church and the Left.* Chicago: University of Chicago Press, 1993.

Osa, Maryjane. *Pastoral Mobilization and Contention: The Religious Foundations of the Solidarity Movement in Poland* (New York: Routledge, 1989), 145.

Repa, Jan. "Analysis: Solidarity's Legacy." *BBC News,* August 12, 2005. Accessed July 28, 2010. http://news.bbc.co.uk/2/hi/europe/4142268.stm.

Sandford, George, *Military Rule in Poland: The Rebuilding of Communist Power, 1981–1983.* London: Croom-Helm, 1986.

Touraine, Alain. *Solidarity: The Analysis of a Social Movement: Poland, 1980–1981.* Cambridge, UK: Cambridge University Press, 1983.

Walesa, Lech. "Foreword." In *From Solidarity to Martial Law: The Polish Crisis of 1980-1981, a Documentary History.* Edited by Andrzej Paczkowski and Malcolm Byrne. Budapest, Hungary: Central European University Press, 2008.

CONCLUSION

The study of insurgency and revolution, like all warfare, is frequently in danger of being too rooted to the past while ignoring or misreading trends and present-day lessons. The purpose of this broad survey of recent revolutions is to expose the reader to the wide variety that such warfare may exhibit while also reinforcing the common elements and frequently observed characteristics particular to this class of conflict. The structure of this Casebook is meant to provide a template for learning and assessing the vital aspects of revolutionary warfare and allow for easy comparison of techniques and trends across the case studies. It should also provide the reader with a scheme for analyzing evolutionary changes to strategies, tactics, technologies, and sociopolitical contexts.

In that spirit, this conclusion lays out observations for each framework section that have been proposed by the research team for consideration, discussion, and debate. We expect that some will disagree and others will want to modify any or all of the below observations, but we offer them as a starting point for a classroom setting, discussion group, or individual self-reflection.

A revolution is *ipso facto* brought about by the desire for change. Therefore, the motivations and objectives of the movement center on the alteration of the government system and/or policies. The appeal of the alternatives that the revolutionaries want to implement is paramount to their success. A clear and ideologically mature replacement system is conducive to the formation of the revolutionary movement and the collection of support and sympathy. It also provides a motivational tool when the inevitable time of despair and crisis occurs during the revolutionary struggle. However, although a strong and clear objective provides benefits, a rigid and firm vision can become a major hindrance at the negotiating table when compromise is expected. Popular support may dwindle rapidly when an ideologically pure but rigid insurgent group rejects a viable political solution to the conflict. The sociopolitical context can evolve past a group's initial vision, rendering them less supportable, less viable, and perhaps irrelevant.

Revolutionary objectives frequently define an identity group, either explicitly or by the nature of its socioeconomic or cultural framing. Defining sides, i.e., "us" and "them," is a key objective in all warfare but is very important in civil or revolutionary war. This partitioning is key to validating claims of grievance, support, and legitimacy. The deeper these categories, especially when based on cultural, ethnic, nationality (in the case of a foreign military), or class divisions, allows

861

the revolutionary group to recruit, frame propaganda, and sanctify their operations against "them." Successful movements often rely on a narrative for recruitment, legitimacy, and support that resonates with a deep cultural, ethnic, or historical myth/memory within the population.

Organization and leadership styles of all types are used in insurgencies, but smaller, tight leadership cadres seem to be favorable in the nascent stages. This allows groups to build a cohesive ideological framework and form strategy and initial plans for the revolution. Command and control is often very centralized and tightly controlled at this stage, but revolutionary movements must be adaptable and flexible when the counterinsurgent effort matures, and must usually change to a less-centralized, cellular, or networked structure in order to defeat penetration, co-option, and other types of intelligence gathering. While a movement may be able to change its structure to something more conducive against government actions, successful movements must reach a point of self-sustainment. They must be able to survive a loss of leadership, in terms of both morale as well as the ability to execute command of the movement. The capture or death of the movement's leadership often critically dampens the revolutionary movement in its early stages or can cripple any group that is still highly dependent on specific leadership personalities.

Revolutions that occur over a long period of time usually have organizational structures in place that allow for mass communication and participation beyond the guerrilla/operations group. An insurgency that has an accompanying overt organization, such as a legal political party, has the advantage of a communications conduit, a negotiating intermediary, and a means to involve people who are committed to the cause but are unwilling to participate in violence. These people may distribute propaganda, run political candidates in the legal electoral system, collect financing, and arrange legal protests or other events to garner attention and sympathy. Having a range of possible support opportunities allows the revolution to broaden beyond those who have been "radicalized" or support violence against the state. It also provides a proxy through which both the government and the insurgent group can communicate, negotiate, and tactfully make concessions.

Active operations within revolutions span the gamut of irregular warfare—such as terrorism, guerrilla warfare, protest, and sabotage— and sometimes blend into conventional warfare as well. Once a group has decided on a strategic approach, access to technology, advanced weaponry and materiel, and funds often provide the boundaries to a group's operations. Beyond the type of actions employed, the sustainment of operations is critical to the continued viability of a movement, perhaps as much as or more than its choice of targets or level of intensity. The ability to sustain operations allows the movement to continually stress the opposing forces and keep the security environment unstable. A movement that can methodically perpetuate operations continues to show its relevance, wears down the opponent, and keeps the population in fear.

Success in recruitment is tied to two crucial factors: a sufficient pool of potential recruits from the core "identity group" to which the movement is tailored, as well as a large enough structure to allow multiple paths of support and membership within the broader population. Large segments of unemployed/underemployed young males are increasingly susceptible to revolutionary activities that are sufficiently organized, while long-standing insurgencies have various overt, underground, and military networks that can accommodate the sympathizers as well as the fervent or radicalized.

These larger networks of support are traditionally based on social networks at their core, whether built around familial, ethnic, religious, or cultural affinities. But business and criminal interests seem to be of increasing importance to the successful sustainment and execution of this type of warfare, allowing quick access to training, financing, and materiel that used to require time to develop. The access to large international networks has provided a boon to nascent insurgencies that do not have ready-made support streams.

The importance of legitimacy in the hearts and minds of the surrounding population has been well discussed and corroborated in recent insurgency warfare literature. Most of the cases in this collection stress the importance of the underlying narrative, especially those that use cultural and historical themes, for a movement's legitimacy. The balance between the military operations of the government and those of the revolution also greatly affects the weight upon which the population is willing to see either the insurgent or the government as more worthy of rule and support. Propaganda and psychological operations have their effect, but deeply rooted messages and actions that correspond to these messages are vital to each side.

Often overlooked in recent discussions of insurgency is the concept of safe havens. Physical safe havens are important in the early, vulnerable periods of an insurgency, when political objectives are still being formulated and a support base is being established. These safe havens generally correspond to an area with a disenfranchised populace or with significant cultural or economic disparities with the greater state. Operationally, the segregated spaces provide a secure location both for launching operations and for training; they also offer a defensible enclave, while making the group more cohesive.

Finally, we must stress that all revolutions, like elections, are local affairs. Although national or international contexts may play a role in setting the motivations, support structures, and other elements of a revolution, local issues such as culture, history, demographics, social networks, economics, and preexisting conditions play an even greater role in the development, execution, and eventual success of a movement. Studying the situational environment before a revolution is necessary to the understanding of any case study, and we hope that this text has shown that those underlying factors are as important to study as the conduct of the war itself.

www.ingramcontent.com/pod-product-compliance
Lightning Source LLC
Chambersburg PA
CBHW052106020426
42335CB00021B/2665